This is
an Open Letter
on the
Veracity of
Gravity

ISBN-13: 978-1499321845
ISBN-10: 1499321848

Information studied by PEET (P.S.J.) SCHUTTE

FROM THE ORIGINAL AFRIKAANS: "MATERIE SE TYD IN RUIMTE BY PEET (P.S.J.) SCHUTTE

© KOSMOLOGIESE EN ASTRONOMIESE TEGNIKA

Announcing this book that reveals Nature's Natural Science.

naturescosmicconcept

With this book and others also already published Science principles changes forever. Now nature is explained, which proves Newtonian science is without a basis of truth. A new approach to science is revealed in the website called **naturescosmicconcept** a website introducing true naturals science as the truth. Now your first thought would be that scientists are doing the flip flop with joy! No such a chance because these findings destroy everything science pretend is true.

Should you wish to find out how nature forms the Universe then this book is written with you in mind. It shows why nature uses the four cosmic pillars

This is to inform whom it may concern about my latest book aiming too bring justice to nature by showing how nature and not Newtonian principles form the cosmos. There are four cosmic principles that nature applies and Newtonian science ignores. Nature form the cosmos by applying the following four principles found in nature:

The Titius Bode law
The Coanda effect
The Roche limit
The Lagrangian points

Have you heard of these cosmic phenomena? These phenomena are the building blocks by which and on which the entirety of the Universe forms and yet science in the Newtonian fashion ignore not only the importance but also the phenomena having any role in physics. While Newtonian science ignores these natural phenomena because it annihilates the way Newton saw the cosmos, we have a scenario where we can apply only one of the two options that are available which can be correct. We can forcibly conspire to make Newtonian principles apply as is the current teaching practise notwithstanding everything pointing to this being wrong or we can look at nature and find the correct way in which the Universe works.

Nature and Newton is not compatible even in the least. We have to distrust science or we have to distrust nature.
I have written many books, each formulated to serve a different intellectual requirement in understanding science. This specific book present the way of how nature forms the solar system and therefore how the Universe forms. This puts Newtonian science flat and corrupted. In Nature mass has no role
You now can for the first time in human history choose to read the truth about the cosmos. For the first time nature is the teacher how the cosmos works in its dictating science.

Synopsis

Everything in the Universe is round. Anything that is round has to apply the value of Π. This is a fact of mathematics but while Newtonian science forever tells the Universe to have "mass" and to use "mass" nowhere in science would one find Π used in prominence. Whatever you may study in astrophysics, go where you wish but never would you find Newtonian science taking the fact of Π into any prominence. When you read any of my books you will see that gravity forms by movement applying Π as a value. I have found the four phenomena that put Π in astrophysics. By valuing gravity as Π therefore the Universe consists of gravity that forms by the working of the four phenomena that Newtonian science hardly ever mention but dubbed "a freak of nature". Science pretends to search for planets far outside our solar system and all the while science has no idea why our solar system functions in the manner that it does. Would it not be much better to first find out how nature applies in forming our solar system before trying to pretend to know about other thought to be planets? Science ignores nature and pretends to know much more while all of science know so little they have not figured out how gravity works…and that I prove! When you read this book you will know much more about the solar system than what the Super-Educated ever knew about science.

The Titius Bode law: The Titius Bode law has been around for centuries and with all the mathematical splendour available there for all to use, all the brilliant mathematicians could never come close to show any ability of understanding any of this very important phenomena. They could mathematically equate the formula the sequence applying as the formula, but then after that their superior human intellect dries up as they hide behind worthless equations.

The Roche limit: The Roche limit has been around for centuries and with all the mathematical splendour available to apply in order to fathom concepts behind this phenomenon, still with all the computing ability of a machine all those physicists with all the mathematical superiority could not touch any understanding about the concept forming the background. Yet when using the truth about gravity in physics the answer is simple; it is that gravity is Π.

The Lagrangian points system

The Lagrangian points have been known to science for centuries and with all the mathematical splendour available not one calculation could ever explain why this event is taking place.

The Coanda effect

The Coanda effec has powered turbine engines and aeroplanes in flight for almost a century and with all the mathematical splendour available to design the most terrific aircraft, not one engineer could mathematically compute one fact to show understanding why this takes place. How sad it is that those claiming of much superior intellect in physics remain just no more than having computing power. The understanding is not complex. I have to warn the readers that the topics are showing a very new approach with no quick answers. Understanding is in the proof and that does not come by reading just a few lines and then forming conclusions. The information is new but not hard to grasp. I did not put these phenomena in place and these phenomena nullifies Newton's correctness and that proof I bring goes beyond any doubt. I prove the Titius Bode law. I have published the Titius Bode law in four already published books but in this one I go deeper than the four already published.

This book clears up many questions that have never been answered such as: where is the centre of the Universe?

Have you ever thought about where to find the centre of the Universe? I found the centre of the Universe and this is no joke! Almost my entire lifetime (since I was five) I thought about this question and after decades of studying astrophysics I can prove where the centre of the Universe is. Does science applaud me...hell no because I prove Ptolemy was more correct than what science will ever dare to admit...well more accurate than Newton. My finding the facts Kepler introduced put their findings in jeopardy! To find out where you can locate the centre of the Universe then read the "more Information" part and you will be well on your way to locate the centre of the Universe.

Look at the cosmic picture that the Universe represent you with and see your role you have within this large concept. Look at every spot of light you can see and see where your position is to where you stand in all of this. Something mighty big is eluding everybody within this picture, something that all the wise in the past missed and the entire mathematical brilliant in the present never came to witness. The largest issue of all issues is to figure out where is the centre of the Universe. Where is the point where everything in the Universe comes together to join in one single spot?

Every light photon come across the sky from all regions there is within the Universe and they all meet at the spot where you are as the viewer of this spectacle. Where you stand in the picture every photon that came from everywhere rushed towards you at the speed of light and not one photon coming from any spot ever missed you at the point where you stand. That places you as the onlooker right in the centre of the Universe.

WHOM IT MAY CONCERN,

I am P.S.J Schutte, nicknamed Peet. Being a white South African my mother tongue is Afrikaans and my second language is English. I have per suiting a new cosmic theory that I partly present in a six part theses, of which the investigating research began in 1977. First I located what was wrong in physics. I compiled my presentation of The theses called The Absolute Relevancy of Singularity and then six separate thesis parts forming the theses published through LULU.com which I saw as way the only manner whereby I could generate funding by which I would be able to have the thirty seven or so books I already wrote linguistically edited and then to have the books published on a Print-On-Demand basis. I compiled a new cosmic theory by which I eliminated all the incorrectness that Newton has burdened science with but with this being my opinion I did not find a garage full of academics supporters waiting to applaud me and to uphold my views on the matter. Yet still I was not going to be ambushed by their relentless stonewalling my efforts and blocking my efforts in introducing both the incorrectness and the new cosmic theorem I concluded. Their blocking convinced me about a Conspiracy in Science in Progress and this spurred me on to tell the entire world about their brainwashing of the minds of students.

This kept me busy for the past going on to twelve years on full time basis whereby I was trying to introduce my findings to many academics without finding much joy from my efforts. This past almost thirteen years plus saw me go without any income as I tried to get my theorem recognised as well as get my warning noted. Going without a steady income left me almost destitute and in order to find a manner to get my theory across to the attention of influential readers, I decided to publish a theses of six books electronically as to try and get around the stranglehold of Newtonian bias controlling science at present worldwide. I decided to publish electronically which those in power do not control.

It is said that when any person is capable of understanding 500 words in any language such a person is able to converse in that language. Well I have not counted the exact number of words that I can use to express myself in English but my guess is that it should be close to 495 words that I have mastered in English and so by reading this book you then would be a witness to this. With my first language not English and the books not linguistically checked by an expert there are bound to be language errors that readers will notice. In the past I tried to check my work myself but after checking say one hundred and fifty pages for language corrections, then after days of toiling instead of having corrected work I ended having four hundred pages of newly written information which is still not linguistically corrected but holds a lot more information. The language and spelling errors compiled instead of reduced. This is because my priorities lie elsewhere. I aim to spend money on correcting the work as far as language goes, as I receive money in the selling of my theses and in the hope that I will receive money. I will have all my work including the one you are reading edited professionally and corrected as I find money to do so...But first I have to get the public aware of the problem to get the academics to appreciate the problem.

This is my introduction and this is my prologue: But before I can commence with that task I have another duty to administer: I am about to warn every person in sight of my work about my preserved slender abilities...pointing out weaknesses seen by experts and it is in response to these acquisitions about my approach not matching the normal norm and by me not conforming to the current accepted norms I am expected to confirm their believes never taken with equanimity. The current norms are correct. Any deviation from such thoughts is unacceptable.

Therefore in the light of what the most respected academic group on Earth accuses me of, I therefore have to issue a most serious warning to any person with the intention of making some kind of inquiry to the content this book holds, then the most concerning matter involving any content within the pages of this book you inquire to acquire that you must please seriously consider that where the stating declares the possibility that the content in this book has been (written by...) then don't take the announcing Written By Peet Schutte (Petrus S. J. Schutte) very seriously for there are grievous doubts leaving considerable dispute about the possibility, which underwrites the authenticity of Peet Schutte achieving the (written by...Peet Schutte) status. Please take note of the following dehortation. In the light of the reference to me serving in the capacity as being responsible for authoring, (written by...) in line of keeping fairness and justice to members of society, where all civil beings should carry reputed honesty, then: Please be warned before any reader starts reading about the following extremely serious admonition: I am bound by my conscience to warn all intended readers that I am placed under caution by the Academics in Physics. Those most esteemed members responsible for the guardianship and maintaining the ethos in physics are of the opinion that I, Peet Schutte, am unable to write any book on the science of Physics as well as Astrophysics. Therefore I, Peet Schutte, must declare that I should be considered as not very able to write anything, because I am incapable thereof. I suppose, I merely generate new information, which I establish as thoughts and then gather as concepts. I further collect the result as words, which I put on paper using alphabetic symbols. I then compile that in a format that others may confuse with a book, but a book it cannot be, since the Masters in science found me unable to write a book. However, as you are about to see in due course that I am warned by the most esteemed academics in physics that I cannot use words to describe physics. But before you go further and follow my arguments, I first have to level with you about how academics view me in the position I hold. Please do not allow me to fool you, for this then cannot be, or represent a book because I use words and words are what Astrophysicists don't use because then they might detect their stupidity.

Science went corrupt in 1705 when Edmund Halley told the world he used his friend Isaac Newton's physics formula to calculate the route and time that the comet that was named after him would arrive. This was where science went crooked and started to corrupt science, a position that went on ever since because that same dishonesty is still present in Newtonians science. Halley calculated the time periods since 1066 at the battle of Hastings and found a comet was mentioned every seventy-six years. This was very ordinary for a man of his class so he had to get far cleverer than backdate history to get a time frame. So he really got clever and conspired with the biggest fraud in science ever since; the man hat stole all physics Doctor Hooks invented, the man that even got Kepler' figures wrong the man called Isaac Newton.

If Halley was honest about tracing the arrival of a comet that was mentioned ever seventy – six years then Halley was no more than bloody ordinary and hat Halley could never be. So to look smart Halley said he used the formula of Newton to calculate the rout the comet took. This says he used mass to calculate how the comet came to the sun. That says the sun's mass pulled the comet and the comet's mass pulled right back and this way the comet came to the sun. I don't go into the comet as such in this book but I do in other books. In this book I show how Newtonian science started to go corrupt in 1705 with one conspiracy to cheat and became the corrupt myth it now developed into. How do I know Halley did not use the mass pull mass idea because if he did then how did he calculate that the comet was cyclic or that it returns every seventy-six years. If mass pulled the comet to the sun what then pushed to comet back into outer space?

His big ambition was to prove the comet comes and goes but if mass makes the comet come what pushes the comet back. You know what is the biggest fraud that came to be called Newtonian science? The most brilliant minds on earth this past three hundred years failed to asks this simple question: if mass pulls the comet closer what pushes the comet away? If mass forms the force of pulling and pulled the comet closer than what pushed the comet back into the darkness of the beyond. How did he know the mass of Halley's comet? Nobody then asked questions. No one asks uneasy question…except I. I show the fake science we have by just questioning science in search of the truth. Newton and Halley got away with corrupt science. Today Newtonians get away with corrupt science. Then those in science question my integrity because I question the integrity of those in physics and in this book you read how a bag of stinking shit flies into the faces of the most holly, the most intellectual mind the world ever produced. Brainwashed as you might be just please try to understand this if you can't understand anything else; those that are too stupid or too brainwashed to see how and that Halley cheated to make him look good and to falsify the truth so that Newton could falsify physics and to make Newton a genius of all times, which is just what Newton became after this dishonesty, then don't by this book because you are brainwashed beyond recovery. I don't want to steel the money of a capitulated mindless person.

However, if you think I am about to praise the stern abilities that the science known as Physics hold while claiming their honesty and purity then you came to the wrong website. To be further mislead go elsewhere. If you think you are about to read an ode to the honour and the ability of the total genius Newton in what his work represents then you better start to think again. If you think this is an endorsement for the incorruptibility of science and the absolute correctness science portrays throughout, then you are in for a shock. I am not going to bullshit you about the sincerity of science or the honour that scientists hold above all else because the public is disillusioned enough with all the sanctimonious garbage science claims. They've been leading the public down the garden path for far too long and it is time to come clean. If you believe in the unwavering honesty of scientists then either brave yourself or page on because if you carry on reading you are going to be very disillusioned about what you are about to learn.

We see their achievements and we gasp breath in admiration. With all the amazing achievements accounted for and when recognising all the things with which that science changed our way of living on the earth and what was achieved by scientists developing this super mentality and in that also giving science all the admiration dually admitted, notwithstanding I am about to dump on you the biggest conspiracy that has ever been presented and that was ever undertaken by any group of persons in the entire human race. Think of anything you might think is big or outlandish by nature and that dwarfs in comparison to what I am about to reveal. If you studied science then your surprise will be that harsher.

It is so large that there is nothing in the past history of man with which one could compare this revelation that I reveal. It involves every aspect of all aspects of the life of every human being and this shadow hanging in our midst covers the darkest secret that was ever hidden from intellectual human view. Those we absolutely unconditionally trust in all aspects perpetrate it. It touched on every individual walking the surface of the earth and that excludes no person of any status albeit it an infant or someone in old age.

According to the Super—Educated I didn't write any books of these books since according to those Masters – of Physics I am not schooled to do so. They are so educated that they apparently are wise enough to judge my work without ever reading my work and then seem fit enough to form an opinion without insight about my work's content. Mot one academic in fifteen years took time to read any of my work. It is my guess that I merely generated uninformed thoughts, which I collected as alphabetic symbols and plotted that in ink on paper. This effort I achieved from harbouring my delusional ideas spawned by a dehumanised brain. It only proves my weak and under developed mentality, due to my lack of an informed insight that is a typical symptom that all those have that is suffering from a disadvantaged past that one can only have when the person obviously lacks a Newtonian opinion. While you are reading the letter deciding to regard or dismiss my work, then also please keep in mind when reading my language used and also please give credit where it belongs…if you do find linguistically improper use of words or misspelling, then remember that I am a feeble minded because I disagree with Newton and not a literal giant. Now I have done my duty in warning everyone and in that, I denounce further participating with any purposive intention to wilfully bring down the crux of civilization by acting unacceptable and irresponsible.

I compiled a new cosmic theory by which I eliminated all the incorrectness that Newton has burdened science with but with this being my opinion I did not find a garage full of academics supporters waiting to applaud me and to uphold my views on the matter. Yet still I was not going to be ambushed by their relentless stonewalling my efforts and blocking my efforts in introducing both the incorrectness and the new cosmic theorem I concluded. Their mannerism in blocking and frustrating my opinion when showing the mistakes in science convinced me about a Conspiracy in Science in Progress and this spurred me on to tell the entire world about their brainwashing students minds. By the manner they selectively withhold information when teaching science, amounts to deliberate brainwashing of students in physics by "normal" education practises.

Trying to convey my message kept me busy for the past going on to twelve years on full time basis whereby I was trying to introduce my findings to many academics without finding much joy from my efforts. This past eleven years plus saw me go without any income as I tried to get my theorem recognised as well as get my warning noted.

Going without a steady income left me almost destitute and in order to find a manner to get my theory across to the attention of influential readers, I decided to publish a theses of six books electronically as to try and get around the stranglehold of Newtonian bias controlling science at present worldwide. I decided to publish electronically which those in power do not control. However to get people to believe me is to change science that everyone believes as culture.

With my first language not English and the books not linguistically checked by an expert there are bound to be language errors that readers will notice. In the past I tried to check my work myself but after checking say one hundred and fifty pages for language corrections, then after days of toiling instead of having corrected work I ended having four hundred pages of newly written information which is still not linguistically corrected but holds a lot more information. The language and spelling errors compiled instead of reduced. This is because my priorities lie elsewhere. I aim to spend money on correcting the work as far as language goes, as I receive money in the selling of my theses and in the hope that I will receive money. I will have all my work including the one you are reading edited professionally and corrected as I find money to do so…But first I have to get the public aware of the problem to get the academics to appreciate the problem. In everyone's mind science is more perfect than religion is.

I discovered the building blocks of nature where my discovery puts all other cosmic aspects of science into science fiction. Those who force-feed non-existing dogma do so to brainwash students to hide the incompetence of "modern science" so they can rule supreme while ignoring the truth that they deliberately hide by concocting a conspiracy. To keep everyone unguarded they practise a conspiracy by which they perform an accepted practise of thought control on students to further the false dogma presently in place. I try to blow the whistle on such a practise but accepting my resolution makes every thesis ever written science fiction. Therefore no one in science dare to read my work leave alone appreciates the revolutionary nature thereof. Whatever now is deemed to be accepted science would then become what is the past tense in science because the flaws that those in power of science principles kept coated for centuries on end as untouchable truth will then be rust that breaks the surface to show the holes! They try to silence me but surely somehow somewhere I have to break through with my massage! I bring you a true form of science as never seen before in all of history and I do that when I dispose of the conspiracy that hides all the incorrectness and the failures that haunts science today. Science is accepted as the most righteous information available to man and that is a scam. It is not what science declares that is important but it is always what scientists don't declare that holds prominence and more so the reason why science keeps a silence about the information they do not disclose. It is never about what they say but it is why they don't say other things they keep quiet about. You will read how they never disclose the entire truth because science is about promoting one-sided and selectively opinionated information forming fraud no less. I have been per suiting a new cosmic theory that I partly present in a six part theses, of which the investigating research began in 1977.

I do find much pride in my status as being Afrikaner and would like to have my names used by pronouncing it in the manner Afrikaans dictates...therefore I would sincerely appreciate the courtesy when readers will take note that my name and last name are pronounced in Afrikaans, which is originally from Dutch and must be pronounced that way. Peet one would pronounce "here" which is the closest English to the pronouncing of the "ee". The "Sch" in Schutte is pronounced exactly as school is where both actually are pronounced Skutte or "skool". By pronouncing my name in Afrikaans you do me the utmost courtesy any one can. Being an Afrikaner is what I am most proud of. I submit article to well known physics magazines but my articles are rejected on the most unappeasable grounds and for the most outrageously ridiculous reasons the Newtonians can think of. I explain how gravity forms but I am rejected because they are of the opinion that my work does not meet. One such an article I may use because I said I was going to use the material as an open letter I gladly show. I submitted an article in which I show what the manner is in which gravity conducts movement by means of singularity. I wish to produce as evidence e- mail response I received when I contacted members of the Physics Academic establishment and show my case while also showing the response I received. Readers will find witness of what I accuse science of and their demeaning attitude from physicist wishing to silence me at any cost.

Have you, the person reading this, ever thought how it is possible to see that much information that you see at night when looking at the sky. Ever thought about how you are able to see when you see everything in the night sky and how that much light information can fit into such a small space as your eye? Have you ever sat back and think what the amount of information it is that you see when you see the entirety of the Universe when looking at the Universe at night and what the size is of everything of that which you are able to see? The one star you see seems to be a near visible dot in the picture while the dot might be hundreds or might even be many thousands of times the size of the sun...and we think of the sun as big. The dot is then that much bigger than the sun because the star we think we see could be a galactica hundreds of times the size of our Milky Way galactica but that shows as in the sky as one little dot and yet that entire structure as big as it is, does also fit into our eye socket. But that is not all...there are trillions of such light images and they all fit into one eye socket. What we see is immeasurable and yet we see it effortlessly in the space our eye holds...how can that be? How is it possible to fit what we see into the space of our eyes we have? Think how much is the entire information that is visible at night and think about how all of that fit into the space your eye holds? Consider how big is what is visible and put that space into the size is of what your eye can hold and ask your mathematically educated Professor in physics to find some ratio between what you observe and the size of your eye. The ratio is astonishing, but more-over what is truly astonishing is the arrogance of man to think of his position, as being important while the space man holds is beyond any comparison in ratio to everything we see in the Universe we see.

Think how small we are when we are able to see the entirety out there! Even if there was other life out there, what is the worth of it in comparison to what there is that we see?

It is said that when any person is capable of understanding 500 words in any language such a person is able to converse in that language. Well I have not counted the exact number of words that I can use to express myself in English but my guess is that it should be close to 495 words that I have mastered in English and so by reading this book you then would be a witness to this. With my first language not English and the books not linguistically checked by an expert there are bound to be language errors that readers will notice. In the past I tried to check my work myself but after checking say one hundred and fifty pages for language corrections, then after days of toiling instead of having corrected work I ended having four hundred pages of newly written information which is still not linguistically corrected but holds a lot more information. The language and spelling errors compiled instead of reduced. This is because my priorities lie elsewhere. I aim to spend money on correcting the work as far as language goes, as I receive money in the selling of my theses and in the hope that I will receive money. I will have all my work including the one you are reading edited professionally and corrected as I find money to do so...But first I have to get the public aware of the problem to get the academics to appreciate the problem.

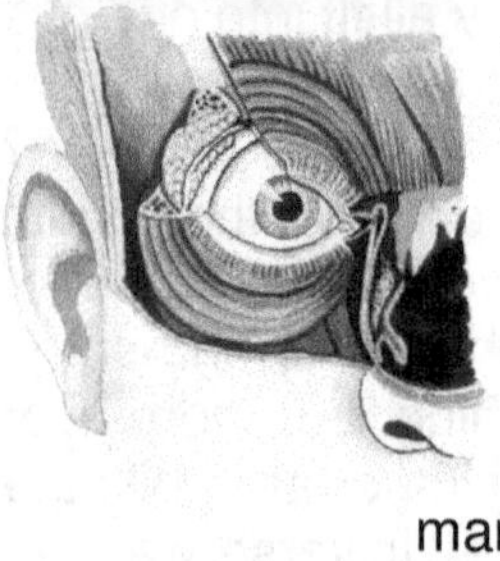

Take every speck to represent a galactica and fit that galactica being the size it holds into you eye. Then go one better. Fit all the space in this picture whether it is light or not and fit that volume of space into your eye and explain how it fits into your eye when you see the vastness outer space holds in the volume of space.

Take every speck to represent a galactica and fit that galactica being the size it holds into you eye. Then go one better. Fit all the space in this picture whether it is light or not and fit that volume of space into your eye and explain how it fits into your eye when you see the vastness outer space holds in the volume of space. In order to understand my presentation of physics go out and look at the night sky and witness the many stars you are able to see in a clear night at an unpolluted spot out in the country. Look at the picture above and see how many dots represents a star or a galactica or even many galactica for all we know every dot and every spot represents a vastness filled with material in space that are bigger and much bigger and enormously bigger than what the sun is that we see during the day. In our small view we have of the Universe the sun is enormous but the sun is so small it is a yellow midget. In context of the picture above one would never spot the sun in the picture because the sun is far too small. Think how big the light dots are when considering each speck might represent a galactica that is even much bigger than our Milky Way is. This is physics in cosmology but physics can't even begin to fathom this notion.

Now comes the question: how can all the light you see fit into an area as small as what your eye is that you use to see with? Can you explain how all this light can fit into the eye by which you see? Better still explain mathematically how does all the light coming from the vastness of outer space that includes thousands of trillions of stars and billions of galactica all then unite into one spot in your eye forming a picture that your eye tells you is the Universe you see. How can you see the entirety representing the picture in 3D as a complete unit? Think of the space the eye uses through which to see and all that space that is seen comes through a port in the eye no bigger than an electron and yet it is big enough to allow all the light coming from the entire Universe to be collected at one point that meets something as small as your eye. Don't let this thought just pass you by but consider it with the devotion it requires because behind this answer is the true foundation of physics.

This is where physics starts and not with some mathematical formula someone devised without having an afterthought or an appreciation about what forms the entire picture where it involves physics. If you are substantially schooled in physics then give me the formula that would mathematically explain how all the light coming from space as vast as I would never be able to appreciate when using something as small as the human mind can fit into something even much smaller like my human eye and then formulate this and let us appreciate your substantial mathematical ability.

See all the light coming from all over rushing to the point you fill in the cosmos and explain that with mathematics.

I take physics into a Universe that was in place before light was in place. I introduce the Universe when only darkness prevailed because light calls for space and in that era of singularity I introduce, space was not even a thought yet. I show why the Universe goes "flat" and in a "flat" Universe only the value of 1 holds measure since singularity is 1. If you can understand 1 or $5^0 x 7^0 x 3^0 = 1^1$ you have all the mathematical skills required to understand the applying concepts. To reach a value of 1 does not require big mathematical equations but to reach singularity requires 1. The collection I named **The Absolute Relevancy of Singularity**: The Theses and the collection as such forms a small introduction to the thirty-two or so books I wrote on various matters concerning physics with gravity in mind, but **The Theses** as such in the entirety of the eight books does not officially even start to introduce the spectrum of every aspect of my work.

Let's run through this again because this statement underwrites physics as much as it disperses astrophysics seen by Newtonian science. This is the point where physics starts and no mathematics can ever begin to exercise what the understanding is behind the concept. The picture below is one small part of what we know is the total entirety of the cosmos that we see.

Every speck of light is many times over the size of our sun. A star the size of the sun would not even feature as a shining speck between every galactica that is shining on this picture.

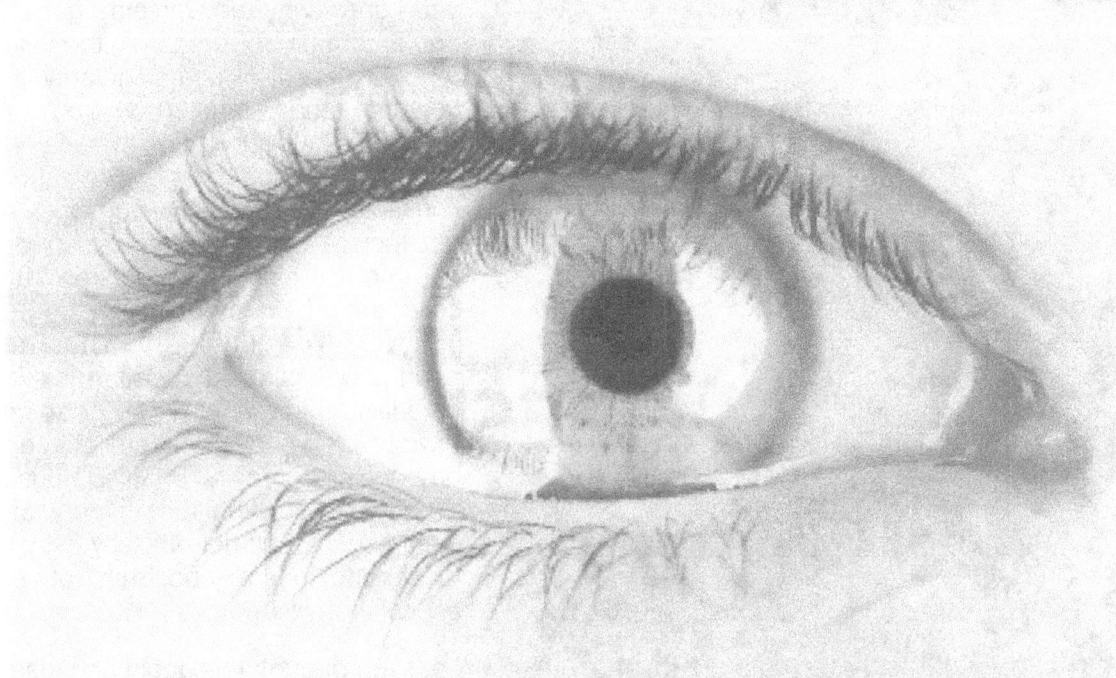

Yet the entire vastness of space we are able to see having all that space coming through the size of a nerve ending. Think of the volumetric space that is evident in this picture alone and try to realize the fact that that immeasurable volume of space held as information can be going through a nerve the size we have in our eye.

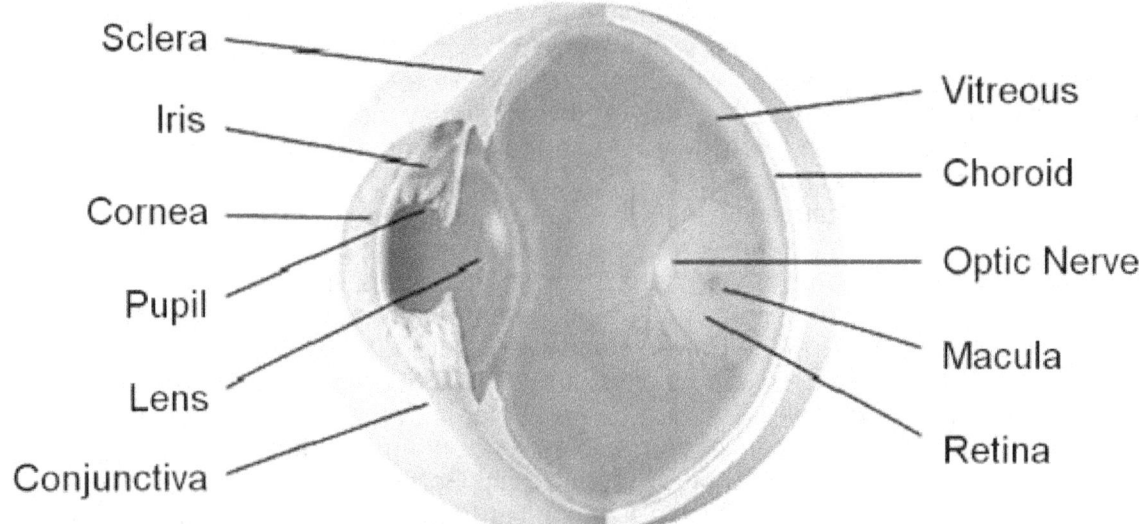

Sclera
Iris
Cornea
Pupil
Lens
Conjunctiva

Vitreous
Choroid
Optic Nerve
Macula
Retina

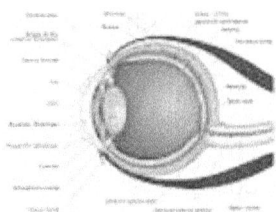

Then go one further…think why would all the light coming from everywhere come straight to you, the onlooker! How is it possible that light that left the original position travelled in some cases 12 billion years come to find you who are looking at the sky? Why did the light cross that vastness of space to find you where you are looking at the night sky? Why did the light not simply just miss you and went past you as it disappeared into the blackness. What makes the position you hold so important that all the light comes rushing to you, precisely to the point where you are regardless of where you are? Why would not one photon miss you as you stand and stare at the night sky? What makes the position you have that important that all the light will come to you from all over the Universe and from wherever the Universe has space and moreover rush to you at the speed of light. Not one photon misses you standing and staring at the night because you see with your eyes the full picture out there. What puts you in the place where all the light coming from all over then will meet at that point? Why would the entirety of all light gather where you are and not one photon misses you? If the light did not meet and cross at the very point you are, the light would miss you by far, and it does not!

This following concept forms the entire basis of everything forming anything in physics that is part of science as the cosmos presents science. Around this following thought and the previous thought I entrusted you with about light fitting into your eyes socket forms the reality on which I base my entire New Concept about physics. If physics as it is presented in the present can't prove these two thoughts then physics is as far from reality as madness can be from sanity. If ever any thought represented physics then this is the most fundamental start of physics. Go out once again and gaze into the night sky. Look at the cosmic picture that the Universe represent you with and see your role you have within this large concept. Look at every spot of light you can see and see where your position is to where you stand in all of this. Something mighty big is eluding everybody within this picture, something that all the wise in the past missed and the entire mathematical brilliant in the present never came to witness. The largest issue of all issues is to figure out where is the centre of the Universe. Where is the point where everything in the Universe comes together to join in one single spot? Every light photon come across the sky from all regions there is within the Universe and they all meet at the spot where you are as the viewer of this spectacle. Where you stand in the picture every photon that came from everywhere rushed towards you at the speed of light and not one photon coming from any spot ever missed you at the point where you stand. That places you as the onlooker right in the centre of the Universe. If all the light crosses at the point you fill then all the light puts you in the centre of the Universe. Not one speck of light could dare to miss the point you fill and that makes you standing on Earth the centre of the Universe. Is this wrong to think that way? …No hell it is the Universe that forces you to think of yourself in terms of filling the centre of the Universe!

We take so much of the characteristics of light for granted while never thinking for one second how impossible our relation with light truly is considering our position we have in relation to the entirety we think of as the Universe. This totally extraordinary relation we have with light must be one of the reasons why we humans put our position we have in the Universe in such a pivotal place. The fact that we as persons which includes all those with the ability to use our hind legs to walk on Earth has this extraordinary idea that the Universe was created especially for us, us being those holding life within the human context. Such an idea is absolutely bizarre to say the least and we all are guilty of having this impression. Thinking in that way would be the same as if the ant in the park thinks a thousand people maintain Central Park in New York with one purpose and that is to please that one ant. ...And yet that is what is happening with the light and us. Every person is standing in the Universe and all the light through out the Universe is directly flowing to the very point any and every person is standing. It happens to all of us. The place where I stand or any other individual for that matter is standing is positioned in such a manner that every beam is directly flowing to that very specific spot. From all the corners of the Universe and every individual spot in the Universe there is one line of light that is especially directed to that location. The light departed from every atom in all locations and from there it is following one direction and that is to the spot where I am filling that spot. All the light in the Universe is coming to me to surrender to me at that location. It is coming straight to me where I am standing filling one spot on Earth. Go outside and see the vastness the light is coming from. It is coming from all over. It is coming from areas so large not even Einstein can calculate the size it has because our telescopes can't reach that far and it is rushing towards me specifically. There is not one ray that is going to miss me by fluke or accident. The light has one purpose and that is to meet me at the point in which I am. Every beam has my name on it and it is coming to meet my eyes. Can any one imagine if a person was standing in a location and found all the persons in that city was running towards him where he is occupying that point, how frightening such a person must feel. Yet it is happening to every one from wherever the vastness of space is situated and is coming across space to that very specific point the viewer is standing. Even if I shift to another position the light will change direction and trace me standing in my new location. Even if my new location is on the other side of the Earth, the light will still get me at that location while tracing me all the way. The light flows to me from where ever and to top that it is also flowing to all other persons during the same instant. That means it is not the Earth that is that important but it is where the location is and that point the observer is using to view from that is that important. If it was only the sun that the light was streaming from that is choosing me as the centre of the Universe it then cannot be that very exclusive. The sun is close and the light is plenty. But it is coming from all over and everywhere excluding no point at all.

That is just one small part of the fantastic affair. Some of the light left the stations they come from some 12×10^9 years ago to meet little old me in this spot I am filling. The light has been travelling 12×10^9 at the speed of light, which I might add is much before my birth crossing space and time, rushing all the way to meet me at this point. No one ever thinks how was it possible for the light to know I was going to stand at this point and wait for the light to arrive but nonetheless it finds me after all the time lapsed. How did the light know I was going to take centre stage at that moment and fill the specific centre of the entire Universe? I have to be in the centre of the Universe because all the light is travelling to this spot filling the centre of the Universe. The light sometimes takes many, many million years and that is when it is coming only from the closest next galactica to meet me here and after all the time it is meeting me in the centre of the Universe. Regarding me in such a position then how important can I ever be when filling the centre of the Universe? Light is coming across time measured in millions and billons of years through space measured in millions of trillions of Yottametres, ignoring all other places it could go to and yet every beam of light came to meet me regarding me as being in the centre of the Universe.

To the light on route time means nothing and space even less. Still light cannot be more motivated to reach me at this point that I am filling at this moment in time. Not one ray is by accident missing me except by my choice prevailing. It flows through the Universe in time and in space in the hope I whom is filling this spot, the spot all light is coming too is graceful enough to notice the light. If I were not in the mood to acknowledge the light, the light would have done all the travelling just to be disappointed by my not meeting the light. An effort sometimes spanning billions of years and an effort stretching trillions of mega kilometres was all in vain because I neglected to acknowledge the arrival of the light. From everywhere the light is coming my way and that miracle is passing me by because I am feeling even more important than as to acknowledge the total importance I have. The light is tracing me specifically at the location I am occupying just to please me and serve me with all the information about the history of the Universe. I can accept and acknowledge the effort or I can dismiss and ignore the lights efforts.

I suppose that will allow me some arrogance and encourage me to think this all was specially created with only me in mind and if I wish to draw a map about the Universe I have all the right in the world (or is it the Universe) to place me in the centre of the Universe from where the all of the everything is meeting. After all, all the light is doing it and so the Universe is telling me it is correct! We know Americans are filling the centre of the Universe but even those not being American has the idea that they are special because they fit bang into the centre of the Universe and all the while it is the Universe telling you and proving to you that you are in the centre of the Universe. Go on and mathematically formulate this truth. Those considering their wits as being superior and above all others go on and mathematically prove how this concept becomes the truth above any criteria they might produce as the truth. These concepts I just mentioned are the realities forming the concepts that rule the Universe and I am able to formulate these concepts mathematically. But as I introduce and formulate the concepts I introduce I show incorrectness in science that all those practising science try to hide because they conceal the anomalies they can't explain …and so they hide the biggest corruption in science.

Take every speck to represent a galactica and fit that galactica being the size it holds into you eye. Then go one better. Fit all the space in this picture whether it is light or not and fit that volume of space into your eye and explain how it fits into your eye when you see the vastness outer space holds in the volume of space. In order to understand my presentation of physics go out and look at the night sky and witness the many stars you are able to see in a clear night at an unpolluted spot out in the country. Look at the picture above and see how many dots represents a star or a galactica or even many galactica for all we know every dot and every spot represents a vastness filled with material in space that are bigger and much bigger and enormously bigger than what the sun is that we see during the day. In our small view we have of the Universe the sun is enormous but the sun is so small it is a yellow midget. In context of the picture above one would never spot the sun in the picture because the sun is far too small. Think how big the light dots are when considering each speck might represent a galactica that is even much bigger than our Milky Way is. This is physics in cosmology but physics can't even begin to fathom this notion.

Now comes the question: how can all the light you see fit into an area as small as what your eye is that you use to see with? Can you explain how all this light can fit into the eye by which you see? Better still explain mathematically how does all the light coming from the vastness of outer space that includes thousands of trillions of stars and billions of galactica all then unite into one spot in your eye forming a picture that your eye tells you is the Universe you see. How can you see the entirety representing the picture in 3D as a complete unit? Think of the space the eye uses through which to see and all that space that is seen comes through a port in the eye no bigger than an electron and yet it is big enough to allow all the light coming from the entire Universe to be collected at one point that meets something as small as your eye. Don't let this thought just pass you by but consider it with the devotion it requires because behind this answer is the true foundation of physics. **Please be aware that I was told so many times that I am not able to _"understand Newton's classical mechanics"_ and therefore I am not able to _understand Newton_ and my inability of _not understanding Newton_ is the result of my compromised outlook on physics because of stupidity. Please gauge me either as you see me by the facts I present or as they see me but then use the facts I present and not physics based on culture taught over millennium without any proof.**

What you are about to read holds the dynamics that would change physics for all time to come. For the first time you will learn what gravity is as you would learn what singularity is as much as you will learn how to venture into a Universe that holds what there is together without mass, but all pulling goes by singularity arresting a bonded "flat" Universe in a state of gravity manipulating singularity. It is singularity holding the Universe together by moving time through space.

Gravity arrests the Universe, but no one ever managed to find the way it does arrest the Universe. If you read on you are about to find out how. I am taking the reader into a cosmos that holds a maximum value of 1 and anything greater than 1 does not fit into the Universe you are about to enter. Al the mind-boggling formulas used to impress has no meaning in singularity or in 1.

The Universe you are about to enter doesn't rely on a mathematical computing skills but an ability requiring human intellect through reasoning and following a line of debating. It requires the skill no computer could produce because it requires intelligent understanding of issues going beyond simply calculating and drawing unconsidered conclusions that is void of any intellectual understanding of cosmic principle, such as for example space whirls. If you have an ability to think and reason and don't require some mathematical disposition to rely on to help you think, then read on but be warned, this might be the highest intellectual level you ever called on.

The first question that would springs to mind when any person reads this work is to ask why I have is attitude in the manner of how aggressive I am. What would spur on such aggression towards the faculty of science and all that administrate all the small parts, as you are about to encounter? Why would any person find it necessary to attack the establishment of physics with such aggression? Why would any body insist there is something missing in physics when physics work so well and no one but me finds physics questionable? It is possible that the entire world find physics flawless except me Peet Schutte. The idea that every human on earth could be wrong and only I, the one named Peet Schutte is correct, such an idea crosses over from mental insanity to clear madness. Madness is when a person finds the world of physics at fault and more so considering that the one person says that that person has the remedy to cure all the mistakes that only that person are able to witness. I can see why everyone thinks I am mad. But also I seem the be the only person on earth that sees through the brainwashing and mind control going on in physics. Now that I have shocked everyone into a feeling of total rejection, I challenge you to read on and in the end, see if I am mad or if I am correct…then and only then form your opinion!

Those Super-Educated in Science tells the Universe what the Universe is but never ask the Universe what the Universe is. In this idea about how you are able to see the entire Universe you will find all the answers to the questions about how physics use time to employ gravity.

Do you know where the centre of the Universe is?
By the time you have completed this book you will know where to find the centre of the Universe.

You will find that there is a conspiracy in science in progress this minute that involves everybody. You are keeping your breath awaiting the punch that makes the money because this sounds big and the punch that is coming is what rakes in the money. Now you think I am going to inform you about another conspiracy that eventually turns out to be as unfounded as all the other untruthful and baseless conspiracy theories you've encountered and discarded as another fable in place only to earn money?

Alas, this conspiracy I reveal is so big and involves such a big part of civil culture accepted to form the basis of human intellect and the foundation of what we think truth is that if I give you the tiniest hint about what it is you will not be prepared to investigate the issue further because you will discard this article and my statements as another hoax because it is too big to be believed. Up to now the overall scale that this issue implicates is so big, it's the size it captures of what social acceptance hold as basis of civilisation that keeps what this conspiracy hides part of human intellect. Believe me this is big. It's bigger than any revelation about any religious misconduct. This conspiracy has been around for as long as the British Empire was in place…

It is bigger than what you could ever imagine because the truth you regard that supports this hoax and the belief you have in that which protects this conspiracy is more rock-fast than anything else you believe in or might trust to be solid truth. All that I ask of you is just go to my website at and see what I explain in a few pages and then decide on the validity of my claims. It will take but a few minutes of your time but it will change your outlook on what to trust for as long as you may live. I show you in the website where this mother of all conspiracies started. I elaborate on this conspiracy and show how it is still in progress even today, three hundred years after it started. It began in 1705 and swept aside all resistance...

...Then it grew into a fashion and intellectual statement that evolved into the mother of all the other conspiracies in science. All conspiracies we have in place about science serves one goal and that is to divert the investigative attention away from discovering the Mother Conspiracy. It is hidden so well that in three hundred years no one that is not part of science got a sniff about what the Mother Conspiracy entails. The conspiracy wittingly or otherwise involves us all, including you and me.

We all commit to the conspiracy and in the beginning it is unwittingly but keeping a blindfold becomes an excuse not to think about what is obvious, which becomes a norm that preserves to keep the conspiracy in place as part of our education in science. By the time that those in science discover the anomaly that science hides, they also have too many years of devoted study to lose to reveal this conspiracy because they already are equipped with an education and working in science and so to protect what they have worked for, they become part of those preserving the conspiracy. Do you understand Newton...and moreover are you proud of this fact; if it is yes on one or both counts don't be proud but be very worried. If you want to know why you should be worried read on and find out.

I am proud to be the only person on earth that disagrees with Newton's science and I'm not joking and neither am I an escapee from some insanity rehabilitation institution. In 1970 I was a student that stared at a handbook while battling to make sense of Newton. I had to learn that a wheel that turns did no work and I clearly saw that was more than obvious total rubbish. If the wheel turned while it did no work then how does a wheel drive a car forward?

It is the circle that forms the straight line.

When the wheel turns the circle pushes the car in a straight line and the circle is doing the job of moving the car. The work was in the movement of the car and to drive the car. The car had to go forward and to go forward the wheel had to turn so the wheel circling was doing the work. I showed a fellow student the mistake.

By turning, the wheel lands the completed circle on another spot. The wheel has gone the circle, which pushed the car the length of the radius but in the drive train it was the circle that placed the movement of r to the car's new position. I was informed to disown and reject this concept about the truth because Newton said so and Newton is physics. Much to everyone's surprise I was not intimidated by Newton's fame.

They saw my unwillingness to agree with their teachings, as proof that I couldn't understand because I was unable to follow Newton and they therefore had the view that I was mentally handicapped because I did not understand Newton or accept Newton.

I know am the only person on earth who think that when one believes Newton's ideas, such a person then have to be part of a criminal gang that fake science and I say that. I believe only those who are the best in not getting to terms with reality and are prepared to bullshit everyone are the very ones stupid enough to understand Newton.

This thinking I not only promote in my books but also I prove this fact in all my work. Because I found out how nature works I now can give you true science as nature and not Newton form science. I studied nature and now I am able to present nature that uses pi as gravity where gravity is movement. Read all about it. This book is most affordable in e-book format to allow all persons access.

I wish to make one fact very clear. I base my work on formulating the working process of four cosmic principles in Nature. These are:

1) The Coanda effect
2) The Titius Bode law
3) The Roche limit
4) The Lagrangian points.

I did not discover these phenomena because science knows about these phenomena for a very long time and in some cases even for hundreds of years. Science knows they apply and where they apply. When science discovered or allocated missing planets they used the law applying such as the Titius Bode law from which they deducted positions that they knew in that circle according to the planetary layout that the law predicts there had to be a planet according to the law. Science did not apply Newton's formula to discover and locate planets but they applied these phenomena and especially applied the law of planetary allocation to discover the precise location the planets discovered after Galileo.

Every one in science knows these phenomena is there and is in place and they rule the orbit set-up of the planets. The solar system functions according to them. These four laws on planetary motion that is used by nature at this moment and has been in place since time began are what apply and they dismiss Newton. If you argue with me about Newton being correct you better take your case to God or the solar system because the four cosmic phenomena is working in nature and nothing Newton said is applying in nature. This is a truth and a fact and a foregone conclusion and can never to be in doubt.

The phenomena are what we find to be used in the cosmos while Newton is in the imagination part of the minds of scientists and nowhere else. If you don't believe me and if you wish to discredit me first find out a little more about science. Then deflate your ego as to what you think you know.

Science never mentions these phenomena because science can't use Newton and explain these phenomena or use these phenomena to prove Newton. These four phenomena that the cosmos uses as we speak have been in place ever since the Universe formed. Since science can't explain the phenomena and the phenomena destroy the credibility of Newton science avoids these phenomena as if it brought the plague.

You can't choose between Newton and me because I did not put the cosmic phenomena in place. All I did was doing a study since 1977 to formulate how and why these phenomena work and how these phenomena keep the cosmos and the solar system working. I am the first in history to show why they work. We are all been brainwashed for centuries to believe Newton. Should your brainwashing kick in and you have an axe to grind with me about what I say, then first prove these phenomena are not in the cosmos and are not applying to form the laws that the cosmos put in place as gravity. I only found out how they work and why they work and I did not make the phenomena work. All those clever stooges that have so much to say even before you read first learn what is in place before getting so opinionated.

I am not fighting science or the credibility of Newton or what might be true or not true but I am fighting centuries of brainwashing and I have to dismiss the brainwashing and the systematic mind control that those teaching science inflicted on us all. My fight is not about what is true or not true but what is accepted as culture and which was not even once been proven by science. Please do read on to investigate before becoming self-opinionated.

Please be warned about aspects of this book some may find offensive. Apparently in the following accusations aimed at scientists I use degrading language associated with persons of crude behaviour but the thoughts are not that petty. With this book I aim to address the public that are not well developed in all the facts just as I was a student but after decades of study I decided that it was time I placed these thoughts by pen on paper, I do this to try and find a means to reach the young minds with young thoughts and from humble and crude ideas I developed a new Cosmic Concept I wish to share. I wish the public to realise what science is hiding under a cover-up conspiracy. I thought it was time to lift the cover about information that I came to realise years ago when I started out thinking. I leave the language unchanged and I know too well that in the eyes of the Most Respected Physicists that the language, grammar phonetics I use is unsophisticated.

To the Highly esteemed well-to-do Academics in Physics all this information I provide in these statements and concept also comes across a well below a standard, which those well-educated and highly developed Newtonians are used too, but also the way they frown on anything regarding me. I am by now use to their frowning upon me.

Hey, for you members of the public I have books written that are coming to press, which they will not want you to read...

This crudeness of my unwaveringly impoliteness is distasteful too the highly educated members of the physics establishment and my writing also lacks every bit of linguistic sophistication and terminology they use to avoid the truth. I saw what no one else sees. I see misconduct that is rife in physics. I accomplished what I aimed to achieve solving misguided concepts and that I produced as I intended when I started out finding a path amongst the maize of corrupted science by using arguments that Newtonians never even bothered before to give a thought. If the language is poor, to that I will admit but the ideas surpasses even the best Newtonian mind because where it ventures as crude as it is according to the uttermost sophisticated members of science, and yet where I am no Newtonian brain even realised the ability to go there.

Dear Mr. Schutte

Having browsed through the cover letter and the attached files, my colleague Alexander Grossman, who is the Editorial Director for our Physics and Maths publishing list, would like to suggest that you submit your work to a journal rather than further considering the idea of a book. This is the usual way in science for new theories or concepts and he is confident you will receive constructive feedback from the reviewers.

He suggests that you submit your work as an original paper to our physics journal 'Annalen der Physik' (www.ann-phys.org)? You will then receive feedback from the Editor in Chief or the reviewers.
I trust this is helpful.
With best wishes
Lesley

Lesley Smith
PA to Mike Davis, Managing Director
Wiley-Blackwell Life Sciences
John Wiley & Sons Ltd
The Atrium
Southern Gate
Chichester
West Sussex
PO19 8SQ

Wiley Bicentennial: Knowledge for Generations
1807 - 2007
Tel: +44 (0) 1243 770110

John Wiley & Sons Limited is a private limited company registered in England with registered number 641132.

Registered office address:

The Atrium, Southern Gate, Chichester, West Sussex, PO19 8SQ.

This above writing with which I open my letter, is the first ever response (positive or negative) that I received in eight years of contacting publishers as well as many, many Universities and to show my utmost gratitude for their help in the matter, I mention the person and the name of the publisher.

I am well aware that I have no hope that my effort in sending you a copy of this manuscript would urge a response from you, but without my trying, I forfeit even the least chance of finding a positive response from your institution, however slim the chance might be and with that I then deny myself of having any chance I might ever have of even having doubt about sending you this manuscript. I am sending a complete manuscript and for that I have a very good reason. A brief proposal about my theory, which I jotted in an essay format, will belittle my work and the essence would be lost. The volume of information that my work undertakes in changing science can't fit in a journal because what little information this book holds alone, will change every aspect currently understood by science, and when I say that it is not a madman boasting, for I challenge any person that are able to read this work, to show that my statement of changes that has to come to physics are not true. However, my claim about this work having the ability to bring change to science cannot come from brief communication about a few topics I lightly touch or a number of thoughts to be studied. When one says a piece can change science that would entail a massive undertaking covering a comprehensive background of detail in a study of gigantic proportions. This is what this is but also this is just a small fragment of what detail the entirety of my wok holds that is consisting of more then forty other books covering the total view. There is no chance in hell that the content containing the fundamental changes required to correct physics as a whole can be dealt with by simply reducing the information that just this book alone presents as little as it is, to be contained by an article presented in a journal. If it is thought that the information presented in the other twenty or more books will also be added to the information in this book, the entire thought of confining the concept to an article in a journal becomes bizarre. The information that just this book holds as it is, presented in this book form alone, will change the understanding of mathematical physics completely, and yet as I say that, the madness that accompanies the words I utter cannot escape even me the person that declares that the changes has o come and even I see how ludicrous it sounds when I say I am the only person on earth hat understands cosmology correctly and the ironic obscenity in such a statement does not escape me but also rings through my mind as it does ring through the minds of any person reading the statement, notwithstanding how true such a statement might be. Even thinking it, the thought alone that I am the only person on Earth that has a fundamental understanding about cosmology sounds like utter madness and when I declare that I am the only person on Earth that is correct about science then I too can see how any one would find me as being the joke of the century by saying this, and yet I dare any person to prove I am incorrect when I charge these claims, that is after reading just this one book. This statement I make about me being able to understand cosmology could only become true and sound in the fundamental covering of the concept, when the entirety of the information that is presented in this book alone, is fully covered by all the detail I present to you. Leaving out even one thought that may be viewed on its own as insignificant when standing apart from all the rest of the other thoughts, be it as fundamental as such a thought might be, then the exclusion of that one thought will possibly nullify a vital part of the entire concept as a whole that I present about the facts that the view I bring forward will bring changes to physics and mathematics. When leaving out one argument that is said without the support of the rest of the content in this book will then sound like the delusions of a drug addict and not finding support of the compliment of arguments without reading the rest of the information anything you are then about to read will sound trivial and even mad. Introducing the following thoughts is but a very few examples I use to try and show my point I wish to make about reducing parts of the concepts and then belittling the entire unit as an entirety. Here are a few thoughts

When I say a line can't start with zero, it sounds as the most trivial statement any one can make, and yet in understanding this concept brings holds the difference between the veracity of a Universe being in place and a Universe being totally absent and being void of ever being established.

When I declare that pi ($\Pi = 3.1416$) is flat before you have read the book, and then you would think my remark is the most ridiculous statement I could ever make and when I proceed to prove that pi is flat, you will rubbish the argument without further thought because you would be unable to see the purpose of wasting time on such trivial information. Putting pi down to being flat makes no sense before acquiring all the information I supply in the book, but after four hundred pages of other arguments that knits the woven blanket forming my entire theory as I use which I make in support of one another throughout the entire book, you will learn that this statement unlocks our fundamental understanding of the Black Hole and therefore the concept we have of space-time. But when putting the argument on its own in an article to be published in a journal would be wasting paper because if it is published on its own, the merit of doing so evaporates, as there is no purpose of putting forward such an argument.

When I say a graph can't have zero dividing the graph on any axis, by my saying that I clash with thousands of years of mathematical culture and who ever is reading this suggestion, is then disinterested and their response is to stop reading the rest of my work because such trivial statements can't interest any one ever…and yet, mathematics has been incorrectly assessed in this matter for as many years as human memory will allow. This idea by its own prevents Science from mathematically going to a pre Big Bang situation because at present science covers that era with a value of zero. Science can't locate singularity because science awarded singularity the value of zero as we find they do when using zero as the axis in the mathematical graph. My suggestion about this matter alone defies human mathematical understanding. Yet, do not underestimate the importance this issue may have and thereby show disregard just because you are showing loyalty to mathematical culture because that loyalty will completely desecrate all mathematical principles formulating physical mathematics. Yet, try to multiply the growth that a graph shows when it breaks through the line of division forming the horizontal axis as it now is thought to hold zero (zero representing the measured value as the line that divides the graph must have when it is zero as science claims it is) and see how any value may progress from any point formed by the horizontal axis. The principle applying stopping this axis from being zero is most basic and yet never realised. By multiplying or extending the progress that the horizontal axis as zero may hold in growth onto a point on the vertical axis, then this value of zero should have in further growth into the graph on the vertical axis. By multiplying any growth with zero will nullify all potential growth because $0 \times 1 = 0$, and in doing that it will have the entire graph form a continuing of zero and then all further reading on the graph must be nullified in value. Multiplying any progress in value that may extend from the zero with which any value flowing from the line of progress must have, must bring more zero to the graph as zero then becomes more because the mathematical principle on multiplying anything with zero insist that the answer that follows must be zero. Anything multiplied by zero becomes zero. No dividing line in a graph can be zero and still this concept has been used for as long as humans used mathematics. The idea of realising this concept sounds trivial but because of this misconception no one ever and that is including Albert Einstein, was ever able to pinpoint singularity, a fact I do on page three or thereabout of this very book. I take you and I show you where singularity is and you are able to visually see the point holding singularity; that is if you will just bother to read my work! However finding singularity becomes a prelude of the rest of other detail as I then from there uncovers more about the cosmos than that which is uncovered to this day. I say this without a shadow of boastfulness: that after you finished reading this small book you will know more about cosmology than any other person does, and believe me I am way past a point of desperation to even consider a statement just to be boastful. Mentioning the zero aspect about the graph not having zero by that fact alone it then changes nothing about the graph by own implication of zero bringing a change in influence of the reading information on the graph, but in the total overview of how I show how everything that then changes altogether, then nothing remains unchanged. That line represents not zero, but singularity.

When I say the Sun is freezing cold at the temperature of 6500° C, I sound like a loony that escaped from a dangerous asylum and yet, if my theory about gravity is correct, about which I have no doubt that even the smallest detail is correct, then the Sun is freezing cold at 6500° C. If I try to convey the theory about the Sun being cold without the support of the rest of my theory that no line can start at zero, that pi forms the curvature of space-time, that gravity is about condensing heat and mass has nothing to do with gravity applying, that gravity is the movement of space in space and through space and gravity has nothing to do with attraction, then without explaining all these issues, everything I say becomes the mumble of a lunatic, and yet, I challenge everyone to show me where I am mistaken even in one thought, but that can only be clearly understood when the total content supports every suggestion. Up to this point I couldn't find one academic with the guts to prove me wrong, but I found many refusing to read my work and thereby avoid the issues I introduce.

If I say that there is no mass found in the Universe and there is not even one shred of evidence to prove that the Universe supports the idea of mass, then everyone concerned with physics rebuffs me as if I have become criminally insane with dangerous tendencies to commit murder. Every reader so far that read this statement I make, when reading my work then in response to what I say threw down the book in disgust at page four without completing the rest of a seven hundred page book…and yet I challenge any person to prove where the Universe does support the factor of mass playing a part in physics! Moreover, I challenge anyone unconditionally to bring any proof that mass does produce gravity and then also show how that is done and indicate the system whereby that is done. Show how mass does bring about gravity as a pulling power! Show me where the proof is that mass is a force of attraction and I will retract every accusation I made towards Newton. I say Newton is wrong and he created mass where mass, as a cosmic factor has no role to play in cosmic physics in any shape or form. The smallest piece of dust in the Asteroid belt spins as fast or as slow in orbit around the Sun as the giant Jupiter does. Mass does not bring about any changes in any particle circling around the Sun and with that being true, then what role does Jupiter's mass play in its circling of the Sun giving it any advantage in size in relation to any smaller particle. If it doesn't show mass then there is no mass present and only human hallucination and mind games then bring about the imaginary presence of mass as a cosmic factor. A giant elephant falls as fast or slow as a mouse does and mass does not play a part in the descending of either. This was what Galileo said and if Galileo is correct about mass not influencing the fall, then Newton is incorrect by putting mass down as the prime factor for being responsible for objects falling! It only shows that as far as cosmology goes, Jupiter and the dust particle has the same mass because they both show the same gravity influences on the attraction the Sun has on both particles and the mouse and the elephant fall down equally evenly. That proves that while falling, the mouse and the elephant are granted the same mass factor because both fall equally…and one either accepts this while accepting Galileo or reject this by clinging onto mass being a presence and then by clinging onto Newton one has to reject reality as well as Galileo. Where is the mass factor prominent in Jupiter's orbit around the Sun when Jupiter spins at the same pace as the tiny dust speck or where the mouse and the elephant falls alike? Gravity does not rely on mass but there are four phenomena working in tandem and in conjunction that forms gravity by measure of the square of Π. The existence or even the presence of the four phenomena are not even entirely recognised by Mainstream Science much less than they are understood by anyone at present holding any position in physics. How can I explain the value of the four phenomena and show how these phenomena interpret singularity without explaining that no line can ever start with zero or how can I implicate the importance of understanding the fact that no line can start at zero and all lines proceed to become a flat Π, if I am unable to extend this understanding to introduce the understanding in supporting of how the four principles form the measured compliment that is behind the movement which is forming gravity. The arguments form a woven blanket and by removing one shred, the blanket falls to ribbons.

I do show where the Universe starts with infinity and where to locate the starting point at infinity as much as I show that the Universe can never end because of eternity holding an absolute relevancy with infinity which then forms time. I show where to find infinity as much as I show precisely where to locate eternity. I show how to find that which can never start and where to locate that which can never end and how to read time forming space between these cosmic limits. I take your finger and place it on the two locations…that are how precise I indicate the positions of each of the limits but that can't be done without accepting that the graph cannot present a line of zero on the horizontal axis as a form of diving the graph. This I can only do when I am able to show where the coldest point is in the Universe and that that point is in the Sun as much as it is in all stars. I have to pinpoint singularity at an infinite position before I can explain gravity forming the Black Hole…and this has nothing to do with mass being a factor. In relation to that I then have to show where the hottest part of the Universe is and that I can only show by showing no line can start with zero because if zero is the starting factor of a line, then pi can't start space-time. I challenge any person to show which part of the overall concept that just this small book brings can be left out as that part does not support the rest of my concepts. I challenge all persons concerned to prove that mass has any function in gravity and then to disprove my proof that gravity is formed by the interaction of the **Titius Bode law**, the **Roche limit**, the **Lagrangian positions** which in conjunction form the **Coanda Effect**, where the **Coanda Effect** is gravity applying. At present in view of Mainstream science policy these phenomena are hidden so deep in Newtonian misunderstanding that most people never even heard of these phenomena being a part of cosmology and with that in mind it is very likely that most academics reading my work might never even have heard of these phenomena. Since Newtonians can't use Newton to explain the **Titius Bode law**, the **Roche limit**, the **Lagrangian positions** and the **Coanda Effect**, these phenomena are hidden so deep that it is very likely that you as the reader might never even know anything about these phenomena, albeit that they play such a most vital part in translating singularity into space-time.

I prove mathematically that everything in the entire Universe we know as a unit starts at a displacement value of 112 (which the displacement value is normally considered as the proton number).

At a displacement value of 112 is also the point where singularity being the heat forming outer space (yes, outer space is formed from singularity also known as heat and not nothing as mainstream Science propagates at present) becomes six sided and double three-dimensional and for that there is a very good and well-defined reason.

At a displacement value of 112 the atom forms a relevancy unifying the proton, neutron and electron as a sustainable unit, at that point gravity becomes displaced as a liquid heat that we also think of as electricity and the only difference between electricity and gravity is the scale of generating taking place and this happens at 112. Also at 112 the Universe becomes a three-dimensional sphere, this I prove mathematically by not using mass. If you read this you will learn how I mathematically prove all this that I claim.

I prove mathematically that everything in the entire Universe as a unit holds absolute relevancy through singularity to everything else forming even the most insignificant detail in the Universe...but if I wish to achieve this statement and show what I say I show, then also every word in this book holds relevance to everything stated by every other word in this book...which brings me to the question I answer in this book and that is how is that achieved, where everything links by singularity to become one united Universe and to do that I cannot reduce any word forming this book because every word supports the rest of the book by linking the entire concept through every word supporting every other word. I do realise it is not contained in the format you may prefer. It is not conducted in the form you insist on, but just read the content, and when finishing it, then you asses which part were not worth your while and what part I should have dumped to suit your required conditional format. Read the content and then you decide what part could be chucked into the waste paper basket because in your opinion that part is old news to you, is irrelevant to the rest of the information and is not critical when seen in the light as to how that information is not supporting the rest of the network of information. When you conclude what it is about which part you decide to excluding because that part that is not worth noticing and is worth rubbish, then please judge yourself why you would think that the part is not worth noticing and by dumping that, answer to your conscience why you think that science would not be poorer for your doing so when you are discarding that information.

I am of the opinion that if you do not read the book I present to you, which I wrote in a format representing an open letter, then you would be poorer for not doing so, but moreover, science in the history thereof would altogether be poorer for you not doing so. This book has not been edited by any linguist because I do not have the recourses or the required financial funding in paying for it, but if that is a hurdle in your opinion as it previously was with many of my other books I sent to academics in the past, then if that is how you feel, let it be a hurdle...but it does not take away any of the authenticity of newly introduced information that I put to view on the science table and it does not reduce the veracity of the information I divulge to science which was never yet before presented, linguistically unsupervised or not. Should you think I am boasting about the content and the new information that I bring to your table, then just go on and read the first hundred pages and see whether I am boasting or whether my warning about science missing a great opportunity when you decide not to finish and if I am just vain when warning that you lose out on your personal gain in science knowledge while a great opportunity to further science is going begging. After you have read a hundred pages, then you judge yourself if my saying all this that I am saying is truly boastful or is an honest assessment of reality?

They swindling by giving a list of half-truths because only science, Newton and God in that order is never wrong and that is religion. Lately someone was found to be wrong and science decided it had to be God that misplaced dark matter in obscure places in order to have the Universe expand instead of contract as Newton said it does. Science decide on behalf of God that God was wrong when the Universe expanded instead of contracted as Newton said it does because Newton couldn't be wrong so it's God that wrong. The way science present a star such as the sun is similar to a furnish burning to boil water. The furnace will burn until the coal is finished and then the star will **_die._** This view is carried over from since a time before they new about electricity or internal combustion engines or air-conditioning or nuclear energy. Science at best can't begin to explain what makes these phenomena pictured above on the first few pages do occur or what happens when these phenomena occur. I can and I do explain in detail why these phenomena happen and I prove what I explain. Then why would science in its current concept not accept my explaining? If they do accept my explanation about why my concepts are correct then they have to rubbish their entirety of everything they think they know and trash every thesis down to the last dissertation they ever wrote on science presenting it the way thy do. This will cost the industry every penny they spent this far on science. Acknowledging my concepts about cosmic physics and cosmology renders all their work useless and moreover it clearly proves how wrong they are about the woefully outdated Newtonian approach. It will become clear how little they understand about science because everything they say they understand is incorrect and applauding my view makes every book published on cosmology instant science fiction. You think they will embrace my work…hell no and now I turn to the public. I let the people decide what makes sense and what is rubbish. Purchase this book and then you decide! You will have to choose between nature and Newton.

If you have never heard of these phenomena it is not that surprising because you either have to believe in Newtonian magic Newtonians call science or believe in nature and nature implementing science because this is what nature uses instead of Newton. These 4 phenomena used by nature in nature disprove Newton completely. Science never shows how these 4 works because they can't. I prove that the Universe consists of two cosmic substances which is material and space albeit either in a denser cosmic liquid such as light and cosmic clouds or less dense cosmic gas such as dark outer space Gravity is the movement of space within space. The entire Universe is relevancies formed by differentiation of density. Movement brings comparable density differences. This concept destroys the myth of mass pulling mass.

A Black Hole is much denser than the sun because it spins faster than the speed of light and the inside of a galaxy is denser than the outside since the inside spins faster. Everything in the Universe moves. By turning material contracts space thus reducing space surrounding objects and this leads to increases seen as space expanding. The faster a star spins the denser is the star and therefore the denser a star is by fast movement the smaller the star becomes in overall size and space. Since the area called outer space does not move the density reduce as stars collect the density by contraction of space. As outer space does not move it loses density in relevancy to material growing in density so it seems as if space expands in relation to material seemingly growing. This is relevancies changing as it seems the earth is getting larger at the circumference and it seems the moon is growing further apart from the earth but it merely seems that way because it is density moving by gravity from space unoccupied to space occupied. But space can't grow because the Universe holds all the space there ever can be and so expanding space is impossible because whereto will it expand; space has no where to go!

I wish to make one fact very clear…the way that Newtonian science shows the Universe work is not working anywhere. There is weight but not mass. Weight weighs and according to science mass pulls. I base my work on formulating the working process of four cosmic principles in Nature. These are: **The Coanda effect**, which is the way, the atmosphere forms or liquids respond to solids moving.
> **The Titius Bode law** is how planets use a very specific ratio to arrange their allocated positions.
> **The Roche limit** is amongst many also the law that applies to what we call as the "sound barrier".
> **The Lagrangian points** are why the different layers that form the atmosphere around the earth.

I did not discover these phenomena because science knows about these phenomena for a very long time and in some cases even for hundreds of years. Science knows they apply and where they apply. When science discovered or allocated missing planets they used the law applying such as the Titius Bode law from which they deducted positions that they knew in that circle according to the planetary layout that the law predicts there had to be a planet according to the Titius Bode law. Science did not apply Newton's formula to discover and locate planets but they applied these phenomena and especially applied the law of planetary allocation to discover the precise location the planets discovered after Galileo.

The response I get from a uniformed part of the public is astonishing. Believe me if there is one thing I appreciate then it is criticism backed by intellect. It seems the less they know the more they wish to teach me on matters about how they were taught how Newton presented cosmology. I am not interested how you think Newtonian explanations work in the Universe because Newton never applies in any form in the Universe. That which you believe explains Newton; it explains your personal brainwashing and how you convinced your mind that Newtonian impossibilities are possible. I am not interested in how you see Newton works in nature because nature has no room for any Newtonian principles. If you think nature uses Newton you better start reading up on what truly goes on in nature. Those that criticize me usually start off by saying "I did not read the entire book" and then goes on to try and teach Newton to me. I know all about Newton because I have studies Newton since 1970. The reason why you did not read the entire book is because the truth about physics goes beyond your mind limit. Don't think I insult you, every Newtonian out there that has a PhD in physics have the same problem; nature is above their understanding and the Newtonian simplicity calling on to enforce magic powers they can manage and therefore they go for Newton...not because Newton is correct but that is all those physicians can understands about physics. Newton saw forces and he was an alchemist by trade and therefore he had to believe in magical forces and powers no one can explain such as gravity. Those Brainy Newtonian Masterminds are fortunate because they can repeat after Newton albeit incorrect or disastrously wrong and that is because they **understand Newton.** I present you with facts from nature and not Newton!

Every one in science knows these phenomena are there and are in place and they rule the orbit set-up by forming the planets' gravity. The solar system functions according to them. These four laws on planetary motion that is used by nature at this moment and has been in place since time began are what apply and they dismiss Newton. If you argue with me about Newton being correct you better take your case to God or the solar system because the four cosmic phenomena is working in nature and nothing Newton said is applying in nature. This is a truth and a fact and a foregone conclusion and can never to be in doubt. Get educated in cosmology before you try to get wise and become opinionated about Newtonian correctness.

Brainwashed as you may be in believing Newton you can't either side with my view or decide on Newton because it is not a case of choosing between Newton and me. I'm out of the picture! It is either telling nature to listen to Newton and change what is in place or read and see what is in place and what is applying in nature all along. The phenomena are what we find to be used in the cosmos while Newton is in the imagination part of the minds of scientists and nowhere else. If you don't believe me and if you wish to discredit me first find out a little more about science.

Then deflate your ego as to what you think you know. Go and study what is in the cosmos and how the cosmos works before giving me a lecture how Newton works. I am not interested in how you were brainwashed and how you now are programmed to believe in the impossible. Go and find out why the rings are circling around the large planets such as they do around Saturn and why the material is formed as rings. If it was "mass" of the sun pulling the "mass" of the comets and therefore the comets are drawn to the sun then go find out why comets don't collide into the sun but circle around the sun and disappear into the darkness of space. Go find out why the moon is departing from the earth notwithstanding that according to Newton the moon and earth must come closer.

I found out why the four phenomena are in place instead of Newton ideas. Science never mentions these phenomena because science can't use Newton and explain these phenomena or use these phenomena to prove Newton. These four phenomena that the cosmos uses as we speak have been in place ever since the Universe formed. Since science can't explain the phenomena and the phenomena destroy the credibility of Newton science avoids these phenomena as if it brought the plague. You can't choose between Newton and me because I did not put the cosmic phenomena in place. All I did was doing a study since 1977 to formulate how and why these phenomena work and how these phenomena keep the cosmos and the solar system working. I am the first person in human history able to show why they work.

We are all been brainwashed for centuries to believe Newton. Should your brainwashing kick in and you have an axe to grind with me about what I say, then first prove these phenomena are not in the cosmos and are not applying to form the laws that the cosmos put in place as gravity. I only found out how they work and why they work and I did not make the phenomena work. All those clever stooges that have so much to say even before they read my work first learn what is in place before getting opinionated. It is not me that takes Newton out of the cosmos and replace Newton because Newton was never in the cosmos to begin with. While the four phenomena are in the cosmos Newton is in the imagination of scientists and in the mind control they placed on students? When you argue with me it only shows how little you know.

I am not fighting science or the credibility of Newton or what might be true or not true but I am fighting centuries of brainwashing unleashed on everyone young and old and I have to dismiss the mind manipulation, the brainwashing and the systematic mind control that those teaching science inflicted on us all. My fight is not about what is true or not true but what is accepted as culture and which was not even once been proven by science. Please do read on to investigate before you have the enormous wealth of wisdom that fires you up and you become self-opinionated. If your brainwashing starts to control your common sense please don't share it with me but go and put Newton in the cosmos and show where you found Newton was. I am not interested in the method by which they brainwashed you. First show me where it is in the cosmos that Newton hides and where it is that Newton's ideas can be located.

Before the French revolution people were force fed information when everything was told to everyone and it was expected of the faceless mobs and working class forming the illiterates to believe what they were told by the educated that told everyone what to believe and what to think. So since then what have changed because after everything the people still can't think using their own minds! When I came along and was not impressed by others telling me and did not take fore granted what others told me what to believe I became part of what was the intellectually deprived because **I did not understand Newton.** When the urge overwhelms you to confront me about the honourable scientists all working in physics and how I am suppose to kneel before their supreme and ultimate wisdom allow any of them to prove mathematically how the four cosmic phenomena work and why do they work. Before you defend Newton, first try to prove Newton because that part has never been achieved in over three hundred years of trying to.

Peet Schutte.

An Open Letter

ON THE VERACITY OF GRAVITY

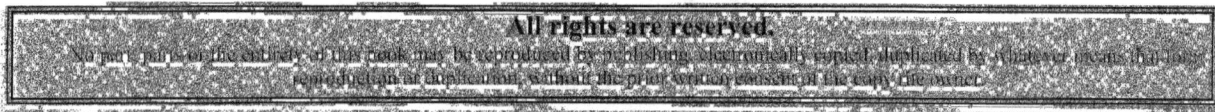

ISBN-13: 978-1499321845 ISBN-10: 1499321848 **BY** PEET SCHUTTE
FROM THE ORIGINAL AFRIKAANS: "MATERIE SE TYD IN RUIMTE BY PEET SCHUTTE
© KOSMOLOGIESE EN ASTRONOMIESE TEGNIKA

TO WHOM IT MAY CONCERN,

I do find much pride in my status as being Afrikaner and would like to have my names used by pronouncing it in the manner Afrikaans dictates…therefore I would sincerely appreciate the courtesy when readers will take note that my name and last name are pronounced in Afrikaans, which is originally from Dutch and must be pronounced that way. Peet one would pronounce "here" which is the closest English to the pronouncing of the "ee". The "Sch" in Schutte is pronounced exactly as school is where both actually are pronounced Skutte or "skool". By pronouncing my name in Afrikaans you do me the utmost courtesy any one can. Being an Afrikaner is what I am most proud of.

I have no idea why I still bother to write this book or why I still bother to think about the subject because just as before with all the other books that I wrote; again there will be no one that will read this book and this I say before starting this book so that I am on record this time confirming the fact that I wrote this book without any clear purpose as to why I do so. What I do know is that somewhere in the far future there will be someone that will eventually read this book and then those that at present did not bother to read this book will become the fools they are for not bothering to read this book. There has only been one academic in the past that was prepared to read my work and that person was also the one that read my first work in Afrikaans being Professor Luth Strauss of Pretoria. I therefore attribute the assembling of this book taken from various other articles I wrote in the past, and dedicate this work to him and I plan to send him a copy as he was up to now the only person being an Academic that treated me with dignity and with human respect in all my dealings with physics academics. The rest of all other academics in physics can go to hell as far as I am concerned and I truly couldn't be less bothered if any of them ever read my work because time will finally destroy their incoherent philosophies as time somewhere in the future will acknowledge my work. There may eventually in the distant future be someone that will read this work in eighty yeas from now and then take advantage in annexing all the philosophies I introduce, facts that I prove and concepts that I now put forward, stealing the content and the impact of the work by giving me the credit in naming something after me as to uphold my memory and then go on to reap all the financial reward. Newton did that to Kepler, Einstein did it when he was a clerk in the patent office, so it must be some Newtonian tradition. While I know before starting this book that no one will read the book, I still do it and attribute the assembling and existing of this book in dedication of and to Professor Luth Strauss of Pretoria.

Dear Professor,

I am Petrus Stephanus Jacobus Schutte going by the name of Peet, the author of the above-mentioned book(s). I hope you find your reading of this book presented as an open letter a most fruitful experience. It is thought provoking because it strays widely from mainstream science and for that there is a very good reason. I researched a most prominent man that everyone ignores. It is therefore almost absolutely realistic to say that what you are about to read was never yet printed. It seems to me that any research predating Newton never came into use or in practise. You are going to find the topic under discussion most controversial since I dispute a part of science no individual since the time of Newton even investigated. In the least it should be thought provoking because it strays widely from mainstream science but for that there is a very good reason. The proof I present is real and carries the arguments in a manner, which is by nature childlike simple to follow, should one allow oneself to have an open mind on the subject under discussion. Tracing the argument in detail helps to uncover incorrect principles currently applied by science and it is on this matter that the controversial aspect about the arguments rests. The accepting of those principles that I view as incorrect depends on your inquisitive nature and your ability to consider dismissing your personal culture bringing about your personal as well as academic bias on issues at hand which to my thinking is solicited in the first place through your historical accepting of facts used in science and not facts backed by scientific proof.

Through my studies on an on and off basis that is going on for several decades (as time and when time would permit), I have realised the reason why by replacing $a^3 = T^2k$ with Π in Kepler formula and too include the measure of Π to his formula makes cosmology much better understood. In this book I prove that the reason why adding Π to the rest of Kepler's formula is vital as it is necessary This addition is because when going one step further in the investigation one will find that **k** and **a** and **T** is symbolising the same value with the only difference being that each one represent a different dimension to our six dimensional or six sided Universe we enjoy. In fact I shall show that by replacing "**a**" and "**k**" and "**T**" with Π brings about the true value that should be in place to indicate the correct value nominating Π. Therefore the formula is in fact $\Pi^3 = \Pi^2 \Pi$ where I shall show that Π represents singularity wherefrom the entire Universe sprang. That will become clear when one dissect different facts coming from studying Kepler. There is need to replace $a^3 = T^2k$ with Π in form because Kepler discovered the ultimate Π in the Universe, the Π giving the Universe form and gravity. When replacing the symbols with Π the Universe facts become self-explaining.

PLANET	PERIOD (Years) (T)	MOVEMENT (T^2)	DISTANCE k	SPACE (a^3)	RATIO
Mercury	0.241	0.058	0.39	0.059	0.983
Venus	0.615	0.378	0.728	0.381	0.992
Earth	1.000	1.000	1.000	1.000	1.000
Mars	1.881	3.54	1.524	3.54	1.000
Jupiter	11.86	140.66	5.20	140.6	1.000
Saturn	29.46	867.9	9.54	868.25	0.999
Uranus	84.008	7069	19.19	7067	1.000
Neptune	164.8	27159	30.07	27189	0.999
Pluto	248.4	61703	39.46	61443	1.004

Mainstream Science was up to this point unable to supply **any answer on** any of the groundbreaking issues about dimensions or any of the following questions, while Kepler answered the lot. Test yourself again and see what your personal answer will be when asked:
…**where is the Universe coming from**……**in which direction is the Universe going**……and most of all…**WHY** is it travelling through time …**what** it is that is travelling through time…?
…all these answers remained unanswered because Mainstream Science ignored Kepler and his wisdom. 'Have you as an academic ever tried to answer facts about the Universe in as much as…
…what is expanding …what brings about the expanding? Kepler answered all these questions…and brought even much more answers to the table than he ever got recognition for! All that is required to translate Kepler's mathematics to English is not to read Kepler's formula but read into Kepler's formula. My investigation of Kepler's work brought about a conclusion that no one yet arrived at concerning the findings of Kepler because no one other than me scrutinised Kepler's formula in any way to this day. Kepler found planets rotating around a centre but Newton saw a

circle and added what is mathematically required to indicate such a circle. Newton added a mathematical $4\Pi^2$ to the formula of Kepler and removed the distance symbolising measure that Kepler introduced using **k**. On the other side Newton changed the symbol of **k** by using G (m + m_p). This is just a longer and probably a more detailed manner of indicating **k** and better defining of **k** but it symbolises precisely to the point what **k** stands for nonetheless and by claiming G (m + m_p) Newton mollified his other presumption of $a^3 = T^2$ with **k = 0**. However, what Newton added to Kepler cannot be added because Newton never had the insight to read what Kepler unveiled. However one may argue and whatever reasons one may present, it is not possible to argue away mathematical values that exist in the cosmos in a ratio we have in space-time. If there is a table with a column presenting values that match a ratio, used in an applying formula then no argument Newton could ever advocate can have the ability to nullify the column, the values, the ratio thee is or the fact that the values are part of a mathematical equation.

The German mathematician and astronomer KEPLER, JOHANNES (1571-1630)
German mathematical and astronomer became Tycho Brahe's assistant in Prague in 1600 A. D. where he undertook to complete the tables of planetary motion Tycho had begun. Kepler first calculated the orbit of Mars. He spent much time trying to reconcile Tycho's accurate observations of the planet with a circular orbit, but concluded (in Astronomia nova, published in 1609) that Mars moved instead in an elliptical orbit. Thus, he established the first of his laws of planetary motion. A theory that the Sun controlled the planets by a magnetic force led him to the second and third of his laws, which were published as part of his treatise on theoretical astronomy, Epitome astronomiae Coernicanae (1618-21). The Rudolphine Tables (named after Tycho's patron, the Holy Roman Emperor Rudolph II) of planetary motion appeared in 1627 and were still in use in the 18th century. Kepler also wrote De Stella nova, on the supernova of 1604 and Diptirce on optics and the theory of the telescope. The overall view followed in this book **Matter's Time in Space** places the true significance of his work in true contents.

In KEPLER'S EQUATION is the equation that relates the eccentric anomaly of a body in an elliptical orbit to its mean anomaly. The equation is $E - e \sin E = M$., where E is the eccentric anomaly, M the mean anomaly, and e the eccentricity of the orbit. It is important as one of the mathematical relations enabling the position of a planet about the Sun, or a satellite about is planet, to be calculated from the orbital elements for any time. However this only relates to the solar system, and KEPLER'S LAWS only apply in the contents of the solar system. The three laws governing the orbital motions of the planets, discovered by J. Kepler is as follows: The first law states that the orbit of a planet is an ellipse with the Sun at one focus of the ellipse. The second law states that the radius vector joining planet to Sun sweeps out equal areas in equal times which as it says refers to time and not the circle. The third law states that the square of the orbital period of each planet in years is proportional to the cube of the semi major axis of the planet's orbit. The first law gives the shape of the planet's orbit; the second describes how the planet must continuously vary its speed as it follows its orbit, moving fastest at perihelion and slowest at aphelion. The third law gives the relationship between the planets' average distances from the Sun and their periods of revolution. Instead of placing, the true value to Kepler's laws I. Newton placed his own interpretation to Kepler's laws, and in doing this, he wilfully destroyed the principle working of the Creation. Through Newton's tunnel vision, he applied his own miss interpretations to the correct presumptions of Kepler. Newton reduced the implication that Kepler findings hold by introducing to the law of gravitation. He then went about and changed it to three laws of motion. I. Newton generalized Kepler's first law, verified the second law, and showed that the third law should be amended to the form; $4 \pi^2 a^3 / T^2 = G (m + m_p)$. In this, the value of T and a are the period of revolution and semi major axis of the orbit of a planet of mass m_p about the Sun of mass m, and G is the gravitational constant. The major aim of this book is to correct these misgivings of Newton. I shall return to the statement about $4 \pi^2 a^3 / T^2 = G (m + m_p)$.

What Kepler saw was more of a dimensional nature than the practical mathematic symbols and values. On the one hand was a value to the third dimension, which equalled two-dimensional values one the second dimension, and one to the first dimension. In the argument Kepler made he had hide so much more facts into one formula than what I think even he realised. Well, it is much more than that the Accepted Policy Protectors Of Science ever came to realise. He officially formulated space-time, he officially coined not the name but the origins of the Universe being the Big Bang and he was the first to put the speed of light in relation to cosmic development…and all of that with his rather simple formula. He said the space a^3 not the circle (a) or the circumference a^2 but in the circle a^3… where such a circle represent a factor in the third dimension.

The formula he compiled was not rather but very specific about the area being a third dimension area and to prove it beyond doubt he placed it in the relevancy of the formula in a ratio of presenting the third dimension in space. He said a^3 is equal to T^2 **k.** Newton and Newtonians came afterwards and played with mathematical toys as to challenge their mental capabilities. Newton introduced a $4\Pi^2$ to indicate the presume circle on the one hand and on the other hand he brought this lot equal to $\{G\ (m + m_p)\}$ which he then presumed to be the general Universal gravity constant (G) and the sum total of the two structure mass. Newton saw a ring circling around a centre having $4\Pi^2$ to indicate such a ring outside a centre and he positioned $\{G(m + m_p)\}$ where the two mass factors combine the gravity effort in the general grand gravity constant in space. I have had so much resistance in the past from all Academics but that is not what I see what Kepler saw. With what I saw what Kepler's saw I shall trace that back even as far as to the centre of creation.

For decades I tried to come to terms with the inability in science to explain cosmology while the understanding and advancing of physics mathematics and chemistry as subjects was flourishing. To name but one example is the formula they say can used accurately to calculate the gravity of a Black Hole. If one applies that formula to all the stars holding a gravity that is well above the speed of light, then something as insignificant as a neutron star, and a neutron star pails in comparison to a Black Hole, has a stronger gravity fields than a Black Hole has. That is rubbish. I saw how little there was available in explaining cosmic phenomenon and how much the other departments in other fields of science such as chemistry and medicine could offer as results coming from research. Even the little explaining that is available is confusing to say the least and at best applying double standards. For decades photographs were the only progress that was forthcoming in the field of astronomy but as such were only photographs carrying beautiful pictures that pleased the less informed, except the photographs did not bring progress to cosmology intellectually by promoting insight to the subject. Science still was more unable to explain what the photographs depicted and the lack in knowledge in astronomy became even more apparent. So it became apparent that the more advanced the picture became he more was it a case of showing what it really did was to underline what lack in progress there truly is in the perceptibility science had about the cosmos.

While such Hubble images might seem to be clear as daylight it was even clearer that there can be no real progress without any one showing science had the slightest clue of what is going on in the Universe when investigating the new images coming from scientifically advanced telescopes. Most images contradicted Newton and for my saying that every Academic I came across in the past ostracized me. That bothers me little! I know I am not the only one with serious doubts because I cannot be the only one with serious doubts when what there is, is so very doubt provoking. Every one able to read mathematics has to realise that Newton suggested the future only holds collisions between cosmic structures and destruction by collisions must eventually come about as gravity erodes the distance separating the cosmic structures at a rate measured by multiplying the product of the mass of both structures and then multiply this value by the gravitational constant which then is divided by the square of the radius separating those that contribute to the calculation by supplying the mass factors. Newton said the multiplying mass of both structures destroy the distance between the structures. The cosmos then must end in a Big Crunch with all material joining together, but that joining is not that apparent or forthcoming at all…and that only indicates how much insufficient understanding there are on offer to explain their field in the ranks of cosmologists.

The picture we see coming from Hubble shows why, in the perfect Universe…the radius parting the sun and its planets would not behave so mystifying…but there is a perfect Universe and it becomes ever more perfect as one learn to understand the perfect Universe even better. But it does need an open and clear mind and it needs no preconception that is saluting interpretations about facts surmised and not carefully studied. It becomes obvious that Newton never gave careful attention to Kepler's findings because Kepler said without using names as the name gravity was not yet introduced, that the structures orbit because there is a space that circle around a centre and this process is keeping the Universe secure and comes about with a process Newton later named as gravity. Then eighty years on Newton named the force keeping the Universe together. He named it gravity. Why he chose to ignore Kepler's findings on gravity we shall never know but why the world still choose to ignore Kepler's findings about gravity almost four hundred years after the fact I shall never know. Translating Kepler's mathematical expression $a^3 = T^2k$ correctly to the verbal statement in English Kepler said that there is a space a^3 which is equal $=$ to the motion in the time duration T^2 thereof between two specific points which holds a relation to a centre where from there forms a straight line k and is located on the spot where space begins the circle therefore that spot has the least space.

By using Kepler's formula I can prove not only what gravity is but also prove:

1) **The location position and value of singularity** as a factor forming **space-time**
2) **Finding space-time,** proving **space-time** and aligning **space-time** with **gravity**
3) **The working principals behind** and manifesting of **gravity** as a cosmic occurrence.
4) **The Roche limit,** and explaining the resulting as a law coming about from **singularity.**
5) **The Lagrangian system** and how and why that becomes the building form of the Universe.
6) **The Titius Bode rule** and how **gravity** comes about from that
7) **The Coanda effect** and the producing of **gravity** through applying **space-time**.
8) **The sound barrier** that is coming about by duplicating relations in **space** and **time**

Before attempting any investigation there must be coherence about what gravity is. Anything we use in order to base a decision on must support the fact that it is gravity that prevents planets from dislodging from the grip that the Sun has on them. If it is gravity we are in search of, then we must look at the behaviour of the structures indicating a response coming as a result of gravity. We have to see why the planets do not reduce the radius as Newton suggested because if anything they are departing as they extend the radius connecting them to the Sun. Only a fool will fully support the total undeniable accuracy the formula $F = G\dfrac{M_1 M_2}{r^2}$ presents in science.

The fact that the cosmos grows everywhere is the way having gravity contributes to the cosmos because it applies through out the Universe. That then is the gravity we have to study to find the Universal enticing by which gravity is holding the Universe together. By close investigation one will find three factors in urgent need of investigation. There is firstly a centre that seemingly is drawing the object closer. If gravity was not drawing the objects circling around the common centre closer, then the object would not be orbiting around the centre and apply circling motion. We are using Newton's $F = G\dfrac{M_1 M_2}{r^2}$ and the entirety out there that is spinning around some centre must come closer according to Newtonian declarations, but that is not happening. In fact everything is departing and not arriving. Even the moon is drifting away from the Earth and this information comes about from the most advanced investigation.

KEPLER'S LAW OF PERIODS FOR THE SOLAR SYSTEM			
PLANET	**SEMIMAJOR AXIS** $a\left(10^{10}m\right)$	**PERIOD** **T (y)**	T^2/a^3 $\left(10^{-34}\,y^2/m^3\right)$
Mercury	5.79	0.241	k = 2.99
Venus	10.8	0.615	k = 3.00
Earth	15.0	1.00	k = 2.96
Mars	22.8	1.88	k = 2.98
Jupiter	77.8	11.9	k = 3.01
Saturn	143	29.5	k = 2.98
Uranus	287	84.0	k = 2.98
Neptune	450	165	k = 2.99
Pluto	590	248	k = 2.99

The **cosmos** spoke to Kepler about space-time coming from singularity. Kepler gave us his findings. It is $a^3=T^2k$ and it is not as Newton later changed it to $a^3=T^2$. The cosmos mathematically spoke to Kepler personally using mathematics as the medium that provided us information about the cosmos.

If Newton was correct in surmising that $a^3=T^2$ I would love to here an intelligent explaining about what happened to all the k factors in Kepler's table. There they are and every one of those factors has a value and not one of those factors carry a value of zero **(k=0)**

It is a matter of seeing that while we use gravity, the using of gravity makes us part of the Earth by mass forcing us onto the Earth as to bind us to the Earth in order to make us become a semi unit with the Earth. Is that truly gravity? Much of the proof about gravity is part of our perception about gravity because we experience certain conditions with gravity. But are our perceptions about gravity truly correct? We only experience gravity as a factor from the position we have on Earth and while we are being forced to be part of the Earth. Kepler was the very first person to mathematically introduce space a^3 singularity k^0 relating with time T^2k. Not only did he introduce space-time $a^3 / T^2 k$ but he also placed space a^3 and time T^2k in a relevancy long before Einstein did and placed gravity in space-time $a^3 / T^2 k$ even before Newton named gravity. Kepler was the person who placed gravity as the ingredient in the Universe that determines space a^3 and time $T^2 k$ and much more. Kepler was the first one that saw that gravity comprises of two factors being **k** or linear gravity and circular gravity or T^2. By **reading what Kepler** said **correctly** so many centuries ago, the effort brings all the answers to the questions we have. But it does not involve looking at what Newton said about what Kepler said… it is all about looking at what Kepler said. To understand Kepler one has to **include the opinion of Kepler** and **remove the opinion of Newton** about Kepler's findings.

If Newton was that much correct in all his surmising as those in Mainstream Physics agree that his work is, then if that is the case then is we could make deductions which at this time we can't do. If Newton's gravitational principle was as correct all this time as Mainstream Physics corroborate it to be since Isaac Newton introduced the concept of gravity being a force of attraction, it then must be possible that we should be able to use this force of attraction as a calculation standard in a way that we could predict the future. We should be able to calculate when the Moon and the Earth will destruct in a spectacle. This will be when this Newtonian gravitational force of attraction is going to bring planets together by the force of gravity through the intervention of mass. By applying Newton, it then would be simple to calculate the time measured in millions that we have left to live on Earth and it could be measured in the number of millennia that there is left in the future. We could calculate how long it will be until such time as that the Earth and Moon are to collide and destroy all forms of life.

This inevitable coming collision between the Earth and the Moon is what then should bring about the end of humanity on Earth! We have the mass of the Earth that is well known and is generally available and we know the mass of the Moon that is well known and is generally available so why don't we use it as Newton intended it should apply. At the very same time the radius between the Earth and Moon is also well measured, so why not do some calculating. We can use this information to our advantage. Then it would be possible to measure what time life still has available before all the planets would unite with the Sun, beginning first with the ending of Mercury, which is the closest and then much later on finishing the solar system by concluding the ending of the solar system in destroying Pluto. Then it would be possible to say what time span this force of attraction would allow the solar system to maintain its orbit around the Sun and what affect the Earth's coming closer to the Sun must have on climate change…that is if Newton' gravitational principles of attracting objects by mass is correct!

If Newton was that much correct since the time Isaac Newton introduced the concept of gravity being a force of attraction and if this concept of being a force of attraction was what brought on gravity, meaning that this force of attraction had the ability of bringing planets together by the force of gravity unleashed by the Sun, and also through the intervention of mass, then it would be simple to calculate what the time period would be that is left until such a time when the Earth and Moon was going to collide and this collision then was going to bring about the end of humanity on Earth! It must be possible to use Isaac Newton' principle formula $F = \frac{M_1 M}{r^2} G$ to see what the time is that there is left for life to be on Earth! Then it would be possible to measure the exact time when all the planets would unite with the Sun, beginning the inevitable unification that will end the solar system and with Mercury being the first to become a Sun spot and ending with Pluto forming Sun material. Then it would be possible to say what time span this force of attraction would allow for the solar system to maintain orbit and what affect the Earth's coming closer to the Sun must have on climate change...that is if Newton is correct! Is there even one professor that can show a class how to go about to use $F = \frac{M_1 M}{r^2} G$ and with that then will set about determining the time life has left on Earth. If your professor fails to commit to a plausible calculated answer, then you should hear an alarm sounding because then your professor admits that Newton's $F = \frac{M_1 M}{r^2} G$ is completely worthless.

Without your professor showing how $F = \frac{M_1 M}{r^2} G$ can be employed to serve a mathematical purpose, then read the rest of this book to see how implausible $F = \frac{M_1 M}{r^2} G$ truly is.

This remark and the viability of the method throws serious doubt on the validity of Newton's gravitational formula used (and also never used) as $F = \frac{M_1 M}{r^2} G$.

Inspired by my belief that I have uncovered a mistake in mathematics led me to a new approach to cosmology. Admittedly the content of this letter is controversial. That is what my views have been blamed for in the past by those who are accepted and self proclaimed members of Mainstream Physics, and that I accept, but by such acceptance I admit that my line of thought is and extraordinary. Since the concepts I follow starts with the start, which, was named as the Big Bang, I start by tracing a new approach and add to the already existing theories. Every one accepts the Big Bang Beginning. In admitting that as a reality applying mathematically, we then can proceed from that point without further arguing or debating about that part. In admitting to such a start we can conclude that with that comes an inconsistency not yet discovered. The official father of the Big Bang being accepted is a person by the name of Father LE MAÎTRE.

Father LE MAÎTRE, GEORGE ÉDOUARD (1894-1966) was a Belgian priest and cosmologist who was the first person to embrace the fact that the Universe expanded from an infant stage. His model of an expanding Universe (1927) was superior to that of W. de Sitter in that it took into account mass, gravitation and the curvature of space. Similar models were proposed in the early 1920s by the Russian mathematician Alexander Alexandrovich Friedmann (1888-1925) but Friedman compiled various such possibilities. Lemaître argued further (1931) that the quantum theory supported an origin in the explosion of a 'primeval atom' or 'cosmic egg' into which was originally concentrated all mass and energy. As modified by A.S. Eddington, Lemaître's model provided the springboard for G. Gamow's Big Bang theory. In the wider picture of science in general a lot changed to just allow such turnabout in thought since the day of Isaac Newton. From Newton's attraction and contraction many things came into place that allowed changed in the most hardened minds. Accepting facts about the Big Bang concept is quite radical.

LE MAÎTRE'S UNIVERSE
A model of the Universe containing a cosmological constant term, named after G.E. Lemaître. In this model, space has a positive curvature but expands forever. The Lemaître Universe is both homogeneous and isotropic. The most interesting aspect of such a Universe is that it undergoes a so-called coasting phase in which the cosmic scale factor is roughly constant with time. This theory was an evolutionary process to act as an alternative term for the standard Big Bang theory. The word 'hot' was initially used to distinguish it from a rival theory, which had a cold initial phase. The existence of the cosmic background radiation requires that the Universe must have been hot in the past if the Big Bang picture is correct.

The Big Bang theory accounts for the expansion of the Universe; the existence of the cosmic background radiation; and the abundance of light nuclei such as helium, helium-3, deuterium, and lithium-7, which are predicted to have been formed about 1 second after the Big Bang when the temperature was 10^{10} K. The cosmic background radiation provides the most direct evidence that the Universe went through a hot, dense phase. In the Big Bang theory, the background radiation is accounted for by the fact that, for the firs million years or so (i.e. before the decoupling of matter and radiation), the Universe was filled with plasma that was opaque to radiation and therefore in thermal equilibrium with it. This phase is usually called the primordial fireball. When the Universe expanded and cooled to about 300 K it became transparent to radiation.   The discovery of the microwave background in 1965 resolved a long-standing battle between the Big Bang and its then rival, the steady-state theory, which cannot explain the blackbody form of the microwave background. Ironically, the term Big Bang was initially intended, to be derogatory and was coined by F. Hoyle, one of the strongest advocates of the steady state.

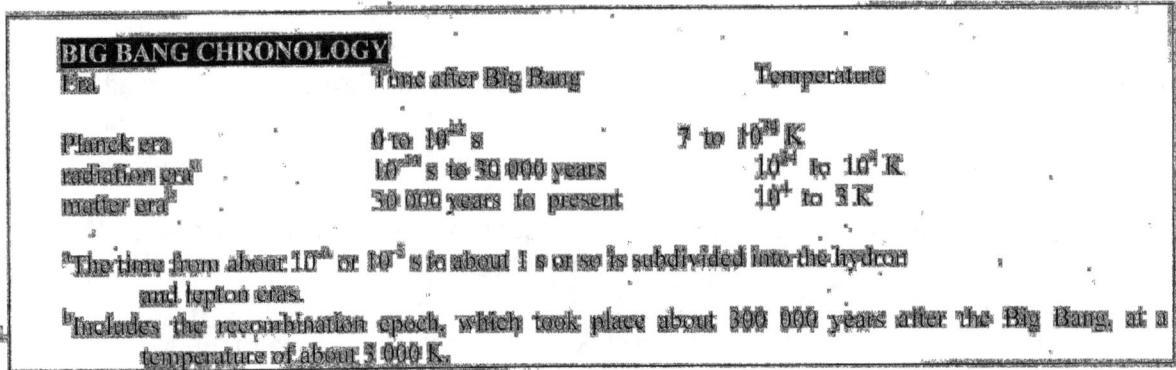

This clearly shows that the Universe came form one single point where everything was concentrated and was dense to the extreme, which disprove Newton's presumption of gravity produced by attraction. We are coming from a single point and the cosmos is expanding and we are definitely not contracting to a single spot as the idea of gravity being presented as a force that produces attraction. At such a point where the Universe came about in such a small confined space present during period the gravity must have been supreme by having the ability to confine the Universe into a space the size of less than one atom. The preference choice of gravity is to reduce space. This statement too may see shocking at first to those pure at heart but a statement I do explore later nonetheless but ask the reader at this stage to accept. Gravity holds the sphere as the preference choice with out a doubt. It means that to find gravity we have to investigate the sphere. Gravity holds the strongest locality and holds the sphere as the strongest form of all forms available. This will become better understood and might then be less reluctant accepted when I indicate the value as well as the Location of singularity. In the sphere where space is the least and that will be in the centre of the circle gravity bonds all atoms together in a unit as well as distribute a specific alliance in shape and form. Gravity is the strongest in all cosmic structures holding the form of the sphere and is in the very centre where space is the least therefore the more any star produces gravity, the smaller the star gets as far as volumetric occupation goes. Take the Neutron star and the Black Hole as an example and compare that with the Sun and the answer are self-proving. I claim that gravity is all about reducing space and not attracting matter but that I explain on another occasion. Therefore gravity must be where space is the least. Looking at a sphere we find that which holds the sphere true to form is in the centre of the sphere that then has to be the most intense point of gravity. By confirming the round shape without favouring any specific point, one can see that the sphere as a form is dominated or controlled from one specific location in the centre. From the centre every part of the circle structure and all structural positions of the circle in all circles refer to the centre in perfect aligning. In the middle the sphere bonds all sides there is possible equal and that then has to be where the strongest gravity can be located. The Big Bang was where gravity held the Universe in the least space there ever was. To find the original gravity we therefore have to reduce the sphere to the circle and reduce the circle from there as far as one can go. In our reducing of the Universe we must first acknowledge that the Universe constitutes many spheres, which is giving the Universe gravity as a combining unifying part of the sphere giving the sphere form (or gravity) the sphere being a circle in many positions from where gravity secures form. If we wish to go back in time and at the same time maintain some coherency we must concentrate on a single circle because a sphere is a circle by millions of possibilities linked together by a name that changes the concept. Going into our past we must take a circle that is representing the cosmos and reduce the circle to see what comes from there. That is what I did and in that I uncovered a hornet's nest. The hornet sting depends on the individual person's view of what a sting is. The sphere is any circle and all circles have two parts. On part is a line that indicates the distance between the two opposing sides being 180^0 apart. The second factor is the pi indicating the result of the square value of the other factor being the straight line which by the square of the shape it indicates...but where as in the normal square, the line matches angle touching lines that connects and in the square on will find the surface in the lines. With the circle the square falls inside the square but the lines cross at an angle and the edges hold the form value of the circle being Π. The circle holds a cross inside and a square never crosses the lines but the touch at angles. With the square we refer to the lines indicating the borders of the square by side we face as in example length and breadth. In the case of the circle we use the lines crossing to the inside on the length they represent being a full line or half a line. In the case of the full line the name used for such a line is a diameter and in the case where only half the inside line of the circle apply we gave it the name of a radius. Being fully aware of the various names I prefer to use r to indicate any and all lines that forms a combination to indicate either surface or volume.

To go back into the past one have to reduce the radius because the radius of the circle indicate the size while pi indicate form. By reducing the circle through the radius it becomes another matter of eliminating form because in such reducing of the radius it then becomes a matter of reducing a straight line. When reducing the circle in size one has to reduce the radius or the diameter because the pi factor is the indicator of the form as being a circle. It then becomes a process where it is just dividing the radius defined by the symbol r by halving the answer every time until there can be no dividing any further and such a reducing cannot end by becoming zero. One may divide by two, halving the result every time, which is a normal mathematical expression and by reducing by half can never reach zero. Allowing zero to be accomplished through any legal mathematical equation is not a mathematical fact and I challenge any person to prove such a feat. The numerical procedure may become tiresome or to small for any human to make sense of the outcome but never can such a dividing bring about zero in the ultimate answer. Zero cannot divide nor can zero multiply should I wish to retrace my steps to where I started. In using the method of the dividing by reducing the answer to half the size it will forever allow such a process to continue without ending in zero because no matter how small, the next value will be dividable by two in that will forever be a value in place. This stands as a mathematical fact and I do not have to prove my statement but those persons in academic positions, which portray me as a person trying to manipulate facts to fit my convenience, must explain how one can multiply distances filled with zero as a concept. This alone is the biggest obstacle why there is so little exploring going about the space expansion present overall. With that remark I am just referring to the work of Hubble but trying to make all out sense of the concept in general.

The man that (to my humble opinion) took cosmology into a new dimension was HUBBLE, EDWIN POWELL (1899-1953), the American astronomer. He first studied nebulae, concluding in 1917 that the spiral-shaped ones (which we know as galaxies) were different in nature from diffuse nebulae, which he found to be gas clouds illuminated by stars. From 1923, using the 100-inch (2.5-m) telescope at Mount Wilson Observatory, he resolved the outer regions of the spiral nebulae M31 and M33 into star, identifying over 30 Cepheid variables in them. This proved that such 'nebulae' were truly independent star systems like our own – other galaxies. In 1925, he devised the so-called tuning-fork diagram of galaxies, dividing them into ellipticals, spirals, and barred spirals, which he believed to indicate an evolutionary sequence. By 1929, Hubble had good distance measurements for over twenty galaxies, including members of the Virgo Cluster. By comparing distances with their velocities, as revealed by the redshifts in their spectra, he concluded that galaxies were receding with speeds that increased with their distance, a relationship known as the Hubble law. This was powerful evidence that the Universe is expanding. The dynamics of his work was so far reaching everybody (including Einstein had to revise their theories to accommodate his findings. His findings are the most disputed, undisputed observations in all of history. The HUBBLE CLASSIFICATION is a widely used system for classifying galaxies according to their visual appearance, illustrated on the tuning-fork diagram. The sequence is based on three criteria: the relative sizes of the central bulge of stars and the flattened disk; the existence and character of spiral arms; and the resolution of the spiral arms and / or disk into stars and H II regions. The system was originated by E.P. Hubble.

The sequence starts with round elliptical galaxies (EO) showing no disks. Increasing flattening of a galaxy is indicated by a number which is calculated from 10 (a − b)/a, where, a, and b, are the major a minor axes as measured on the sky. No elliptical is known that is flatter than E7. Beyond this, a clear disk is apparent in the ventricular or SO galaxies. The classification then splits into two parallel sequences of disk galaxies showing spiral structure: ordinary spirals, S, and barred spirals, SB. The spiral types are subdivided into Sa, Sb, Sc, Sd (Sba, SBb, SBc, SBd for barred spirals). With each successive subdivision, the arms become less tightly wound (but more easily resolvable into stars and H II regions), and the central bulge becomes less dominant. Two types of irregular galaxy are defined. Irr I galaxies show rather amorphous, irregular structure with perhaps a hint of a spiral arm or bar, and can be placed at the far end of the spiral sequence. Irr II galaxies are sufficiently unusual to defy assignment to any of the other types, although this category encompasses only about 2% of bright or moderately bright galaxies in the nearby Universe. The original, erroneous idea that the sequence might be an evolutionary one led to the ellipticals refers to, as early-type galaxies, and the spirals and Irr I irregulars as late-type galaxies. Colour and amount of interstellar material vary systematically along the Hubble sequence: ellipticals are red and contain little interstellar gas or dust, whereas late spirals and Irr I galaxies are blue, with significant amounts of interstellar material.

The relatively faint dwarf spheroidal galaxies were not recognized as a separate type in the Hubble classification. Some variants of the Hubble classification use plus and minus signs to subdivide classes, so that Sa^+ is later than Sa, but earlier than Sb^-. The importance of the HUBBLE CONSTANT is still to this day, underestimated. This "constant" is well explained, for the first time, I might add, in this book. The Symbol H_o is the figure that relates the speed of an object's recession in the expanding Universe to its distance in the Hubble law. It represents the current rate of expansion of the Universe. This important cosmological parameter is usually measured in units of kilometres per second per mega parsec. In the Big Bang theory, H_o varies with time and it is therefore more properly known as the Hubble parameter. Its value is not accurately known but is thought to lie between 50 and 100 km/s/Mpc, recent research tending to favour values towards the lower end of this range. In the HUBBLE DIAGRAM, a graph plots either the red shift, or velocity of recession of galaxies against their apparent magnitude or distance from us.

The Hubble law appears in the form of a straight line on such a plot. The original diagram, presented by E.P. Hubble in 1929, was the first indication that the Universe is expanding. The Hubble diagram is mainly used to test the geometry of the Universe, since at large distances any departures from the simple linear form of the Hubble law should show up as a curve. The HUBBLE FLOW is the general outward motion of galaxies resulting from the uniform expansion of the Universe. All motions lie in a radial direction from the observer, and the velocities are proportional to the distance of the galaxies. The real pattern of galaxy motions is not exactly of this form, particularly close to us, because of the mutual gravitational interaction between galaxies; some nearby galaxies are even moving towards the Milky Way. At large distances, however, the discrepancies are small compared with the Hubble flow. All these findings are incorporated in the HUBBLE LAW, which is the mathematical equation of the principle law that governs the expansion of the Universe. According to the law, the apparent recession velocity of galaxies is proportional to their distance from the observer. In mathematical terms, $v = H_o r$, where v is the velocity, r the distance, and H_o the Hubble constant. The law was put forward in 1929 by E. P. Hubble.

The HUBBLE RADIUS is a distance defined as the ratio of the velocity of light, c, to the value of the Hubble constant, H_o, This gives the distance from the observer at which the recession velocity of a galaxy would equal the speed of light. Roughly speaking, the Hubble radius is the radius of the observable Universe. Depending on the precise value of the Hubble constant, the Hubble radius lies between 9 and 18 billion l.y. This data is the basis on which the age of the Universe depends and is the HUBBLE TIME. The time required for the Universe to expand to its present size, assuming that the Hubble constant has remained unchanged since the Big Bang. It is defined as the reciprocal of the Hubble constant, $1/H_o$. Depending on the precise value of the Hubble constant, the Hubble time is between 9 and 18 billion years. In the standard Big Bang theory, the actual age of the Universe is always less than the Hubble time, because the expansion was faster in the past.

The distance between the Sun and Pluto is roughly one hundred times more and if the distance between Mercury and the Sun, but both has nothing between them and the Sun. If space comprises of nothing as Mainstream Science advocates, then could anyone being sober and of a sane mind explain how nothing can become plural forming more than nothing, or be multiplied as to give a multiple number value. If a line started from a single point and went on to become one hundred, then the one from where the start originated cannot contribute to a progressive value if science is correct in the surmising that lines start from nothing. To grow into a multiple value, the starting point of any line then must be part of something of value to use as a measure in continuing use. If the one substituted the nothing, all laws of mathematics will go in disarray because when one multiply any number by zero it becomes zero and if it was true that there was nothing in outer space, then it is nothing separating Pluto as well as Mercury from the Sun. With nothing being in between Pluto and Mercury and the Sun, this would place both planets in the centre of the Sun. By excluding nothing from the equation, space becomes something bringing in a value lying inside the realms of the infinite that must form singularity.

Mercury has 58×10^6 km and Pluto is 5900×10^6 km space between the Sun and the planet. That indicates a distance and a distance comprises of something, for if was nothing then both would have equal nothing and be next to the Sun. I repeat, having any indication of a distance being in place, the distance indicates something because nothing would place them both in the sun. The problem is identifying something from nothing that defines the difference there is in science. I cannot see how nothing can become plural or become more sometimes and in other cases indicate a value of zero.

Taking that into account it is important to recognise that notwithstanding the size of a line, there is another line (or dot) eternally bigger as well as eternally smaller than the line in question. We can never grasp the size of a line that forms the utmost or the least of possibilities and therefore size belongs to the human mind forming conceptions of big and small, but it has no place in the cosmos at large. This concept not only applies to size, but to all limits and divides we wish to create forming borders we can appreciate. When looking at the circle in the conventional manner, we persist with errors brought about in culture and not by applying some significant modern logic.

From the smallest ever possible dot will grow a line in every imaginable direction relating to a prospect of Π where the value of Π will not favour any one direction. This reality puts all directions at equilibrium meaning that any form of what ever might develop from such a spot, will have the end and the start being in the same position, which will also have to be a sphere as the flow outward will be equal in all directions. This reasoning prompted me to look for singularity in such a spot because if the prime spot from which all came was a spot holding all, then the spot must hold the shortest line but more prominent it will hold the smallest form including the smallest circle or for that matter the smallest sphere. One possibility that the shortest line or smallest spot can never have is having a starting point on the zero mark. If the mark of zero holds the start it must also hold the end because the end and the beginning have the same position. If the position of zero then is the beginning, the end will also be zero leaving the line or spot without an end as well as without a beginning. Such a spot will constitute all through the distance covered by nothing, which have a total of nothing. Any line starting from zero would inevitably start from a point where it ignores the zero mark because the fact of zero being present implicates the absence of any possible starting point and zero by value does not implicate a start or a size of value, but only the not being there in that position. All lines would form a duplication of another line sharing value since there will always be a possibility of yet another line in the realms of singularity lying between the two lines in question reducing the size infinitely to either side of the divide we humans create. Boundaries therefore are human and as man made substances it does not belong to the cosmos outside the influence of man and must in such a case be discarded.

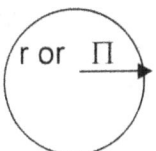

•$r/2$ •$r/2$ • $r/2$ dividing r reduces r to infinity but not Π as Π remains stable, protected by the rotation of matter forming a circle around singularity.
When **the circle reduces**, the depreciating **value** allocated to **r** as the radius will become implicated because **r determines specific size** of the circle that reduces. That **doesn't apply in** the **case of Π, because** Π in the true sense only **indicate to**

form bringing differentiation between the circle and a square where the circle is without corners and therefore Π **dictates form and not size**. By **reducing size** only **r comes into question** and only r will be affected by such reduction. By **reducing** the circle in depreciating the **radius r by half continuously** will lead to an **infinite small circle** but Π **will remain because the circle as a form remains** even being infinitely small

In any circle or sphere the size only depends on the fluctuation of r to the square value, as a component to the circle or sphere but that does not affect the form by indication of Π in any way there may be. The conclusion from this is that no line can start at zero because that will be a mathematical impossibility. A line or spot starting at zero would therefore be shorter than the shortest line possible. For obvious reasons can no line, or any line grow or extend from zero because such a line must then quit zero and become something, thus abandon its original value. That would mean when using zero as a starting value, the start of the line has a different value to the end and his is not possible as a line holds conformity through out. When any line is starting from point zero, it can never leave zero because of the influence of being zero disqualifies any possibility of growth. If the line then had to grow in all directions at the same pace the line must therefore be a circle or being three-dimensional, a sphere. Flowing from this fact is that in the Universe there can be no zero point or unfilled space. In the case of the growing sphere, the value of the circle is Π, and that is where creation started. That gave me the clue where to start looking for singularity. One would find singularity in the value Π and the value Π will be in all things rotating in a circle. You might wonder how does that apply to the cosmos and moreover to gravity since the only value one can attribute to singularity must be 1?

By reducing r indefinitely to the tune of half each time, r would become infinitely small, beyond human calculating means, however as mentioned in the case of the smallest dot holding one spot, r would become insignificant beyond human comprehension even, but never reaching zero and still Π would remain intact and  dictating form. Even when r loses all value to indicate any distance of any sorts, what thereafter will remain is Π. An observation coming instinctively to mind one may recognise is that the form with r reducing, this reminds rather explicitly to the form associated with natural phenomenon such as hurricanes, water whirls and even the shape most commonly favoured to express the cosmic object referred too as a Black Hole. The similarity may be more than coincidental. Let us consider the statement in the reverse.

Anything occupying space in the cube will apply r, notwithstanding the name used confirming the shape or r named as length width or height, it is all just a straight line bringing about the cube with all its other names that may find attachment to specific form but nevertheless still remains only a six-sided cube with connecting lines applying different angles that changes in some cases. The normal perception is that any circle growing spontaneous would grow by the radius, which is r. That cannot be the case because r is an indication of a straight line. By growing with the aid of a straight line that is running from the centre to circle's end, the influence that that would have on the circle would result in many circles following one another and not be indicative of a continuous growth. Gravity is the dimensional changing of space holding r as reference in the cube formed as a circle to then form a sphere holding Π as the reference. In order to generate spin producing time in matter occupying space, therefore creating dimensional change Π has to be a factor indicating the possibility of spin because implementing Π the circle sides will follow one another without establishing separation. The answer must be in finding Π and thereby locating singularity.

Locating and finding **Singularity**

In the **precise middle** of all **objects in rotation** is a precise centre dividing the object in sectors that will **start the spinning initiation** from that centre point. Thus, the spinning object **will have a middle point**, a very specific **centre point that does not spin** and only holds Π as a specific value. One value such a line **cannot have is zero** because **zero does not start any** line and therefore the **value of the line must be infinite,** just as described in **accordance** and by **the definition of singularity.**

When the top is not spinning, there is no such line detected, but when the top spins, the line forms from the top to the bottom running all along the centre. This line divides the top into four directional sectors that opposes in direction of rotation. The line's worth has gone undetected since Newton announced gravity and whereas the line is the most important aspect concerning physics, yet the existence of the line was previously never noticed.

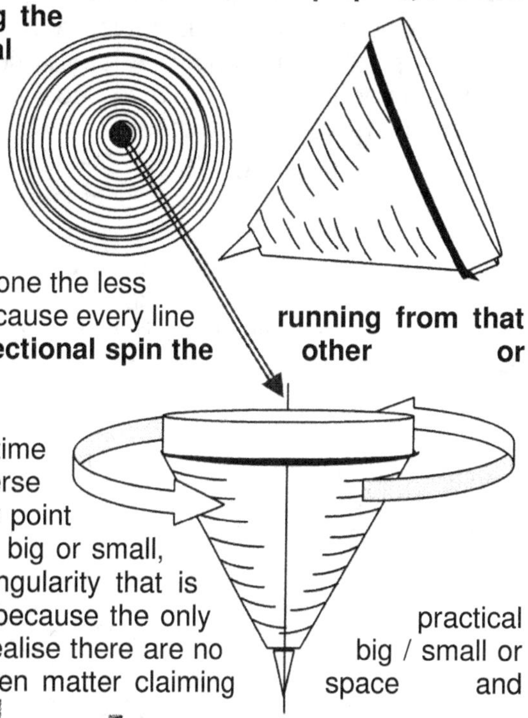

That point albeit hypothetical, is also as much a reality none the less and is placed where that point **must be standing still** because every line **point** in **opposing directions** are also **in opposing directional spin the opposing side.**

running from that other   or

Arriving at the question about locating the space and time forming the centre of a circle, or the centre of the Universe one has to realise the centre of the Universe is in every point forming where singularity is forming matter weather it is big or small, size carries no significance. It is the impartiality of singularity that is claiming the value and not the differentiation of matter because the only value one may contribute to singularity is 1. One must realise there are no hot /cold or near / far; these are all relevancies between matter claiming

practical
big / small or
space   and

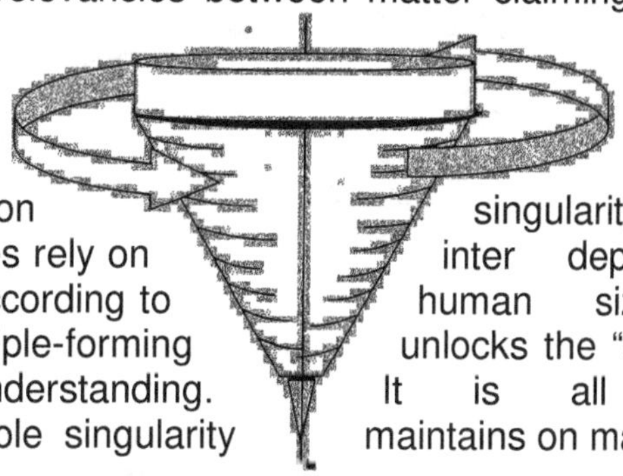

space is heat in a turnabout manner. cosmos where differentiation would Universes, sealed off from other inclusive or exclusive depending on relating to one another. The relevancies rely on linking, but there are no differences according to Accepting these facts as being principle-forming the Universe and brings about clear understanding. acknowledging and interpreting the role singularity

Every aspect in the apply is locked-in Universes and is singularity holding relevancies inter dependence and inter human sizes or standards. unlocks the "so called mysteries" of It is all about accepting, maintains on matter.

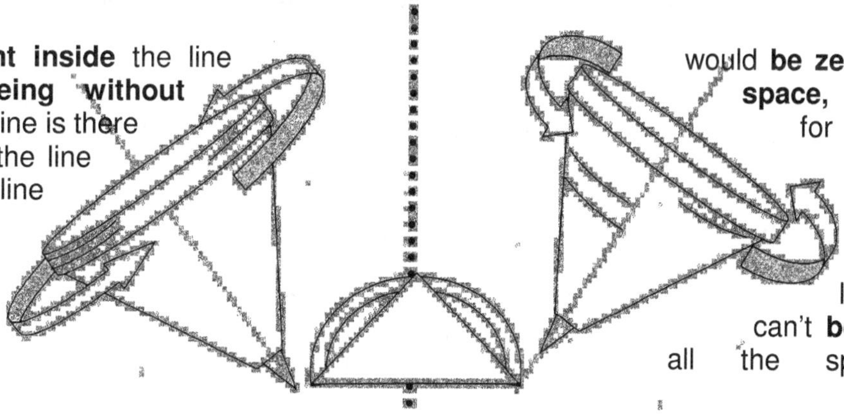

Spinning or movement inside the line the line, although **being without can't be zero** since the line is there see**.** The movement in the line **zero** and the space the line **is zero** and the line holding a value in might **be zero,** but the **a cosmic reality** just since the line controls taking place.

would **be zero,** but **space,** also for all to **is** uses

size line as can't **be zero** all the spinning

From this centre line that is only theoretical definable, but is still there all the same, an opposing value always form from a previous turning position to the next turning position that becomes real and distinct when rotating, but loses its distinction when not rotating because then all traces of the line that is not there is lost as the line disappears. As the line disappears the value of the line not being there changes from being noticeable to zero, and as the line removes from having a notice ability in securing a value by spin to then when not spinning have a value of zero, where this zero value then replaces the most original value it had. When not rotating, zero removes the line from a position it never held in space previously. When rotation begins, the line forms and are only backed in value by having only a

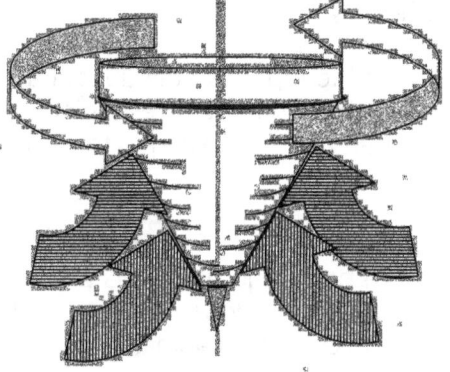

hypothetical position claiming zero in spin and in space but not in presence. Being without space doesn't make that the line is not less distinct but the line is more distinct than any other part that in reality does hold space and therefore participate in spin because from that point every rotating piece of what ever is, then will spin around this line that is not there to start with and such spinning will clearly carry from where the line only has a distinction value in the singularity to carry on with a value of Π implicating rotation. The line forming hold 1 in singularity and from where such a line ends, only there does the circle value of Π start.

If the spinning top is all the evidence any one needs to come to such a conclusion what will bring any proof that the singularity governing the top connects too anything anyway we will then find it when studying the spinning behaviour the top represents. Placing singularity in a location not being present in the Universe is fair and fine, but what will the evidence be in proving its activeness as part of the creation at large?

The reason why we can be sure it is active is that when spinning, it shows borders implicating restraining of further set limits. It is the fact comes about when triggers the spinning too fast because the upright the top the airborne and retaining. When the top movements outside the that the same affect spinning too slow that questions. When the top fights something alignment keeping it starts to tarnish. When spins veraciously in speed, the top clearly starts to fight the restraining of gravity as top tries to become lift from the Earth's spins to slow, it fights desperately hard against the restraining of gravity trying to retain the top's position accordance to having no motion or having mass. The same apply when the top is spinning too slowly as when the top spins too fast and this indicates clear borders in which the top can conduct its spinning performance.

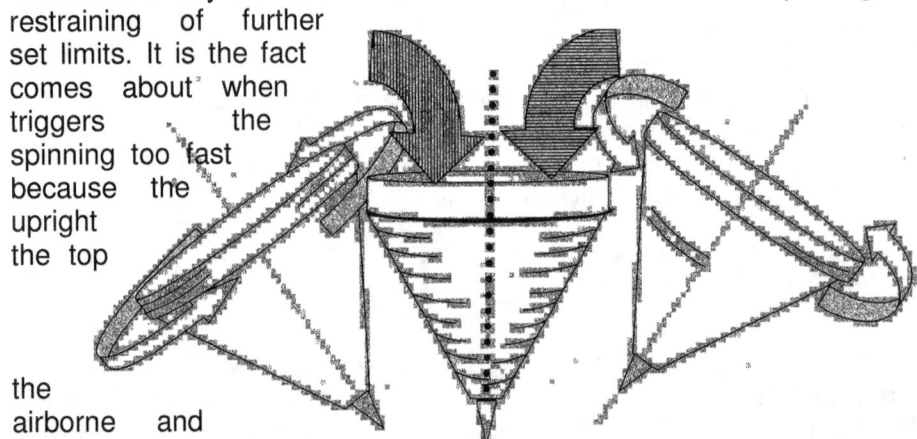

By going faster (past the upward border) the spin goes oblong where it actively tries to change the position the top holds in relation to the Earth and moreover in relation to the surface of the Earth. By going too slow it once again shows identical characteristic behaviour. When going too fast it indicates an attempt to rise into the air, therefore relieve its singularity in an effort to part with the Earth's by fighting off the Earth's restraining gravity.

It shows unmistakable characteristics of trying to become airborne securing an independent position from the Earth, which holds it down. At the bottom we surmise correctly that it wishes to fight the coming restraining and prevent the inevitable toppling over and falling down.

Of course the bottoming out shows the same characteristics whereby we gauge that to be the normal process of falling down, but it is a clear fight to survive that is going on. If the bottoming is also relative to the Earth's gravity placing limits on the speed in which the top may stay upright and if we recognise the process as being normal, then the top trying to fight of gravity by escaping the Earth's boundaries where it reaches the upper speed limits should then also be adjudged to be just as recognisable with the view of normality.

In determining this behaviour while reconciling it as forming part of a cosmic process where matter interact with matter in an laid down set of rules, we should once more be asking questions and this time it is whether the top will show the same behaviour in outer space as it does on Earth. With giving a reply of "no the top would not behave in the same manner" also comes an admitting that the process involves the interacting of singularity of the Earth with the singularity of the top where the spinning created independent singularity, making the top's gain in singularity being as valid as that of the Earth because the Earth has a role in sustaining it or destroying it at the border ends.

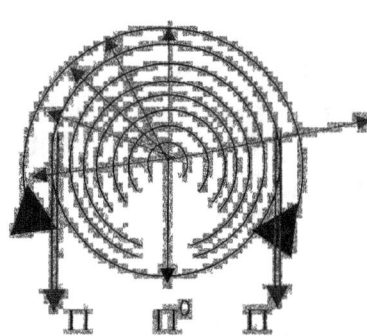

in constant directional change as time flows through rotation

In the centre is a spot that is not there because although being there in having a presence the dot has no space and no part of the Cosmos.

Pinpoint positioning of singularity Π^0 with Π positioning space to either side forming the border set by singularity

While spinning the movement will create that the new direction is also pointing to a new location in relation to the previous point and the following position will oppose the previous point it had in relation to direction considering the centre point...everything is always changing while nothing remains the same.

In the sketch the circle to the right would come about from a straight line r growing by influencing the appreciation of Π, but to influence Π would lead to a breakdown in r as Π and r are different entities. The circles to the left shows a continuous growth by extending Π every time and since Π is the same part as the previous Π, only extending that billionth of a millimetre each time, the circle will be truly continuous without any signs of a break as 3ould be when identifying a new location that serves a new r every time.

Looking at the affect of gravity it shows the precise quality of no distinctive point, as gravity never seems to end at a point but flows all over affecting all that holds a position in its sphere of influence. The gravity coming from China meets the gravity coming from America at no particular spot but intermingles without distinction.

Every person that is the least familiar with mathematics would know that a straight line, and a half circle as well as a triangle are all equal to 180°. Ask the greatest mathematical minds why this is the case, while the form attached to every individual shape varies so much, and not one would be able to answer in any specific detail. I found the answer to be in singularity and as the book progress I will show how this comes about.

The triangle, the half circle and the straight –line has two things in common, they share 180^0 as a mutual value and they are part of singularity because the share a value and not any form.

Using the concept that gravity applies Π as the circle factor Π as well as Π^2 replacing r^2 the replacing by Π brings two values as Π and Π^2. That I found is the case with gravity and will be apparent when explaining the sound barrier as well as the Four Cosmic Pillars. In order to create a distinction I remained using r as the indicator of the cube or non-circle that has vacant space and by vacant space I refer to non-solid structures. In the solid structure I use Π as a value for reasons that will become apparent in due time.

Space-time is a four dimensional position of the Universe where the position of an object is specified by three coordinates in space and one position in time.

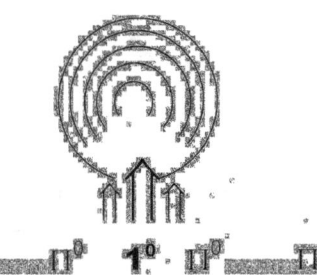

With singularity placed in infinity within the centre of every rotating object every atom and its relation to its surroundings including other

atoms form space-time diverting from the point holding singularity as far as rotation goes because every object holds three relative positions in as far as where it was, where it is and where it will be in relation to singularity providing time. I elaborate on this else where.

Due to the spinning nature of such a point with all surrounding the point will be alternating direction favouring change every second and in that the value to such a point can only be Π because of its constant changing. Using r would specifically oppose another r from every angle because the use of r will bring about a static relation to the previous and following instant and therefore it will cancel the constant spin flow. In the circle using $r^2\Pi$ the r has to have distinctive qualities placing it as a factor apart from Π. Where the growth shows no separate distinction but a continuous flow from the precise centre to the precise edge the flow would become in relation with Π depicting the circle and Π replacing r as reference to any point on the circle. By using r as a distinction in the circle division is possible but by using Π there is no distinction possible making it a solid flow and that is one of the main qualities we attribute to gravity. In considering the spinning motion in the fraction of time in the detailed instant every aspect of rotation will turn in every instant of change in time. Although the points had the same characteristics only seconds before, they oppose the characteristics it had just before and just after the very second in which they are and to which they relate by similar points also in rotation. The fact of the graph proves my point in quarterly opposing dimensions and values,

Any point will be opposing itself within the **rotating of 180°** where it **then change every aspect** of its **previous flowing** characteristics it had or **will once again have** in **360°** from there. While in rotation from the view point of a bystander it all may seem static and never changing but to the object in spin every next instant in time will be diverting from every aspect it had every second passing, and the direction it held in relation to the direction it held the previous mille, mille second will totally be incompatible with the direction it holds the very next mille, mille second of rotation. This is why we can use degrees measuring the circle by (6^2) (forming the square relating to matter through singularity) X 10 (square if space) = 360° however it is always in motion. That proves no point can be static or constant, though it may seem that way to outsiders. Although matter is matter, matter can also be anti-matter and moreover form its own anti-matter at the same time, just by the **rotating of 180°**. This degeneration of structure is very likely to occur with overheating. Revaluing Π to Π^2 will bring about a new contact point where Π meets **r** forming another relation in Π^2 **Time is** the **changes in relation** where Π **contacts a different r** not withstanding the many r points there may form because **every r constitutes a different value** to the Universe through other ratios and relevancies brought about **by heat and light. Time is the duration it takes Π to rotate between any two given points of r** and therefore must always amount to **a square (T^2)** moving from point to point through the **cube of space (a^3)** in that **duration of time (k)**. With that it proves **Kepler's a^3 (space) $=T^2$ k (time in the instant of motion)** but motion must continue through a specific value in space where the space-time is maintaining relevant equilibriums throughout singularity connecting.

Since occupation may or may not be placing the factor in infinite, the space therefore holds the premier singularity of infinite from which all included in the Universe has come. When the top starts spinning in a specific position the top merely executed the option to fill the premier singularity at that specific point. When it moves it may take the premier singularity with to the new location it moves through spin or it may fill yet another position in singularity as all is the same.

The influence immediately above the circle will have the biggest influence and reduce gradually as the value of Π reduces in the leverage that the space has on Π and a gradual but definite change from Π to r will affect the extending of Π progressively more. The decline of Π will follow the same contour of the circle at 7°.

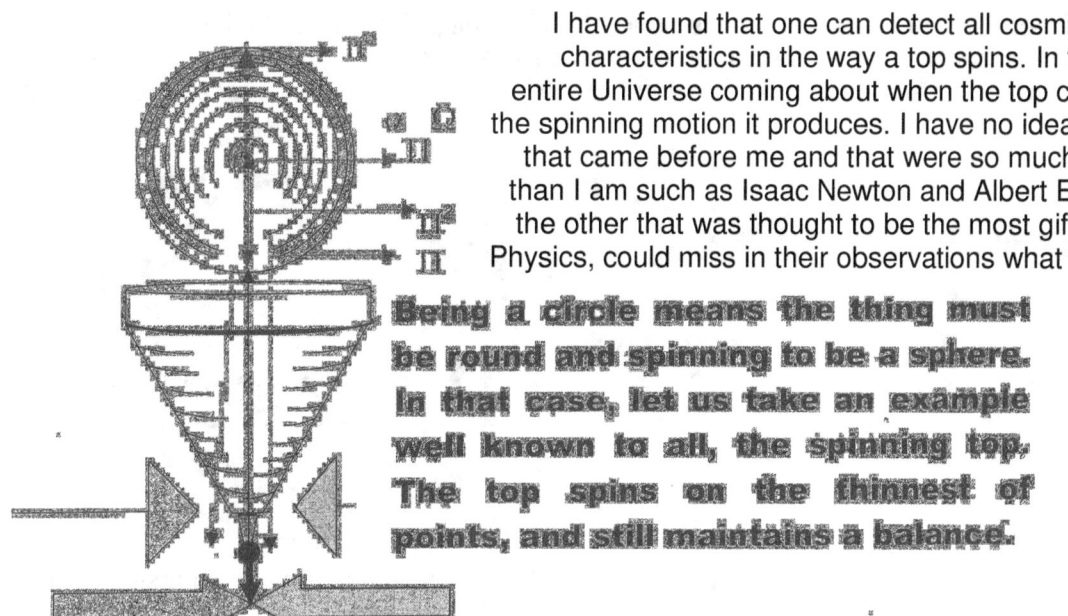

I have found that one can detect all cosmic gravitational characteristics in the way a top spins. In fact I found an entire Universe coming about when the top comes erect by the spinning motion it produces. I have no idea how all those that came before me and that were so much more brilliant than I am such as Isaac Newton and Albert Einstein and all the other that was thought to be the most gifted masters in Physics, could miss in their observations what I saw and that which discovered. I discovered that the top adheres to every gravitational principle and the top revealed so much more than science ever knew before the top was investigated.

Being a circle means the thing must be round and spinning to be a sphere. In that case, let us take an example well known to all, the spinning top. The top spins on the thinnest of points, and still maintains a balance.

Looking at the top one is not struck by awe in regard to mechanical complexity or operational brilliance, but in closer inspection the top proves gravity far more convincing than does any argument that came from any gifted wise before. There are so many names attached to the discovery that Albert Einstein came upon during his unveiling of his theorem on the Special Relativity Theory with the Universe going flat at the point gravity is strongest and with the curvature of space-time is coming into effect at such a point. In Einstein's arguments the grandeur and the utmost splendour he showed the cosmos has reminds of a fairy tale land hidden below dimensional patricians. This breathtaking mystery world or mystery Universe had principles hidden from sight that struck us all with a feeling of bewilderment. In the arguments that the Master made, he portrayed a part of the Universe that was seemingly magical and beyond approach by the ordinary human such as me. The knowledge preserved about that part of the cosmos was only reserved for the Super Beings that had education more than I had breath. Those person's who carries knowledge in their veins where us mere mortals use blood, are the only Super intellectuals to claim the rite to understand aspects of this secret world in the cosmos hidden below dimensions which Einstein brought into the open. But to view it was preserved for the very few that had the mentality to understand the going on in that dimensionless shady world where light travels through darkness.

Then, Professor I found a way to make all that marvellous information that only the most brilliant intellectuals could grasp become so simple that a child's toy brings all the hidden facts to the understanding level of the intellectual commoner such as I. May God forgive me that I had the audacity to break the bubble of the mysterious world of science and show that what Einstein thought was special relativity, is common, everyday relativity and which is what the cosmos uses in all events. I made what was only reserved for the special and the privileged in science, become common and everyday facts. That I did by studying the most unlikely machine devised by man. What is most shocking is that a person with such meagre abilities such as I have, now has the capability to see what Einstein and others never could see. I made a search and I found singularity, not only located where Einstein said it was, but everywhere and in general use. If I had the tenacity, I would even call my theory the Relativity of not even General Relativity but Relativity in General common use found as singularity and such a relativity is found and in use all over and being in everyday use in comparison the Einstein having singularity only found in very selectively locations such as Black Holes and flat Universes and that makes his relativity extremely special when compared to mine being very commonly distributed. I burst the Newtonian bubble of placing singularity outside the realms of the general population and in a magical mystery Universe where gravity pulls the lot flat and so many other mystical things happen to out Universe. Also I burst the bubble on Newton's magical mass with such enormous powers as that it can pull a Universe all over and still get nowhere with al the pulling that is done. But I had help. It came in the form of a Man that measured the solar system and as he did so, the Universe let him in on the secret recipe by which the Universe is formed. The man is Johannes Kepler and his secret formula is space-time $a^3 = T^2k$

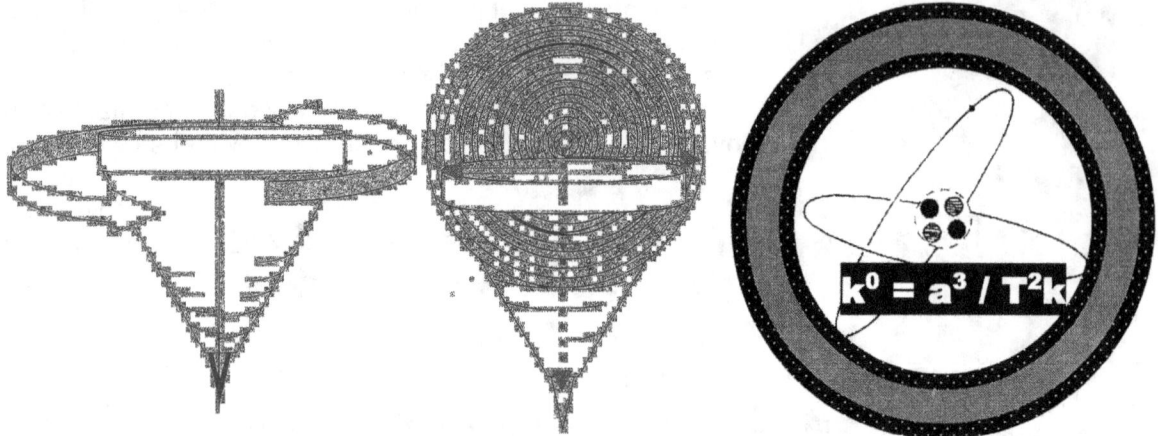

$$k^0 = a^3 / T^2k$$

It is known to science that a Polar Bear can smell a seal through an ice sheet that is up to and bigger than two meters of ice. When hunting for fish seals will be under water and come up for air, but they will hide in pockets of air captured between the ice on top of the pocket and the water line below the air pocket. The air they use may be in captured pockets, which are retained in holes formed under the ice, but the pocket is not connected to the surface by any measure. Between the captured air in the pocket and the air in the sky above the ice sheet there are meters filled with highly dense ice that is about as dense as concrete. The pocket might be secluded from the air and the top of the ice sheet where the bears hunt by a layer as thick as two meters. The sheet of ice might be twenty kilometres long by thirty kilometres wide with no hole in the whole sheet. Yet, the bear can smell the seal and can start digging exactly on top of the air pocket where the seal is hiding in air sanctuary. The ice sheet is as solid as concrete or metal. The ice sheet spans across six hundred square kilometres. There is no interrupting of the solidness of the ice and the top is completely secluded from the bottom. Yet with all this secluding of the air pocket the bear smells the seal. For the bear to be able to smell there has to be a release of tiny particles of the material being smelt which evaporates into the air. Through the nostrils the smelling organs pick up these tiny particles and analyse it to supply the information to the brain about what it is being smelt. The particles are microscopic in size, yet they are solid and they have to come from the smelly to the smelling and have to be carried by air. They are quite independent and float in the air as an air born substance. Only by using these microscopic particles can the bear find the ability to trace the seal. Yet the bear goes to the exact location and start digging. The unsuspecting seal has no idea the bear is on to hunt him because the seal are unable to smell the bear. Eventually the bear secures its meal and the fish has one less seal to fear. Yea sure no fine, so what's this got to do with Einstein's theory on relativity, I hear one ask?

The story will tantalise any child and pass completely by most grownups because it is an everyday story leaving not much to think about...ore does it? Let's see what part of cosmology is it that the average thinking person missing... The ice is as solid as rock. If it was water that was dividing the two animals then one may presume that some of the particles flowed in the water with the water to the bear and the bear tasted the particles in the water vapour... but it's not water it's ice. The density of the ice is so solid the ice can crush the hull of a ship with a small effort. Yet the particles run through the ice to the nose of the bear on top of the ice. Not only that, but the bear are able to smell the spot from quite a distance because the bear follow the trail across the ice to the precise spot where it starts digging. That means there is one very precise spot the microscopic particles leaves the ice. It is right on top of the seal. The particles leave the seal and run through the solid ice up to the top of the ice sheet from where the particles then distribute in all directional, as they are air born. That is where the scientist leaves the story. The bear finds the seal and the seal becomes a meal.

That is where the story starts. How did the bear smell the seal? The ice is solid. The ice is six hundred square kilometres of an ice blanket that covers the sea. There is no chance the air went out on the rim of the ice and by that allowed the bear to smell the seal. The ice is two meters thick and dense enough to sink the unsinkable Titanic. Yet it is porous enough to allow a bear to smell a seal lurking within the air sanctuary. How does that happen?

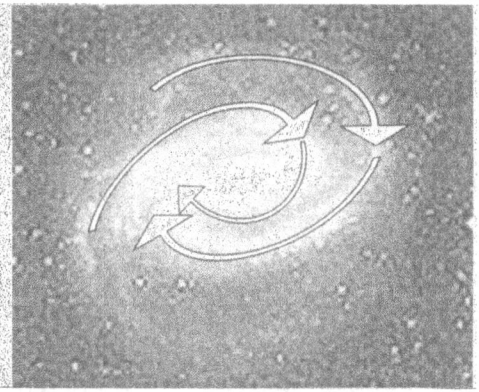

That is where the air being liquid can transport the microscopic particle being microscopic and therefore also liquid through the solid ice. That is the Coanda effect and that is the basis for all gravity where gravity is time moving space. This raises the question of motion. How does an object come into motion? How does an object displace the space it retains to another spot where it claims new space to occupy and in that motion it displaces the space it retains as well as the space it exchanged to occupy? Motion is the most complex manoeuvre of all parts of the cosmos and yet all in the cosmos must be in motion or it cannot be in space. Kepler showed that space is the motion of the space in time. This leaves the question about what is space and what is motion and what is time. It is easy to refer to space-time and it is easier to be Newtonian and have no idea what space-time is except by accepting it is a mathematical statement Einstein uncovered with his mathematics but even Einstein never understood what it was. I do not wish to belittle Einstein for he was a wonderful mathematician but after all was said, in the end and after he died, that was what he was, a wonderful mathematician.

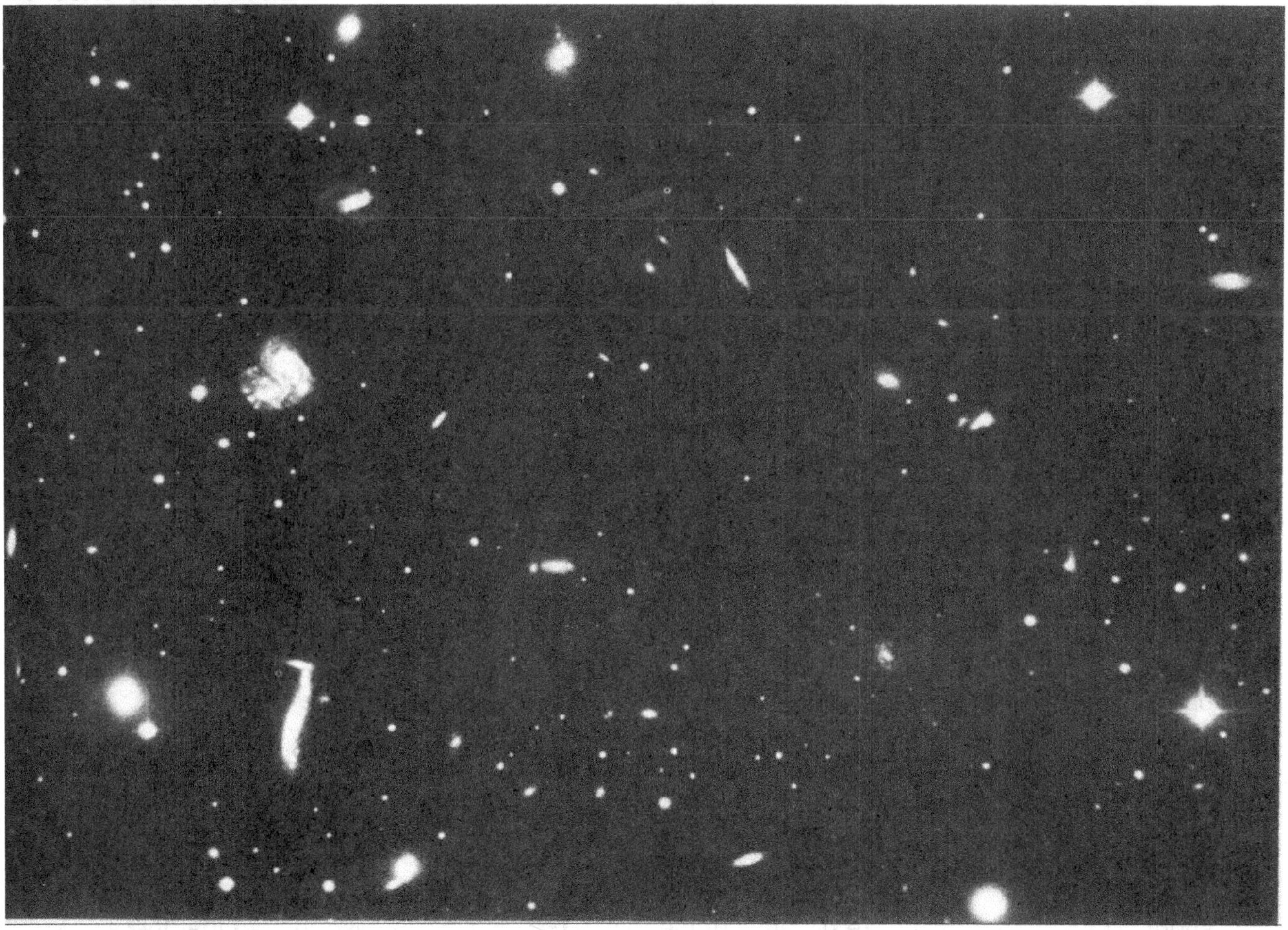

Johannes Kepler's secret formula is space-time $a^3 = T^2k$ and this means that everything is on the move and always encircling something of greater importance. A top can spin but the parameters of its spin are limiting the motion it can apply. By not spinning the top is still spinning as the Earth is doing the spinning on its behalf. When looking at the cosmos from whichever angle indicates the fact that the cosmos is moving. It is forever spinning and it is all going towards as much where it is coming from. When looking at the cosmos from whichever angle indicates the fact that the cosmos is moving. Everything is moving in respect to everything else. It is forever spinning and it is all going towards as much where it is coming from. The cosmos seems to spin in relation to a centre point being everywhere.

What would all this information leave us with I wonder. Please allow me to show what I found by studying all this information and allow me to share a thought that all this information left me with. My new cosmic theory arose from two questions that I asked myself and which I thought must be answered if I truly wished to know what the known Universe comprises of.

The first question is how is it possible for me to see nothing or to see the black sky at night. Both questions I answer with much detail in **an Open Letter on Gravity volume 1 and 2 part 1 and 2**. The second question is this: Looking at the cosmos at night I receive a picture that covers just about the entire sky on the one side of my viewing direction. The other side is also there for me to witness and I should be able to see it, if I had eyes on the back of my head. I see such a wide variety of stars through something as small as my eye socket. I do not see one star or a small sector of one star and then if I wished to see another star, I then have to shift the angle of my view to focus on another sector of night sky representing a part or fraction of the same star, or a tiny part of another star, but my vision allows me to view the entire view as one wide angled night sky. If this is possible to view everything with one glance, then the question to answer is then how is this possible? How can I see everything as big as everything is that is out there representing the cosmos through something as tiny as my eye-nerve? How can one photon carry the information not only coming from the atom that it left, not only from the layer in the star that it left, not only from the entire massive star that it left, not only from the region it represents, but it consolidates information coming from all across and everywhere that is covered by the direction in my view. One photon consolidates information coming from $180°$ that is seen by me as that which is on my left, that which is on my right, that which is in my centre and that which is top and bottom of me. This one small messenger, so small it is not humanly possible to measure and yet it is packed with information that underwrites all knowledge we could have about the Universe because in this hides all the knowledge of space in that instant that we could possibly know about the Universe.

When standing outside the night sky with one of my boys one night, I looked up at the sky. When they were still small, I often took my boys out into the open night sky on my farm and explained to them the constellations and the significant of stars they view. My farm that I had is situated in a semi desert and only those persons blessed with utter stupidity and that lacked the brainpower to realise there has to be other places one can fair better. Only those with enough self-hate were simple – minded enough not to see how much they suffer in the area where we farm. Only those such as I that is bent on seeing how much any human can suffer, would think of farming in the region my farm are situated since we only farm with drought because in the climate we have, the area could produce no other commodity. Droughts are ongoing events that are interrupted by the occasional all-destroying floods and in-between the monotonous of drought is interrupted by events of all-consuming insect pests that can only be compared in proportion to events reminding us of Biblical times. To farm in this area, one has to have the stupidity of a brainless zombie and the resilience of one that is too stupid to be capable of thoughts as well as having no money to go and buy any other place on Earth where conditions are more suitable for life.

 When God created this area, I suppose God was almost done for the day and night was setting in, so God left out the possibility to establish conditions that could be suitable for human life to become part of the area because it became dark and that meant it was the end of the day of creation. However, during the night and by candle light (I suppose), God made persons as stupid and as thick-headed as I am with the tenacity to fight for myself against all odds and fend off all the shortfalls that come as a prize because of the things God left out in this area, so that there at least was one place so close to hell, that when we lot get to hell we can take charge of others not used to such conditions because we then were already used to living in hell. I suppose our blessings will come in our afterlife when the toughness that hell brings along, would seem as everyday occurring events again. Our suffering could after all and eventually then be beneficial for other humans too.

In our part of hell daily temperatures in the shade in summer could be on average 42°C while not letting off that much at night and on average in the decade of the eighties my average rainfall was often below fifty millimetres for one entire year. Then, if it rained, it poured for a few seconds and that unleashed floods that brought on swarms of insects capable of destroying any conceivable form of plant life. However, in these unbearable weather conditions, God gave us one substitute that makes up for all other disadvantages, and because night fell while God was creating this place, God remembered to give us a night sky worthy of His Creation. It is like no other view on Earth could be because the night sky is almost never cloudy and is always crystal clear and using these conditions it gives a view with which it is possible to see forever in three dimensions.

The night sky seems as if it is three dimensional and as pure as it can be. One can see double stars with the naked eye and colours of stars are visible when looking at the star with unaided site. The brilliance is breathtaking and if it wasn't for the many droughts during the time I farmed or tried to farm and the droughts taking all my time all the time, I would be occupied with farming and would have worked day and night leaving me no time to ponder on physics. But as it happened, the droughts and the leisure time that it presented, gave me a lot of vacant time to study cosmology. It gave me the opportunity to think about other things, better things than to worry about my incomprehensible financial position and by escaping from my dismal living conditions, it helped me to use time in which to think about questions and to try and find answer about physics. If it wasn't for the unbearable nature, I would have been busy farming while trying to get rich as all humans desire to do and not kept myself busy contemplating about unanswerable question Newton left us in physics. By not facing up to my unsolvable situation, I escaped into a wonderful world of solving unanswerable question about cosmology and in the same time solve problems about physics that no other person apparently observed. That left me time in which to view the night sky in detail and one question came back every time to confront me. It is a question of size and space in space.

If light came as individual streams of photon flurries our visage would translate that as such. We would view the sky in that manner and see individual single items not forming a picture as a formation or a unit. It would be a picture of individual items standing far apart while being unconnected, bringing across some photons in the manner where every object stands apart not being related in any way and that will be what we see, if it is anything that we see. That we know is not the case and that means geodesic zero is as much rubbish as anything Newtonians regard with simplicity and with careless thought. This thoughtless attitude runs as wide as science goes, as deep as science take matters and it comes from the most basic we can find in science. Please allow me to indicate how basic and how enormous the thoughtlessness truly is. It comes in gravity and how much more basic can science go than the manner in which Newton simplified gravity.

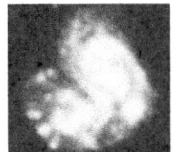

If light was simple straight-ahead flowing photons that was unconnected since the photons were individual particles, then every frame would be seen as either one of the frames and that will leave out the other frames not seen because of the photon flurries being unattached single beams, and that meant this would leave us to see being aided by non-connecting frames of light we had to employ in order to see. But because we see the entire picture as one structure, the photons must in some way link and connect to mingle in such a way that it will enable us the full view by using only 0-a few photons. By thinking that the light flow via outer space in one direct line from the source to our eye cannot bring a picture of even a part of one star, let alone the picture compliment coming from one spectrum of the Universe. Such obvious complexity goes completely unnoticed because the atheist mind of the general Newtonian intellect always present the complicated as simple and that makes the simple complicated so that the atheist Newtonian would seem wise and knowledgeable while they only know about total insane rubbish that they could think of as if it is complicated.

The idea might sound bizarre but it is not our fault we have such an idea because that is happening with the light travelling through the cosmos to us and to serve us. Every person standing in the

Universe is under the illusion that all the light throughout the Universe is directly flowing to the very point the person is standing. It happens to all of us. The place where I stand or any other individual for that matter is standing is positioned in such a manner that every beam is directly flowing to that very specific spot. Every beam is coming at the speed of light through the entire Universe to locate such a person with that magnitude in honour and glory as to fill the centre of the Universe.

From all the corners of the Universe and in specific every spot imagined one line of light is especially directed to go to that specific location used by that specific person for that specific instant in time. That line that is coming from every possible spot there may be is competing for the honour to reach the person filling that centre spot...and only the toughest will endure. It is sent from that spot as vast as a galactica might be or as huge as a star might be and this line is delegated to represent that huge area at the point the person stands, with the one lucky ray that finally made the journey successful. One very important human being is filling such a location of absolute splendour where all the light travelling from all over the Universe will meet to honour the person's presence. The light departed from every location in all points throughout the entire Universe, stretching further than the mind can admit there are such places, directed by a purpose to succeed and was determined to remain on course, notwithstanding any difficulty such a journey might represent, just to meet the person in that centre spot. We are all in such a centre spot. The light followed one after the other, dedicated, motivated by the idea of meeting such a person filling the centre spot that the rush to meet the person is coming in a stream of innumerable number of photons crossing innumerable number of millimetres and travelled unabated for billions of years to flow in the direction and directly to that spot, never diverting for one instant, to come to where I am filling that spot in that centre of the Universe. Thinking about that makes me seem rather important, don't you think!

All the light in the Universe is coming to me. It is on route straight to me where I am standing filling one spot on Earth. If there are those that do not believe me, well those I challenge then to go outside at night and see the vastness everywhere from wherever the light is coming from and see how all the light is heading precisely to where that person or me is standing. It is coming from all over, dedicated to find me where I am standing. It is coming from areas so large not even Einstein can calculate the size or content of such measure and it is rushing towards me specifically with a single-mindedness no other can match. It was one purpose to fore fill before ending its life long task and that is to locate me and meet me and inform me about the region that that light represents. There is not one ray that is going to miss me by fluke or accident. The light has one purpose and that is to meet me at the point I am presently located. Every beam has my name on it and it is coming for my eyes. Can any one imagine if a person was standing in a location and found all the persons in a big city of millions of soles within that city was running towards him where he is occupying that point whereto everyone is heading, how frightened such a person must feel? Yet, it is happening to every one with light coming from where ever the vastness of space is situated and is coming across space to that very specific point the viewer is standing.

Even if I shift to another position on the other side of the Earth or to the moon, the light will change direction and trace me in my new location. Even if my new location is in a camera and the camera is in a vehicle in the centre of Mars, the light will know that I am using such a point and trace the camera so that I may still be in the centre of the Universe. Wherever I might be, the light will still get me at that location. The light flows to me from where ever and to top that it is flowing to all other persons in the same manner and moreover under the same pretences. This puts every individual in the centre of the Universe without letting any other person in on the secret. In doing that (putting us all in the centre of the Universe) and by not telling others that we all each one) fills one dedicated centre spot, it is putting everyone in the centre of the Universe, and that is having everyone getting this idea that everyone walks around with this the knowledge that only he or only she is in the centre of the Universe. That is why there are all the wars and conflicts going on. That is why everyone thinks his view about God is correct and no other view can be correct because everyone has this bolstered and gloated idea whereby everyone thinks of his person filling the centre of the Universe. That means it is not the Earth that is that important but it is where the point holding the location is and that point which the observer is using to view from, that is the most important place there could ever be. If it was only the Sun that the light was streaming from that is choosing me as representing the Universal centre being the centre of the Universe it then cannot be that very exclusive because in all fairness and in contrast to Newtonian opinion, the Sun is not that big issue and there are much bigger stars going about in the Universe. The Sun is close and the light is plenty. That, which I am referring to, that it is coming from all over.

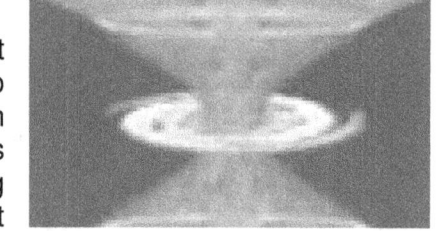

That is just one small part of the fantastic affair. Some of the light left the stations they come from some 12×10^9 years ago to meet little old me in this spot I is filling. The light has been travelling 12×10^9 at the speed of light, which I might add is much before my birth, it came crossing space and time, rushing all the way to meet me at this point. No one ever thinks how it was possible for the light to know I was going to stand at this point and be here the moment the light arrived. How did the light know I was going to take centre stage at that moment when it left so many billions of years ago and came with one purpose and that is to meet me filling the specific centre of the entire Universe? I have to be in the centre of the Universe because all the light is travelling to this spot where I am standing and I am filling the centre of the Universe without one straying off course and missing my spot. The light takes two million years coming only from the closest next galactica to meet me here, if taken into account the prefect timing it applied after all that travelling this far just to be in time just to meet me in the centre of the Universe. How important can I ever dream to be? Light is coming across time measured in millions and billons of years through space measured in millions of trillions of kilometres, travelling at the maximum speed the cosmos will allow, ignoring all other places it could go to and came to meet me in the centre of the universe.

When we look at the night sky we see images of stars. I am of the opinion that our vision of stars and our interest we show in stars is just what sets us apart from other species. We are able see what we never can touch though we can appreciate what we never can have. We interpret what we see without ever making contact to confirm and that gives us external knowledge and insight. Our vision about that, which we see tell us that there is more than the animal's concept of a plain survival on Earth. The stars render us the knowledge that there are far more to life than trying to secure survival. The cosmos is about more that being where you find sanctuary in the knowledge that you can eat or you can be eaten. Fathers show their children the constellations and although we no longer attach religion to our stargazing, it never subdued the bliss we find in our astonishment about stars.

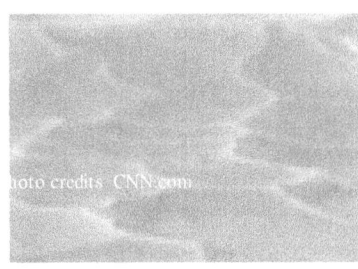

photo credits CNN.com

The star that gives us our greatest wonder is a star we cannot see. Every one stands amazed at the fact that there can be a thing such as a Black Hole. There is so much to ask and such a lot to wonder as to why and how and where and which…yet, we cannot see any that we interpret. We see but we cannot see and that makes us wonder what it is we cannot see and what it is we wish to see. The fact that our view is obscured by the fact that our view is obscured dramatizes our sensation of wonder many fold. That is human and that is why we are what we are and why we are in terms where we are.

It is part of the human concept to believe your eyes. In time gone by if a person saw a lion coming there was no time to look for other confirmation, as there was no time to do that. What your eyes tell you is what you believe and that is how we stayed alive. Seeing the sand dunes on Mars is equivalent to seeing the sand dunes on Earth whether it is seeing a picture in a book or by means of the television media and what you see could just as well be a picture presenting an image of dunes in the Sahara. The Sahara is a place we can go and visit, should any of us wish to do so, but the dunes on Mars are another problem. I can cross the space by land or see and travel to the location where I then am in the Sahara. Visiting and confirming what we then see is not that simple to accomplish. It will require that we cross the space. The Martian dunes are not only space away, which means I can cross the space in time and visit in the manner that I can visit the dunes of the Sahara. The dunes of Mars are not even space away, but are time away. There is no way I would ever cross time to see for myself what there is to see while standing on the Martian dunes. This is the part our Newtonian space travellers do not begin to understand and that is if they ever understand anything!

That is what is wrong with science, amongst others. Science is of the opinion we see space. We do not see space. We see time, where light used time to cross space to get to where I am, but then it is not time we see, it is the distortion of time that we see, which we think of as space.

We use the most infinite to view and formulate what we think is going on in the Universe. Being Π sets us in the centre of the Universe. We take so much light for granted, never thinking for one second how impossible our relation with light truly is. This totally extraordinary relation we have with light must be one of the reasons why we humans put our position we have in the Universe in such a pivotal place. The fact that we hold a centre spot with light confirming our idea as life is carrying individuals. As we see us filling the centre of the Universe (and everyone is subconsciously having this idea) especially where we are in person in a position that confirms this idea every second that passes, that we are all blessed with the ability to only use our hind legs to walk on, and then walk upright on the surface of the Earth. We are the only ones that are able to do so and that also adds to our feeling of superiority. With all that wonderful conformation about how special we are, it is not that highly surprising that we have the idea that the Universe was created especially for us, us being those holding life. Think again and such an idea finds support from everything everyone thinks of and whatever is there, is confirming the idea about his own importance. Then on the other hand thinking that the Universe came to be what it is, that the Universe came to be dedicated and devoted to the cause of life is human, is Newtonian as well as us having such an idea is absolutely bizarre. If there is no more life there is still a lot of Universe going around. If there is no Universe, there is no life left and it is time we start realising this. The cosmos is about confirming time and not supporting life.

The "further away" we look the more time we see. However it is not time we see. It is the distortion of time that we see. It is the history that time left us in the form of space that we see. The further something is away, the more it is in the distance, the longer it will take the light coming from the object to reach us. That means the longer it takes light to reach us, the more time is distorted to put distance between what we see and what there is to see. It is not space that we see but the distortion or the compromising of time. It is the time delay between here where we are and there where the object is that we see and we do not see the object or the space the object has or even the space between us and the object. We see the time delay there is between the object and us. We see what was there in time gone by, however, we do not see what is there and we see space for what space represents to the Universe. We see space as time delay, time slowed down. That is what space is, space is time delay. That concept urged me to go and look for the beginning of space and the beginning of time and the origins of the concept space-time. Please allow me to explain the beginning of space by measure of time delay. At the time of the Big Bang everything was small...not

so...it was as big as it is today. If the Universe was the size of a neutron, then we had no size at all. One cannot compare apples with oranges and see bananas. The space we see is the distortion time has to separate points of comparison.

In order to understand what I am trying to say I have to use a picture that is most probably not a true event. What we think we see is space. It cannot be space that we see. In the forefront we see a light and that is a result from that we have this concept that that star, although much smaller must be much closer. Then there are pixels indicating lesser star structures and some clear dots indicating stronger light spots, which would personify larger stars being present in that direction. The rest we see is the black of night. If the Big Bang theory is correct and to my thinking there is no doubt about that, then not to long ago there was a lot less space between the objects than is the case at the present. The space was less. That cannot be the case because if the space was less it would then take the light much quicker to arrive at the spot we are at present. The light coming from what should be the inferior but closer star is relatively quick in reaching my location while there may be some of the faint dots that have light travelling a considerable time to get to me. I presume the star in the forefront is closer because of its allocated position in relevance to where I am, and on those grounds I presume it is closer.

Looking at the image of a roving planet it shows a structure filling space at intervals. When one picture is taken every night at the same time we see a planet shift as if it is moving at a certain velocity. The space it fills is a constant because the space does not change in becoming bigger or smaller. However, the space it is moving through appears to grant the roving planet another position every time it is photographed. It is in the terms of time that the answer is. It takes a different period to position and obtain the light coming from the different position where the object is located. This buggy carrying the Martian camera is not a distance way, but it is time away because whatever we accomplish with the photo that the camera takes, we will only be able to see less history of time, but we will remain seeing the history of time because we might bridge space, but time we cannot cross. There will always be a time delay between what we see and when we are able to see what we see.

If the prime object were the space we wish to have photographed as it is in the case of the space serving to fill the roving planet, then no changes would come about to the space. If Einstein's novel idea were correct that the speed of light is equal to time then it would take as long to fill the space between the object and where I as viewer am in position as it would be for me to see the object. If light were one with the travelling of time, light would cross the space in using no time to do so because then time and light travelling in time through space in time would be one. That is clearly not the case. It does not form a viewer's instant portray because the motion that the light has to endure is longer by time duration to cross than the shorter period to fill the picture with pixels. It is the space that is constant, yet the time to travel varies. It takes time to cross the space  whereas the space holding the object remains filled at an even volume every time. In the case of the planet that fills space with material by gravity is the same space as it always are and is always filled without changes entering the picture. In the case of the dark space, the black stuff Newtonians think nothing of and think of as nothing, that space is putting time at a different duration to reach the location where I am standing at a distance and since it has the ability to change the time it takes light to travel from wherever to me being here, therefore that black space can't hold nothing, because nothing cant alter time.

It is part of the human concept to believe your eyes. Seeing the sand dunes on Mars is equivalent to seeing the sand dunes on Earth, be it by means of the television media or using your senses directly at the location and when looking at the Martian dunes it could just as well be as if it is looking at the Sahara dunes, because we believe our eyes. The Sahara is a place we can go and visit should any of us wish to do so, but the dunes on Mars holds another problem. Visiting and confirming what we then see is not that simple to accomplish. The Martian dunes are not only space away, which means I can cross space by means of travelling in time and visit and where I have the ability to do so, because between Mars and me there is not space forming the black stuff the Newtonians think nothing of and think is nothing to think of but it is time which I have to cross to get there. The dunes of Mars are not even space away but are time away because I have to beat the time the Earth's gravity constrains me with to get there. I have to beat gravity and I am going to show that gravity and time is the same thing. There is no way I would ever cross time to see for myself what there is to see while being on Martian space and all the future prospect in "space-travel" is hampered by this. Gravity on Mars is time on Mars and my senses will not function in the same manner as what I am accustomed to when being on Earth in Earth –time.

Again I have to repeat what I previously said, because in this idea the entire Universe comes across differently. Because Einstein saw the Universe not for what the Universe is, therefore Einstein theoretically was insane while he was more correct than any other person before him ever was. The Universe that Einstein saw and that Einstein said is going flat and the Universe Einstein thought of, as being the Universe that is supposedly going flat is not even remotely the same place. The Universe that Einstein saw was going flat group together to form a conclusion that time then presents as forming a unit and what Einstein thought of as going flat was the unit that formed while that what Einstein saw was going flat forms the Universe that Einstein thought was going flat. That is what is wrong with science, amongst other things being incorrect. Science is of the opinion we see space when we stair into the black of the night. We do not see space. That which we see as blackness, that is time, but it is not time we see, it is the distortion of time that we see formed as space to become the historical overview on time that lapsed. The "further away" we look, the more time we see being in a historical perspective. However it is not time we see but what time duplicates as the past when time takes the future onwards to form the past. It is the distortion of time that we see of what was when that was in the present. The further something is away, the more it is in the distance, the longer it will take the light coming from the object to reach us, therefore the more time there is located between there and here in relation to then and now. That means the longer it takes light to reach us the more time is distorted to put distance between what we see and what there is to see. It is not space that we see that is what we think we see, but what we see is the distortion or the compromising of time that in the meanwhile changed that which we think we see and the space that we think we see is totally different and on another location than where the images are at present that we think we see of what we see. We see the difference of what is there in relation to what there was in relation to where we are in relation to where we were when that we see was what was there where we see it was. It is the time delay between here where we are and there where the object is that we see and we do not see the object or the space the object has or even the space between us and the object. We see the time delay there is between the object and us. We see what was there in time gone by, however, we do not see what is there and we see space for what space represents to the Universe. We see space as time delay, time slowed down, time reflecting on the past. That is what space is, space is time delay, the paper on which time writes the history of events gone by. That concept urged me to go and look for the beginning of space and the beginning of time and the origins of the concept space-time. Please allow me to explain the beginning of space by measure of time delay. At the time of the Big Bang everything was small…that is not so…that is untrue…it was as big as it is today and that makes a case for everything then being as big as everything is today. If the Universe was the size of a neutron, then we had no size at all because everything in that neutron had an allocated relevance to everything else that is today; back then also shared confinement in that neutron in that time. Since nothing can add and nothing can remove, then therefore what is today was also present then. If the Universe then fitted into a Neutron, the Universe still fits into a neutron because it is impossible for anything to add to the Universe, except time. Time is that part in the Universe that eternally adds the past to the Universe as forming the future which is the adding to the past that it presents. Whatever was, still is and if that which was fitted into a neutron then today nothing can be bigger or more because what was can only still be. One cannot compare apples with oranges and see bananas. The space we see is the distortion time has to separate points of comparison.

In   order to understand what I am trying to say I have to use a picture that is   most probably not a true event. What we think we see is space. It cannot be space that we see. The fact that I wish to make is that if the Universe can be compressed back to the size it had at the point of 10^{-38} seconds after the Big Bang temperatures of 10^{27} K will come about once more. I conclude that the expansion was the result of compressing heat because of the insufficient space prevailing at the time. If the Universe was in a vacuum as big as being available now, but only filling the smallest fraction of such a blown up Universe as it did then in comparison to space being available at present, then what was the temperature of the vacuum that was not filled at the time during the initial Big Bang encounter while it was empty as before material filled it later. This presumption works on the Newtonian suggested conclusion that there was unlimited space available during the Big Bang because the hollowness in the Newtonian Universe filled in time laps as the space was nothing and presuming the space in outer space is still nothing, then I presume there was as much vacuum available then as what is presently available today. I am not suggesting this madness but it is a Newtonian conclusion I am putting to the sword, so please don't think that I am off my mind thinking this way…it is a typical Newtonian way of arguing which I now wish to question. Why was the vacuum that was present there at the time the bang came about as it is now still is present because it is vacuum or still being "nothing" in the present, not filled back then as it has been filling from then to now. If the Universe then employed the space of say one atom, the impression comes through that from edge to edge and from Universal border to border the space that was occupied at that time was the same as one atom will claim as personal space in our present day and age. Normal gravity started at 10^{-43} seconds. The Universe was the size of a neutron or somewhere in that vicinity according to the big Bang theory. It is not the Big Bang theory as such that I dispute because the Bible confirms the Big Bang theory and that I prove! It is Newtonian simple-mindedness that I try to demolish. The Big Bang began and GUT or the grand unified theory produces the attempt to describe the strong and weak nuclear forces and electromagnetism in one single mathematical theory. Somewhere before 10^{-12} seconds of counting the Universe cooled to about 10^{15} K, which was when the electromagnetic and the weak interactions acted as one single physical force. Science recon that unification may come about at temperatures of 10^{27} K, which was the temperature of the day at 10^{-38} seconds after the Big Bang. This statement echoes my viewpoint but one has to look carefully for that to surface.

In the suggestion the presumption reads that all the space that the Universe made available at that time was the total space one atom might take up today. Or was the rest in a vacuum chamber that is still filling at present? If the vacuum was there at the time then the vacuum is still here at this present time because what we find in the Universe at the time still must also at present be available. There is no possible adding or removing in the Universe then and now. However, with the Universe at the time having only that one tiny hot spot filled with huge volumes of heat with the rest of space being unfilled while the unfilled was also being there all along but filled with emptiness, then where is the rest of the available vacuum now? Then that statement suggests that in this hot Universe there were light-years upon light-years of vacuum waiting to be filled by the intense heat souring in the smallest spot one can imagine. Was the Universe overall cold, locked in with one spot of the vacuum filled with temperatures so hot we can only produce it in numbers that has suggesting values but we can never claim to digest the reality thereof? Was the space available at present available then or was the hot space the only space available at the time. If so why did the heat not instantaneously fill the eternally cold vacuum, as it should?

The answer to that is absolutely crucial because how did the Universe decide to fill. If that is true that the filling came by emptying the emptiness of the unfilled nothing, why did gravity not prevent the vacuum filling? If gravity contracts, then back then gravity had the opportunity that will never rise again to contract the lot and stop the expanding thereof…that is if the mass was there as it should be because nothing can add and if mass does produce gravity by contraction!

This question seems small-minded belonging to a child with not much factual appreciation developed yet. Please do not see it that way. If the space is nothing then the space was as large as it is at present and then there was no need for such a small area to fill since all the space we know about was there for the taking. The vacuum should then draw all the heat and expand even further many times over, while gravity should get hold of the material and compress this lot into a lump so dense we would never find a name to suite the occasion…and Newtonians are notoriously famous for devising names! That means the Universe should then have instantly develop from terribly hot to terribly cold in less than a thought and the expanding should therefore far exceed the speed of light, while the contracting should match the expanding speed but just contract in the other direction. With all the heat and all the temperature available to expand in one side and all the spaciousness and cold holding the rest, the heat and the space that the vacuum holds should grow into one in less than time can measure. How long can it take for such a lot of heat to fill vacuum while material is creating more vacuum by contracting into a lump of thickness?

A very close friend of mine one night asked me about space and where it must end. He said in his argument correctly that if it is space we see which is what we see, then if it is space that space has to end because space ends. That is true but I tried to tell him we are not looking at space, but at time forming that results in space forming as the history and the story told by time. I tried to answer him but saw the argument was too complex for a mere informal discussion, so I thought it best at the time to write my thoughts down so that Johan Boonzaaier (the friend I am referring to) could read my ideas at leisure and have time and chew on it. That led to the first book written by me (in Afrikaans my native language). If the Universe was the size of say even a tennis ball with only the size of a tennis ball being the very all of space there is available, then yes, it must take time too expand from that having the excessive heat there was back then to all the space we have at present. It then is converting heat into space bringing about the expansion. But the space then also developed as the Universe developed and if space developed then that blackness out there couldn't be total vacuum filled with nothing because "nothing" cannot develop. You, the reader must judge between my view that space developed with the Universe expanding as part of the Universe and reject the official view about space being nothing or otherwise you, the reader, must then decide that I am wrong, but should you do that, then find a reason why the Big Bang started out small and filled all the available vacuum we have with the motion of time. Then you have to show what is big in the Universe compared to the greatness of the Universe.

Coming back to the concept of seeing with my eyes while believing wit my mind and I believe my eyes by using my mind while this is where intellect now has to start to override culture. Is the moon some distance or some time away?

If the moon was space away, then I would be able to travel there by ordinary means of transport. But such as it is travelling to the moon requires breaking through the time zone the Earth olds. In order to be in outer space there has to be a relevancy in place with the Earth formed at an orbit value of $(4\Pi^2)$. This $(4\Pi^2)$ is a requirement to be able to orbit the Earth. Then we have to increase the Earth's rotational gravity $7(3\Pi^2)$, and this is the velocity that all things fall by. Further we have to get to outer space and when doing that we have to bridge the Roche limit be half $\Pi^2/2$. This is how much we have to increase the speed with which we travel. Then to top this we have to overcome the constraints the Earth's atmosphere places on the movement we have to overcome the time division there is in place between the Earth (6^2) and the Sky (10^2).

$$((4\Pi^3)7(3\Pi^2))\Pi^2/2 + (6^2 \times 10^3)$$

Since we require heat in an impulse to overcome the barrier of time that the Earth holds us within the time the Earth has and that puts us in the space captured, we are within the time the Earth has which makes us space-time within the Earth.

If time was where the Newtonians put space, then space as time can grow because time moves on and time becomes more. However if it is truly space as space is seen as substance filling outer space by not filling such as in the case of where they say space is serving to fill the distance between the roving planet and the Sun, then that must be a constant that can never change to the volume of space. Again I state my case that what is in the Universe remains in the Universe with no possibility of any adding or removing of anything which already is part of the Universe or which is not part of the Universe.

If the Universe did fit into a neutron at one time, then the Moon at that time was much closer than the Moon is today. Let's forget this Newtonian madness about planets forming from dust to dirt and then to rock because that line of thought is as Newtonian as the thought of mass performing magic that will draw the Universe together. It is not happening the whole Newtonian concept is not realistic. So the material that now forms the Moon was much closer to the Earth sometime in the past than it is today because the Moon and Earth are still drifting apart.

Let's take light as the criteria and when doing so I am not admitting for one second that Einstein was correct when he declared that time and light is equal. But since the GUT period we see that light forms the basis on which all other objects form a relevancy to time applying in relation to movement. The "further" the object is in distance from where I am, the more time it would take the light coming from the object to reach me. The object will appear smaller as the distance increases but I know the space the object holds is filling the same volume as it does when being close to me. The space I now refer too is true space and that means it is that space, which is what material, is holding captured as occupied space-time. One can't place that which fills the Moon in the same category as the space I see between the Moon and me. There has to be a different concept since the two factors are obviously not the same. In the case of the space filled, that space appears to change but that space is filling a volume at a constant, which depends on the amount of material used to form the object. It is when the space increases in which the object moves, and then the apparent space that the object holds diminish, as the object seems to get smaller. The object does not reduce in size and that much I realise, but with time becoming more, the size of the object seemingly reduces. The space that the light has to pass through to bring me the picture of the object increases, but this happens when time increases and therefore the connection is in the time factor

that increases. The space filled with material remains the same although the space I associate with time increasing becomes more and in doing so it seems to reduce the space material fills. That reducing cannot be true because the space is filled evenly and on a constant basis by material occupying space in time all the time by the same margin. That which becomes more is the part being associated with time increasing. It is the time the light takes to bring me the picture that increase and it is that light that shows me a diminishing space, which is filled with material. It is not the space between the object and my location that increases, but the time that increases because the light uses more time to reach me and by allowing the time to increase, I allow the space of the object to appear to become lesser. The space the object holds has to remain the same and the space between the object and me cannot change by motion. It is filled by volume that motion cannot change.

There are arguments that the space too becomes more as the material in the space the size increases, but that I show in another book as also being a time issue that Newtonians never compensate for in the views they hold about the Universe. But this argument about time increasing material in size I bring in as part of another more comprehensive overview, which is in another book. But if I start to dabble in that argument too, then this introduction will truly become confusing to follow. One first should find an understanding of what comprises of space and what comprises of time to understand that gravity forms by the interaction of space-time and in that there has to be clarity about what space is and what time is. Only time can be affected by motion and since it is motion that is changing by forming time, it can only be time that the motion can change. The slower the motion is of taking time to run through space, the longer will the time be that it takes the motion to negotiate the space. That stuff forming black of the night that I see, is not space that I see, but is time that I see and the space I think I see is the retarding or slowing of time that I see. Outer space is not space but it is time that space retards and therefore the space we regard to be space that space is not space but a retarding or a distorting of time. That means that which we see while thinking it is space that we see is all the time, time that we see and being time it has no outside because time is eternal. Time can never end while space has to end and therefore what we see filling that which we see is between objects are endless and is therefore time. Space, being infinite interrupts time to give time in eternity duration value. I come back to this argument later on...

The relevancies we are about to address are about form. The concept that there was a Universe without form and before form entered the Universe is substantiated by the bible and I am going to show just where that still is forming part of our Universe today. It takes us into a Universe back in time to when a line had the same value as what a half circle has and also the same value as that value which a triangle has. That proves mathematically there was a time when form wasn't part of the Universe. It takes us beyond space to a Universe when time formed no space. It puts the Universe beyond distance. It is what came about when space interrupted time to deliver us the black of night, which we incorrectly think of, as space. The Universe did not start small and then grew bigger because we now sit with an enormously big Universe but it started back then as being outrageously big. The Universe is not expanding as much as it is reducing. When the Universe started there was no outside to that which started because if there is an outside then what is on the outside of that which started. The problem is when comparing the size of the Universe back then with a neutron because in our minds a neutron or an atom has a limit and an end. This is what the Universe never had nor can it ever have an end. The Universe has always been with only one side, which is an inside, and the only direction in which the Universe can expand is to the inside by going to where it goes smaller. The limits grew smaller not bigger because the only limits there are, is on the inside and not on the outside of the Universe because they're never can be an outside in the Universe! The initial start had no limits, which means at that point there was no outside. The only limit is on the very inside of anything spinning and in the circle formed by the spin is the inside of everything spinning, which is an end to space forming. That which we think of as so small and tiny, so small it has no sides and is also so big it cannot have sides because it is too big to have an outside and all we see and all we cannot see what fills the inside.

Where we are now in the Universe, we are so much smaller than what was when the Universe was the size of a neutron or whatever it was. If the Universe then was one block without limits, which had no outer limits and was the size of a neutron, then it grew smaller because what was our size, we on Earth and the Earth included, when the Universe contained all it had in a neutron. If the Universe could fit into a Neutron, how small was the Earth and therefore how small were we. We were so small that when the Universe was a neutron, we were not even a thought. It is easy to lose perspective but perspective is all we dare not lose. That which took all the space a neutron could offer back then has no limits now and has no boundaries. That which fitted into a Neutron is today too big to be cooped up by limitations and boundaries. We with limits and boundaries now have measurable quantity to calculate, but what was the Universe then has no calculations art present.

Where there is no boundary to shift what shifts then and yet they say the Universe is shifting its boundaries because the Universe is expanding therefore it shifts! Where no growth is possible because it has no limits to extend since it captured all the possible growth at the beginning, then the question to answer will be where too can it grow if it already grew all it could? Where will the end be located of that which can't shift to an end? If there is no end to such a shift of that what cannot shift, then how can it shift too when there is no place available to go. This location that can't be, because there is no space allocated to be, then where no shift is possible is then designated to eventuality become what they named The Big Crunch even before locating the Big Crunch. Plainly said, with no outside available, where will the Universe end since the Universe only has an inside in which it can end? The idea of a Big Bang puts limits on what has no limits and then those wishing to limit the limitless, wishes to end that which has no end ever, in what would be a Big Crunch that can never be because there is no outside to end by having an outside end on an inside. There is no Big Crunch possible because where would the Universe find an outside to unite with an inside if it never can have an outside to end. Still the Brainy Bunch Newtonians were masterful enough to name what apparently can never be.

It is like naming a baby even long before knowing how the procreating is taking place that will lead to impregnating of some member of the specie (which member it will be is still unclear at the time the name giving ceremony was undertaken) where it later on will lead to conceiving the baby ... that is the manner in which science dogma is enunciated but that is how clever those mathematicians are that knows everything there is to know on science. They can name a baby before even knowing what procreation is and that they do by calculating what they don't know anything about... like procreating the baby! It seems more likely that that which has no prominence finds prominence, which means the lot is shrinking. The Universe is surely shrinking to give us space in which to be. If the Universe still were the size of a Neutron, then we still would not be the size of a thought. With us being where we are, things are getting smaller just because we can be where we now are, and to fit an entire Universe into a neutron would not leave much room for anything else. This is not the case and it only means that according to calculations. All the atoms that is in the Universe, by then developed to the Neutron level, where the Neutron acted as a liquid substance giving the entire Universe buoyancy. That is the reason why the neutron holds no mass and the proton as well as the electron does hold mass or shows a resistance in declining in space-time values. As the neutron is forming the part of being cosmic water in relation to the proton being the cosmic solid, gravity by way of the Coanda effect applied at the earliest stages. With the protons being Earth, or solid or material that makes the neutron a type of cosmic water or a fluid it gives the entire Universe buoyancy in which the Universe floats. That is the entire crux of the Big bang theory and that is what the Bible explains happened during Creation, but I am getting ahead of myself. .

When we look at the images of the two solar objects it is so easy to put them out of perspective and in the same size, although we know they are not the same size. All one needs to do is just play with the dimensions and find the results. One change the space they have to match and they are equal

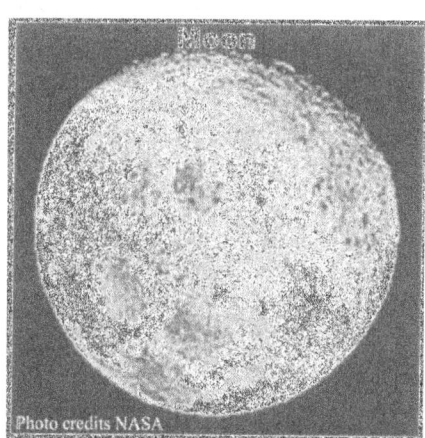

in size. That we can accomplish by fiddle ling around with a piece of glass called a telescope. In cosmic reality and the reality we as humans culturally experience we find the two concepts are very much substantially different.

When we put our hand out we are able to touch...say the door which is what we

are looking at, at that moment, we are immediately in contact with the door which we see where our eyes places it in front of us.

It is the door we touch because it is the door we see we touch. Moving back one meter we find we are no longer able to stand upright and touch the door because we are one meter away from the door. We are one meter away because we can see we are one meter away from the door. We grew accustomed to this thought because Galileo's pendulum shows we are in time in space in the Earth time in space. The time we will take to touch the door corresponds directly on Earth with the distance there is. We developed all our skills as being part of the Earth space and living in the Earth's time or gravity. Things change drastically when we leave the Earth or when we view object not confined to the Earth as we are. The truth is we are accustomed to think we are one meter away from the door and in that idea we are therefore able to touch what we see when we lean a little forward because we think we see the door is one meter away. However that it is not the door we see. We cannot see the door, because the door is not there for us to see. What we see and what is there is connected in brain association and in nothing more. We see light banging on the door and as the same door, that rejected, rejects the light is what we see as the light that comes flowing to us. We see the rejected light bringing an image of the door we cannot see. It is light we use and that image we are used too associate with what we see and the image, not reality, is what we are accustomed to use to confirm what we see, but such confirmation is what makes the most intellectual stumble when extending our culture (not our intellect because we can't be that stupid and still be what we think we are) as we take this what we culturally believe is true outside the Earth and into the realms of outer space.

In quite the same manner we see the darkness of the night and observe such darkness as darkness. By darkness we interpret the meaning as that which we cannot see or that which we are unable to see. Reality tells me that the darkness is light that is too bright for us to see. Take an image of Mars with a close up view. Then reduce it and go on reducing it until it is so small it becomes invisible. The space filling darkness is not darkness filling space because the ratio of darkness increases as the ratio of light in comparison to the darkness reduces. The object does not go dark by moving back. It rather becomes more of the same when it blends with the darkness, which proves the darkness is not darkness but it is light. By reducing the space an object has the darkness becomes either more or less but the darkness promotes the object or reduces the object. The fact that large objects are close and small objects are at a distance we on Earth relate to more space and less space. The only factor that can produce more space and less space is time because time is irremovably connected to developing what already is the past-time or the history of time that became what we think of as space as time is progressing to become more of what was forming the past and build on that in the future. By reducing the share of the combination of space-time, time must reduce or increase to allow space to do the opposite.

Even the fact that the objects seem to be as near or far is defined by Kepler as a fact in $a^3 / T^2 = k$. This answers the question, which a friend of mine Johan Boonzaaier asked me why would space not end and is was that question to me that set me off to start writing these books...It was his question about what space was and why space in outer space can never end that started me off on this journey of investigating possibilities in how to answer him. Now, even eight years later their still is no one out there that is prepared to read what I say about what I write my work, just because I go against the grain if Newton. Johan, that which you asked me what it was out there in the blackness of the night, that is not space, it is time. If it was space it was defined by limits, but because it is limitless it is time eternal. Borders will always define space being infinite but time is forever eternal.

The realisation that only time can affect space by the measure of what could bring about an apparent increase to space, and what forms the appearance of space becoming more in relation to what was and what will apply in the future, this realisation is a huge step in the right direction. Everybody always were searching for a White Hole in relation to a Black Hole but no person ever connected time in progress to form the so-called white hole. In the same manner as time dumps space down the Black Hole funnel, time dilates the density of space to make it seem to become more. Space is a constant therefore time has to influence the appearance of space to become apparently more or apparently reduce to become less. Being big is a sure sign to the brain of an object being close. That would then appear as if there is little space between the observer and the object in observation. That is culture talking because space may appear larger or smaller but it can only be a medium of space that may allow space to appear. Space as such has the same measure and has the same prominence when measured. Time is the factor that allows space to reduce and even to reduce to the obscure.

This is the best example we may ever find of space-time. The time factor reduces the space factor as it divides the space factor into smaller parts of time holding space in eternity. That proves that all containing space is in fact time holding space. $a^3/T^2k = k^0$.

The more time develops and time pushes the object "further away" the smaller the object will seem in relation to the containing space or time. Outer space is not space but eternal time holding space. In this concept is vested every mathematical aspect we find gravity has without

Photo credits NASA

us getting silly about mass becoming magical with pulling powers and full of forces as well as other ghosts of a divine nature.

In the past we developed a culture that placed all emphasis on space since we all ran about the same speed. To get to the dear we wish to hunt, was measured in distance because everyone ran approximately the same, so the time fluctuation was not important and in any case we all carried the time component the Earth holds us by. The space aspect, as in determining the distance the dear was away from the hunter that was the all-important factor that determined success or failure. We had the same time and the same abilities so the distance our ability to measure, determine and compensate for distance fluctuations brought success or failure.

When we altered the size of the moon in relation to the size Mars has what we did was change our relevance to that of Mars. We first brought Mars on a time line as close as it would be if it were hanging around in the space the moon has at present. Then we moved the time line back because it takes time to travel to the structure. Pushing Mars back does not increase the space, because eventually Mars fills the same space. It increases the time duration between Mars and us. It is not space we cover. If it were space then the time would be equal for light to complete the journey in the same time. By changing the time the relevance change as to how long it would take to get from they're all the way to here.

Therefore moving the object that fills in our vision backwards by reducing the optical value does not diminish the space the object has. What does increase is the time differentiation between the position I hold and the relation of that to the new allocated position the object holds. Should I wish to reach the object at the speed of light, I then will have to diminish my space to that of a photon to be able to reach the speed the photon has. This is what Kepler's formula says taken as Kepler gave it before Newton raped it. $T^2 = a^3/k.$ The speed T^2 that the object travels reduces k^{-1} of the space a^3 the object holds. Life can't remotely stay alive under those conditions or even operate close that, because of what the heat levels will rise to when reducing to that size and the decrease in structure size that will come about from that. However, in the event where I do not wish to surrender any space because I as do not wish to relinquish for reasons of the survival of life, I then would have to travel taking much more time to reach the object at the location the object holds that what time it would take light to cover the distance. Then we get to the question of what did it take to shift the object into the allocated slot it now has in the distance it now has. It took time to shift because it did not take space to move. The time it took the object to move to a new location was added between the moving object and me. It would take the same time but in a different ratio of moving to retain the time should I wish to move to the location where the object filled the same part of my view I have of space with the object in that space. That means since the Big Bang there was no space added but only time increased what parted the objects since then. Space is the retarding of time.

The big issue Newtonians miss about all of this is so apparent their missing the point makes their observation abilities seems as if it is childish.

If it is possible to move objects back and foreword by the relation we can manipulate and change by having two pieces of polished glass change the flow of light, and the images become brighter or dimmer because the images is seemingly smaller or larger, then the night sky must have the same qualities of a lens. The sky must be a lens. Only a lens will act as a lens when this lens encounters yet another lens that can bend the light coming from the sky back in relation to what the sky bent the light in applying time. If we see the sky more clearly by suing lenses, then we can counteract the distortion time brings about to the view we have of the sky by bending the light back with glass that also bends light. Every picture Newtonians show of the overall view they have of the cosmos as the cosmos supposedly grows from the Big Bang shows a form of a circle or a sphere. Yet, when Newtonians put this interpretation into practise, then the nothing they have filling the sky, equals the nothing they think about when investigating the cosmos. Instead of looking at the cosmos for what properties they could find, they give the properties which they have in their heads to the cosmos and pass on the nothing they can think about by placing nothing in outer space. If the cosmos is round as they wish to interpret and which they so frequently depict, then the cosmos must have the qualities one would associate with a round lens because we see through the cosmos as if it is a lens and then we manipulate the distorting properties of the sky by correcting it with sing a lens. However, when the arguments come about what the sky is. Then true Newtonian mind overrules logic and they put Newton's nothing in place to fill an entire Universe! Then what they think of in terms of logic falls down to nothing! Space is as much water and that is why space distorts light as water or glass does.

Space is nothing less that water or liquid or something with buoyancy qualities. Space holds buoyancy and therefore birds can fly by flapping their wings. Man made his first attempt to fly by fixing flapping wings on his body with ill results because the flapping was not the reason for flying. It is not the flapping motion that acquires flying, it is the buoyancy about the body relating to space holding heat (time) that brings about flight. In other words, the movement decreases the density of the moving body in relation to the liquid substance in which it is. Skiing on water is the same as flying on air.

The liquid used has only a different relevancy or density, but still has the same properties about buoyancy. I am not going to make an issue about this issue at this point, since that I do by explaining the following argument very extensively in other books, but I merely wish to point out that a body loses it mass qualities when floating not on the water, but in the water. When another body, other than water floats in the water, that body in the water can just as well be water because it takes on the same density factor as water does because the water and the object in the water holds the same relevancy to space-time unoccupied and that puts the body floating in the water in the same density level as the water has. While this happens with the body floating in the water, the body has lost all mass in relation to the density of the water and the air or outer space or sky or atmosphere or whatever name goes with the concept you have about the blue stuff above your head, that stuff is then same stuff as water is, except for the fact that it is about 6×10^2 times less dense than what water is. If that stuff above your head has liquid properties, then the liquid properties go on and on, indefinitely except that it becomes less dense in structure, because there is no point that one can say that sky or atmosphere stops here and nothing follows after that point. In all I Have said, mass has no input, except when the object hits the ground and then by receiving mass, finds it then has a position with the solid and is then no longer part of the liquid that descends towards the solid.

Then only does Newton's mass find circumstances granting mass a valid ness to become a factor but mass does no bring on gravity, where gravity by the value of Π^2 forms what one then may read as being mass. I am going to use a sketch in order to show that mass has no validity…not in cosmology as far as gravity applies to cosmology, because the reference to gravity comes as part of movement and there is no gravity coming by from attraction, sucking, pulling, pushing or any magical connection between what will have mass if it stops on the Earth surface and the fact it will somewhere presume with mass while it is falling with space. Newton incorrectly concluded that

$$\frac{dJ}{dt} = 0$$ whereas it should be $$\frac{dJ}{dt} = 1^0$$ and I am going to show that that is singularity and also I

am going to show where one may find singularity. What this says is that rotary motion does not

eliminate space-time $$\frac{dJ}{dt} = 0$$ but in fact installs space-time $$\frac{dJ}{dt} = 1^0$$

Only water is a natural fluid because becoming solid it does not destroy space but rather accumulate space. In outer space objects float as objects float on water because of the similarity existing between water and space-time (heat in space). While there is a gravity thrusting the object downwards, this implication has no bearing on mass at all. However mass becomes a factor derived from this movement of thrust of space moving down by time or by gravity, one may use the any of the two names since both names are the same but for the name. When "gravity" gets hold of you, the body is no less buoyant than what it was in space. If there were a certain speed attached to any object and such speed can maintain a high enough speed, it would float in space as if it is buoyant. This we call being in orbit or holding a satellite position. The object then does not have minimum gravity as Newtonians think, but it has minimum mass because
with such reduced mass it could maintain the orbit while it
has maximum gravity drive.

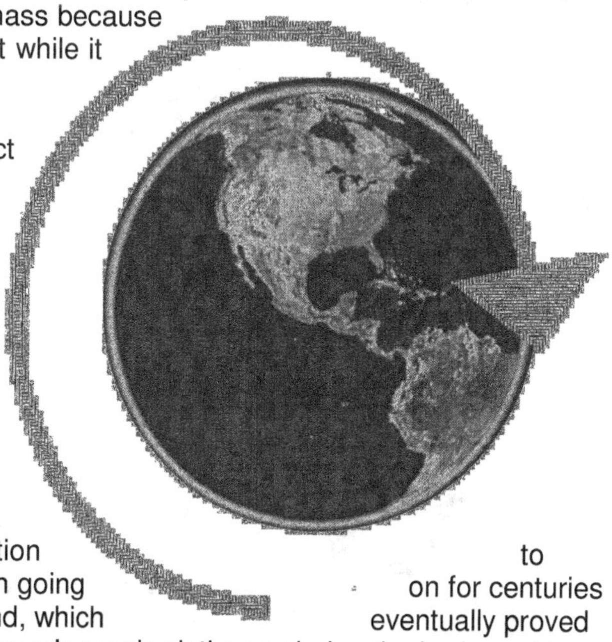

I am well aware of the fact that Newtonians connect
the idea of having mass too the idea of a body's
ability in forming a solid structure, but that is part
of the lie that
Newtonians
repeated
from

generation   to
generation going   on for centuries
without end, which   eventually proved
to be the Emperor's magic clothes as being the brainwashing
part in culture they presumed to give Newton's view legitimacy. I
am sorry, but I am not taking part in that sham just to make sense of
what clearly is senseless.

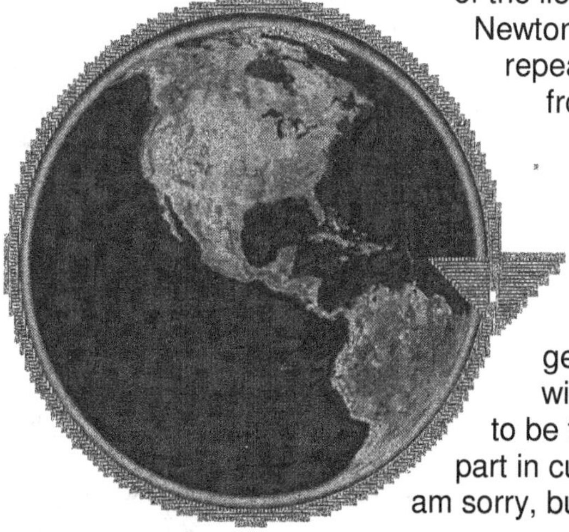

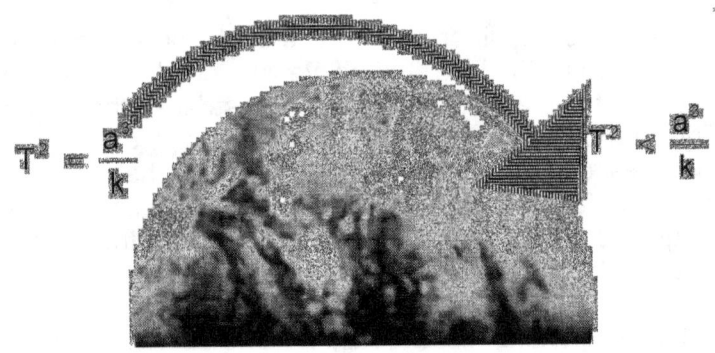

$$T^2 = \frac{a^3}{k} \qquad T^2 < \frac{a^3}{k}$$

Up there floating in the sky or in outer space, the body is equal to the air or space and has no mass because it resumes its place in space as the maintaining of the correct velocity grants the object the density that is equal to the space in which the body floats. Up there the body is just more air. Only if the movement becomes too slow for the have the object to maintain the speed in which the density is in equilibrium with the air, does the object lose this gravitational buoyancy. The downward thrust then overcomes the rotating thrust of gravity and when that happens it will move the space in which the object is moving too slow to keep it that a certain height, downwards. The forward thrust of space will be overcome by the downward thrust of gravity and this will redirect the body from going sideways to going downwards.

$$T^2 > \frac{a^3}{k} \qquad T^2 = \frac{a^3}{k}$$
$$T^2 > \frac{a^3}{k}$$

It is at that point that I make an issue about Newton not holding the same view as Galileo. Falling to the Earth comes as a result of the body becoming less buoyant because there is then too small ratio between heat or space and the density that movement allows the body to have. MASS PLAYS NO PART in the falling or the floating in space, just as Galileo said when he said all bodies large or small (in other words mass plays NO PART in the falling process) will fall alike and at an equal rate. What makes the body fall is if the body gets stuck in space not going sideways and becomes part of space going down and then goes down with the down going space as part of being one with the space. It is not the body that falls but the space within which the body is and in that the body density as a relevancy to the distribution with the volume of air and this density in relevancy play a part and not the mass of the body. In the cosmos everything is space. There is no mass given to some lump of compacted space and where there is no lump of compacted space there is nothing as Newton suggested in his declaring that the spinning motion absolves the turning material when he said $\dfrac{dJ}{dt} = 0$ and from that he assumed that $a^3 = T^2$ because $0 = k$ because that is bizarre. Kepler proved mathematically that $a^3 = T^2 k$ and one can't fiddle with mathematics to prove a point. Once mathematics prove the point all other arguments to the contrary seize having validation. That is the purpose of proving by using mathematics. Saying that $a^3 = T^2$ is saying a person in three dimensions can climb into a mirror and presume the status of the two dimensional image in the mirror. It is corrupting science! In relation to the material where more material adds to the already moving material, then the factor of mass could be contributed but only as a human concept and not as a cosmic institution. Again I repeat that in cosmos everything is space and only by some space spinning the space condenses space to condensate space into becoming concentrated space, which we know as material

When hot air is over concentrated in a confined space by pushing in more hot air, increasing the existing ratio would bring about that ratio between material and liquid air tips in favour of air and this rebalance in the ratio of air to material will cause air to stop descending. Such tipping of the scale will lead to material moving up into space instead of moving down the space because of gravity reducing space which I may add is then anti-gravity. It is the heat or space or gas or liquid (water) that moves around the Earth that becomes more compact or denser because of the Earth rotating.

As the air speed reduces this brings about that smaller sections of air move slower and therefore moves down to the Earth. This has nothing to do with mass acting on behalf of attracting or instigating gravity to start pulling. By the rotating of the Earth space contracts and thereby the space surrounding the Earth becomes denser, the closer it gets to the Earth. In other words, the sky is falling and the closer the sky is to the Earth, the more compact will the sky become as the falling process increases the density of the sky. By blowing hot air into the balloon bag, the density of the air reduces and with the expanding the hot air acts as anti gravity. This density increase has a Newtonian name by which it goes by and is called ...the atmosphere! Whether the object is solid or not, it is not the object holding mass that falls because only when the object lands on Earth does the density of the object serve to give the object a factor Newtonians thought to name as mass. However, as Galileo said when objects fall, all objects fall equally and this illuminates discrimination that mass brings into the equation. In the sky objects either move faster than gravity or gravity pulls the object by the compressing of the space in which the object is descending to the Earth by the rotation or gravity Π^2 of the Earth. It could also be seen as the body is still floating in the tub, but something pulled the plug in the tub and now the body is not able to float on the water (heat) because the slowing of orbit speed brought the reducing of density and this gives gravity the declining direction (the compressing of space around the Earth) which leads to space moving towards the Earth centre while it is the draining of space that increases the levels of the density by the reduction of movement in the direction of the cosmic atom (Earth). However, the body still holds buoyancy to the surrounding space being weightless or mass less and the body joins the flow of the heat as it does not fall by it being heavy or laden with mass, but it falls by the space or the heat compressing denser.

 Having mass or falling without mass is the changing of relevancies. When falling the object is part of the liquid ($k^{-1} = T^2 \div a^3$), therefore is pure moving space and becomes part of that which is contracting. The instant the object touches the surface of the Earth, the alliance the object holds, changes entirely and the object then associates with the solid by spinning with the solid ($k = a^3 \div T^2$), hence it becomes part of the Earth $\dfrac{dJ}{dt} = 1^0$.

$$k^{-1} = T^2 \div a^3$$

$$k = a^3 \div T^2$$

The compressing of space is the result of the Earth rotating and in this space that never moves but for expanding, (which is the outer space Newtonians think of as nothing and therefore thinks nothing of) suddenly moves by the orbit that the Earth (or any other cosmic body) introduce by the spin thereof. It falls through the movement of the Earth and when a body moves slower that the speed the Earth rotates, the matter occupying the share of space it holds, moves as space, not as matter or material particles, just because it is material and therefore because it must have mass. It is not because it has mass that it falls, but it falls due to the loss or the lack of rotating speed, that then is unable to support the object's ability to move about in suspended air at a bigger rate than the Earth suspends air, and in that becomes attached in the falling process of air. If the orbit speed is maintained at a precise level, then the density factor does not become an issue. That then becomes another issue to explain involving the Lagrangian points system, which is one of the four cosmic pillars or cosmic phenomena where the four phenomena interact to form gravity. If the body were plummeting down on own steam, mass would matter because the body then by virtue of its composition is the subject falling, but then all bodies will fall at different tempos because all bodies would have different composition in material lending each object different mass that will lead to different speed of descending. This is not the case and this part of physics was accepted even before the birth date of Isaac Newton! Since the body falling can only reach the same falling speed as space or air or sky or heat can do in the falling process, the body will not exceed the heat's plummeting speed, and therefore all bodies fall alike, just as Galileo said. Wind resistance HAS NOTHING TO DO WITH THE WHOLE ISSUE because when a body exceeds the heat's dropping speed, the body falling will divert in direction by falling at an angle to the singularity point of free fall. In the compensating of direct fall, the body also compensate its singularity or density from singularity should the singularity prove to be sufficient to hold steadfast. In gravity the forward motion T^2 is directly linking downward motion **k** to evaluate the density of space a^3 again Kepler of coarse and that is the formula for gravity $a^3 = T^2 k$. This is evident in all walks of physics but Newtonians has corrupted true physics with Newtonian hogwash to a point no one is able to read the truth from fallacy. I am about to try and explain if you would permit me getting all-artistic on you.

Let's pretend this is a boat. I have never boasted about my artistic skills because there is not such a thing in all of my credentials, but if you have as bad an imagination as I have, then it would be wise to advise you to the fact that this sketch represents a boat.

To help you solve my artistic riddle I wish to show you some pointers that might guide you to a better understanding of my artwork. The boat being a triangle has a sharp edge front and a straight down back and a line of arrows representing the waterline in which the boat floats.

The story holds a boat with material and water giving the material buoyancy. Since the boat is at a certain level in the water the boat clearly has mass. The boat having mass is sunken into the water up to a point and this point is called the water line. This is the equivalent of the mass of the water that the boat displaces. In this picture there are two relevancies forming factors, which represents gravity. Mass is a Newtonian daydream and mass has nothing to do with gravity, in spite of everything said by science this past three hundred plus years. The boat holds space in relation to the density of the water.

There is a specific dynamic between the water holding the material and the material substituting the space that the water would claim if material were not present. In this we have the three components Kepler introduced which is space in volume a^3, movement of space downwards **k** and movement of space sideways T^2, and to this end the cosmos unveiled to Johannes Kepler the formula to space-time and the mixture that represents the forming of space-time as $a^3 = T^2 k$. This Kepler did not do because Kepler read only what the cosmos mathematically gave Kepler to read $a^3 = T^2 k$.

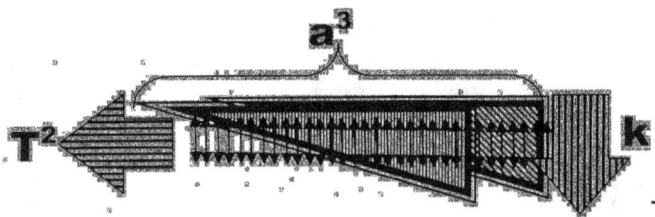

The boat applies to three factors. There is the downward motion **k**, which is the movement associate with what will lead to the object getting mass. Kepler gave that a symbolic value of **k** in his equation. This is the part Newton illegally chucked out $\dfrac{dJ}{dt} = 0$, but I have written so many books on that issue that I am not spending more time trying to disprove this. To summarise my findings on the matter is that the value applied to space – time $\mathbf{a^3 = T^2 k}$ is that space in volume in ratio depends on the motion of downward thrusting **k** while adhering to sideways thrusting $\mathbf{T^2}$ and this ratio will give space $\mathbf{a^3}$ validity according to gravity. This is where Newton's mass comes in, but again I have to stress that mass is a human thing and does not apply to the cosmos as a cosmic reality.

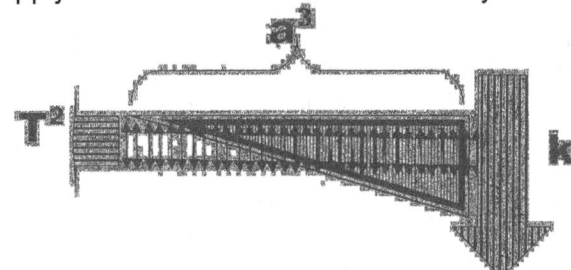

When the boat suffers from mass, the boat relies on the forward motion the Earth provides and in that ratio the downward motion that the Earth provide will also affect the space-time displacement. When the boat is motionless, as Newton wished to see things, the boat is not motionless because nothing in the cosmos can ever be motionless. The boat is moving in terms of the movement the Earth provides and in that sense the relevancy applying to the forward thrust in relation to the Earth providing forward thrust is 1. Having 1 as a thrust, this applies motion in terms of $\dfrac{dJ}{dt} = 1^0$ and in that the thrust is matching singularity, which is what I am going to explain in a short while.

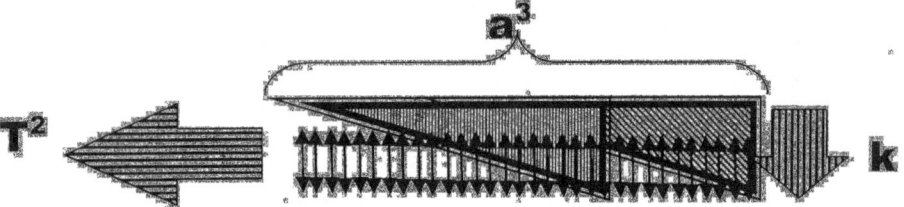

When the boat starts moving in relation to establishing a forward thrust, $\mathbf{T^2}$ such movement can only come about as time distributes the boat over more space and in terms of that the downward thrust decreases as the sideward thrust increases. The density of the body reduces in terms of the density the water holds and this comes about as the body spreads thinner over a larger area of space.

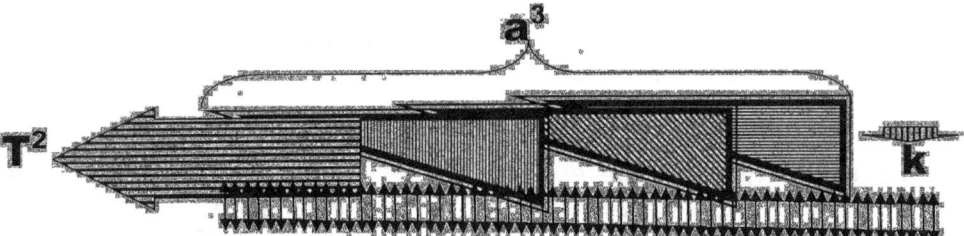

At a speed of $7(3\Pi^2)$ the body of the boat will start to lift completely out of the water and start to get airborne. This has nothing to do with air going underneath the hull as Newtonians explain the situation, because at that time there is so little of the boat touching the hull the boat almost lost all contact it has with the water. The forward thrust is reducing the downward thrust and that gives the boat such a large area of water to relate to that the hull spreads so thin the density of the boat begins to equal the density of gas. The downwards thrust has so little time to effect the space because the sideways thrust is propelling the boat along that it allows very little time for the boat to remain in one position as to give the downward thrust any time to be effective.

If one would put rocket propulsion on this boat, the boat will become airborne and lift out of the water and into the air completely. Then in that case, the boat's density exceeded the mass component completely. This is what happens with racing cars and airplanes and all things that fly. The motion of the craft exceeds the rotation displacement value of the Earth and the density of the object in relation to the air surrounding the object becomes so small. It makes the moving object as tin as air.

The previous factor that was explained was the effect that the negative flow of space-time (contraction) has on matter whether it is the Universal expanding negatively or plainly space-time displacement it still is the same. We have to look at the contraction (negative space-time displacement or negative Universal expanding) as just being more expanding going in the opposite direction of normal expanding as in time forming space, because the more development that goes in the direction of contraction (negative expanding) the less would go in the direction off expanding (positive space-time displacement or positive expanding). The negative growth or negative expanding or then contraction of space-time by displacement is the result of atomic spinning at about the value of C, be it an electron beaconing the atomic border or whether it is one single photon forming a cosmic projectile that is hurtling through space-time at the value of C. This restraining of the growth in the cosmos by the speed of light connects the entire Universe in development in both directions to the speed of light or C as it is better known. C forms the limit of heat moving in space and though space and is therefore forming a border for cosmic development in positive expanding or in negative expanding. In this event, the prime value has to be the speed of light connecting the unoccupied space-time value to the occupied space-time value within the atom.

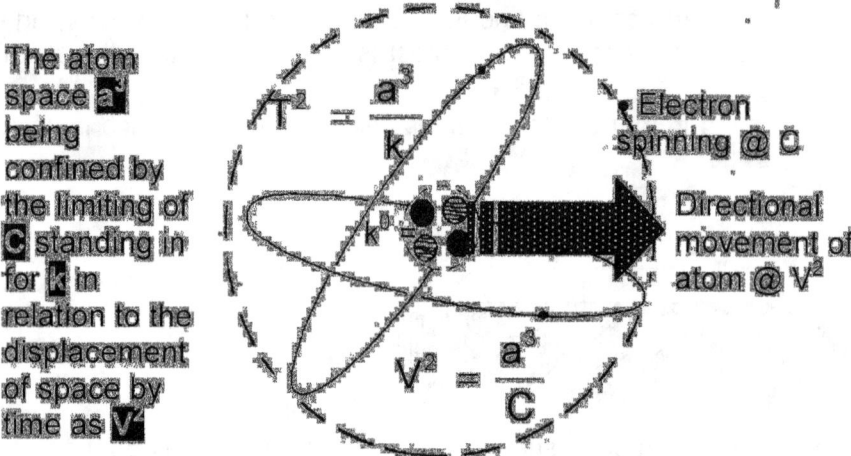

The atom space a² being confined by the limiting of C standing in for k in relation to the displacement of space by time as V²

$$T^2 = \frac{a^2}{k}$$

Electron spinning @ C

Directional movement of atom @ V²

$$V^2 = \frac{a^2}{C}$$

As already explained this brings a further implication. Once the positive space-time is well defined and its prime value lies within the value of the atom structure, the negative space-time displacement is accordingly accelerated to reach the ultimate value of C. In this there are two determining issues within the atom that will establish the atomic size prevailing in the star.

There is the spinning of the star that thrusts the atom in a forward direction which is represented by the symbol V²...And then there is the gravity or downward thrust of the contracting of space-time or negative space-time displacement of the electron freezing the space-time to the value of C. In time there are always these components being responsible for the movement of the atom where the atoms in total combining in a united effort to bring about the movement that establishes the gravity or the movement of the star or planet.

The only place in present time where light stagnates in time is within the Black star (Black Hole). What is happening inside the Black Hole goes far beyond our limited understanding because in the Black Hole, the atom gets drawn back into singularity. The Black Hole return the atom to the point where the Bible describes where the atom started: in a place where material (Earth) is without form and void, with darkness over the face of the abyss (singularity) and a mighty wind (space-time displacement) that sweeps over the waters (the neutron), but I am getting to that in a short while albeit very briefly in this book. Can anyone ever imagine a better description of the Black Hole and can any one ever imagine a better description of conditions prior to the Big Bang? The effect of time development in the geodesic sense can only be described in the fact that as time slows, space is moving negative to a point where time stands still with the total annihilation of space because there is no possible direction left for space to develop in time. No sketch can show how this concept is working because like the time value of the light wave, we can only follow in understanding when space-time displacement insist that the expansion is spherical in all directions, whether it is in positive developing or negative developing. At the point where the Bible describes a pre Big Bang applying when conditions was such that only atoms formed with no light being possible to be in space, and then when the Creator gave the command "let there be light, this is the best description any one can possibly give to conditions that came in place at the point where the GUT started the Big Bang that Newtonians are able to see. This is where the electron came in place that secluded every proton from the rest and gave atoms confinement in which to be in by excluding outer space-time (unoccupied space-time) from inner atom space (occupied space-time). The Bible is a wonderful scientific book but demands intelligence when people read it and where there is an obvious shortfall of vision as all Newtonians suffer from, no reading of intellectual material will stimulate intelligent thoughts in the mind of the reader as to spawn understanding of such deep intellectual concepts as what the Bible describes. One can hardly expect a person that believes that mass is responsible for gravity and with that show childish belief in the unnatural occurring of magical forces creating an even more magical concept of gravity whereby they are willingly and with no explaining to underwrite the concepts, can believe that in spite of having clear proof thereof, how or even that these forces pulling, is happening and therefore they resort to creating magic –like fairy-tale substances called gravitons which is also very unexplainable: and furthermore their childlike mentality can have them believe that there is force of various descriptions able to and therefore that can pull on something that is at a far distance while also believing vividly and literally that this pulling can occur while nothing is filling the distance that is between the objects pulling on one another. Those persons believing in such events and clearly underwrite the mystery of magical forces cannot be expected to understand the true implications of the cosmos forming as the Bible describes events happening and show intellectual concepts when the Bible uses physics of a proven and practical nature to explain how Creation progressed.

Behind everything said and to this very end there is the cosmos expanding into the future as well as negative expanding in reverse where the effect of this is understood as the positive space time displacement value, which in contraction is also negative in direction. However, we must understand that time is a factor that adds space to the concept we have as a Universe and time adding space places the future in expanding in all possible direction, even in the direction of what we would think of as being negative or confining.

This value of time can only be determined because time has an equal value to all matter in the Universe. We call that value space. However, as already explained this is not the case, because of movement dictating different expanding taking place. Time has a proportional equal value throughout the Universe which one may link to C when applying light standards or singularity when applying pre light standards. In both cases there remains the geodesic space-time value to all positions of all matter at any given instant of time in space according to each aspect of the position in space-time. However, one may not mix the two standards applying.

The value of the space-time position can relate to many factors, ranging from orbital positions of structures sharing space-time, as in the case of the Centauri triple star system. This relates to a certain group of stars sharing space-time in a particular part of a galactica; or it stand in relation to the whole galactica space-time. Then the space-time value relates to a group of galactica that once formed a unit but only share space-time and so the relevancy applies by circling out to a broader spectrum until one reaches the universal geodesic space-time value that applies to all matter in the Universe. Here I must stress that there seems an infinite value to both the smaller growing factor, as there is an infinite value to the larger growing factor. However in all circumstances, the use of ordinary mathematics in terms of what applies on Earth must be avoided. We might see the Sun as a sphere and therefore wish to put the calculating formula we apply to measure a sphere in terms of trying to measure the Sun. This does not apply and normal mathematics just doesn't work in cosmology.

So the value of the geodesic space-time has to be $a^3 = T^2k$ which stands in terms of the atomic value Einstein gave to nuclear gravity, which is by formula $E = MC^2$ or more correctly stated as being $E^3 = MC^2$ in terms of space-time expanding. This is only applying Kepler's cosmic formula in terms of an atom performing in singularity terms on Earth. But it has no applicable connection to the formula, which we put in relevance to spherical dimensions as having the value of $\dfrac{4\pi r^3}{3}$ which is a spherical value. This just doesn't apply because most of the size is material retained by the movement and not the mass density. In cosmology this means nothing because the size has not

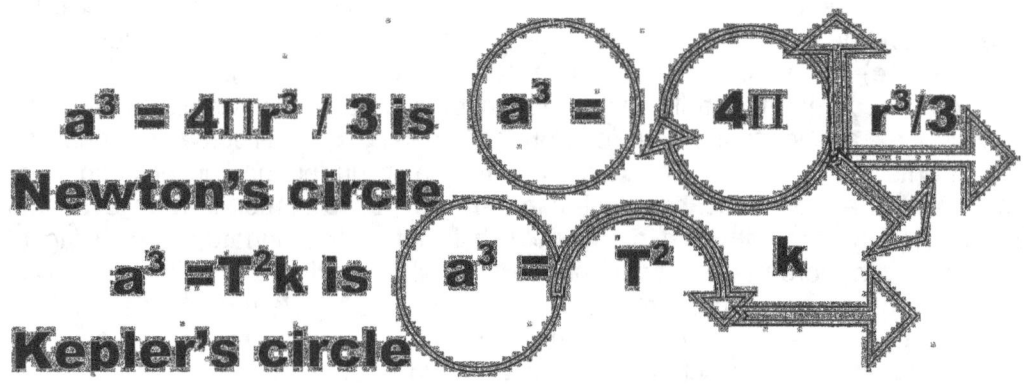

that much implication on the size or the gravity. The rotational speed has as much to do with the gravity as the distance in meters from centre to the side. This rotating speed is responsible for the time applying within the star. There are no two places on Earth where time is equal, yet Newtonians have the Earth's rotating time around the Sun as a fixed unit applying to time even at the beginning of the Big Bang. Time is the relation there is between what is heat and what is cold keeping in mind that motion is the result of heat where in the case of life; we humans associate movement with the attribution of life interfering in cosmic affairs. Life can interfere at a very low level and under very slim conditions that will have no Universal effect in cosmic terms.

Some aspects of the Universe go beyond mathematics and some even go beyond words. It is our task to find space, to find time and moreover it is our optimal task to find the Universe. This line of thought is in concept a joke but as much as a joke it may be it is the truth as no other. It also is the scientific Newtonian inspired approach to science that brings the thought pattern of truth in all people to mind. Because we use the Earth and on the Earth a meter is a meter in a second, we believe that meter per second will also be a second in measure of all meters everywhere and the distance duration will be the duration of all seconds per meter everywhere. Let's see how much this joke is a reality in the minds of Newtonians. What is gravity, asking that question it would be more correct to start by asking why is gravity there in the first place, other than to give Isaac Newton a permanent prominent position in history. What would be the reason why there is such a thing as gravity? To find an answer we have to look at what stars are because stars provide the Universe with gravity.

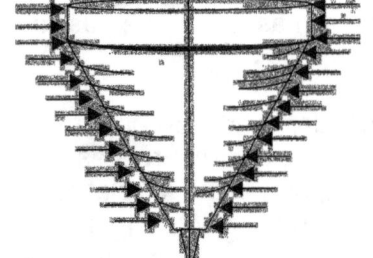

There is space on the outside, which seems unseen that is in support of the space holding the material erect and in a balance. The faster that the top spins, the higher is the intensity will be at the point where the unseen space connects with the visible

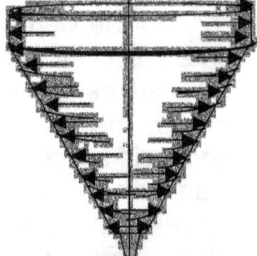

space.   There is space spinning and is therefore supported by the space on the outside of the space holding material and the support is what is in place to keep the top erect and upright. Mass could have no part to play

in the top being uptight or lying down. Where the two substances meet there is a field formed and it is the strength of this field that enables the top to maintain an upright stance.
There is no mass in the cosmos. There is no evidence of mass being present or being presented in any way except in Newton's profile of physics. If you think of a cosmic object filled with material floating in outer space and by the reason of being filled with material therefore having mass, then no there is no mass. There is space-time that is occupied by space being in a movement of spin and this movement places the spinning space in a relation to space-time not occupied and are therefore in relevance to the other party, not spinning as much. There is no mass because there is only space, either filled while spinning and through spinning it is cold while there is other space not filled and only moving by measure of expanding and is therefore overheating, but no mass as a conclusive factor, that there is no such evidence thereof. There cannot be mass just because there is only cold and hot in relation to preservation of singularity. However shocking this is to the Newtonian that was taught for decades how to appreciate the mass being everywhere, there is no mass and the above is true notwithstanding the fact that it destroys Newton's entire physics reputation. The two things science cannot work without, while it is completely Newtonian fabricated and therefore are created by man with no hard evidence to back any claim of any object having such a factor in the cosmos.

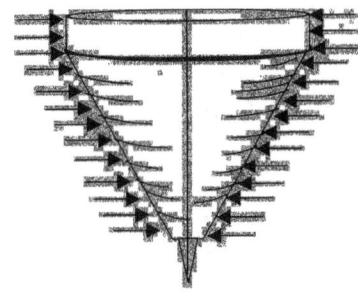 **There is space not spinning and is therefore not occupied with material.**

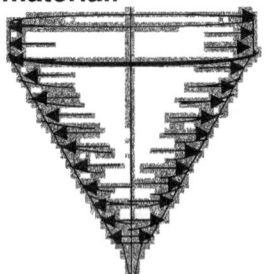

There is space spinning and is therefore occupied with material. That is it. That is what there is in the cosmos.

There is no mass in the cosmos. There is no evidence of mass being present or being presented in any way except in Newton's profile of physics. If you think of a cosmic object filled with material floating in outer space and by the reason of being filled with material therefore having mass, then no there is no mass. There is space-time that is occupied by space being in a movement of spin and this movement places the spinning space in a relation to space-time not occupied and are therefore in relevance to the other party, not spinning as much. There is no mass because there is only space, either filled while spinning and through spinning it is cold while there is other space not filled and only moving by measure of expanding and is therefore overheating, but no mass as a conclusive factor, that there is no such evidence thereof. There cannot be mass just because there is only cold and hot in relation to preservation of singularity. However shocking this is to the Newtonian that was taught for decades how to appreciate the mass being everywhere, there is no mass and the above is true notwithstanding the fact that it destroys Newton's entire physics reputation. The two things science cannot work without, while it is completely Newtonian fabricated and therefore are created by man with no hard evidence to back any claim of any object having such a factor in the cosmos.

Let's go on a search for of mass being present in out solar system in order to find evidence or even traces of mass occurring readily where the evidence of mass would support Newton's theories on mass. If mass was something the cosmos applied, then the cosmos has to bring mass in as some guide to the way planets orbit in ratio to mass, or in mass contributing to a faster or a slower rate of travel or determine the orbit length of the planet as the planet goes around the Sun. There just has to be some indication that it is not only Newton that saw mass fit for purpose. If there is no indication that the cosmos use mass in any way, then we have to surmise that space forms as space and that which fills space has no influence on the space being in place.

Every planet orbits at a ratio set at a relevance of ± 0.986 or they're about and this the tables Kepler calculated years before Newton formulated his theorem of gravity based on mass. The discrepancy detected from Kepler's table come from the fact that Kepler had to work from the Earth to calculate and the Earth does not form the pivotal centre of the Solar system. This would inevitably lead to figure not conforming but as I will prove, when the centre is correctly aligned, the ratio would have to be 0.986.

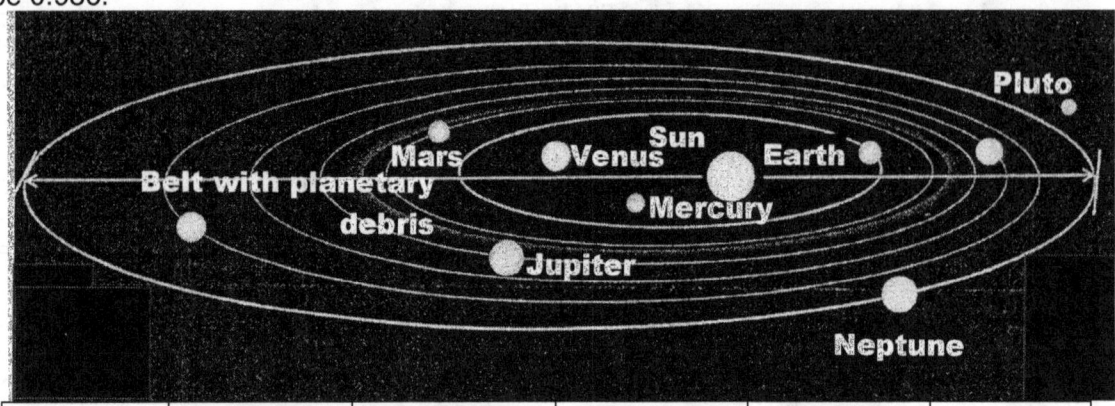

PLANET	PERIOD (Years) (T)	MOVEMENT (T^2)	DISTANCE k	SPACE (a^3)	RATIO
Mercury	0.241	0.058	0.39	0.059	0.983
Venus	0.615	0.378	0.728	0.381	0.992
Earth	1.000	1.000	1.000	1.000	1.000
Mars	1.881	3.54	1.524	3.54	1.000
Jupiter	11.86	140.66	5.20	140.6	1.000
Saturn	29.46	867.9	9.54	868.25	0.999
Uranus	84.008	7069	19.19	7067	1.000
Neptune	164.8	27159	30.07	27189	0.999
Pluto	248.4	61703	39.46	61443	1.004

Looking at a picture that should represent the solar system we have the smallest planets in the most inner orbit and the most outer orbit. This leaves us with no distinct evidence of mass being present. The further on we have two almost equally sized planets orbiting in the second (Venus) and the third (Earth) orbit rings and that is also no proof of mass. The two planets are very similar… And notwithstanding the similarity, the two planes don't nearly share a common distance from the Sun. However, both are orbital placed in true alliance with the Titius Bode planetary formation layout. In any of these cases mentioned this far I was unable to see mass bringing any inspired input into any evidence we see.

Then we reach Mars that is much smaller that the Earth but are in orbit at a same rate as the Earth orbits and is in different orbit while Mars is such difference in size. This could be proof but with less mass and even a bigger distance why is Mars orbiting also at the same pace as those planets orbit that are much closer while being bigger. With much discrepancy in size as well as distance in orbit in relation to the Sun the two planets share a common velocity of travel around the Sun, which does no bring mass into consideration as a factor. Then we go further than to the gas giants that are beyond Mars, but before we get beyond where the true giants lurk… we have a ring filled with fragmented rock particle with an array of different sizes while the lot is following in the same order as it has done since the rocks started such an orbit These small pieces of crumbled rock is where we find a ring of debris Newtonians say is evidence of the mass not combining the material into a constructed planet…what mass where? This is my problem I have with Newtonian punch-happy madness since the mass they secure, as a factor is never proven to be a factor. By presenting mass being a factor is either coming from the imagination of the Newtonian dreams where such a factor obviously belong or if it is in the reality of the Solar system and not only in the minds of Newtonians, then where is the proof that there is mass applying as a factor? The debris floating in space is scattered in random. There are no smaller parts collected by mass in some areas and the bigger particles gather at another location because mass contributes to greater gravity achieved. For billions of years the similar particles followed the bigger particle that took the lead from other smaller particles and this lot did not pass or speed up in relation to mass discrepancies but are following in orbit in a fashion that disproves mass as a factor. If there was mass applying, these fragments would either group in a fashion where mass collect the smaller parts while mass group together the bigger fragments and the smaller would travel faster or slower than the bigger particles. There is no evidence of mass anywhere as the bigger pieces move at the same rate as the smaller pieces and while the bigger pieces did not venture closer to the Sun as the smaller pieces then in the light of all of this how the hell can they award mass!

Just one orbit circle on we have the real big one, the giant, the massive, mass collector…and if you guess that it is orbiting at the very same pace as all those tiny fragments do, then you have guessed well. If any Newtonian could just once bring proof of mass being a factor in cosmology, then I would accept the fact that there is mass.

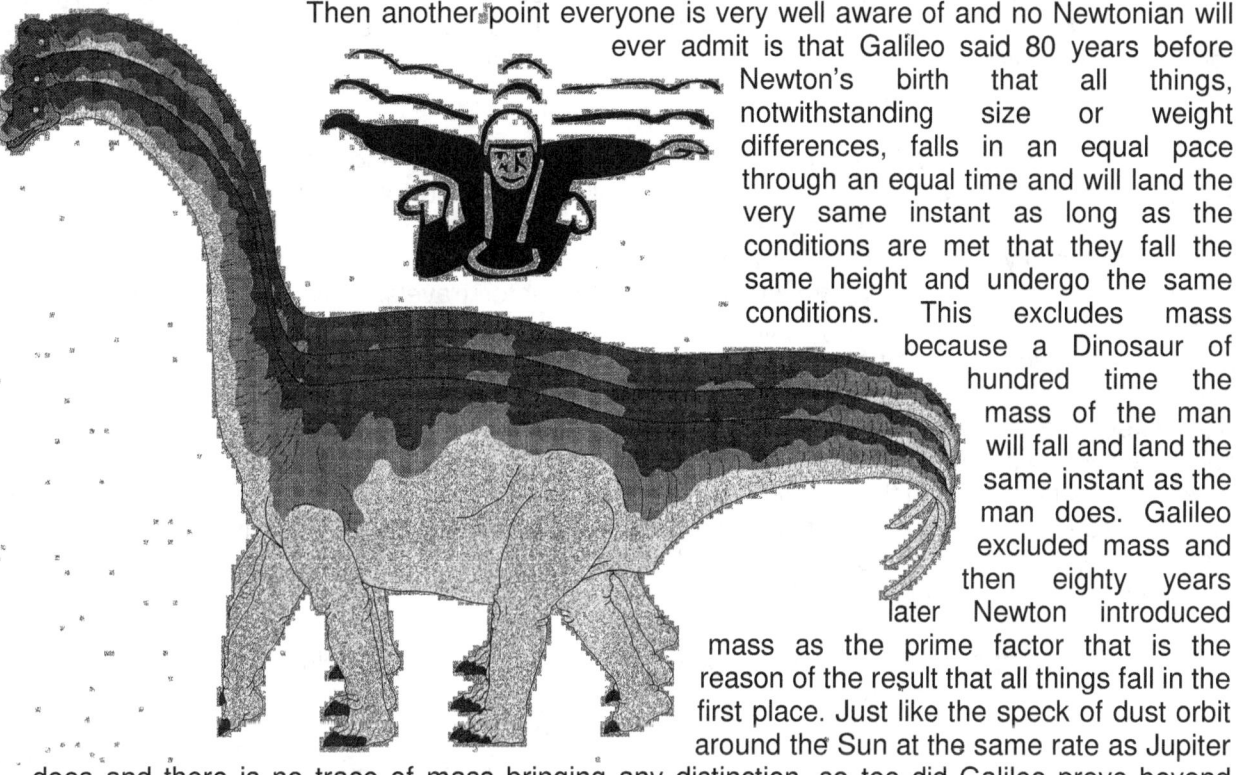

Then another point everyone is very well aware of and no Newtonian will ever admit is that Galileo said 80 years before Newton's birth that all things, notwithstanding size or weight differences, falls in an equal pace through an equal time and will land the very same instant as long as the conditions are met that they fall the same height and undergo the same conditions. This excludes mass because a Dinosaur of hundred time the mass of the man will fall and land the same instant as the man does. Galileo excluded mass and then eighty years later Newton introduced mass as the prime factor that is the reason of the result that all things fall in the first place. Just like the speck of dust orbit around the Sun at the same rate as Jupiter does and there is no trace of mass bringing any distinction, so too did Galileo prove beyond doubt that while falling all things fall equal and this eliminates every and all possibilities Newton later claimed about mass being the factor that initiates gravity. Newton's View can never in a million years compliment that of either Kepler or Galileo.

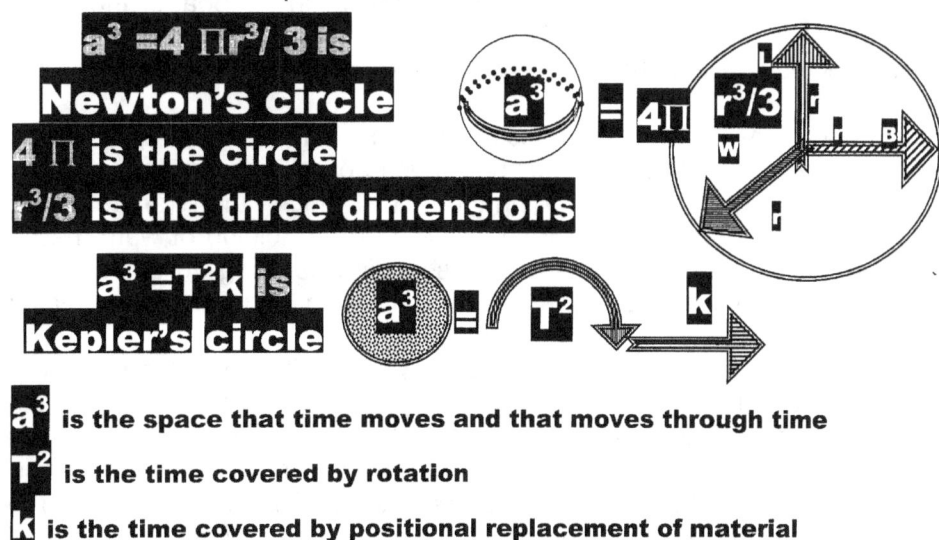

$a^3 = 4\Pi r^3/3$ is **Newton's circle**
4Π **is the circle**
$r^3/3$ **is the three dimensions**

$a^3 = T^2k$ is **Kepler's circle**

a^3 is the space that time moves and that moves through time

T^2 is the time covered by rotation

k is the time covered by positional replacement of material

Let's investigate what happens during the falling as such and see what happens as I go into this fall. The Dinosaur falls at the same pace in which the skydiver falls, which is the same pace as that which everything big or small falls, and on that I have a giant such as Galileo supporting my view. If the Dinosaur falls at the same pace as the skydiver there then has to be a common denominator in this process and since the common denominator eliminates size, form and shape, we can eliminate mass. Mass brings distinction and the falling eliminates any form of distinction which puts all equal. Let's put me personally in the picture and see where I personally will fit in.

I row a boat and never see or hear a waterfall nearby. By the time I realised there was trouble, I was too late to respond to avoid the coming fall and I had to sit back and enjoy the oncoming lesson in physics. When I fall down a waterfall with a boat I travel the same pace, as does the boat. The boat, the water and will descend as if we are inseparable because the boat, the water and me comes down as if the three of us are roped together. Maybe we are roped together because we are falling at the very same pace. That could be because I am fixed to the boat by sitting in the boat. But my sitting in the boat has certain condition and one is that I can remain sitting because I fall the same pace as the boat is falling and all the wile all the water is falling exactly as fast as the boat falls with me in the boat or outside the boat.

I fall down this waterfall with the boat and the boat and me are forming a distinctive unit that falls at the same pace as the water that forms the waterfall falls. Should I at the time of my falling hold an empty mug in my hand and I wish to fill the mug with water, and then I will have to move the mug upwards to the sky and against the flow of water streaming down the waterfall because the mug will travel as fast as I travel and I travel as fast as the water travels. I will have to thrust my mug upwards at a faster pace than my descending is while casting the mug up into the air and therefore while I accompanying the mug down the waterfall in tandem with the water I then must act against the grain of gravity to have water (mass) filling my cup (space). My mug will not automatically fill with water or if there was water in the mug my mug will not automatically empty with water just because even with the mug being filled with all the emptiness filling the mug the emptiness will travel at no different speed than would the water travel if there was water inside my now empty mug. Should I wish to fill the mug with mass I will have to go against the grain of gravity to get mass into the empty cup and in doing so my acting will have to be at a different pace than the content that is otherwise descending with me while not filling the mug!

The mug being empty falls as fast as the boat and I. The mass less ness inside the mug is not filling by the mass of the water just because the water has mass and the mass is attracted to the emptiness of the mug not filled by water and the emptiness can only fall if the emptiness was filled with mass because everything can only fall by mass. This shows that anything not filled with mass is falling as fast as everything that is filled with mass. This states the Galileo is correct and Newtonian presumption of mass creating a falling goes to the toilet. The empty space in the mug is falling as fast as the mug will fall when the mug is filled to the brim with what ever can fill a mug to the brim. Notwithstanding the content within the mug or the content within the boat or the content within the water being within the waterfall, the very lot is falling at a similar pace. By lifting the cup up into the air while falling with the cup down to the ground is the only way the cup will fill with water.

If I move the cup skywards while the lot of us is going down, I am not putting the water into the cup but I am exchanging the space that the water holds with space that the empty cup holds and my action in truth has no bearing on the water filling the space, which I then transfer into the cup. I am filling the cup with space that at that point holds water but the holding of water has nothing to do with the transferring of space or with the space descending as filling an empty cup or the space filled with water and as space filled with water will then fill the no longer presumed empty cup.

If I leaped from the boat and fell, I would fall alongside the boat. The boat then after my jumping out of the boat will be empty, but will fall at the same pace and using the same space as I fall notwithstanding being empty. The mug being empty will fall at the same pace as the boat being empty which will fall at the same pace as the water in the waterfall and this lot will fall at the same pace as I would fall, where I either am filling the boat's emptiness or falling next to the boat filled with emptiness. This brings the common denominator in the entire process in hand.

The space in the boat, which is empty if I do not fill the space, will fall at the same pace as the empty space, which fills the mug, and the mug will fall at the same pace whether the space in the mug contains or doesn't contain whatever can fill a mug. The space filling the mug is falling at the same pace, as the water that would fill the space in the mug should the mug be filled with water.

The space in the boat is falling at the same pace as I would fall, whether I am filling the vacant space in the boat or otherwise filling the vacant space next to the boat. It is the space that falls and not the object filling space while falling through the emptiness of space because it is clear that it is the space that is falling with mass in it or not being filled with mass. It is the space that is filled or not filled that is dropping down because the space being filled is in decline and holding mass or not holding mass has no relevancy on any movement forthcoming from the falling action. If it was not the space that fell, then the space within the mug would fill first as the mug and the boat fell because the empty space would first fill before it could take anything down. But since the boat falls as fast as it would whether it is being filled or not, we can assume that the space which the boat fills or does not fill is falling as fast as it would fall whether it is holding the boat or I or holding the boat and I. The space not filled by mass also moves just as fast as space filled by mass.

When the object, such as the mug or the boat or I connect with the Earth and then in touching the surface of the earth and by doing that is then awarded mass as a factor, the Earth disallows the object further freedom in motion which the object enjoyed while falling. While moving in freedom, the object claims more space in ratio than the earth time will allow. By falling and thus moving, the falling object takes in ratio to what the Earth would allow, more space than the object has in ratio claim on in relation to everything that moves in accordance with the centre of the Earth. The object now has to vacate the space it had claimed when it moved freely and take on new space that the Earth allows the object in relation to everything moving. It is all space in movement allowing claims to movement in relation to what flows by contraction to the centre of the Earth. In forming a blocking of further descending movement, the Earth resists the flow or the gravity or space lining up with the centre of the Earth. The flowing of space by contraction is gravity but the object being in the space that flows becomes and obstacle through which the oncoming space must drag in order to flow to the centre of the Earth. The space flowing is as much a substance as the material blocking the flow. It forms resisting of allowing space claimed to release to the normal flow when the object will not relent form in favour of gravity. This resisting such relenting of form and consequently forming a frustrating barrier that blocks the free flow of space towards the centre is time displacement of space and this relenting of space-time flowing freely becomes the mass factor. The density and the resistance that the particles show forms the mass that implicate the degree of the frustrating or preventing or disabling of such free flow of space through time and the displacement of space during time is space-time notwithstanding what ever irrational connection Newtonians wish to add too space-time. Allowing space to displace through time to form time is space-time and that is gravity. How bright am I to figure this out? Well I am as bright as persons were two and a half thousand years ago before Newton mesmerised science by adding fraud to science. Newton should have been aware of Galileo's finding and Newton was well aware of Kepler's finding, yet Newton chose to ignore both Maters' conclusions and introduce completely fraudulent arguments, which the world was very willing to accept because accepting Newton's fraud gave science the opportunity to play God with mathematics. All I did was not believe Newton and by realising I don't accept Newton I went in search of what was assumed as truth before Newton raped science and misplaced the truth.

All this is not new and I am not the first and the big genius that thought this out for the very first time since Eve had a bite on the forbidden fruit. In around 450 BC a man going by the name of Empedocles killed the myth that it is nothing that fills the bowl when water runs from the bowl. The bowl in use was named the clepsydra meaning water thief. Empedocles proved at the time that something other that water fills the container while the container is emptying of the water.

Empedocles' Clepsydra of 450 BC

Connected pipe allowing filling of bowl by water

Round Container Filled with water

Water running from outlet at the bottom

The process worked as follows: the clepsydra was filled with water by dipping and then submerging the jar into the water. The jar had a number of small holes in the bottom and a pipe is extending from the top upwards. When the jar is filled with water, the person places a finger on the top of the pipe as to block the opening of the pipe on top of this jar. By blocking the pipe entrance this action prevented the water from flowing out that the many bottom holes.

The water will start running only when a finger lifts from the pipe where the finger before the time blocked the intake of the pipe and therefore prevented something from entering and therefore releasing the water at the from the container. This Empedocles bottom end interpreted as being that the water was not running out from the container but was being pushed out from the container. The container was filling with something as it was being emptied of the water by something being anything other than nothing. This experiment was done some almost two thousand five hundred years ago and still Newtonians have to find a manner in which their grasp will accommodate these facts. It is the air filling the space that pushes the space filled with water from the bowl and the filling process of either space filled with water or space not filled at all that is named gravity. How difficult can it be to grasp an experiment that was understood two thousand five hundred years ago?

The space that holds the water is running down and away from the bowl and the water thief or clepsydra is stealing not the water but the space that holds the water. Once the water is out we can presume this displacement of space continues because there is no reason to think that the factor of nothing suddenly enters the scenario and the process stops because Newton saw nothing where something had to be. The space keeps repeating the process of displacing the space it follows as it is displaced by the space filling it. It is a continuous cycle never ending and the space flows notwithstanding it being filled with what Newtonians can understand or the nothing they do understand. They understand nothing so well that they used nothing to fill the entire Universe and in that they found somewhere to go with the entire nothing they do understand.

It is so clear that it is space that is moving and the fact of being filled by material or not filled by material stands no ground in the process. It is not the apple that Newton saw that fell but the space he saw as nothing it was not nothing because that which fell either was filled with an apple or empty from an apple and the space that was falling took the apple with while the space that followed the space that carried the apple still fell, whether it had an apple to accompany down or was filled with a Newtonian filling of nothing. The space holding the boat being next to the space not holding the boat and both space holding a boat and space not holding a boat is falling as fast as the space holding me with my mug in hand and the space next to me with my mug in hand as is the space holding the water which is not holding me with my mug in hand or is the space holding the water where this lot is falling just as fast as the space holding the everything or not holding anything. The space is falling. The space is falling whether it is filled or whether it is empty and that means the mass has as little to do with the falling, as the colour of onions has to do with the depth of the sea or the temperature of the shining Sun.

Newtonians should start to expand science and not the Universe for the Universe is the only aspect that has not the ability to expand. I challenge all of you Newtonians to prove $\dfrac{M_1 \times M_2}{r^2} G$ and not just go about as if it was declare proven because it is in use since the Dark ages. Expand your mind and double check the formula you all so vividly underwrite and support. Prove why you support the formula in a modern and a scientific era. Explore the correctness that this formula $\dfrac{M_1 \times M_2}{r^2} G$ underwrites. Be a true exploring scientist and journey with me through the following pages while we venture on the quest to find and vindicate my incorrectness by proving the truth vested in the formula $\dfrac{M_1 \times M_2}{r^2} G$ that carries the entire physics everyone uses. Let us start where the lot should start and get two Masters together on one point of argument. Galileo said all things fall equal. That says all things fall alike. The first thing anyone brings in is the vacuum bit with the feather and the hammer and since we do not live in vacuum there is no chance of finding a feather that will fall as fast as a hammer. Since the feather does not fall as fast as the hammer does we immediately jump to the conclusion that there are falling disparities because of the falling discrepancy we find between the hammer falling and the feather falling in the surroundings that is lacking vacuum as an active part of space. Then what would give the feather the time to fall longer than the hammer does. Everyone concludes about mass coming into play and they are correct. But they are half correct while Newton still is completely incorrect by attaching mass to the entire idea of falling. Take away the resisting of the feather and replace it with something far less air resistant and one will come to a different conclusion.

We have to dissect what factor consists of gravity and what factor represents mass. Then we have to dissect which part does mass play and what part does gravity play. The falling object experienced no mass while falling therefore the falling or moving must be gravity's contribution. While objects are in motion those moving objects is experiencing gravity.

When restrained by experiencing mass the object shows mass as a vivid factor only when the object has a tendency to move but the motion towards the centre of the Earth no longer takes place. That means mass is the restraining of the motion or is that which prevents the motion or gravity taking place. On Earth, objects experiences mass only when movement is restrained as by blocking or restricting gravity or motion with the Earth giving mass but taking away free motion. By giving mass the Earth forces the object to become one with the Earth and move with the Earth as forming a united part of the Earth.

Persons falling will experiences weightless ness while falling and they have a weightless state while falling. One cannot then go on to declare that the factor, which prevents motion, is the factor that causes motion because that is totally contradictory. The motion takes place without the presence of mass because the boat and the water are falling equally fast. When landing the motion of the boat and the motion of me ends. Then the two have very different mass values because neither shows the ability to break from mass and move further towards the centre of the Earth. Kepler said the space a^3 is equal $=$ to the motion in a line k as well as a circle T^2.

While experiencing unrestricted gravitational motion a body a^3 is equal $=$ to the motion T^2k as Kepler said gravity is: ($a^3 = T^2k$). When motion stops, then blocking of further movement restricts motion and weight or mass forms as a result. While falling we find that gravity applies as individual separate space is moving and putting time in relation to the distance that the falling object travelled. That makes the falling factor the part that is the motion that confirms gravity. In the motion or movement we find the gravity still attempting to move because that part forming gravity even remains as being a permanent attempt to move. Even when mass comes in as that which results in the ending of the gravity and in that gravity as a term is also forming the motion factor, gravity being present still remains as an attempt to move. Then while moving Galileo proved mass is not present because all things fall equal. Mass comes in when movement is retained and although the mass is present as a factor that factor that mass represents is what produces restriction of such a movement and not resulting in such a movement. The factor that mass represents is the containing of further downward movement. Looking at the factors separately it is obvious that mass as a factor cannot produce gravity. Mass is the restraining of motion that leaves gravity as intending motion. Mass occurs only when motion is prevented and when mass prevents further motion resting objects leans against each other. When objects rest against each other they restricts individual gravity motion. Mass is a substituting factor, compensating for motion loss. When mass restricts motion gravity becomes the tendency of motion. Mass counters motion when the Earth restrains further motion of falling objects. When motion seizes, falling objects remains individual while still tending to move. The Earth resists further movement of falling bodies' movement by restricting motion individuality. Having mass does not bring about gravity but it does restrict gravity's motion, which is what brings about mass.

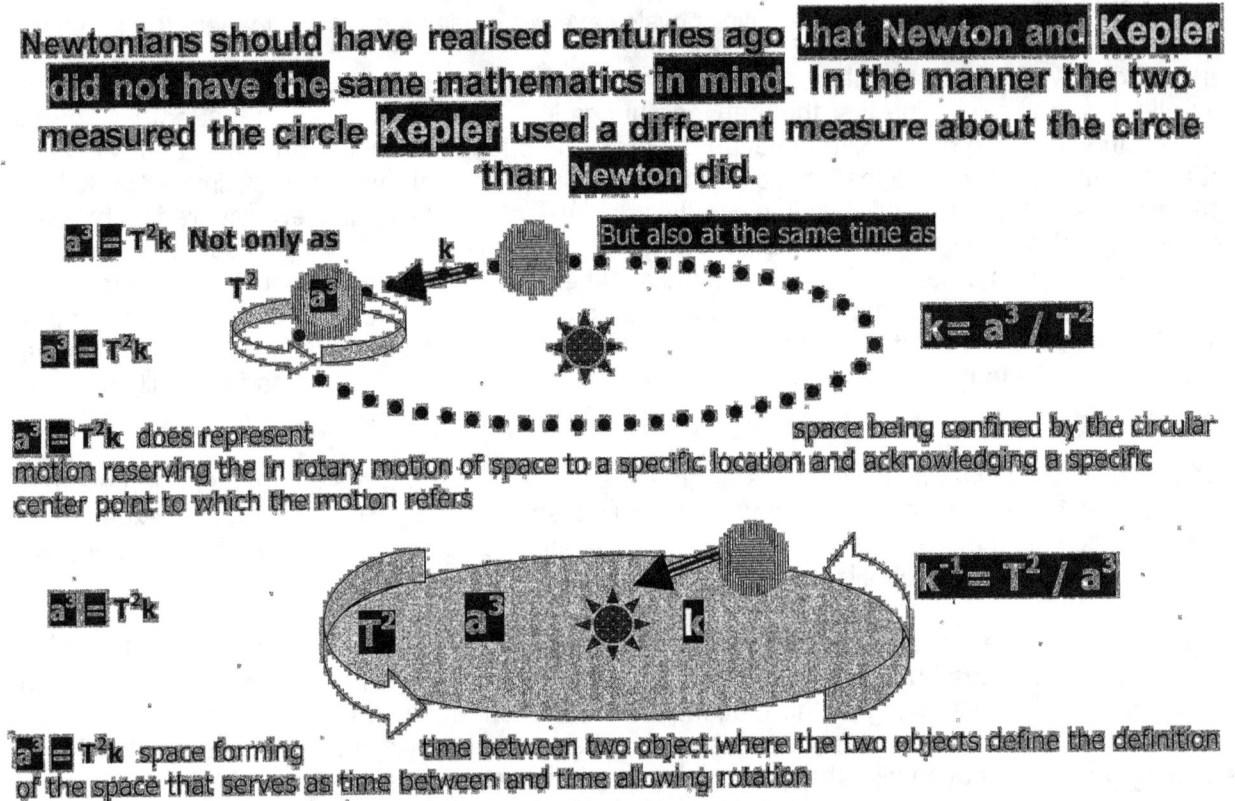

As I have indicated the one factor that is no-where to be found is mass except with an object forming part of the Earth's movement. The other thing not present is pressure. That too, is man made. Mass is the result of let us call it "stationary friction" which is the relation between two cosmic objects having motion equality while still sharing space within space. When serving under these conditions we have given a relevant value as if the lesser object holds mass while the prominent object supply movement and direction. Man has created mass and then for the sake of being clever, supplied this factor to anything and everything we could find. We gave mass a value and take fore granted that this value holds equal measure everywhere.

Mass is a thousand kilograms held by one cubic meter of water while travelling one degree in space during a specific proportion of 24 hours. Mass has to have the object having mass move with the Earth at the very same rate as the Earth does. The object must not move on own accord but must be very still and in no way fight the friction by which the Earth is holding the object in mass at the same rate as that of the Earth. By creating unequal movement while touching the object exerting mass on the Earth will cause friction. The friction comes about in the form of heat and the heat results as unequal movement apply between any two objects moving at an unequal rate. The bringing about of any form of motion in discrepancy between objects in any such a test performed, such friction coming about will produce heat and the heat will result in space forming. In such contact between objects in different speeds that such motion discrepancy produces to cause destruction of matter in space and heat comes about. In that the net result eventually leaves space created when overheating material no longer fills the space after cooling sets in.

The cracks showing in the cooled material afterwards is a result from the overheating of the material that created the extra space and then reset the occupation to what it was before. It is the result of material growing more when artificially heated that normal. The material that is reducing from retracting used space with the becoming colder again leaves cracks behind on the surface that proves that there was more space filled when the material was heated than what it had before the material was heated compared to the decrease in space after the material; again cooled down back to what it was before the material was heated. Evidence of this is evident in all supersonic aircraft as the fuselage forms cracks in the body structure of such an aircraft. This underwrites my claim that heat feeds singularity and where more heat that is delivered to an object the space held by the material and in particular metal will increase.

The outcome of this heat levels increasing is that when cooled, the material occupies slightly more space than what space it occupied before it was extensively heated. The heat caused the material to grow and after it shrinks back to the levels of heat it had prior to the overheating, there will be evidence of cracks that formed on the surface of any metal that was heated extensively. The grown space then tries to fit into the area it did fit into before the heating but with the extra space growth that came as a result of heat increasing extensively, the extending of space used to house the same material always show signs where the metal cracks.

The cracks are an over fill of previous space it once filled before the heating started. The metal uses more space than what it employed what it previously had, before the heating process started and during the heating process the volume of space occupied grew extensively with the adding of heat...therefore adding heat increases volumetric space. Then when the cooling came about the heat levels fell and with it the volumetric space that the metal held during the heating process also reduced substantially but when the cooled metal return to previous heat levels as it was before the heating began, it shows signs of metal growth indicated afterwards as cracks. This takes us back to what Kepler said. In Kepler's formula it is the extending of the distance **k** that influences the time aspect T^2 which the supersonic aircraft does by its going supersonic and by shifting **k** from the previous location the Earth prescribed to the new **k** the aircraft implicate in according to the **k** that comes about since the distance in affect becomes longer.

The aircraft produce a new time value T^2 in accordance with the Earth time factor T^2 because of the fact the Aircraft still shares space in space of the Earth with the Earth. The aircraft now has a bigger time vale T^2 in the space a^3 of the Earth using the Earth time so therefore the fuselage of the aircraft has to reduce its space a^3 it claims compensate for the extending of **k** which it does by going faster as the extending of **k** will introduce a bigger time factor T^2 that will reduce the aircrafts occupying space since the Earths atmospheric space will not compromise and the aircraft still remains in the space of the Earth. I have explained this with my marvellous exhibition of my artistic skills when I created the masterpiece I call the boat in the water sketch. The space the aircraft takes up increases in relation to the Earth's atmosphere that remains the same and by applying a bigger ratio the density factor of the aircraft reduces in relation to space in the atmosphere. A man fell from the Niagara Falls and that made the Engineers perform testes. They reasoned that by the water churning the water filled with a lot of air and this changes the buoyancy ratio between the man and the water, which to my thinking is precisely what I can detect when applying Kepler's formula. The mass the man had changed considerably which allowed the man to retain a floating stance and prevented the person from diving to the bottom of the river.

By heating any metal surface the heat level raises the volumetric space object that the the heat acquires. As increases the ratio artificially the heat as volumetric space that material holds becomes visually more and claims a much bigger area. Then when cooling brings the heat levels back to what it was before the artificial heating, the material occupying space in all atomic structures claim more space used, By trying to fit more material into the same volume of space the material had prior to the heating, the object shows surface cracks appearing.

By having more heat per volume in ratio, the material will claim and introduce new space that formed and recognising this while appreciating the importance of this aspect is much more preferable in cosmic science than placing a value on an artificial concept such as awarding non-existing mass. This new space is included in the occupied space and it has been accommodated by the atom structure. Heat establishes space that expands. This truth science does not recognise. The claiming of more space and disposing of the space after cooling shows new space formed in the material occupying space being the result of the process of heat multiplying where there were no space before that was claimed by the material. The Concorde supersonic aeroplanes' wings and other body parts were very indicative of this structural depreciation of integrity depreciation

There is now after the cooling followed the process of artificial heating material filling space that was not previously filled. The material fills more than what the case before the heating was started. There is obviously more material filling space when the material arrived back in the cool state after the heating and cooling came about and afterwards when the cooling process was introduced the space that then holds the material cannot accommodate the grown quantity in terms of space required for the claim of the material in relation to what was the case prior to the heating. This is not because of the void that came as a result of the material getting cold that brings about the contracting space or reducing the space when the heat levels dissipates. This is extra filled space even when temperatures return to levels to what was applying prior to the heat. Even after cooling the occupied space was becoming more volumetric as the space filled when the material was overheated and the reducing did not return to what was used before the heating started. I am persisting to explain this issue because behind this evidence we find what drives the cosmos since the Big Bang commenced and in Newtonian science this does not ever receive a thought far less than receiving the prominence it should receive. If material employs this as a basic technique today by increasing volumetric space when heat levels outside the atomic structure reduces, it then was also a basic technique that applied back during the Big Bang and consequently applied ever since. That evidence we can see when material having a heat level that is amplifying upwards when motion difference brings on friction and such friction brings on heat.

Two opposing issues came about, but both opposing issues are still present in the cosmos somewhere in a place where Newtonian science are missing the presence thereof. Material is composing of energy and energy is indestructible. Please take not that I am very reluctant to use the word energy because Newtonians use the word energy when they admit in silence that they have no bloody clue what is happening around them in terms of understanding science. In order to hide their misunderstanding they use pressure, mass and energy and then everyone knows that everyone has no idea what the hell is going on. But at this point I'll use energy in order to stoop down to the levels Newtonians work on. However energy can change form...yes that we all know and energy may even hide appearances. I am going to show that it is a motion discrepancy that produced matter and if there was something such as anti matter, (whatever anti-matter might be and I say that because I am not convinced that the substance thought to be anti matter truly exist). If there is a positive then for the value being present its presence must produce a negative proton and such a performing sub atomic structure cannot be possible. If something is positive, then what makes the something positive? To be anti it must be negative to what is positive, so what makes something positive? If there is anti matter and anti matter attacks matter by eating up all the matter and still remains anti matter not found anywhere we have to start to define what would anti matter compose of. The story sounds much more than Goldie locks eating the three bears' porridge. Let's place this matter being positive and anti matter then being negative in perception with the proton performing gravity. The proton instigates gravity and how and why the proton achieves forming gravity at this point is not the issue. The proton being considered as positive is what we try to establish and from there see what Newtonians regard as being negative or anti or whatever it is they say anti matter is. If being positive the proton is attracting then by the same token the opposite must apply in allowing the proton to changing legions and then getting all negative the proton must then perform gravity by rejecting material or if I am correct, producing space! I am about to prove that antimatter is in fact a process where the heat that became formed heat, which forms space, and therefore space has a valid substance other than being nothing as Newtonians propagate we have filling outer space. The motion between particles in a cramped space as the case was during the initial stages of the Big Bang would have brought on friction in space we cannot even calculate.

There is space not spinning and is therefore not controlled by directional planned movement that is forming material. To our view the space is standing still and therefore the space is expanding

There is space spinning and is therefore the spinning movement controls the directional flow of the heat and by that allows compacting of space where the compacting within material causes contracting. Again I repeat what I said before…that is it. That is what there is in the cosmos.

It is the combination of what controlled space is and what we might view as uncontrolled space that forms gravity by motion. The result is that in the very beginning, some matter particles produced gravity in their sustaining of independent singularity by applying motion and by applying motion it creates contraction. By forming movement conformed density into solidness but everything can't be sold because that would lead to friction by solids coming into contact causing friction and friction would result in heat rising. If this occurred then in some cases this would have lead to the demise of about the half of what was solid overheating to become liquid where some by liquefying formed space-time in converting the solidness it had into liquid where some then compromised solidness to overheat through friction and became liquid in the process.

This route the one side took resulted in plasma forming on the one side and

Maintaining a^3 to stay a solid T^2 by keeping space integrity

Expanding k into liquid T^2 by losing space integrity

material on the other side. It is clear that by Newtonians regarding outer space as being nothing then also see all the liquid that forms outer space disappearing into nothing. There is no say what Newtonians might come up with next so we have to consider that Newtonians saw the liquid forming plasma as becoming the nothing that holds the something that became matter.

This was done because there was less control of movement in one part of space as that part in space slowed down due to friction tension conforming solidity into a liquid that confirmed the space and the volumetric space grew in the part overheating. By expanding through loss of density that results from overheating is then having some part

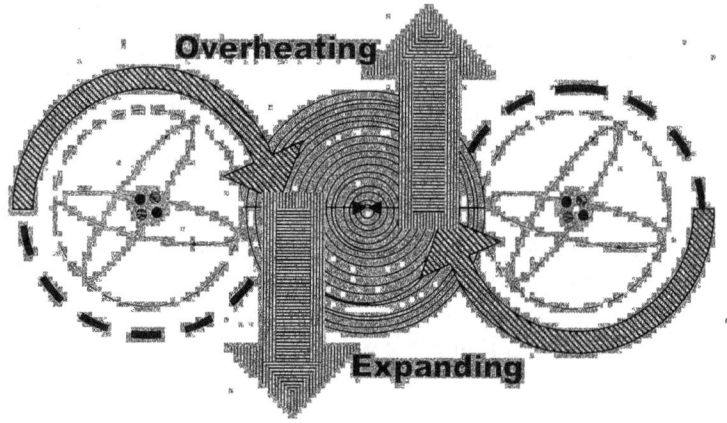

in space being softer than the other harder ones. The harder space was controlled and became solid while the other part became the softer space where that one became a liquid. The notion or defining of a liquid is very relative because as solid as the Earth seems in and during an Earthquake I saw on a moving picture how the Earth starts to vibrate as a seemingly solid then turns to acting as if it is a liquid during an Earthquake. It forms waves many meters high just like it would be a liquid like the see. Afterwards when those in charge of damage control come to assess the damage it is hard to digest the destruction and damage because all liquid-likeness disappeared and the Earth returned to an absolute solid state. Then how and where did this lot start?

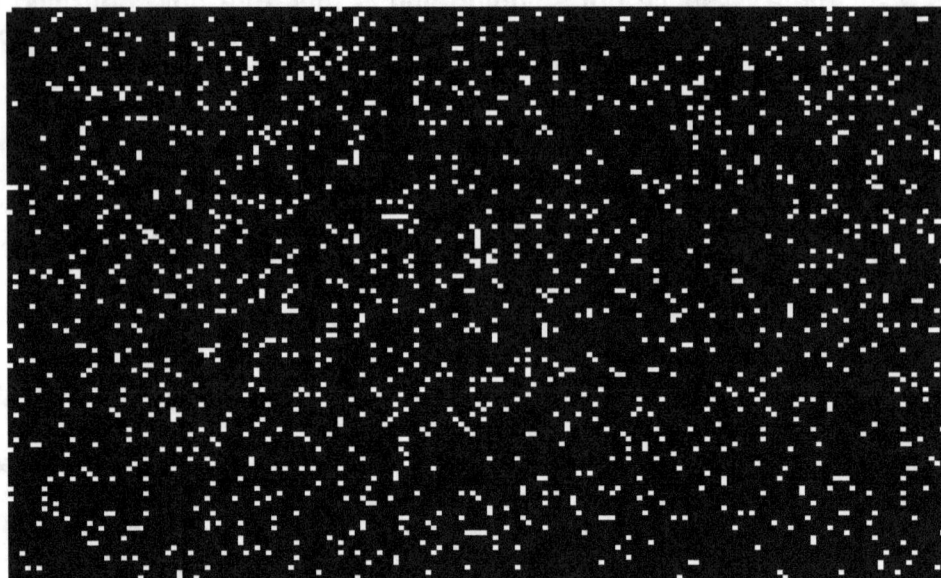

Every dot is a line connecting that dot to the point where I stand and therefore every dot is actually a line crossing space and time.

It started off with one dot so small eternity met infinity within on that spot. Then came one more dot, and with the one came an immeasurable many dots being one as they continued coming until there were a countless number of dots. The accumulative size of the dots were the same size as one dot because in the true Universe big and small plays no part. The dots were infinitely small and eternally big at the same time because size is a relevancy and without differentiation with everything being singularity and singularity being equal to one, the lot was the same size because the lot was one and one representing all there was no other size. So in the true perception, there is no difference in size. That is what still is filling the Universe. The Universe is filled with singularity being $1^0 = 1$

When the cosmos came to motion, motion was not yet defined. When the cosmos brought about motion, the first motion was relevancies.

Cold parted from hot. Eternity parted from infinity. Motion parted from motion absence. Infinity broke the laboriousness of eternity for the duration of infinity. The spot became and grew into the dot.

From what the spot was to what the dot now is might be just a mathematical implication of going from 1^0 to 1^1 but in reality that first motion was the creating of and establishing of an entire Universe with all possibilities now in it. Never again can that much growth become a reality, although to us the growth is beyond what we ever can notice. But it is because the growth is so massive and we are so small that we are unable to notice such almighty growth.

When the spot Π^0 became functional and established all relevancies possible, heat parted from cold as eternity parted from infinity. The expansion was not clear motion but more a parting of relevancies where a centre formed a relevancy because the centre could not provide motion. Without being capable of motion, the centre established four points, which also served singularity. From the inverse square law we know that the centre doubled by producing the four points holding singularity.

By exciting the centre spot, the centre spot came to be because of the heat that formed in relevancy as heat parted from the cold bringing about the division that followed and that was the motion that formed. Therefore the heat had to move but being singularity it could not get singularity to move. In an attempt to establish growth, singularity activated six spots of which four was having motion drawn into relevance four spots that was providing what was to be motion and three that was to be securing the position the centre holds. There were four forming a ring around singularity with two forming in locations we will refer to as above and as below or north and south.

It started with the fact that there is no place or part in with which one may associate zero or nothing. There are no room for a number such as nothing. The dot is eternally bigger than the spot and yet to our understanding there is no parting between the spot converting infinity to the dot representing eternity. Later on I will show precisely where both values today still are but when realising that, we must then acknowledge that the Universe is shrinking and not expanding as Newtonian culture promotes. The spot developed everything within the dot and the dot grew from the spot and we are all on the outside of the spot but on the inside of the dot. Next to the one dot (infinitely close) one will find the next dot, and if nothing was a factor then that is precisely what one will find between the two dots. Nothing of space, a non existing entity, taking up no space, and much more important, no time, therefore the dots are infinitely close to one another, being the same space, eternally big as much as infinitely small. If we as humans cannot find a manner in comprehending this notion, there can be no manner ever understanding the cosmos as much as the start to the cosmos.

Every dot was a Universe in its own and the accumulation was a Universe. The Earth in itself is a Universe as the moon is a Universe, as every single atom forms a Universe because rules applying on Earth do not apply on the Moon and visa versa. When in the ocean another set of rules apply, therefore being in the sea places a body in another Universe. The number of Universal entities is still countless, as much as it was in the beginning.

Every dot insignificantly small as it may be, is a part of another Universe as much as it is part of the accumulative Universe and every dot in the infinity holds singularity, which we translate as " nothing" being " darkness". There cannot be "nothing" just as much as there cannot be "darkness". There cannot be something big or small, but when placed into relevancy of perception, and then the relativity of perception becomes the question. There cannot be hot as much as there cannot be cold. The sun FREEZES hydrogen to a liquid at six and a half thousand degrees Celsius and Universe boils over in the form of the Hubble constant at the temperature (we presume from our vantage point) at minus 273 degrees C. If we Humans cannot or will not abandon our human perception and our manly perspective, we may as well return to astrology for all its worth.

Every point in the infinity we may observe, is not merely part of the Universe in not being anything, but is the point where the Universe started representing singularity. It is the very first point where everything began so many eternities ago, because after all, how can we ever determine where the first point was, as they were very much equal and alike at the beginning. Every aspect of the Universe started with the fundamental fact that no point in the Universe can represent "nothing" as a number, because every aspect in the Universe represents singularity in what ever form it may hold in that specific spot forming space-time. If man does not reach a conclusion where that conclusion is matching the Universe and stop to match the Universe with man (and man's incapability's and failure to understand that the Universe is too big too ever be understood by man), we may all go back to caves and become starving hunter-gatherers again, because we will never find a way to progress to the ultimate understanding of the Universe.

The First question is where the cosmos starts because the cosmos has to start somewhere. If it is there and we can see the cosmos is there, then it started somewhere. Einstein reckoned the cosmos started from a point holding singularity and with that I couldn't agree more. However, looking at the cosmos, the cosmos is innumerable lines running all over and connects everything to everything else. That then is what the cosmos are…it is lines. Einstein said that the cosmos started or starts from one point holding singularity and this means that singularity has to progress to a line and since the line is always the radius in the circle that forms the cosmos, therefore the line being the radius has to start at singularity r^0…that is if Einstein is correct and I have no doubt that he s correct.

Having r^0 as one can be any length. However Π is the form indicating a circle, and about that there is no question. With the radius starting the line as r^0 that only means that Π^0 established units of one because $r^0 = \Pi^0$. What and where ever that one was I do show later on. That brings us to the point from where the cosmos started and that point no Newtonian ever found however I have located the point and also gave it the same name as the name Einstein gave it, but to get there we have to start with a line. Every Newtonian I ever came across says a line starts with zero and that is impossible.

Lines mathematically cannot start at zero because there is no evidence of zero as a factor in mathematics. In arithmetic it could be, yes but not in mathematics. The shortest possible line (hypothetically) must be so short it must have an initial and ultimate point sharing the same spot. There has to be a point but the point has to start and has to end at the very same place while there is a point there. Any theoretical line being the shortest possible line cannot have the line holding the initial starting point at point zero and advance from there. If it used zero as a start, the zero part would not count, because the line will only start at a point past zero where the line then will start at an actual value, since zero holds no value. When the line has a beginning and an end at the very same spot and it wishes to extend the position as to further the possibility in exploring all growing it has, which direction should it favour? Extending the line in any one direction will favour one direction without any clear reason why the line is not extending in other directions as well. The only option about extending will be in all directions equally in order to give a meaningful non-bias flow of

Starting point ▸ **Extending from start.**

Zero point

mathematical equilibrium.   The shortest line in the realm of possibilities must have a start and finish holding one spot and such a line will also be a dot or a circle. Not favouring one direction puts all directions at equilibrium meaning that any form of what ever might develop from such a spot with the end and the start being in the same position also has to be a sphere. This reasoning prompted me to look for singularity in such a spot because if the prime spot from which all came was a spot holding all, then the spot must hold the shortest line but more prominent it will hold the smallest form including the smallest circle.

Again I have to repeat that the one possibility that the shortest spot can never have, is having a starting point on the zero mark. If the mark of zero holds the start it must also hold the end because the end and the beginning have the same position. If the position of zero then is the beginning, the end will also be zero leaving the line without an end as well as without a beginning. Mathematically multiplying any number with a zero factor equates to leaving another zero as the total gained.

The conclusion from this is that no line can start at zero because that will be a mathematical impossibility. A line or spot starting at zero would therefore be shorter than the shortest line possible. A line growing or extending from zero can never leave zero because of the influence of being zero disqualify any possibility of growth! If the line then had to grow in all directions at the same pace the line must therefore be a circle or being three-dimensional, a sphere. However, if the line started at a point (1) and ended at a point (1) then the line has to equate to $1 \times 1 = 1$ and therefore, mathematically, any line has to start at one which can only represent the value of singularity 1^0 to 1^1. If this is taken further we find that when using the equation of measuring the circle (1^0 to 1^1 can only represent a flat circle) the equation will be $r^2\Pi$. With r being the line from star to end and we wish to produce the smallest circle possible, then the radius would be 1 in the square $1^2 \times \Pi = \Pi$. Flowing from this fact is that in the Universe there can be no zero point or unfilled space. The value of the smallest circle is Π, because of singularity influencing the outcome the smallest circle would start with $1^2 \times \Pi = \Pi$ and that is where creation started. That is where the Universe starts…it starts with $1^2 \times \Pi = \Pi$.

That gave me the clue where to start looking for singularity. One would find singularity in the value Π and the value Π will be in all things rotating in a circle. You might wonder how does that apply to the cosmos and moreover to gravity? Gravity is the dimensional changing of space holding r as reference to the sphere holding Π as the reference. Heat occupying space has the cube that can apply r, as a straight line bringing about the cube with all its other names that may find attachment to specific form but nevertheless still remains only a six-sided cube with angles changing in some cases. In the sphere there is no radius but only the extending of Π from the centre Π in six opposing directions relating to one another by the square but remaining Π because of the unity the matter holds in relating to space. It is not possible to draw a precise line that would form a precise ring and not cut some atoms in parts.

There can never be an absence of a line or have a situation that does not present or prove that there is no line at all because if there was nothing there would not be any possibility of a line ever being at that point as nothing can hold no line. Since everything in the Universe can only be while being a line, therefore if there is a Universe, no matter how small or how large, then there has to be a line. Because there is a possibility of a line forming being the shortest line possible, the line is infinitely small yet the possibility does not exclude the line in totality as would zero do. Therefore the line is there, albeit only in the possibility of a line being there $1^2 \times \Pi = \Pi$.

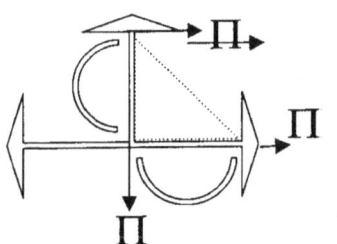

That may be one solution except for the fact that only nothing is that simple. The rejecting and / or the accepting are correct, but the spinning part just comes across a tad too simple to make sense. To find substantiation one has to find the manner in which light connects to singularity because everything connects to singularity. Since light holds the most basic evidence of when movement started, we will have to go there where singularity becomes gravity by applying the very first indication of movement.

With the establishing if the value Π and identifying r, one has to distinguish and define each item in order to bring comparison. No line can start at zero but can only as the smallest dot one may imagine, bringing about that the line will start with the value of Π and proceed and proceed following the same value. The value has to be an extending of the original value because no evidence indicates changes that may take place.

By extending Π, a continuing of The form we multiplication is a brake anywhere that Π will because r = 1 therefore only Π because $r^2 = 1^2$ and that proves a stationary position, it will alter position to comply with the new standards set by shape.

the dot • to the value of the next Π will flow in all directions evenly. sequestrate from this sphere, but when the sphere showing a flat surface, we find continue going on to the next Π, and therefore $r^2 = 1^2$ and will progress onwards. With only Π progressing

In that manner Pythagoras becomes defined because the straight line and the triangle hold equilibrium in the half circle. By extending the value of Π to the next position that Π maintain in the time, a position Π^2 becomes available where the Π^2 indicate time (or the which follows from the past to the present) and the square of the triangle indicate space developing will establish the square of time.

From the governing singularity comes the major singularity providing Π^3 as space, which time claims $\Pi^2 + \Pi^2$ and space holds time to control space $\Pi^2\Pi$ from where singularity influences space by the three sides a cube shows to one side of the universe using r as a means of doing so. I explain this in much more detail in other more technical orientated books, but I try to avoid getting overbearingly technical in this book.

From this (I suspect) there is a value difference where the one singularity holds a factor of 1 in the 21.991 / 7 and the other has a value of .991. Where $\Pi^0 = 1$ and $^{\alpha}\Pi^{\Omega}$ has to be smaller, it holds the value of .991 in some cases and other cases the value of .91. Closer than such an explanation I could not come without rambling on for bout three hundred pages in how this happens. In ***An Open Letter On Gravity Vol. 1 and 2, Part 1 and 2*** I am much more explicit but have to frank that when doing so, the reading becomes rather elaborate. I have to name the opposing linking singularity although to my mind humans use the naming of objects as a shield to cover their poor concept of understanding. The singularity Π^0 I refer too at times as the governing singularity. In another part I shall show why light uses $^{\alpha}\Pi^{\Omega}$ by means to travel as a conductor, but that comes later. I mention that fact because the singularity from where everything came is still in our midst and used by all with vision. My dismay in naming is clear as through out the book there is clear evidence of how names had replaced recognising of various factors and through naming what ever differently; humans go at fault in recognising what the product truly represents. The governing singularity is the unseen line that all matter refer to using the space provided by the governing singularity and I sometimes refer to that as the major singularity. Where is singularity…?

There is no need to get all-Bohemian and mesmerised to locate singularity. Locating and finding Singularity in places other than Black Holes must be our next quest because to my view everything connects and that view places a Black Hole in the position of being just another star.

The location of singularity is so obvious it is even slightly funny. In the precise middle of all objects in rotation is a precise centre dividing the object in sectors that will start the spinning initiation from that centre point. But the spinning object will have a middle point, a very specific centre point that does not spin, that can never spin for it has no space within which it could spin and holds Π as only value. This is so new I have to repeat this discovery many times in this book and yet this concept is as old as the use of the wheel.

That point, albeit hypothetical but it is also as much a reality none the less, is where the point is that is holding singularity because is standing still. Every line running from that point in opposing directions is also in opposing directional spin to each other. Another aspect making such a line prominent is the fact that such a line is eternal, while being motionless at the same time and that is precisely the requirements to find the line from which all known and not known came about.

When one goes about drawing a line, the first motion will be to place the pen on the paper. The paper is blank but the blank paper does not represent zero as an option. If it did, there should either be no paper at all or no pen or both will be missing. But that does not bring about the option of zero either because a possibility may arise where one will obtain any instrument with which to draw the line. It could be a stick or even a finger and the paper may be sand or wood to carve on. The main issue is that there are always many possibilities whereby a line may form and that exclude zero as a possibility. Even by the two stars lining up brings about the forming of a possible line. Any space between whatever and me might be in my sight, that space holds a line and even if it is not in my sight, there remains a possible line. The shortest line in the realm of possibilities must have a start and finish holding one spot and such a line will also be a dot or a circle. When placing the pen on paper or the chisel on rock or whichever device one may choose to draw a line, the line will start with a mark. The mark may be a sizable hole, or a chip so small it is barely visible to the naked eye, but it is there, even if it is so small it is only visible by the use of an electronic micro scope, the dot is there.

When a human draws such a line, the human will tend to favour one specific direction to take the line from the position of the dot, but humans are not creating, we are merely duplicating what is already about. **Singularity is a mathematical point at which certain physical quantities reach infinite values for example, according to the general relativity the curvature of space-time becomes infinite in a black hole.** If singularity is a mathematical point we then should discard man's options and look for the option where mathematics will bring about a non-bias flow. Mathematic has no pre conception but man-applying mathematics does and that is where mathematics goes astray. We also have to presume that space-time came from the first dot because where else did it come from? One thing about the Universe is that it holds conformity going out in all directions even-handedly because if it did not there has to be borders, which there are not.

Not favouring one direction puts all directions at equilibrium meaning that any form of what ever might develop from such a spot with the end and the start being in the same position also has to be a sphere because the flow outward will be equal in all directions. In this there are directions north, south, west, east, back and front. This reasoning prompted me to look for singularity in such a spot because if the prime spot from which all came was a spot holding all, then the spot must hold the shortest line but more prominent it will hold the smallest form including the smallest circle or for that matter the smallest sphere. One possibility that the shortest line or smallest spot can never have is having a starting point on the zero mark. If the mark of zero holds the start it must also hold the end because the end and the beginning have the same position. If the position of zero then is the beginning, the end will also be zero leaving the line or spot without an end as well as without a beginning. Such a spot will constitute all of nothing.

The conclusion from this is that no line can start at zero because that will be a mathematical impossibility. A line or spot starting at zero would therefore be shorter than the shortest line possible. A line growing or extending from zero can never leave zero because of the influence of being zero disqualify any possibility of growth! If the line then had to grow in all directions at the same pace the line must therefore be a circle or being three-dimensional, a sphere. Flowing from this fact is that in the Universe there can be no zero point or unfilled space. The value of the circle is Π, and that is where creation started. That gave me the clue where to start looking for singularity. One would find singularity in the value Π and the value Π will be in all things rotating in a circle. You might wonder how does that apply to the cosmos and moreover to gravity? Gravity is the dimensional changing of space holding r as reference to the sphere holding Π as the reference. Heat occupying space has the cube that can apply r, as a straight line bringing about the cube with all its other names that may find attachment to specific form but nevertheless still remains only a six-sided cube with angles changing in some cases. In the sphere there is no radius but only the extending of Π from the centre Π^0 in six opposing directions relating to one another by the square but remaining Π because of the unity the matter holds in relating to space. If closely inspected one may recognise Einstein's curvature of space-time following the form of Π surprisingly close. Might that be a coincidence since Newtonians love to put whatever the never can't understand down to being coincidental. However taking the line one step further, it is not possible to draw a precise line that would form a precise ring and not cut some atoms in parts.

From the possibility of the line being, the line may have the space of the premier singularity from which all of the universe arrived, but remains a possible factor and as such then also remains a very plausible and possible factor one might discover at any point throughout the entire Universe. The fact that a line is not excluded as a possible factor presents the presence of a line notwithstanding out visual competence to see it or not to be able to see it. In the case of zero the factor becomes a definite excluding factor whereas without zero in infinity the line becomes a definite included factor, as it would be most incorrect to surmise no line will ever form at such a point. The big picture in the Universe is that any point can hold every point on and any in any form that material may take on or not take by never excluding the possibilities to represent point holding whatever material; the Universe is about inclusiveness and never about exclusivity. Zero brings conclusive exclusion while infinity brings conclusive inclusion and as such the two values oppose the concept surrounding the line. It is moreover the individual singularity in the major singularity, which sustains maintaining the governing singularity providing equilibrium in space-time.

Seeing our spinning top from the top, there are four quarters opposing each other and by that opposing one another.

Any object in rotation will have a middle point, a very specific centre point that does not spin. That point is once again albeit hypothetical, but none the less must be by the mere location thereof, in place and also be standing still because every line running from that point in opposing directions are also in opposing directional spin to each other. If the one forms matter, the other would then provide anti matter and if the one is gravity, then the other would at that instant be anti gravity.

From such a point every other point will be opposing any other point not pointing in the direction to which the first point is pointing, whereby it extends the direction it holds. No matter what the point is or where the point leads, such a point holding a specific direction will be unique in the direction it is rotating because at that or any other specific point wherever, it will be directing not in the direction it spins but in the direction flowing from the centre point outwards.

Any point will be opposing itself within the rotating of 180° and by doing so it will be changing

East going west

West going east

every aspect of its previous flowing characteristics it previously had or will once again have in 180^0 from there. While in rotation from the point of an outside observer all in the visible Universe may seem static and never changing but to the object in spin every next second will be a diverting from every aspect it was in every second passing, and the direction it held in relation to the direction it held the previous mille, mille second will totally be incompatible with the direction it holds the very next mille, mille second of rotation.   That proves no point can be static or constant, all though it may seem that way to outsiders.

This argument I raise must be very, very well understood. In this is global warming. The Universe and everything the Universe holds are subject to cyclic inter action by progressive changes opposing its position from what it is to what will be in development. In concept is ice ages and drought name I improvise because we yet invented a name for the we are going too end in just everything is once again going we are heading for the following

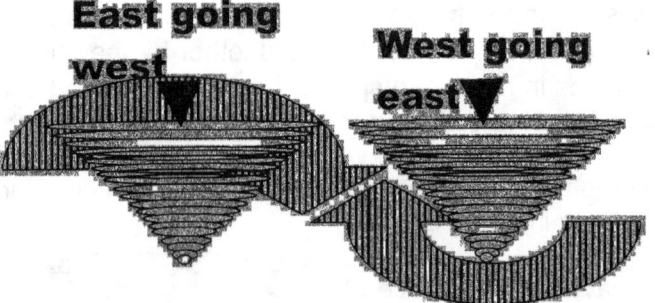

this ages (a have not cycle that before to cool while ice age where the turn around from cycle to cycle will last

$10\Pi \times 10^3$ years going from ice age beginning to ice age ending, but we only have a name for the cycle from where we are coming and that was the last ice age). In this are summer and winter and not the angle that the Earth has with the Sun. In this we find that the same gravity that pulls everyone down in England pulls everyone up in South Africa. In this is hot and cold, summer and winter, the fact of night and day, seasons forever changing, the long Antarctic days while at the same time the Arctic enjoys long nights. In this we find the reason why the polarisation of the Earth swaps when the North and South Poles change and alter their magnetic fields in relation.

Whatever is now summer, will again become winter when the cycle has gone through half a circle. What is now positive will be negative the next time around. This is the biggest issue to recognise when researching the cosmos. Wherever we are heading too, we will return from there half a cycle later.

Because there is a space that may not be occupied by a particle does not exclude the possibility of a particle sometime to the future occupying that space. If the space was nothing all possibility of future occupation will become excluded by the presence of zero that is unable ever to include occupation

From the centre of the top runs the premier singularity and as the top starts rotating the top's rotation bring about the sides to singularity, which too was present all the time but also forever changing by the rotation of the top. This is the essence that keeps the top erect and stable. Let's put it this way, the lines were there whether the top chose to fill it by moving in it and with it or whether to top chose to have

mass

and not manipulate the lines holding the top erect. The movement of the top may charge the line that is already in place but by charging the line that is in place, the top promotes the line to serve as a premier singularity.

Since occupation may or may not be placing the factor in infinite, the space therefore holds the premier singularity of infinite from which all included in the Universe has come.

When the top starts executed the option When it moves it location it moves singularity as all is

spinning in a specific position the top merely to fill the premier singularity at that specific point. may take the premier singularity with to the new through spin or it may fill yet another position in the same.

The sectors provide in sustaining which provision maintaining singularity spin

individual singularity a governing singularity comes through governing the required in

means by

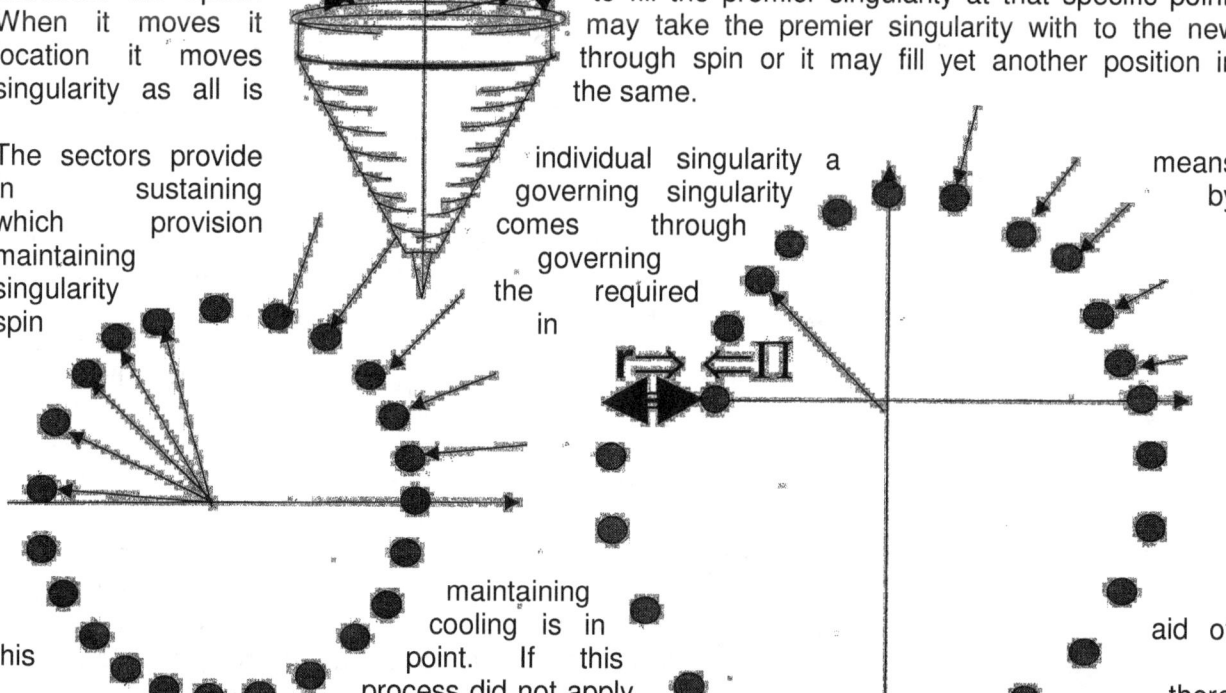

maintaining cooling is in point. If this process did not apply,

aid of

there singularity to singularity, then the governing

this

would be no connecting individual major singularity. If the Sun forms the governing singularity, then the top will hold the individual singularity. If the Earth provides the Earth will hold the individual singularity. If the Earth provides singularity, then the top will hold the individual singularity. If the Milky Way holds the governing singularity, the Sun would hold the individual singularity. There will forever be a reference forming from which a governing singularity supports and guides individual singularity. It is on this bass that the Titius Bode law doubles its distance in relation to the Sun forming the centre point of rotation to all the planets, but this I explain much further on. By claiming the position held by singularity premier as a vacant spot until the arrival of the top, the singularity of the top divides the point flowing from singularity into four sectors holding two half circles

From such a point, every other point will be opposing any other point not pointing in the direction to which the first point is pointing, whereby it extends the direction it holds. No matter what the point is or where the point leads, such a point holding a specific direction will be unique in the direction it is rotating because at that or any other specific point wherever, it will be directing not in the direction going straight from where it spins, but in the direction flowing from the centre point outwards. Recognising singularity requires the realising that any object in rotation will have a middle point, a very specific centre point that does not spin. That point is as said before is said once again to be totally hypothetical and seems completely undetectable, but for human intellect providing the ability to interpret such a point as a secured reality, but none the less its visibility or invisibility, it asks for the understanding that that point must be standing still because every line running from that point is running in opposing directions and are also in opposing directional spin to each other, notwithstanding the point not being part of either space or time.

Although the points in rotation had the same characteristics only seconds before as it was spinning, they oppose the characteristics it had just before and just after the very instant in which they are and to which they relate coming from the past or going into the future location by similar points also in rotation. Due to the spinning nature of such a point with all surrounding the point very varying second, the value to such a point can only be Π because of its constant changing. Using r would specifically use a varying value that is in opposition from one another r from every angle. This value holding singularity can only be if Π forms the progress while r = 1 or then refers to singularity.

An object maintaining time will have Π confirming the next Π in the same positing in relation to r = 1 as the previous Π. By maintaining Π in relation to the same position Π had, then following Π to develop to the next Π will produce a value of Π^2 that confirms time. By maintaining the value of Π in relation to singularity as Π^0 it also confirms all other aspects of singularity in as much as $\Pi\Pi\Pi$ and Π^3. In that way singularity secures its relevancy to matter as much as matter secures its relevancy to the point serving as singularity Π^0 and is singularity. When Π crosses over from the one quarter to the next quarter and it does not maintain a constant position in Π^2 forming a relation with Π^0 the relevancy will change as time no longer can apply a true value. When time diverts it will also affect density and the standing the particle has to surrounding matter.

By applying a different position Π becomes r in relation to the previous position Π held because the circle now has to introduce a line in support of the new circle. The loss to density through the application of a new time relation will be suspended matter forming in a heat release.

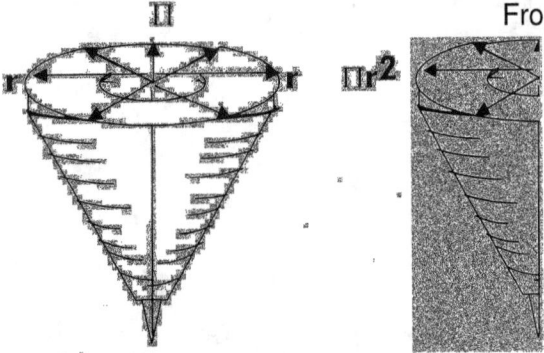

From the centre of the top running outwards is the radius extending by the square. The radius will be balanced performing as Π as long as particles remain connected to and maintaining singularity Π^0 which then is indicating a solid substance. The formula that uses the radius by the square in conjunction with Π is Πr^2. Because it is in the square it has to be from one side to the other side and using Π to improvise the lack of being the square as the square will be in the cube. The formula is in use for many years and used extensively. From the centre is the same radius that Newton used for his just as famous formula $$F = G\frac{M_1 M_2}{r^2}.$$ This formula is a lie while gravity is the truth. Newtonians are experts in mixing the truth with a lie so well no one can separate the two. It is like putting milk in coffee. They are presently doing it with global warming. The truth is that there is global warming but it has been going on ever since the last ice age and the lie is that carbon emissions are responsible for global warming. The same lie goes back to gravity and mass where it took me twenty-seven years to separate the truth of gravity from the lie of mass.

The size or mass only depends on the fluctuation of r in the square as a component to the circle or sphere but that does not affect the form by indication of Π in any way there may be. The conclusion from this is that no line can start at zero because that will be a mathematical impossibility, because $\Pi r^2 = \Pi 1^2 = \Pi$ but $\Pi r^2 = \Pi 0^2 = 0$ and notwithstanding the correctness about my argument, my work was rejected by three University Professors at three institutions because they said my statement that zero is not a factor in outer space was being incoherent on my part. A line or spot starting at zero would therefore be shorter than the shortest line possible. For obvious reasons can no line, or any line grow or extend from zero $\Pi r^2 = \Pi 0^2 = 0$ because such a line must then quit zero and become something, thus abandon its original value.

That then in such an event would mean the start of the line has a different value to the end and a line holds conformity through out. When any line is starting from point zero it can never leave zero because of the influence of being zero disqualifies any possibility of growth. If the line then had to grow in all directions at the same pace the line must therefore be a circle or being three-dimensional, a sphere. Flowing from this fact is that in the Universe there can be no zero point or unfilled space. In the case of the growing sphere the value of the circle is Π, and that is where creation started. That gave me the clue where to start looking for singularity. One would find singularity only associating with the value Π and the value Π will be in all things rotating in a circle. You might wonder how does that apply to the cosmos and moreover to gravity? Let us find the answer from what we know and work our way back to the infinite where we also know we will find singularity as defined by the Brainy Bunch.

Using: r as reference or using Π as reference only changes form.
By reducing r indefinitely to the tune of half each time, r would become infinitely small, beyond human calculating means, however as mentioned in the case of the smallest dot holding one spot, r would become insignificant beyond human comprehension even, but never reaching zero and still Π would remain intact and dictating form. Reducing r by half would cut the circle by four but that does not mean that Π as such reduces. It is r going in four directions that reduce leaving Π to maintain form. It means the reducing does not affect Π but merely influences Π.

When the circle reduces, the value allocated to r will become implicated because r determines specific size. Not so in the case of Π, because Π in the true sense only indicate that the circle is a square without corners and therefore Π dictates form and not size. By reducing size only r comes into contest and will point to such reduction. By reducing the circle radius r by half continuously will lead to an infinite small circle but Π will remain because the circle as a form remains even being infinitely small. Where singularity develops into space-time for the first time it will only do so keeping an interest in Π. That then brings about infinity reaching space-time but leaves out zero as an option because by applying infinity, infinity can never be reached because of the qualifications applying to infinity $\Pi r^2 = \Pi^0 1^2 = \Pi^0 = 1^0$. Being at infinity takes us into the heart of singularity and singularity must therefore be in all round objects since the argument concerns any and all circles excluding not one.

An observation coming instinctively to mind one may recognise is that the form reminds rather explicitly of natural phenomenon as hurricanes, water whirls and even the shape most commonly favoured to express the cosmic object referred too as a Black Hole. The definition of singularity specifically places singularity in the centre of the Black Hole and by reducing r we obviously will reach a line of infinity where singularity is only by definition thereof part of the equation. The similarity may be more than coincidental. Let us consider the statement in the reverse. Even by reducing r to the point of hypothetically not existing any longer, Π still remains a factor because r becomes the implicated factor and Π merely influenced.

Anything occupying space in the cube will apply r, notwithstanding the name used confirming the shape or r named as length width or height, it is all just a straight line bringing about the cube with all its other names that may find attachment to specific form but nevertheless still remains only a six-sided cube with connecting lines applying different angles changing in some cases. This proves that applying Π introduces stability in form while in the absence of Π structural weakness will constrain stability. This has mist prominent implications on the function of gravity. Gravity is not as simple as awarding mass or not awarding mass and the implications of having such a simplistic view truly proves to be outdated.

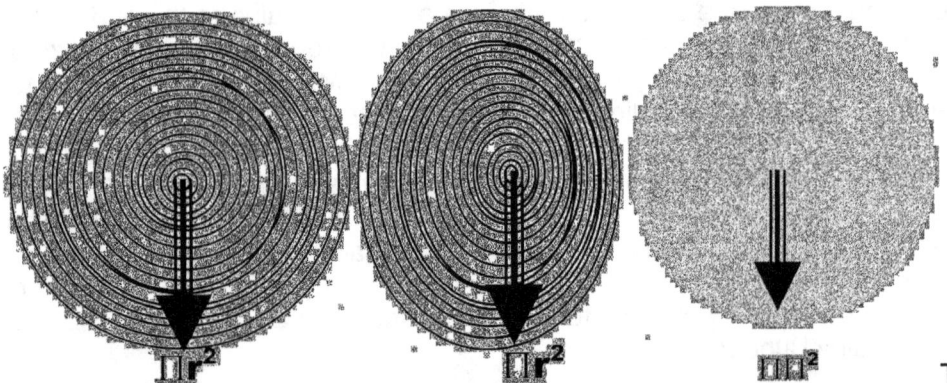

The normal perception is One of such consequences the comes from this revaluating gravity, is that science should start with the recognising that the star uses hydraulic power and not compressible pneumatic power one must accept that r has little use within the star. This is only one of the points of influence that changes the entire evaluation of cosmology. In the star the heat surrounding the particles have almost more density than does the particles themselves, while with applying pneumatic principles gravity then would have to rely on using magical mass to interpret all the magic wonders we find in a star. Where gas breaks down with heat and under severe pressure, we find that hydraulics become ever stronger as heat levels rises because such increase in heat will sustain the power that drives hydraulics, especially in the total absence of viable space as we would find in a hydraulic operating star.

Hydraulics will charge the use of a constant uninterrupted Π because of the unflinching characteristics hydraulics show while pneumatic power will employ the use of r as pneumatic stability leaves a lot to desire.   Matter will even crack and break under hydraulic pressure, but where work sufficiency is called on, hydraulics only increases durability as heat increases and since the liquid is heat, it only brings about that the fluid will become more of what it already is when heated. But when saying pressure it has again more to do with culture in science than with the fact of the matter because heat will flow from hottest to coldest no matter what and therefore the pressure must relate to cold versus hot in relevancy of singularity holding different positions within the stars.

 that any circle growing spontaneous would grow by the radius, which is r. That cannot be the case in singularity because r is an indication of a straight line. By growing with the aid if a straight line the influence that would have on the circle would result in many circles following one another and not a continuous growth. Gravity is the dimensional changing of space holding r as reference to the sphere holding Π as the reference. In order to generate spin that produces time in matter occupying space, therefore creating dimensional change, Π has to be a factor indicating the possibility of spin. The answer must be in finding Π, and thereby locating singularity.

Π=r in constant directional change as time flows through rotation
Pinpoint positioning of singularity Π^0 with Π positioning space to either side forming the border set by singularity

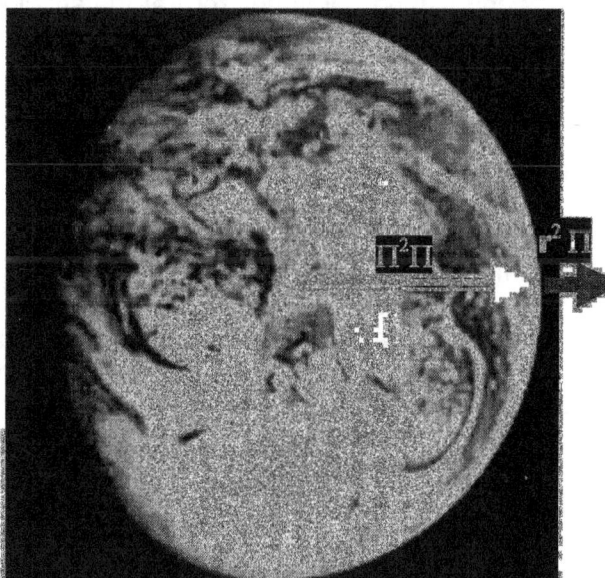

The new direction pointing to a new location in relation to the previous point will oppose the previous point it had in relation to direction considering the centre point.

Looking at the affect of gravity it shows the precise quality of no distinctive progress between points, as gravity never seems to end at a point but flows all over affecting all that holds a position in its sphere of influence. The gravity coming from China meets the gravity coming from America at no particular spot but intermingles without distinction.

Using the concept that gravity applies Π as the circle factor Π as well as Π^2 replacing r^2 the replacing of Π brings two values as Π and Π^2. That I found is the case with gravity and will be apparent when explaining the sound barrier as well as the Roche lobe. In order to create a distinction I remained using r as the indicator of the cube or circle that has vacant space and by vacant space I refer to non-solid structure. In the solid structure I use Π as a value for reasons that will become apparent in due time.

In this difference between whereΠr^2 is applied we can see the difference there is between outer space and what the star such as the Sun represents. The Sun would apply an impenetrable $\Pi\Pi^2$ whereas in outer space one would find the mathematical application for Πr^2 but only up to a point.

In a planet just as much as in a star the relevance that changes when going from outer space to the star is also changing the dynamics by going from Πr^2 to $\Pi\Pi^2$. It becomes a space to move in and move through and a space not to penetrate because of density connecting to singularity. The solid state a cosmic object provides lend the cosmic object the value of having $\Pi\Pi^2$ which is impenetrable while in outer space the dynamics might apply as Πr^2 however, that is only up to a point because singularity might be slightly detached, but singularity still always apply notwithstanding conditional freedom and some leniency in time provided movement. This became a factor when light brought about space. However this is an applying relevancy because on the Earth the atmosphere becomes $\Pi\Pi^2$ giving the Earth a value of $\Pi^2+\Pi^2$. Going above a motion limit in outer space will bring about that the object in motion will return from Πr^2 to $\Pi\Pi^2$ when certain time conditions are breached. This means no object can escape from the solar system and fly to other galaxies and certainly no cosmic aliens can come and go, as they would please. The limit is in the atomic relevancy of $(\Pi^2+\Pi^2)(\Pi\Pi^2)$ 3 = 1836.

In considering the spinning motion in the fraction of time in the detailed instant, every aspect of rotation will turn in every instant of change in time. Although the points had the one characteristic only second before, the next instant will oppose the characteristic it had just before and just after the very instant in which they are and to which they relate by similar points also in rotation. The fact of the graph proves my point in quarterly opposing dimensions and values,

Due to the spinning nature of such a point with all surrounding the point will be alternating direction favouring change every second and in that the value to such a point can only be Π because of its constant changing. Using r would specifically oppose another r from every angle because the use of r will bring about a static relation to the previous and following instant and therefore it will cancel the constant spin flow.

Space-time is a four dimensional position of the universe where the position of an object is specified by three coordinates in space and one position in time. According to the theory of special relativity there is no absolute time, which can be measured independently of the observer, so events that are simultaneous as seen from one observer occur at different times when seen from a different place.   From this view and the knowledge that all things being the atom or sub atomic particles of the atom are in motion, therefore all of the universe are in motion and thus qualifies as space-time.

When taking the radius back towards the centre, the radius has to stop somewhere to that end there is no argument. Where the stopping is of the radius reducing, well that issue becomes another bone of contentions. The spinning top resembles the perfect scenario to apply Kepler's $a^3=T^2k$, where a^3 is the surface of the top, and **k** is the radius. With **k** being r then what will T^2 represent? Reducing the radius means reducing **k** and then **k** has to start at some point and such a point cannot be zero because of the arguments already presented I if started at zero 0 then multiplying the length **k** will be will be multiplying the formula by zero. In the formula **k** has to be one point in the formula and cannot represent 0 or else $a^3=T^2k$ has to be entirely zero. If **k** is one then a^3 will be a^3 and T^2 will be T^2. If **k = 0** then $a^3 = 0$ and $T^2 = 0$ making the formula invalid. In affect the validity of the formula can never disappear because it can only grow and extend if it is not zero at the smallest point of existing.

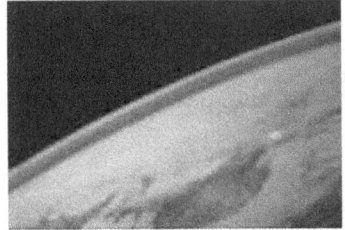 From the dynamics of the Earth and the atmosphere as well as the outer space there is by now given sufficient evidence about the dynamics of dimensions. Coming from outer space towards the solid Earth downwards Newtonians hold the view that it maintains all the same dimension that is coming from outer space and entering the Earth. If there is no dimensional differences then explain why Challenger exploded into pieces when entering the atmosphere incorrectly. At such a point as where Challenger met the unfortunate it did, is a point of enter that demands precise preconditions or total disastrous rejection. It is the same as landing, and the surface of the Earth is definitely another dimension to enter the surface of the Earth, any person either digs a hole or tunnel, or entering is forbidden.

It is a divide. It forms a solid wall where conditions change to new rules. The loss of the Astronauts killed in the disaster is proof of that divide cannot ever consist of being nothing! The same applies when taking **k** down to the very end and finding out there is a divide, but behind the divide **k** continues in another state. The form **k** would take on will bring about a position where **k** is dimensionally less than one. Bringing about that a^3 is less than one and T^2 is also less than one. It is a dimension beyond our understanding, but is a Universe fitting in a human fist as we had during the Big Bang not also beyond human understanding? Reducing **k** or r or whatever symbol connects to the radius, it will continue indefinite going smaller all the time. If a^3 was the size of a man's fist, and $T^2 = 10^{-43}$ what was **k**, with **k** not being in the infinite. The factor of **k** was even less when $T^2 = 10^{-100}$ or $T^2 = 10^{-1000}$. The Big Bang moment of start did not come from zero and therefore **k** was infinitely smaller when a^3 was infinitely smaller and T^2 was infinitely smaller **k** had to be in a dimension of not being zero because the other factors were not zero. That brings us back to the culture of matter and anti matter and the one having the ability to eat up the other one just as Josef of Egypt Biblical fame saw the lean beasts devour the fat beasts and still remained lean.

Newtonians should strongly consider that The Creator used mathematics to create the Universe. Mathematics was just the tool the Universe imprinted as the Universe developed. The Universe installed mathematics. Mathematics is not the god that created the Universe. The Universe laid the foundations and then mathematics followed by leaving the trail that indicated the path that the Universe followed as it developed.

Those highly educated members of the Newtonian Brainy Bunch Establishment are far too highly educated to translate mathematics and see the manner in which the prime of mathematics was set in print during the foundation of the Universe.

Think how far Einstein went in calculations just to establish the fact that singularity is a fact in the Universe. ...and hear is little old me as uneducated as they come, just another illiterate motor mechanic and I show where the most fantastic factor in the Universe is. I use a top, which is a toy that all children, you and me included played with when we were children. In the top hides the most elaborate and most looked for factor there could ever be in the Universe, which is the point where the Universe started 1^0 or going by its other name used as singularity.

If we take the formula used to calculate the surface (the size   or the applying space used in time) then we would find it to beΠ as used in the most basic calculation of the circle formula Πr^2. That is the formula with which the Universe started and from that I find singularity (1^0 extending to 1^1) and how singularity Π^0 developed the curvature of space –timeΠ, which is so dramatic in terms of the Einstein method Newtonians use to embroil their graphic knowledge of mathematics, they portray it as a fairy tale wonderland mystery. I plan to elaborate on the Einstein flat Universe in some detail further on because it is here we find the flat Universe!

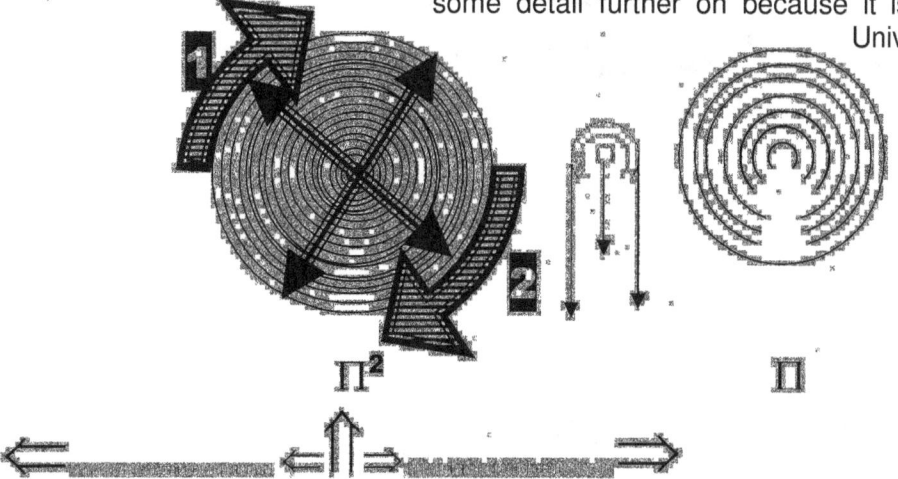

Fortunately I have not such elaborate education such as the most esteemed academics in physics possess and with my simple mind and my simple methods I used my most basic knowledge of mathematics too solve the greatest riddle in the Universe...I solve the mystery behind the Black Hole and showed where to locate singularity.

In exploring the features that provide singularity its most basic characteristics, we find the formula used to calculate a circle by using all four curves of time. That then would be $\Pi d^2 / 4$ but since d crosses the line of divide by four curves, this method then employs the dividing of the square of the diameter or when compiling the full circle one has to calculate the entire curvature ($4\Pi^2$) because in the most basic circle d = singularity equals $1^2 = 1$.

We can find the basic development from of the Neutron as well as the proton masked by these formulas. The value of the neutron is Πr^2 but since singularity applies to r and Π takes over the responsible value of movement across half the circle (r only indicates half the movement or then a straight line being equal to a half circle being equal to a halve square which forms as a triangle and has the directional value of all 180o we replace r^2 with the value of Π^2 and we have the value of the neutron at $\Pi\Pi^2$.

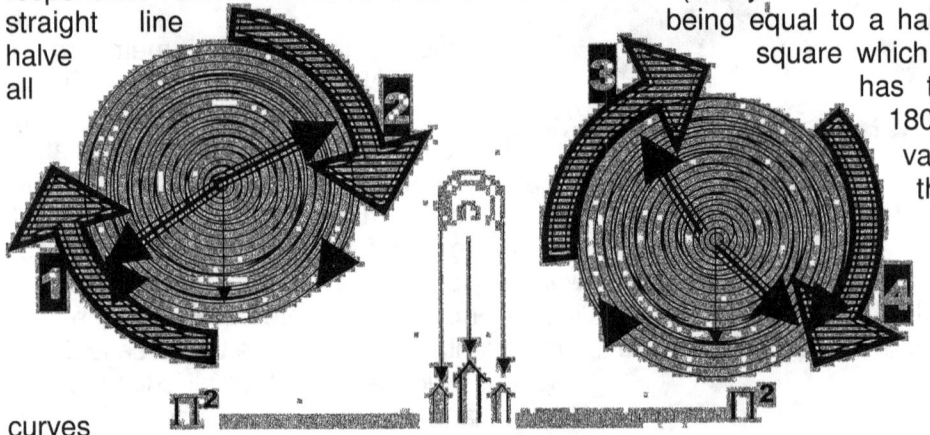

In the proton we find the same formula applying but since the proton always cover all 4 curves we have the double value of half a straight line or half a circle or half time two of a triangle which would then be interpreted as replacing singularity Π with the value singularity holds 1^0 and form there place twice the half value of the radius being touching in all four curves which then add to form the proton at $\Pi^2+\Pi^2$. Then the four curves are eliminated by the Roche limit coming into play in gravity I am not at this point going to prove this statement mathematically because I do so in other books amongst which are the four volumes in two parts of _**An Open Letter on Gravity,**_

Where singularity forms space-time and we can only find the flat Universe Einstein did not find it is there where all the most exotically complicated mathematics goes begging.

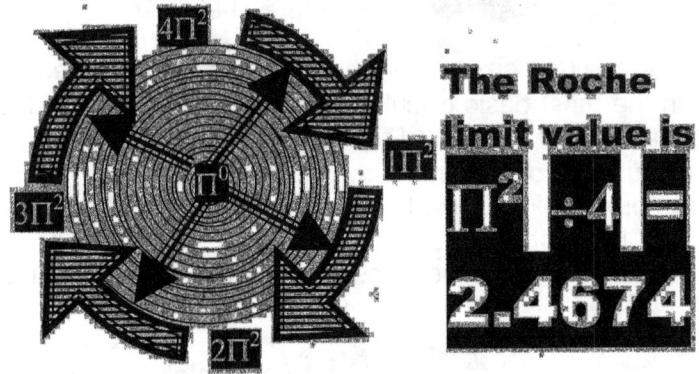

I have not even touched the Titius Bode law and I am only mentioning the Roche limit, at this point so there is no chance of explaining mathematically how the Titius Bode law takes Π as well as seven to form time.

The Titius Bode law as well as the Roche limit and the Lagrangian points combine to form the Coanda effect and the Coanda effect is using Π in tandem with seven to form gravity.

The phenomena I just mentioned being the Roche limit as well as the Titius Bode are two of the four

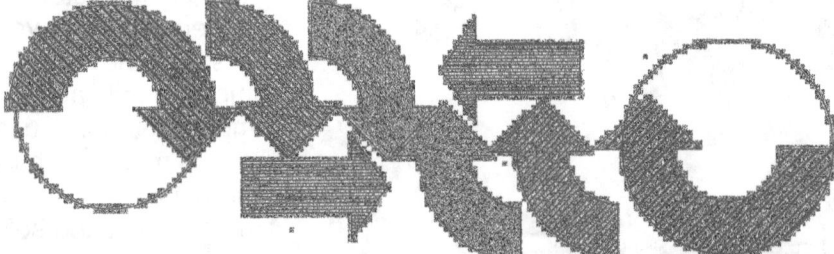

cosmic pillars Newtonians love to hide because while hiding the four principles, with that they are hiding their own short-sightedness and lack of understanding cosmology. With the Newtonian's wisdom applying things can have three values, one is mass which is a non existing factor in the Universe, then there is pressure which is another non existing factor totally fabricated by the Newtonian's shortfall to understand cosmology and thirdly if there is no mass and there is no pressure like we have in outer space, then there has to be nothing which again is a fabrication of the Newtonians poor insight into cosmology.

The Roche limit is the closest that two stars would allow one another to come and the limit applies

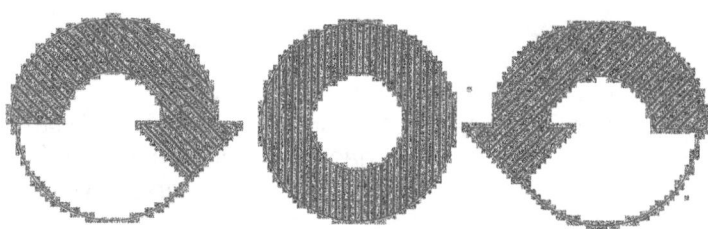

at a distance of 2,467 times the diameter of the larger star as the limit after which if the smaller star is closer, the larger star liquefies the smaller star and then absorb the liquid into the larger star's space-time of cosmic body.

In the Roche limit we find the secret behind the so though of matter and anti matter saga where the one ate up the other one and then had it disappear from the entire cosmos. The four curves counter acting spin holds the secret of all spin. When one take the spin that comes from synchronising time between particles the overheating pattern can come about when spin allows the prevention of overheating. When there is synchronising spin composing opposing spin, gravity will secure the flow of space keeping matter apart and ventilated. Before the initial Big Bang moment which was the point where light formed, there were particles touching and forming $\Pi^2+\Pi^2$ with other of less intensity forming $\Pi\Pi^2$. Then

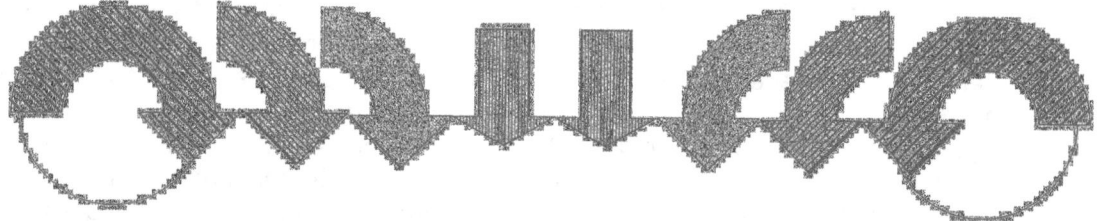

movement brought about space and indescribable friction brought heat discrepancies. This forced some particles to expand uncontrollable in heat (exploding as if banging big) and from that the light or outer space Πr^2 developed. At this point outer space holds the value of Πr^2 while the inner space holds the fluid value of $\Pi\Pi^2$ and the materials within the star holds the proton dynamic relevancy of $\Pi^2+\Pi^2$. These are the same values in relevancy that attach to the atom ($\Pi^2+\Pi$ forming the proton relevancy with $\Pi\Pi^2$ forming the neutron relevance and then outer space has $\Pi^0+r^0+r^0=3$). However if opposing spin comes about, gravity will remove space that is keeping particles apart with diminishing ventilation and this will generate overheating since when exceeding the Roche limit of $\Pi^2/4$ no space is generated that will bring about sufficient cooling. By removing space and not drawing mysteriously on material with the power and the aid of mass intervening in science, gravity holds a balance in keeping space that cools material and holds overheating at bay.

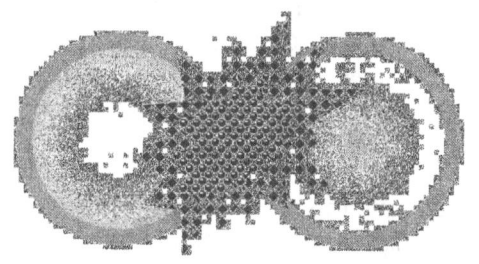

When not enough space flows in

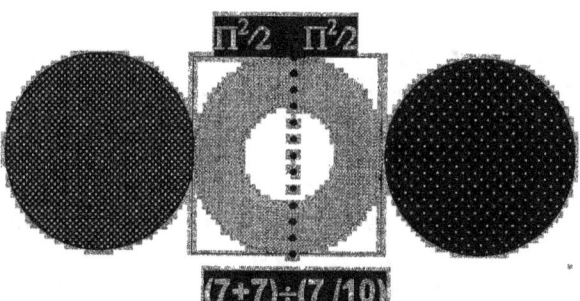

$\Pi^2 2$ $\Pi^2 2$

$(7+7)\div(7/10)$

between the two particles the overheating must
come as a natural result and by that destroying matter through overheating. To prevent this, the Roche limit comes in and that stops the particles closing in on one another where it will overheat from that.

The position of the Titius Bode principle prevents the contact that may come about and on a later occasion I shall mathematically prove how this becomes the value of gravity.

Is the entire argument I put forward about solids going into liquids because of a density crises during the start of the Big Bang not sounding more fundamentally sane than hearing the fairy tale of matter and anti matter (while never getting around to explain exactly what anti matter is) that goes out to war where the one eats up the other one and both disappear from the Universe. One thing Newtonians have to learn is to at all time keep in mind that once anything is part of the Universe that can never change or go away. What is part of the Universe remains part of the Universe until such a time as where infinity unites with eternity.

This electroplating process is forming by heat going into motion and the validity of this entire procedure is possible since electricity is gravity to some intense extreme. Electro motion or electro flow is the concentration of gravity to the limit where we will find gravity has the same intensity in the centre of the Earth as electricity has out in the open atmosphere conducted by conducting wire. By removing material from the less dense and electroplating that which is removed from the less dense and then galvanise that softer material onto the harder material (which by the way is a very natural process taking place all around as a corrosion) the density of the liquid will demise in the liquid sector and the material will grow in the solid sector. This process depicts the Coanda effect and I am about to show in this letter that gravity as well as electricity conductions are the Coanda effect. I believe even to this day and throughout the rest of the Universe wherever there is space, then such space has to have motion and space cannot be anything that it is without having motion. With that in mind that is space-time. Space-time is space flowing on through time creating space.

Where there is motion in space, the motion through of all space is carried along by time in space and time is duplicating space from what was onto the next position where the future will take place. Such movement can only take place where one factor moves, (changes positions) in accordance with other non-similar space. Where a solid rubs against a solid the friction will produce a liquid and where liquid rubs against liquid the two will mingle without discrimination taking place. The plasma is transforming to material through the motion we named gravity and while this happens the process mimics the process whereby liquids are being electroplated onto material. By duplicating space in the process of establishing gravity which is the very same as electricity, the object does not reduce to a standard in occupying space that it had before the motion took place but by placing liquid heat into the form of solid matter the matter use the newly acquired heat through which to cool material as to prevent overheating taking place. This is the cosmic recipe for gravity applying in the small! The process evolves by which solids absorb the liquid heat as another form of material and by boding onto the solid material it uses a process of freezing it into a solid to secure more material in the fight of combating overheating. In other words in the present time in our Universe gravity is freezing space to first become dense and form a liquid after which it then solidify the liquid heat by freezing the liquid into a solid state within the substance that is the atom. However, I do prefer to use heat as the term of choice and not plasma. The process I just explained was the manner used by the cosmos as the cosmos came about and this is the manner that will repeat until such time as will the cosmos conclude its final motion. I believe that the first motion came about as singularity was without space and found irrepressible heat levels rising. This was a product of material being to dense and then starting friction as the lot suddenly began to move by time applying. By overheating it moved into space that was still non-existing and that had therefore produced motion to rebalance the heat. From this I also believe some material that came about from singularity overheating remained as particles forming atoms where there is this relation between the solid proton, the liquid neutron and the gas electron.

The development of space from liquid heat in terms of what we now call the neutron, but is cosmic water or cosmic fluid, such fluid was after the GUT stadium of developing moving from the neutron's water density to becoming a gas that is forming space taxed with the ultimate gravitational relevance that space can carry. This was all contributing to the lack in contracting as the then applying gravity was prompted to enlist expanding gravity. The overeating was inflicted on to those particles that applied lesser motion helping the extending of space to turn into heat that again turned into space. In this release of uncontrolled heat performing as softer space-time, such release of space-time is destroying the applicable density of singularity secured in space, which again I believe (within reason) I do prove. I show that on the one side singularity introduce space-time, which confirms singularity and space-time makes contact with space-time not directly controlled by singularity or that, which is directly confirming singularity. This I conclude from studying Kepler's formula. I believe heat is used as a substance that form material as space controlled by movement or as space not controlled by spinning movement and is therefore expanding and by not controlling the heat density level with movement it becomes destructed density that came as a result o the atom (controlled heat displacement) which became overheated and the confirmation about this is found as evidence in the atomic thermo explosions. But to realise that we must beforehand find what any and all space is and we have to accept that space is made of something instead of Newtonian nothing.

When one applies heat to an object, it expands. That is primary school science. This states that more heat applied to an atom leads to more space acquired by the heated object. In sharp contrast to this is the decline in space when heat levels demise and so freezing brings about the opposite result. When I freeze an object that object reduces its occupied space as it shrinks. That is a fundamental law of nature. Removing heat reduces space. The reducing of space claimed by an atom comes directly as nature responds to the removing of heat and I can prove that easily. By expanding it accumulates space to increase the improving of the size of the material. The increase in heat accumulates the heat in the material for the sake of securing singularity, whereas the freezing tarnishes the overheating symptoms by the removal of material in unoccupied space using a process where heat is exerted from inside the atom to the outside of the atom taking internal matter external. This process works on the principle of setting motion to the exerted liquid material until the solidness contracts to allow the liquid to become a form, which we see as visible heat. The heat is in the form of dissolved singularity that became material as material used it as growth. That is why by freezing it will diminish the space as to accumulate the heat absorbing into the heat into the material to maintain the equilibrium needed in space. Bringing cold can only be the result of movement and overheating can only be the result of being static.

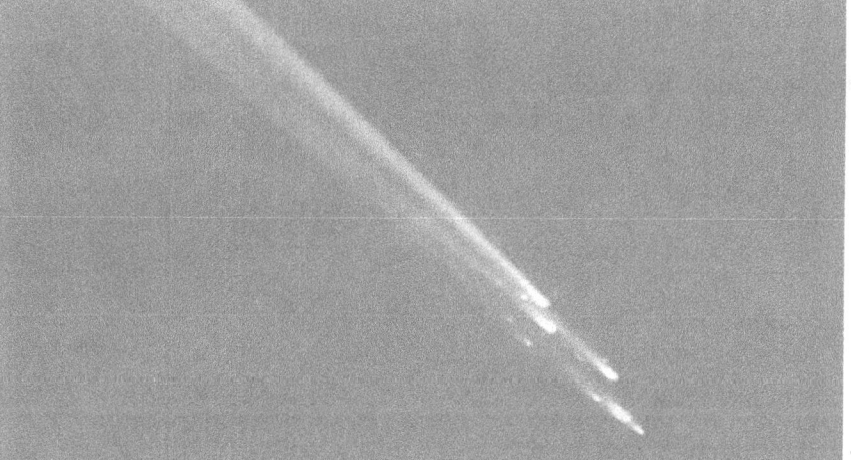

The burning rubble is left over traces of what Muir once was that is entering the Earth's atmosphere. Normal Newtonian disillusions state that this is caused by friction as the particles in the air rub on the particles of Muir and this allows the flames to burn the structure of Muir.

The Newtonian confusion called science first puts in the air down as a grand total of nothing that is keeping the most volatile particles afloat in the least dense area found in the Earth system. The density of the particles is so spares that a human would not be able to breath, ye it is so intense that it could burn metal to cinders. In nature the particles can only naturally be in a gas form on Earth, that is how little dense and how volatile the gasses are. Yet with that little density being that critically low and considering also the volatile nature of the gas particles, Newtonian confusion still explains this as nothing holding gas particles that is very sparsely distributed, rubbing on the structure composed of metals so violently that the structure breaks up in heat and completely disintegrates as a result of having friction with air. As one can see looking at the picture it is clearly to see the heat that is surrounding Muir as it entered from its orbit coming from outer space into that Earth atmosphere. If one would care to interpret this in relation to Kepler's formula it would be as follows:

The reducing must follow the boundaries of Π

When spacecraft renters the Earth the same applies but the reversing of the same applies. The re-entering must come about following the boundaries set by singularity because the reducing will be $k^0 / k = T^2 / a^3 = k^{-1}$. Every time space duplicates T^2 by contraction a^{-3} it releases k^{-1} to reduce heat inside the atom as the space claimed reduces and on the other side outside the atom the very opposite must happen and therefore $k = a^3 / T^2$ leads not only to the confirming of the time component but the time component is forcing the space reducing a^{2-3} because the time factor T^2 becomes the constant the Earth prescribe in the Earths relation to space –time $a^3 = T^2 k^1$, but the space duplication must then accept the ratio the Earth dictates $T^2 = a^3 / k^1$.

By entering the Earth atmosphere the duplication of space of the object entering relates exclusively to the duplication the Earth and that in turn complies with in the specific ratio demanded by the Earth. If space duplication is within the Earth space duplication a comparison must come about where the objects space-time duplication stands directly related in relation to the time component that the Earth holds as a claim to space in regard to the singularity the Earth sustains. That might bring about that a factor such as mass may be presented as the Earth's body restrains forming mass when further movement, but while falling there is not yet mass. It only follows the indicator that will eventually indicate mass if the object can sustain the space diminishing to heat that the Earth will introduce to the incoming object by restricting further descending. The flames surrounding Muir have nothing to do with the incoming objects materials forming friction because there is no friction.

$$k^0 / k = a^3_2 / T^2_2 = k^1_2$$

$$k^0 / k = a^3_4 / T^2_4 = k^1_4$$

Taking this equation of nature that cold contracts while heat expands into outer space, and then we find Newtonians seem to get all confused when applying this same most natural law in the cosmos. Newtonians declare that with outer space being as expanded as anything can ever get, according to Newtonian custom and tradition we have to regard outer space as being incredibly cold because some Newtonian invented some thermometer that declares according to the thermometer scale reading that outer space has to be the coldest place there can be…can you fathom this eccentric Newtonian stance…telling the Universe to ignore cosmic law because some Newtonian self-righteous upstart thought up a scale that then according to that Newtonians' esteemed view the Universe in outer space has to be cold! As heat sets in, the normal flow will bring about expanding because of heat coming into the form we think of as space. By countering this, the cosmos brought in movement and that limits the heat overheating. Outer space is the very edge of expanding of space where heat cannot expand into space any more, but needs time to introduce movement that we see as expanding. Outer space is the limit, the epitome of expanding where heat meets space at the edge of all limits once more. Therefore being the representation of the very limit of expanding outer space has to be the hottest place there is. By applying heat to a kettle holding water, the adding of heat manifests as steam and steam is hot water that traded heat as it evaporated into more space. By allowing the receiving of the heat to continue the container will let loose steam in order to match the contributing of adopting more space. The manner in which heat expresses itself when confronted by overheating is to provide additional space through the expanding of space. Outer space is outer space because outer space has expanded all it can while also it is still expanding to the speed of Hubble's $1/H_0$ which inevitably does not only affect far-off places where we cannot be, but affects us on a daily basis. As outer space is stretched to its limit, its limit will continue to stretch but while it is stretching it has to have more than it had before in that outer space holds the limit of what are the expanding possibilities of heat. Singularity has been expanding since way back when, but that means singularity is still releasing heat as space-time that turns out as space in the universal time of outer space. In outer space heat cannot expand more therefore except for the continual growth that benefits all singularity throughout on continuous bases concerning all outer space. We learn that heat always move from a hotter area to a colder area.

Well there is another way of thinking about the issue, which is more accurate and much more scientific correct in explaining. When an object is heated the object expands and when the object is cooled, it shrinks or that is what we are also been told.

Air passing over matter cools the material by spreading the heat over a larger area that the air is in contact with material. The largest majority of all engines in service if not all engines in service are cooled by means of air passing either over a radiator where the engine heat is transferred by water from the engine to the radiator or the air passes over the engine directly. The airflow is responsible for transferring heat from the source to space. The greater the airflow is the more is its ability to cool the surface of the engine. Air moving over material cools the material under Earth conditions. This is the method that also applies when orbiting objects enter the atmosphere of the Earth. If it depends on the flow of air, the object then has to cool to a point of freezing, but it heats to a point of destruction. The air that should cool the object by passing over it when fanned, as it normally does cool when the air is blown, just can't possibly friction about that would bring flames

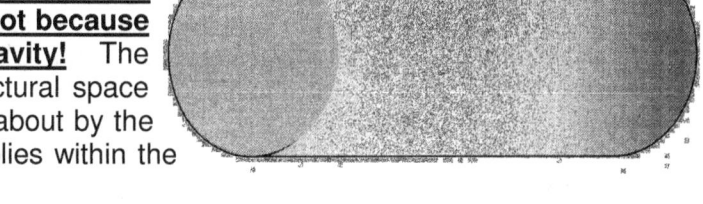

My theory is that gravity is the directional flow of heat caused by the circular movement of material that produces contraction which forms cooler space

bring around the aircraft. With this being an engineering reality then how on Earth can any person in his right mind then suggest that the flow of air over metal can cause friction and even to a point where the material overheats and thereby disintegrates by burning to cinders.

Here is the reason why **objects are larger in outer space** than they **are on Earth, and it is not because the objects are dragged down by gravity!** The material has to comply with its new structural space because of the difference that is brought about by the concentrated time-value (gravity) that applies within the boundaries of the Earth's atmosphere.

Let's take this scenario to the arena of pressure because as in the case of using mass to convince others, I have never seen proof of pressure in the cosmos. Pressure is another Newtonian idea that they implemented to confuse those that confuse gracefully and with style. Due to my limited education I do not confuse easily because I know to little to get confused. Movement brings a relevancy of having a denser inside and a more expanded outside but the denser inside is the result of space moving faster and the expanding outside is the result of space moving slower. That is motion discrepancy because of heat intensity differentiation occurring.

Let's get Newtonian pressure sorted out! If we pump air into a compressor the air gets more inside the compressor. The flow of air is artificial because life intervened and this is causing heat to flow in a specific direction from one specific point in a specific direction through a specific pipe that guides the flow of space to the confining container. There is nothing cosmic about this method of pumping air into a cylinder. In cosmic terms the container is filled with innumerable pumps called atoms that regulate the flow from any point through any space into the un-walled container where the flow is regulated by the innumerable pumps on the inside and not by a all covering all surrounding metal container wall.

The compressor gets hot while the air gets more, which is a clear sign that heat is transported. In order to contain the pressure there has to be an artificial metal wall the confines space being pumped to the inside and this artificial barrier will prevent the perused, pumped air to escape and by increasing the air, the space inside is getting denser and since it is heat that is getting denser, we find the heat levels rise when we touch the cylinder wall. No star or planet is covered by a wall such as the air compressor tank has and therefore in cosmic terms the detention of the pumped space is limited by a specific artificial barrier. This is thought to be pressure increase within the star as it happens inside the

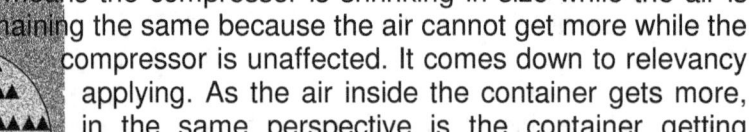

compressor but under closer scrutiny there are no similarities anywhere. More air is pushed inside the container which in terms of reality actually is a case where the size or the compressor remains the same while the air gets more inside the container and that means the compressor is shrinking in size while the air is remaining the same because the air cannot get more while the compressor is unaffected. It comes down to relevancy applying. As the air inside the container gets more, in the same perspective is the container getting smaller.

In relation to the heat, the heat gets more because the surface of the compressor remains the same while the size of the compressor within the relation to the volume applying on the outside of the container the container shrinks in size having the air remaining the same. By increasing the volume of air, the effect is that the space inside the container subsequently reduces. This will lead to a revaluation in the balance of air on the outside of the compressor in regard to the air inside the compressor and quantities will search to find equilibrium that matches. Because it is actually according to nature that the size of the compressor shrinks, (there is no pumping in the cosmos going about except for atomic gravity) the space outside the compressor has to accommodate the flow of heat because equilibrium has to be re installed.

As the pumping continues the relevancy applying is that the size of the container will remain reducing up to a point where the compressor is just too small to accommodate all the air which heat and it is too small for all the accumulated heat it is required to hold. We see it, as the air is getting hot while it is that space that is getting too little that becomes problematic. At a point where the container walls are just too small to accommodate the volume of space less air inside the container, the air will expand the container walls. There is no pressure according to nature but only a reduction in size of the container due to insufficient movement of air inside the container. To get the air moving from the outside to reposition the air inside the container, the air firstly had to cool because of the movement.

By moving the air it reduced the this action increased the space quantity of air and order to accommodate required by the expanded) air. The air the space increased movement. After the air heated a lot lot since the movement conditions the air then will always expanding from the time the compressor inside became air because of movement and required for that measured therefore the heat reduced in the increasing of space moving 9stretched or cooled by movement because with the application of landing inside the container because the space reduced a virtually halted. Under such expand. However the air was that pumping started because smaller as the heat levels rose. This brought on that the air expands the size of the inside of the compressor and that brought about a mismatch to match the air on the outside. We have to think about what pumping constitutes in nature. It is movement. The rotating of the electron increases the moving speed off air from what would be outside the atom to being the speed of light at the point of the electron.

No person that is serious about science can deny the (above mentioned) reality. Through inspecting science in all its forms as all things relate to one another, and affects the position too both space and time in the universe, one has to conclude that there is a definite link between space and heat. As the one all- important factor about heat is that it always flow from the most abandoned to the least abandoned space, and the universe is growing colder since the Big Bang, heat and time is connected. The time within the container holding the compressed air is much higher than the time to the outside, time will flow from inside the container to the outside of the container. For that reason it takes time to cool the air inside the container to match the air outside the container. There is an undeniable, unquestionable link between space and time.

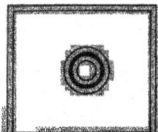
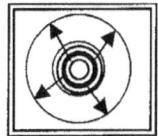

The same goes for material blown by wind to reduce heat. The object has an initial size to start with. Then we put heat to the object and the heat makes the object increase in size. That is hardly the increase worth noting because the relevancy of heat in the air to the heat in the

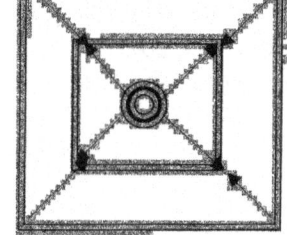

heating object goes array. The heat has to increase the size of the object in relation to the match it has to find in the space it is within. With the heat coming into the object the relation the object has with the heat or air outside makes

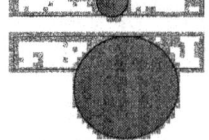

the object that many times bigger because the ratio in the heat balances is disturbed. If we blow air over the object we increase the size of the object by allowing the surface of the object to make contact with much more air in the same period of time, which will reduce the size of the object because the heat then spreads over a larger area in contact with air in relation and considering the contact with air the object expanded by the motion of the air being in contact with the object.

In the normal flow of time the object has a heat to relation set by the time Earth dictates. Then we increase the heat on the object artificially and in that event we actually body has in relation to the heat in the the body we increase the air and the size of the body in the same

Same time frame

the space the go and object increase the size the air. By blowing air over therefore we increase period of time. There is now a dispensation of many

times the body carrying more heat and contacting many times the wind or air which bring the equilibrium back to normal. There was a body size and by applying wind the balance shifted to the reducing of the body size in relation to the wind making contact. The body then had to expand in size with heat applying because the body was too small to incorporate the large volume of heat that was added. Blowing the air over the body increase the size of the body and heating the body decreases the size of the body in comparison with the air it comes in contact with. The body is either expanding or the body is redefining and the balance in heat places the body in relation to either gravity cooling by contraction or expanding by overheating. The very same principle applies in the sound barrier, but I am not getting into that.

A part of this introduction relates to Einstein's theory on the relativity of time in space. Einstein introduced the relativity between matter and space-time where the relative motion lies with the matter factor, as the space-time is regarded to be motionless. In this, the accepting of this statement again refers to the incorrect Newtonian presumption that everything is in a state of rest until a force intervenes. That is as big a lie as mass and pressure. Nothing in the cosmos could ever be motionless or be in a state of rest. Nothing is motionless in the Universe. Therefore, everything in a geodesic sense is in motion to all other factors throughout the Universe.

In reality the geodesic space-time, which includes the stars, is not motionless, but has a geodesic motion factor of one. If the space is not moving by contraction, then the space is moving b expanding. That means there is not one thing that can be regarded as being motionless but everything has a relative slower or faster motion than the rest of all objects travelling through space-time. This motion is the universal expansion rate, amongst many other values. In this, the second value of this is realized where the value of motion lies with space-time expanding and the matter complies with contracting motion but up to now the expanding was regarded as motionless. AS TIME SPEEDS UP, SPACE WILL REDUCE ACCORDINGLY. If time were effected, so would space be affected! The very instance that time increases, that very same instant, space proportionally reduces. Einstein had already explained about one hundred years ago, the volume that a structure holds would be proportionally reduced in the increase of gravity. There still is no understanding in regard to gravity affecting space that is affecting time.

Since time began outer space is expanding and it has been expanding ever since the first instant. If singularity expands when heated and there is a limit to the point it can heat in the moment of time, and that point of maximum expanding has been reached through the unleashing of heat at ever moment in time, and the moment in time depends on them heat which is turning into space, we can with great confidence declare that the area thought of as outer space as the hottest place there is. Whatever expanding that is possible was done this far to secure the cooling and all cooling that can be introduced to bring about further cooling was, is and will be performed as outer space stretches the level in decreasing density, then where the limit of density reduction is, must be that place that is the hottest place there is just because of the shear implication that it can cool no further at this moment and is therefore as hot as it gets anywhere.

If that is the case then it is safe to say that galactica then is freezing cold notwithstanding our concepts of heat and space and heat in space given to us by our collective culture of accepting the accepted in the past and not by our ability to reason. It has expanded to the maximum that this moment in time will allow, yet applying Newtonian culture, we think of not only as outer space being nothing, but being the coldest nothing that nothing can be...that is if nothing can be cold? Newtonian thinking regards that nothing in outer space to be as cold as it gets when everyone also know that outer space has reached the extreme expanding there is. Never is there a connection made by Newtonian science in regard to heat expanding. At the inner core of a star all space shrinks into the oblivious at the centre of the Sun, but we consider that spot in the centre of the Sun to be the hottest spot in the solar system. Thinking of it as the hottest spot while Kepler proved what forms space in the solar system is going that way. The space inside the star is shrink to the minimum there can be and that tells us the space has to be cold because of the shrinking took the space to a position where no space can shrink anymore. Then the space outside the star is expanding and this obvious connection between what is expanding and what is contracting passed by every Newtonian unnoticed for the last almost four hundred years while everyone was telling every other one about the emperors magic clothes being called mass and pressure and only the very stupid will not see the presence of mass and pressure.

That shrinking of contraction to where there is no more space can only be inside the inner star and it also is in that region where gravity is at its strongest. With outer space as expanded as nature may allow the space that grew could only grow in conditions of heat because heat produces expanding and expanding is the result of heat coming about. Space shrink because it is cold: that we know and taking this law to the star centre it means regardless of our interpretation of hot and cold, that area in the star centre is as cold as it can get notwithstanding what our Newtonian culture may tell us. Then obviously the same must apply to outer space for precisely the same reasons because it is so hot it can expand no more. We look at the hotness of space and the coldness of space but it is the relevancy to the solidity that forms the actual heat and cold limits. It is so hot no expansion can produce more space in outer space, as the outer space seems hot and quite the opposite reveals the true scenario inside the star in the centre of a star structure.

That means the number of protons in motion has a lot to do with the cold and hot scenarios because where the protons are most dense the cold is in extreme. Only in the absence of space can so much heat gather in excess and the opposite is true about outer space where the least denseness found brings about the space in heat found in outer space. Our human selecting of hot and of cold and what is hot and what is not prevents us the clear vision we would have when truly understanding the applying temperature. Temperature comes about from spin and the smaller the spin density is the colder the space becomes because the more duplication produces the most cold. We think of outer space as 0^0 Kelvin but in fact it is as hot as no other place can be in the Universe. The coldest is where material is freezing solid as material does when frozen solid and the hottest is when by boiling the material is going into a gas with liquid being the intermediate position where heat acquires the space to perform as a flexible substance.

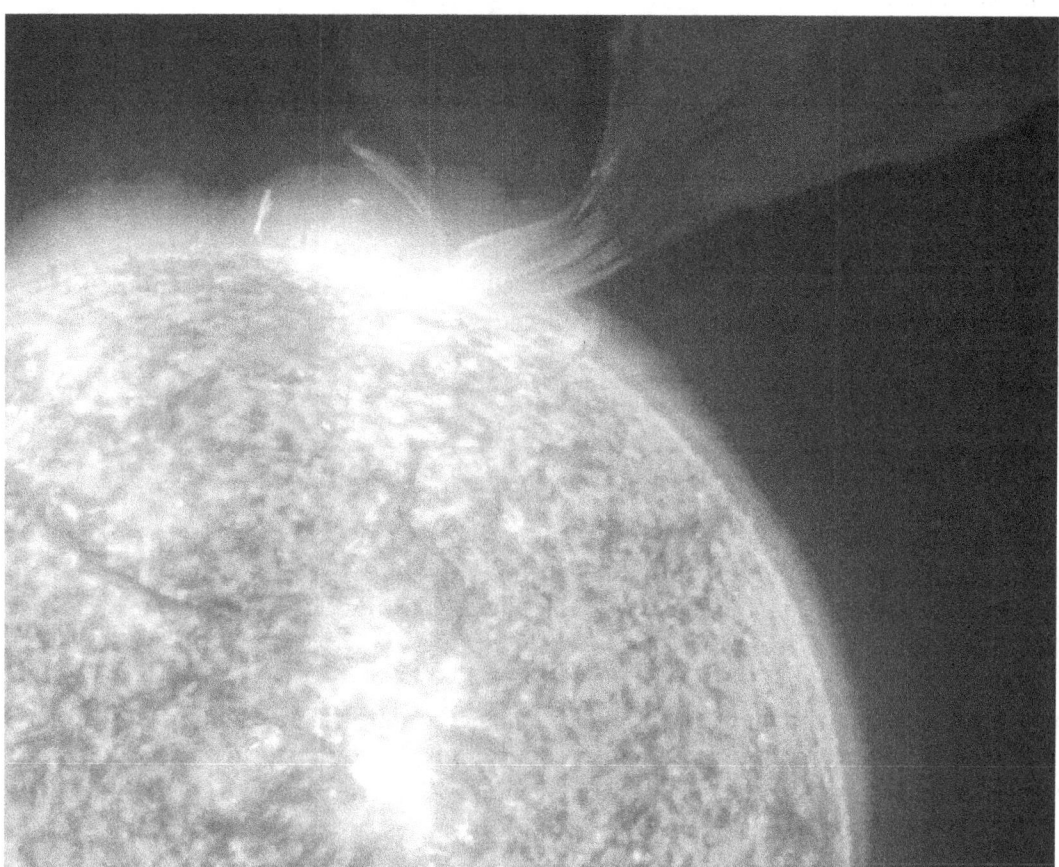

When we look at particles in outer space we see the particles being frozen but in fact the density allows the gasses to be as wide apart as no other place could ever be. It is because there is such a severe contrast between the particles and the environment surrounding the particles bringing the gas formation and not the particles that is so frozen. The particles are in a gas state because the particles do not form a part that is part of the space unit. Hydrogen clouds of hundred of light years in diameter are a common sight in outer space. If outer space was cold, as Newtonian culture would suggest, then those hydrogen clouds should freeze into solid dense hydrogen balls. Hydrogen should freeze solid at 0K. The heat we find filling space is not part of the space but like the particles the heat is a separate issue. That heat filling the space is another form of material that could conduce by diverting from space or marry the union of space by becoming more space or condense material by freezing out all heat. In other words if there is a lot of heat, the solids turn liquid and gas and when there is total absence of heat the gas molecules turn solid. If it were that cold which we think it is, it would not have expanded into such a massive cloud but would have contracted forming a cube of frozen hydrogen. But as we can see the cloud expanded the gas as far as the gas can expand.

It is said by the all-conquering wise called the Newtonian Brainy Bunch that the Sun is a lump of hydrogen gas. Newtonians support the idea that the Sun we see is a gaseous lump floating in space, while squandering its energy formed by the magic of "mass" on soliciting large volumes of magic called the force of gravity and by that attracting whatever needs attraction in order to honour the memory of what Newton brought to magic. See that stuff squirting from the Sun…that stuff Newtonians call prominence because they have no better name for something they have no idea what it is. That stuff squirting from the Sun is fluid heat. It is heat that is frozen by the movement of the Sun at a temperature of $6500°$. But thinking in this manner would not fit the Newtonian wisdom because after everything that was said and done, in the Newtonian mind the Earth still holds the Universe hostage by having all there can be, applying to Earth standards and the Earth still set the rules to what applies in the Newtonian Universe. If $6500°$ is considered hot on Earth, because it fits life's idea of what is hot and what is cold, then that has to be interpreted to fit every spot in the Universe. The Newtonians can only judge a cosmos that supports all the criteria that will support life. Whether they accept it or not accept the following fact, but that stuff that the Sun ejects into space is liquid heat, cooled by the movement of the Sun where the heat that forms outer space is condensed into liquid. The Sun freezes the heat in outer space into a liquid and when frozen even more, the tiny particles are then called electron and photons. Electrons and photons are heat frozen from a gas into a liquid substance.

See the fluid push out of a bowl of liquid, spilling both sides as it falls back into liquid of the Sun. The inside of the Sun is not gas but it is fluid where heat is frozen from the gas it is in outer space to the liquid it is in the Sun.

In all of nature there is no NATURAL GAS as much as there are no elements that are in a NATURAL LIQUID because all elements are in a natural SOLID state. No element is either a gas or is a fluid or is a solid. We arrange the elements in such a manner, but that is only applying to the situation the Earth grants the elements. This standard we use that applies to conditions befitting Earth does not fit the Universe.

When an element freezes it is solid notwithstanding...

When an element melts it becomes a liquid because the atoms are further apart

When an element boils it is a gas again notwithstanding the name used to associate what we think the classification of the element is on Earth elements become gas when there is an increase in heat that surrounds the solid atom. Gas being formed is due to the fact that there is more heat separating the atoms.

Going from a gas onto a liquid and hen to a solid the heat to material level increases in each case. The ratio between what is naturally solid (heat contained in atoms thought of as elements) that is surrounded by a liquid substance in which the presumed solid is floating in a space that is heat, which is uncontained by rapid movement. Material in atoms is solid because movement contains it.

Hydrogen 1	melts at -259^0 C,	boils at -252^0 C,
Helium 2	melts at -269^0 C	boils at $-268,9^0$ C
Lithium 3	melts 180^0 C	boils at 1300^0
Beryllium 4	melts at 1287^0C	boils at 2770^0C
Boron 5	melts at 2030^0 C	boils 2550^0 C
Carbon 6	melts at 804^0C	boils at 3470^0 C
Nitrogen 7	melts at -210^0C	boils at -195.8^0 C
Oxygen 8	melts at -218.8^0C	boils at -183^0 C
Fluorine 9	melts at -219.6^0 C	boils at -188.2^0 C
Neon10	melts at -248.59^0 C	boils at -246^0 C
Sodium 11	melts at 97.85^0 C	boils at 892^0 C
Magnesium12	melts at 650^0 C	boils at 1107^0
Aluminum13	melts at 660^0 C	boils at 2450^0

Hydrogen is as much a liquid as iron is a gas and neon is a solid. It depends on the element relating to the space/heat in the circumstances surrounding the substance at that very precise instant in time. We have to stop telling the cosmos to show us what we wish to find and start accepting what the cosmos is telling us to find. The culture that I am referring to is all about **nothing.** At present we find that there is something we think of as nothing in outer space. Because nothing is what Newtonians wish to find and nothing is precisely what Newtonians are getting because Newtonians think of outer space as nothing. If you accept the cosmos to be nothing, then please define nothing to yourself and find the definition in the cosmos.

Realising what I have stated must bring about some awareness that we have no scientific view

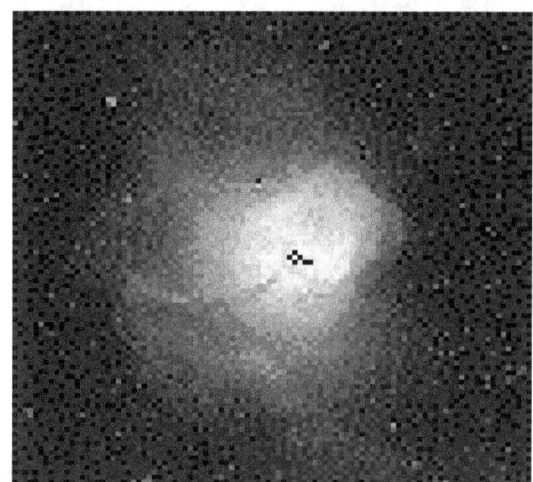

about the cosmos but only a view that came about from culture that we inherited since mass was formed in a mind as an opinion. We look at elements and wish to see solids, liquids and gasses as factors of grouped as natural solids in the one class and liquids in the next class and then solids rounding off the grouping. This is not the case and as long as we are unable to see the incorrectness about this line of thinking, our view on cosmology may just as well return to where we held scientific records preserved as cave paintings.

Forming a solid is that the solid is something impenetrable and forming a liquid means the something is penetrable and forming a gas means the elements hold far more space than it holds density.

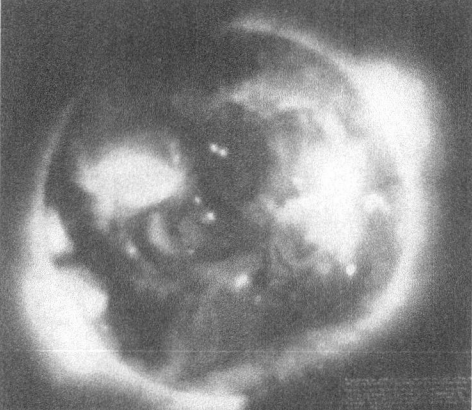

Another point I question about the Official Policy is that they as I am are in agreement that the heat melted particles onto particles and in those joining better combinations of particles came about. How it happened is another bone of contention but more about that a little later on. There was heat on the outside and there was matter on the inside. The heat was liquid because the Sun and other stars still indicate masses of liquid fluid inside. I can only imagine that that liquid inside the Sun holding temperatures as low as 6500^0 K and up to 1.8×10^6 K the heat already is in a molten form. What about the heat then when the frozen outer space was 10^{34} K and such temperatures were the general order of the day back then. If the Sun is liquid now then those temperatures raging back then must put the heat in form available in outer space at the time as thick as mud.

From the outside drawn onto the particle inside the blanket of heat came a flow of soup that became matter. That much I do understand. This carried on until...when? When did this stop? When did the Universe run out of heat? When could one consider outer space as the coldest all around? Where to did the Universe dismiss the heat that was once there but now is empty? How did the process stop of bringing from space intense heat and from that particles grew? When did it stop affecting the growth of particle, the growth of space, in fact the growth of everything that grew came from this first growth. What you see or do not see grew since it was part of the Big Bang and everything in the cosmos at present was part of the cosmos during the Big Bang. I say this process of collecting heat from outer space never stopped but is an on going process we now gave a nice name calling it gravity. Outer space never became empty and void but relevancies changed concepts where centres form that should not be as it then interferes with concepts about relevancies Gravity is not and never was about particles pulling each other closer. If it was, no Big Bang was possible. Gravity is about turning space, which is released heat back to heat and concentrate the heat where gravity is the strongest and heat is the least. Space is the transverse form of heat and visa versa is also true. Should any one not believe me try a bicycle pump by compressing the plunger while blocking the valve but. The heat will burn your finger to blisters if the force on the plunger is strong enough, the plunger seals enough and your ability to withstand pain can last that long. Then answer your own question about where the heat came from because sure as hell is hot, it did not come from friction with air particles such as oxygen and nitrogen escaping through the valve bit. Heat is unleashed space and space is concentrated heat. Reducing space to heat is gravity and antigravity is expanding from overheating blowing into space accumulation.

 When looking at a sphere the inside has always (in a cosmic relevancy) the location with strongest heat also always has the strongest gravity in any given cosmic sphere. The centre of the sphere clusters the combination of particles forming the sphere into unity. By holding a specific centre the sphere becomes the strongest form any object can be. The sphere is without any doubt the favourite choice in form of gravity. Where gravity has the last say without other influences changing possibilities as collisions leaving debris in space or natural out burst like Super Nova explosions, gravity will enforce the sphere to be the form taken by the particle. But there is no evidence of particles of similar size joining in matrimony through gravity being the shotgun at the wedding. In cases where there is a mismatch of size outside any proportions of equality then there is a contracting of the lesser by the greater. In such cases the lesser is not qualifying as material (and that I prove later on) but the greater consider all the lesser to be heat. It is humans bringing distinction to matter in form.

That expanding is indicative of the characteristics that only heat displays and as such this characteristic is extremely important in the process of gravity applying. Only the development of heat can bring about expanding when there is an oversupply and contracting when there is substantial reduction. Only heat can bring about movement and notwithstanding all the magic Newton had attributed to the pulling powers of mass, if gravity is movement it can only apply because of heat interacting. Because outer space is completely overheating the condition it has in support of the particles makes the particles appear to be in a state of freezing but the particles is counteracting the heat limit it meets. However, the particles do not contract because the density required causing movement by which to establish cold enough conditions that would permit space reduction the heat is insufficient. The atomic density brings about the cold conditions and by spinning the atoms pump space into a liquid as we can see from the Sun and in even much bigger stars, the atoms pump the hear coming from outer space directly into a freezing solid state.

The atom must be the utmost coldest and the proton is even much colder because when that cold escapes through nuclear energy release as an atomic bomb, it turns to heat forming space that no one human could truly understand. When nuclear overheating reduces the spin of the atom, this expanding of heat makes the atom expand because the atom is heat and the cold the atom has, is heat captured and controlled by the spin that the atom has within the space-time that the atom has. This expanding slows the spinning down because the electron spin circle increases exponentially and this allows the cold of the atom to escape into forming slower moving hotter space and by suddenly widening the spin value of the atom, the control is suspended. This compacted heat is released. The heat the atom had frozen into space surrounding the atom since time began escapes from a solid to a liquid to a gas and the release of such singularity is beyond what man could even understand. This rapid expansion is the release of time formulating trillions upon trillions of years of controlled, accumulating heat as time and through time forming controlled space. When this heat releases from the containing form of the atom, it brings about much more consequences about than the Human mind can cope with, and this remark is a warning.

We have to acknowledge that there are two forms of substance and only the two forms of substance. The one is material in a controlled state of movement and the other is liquid that can be gas if and when the density reduces the substance. One may not look at the material and judge the surroundings. The fact that hydrogen remains a gas and so does helium in outer space must serve as enough proof that outer space is hot, regardless of our interpretation of the temperature gauge telling us what we wish to hear. One must look at outer space and judge outer space from the findings by only considering outer space in the character it presents. If helium remains a gas it is hot. The removing of heat by creating intense spin makes the centre of the Earth cold although we see it as being terribly hot. That heat we see is that is in place in the centre of the Earth we judge by the intensity it has and it is just because of such intensity increasing the density that makes the liquid heat cold. The atom froze in size due enormous spin applying which makes the atomic material to being much colder and in that to allow such cold conditions where the liquid heat density can rise to become a water substance while it surrounds the atom. However, being in a more concentrated state while forming a liquid, the liquid has to be colder than the gas it forms in outer space because hot and cold is the result of density and not some Newtonian's scale measurements. The only reason why it can seem to be intensely hot is because it is in substance frozen cold and in such a cold environment the heat can gather and the density in space can collect heat intensity because the particles spin fast enough to form the intensity created by the heat density within the surroundings being extremely cold.

The cold in the Earth centre causes the concentration of heat by reducing space through intense spin, and the reducing of space is as a result that all cold surfaces tend to decrease volumetric space. If it was hot the space within the Earth would expand and the space within the Earth where we think so much heat is concentrated, does not expand therefore it must be cold. To gather and accumulate the space in a liquid means it became much colder being a liquid. Finding the surroundings terribly cold will allow the heat to gather and not expand but when the surroundings are hot it will not tolerate more concentration of heat and thus will bring about expanding of heat into more volumes of space in order to get rid of the imbalances due to excess heat within space.

Look at the Sun and see how the Sun turned the hydrogen to a freezing cold liquid at 6500 K. Hydrogen is in a fluid state within the Sun and is colder than the hydrogen that is in a gas form in outer space. The density factor in both cases give me the guideline to use and not some clever Newtonian that made a machine he called a thermometer. The Sun is the coldest place in the solar system and my eyes tell me that. That is when the protons oversupply the removing of space to produce the cold that is so apparent. By the reducing of space it can concentrate heat to a fluid state by producing the opposing cold that finally freezes the heat to a solid state.

The expanding of space is a way of duplicating space without reducing space and by duplicating in the form of expanding it becomes just the opposite to duplicating by motion therefore reducing space by halving space in time. That is what gravity does. By motion space duplicates and by space halving it removes heat in space as well as by dismissing space. In all the applying of gravity space either bites the dust (contraction) or multiplies (expanding). The density of the protons brings about space dense enough to harbour the heat in such densely packed quantities and visa versa applies in outer space.

We have to accept that the coldest place in the solar system is in the very centre of the Sun because there the most number of protons sharing the least amount of space producing the coldest area that can allow therefore the hottest density of heat within the cold environment. Later I will show why the star is so extremely cold and outer space is over boiling with heat expanding into more space. We have to see what forms space and why space can be the absolute basic container through which gravity can relay the influence it carries.

We must come to realise that whatever forms of holding space to form, space has to be composed of that same ingredient which also is the basic component that forms the lot of everything in the entire Universe. When particles heat up the particles expand by claiming more space that the particles hold and this to limit the heat rising by reducing density intensity. The particles claim more space when heated to preserve the cold it received from moving. The claim to more space is a product of more heat in space and by expanding space the movement reduces the intensity leaving less heat density distributed through more space. Such expanding brings about cooling. When particles heat or cool it can only be the result of motion applying in some form. Motion started at a point when the Universe was extremely hot and there was no space. By introducing motion space formed and the lack thereof produced friction that became heat that became space.

The application of gravity is the movement that condenses space and the lack of movement or gravity is bringing about heat levels rising. By applying movement there is compressing of space and that is how we apply artificial human initiated movement Newtonians love to call energy. What we have accomplished we have done so in the way we go about tapping into the energy that nature provide. We humans manipulate the interaction there is between heat levels and space abundance. Internal and external engines combustion engines all rely on this application for harvesting motion by driving power. Compress space even today with a piston in a cylinder and then pump the compressed air into a container and such confining of space will increase the heat by the piston effort to reduce the space brought about in the container. The heat coming about inside the cylinder has no relevance to particles colliding because all compressor cylinders cool down colder because when that cold escape it turns to heat as the heat releases from space forming a secondary form of material forming space that no one can understand when the spin of the atom allow the cold of the atom to release into uncontrolled space. This release and unifying with space that heat does is the heat it had frozen heat because of the motion of spin to space that the atom holds remains in a frozen state under the guard of the spinning electron. But when this heat releases from the containing form of the atom frozen by the spin of the electron it brings about much more heat than the Human mind can cope with.

There is this far in the Newtonian recognising of principles in natural physics not one single reference made to prove their appreciation of this matter albeit the most acknowledged fundamental fact in science. They are bent on particle colliding. When particles collide, such collision forms an atomic thermo release and that action we call an exploding atomic bomb. What principle this argument about particles colliding ignores is that all atoms use negative charged electron forming the atomic limit on the outside forming a definite border to the boundaries of all atoms and in both electrons from different atoms are being negative charged. The closer they come the more violent the rejecting will be and such rejecting is the production of heat that will turn to space. The electrons repel other negative charged sub atomic structures, which the electrons are that form the outer borders of all atoms. With all electrons highly negatively charged (being as negatively charged as any possibility will allow matching the utter extreme) such electrons couldn't touch. Newtonians form the impression that a solid structure has no space parting the structure but a most ordinary child-like argument denounces any validity such an argument may have. Electrons carry negative charges and will never have another electron sharing close space or being a close neighbour. Electrons would violently repel one another but will never touch.

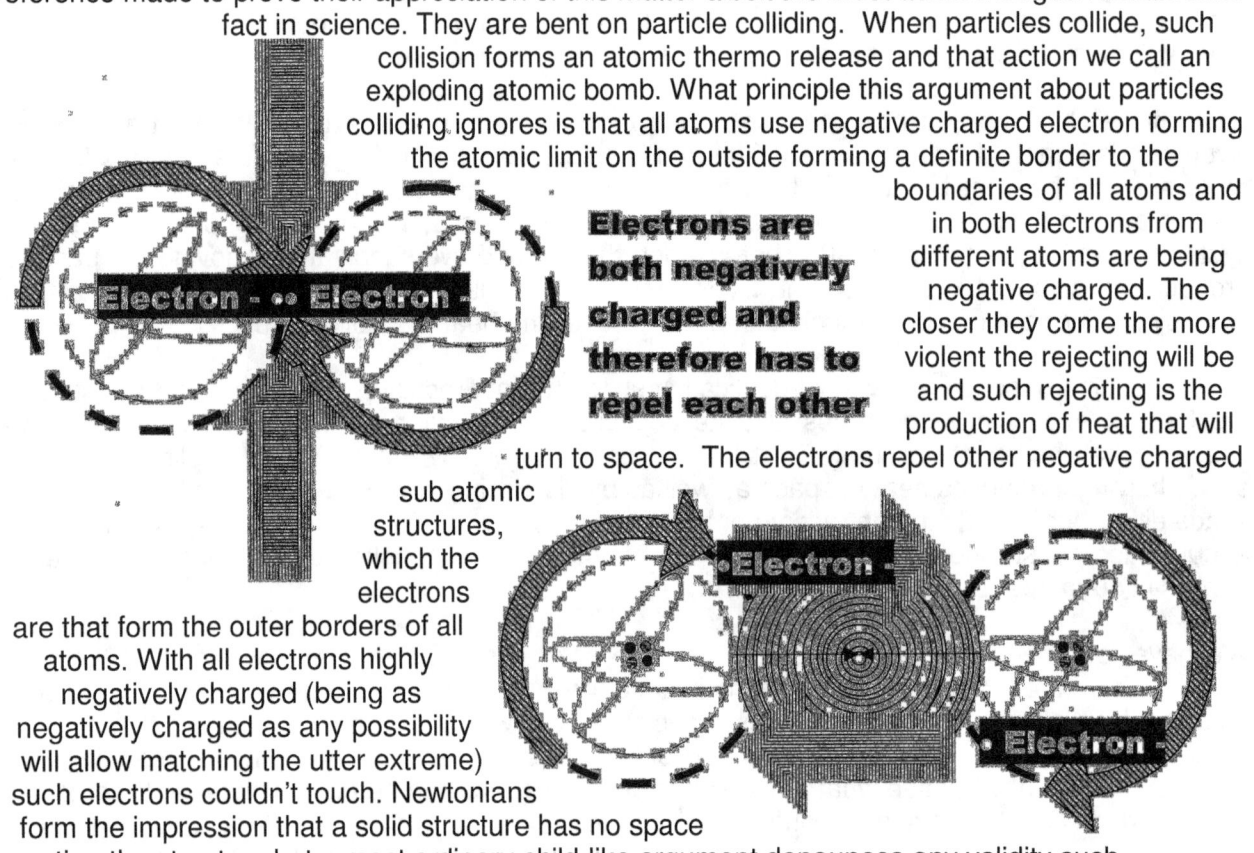

Electrons are both negatively charged and therefore has to repel each other

The balance at first favours the forming of heat from the space coming in and being reduced in the containing size they are squeezed into is reducing the space from what it was on the outside. The space distribution inside then has changed considerably and reduced a great deal compared to conditions outside and with the decrease of the space distribution that space then becomes excess heat on the inside.

The electrons will disallow any contact directly between atoms. It is because of that contact rejection electrons bring about that science has to use an overload of neutral neutrons putting them in the atom nucleus to fake a complying of charges that will eventually lead to atom touching each other in an artificial nuclear release we call an atom bomb but that is through enticing a neutral stance which is enticing a positive overload for a short while. When the touching of electrons does take place the event is called a thermo nuclear reaction where heat is released in unmatchable quantities and the atoms in reaction dissolves all forms of solid heat into a liquid heat. This ratio of heat reduction is time connected as much as it is motion dependent. Motion reduces space by expansion as much as time contributes to space distribution by allowing the flow of heat.

It is said by the Newtonian Brainy Bunch intellectuals that particle collide with each other in a compressor air container and this bumping of atoms form friction and this friction then becomes the heating of the space within the container. Good God, can those with so much intelligence also be that stupid...? When the pumping of air into a compressor air container stops, the heat immediately starts to reduce by escaping through the container metal wall. There are no electrons escaping or particles reducing, but only the concentration of heat in a density reduces as the heat moves through the walls of the container. Most important is the realising that every atom constitutes of two parts. On the outside there is heat in liquid surrounding the inner material part that forms the solid within the sphere and distance of the inner material that is divided by the movement of either being faster than the speed of light or slower than the speed of light. This division parts a solid and a liquid from gas and a gas from a solid and this forms space in dividing atoms from the space between it and the next atom. The electron forms the division between heat uncontained and heat contained. There can be friction between particles in reduced space under controlled circumstances where such particles are grouped together in a unit and as a unit elects a group singularity forming the centre of the chosen form of the unit.

At this point I wish to introduce Kepler's formula in relation to everything I have said before.
The Universe separated heat from material by covering the exterior of material with heat that forms space. Newtonians saw this expanding of heat forming and providing space as the Big Bang. Some material became softer by uncontrolled overheating while others remained more solid by containing form through controlling the overheating as they applies circular movement. On the outside of all elements there are a layer that is the heat the element uses in relation to place relevancies between such an element and the rest of the cosmos. The impotence of this we find in spacecraft coming into the atmosphere at higher than the speed of sound. Supersonic aircraft flies in an envelope of heat that engulfs the structure of the aircraft. The space surrounding the craft becomes liquid as the heat within the space becomes more intense and heat levels rise as the space reduce in concentrating the substance within that forms heat. This is because the supersonic movement that the aircraft moves at exceeds the sound barrier and in that it actually exceeds the limitation gravity on Earth caps on all movement. There can be no particle having friction by flying through air and even more so way up they're in the atmosphere at the altitude where the cosmos meets the atmosphere. This is just because the particles up there are so sparsely distributed in that part of the atmosphere there aren't any to rub against. Above and beyond this lies the fact that all the so called air particles are very volatile and excitable by nature and they are known to turn the slightest heat into rapid motion thus establishing a scene where the particle that supposedly are in contact with the aircraft sheeting will move away from the hot incoming aircraft. If then not for any other reason then it is because the particles are highly volatile and acceptingly sensitive to heat. Airborne particles are prone to motion just because it is the airborne element nature to change heat into motion and the motion comes about from their sensitivity to duplicate. No particle in the air being part of the space we call air which is in a free floating in that air can produce friction because of the volatile nature those elements have. The craft's coming into the atmosphere produces a point where $a^3 = T^2 k$ changes to $k^1 = T^2 / a^3$ (the explanation is forthcoming a little later on). In truth the correct formula to my thinking would be $T^2 = \dfrac{a^3}{k}$ and brought in terms with the atom, then it would be $V^2 = \dfrac{a^3}{C}$. The distance separating the incoming object from the Earth centre reduces rapidly therefore the object start to descend towards the centre of the Earth.

That point will rapidly increase the time factor where the incoming object crossed such a very visible border. By the reducing of distance k space a^3 will have to compromise in the relation of all the factors forming the equation since T^2 will very suddenly grow more acute. What happens is that the applying gravity reduces the space a^3 and the compromising factor comes about since the time factor T^2 moves back to a time where outer space was as dense back then as the density we now have within the atmosphere that then became as the Earth atmosphere. It is outer space that remained denser that what the outer space currently is. I am now referring to a process that I introduce as this letter unfolds which is by nature completely different to what is accepted by mainstream science. That which I refer too came about at a point just before the Earth established an atmosphere that grew through gravity and by the measure of the Earth gravity became separated from the atmosphere. While the gravity of the Earth contained the space surrounding the Earth in a much denser packed envelope the area not under the direct influence of the Earth governing singularity became more spacious.

When the atmosphere grew apart from the outer space there are two ways of looking at the event. One can think that outer space expanded by the implication of the Hubble constant because the density of outer space reduced as the growing gravity of the Earth contracted the atmosphere in placing a limit as a denser unit or that gravity withdrew the atmospheric space of the Earth from outer space by the contraction of gravity at the time that the parting of space came about. In other words one may see that outer space retreated by overheating into becoming thinner, or that the atmosphere parted from outer space by becoming denser and thus thicker. But however you look at it, there was a time when both outer space and the Earth's atmosphere shared equal density as we find it still applies on the moon and on Pluto. The space component is reducing the time component by compacting space to alter the space – time ratio.

The contained Earth atmosphere grew denser as the solar system developed into what it is today. As the atmosphere released from what we think of as outer space that release from outer space made the atmosphere much denser and the space above the Earth which is using a reducing time factor and that makes the Earth more compact. That established the T^2 factor to be that more condensed when one compare in ratio the density with outer space. The density at the time there was when the separation came about in outer space at the time of such parting outer space allowed objects to move away. This parting brought a barrier that is in place between the Earth and the outer space and any object coming from outer space into the Earth's atmosphere. The incoming object then would have to reduce the measure of the space the craft holds as the containing singularity set new standards applying to the incoming object with which the craft then needs to affirms its form and its status within the contained space of the Earth. The reducing will then suddenly no longer use space as the compatible factor but the focus will shift to the time factor that dictates to the space what the space can be. Such reducing comes from the switch there is in space – time where it was in outer space performing as being relative $k = a^3 / T^2$ to what it has to be within the Earth atmosphere $k^{-1} = T^2/a^3$.

Stars can and stars do **overheat**, sometimes and the **Polar Regions** where **the Titius Bode matter-to-matter applies** holding the square matter (7+7) in relation to the square of space (10) and **other times** in a double relation to the **square of space** 10 to that of matter in a half square (10/7). When saying that, one also has to differentiate between the liquid heats gathered or heat accumulating and overheating because a star represents the coldest space in the Universe and not the hottest space as Newtonian religiosity professes. By being cold, a star gathers heat. While gathering heat, there is a strict balance by which the gathering takes place and this gathering of heat in a local space filled with small centrifugal pumps called atoms, the atoms pump in liquid to maintain the heat balance within that star. This pumping process of atoms is called gravity. That is why there is gravity. It is to maintain and to balance the levels in the stars. **Heat and cold are relevant dynamics** forming poles **in appreciation of singularity. The Sun is the coldest place in the solar system** and that is fact. Looking at evidence the Sun provides contradict everything science wishes to believe about cold and hot. Science wish to see the cosmos through the eyes of what fits the needs sustaining life on Earth and only take into account what benefits maintaining surroundings in support of life as one find life's needs on Earth whereas life has no part in the cosmos except for being on the speck of dust we call Earth. Looking at the cosmos impartial to life the evidence supports another view. Every aspect in **the cosmos is the very opposite of what science believe** it is. The Sun is **not a ball of gas but** a **giant sea of liquid**, frozen **without any** form of **gas or air** in the interior. Having a liquid interior **the Sun** has **no pressure** but has the **very opposite of pressure** to which there is yet no name given. **The liquid comes from singularity freezing** space-time within the atmosphere of **the Sun**, and such is the case with all stars still in the shining phase. **Stars that has gone further on the path of developed than the Sun, is frozen solid causing material to freeze because of the cold conditions applying within such a star.**

In **the picture to the left** we find not withstanding whatever name we attach to the **red liquid substance flowing from the Sun** into space and back to the Sun, **that liquid is heat** in a very direct form. **If outer space was the coldest place in the solar system** the heat **should** immediately **escape to outer space in the form of gas or be at least steaming into**

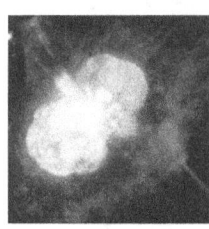

space and **not return to the Sun** in a clear liquid form as it clearly does. If the Sun was hot, then the Sun should have the liquid steam into **outer space. Once in outer space it should then evaporate in a cloud of mist and not flow back to the Sun.** Even if you are as gullible as only Newtonians can be, one can't believe that the liquid will squirt out of the Sun and then have gravity pulling the liquid magically back to the Sun, even if you do believe in magic as Newtonians do. No one can be as stupid as that… really no one can be that stupid as to think overheating lava would rush back to the Sun from which it escaped because it overheated within the Sun. That just can't be a cosmic possibility, even coming from a Newtonian. Once the liquid is in outer space where it then escapes from the extreme hot Sun, it should keep on escaping as far as the liquid can go but it will only return to the Sun if the Sun was colder than outer space is and only then with the Sun offering a colder environment would the heat return to the Sun.

All elements forming matter that is on the periodic table is as much being in a **liquid state as it is it could be in a gas and in a solid** form The heat which is forming **an atom is forming material** by having the **spin with which it controls the heat within the atom exceeding the speed of light and by exceeding the speed of light the heat secures compactness to restrict heat in the atom. The atom is the way heat is contained by contraction in the manner that centrifugal pumps work. There is no hot as there is no cold. It's about storing energy in space or in heat, which is another cosmic equal being opposing similarities.** Hot and cold is in each case a **terminal pole** forming specific **directions** of **cosmic expanding** or **cosmic movement. Hot and cold** are **relevancies brought about by singularity valuating space-time in differentiation between what can move and which point can never move** and during **the Big Bang** the Universe was **freezing cold** at 10^{37}K.

If there was no cold, then there was no hot because only coldness can grant heat a position. It is the relation matter has with heat under the guidance of singularity that provides the form the particle has at that present. The increasing or decreasing the heat will alter the form of the atom and the element proton configuration holds the key to that. Therefore, all elements forming **matter is as much a liquid or not than it is a solid or a gas, although it has to be in one of these forms at any given time**. Being in any one of the three forms is indicative of the ratio there is between what is solid and what is liquid. **It is the space surrounding the atom which forms the liquid factor that provides the form in which the atom find its relativity at that moment to be to the rest of the atoms it share space with**.

Hydrogen is as much a solid as tungsten is a gas depending on the heat in relation to the space matter is within. Should **you reply** that it is **the gravity pulling the heat back to the Sun,** then that **confirms** my theory that **gravity is all about collecting heat onto matter by providing a cold secluded environment where liquid will remain liquid and not turn to a hotter form which is gas. With the Sun obviously holding the liquid and with outer space obviously forming a gas, there is no argument about the fact that** outer space is not the hottest place. **It is the concentration of heat in space being relevant to form. When overheating a star turns its liquid to gas whereby it merely transforms its interior to a relevancy it has from pre- to post-Big Bang, it goes Super Nova. The Super Nova star turned the inner liquid to form gas and gas is only possible to form in outer space where the environment is hot enough to sustain gas.**

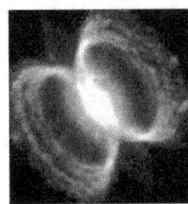

We humans on Earth think that hydrogen is a liquid at -259^0 C but that only apply to conditions matching the Earth. The picture clearly shows the **heat in a liquid** flowing **from the Sun** and **back to the Sun**. In the **Sun the hydrogen holds enormous quantities of heat in a liquid at a temperature of 6500^0 C.**

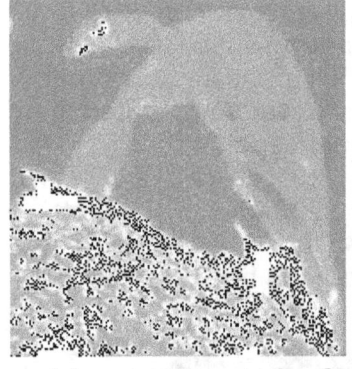

When a star has its singularity secured the star is bitterly cold because it has heat in a liquid form flowing back to the point of singularity although we may regard the star to be rather on the hot side. By applying gravity or as the case is cooling off outer space by applying movement, the Sun (fore instance) freezes hydrogen to form a liquid at 6500^0 C. If hydrogen remains a liquid at $6500\ ^0$ C, just think how cold it must be as the star's interior approaches the point of singularity centred in the middle. Therefore fusing protons comes from cold and not from heat or pressure.

By allowing the singularity to overheat the star overheats and heat within the star flows from singularity to outer space freely. The heating we observe on the edge of the Sun is not coming from or being material connected to the Sun, but is heat turned from gas to liquid by the spin of the Sun and the photons we use to see with, are particles created by condensing gas to liquid on the edge of the Sun. These tiny particles known as photons, escapes the cooling and collecting by freezing gas to liquid because the particles are small enough to remain a gas. In such an event where gas-photons form, outer space is then hotter than the star because as gas, the heat releases to outer space with no intention of returning whereas when preserved in liquid form within the boundaries of the Sun, the liquid returns as soon as it leaves. There are two ways to reduce heat; one is to bring about expanding space, as the photographs clearly show. The second one is where heat will reduce when an object duplicates a position in motion by spin. When there is any withholding or retarding of motion, then the result of that is that matter will overheat. Gravity is the motion of occupied space that condenses unoccupied space through the dimensional transformation to form occupied space.

Motion and space therefore is the anti-, the opposite, and the negative to heat being the positive. With singularity overheating the expansion of the singularity drives heat into space, and by expanding such movement distributes the space more and in doing so, it is creating space to compensate for the overheating that is taking place. **That is a natural phenomenon**. The only reason why **heat will** rather **flow back** to the star than **escape to outer** space once the star released it into outer space is **if outer space presents a hotter environment for the heat than does the star,** because **heat always flows from hot to cold** no matter what influences may arise. This is due to heat forming space and cold reducing space and in that there is a natural direction of flow established. **Outer space must hold more heat than does the star because it holds more expanded space than a star does but the accumulation of space in relation to heat makes it seem colder bringing expanding of heat to become condense in space.** <u>**Space and heat directly relates being the one form of the other**</u>.

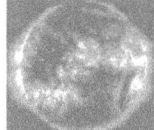  **When any star just as much as it happened to the Sun, came from the galactica cocoon, which is the cradle for all future stars, the Sun started freezing outer space at 6500° C. Then as the part forming outer space further overheated and expanded,** the Sun's governing singularity kept that part of space that the Sun captures and condensed as occupied space-time frozen at 6500° C. The cosmos is all about **converting space to heat** which we see **as gravity** and **returning heat to space** which Newtonians see as expanding space and this is in place as a **control mechanism** always **keeping** a very delicate **balance** which we see as **a star shining or being normal.**

The purpose of the converting of space to heat is to supply the core where singularity is, with heat by regulating overheating of singularity. **It turns space to heat** sustaining matter but sometimes singularity overheats by insufficient particle movement and then matter converts heat back to space allowing heat to convert heat resulting space forming. That we call many names amongst others exploding into Super Nova. Whatever the names used is less important because the **process rests on space and heat interacting to form energy.** The main issue to consider is that space and heat are the flipside of the same coin. Heat is space and space is heat. That was what **the Big Bang** was and **the Hubble Constant** is. In every event it is all about where **matter converts heat to space and condense space to heat.**

I show that **space and heat Is the very same thIng** and there **Is no such a thIng as pressure** In the cosmos but releasing **heat produces space** and **concentrating heat reduces space** with the two interacting on the demand set singularity setting time to space with time being the spin or motion of heat in space. **Heat and space form the second condition holding singularity** caused by the **fragmenting of singularity to compensate overheating during the pre-** Big Bang matter forming era. That is what we see as **light and space,** which again is the **same thing and is fragmented singularity forming radiation and heat, where the star re-transfers heat back to space due to an overload.** Everything there is, is heat or is singularity being at a value of either 1^0 or 1^1 depending whether the singularity at that point reserves relevancy with eternity that always moves or infinity that never moves.

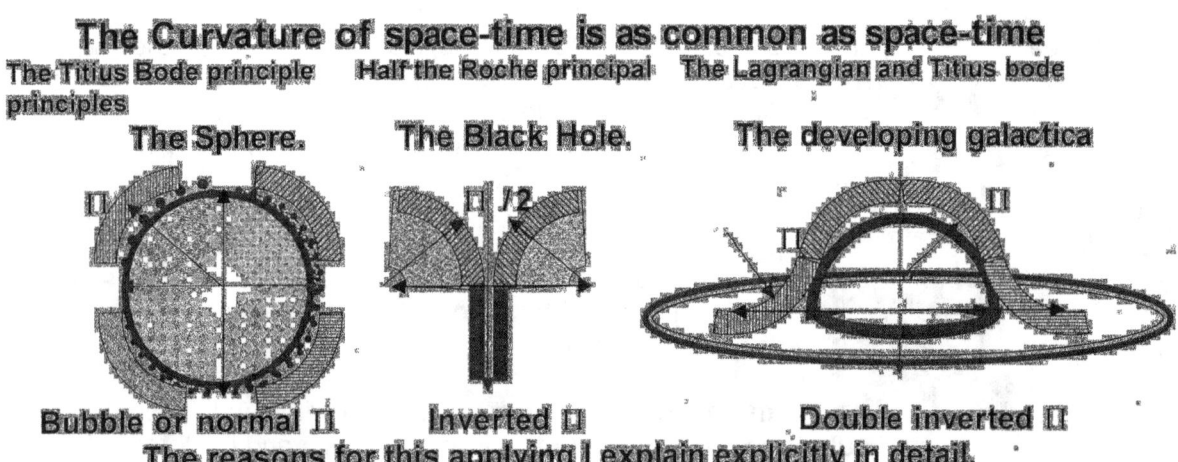

The Curvature of space-time is as common as space-time
The Titius Bode principle   Half the Roche principal   The Lagrangian and Titius bode principles
The Sphere.   The Black Hole.   The developing galactica
Bubble or normal Π   Inverted Π   Double inverted Π
The reasons for this applying I explain explicitly in detail.

This comes about through the overheating of singularity (7 + 7)/10 (top and bottom) or layer overheating 10 / 7 (centre). The fact that stars overheat is evident throughout the Universe and the consideration coming to such a conclusion as such should not be surprising. The reason why they overeat is very simple and not very surprising.

With singularity being where it is and in a common place as it, forming everything there is, it results in presenting the curvature of space-time, which by implication of singularity coming into form, also becomes a common factor, but accepting this idea is not yet commonly distinguished. The Curvature of space-time is the result of singularity changing the profile of being single dimension to being three-dimensional too suits space-time.

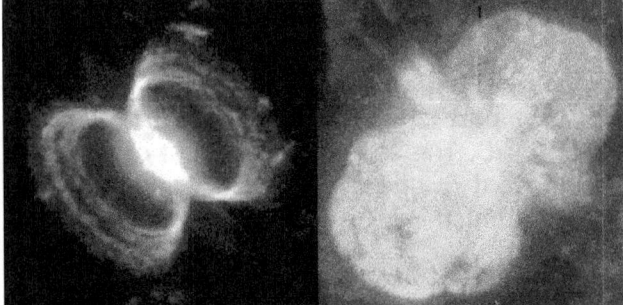

When a star has its singularity in tacked the star is bitterly cold because it has heat in a liquid form flowing back to the point of singularity although as humans with a human mentality, we may regard the star to be rather on the hot side. The Sun (fore instance) keeps hydrogen in a liquid form at 6500 ^{0} C. If hydrogen remains a liquid at 6500 ^{0} C, just think how cold it must be as the star's interior approaches the point of singularity.

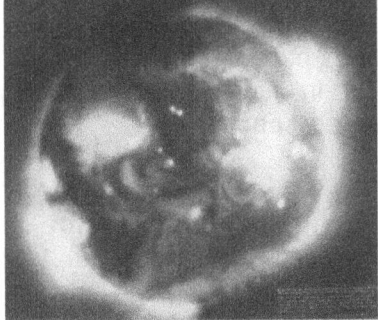

By allowing the singularity to overheat the star overheats and heat within the star flows from singularity to outer space freely. The only way to reduce heat is to bring about expanding space, as the photographs clearly show. . Heat will reduce when in motion.

Heat and cold are relevant dynamics forming **in appreciation of singularity. The Sun is the coldest place in the solar system** and that is fact. Looking at evidence the Sun provides contradict everything science wishes to believe about cold and hot. Science wish to see the cosmos through the eyes of what fits the needs sustaining life on Earth and what benefits maintaining surroundings in support of life as one find on Earth whereas life has no part in the cosmos except for the speck of dust we call earth. Looking at the cosmos impartial to life the evidence supports another view. Every aspect in **the cosmos is the very opposite of what science believe** it is. The Sun is **not a ball of gas but** a **giant sea of liquid**, frozen **without any** form of **gas or air** in the interior. Having a liquid interior **the Sun** has **no pressure** but has the **very opposite of pressure** to which there is yet no name given. **The liquid comes from singularity freezing** space-time within the atmosphere of **the Sun**, and such is the case with all stars still in the shining phase. **Stars more developed than the Sun is frozen solid causing fusion.**

In **the picture to the left** we find not withstanding whatever name we attach to the **red liquid substance flowing from the Sun** into space and back to the Sun, **that liquid is heat** in a very direct form. **If outer space was the coldest place in the solar system** the heat **should** immediately **escape to outer space** and **not return to the Sun** as it clearly does. If **outer space were colder the heat would not return to the Sun**.

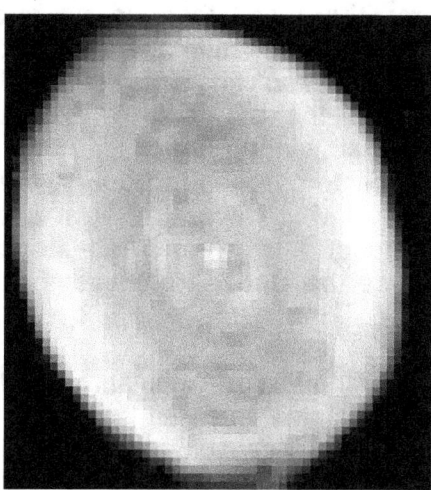

We humans on Earth being surrounded by Earth that provides for life being possible, hydrogen is a liquid at – 259^0 C and to us our Universe on Earth being the Earth but conditions on the Earth. The picture **heat in a liquid** flowing **from the Sun** and In the **Sun the hydrogen holds quantities of heat in a liquid at a 6500^0 C.** When a star has its singularity singularity develops space-time within the is bitterly cold because it liquefies heat to space even becomes frozen solids. The freeze hydrogen in a liquid form at 6500 ^{0} the conditions on we think that being confined to that only apply to clearly shows the **back to the Sun. enormous temperature of** secured and star, then the star the point where Sun (for instance) C.

If hydrogen remains a liquid at 6500 ^{0} C, just think how cold it must be as the star's interior approaches the point of singularity. This the star accomplishes by producing rapid movement and movement condenses space or heat. Therefore fusing protons comes from cold and not from heat or pressure. By allowing the singularity to overheat the star overheats and heat within the star flows from singularity to outer space freely. In such an event outer space is then colder than the star because the heat releases to outer space with no intention of returning whereas in the Sun it returns as soon as it leaves.

There are two ways to reduce heat; one is to bring about expanding space, as the photographs clearly show. The second one is where heat will reduce when in motion by spin. When withholding or retarding motion, such action will lead to material overheating and this is called a Super Nova concurrency. It has nothing to do the intellect of gravity or the wits of the star retarding because as scientifically enormous it may sound in the ears of Newtonians, there just is no possibility for gravity to go mad, as Newtonian so masterly explain the process! Gravity is the motion of occupied space relocating to unoccupied space through the dimensional transformation to occupied space. In a Super Nova some layers slowed down their motion and this formed heat and while other still spin at the correct speed, the layers that moved too slowly overheated and expanded by exploding. However superior cosmologist might think stars and gravity intellectually may seem gravity still has no intellect and therefore the gravity in some stars just can not go mad or go bananas, as scientifically superior such explanation concerning physics might be.

Motion and space therefore is the anti-, the opposite the negative to heat being the positive. With singularity overheating the expansion of the singularity drives heat into space, creating space to compensate overheating. **That is a natural phenomenon.** The only reason why **heat will** rather **flow back** to the star than **escape to outer** space once the star released it into outer space is **if outer space presents more heat than does the star,** because **heat always flows from hot to cold** no matter what influences may arise. **Outer space must hold more heat than does the star but the accumulation of space in relation to heat makes it seem colder bringing expanding of heat to become space. <u>Space and heat directly relates being the one form of the other</u>**.

The cosmos is all about **converting space to heat** which we see **as gravity** and **returning heat to space** as a **control mechanism** always **keeping** a very delicate **balance** which we see as **a star shining or being normal.**

The purpose of the converting of space to heat is to supply the core where singularity is with heat. **It turns space to heat** sustaining matter but sometimes singularity overheats and then matter converts to heat allowing heat to convert to space. That we call many names amongst others exploding into super nova.

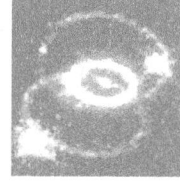

Whatever the names used is less important because the **process rests on space and heat interacting to form energy**. That was what **the Big Bang** was and **the Hubble Constant** is all about where **matter converts heat to space.** I show that **space and heat is the very same thing** and there **is no such a thing as pressure** but releasing **heat produces space** and **concentrating heat reduces space** with the two interacting on singularity demand setting time to space with time being the spin or motion of heat in space. **Heat and space form the second singularity** caused by the **fragmenting of singularity to compensate overheating during the pre-** Big Bang matter forming era. That is what we see as **light and space,** which again is the **same thing and is fragmented singularity forming radiation and heat, where the star re-transfers heat back to space due to an overload.**

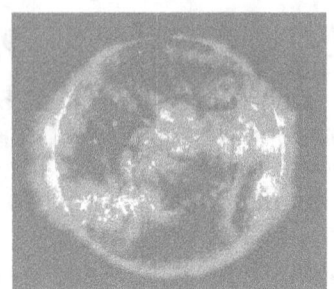

Every position in the Universe holds singularity albeit it in relation to form or as form by the curvature of space-time, which in that case also relates to singularity. There can be no position not being singularity and also there can be no point in the Universe being unrelated to singularity therefore every aspect of the cosmos is space-time in various forms under the provision of singularity connecting to form. Matter cannot be, if not being a point representing singularity where such a point connecting singularity is also singularity in motion that is surrounding singularity which is stationary. Movement is the relocating of

material by duplicating material from one position to another location. That is the function of time. That is the function of gravity. There is space-time not filled by material, which I named unoccupied space-time, and then there is space-time filled with material, which I named unoccupied space-time. Material always represents space-time that forms heat which is under control of the moving of heat that is surrounding singularity and on the outside of the atom there is heat formed as singularity that is moving by expanding in time because the singularity no moving in the unoccupied sector is not under control of the movement of singularity.

Looking at the Sun from the stance the human has to take as we are in the glow of the Sun's light, we are from our perspective looking at what the Sun discards. We are not seeing what the Sun collects because what the Sun collects it holds inside. We se and form an opinion on what the Sun rejects. Science is quick to point out that when we see a yellow flower the flower is every colour but yellow and this statement rings very true. However, when a Newtonian looks at the Sun and sees heat, that Newtonians believes rock fast that the Sun is a pool of heat and light. If heat flows from the Sun in large quantities and huge streams of liquid heat, then that proves that there is more heat than which the Sun can absorb.

If we witness streams of light coming from the Sun, then obviously the Sun on the inside is too cold to cope with that magnitude of heat. The Sun is not absorbing the light that we see. It is rather meaningless to look at the Sun from the perspective off what the Sun rejects and then form an opinion to the fact that that what the Sun doesn't want and rejects is then and therefore the very thing the Sun is made of. Looking at the Sun we find light and dark spots covering the entirety surface of the Sun. The dark spots is where the Sun absorbs heat and light and those spots are truly hot because the Sun absorbs the heat inwards whereas the luminous spots are where the Sun rejects heat and discards heat and they might seem hot on the outside but towards the inside those areas are the lesser hot areas. The dark areas contracts heat and contraction proves gravity is more

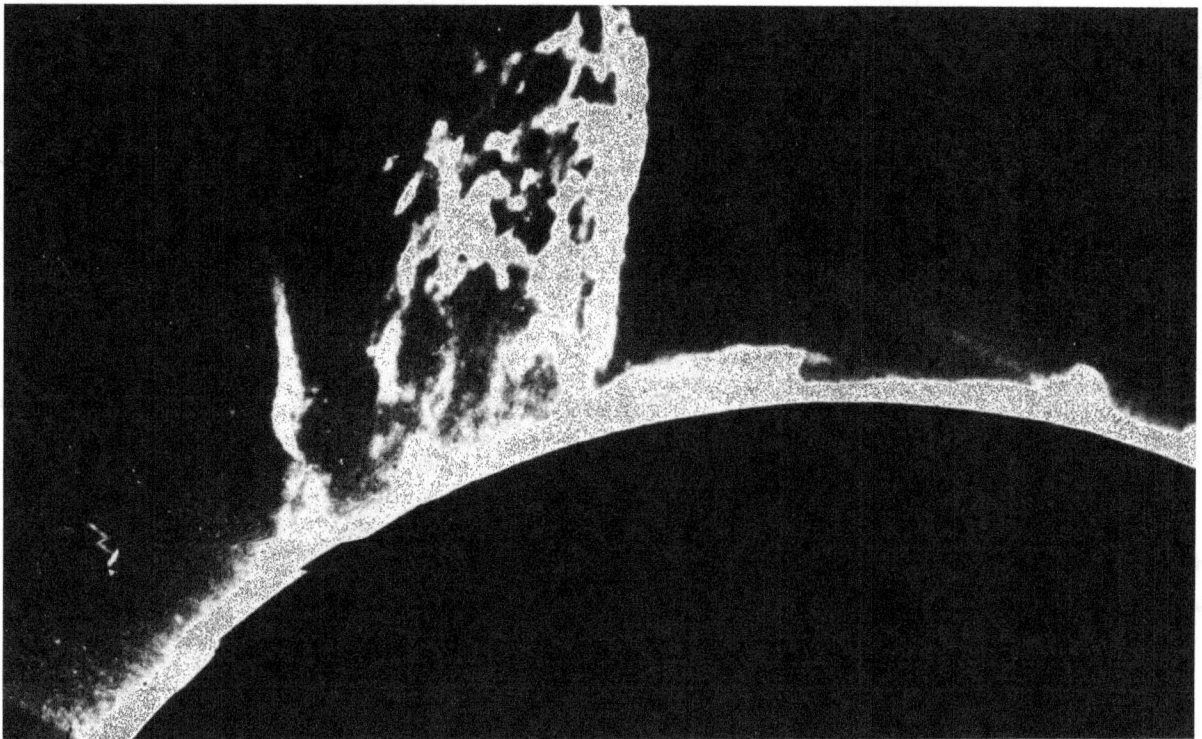

prevalent while the lighter luminous area expands while with this action it is discarding heat. Mainstream physics ignored this visible and most obvious clear connection completely, notwithstanding it being so very obvious. There is this far in their recognising of principles in natural physics not one single reference made to prove their appreciation of this matter. They are bent on particles colliding forming pressure by mass and just the same as in a compressor, the pressure heats the inner star. That is Newtonian wisdom but unlike within the compressor, the heat does not return to what it was, but maintains the heat on the outside. The heat is the result of particles colliding as it is to their opinion the same that happens inside the walled air compressor container.

When particles collide, such collision forms an atomic thermo release and that action we call an exploding atomic bomb. This is when singularity expands from the controlled environment within the atom to the uncontrolled motion outside the atom. Then the atoms destructed ands unlike us breathing oxygen while the oxygen remains the same, the atomic particles can never again form an atomic unit. The atomic constructed structure is dissolved until eternity unites with infinity. That is not the function of a star. A star is there to accumulate and build atomic structures as to unite heat and cold. What principle this argument about particles colliding ignores is that all atoms use negative charged electrons forming the atomic limit on the outside forming a definite border to the boundaries of all atoms and in both electrons from different atoms is negative charged. In being negatively charged, it means both will come out and totally reject the other. The closer they come the more violent the rejecting will be and such rejecting is the production of heat that will turn to space. The electrons repel other negative charged sub atomic structures, which the electrons are and in that the electron forms what the outer borders of all atoms consist of. With all electrons highly negatively charged (being as negatively charged as any possibility will allow matching the utter extreme) such electrons could not touch. It is at this point that a principle applying throughout the cosmos comes in to prevent the touching of particles. It is called the Roche limit and it holds a measured value of 2.47 times the radius of the object as a line from where no cosmic particle would come closer. Should by chance the particles come closer, the lesser one of the particles would turn to liquid as the Roche limit of 2.46 times the radial distance of the particle takes place. How does this come about, you may ask? In the centre we have singularity. The directional changes connected to rotating around singularity come in affect by the value of 4 every cycle the rotation takes place. The rotation is the movement of Π and that value is then gravity or Π^2. When dividing gravity or the rotating of the cycle by the four changes taking place when the rotation direction redirects in order to serve singularity Π^0, we have to divide the rotation Π^2 by the alteration 4 and that places gravity Π^2 in division of 4. Gravity in division $\Pi^2 / 4 = 2.4674$ which then becomes the radius (Π^0 extending to Π) multiplied by gravity redirected $\Pi^2 / 4 = 2.4674$.

It is about time scientists start looking with their minds and not their eyes at the Universe and see what truly is out there for us with intellect to see. All the difference we find is seated in the human mind.   We   humans set differences because we look at the cosmos by placing   humans and the life we find on Earth in a pivotal centre in   the   cosmos instead of placing singularity in the centre and   we have to make an effort to put life where it correctly belongs; which is only on Earth because life is only found on Earth. Einstein proved mathematically that in the presence of a strong gravity time slows down. Surprisingly with that evidence being around this long nobody in science since Einstein's   discovery   took   those statements   and   made   any   further progress from that. It seems to have been left in some drawer to dry or rot. Science still sticks to the opinion that time did not change, not even slightly, since the beginning of the time and holds the same pace ever since the start of the Big Bang notwithstanding the implications this concept carries. Since heir calculations show that the

Earth   turned in orbit around the Sun 13,5 x 10^9 times since the Big Bang,   that is the age of the Universe and to hell with the fact that most of the time   there was no Earth in place to orbit the Sun.

Before the Earth took one year to circle around the Sun and even before the Sun was there to orbit the Milky Way, according to the Newtonian Brainy Bunch, a year was still the same in duration as the duration of one year presently is with the Earth nicely taking 365.25 days to orbit the Sun Those Newtonians still put the Earth in the centre of the entire Universe by placing the entire Universe at the mercy of the Earth to provide the Universe with applying time. Some things just cant change in the Newtonian mind and one is the Earth forming the centre of time. How odd... don't you think ... that the only aspect in the entire Universe that is beyond change is the aspect of time? With the entire Universe including all the gravity now present and not excluding one Black Hole or dust speck pressed in such an area that was possibly the size of a lepton even then the gravity extending from that circumstances must have been beyond what words can ever describe.

When everything was that small at the time when the Big Bang took charge, the gravity applying at the time must have been beyond C^2 because it froze light, and there was such a time because even today in the Black Hole the gravity is beyond the speed of light. If the gravity was that high and Einstein already proved that strong gravity slows time down, then there is one logical conclusion and that is that time was in fact at the time of the Big Bang just about standing still. Mathematically it is incorrect to allow gravity to compress the Universe into a spot smaller than an atom and then exclude any other factors and relevancies to change. If the Universe grew, the duration of time increased and the gravity reduced in intensity, which allows time to increase by deflating gravitational intensity.

If we think of the Big Bang in terms of having a day temperature in outer space at 10^{37} K in the shade of outer space then again there was a time when the Sun was the same frozen state as Pluto now is and was even much more frozen solid. It is obvious that at 10^{37} K something being 6500^0 C is freezing anyone's sock off. If 10^{37} K was the temperature of outer space where we now find 0^oK, then that meant a temperature of what we now consider to be 6500^0 C as extremely cold in comparison to what was applying at that moment throughout the entire outer space. In respect to what was outer space the Sun's temperature at 6500^0 K was a pretty cold place and in that, such a difference was keeping the

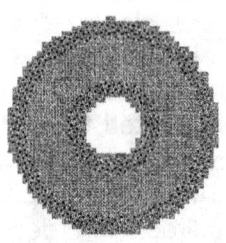

surface of the Sun nice and icy. It is all locked in relevancy because what is freezing to one is glowing hot to the other. At a point 6500^0 K was many times colder than was outer space, but outer space represented one end and matter indicated another end of the relevancy there was between what is hot and what is cold and only human mentality is applying standards of what is lying between what we measure as hot and cold. It is apparent that in the Sun hydrogen freezes heat to a liquid state at 6500K and with oxygen being in a layer found further down towards the colder centre parts of the Sun, it is any person's guess at what temperature oxygen will freeze to a solid state in the Sun. We also know that deep inside the Sun where things really get cold at 18×10^6 K, it gets so cold a hydrogen proton freezes to anything including an iron atom cluster. The name given for this is said to be fusion, although I have my doubts about the reality of this process applying. But with oxygen freezing to point of fusing, the conditions are considerably different with the governing singularity setting other rules as to what we with life can understand.

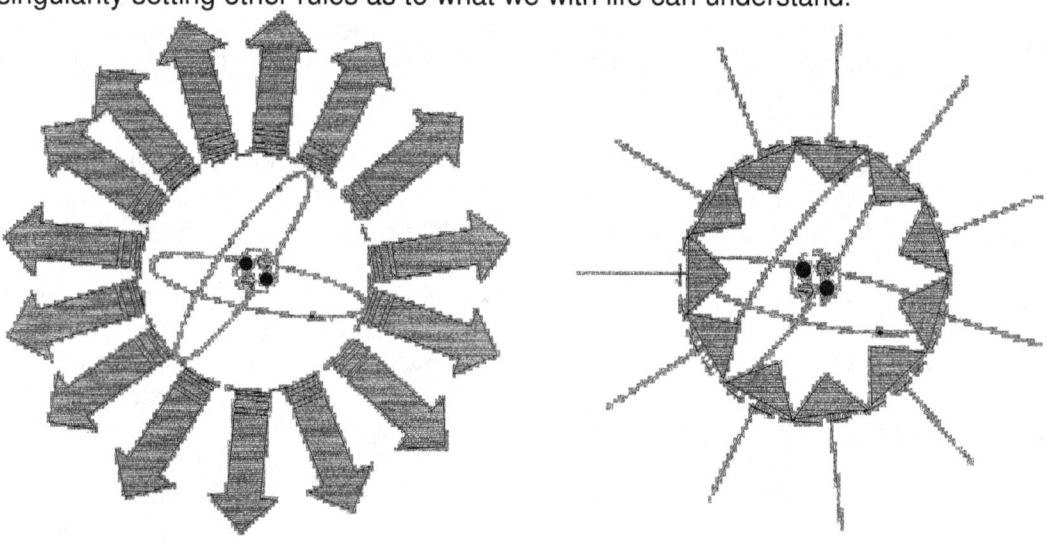

Gravity is movement duplicating material and in effect cooling material by duplication of the location of material by becoming relocation. By moving the space that material holds in effect cools down that location and when the space cools down the volumetric size of the space used by material at that specific place concentrates or shrinks. Gravity is a relation between what is hot on the outside of the atom and what is cold on the very inside of the atom where things move much more rapid than the speed of light. The movement of the material relocating will always be at a disadvantage (slower) than the orbiting of the electron which travels at the speed of light of at C and that proves that the Universe would come to a conclusion by fading out into space unoccupied leaving space occupied to shrink into singularity. The one side of the Universe is expanding into the oblivious where the density later on would be of such a poor quality it would contract into singularity as it does in the case of a Black hole. Then on the other side of the atom which is the inside of the atom, the space inside the atom would become so dense and so rapid moving that the condensation of heat will freeze all heat into infinity, as it does inside the Black hole. That is the route the atom is taking the Universe. Below the atom, the movement is cooling heat and therefore there is contraction of space-time and bringing about a flow of heat to concentrate the heat progressively at the centre. Outside the atom's sphere of influence there is a lack of movement of heat, which forms time, and by heat not moving, the heat further overheats and time moves the heat by introducing expanding

When an object is in outer space that object, encounters a specific relation with what we presume is space. This comes about by motion and through material volumetric size. The space the object encounter by moving through outer space puts a value of a ratio between the space it moved through and the space moving through which Kepler introduced as $a^3 = T^2k$. That means there is a contact ratio between space containing and space contained by.

When the atom is in outer space the atom is surrounded by a temperature of zero Kelvin and that is because zero Kelvin is the presumably the coldest any temperature can get. Being zero Kelvin on the outside and with zero Kelvin being the coldest temperature there can, it would make the atom also zero Kelvin on the inside since there can be no colder than that. That would mean the entire atom is then zero Kelvin.

However applying motion reduces temperature and there is much motion going on inside the atom. That means the fact that zero Kelvin produces the coldest there can be makes a little nonsense of such a statement. When the atom is at a temperature of 40°C then the outside of the atom must affect the atom inside of the because we see expanding as well as atomic shrinking taking place in relation of what the Ballmer and the Lyman series would represent and that proves that the outside temperature of the atom does influence the inside temperature of the atom.

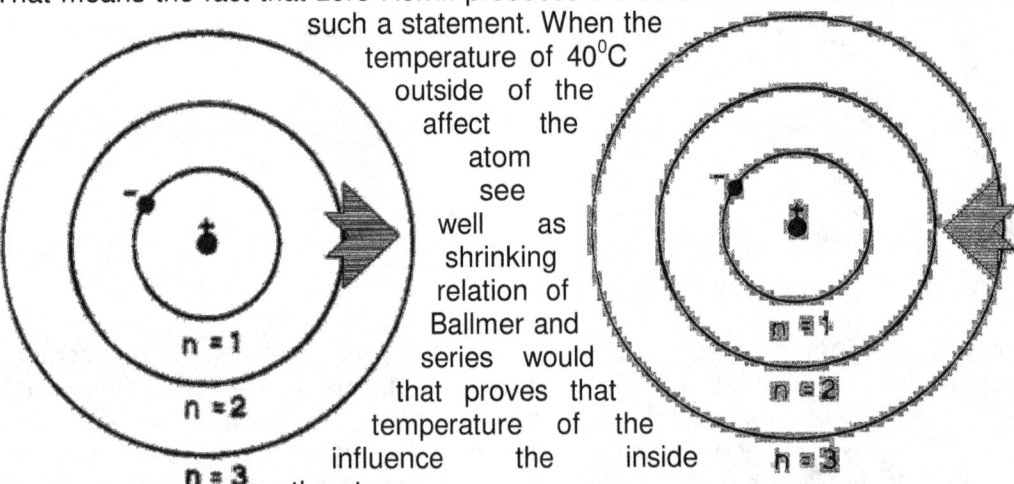

The normal summer's day temperature on my farm is 40° C normally in the shade because at that temperature little is loony enough to venture outside the shade. We consider that the atom must be 40° C because that is what the daily temperature is outside the atom on my farm. We feel and experience the 40° and we presume that all around is suffering from the heat of 40° C. That means the area the atoms on the outside of everything ids in contact with, holds a measured value of 40° C. We know that the inside of the atom is more contracted in heat than it is in cold because of freezing conditions. If this lot was in outer space, this lot was frozen into a solid state notwithstanding what it is that is frozen

We know that the action brings about a reaction and the actions leads to a response. If the atom heats on the outside by measure that it finds a need to reposition the electron by one band, then also the inside got smaller in relation to the growth or space expanding by one band. The relocating of the electron into a new position where the electron jumps a band is done by implication of the Coanda effect and this I prove by proving gravity is the Coanda effect. From the

Coanda effect we know that the liquid attaches to the solid using the formula $T^2 = a^3 / k$ where as space identify new boundaries by identifying the allocated boundary set by the liquid as $k^{-1} = T^2 / a^3$ where the space then forms the limit at $k = a^3 / T^2$. Every time the motion of the liquid intensifies the motion will attach to the solid by applying a new relation, which alters the relation of the solid by extending the space the solid has differently.

We associate such repositioning with the heat on the outside as that it amplifies or reduce space. However the adding of heat brings on a faster flow of liquid, which results in higher motion and it is in the motion that we find the answer to the cosmic principle applying. In the cosmos there is no hot or cold. There is higher or less motion.

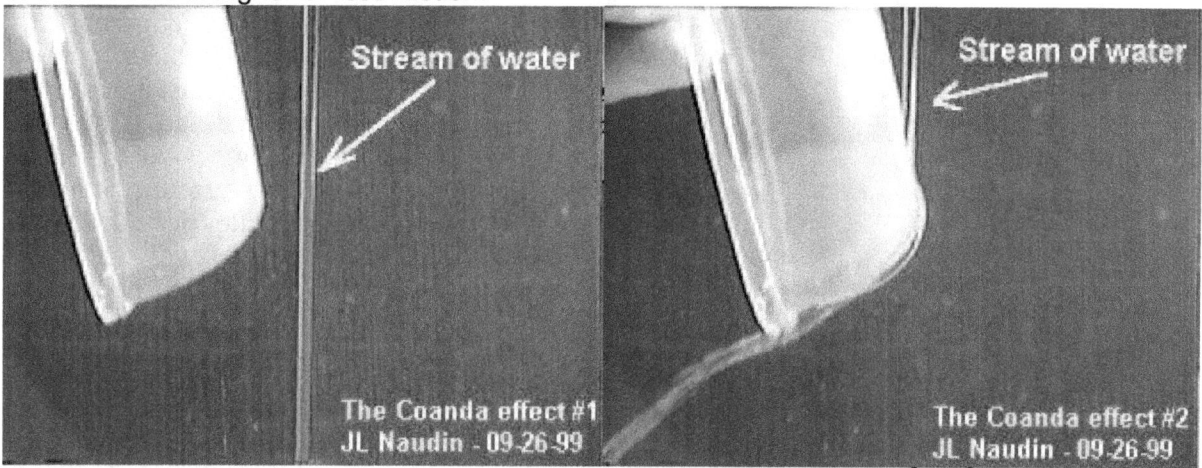

The Coanda effect is gravity forming by implementing Kepler's formula $T^2 = a^3 / k$.

However, we must not lock our focus on the heat but we must refocus on the motion that intensify or weakens. We must gauge the heat as liquid and see the atom as the solid. It is the motion that produces the new electron allocation and the motion produces a heat that establishes a cold. In this we must not look at heat in human terms as being degrees in Celsius or Kelvin, but we must look at the differentiation there is between what we regard as cold and what is hot. It is the difference and not the measured value that is vital. The focus is on the motion because the motion brings on accelerated duplication and accelerated duplication produces cooling that brings on a relevant cold within the atom.

If the temperature on the outside of the atom changes from zero Kelvin to 40°C it is not the temperature that changes but the atom is responding to higher motion applying and setting in place different relevancies.. With the atom in outer space the atom is subject to lesser motion since the atom is still only in distinct and personal orbital motion in relation to the Sun.

That is why the atom can be subjected to zero Kelvin. When the atom is within the boundaries of the Earth and circling around the Sun in a location set by the singularity of the Earth, the motion is distinctly more than what it would be if the atom were located in outer space.

The outside of the atom calls for a direct response to condition inside the atom since the outside can change very little if the inside does not respond in an opposing manner to what the outside produce. In such a relevancy there are always three factors performing as gravity and in that is the Coanda effect in charge of committing the standards by applying the gravity or the motion in relation to the solid.

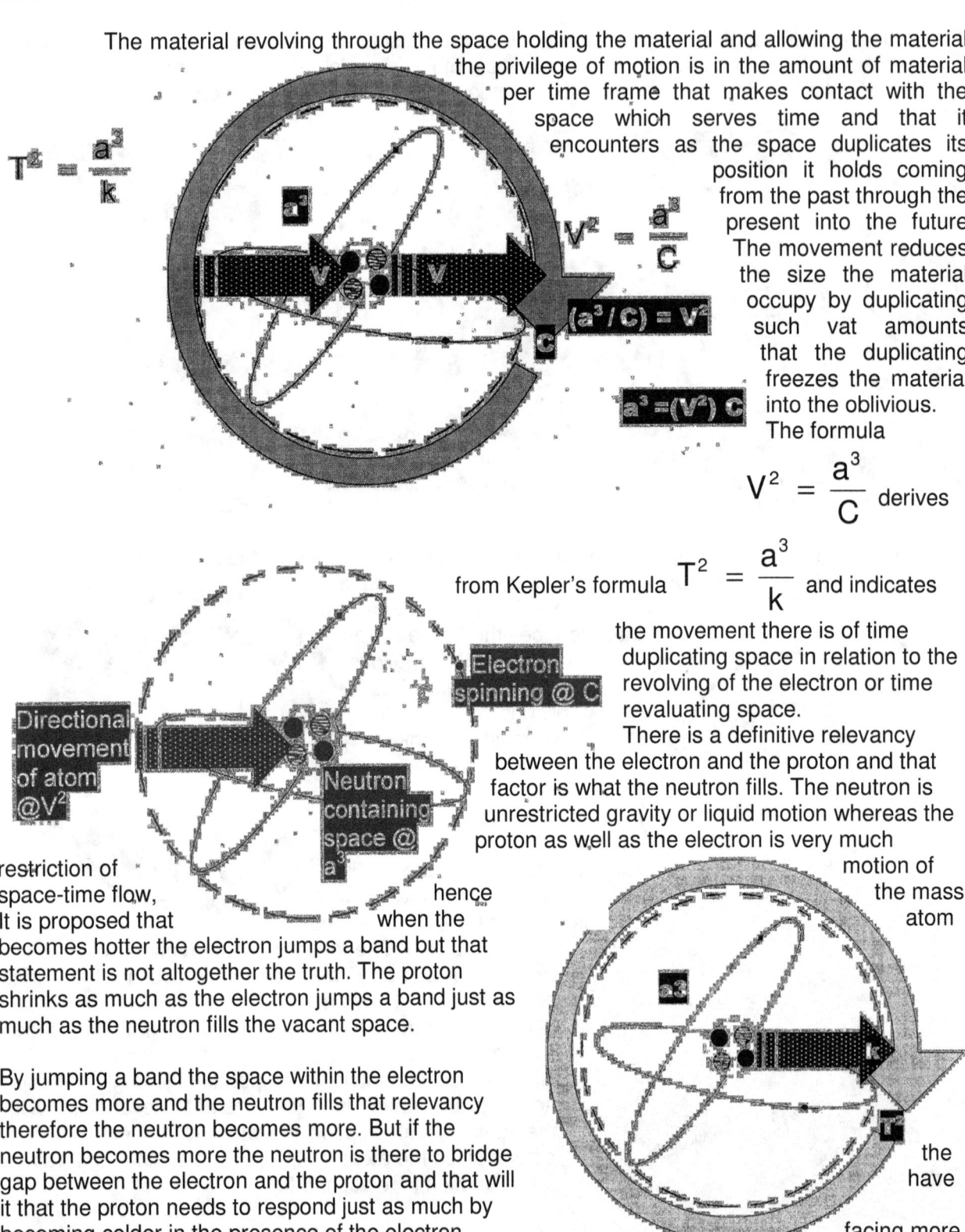

The material revolving through the space holding the material and allowing the material the privilege of motion is in the amount of material per time frame that makes contact with the space which serves time and that it encounters as the space duplicates its position it holds coming from the past through the present into the future The movement reduces the size the material occupy by duplicating such vat amounts that the duplicating freezes the material into the oblivious. The formula

$$V^2 = \frac{a^3}{C}$$ derives

from Kepler's formula $T^2 = \frac{a^3}{k}$ and indicates the movement there is of time duplicating space in relation to the revolving of the electron or time revaluating space.

There is a definitive relevancy between the electron and the proton and that factor is what the neutron fills. The neutron is unrestricted gravity or liquid motion whereas the proton as well as the electron is very much

restriction of space-time flow, It is proposed that becomes hotter the electron jumps a band but that statement is not altogether the truth. The proton shrinks as much as the electron jumps a band just as much as the neutron fills the vacant space.

hence when the

motion of the mass. atom

By jumping a band the space within the electron becomes more and the neutron fills that relevancy therefore the neutron becomes more. But if the neutron becomes more the neutron is there to bridge gap between the electron and the proton and that will it that the proton needs to respond just as much by becoming colder in the presence of the electron heat. The heat is not the factor but the motion

the have

facing more contributed by the

heat is what brings about the larger jump in spin. It would be advisable to see what every sub atomic part supports in the maintaining of the Universe where the Universe is the atom. What Newtonians see as the Universe is time holding innumerable Universe in a state of historical fact and that history of events forming facts s what Newtonians consider too form the Universe. It is not the Universe but it is time holding a picture of what has gone by.

I wish to return to the most scientific accurate book that was ever written which is the Bible. According to the Bible the atoms first came about, as at that point there was only movement but no space in which to move. The Bible says there was darkness over the abyss and strong winds over the waters. The atom is the darkness, as it still is darkness in which light disappears and the neutron being void of mass is clearly a liquid or a form of water in which the proton holding the abyss or singularity is in mass. The atom formed before light formed and the atom was covered in darkness and by darkness whereas when God created light as he said, the electron and the photon came into place to form light. The atom is the Universe and when the Universe formed, time as light was instated where the presence of light forms time. There first was singularity being void and without form, which clearly can only indicate singularity, with darkness over the abyss, the abyss being singularity and the proton forming darkness because the proton absorbs light as the prime function after the winds the bible refer to dissipated. Even today when science accelerates the atom to fit more neutrons as they do in nuclear fuelling, they create winds as closer to the speed of light whereas the winds that the Bible refers to was well above the speed of light and that kept the proton cold. The atom acceleration is exactly how the bible describes the Universe forming where the Universe is the atom that formed. The star constitutes only as a result of the accumulated behaviour of the collective effort of every atom forming gravity in the star. The star gathers atoms and in that gains heat where it later discards the structure that represents the atom when eventually only containing one single point in the abyss forming singularity.

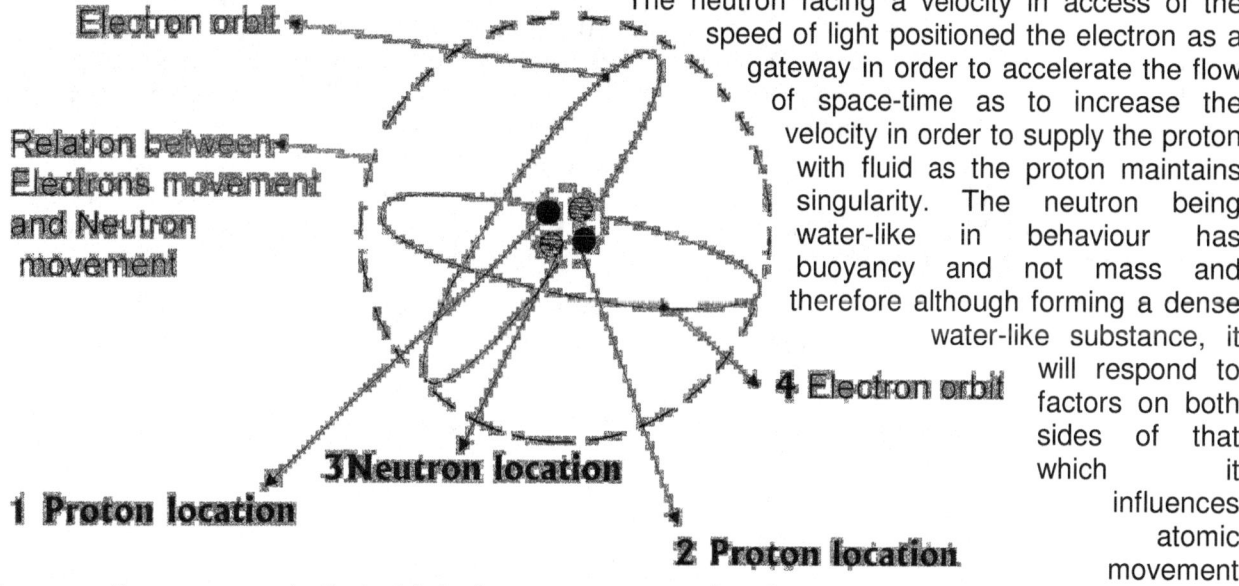

Electron orbit

Relation between Electrons movement and Neutron movement

4 Electron orbit

3 Neutron location

1 Proton location

2 Proton location

The neutron facing a velocity in access of the speed of light positioned the electron as a gateway in order to accelerate the flow of space-time as to increase the velocity in order to supply the proton with fluid as the proton maintains singularity. The neutron being water-like in behaviour has buoyancy and not mass and therefore although forming a dense water-like substance, it will respond to factors on both sides of that which it influences atomic movement because the response is that of bringing over more motion through the electron to the proton. As water does, the neutron also does not restrict the flow of space-time and therefore does no contribute to restraining the flow of heat to the proton and without showing resistance to space-time displacement the Neutron can't have qualities Newtonians attribute to mass. One cannot gauge the electron's behaviour without extending such behaviour qualities to the reaction that the proton would have since the neutron fill the gap between what is material and what formed as time and while linking eternity to infinity the neutron also provide the response on both sides. By the changing of eternity to infinity, the neutron contributes in the way of forming a mass less liquid state (or water as the Bible describes it) by suffering the greater discrepancy coming about as heat changes in form from a gas through a liquid and onto the solid state. However, in the ratio or relevancy there will never be any change in this formulation while time remains a factor within the star. The changes in the star comes when form in the atom starts to disqualify space within the atom holding a position in time where this leads to the demolishing of the atomic structure as it normally come in the form of an atom. The neutron is there solely for the amplifying of the motion, which is a relation the space has with time. When an object is in a location where there is little motion applying, the slack in duplication or reduction in motion present a gain in heat levels as the distribution of space reduces and a lot of increasing of heat results. This rising of heat will take the form of larger space occupied because the distribution of the heat over the space in duplication has very little possibilities of spreading the overall heat over a wider area. The motion of something as small as the Earth will confine the atoms into a relative hot area since the space in duplication does not reduce the extent of the heat by distributing the heat over much space.

In a structure with the size of the Sun the motion of space turning in time has enormous implication of cooling outer space by the sure quantity of space in need of duplication. Shifting that volume of space needs duplication that is millions if not billions of times more extensive than what the Earth may produce. By duplicating such a vast area in a period reduces the individual atom to a fraction of what the situation on Earth would allow. The more the spin of the solid is in relation to the gas is condensing into a liquid state space the ore space holding heat in gas reduces the space to liquid and extends the material in quantifiable measure many billion times over to what conditions in smaller stars will be. It is not the space that holds the matter but it is the spin in relation to what the matter holds that puts the relevancy of hot and cold within the star. The more there is cold because of the more liquid heat making contact with the solid and thus brings about motion, the colder would the atomic material be and the higher the relative contracting gravity that the star produces. This we see in the admitting of Mainstream science confessing that the reducing of space produces an increase in mass and because mass is the frustration of material unable to move by initiating independence in singularity, and in that they admit to the fact that mass in volumetric size has no influence on gravity.

The physics we encounter on Earth allow us to use a common and a constant, a fit all and an all-purpose because we find us captures by the Earth singularity. The Earth provides the space we may claim as well as the time in which such material duplication will take place. The function of stars is to cool off the overheating cosmos and while stars freeze the overheating outer space, stars can also fall victim to overheating. Evidence of this is numerous all the through the Universe. When one applies heat to an object it expands. That is primary school science. This states that more heat applied leads to more space acquired by the heated object. In sharp contrast to this is the growth in space when heat levels rises but freezing brings about the opposite result. When I freeze an object that object reduces its occupied space as it shrinks. Removing heat reduces space. That comes directly as nature responds to heat and I can prove that easily.

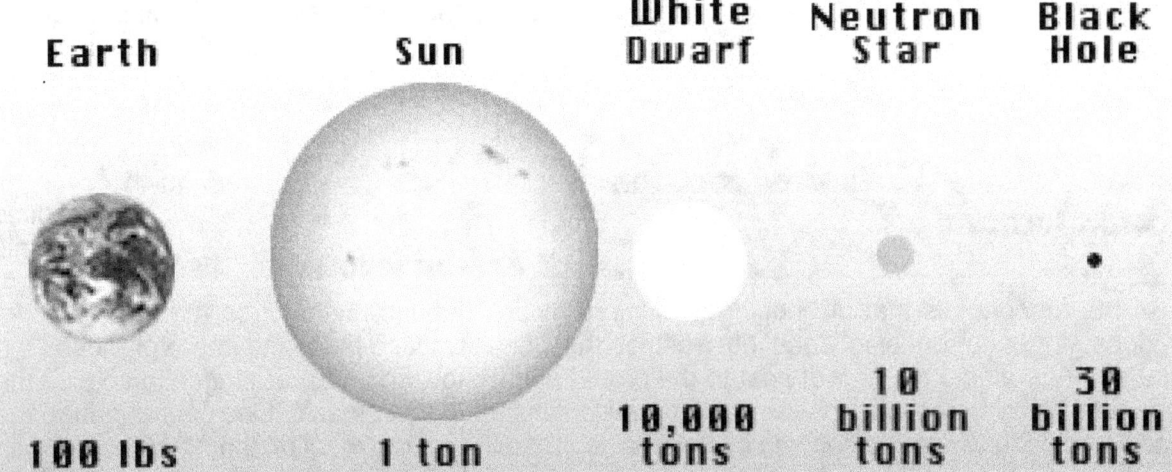

Earth	Sun	White Dwarf	Neutron Star	Black Hole
100 lbs	1 ton	10,000 tons	10 billion tons	30 billion tons

By expanding it accumulates space to increase the improving of the size of the material. When increasing space as heat levels rise, this is done to spread heat out into a larger area. The accumulating of heat is for the sake of securing singularity, which accumulates the heat in the material whereas the freezing tarnishes the overheating symptoms by the removal of material in unoccupied space using external matter and setting motion to the material until it contracts into a form which we see as visible heat. The heat is in the form of dissolved singularity that became material as material used it as growth. However, points holding singularity also unite by joining several points holding singularity within the atom. That is why by freezing it will diminish the space as to accumulate the heat absorbing into the heat into the material to maintain the equilibrium needed in space. Singularity can grow in two forms. As it does in outer space, singularity can duplicate by increasing existing points forming space by dividing space into more points holding singularity. This is the manner in which space duplicates by increasing. Then the other way is to grow by duplicating singularity in uniting two points where the two points duplicate as one and in that way singularity grows. This is following contraction by cooling. The two spots then share one point and the growth is not in heat but in intensity within singularity. This happens within the atom.

The atom is to optimum proof of the statement. The atom is the absorber of heat as well as serving as the release valve of heat. The atom regulates heat in relation to space acquired as well as space acquired. The atom is as much about heat as controlling heat and when the atom expands space it accumulates and store heat. When it cools it reduces and absorbs space. The cosmos is the atom and the atom is what the cosmos use to regulate heat.

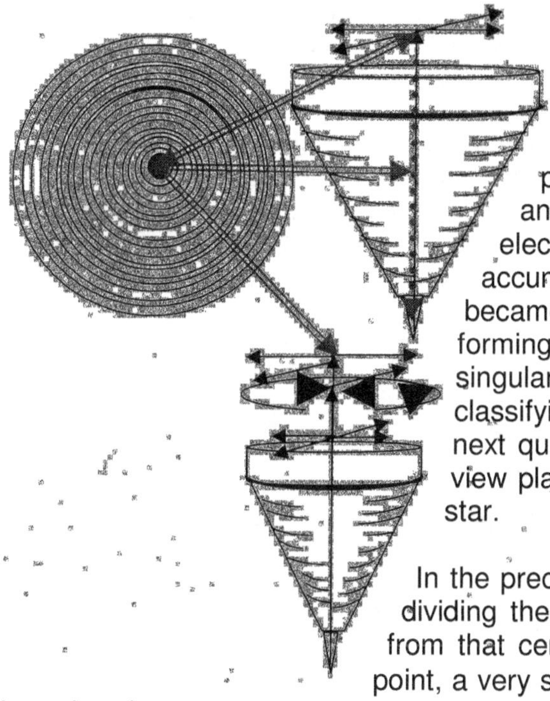

Since the end of W.W.1 everyone came into the knowledge of the singularity factor as being present in the cosmos, but no one new where to look for it. The reason I bring all the deliberation about is to show the cosmos came from singularity that was covered by protons (Earth) that was covered by neutrons (heaven) and finally as a last resort came light (the GUT) or the electron phase, just as the Bible says it did...The accumulation of protons was the universe. The Big Bang became all about forming protons in relation to neutrons forming liquid, NOT SPACE. Locating and positioning singularity accurately in the correct places other than classifying singularity as hiding in Black Holes must be our next quest because to my view everything connects and that view places a Black Hole in the position of being just another star.

In the precise middle of all objects in rotation is a precise centre dividing the object in sectors that will start the spinning initiation from that centre point. But the spinning object will have a middle point, a very specific centre point that does not spin and holds Π^0 as the only value.

What this does is that it puts a Black hole not in a fairy never-never land to be called upon when Newtonians try to become philosophical and mysterious at the same time but a Black Hole is waiting to be uncovered by time within every atom. The atom spins around as the atom has one function and that is to part singularity in infinity from singularity in eternity while controlling the systematic reunification of the two time factors once again. The atom is a cosmic regulator determining a relevancy between what hot (eternity) is and what is cold (infinity)

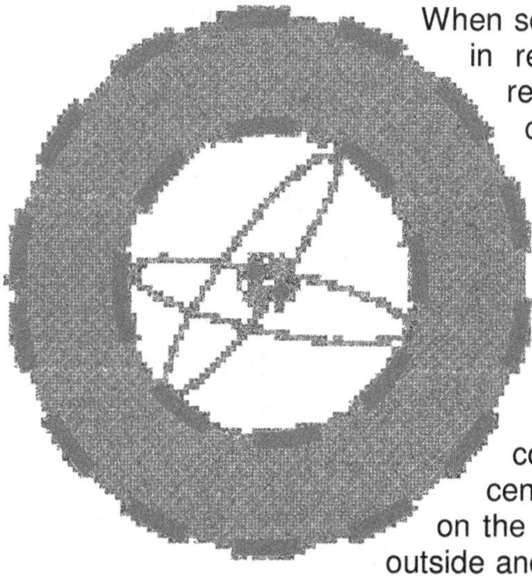

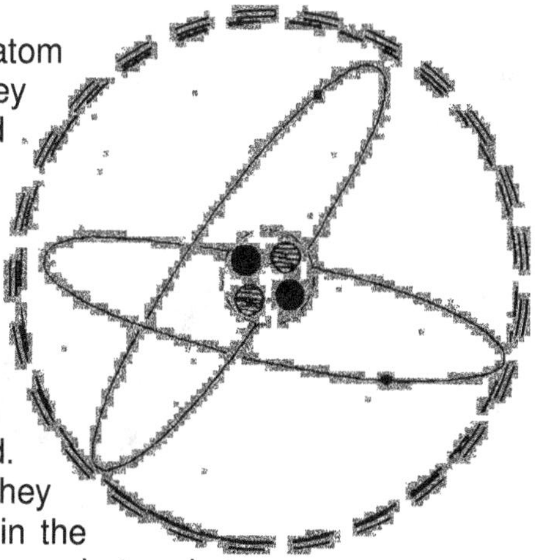

When science look at that atom in relation to what they regard to be hot and cold, science look at the heat surrounding the atom and from that determine whether they regard the atom as hot or as cold. They will say that they consider that atoms in the centre of the Earth are very hot and atoms on the outside of the rim of the Earth are cold. They look on the outside and determine from what is not part of the atom as to classify the atom as hot or cold. That which is on the outside of the atom is time and not space. The time forming heat covering the outside of the atom is not part of the atom but is what space by which the atoms will grow is. That which is inside the atom determines whether the atom could be classified as being in a state of hot or cold.

Taking this equation of nature to outer space we seem to confuse the natural law. With outer space as expanded as anything can get we regard outer space as incredibly cold. As heat sets in the normal flow will bring about expanding of heat into the form we think of as space that limits the heat overheating. Outer space is the very edge of expanding of space where heat cannot expand into space any more. Outer space is the limit, the epitome of expanding where heat meets space at the edge of all limits once more. Therefore being the representation of the very limit of expanding outer space has to be the hottest place there is. By applying heat to a kettle holding water, the adding of heat manifests as steam and steam is hot water that traded heat as it reviewed space. By allowing the receiving of the heat to continue the container will let loose steam in order to match the contributing of space.

We must see the atom as a centrifugal pump, pumping heat Liquid from cosmic time to the star via gravity and into the star via atomic spin. If there was mass on a specific star holding a specific relevancy to the light that the star forms by spin, then the atoms within the star would compress and concentrate space and as a result of this intense condensing that is the concentration of space, the atomic size would reduce as I picture with the tops reducing, but the dynamics of density increasing would have that mass increase in accordance.

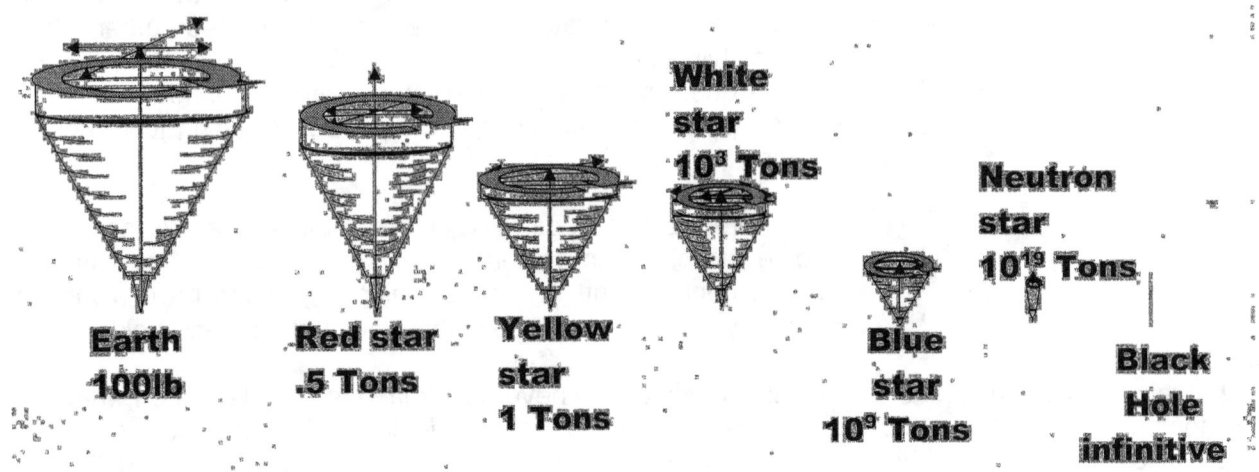

Remember that I say mass is the result of time compressing space by disallowing movement and there is no such a factor in the true cosmos. Mass is just a devise that Newtonians create in order to play mathematical games and while measuring with mathematics a Black Hole, the answer thy gain is with no true connotation to any meaningful proof whatsoever. The question to answer is where does the mass of a mountain starts because one would need life to tell where that person wish to place the divide between the rest of the Earth and what the person says the starting line is that the mountain has,. In cosmic reality, the mountain runs all the way to the centre of the Earth.

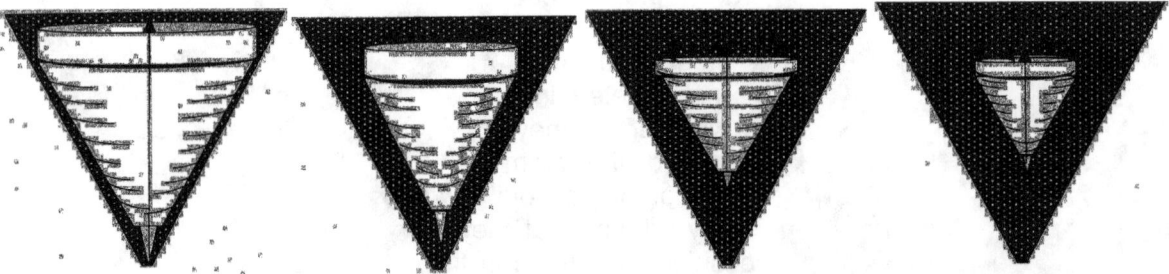

If we put the tops in ratio to what the atoms would be in the star, we would find that the material is progressively shrinking, (not really but seemingly) in relation to the relative time or heat or cosmic space in which the sp[inning material holds a relation. The density of time (space in heat) within the star or for that matter in the atom as well as the liquid outside the atom grows denser while the movement speeds up to make more contact with much more space during one cycle of time.

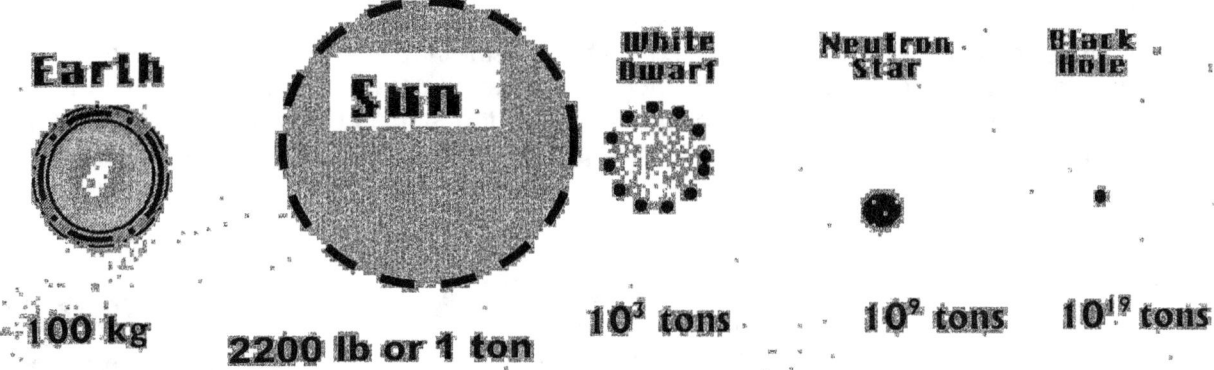

What I am grateful to see as that some Newtonians came around and got something right eventually. They see that it is density bringing about intensity in gravity and large stars grow smaller in size. They no longer connect massive gravity with massive stars having massive sizes and contain massive space. The Hertzsprung- Russell diagram (HR) is a joke that should be shelved alongside Ptolemy's Universe (and Newton's mass in the Cosmos). It is most senseless in the modern age where we begin to understand what the cosmos is all about and we should start to through Newton's magic – in - mass out of science. I put a challenge to the Brainy Bunch: show the world that the Newtonian mentality of mass is correct by proving where it is and that the cosmos applies mass as a classification devise or where the Universe arranges stars (or planets) according to mass or size.

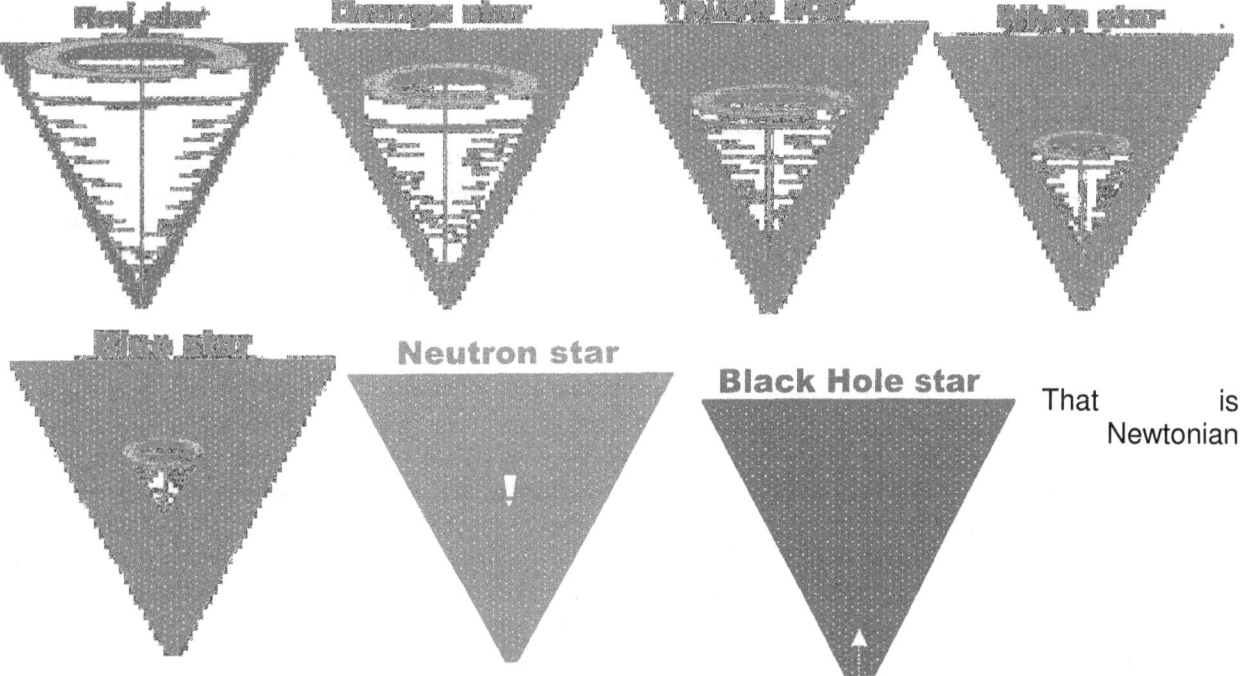

That is Newtonian

mythology...dedicating all aspects of the Universe to one non-existing aspect: having large volumetric space results in much mass and not having mass is contributing to being small and weak and coming across powerless, shit, when thinking about it the only bigger invalid human aspect they can attribute to stars having human characteristics is that the weak star then dies (become dead...passes over into not existing, not burning the bond fire any longer...can you believe it!) They believe that the star gets killed as soon as the star runs out of fuel, because then the star therefore must look small, meaning it is without mass or not that big and looking sadly unimpressive it then also must be without mass and has to be unimposing. How is it possible that so many just can't think and when they are thinking, then they are thinking that having lots of mass introduces large gravity in big stars while reducing the mass then leads to the star getting smaller and weaker and more sickly and then without finding the proper care, the star will die a horrible death by succumbing to malnutrition and become a dead piece of rock floating in space forever and ever... That my friend is Newtonian wisdom...that is the mentality of those that thought they were too smart to be bothered with reading my work because I had the audacity to question Newton!

If any of the tops represented a star, the space held by the star in relation to the space in which the star spins would remind us of what is in the sketch. The reason I press this issue is because this is absolutely vital to understand the working of gravity within the star and how time plays a part by contributing through allowing phases through which progress is made in the development through which the star goes.

The essence of the Sun in comparison with all other objects in the solar system is that the Sun captured time while all the rest are still growing by the measure of the Sun in the time that the Sun dictates. The Sun has not developed the Iron $_{56}$ core as a dominating factor yet, but in spite of Newtonian wisdom, the Sun could not function by having gravity if it did not yet have a gravity producing Iron $_{56}$ core. This enables the Sun to produce gravity much in the same manner as electricity generators charge electricity. By charging gravity, the Sun freezes heat from a gas to liquid by condensing space. In that way the Sun captures time and locks time in outer space in the zone around the Sun to a freezing cold 6500°C.

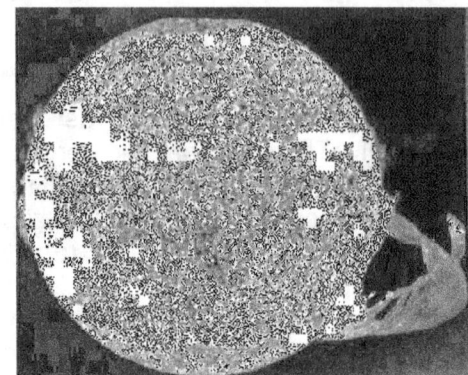

Within the inner part of the Sun the liquid is at a freezing cold 18×10^{60} C. As the space grows bigger in outer space in line with Hubble's findings, outer space grows by expanding. This allows the Sun to accumulate more time and the Sun seems to increase in size. How much growth in time there is, I would not know but as outer space is increasing in expanding, by that same margin we find the Sun is freezing the part around the Sun to what it was back when the Sun became an independent spinning star. But in relation to the Universe expanding, the Universe is also just as much standing still while the Sun is shrinking in size. In time the Sun will become a white star and then a blue star and go on to form a neutron star and eventually end up as a Black Hole while progressing through a process which all the stars follow that forms part of the cosmos. The process through which the Sun will develop is a time-related issue forming a relevancy with space. This is when the relevance in outer space expanded to measure the freezing of time which the star uses as progress.

The manner in which heat expresses itself normally when confronted by overheating is to provide additional space through expanding of space into creating more space. Outer space is the region we consider that holds tarts in place In spite of all Newtonian assuming outer space is the hottest there could ever be, I make this conclusion on the grounds that heat expand when in excess and since outer space could never stop expanding, therefore all heat found in the Universe is accumulated mainly as time in outer space. Outer space consists entirely of heat in a form of gas that has been expanding by the process of overheating. The region thought of as outer space is the hottest anywhere because outer space has expanded all it can it is still expanding which in terms of Newtonian thinking is at the speed of Hubble's $1/H_0$ but again I stand to differ about the validity of this presumed rate of expanding and because this expanding is heat counteracting overheating. Where this expanding is taking place covers whatever vastness covered by space in time that can come to mind and even where such vastness does not come to mind because of the vastness the concept. Newtonian taking the measuring value of $1/H_0$ sees this expanding not affecting us on Earth. However, the overheating and then expanding inevitably does not only affect far-off places where we cannot be, but affects us on a daily basis even in our healing in human body tissue and dictating the rate that our hair and nails grow.

As outer space is stretched to its limit, its limit will continue to stretch but while it is stretching it has to have more than it had before in that what outer space had and as outer space holds the limit of heats expanding possibilities, we have to figure out what is becoming more since the Universe holds all and nothing can be added or taken away. Singularity has been expanding since the time of way back when but that means singularity is still releasing heat as space-time that turns out as space in the universal time of outer space. In outer space heat cannot expand more, yet it continuously does expand more therefore we have to find what becomes more and then forms the part that has the continual growth that benefits all points in singularity throughout on a continuous bases concerning all outer space. This growth is present since the Big Bang and it is there where it started that we should start looking.

As it became evident, we find Time in the atom adhere to two factors that establish a presence that forms space. In normal everyday time as Kepler gave is in his formula $a^3 = T^2k$ we can calculate from that $T^2 = \dfrac{a^3}{k}$ which then in terms of the atom time becomes $V^2 = \dfrac{a^3}{C}$. That means the rotating spin of the star relates directly to the movement of the star forming occupied space-time a^3 in terms of space-time unoccupied through space-time unoccupied V^2 which then must set a spin on heat in terms of the speed of light at C. This comes from the manner in which the atom stands in relation to movement as well as the speed of light. In that way we can determine the colour of light that the star produces which is directly in terms of the velocity of spin that the star has that creates the spin of light as it condenses unoccupied space-time. By the spin V^2 in relation to C we find this ratio produces a value of **k** and that spin sets a matching standard in relation to the light C that the star reflects.

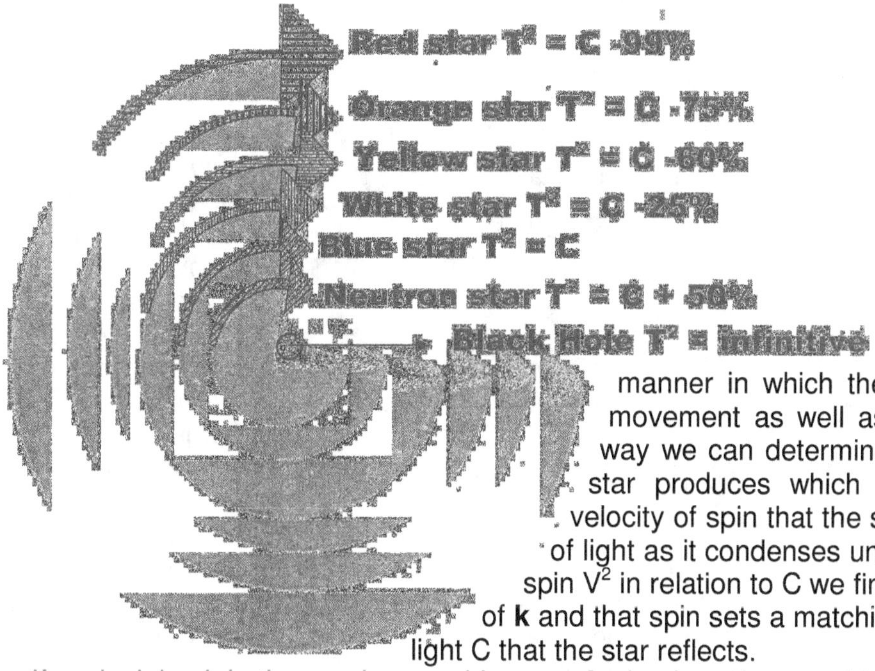

Red star $T^2 = C$ -99%

Orange star $T^2 = C$ -75%

Yellow star $T^2 = C$ -60%

White star $T^2 = C$ -25%

Blue star $T^2 = C$

Neutron star $T^2 = C$ + 50%

Black Hole $T^2 =$ Infinitive

If we look back in time and we could see as far back as time would permit, we would see the Big

Taking the present date 10^{12} years back into the past

Taking the present date 10^{12} years back into the past

Taking the present date 10^{10} years back into the past

Taking the present date 10^{9} years back into the past

Taking the present date 10^{6} years back into the past

Taking the present date 10^{3} years back into the past

Present date

Bang banging big-time. That is as far as we would see, or so we think but if we think that we are thinking with Newtonian restraining. The question would be as to what we would see. Going back that far (going back in time can only be by looking through a telescope and I say this before any clever Newtonian start designing a space whirl to take him back that far) even if it was possible to go back in time that far, we would only see neutrons spinning because that was as big as it got when the lot got started. When the lot started, the Neutrons grouped in bundles where they associated with protons to form atoms and atoms grouped to form stars and stars grouped to form galactica and here we are located in a galactica we call the Milky Way. If we look back from one angle the lot got very much bigger and if we look from another perspective this lot got much smaller. If we look at a Black Hole we see a giant bigger than our mind can cope with also being smaller than our eyes can cope with.

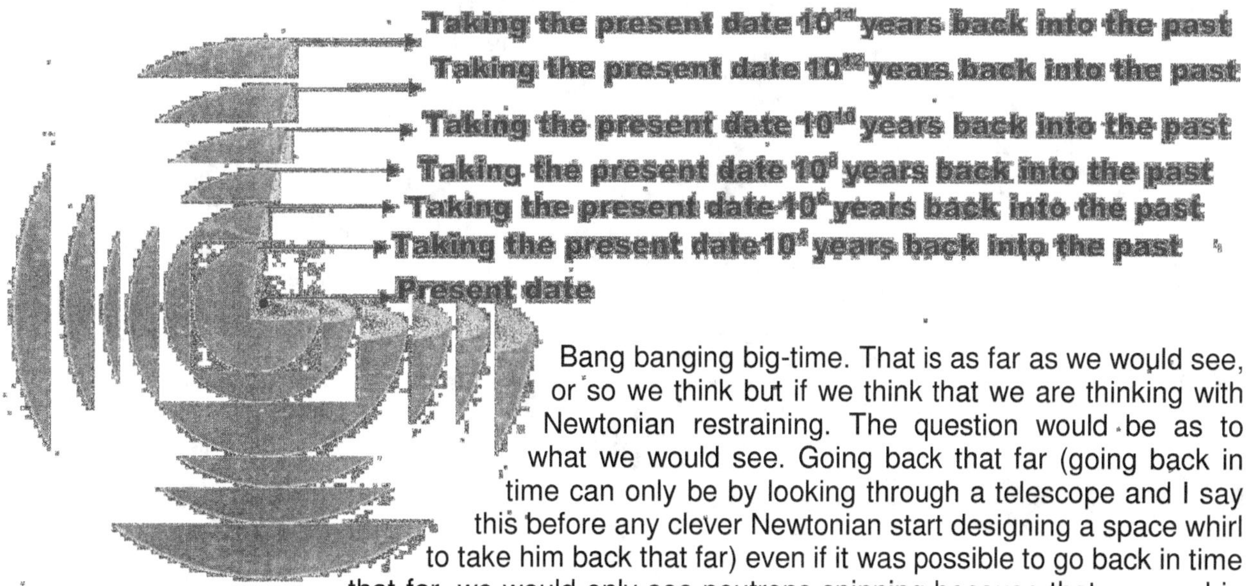

When a star starts developing, it starts to charge gravity by movement coming from the most inner core. The atomic relevancy applying at such a point will be what is required to concentrate heat which is $7/10(4(\Pi^2+\Pi^2))$ =55.2 and this places gravity being the very same as electricity which is heat condensed and which is the gravitational displacement of iron$_{56}$ and this is in relation to $10/7(4((\Pi^2+\Pi^2))$=112.795 which is heat totally expanded at a 112. This inner core action sets the Coanda effect in progress where a flow of liquid starts pumping in relation to the solid atoms and this establishes movement that initiates gravity applying. This then results in movement coming in relation to all the layers in the star and sets a trend by which development will progress. Looking at stars takes us on a journey back into the past where mass fortunately has no role to play. The star captures a time position in space and locks that space into that time. The Sun froze outer space at a relevant position in time when outer space was 6500°C while when the inner-core singularity relevancy was activated the heat was 18×10^6 o C. As much as the time element expands, so the material element shrinks because no one of the two factors could be looked at being apart from each other and by freezing the overheating time, in that manner can a star reduce in relation to cosmic expanding development. In a fully aged developing star the star is still accumulating. But not in claiming more space. Then the star still grows but it removes space from the growth it has by the light colour it sets. This applies to stars and where the planets in the solar system is concerned, not even one started to effect outer space in terms of light production, and that is notwithstanding all the aid provided by the Sun in gravity as abetment coming from the Sun.

When the Sun started evolving inside the galactica cocoon the heat in outer space was the equivalent of what would today be measured at 18×10^6 ° C. That is when the inner core of the Sun within the centre of the star started to thaw. It locked heat levels at 18×10^6 ° C. Then as outer space developed, grew and expanded within the frame of time, the rest of the Sun started defrosting and this came as the overheating expanded in outer space.

All the above may bay be the case in a star including a small star such as the Sun that is a yellow dwarf, but within the limits of the Earth (and all other the Sun) everything is takes the objects sharing Earth is still expanding.

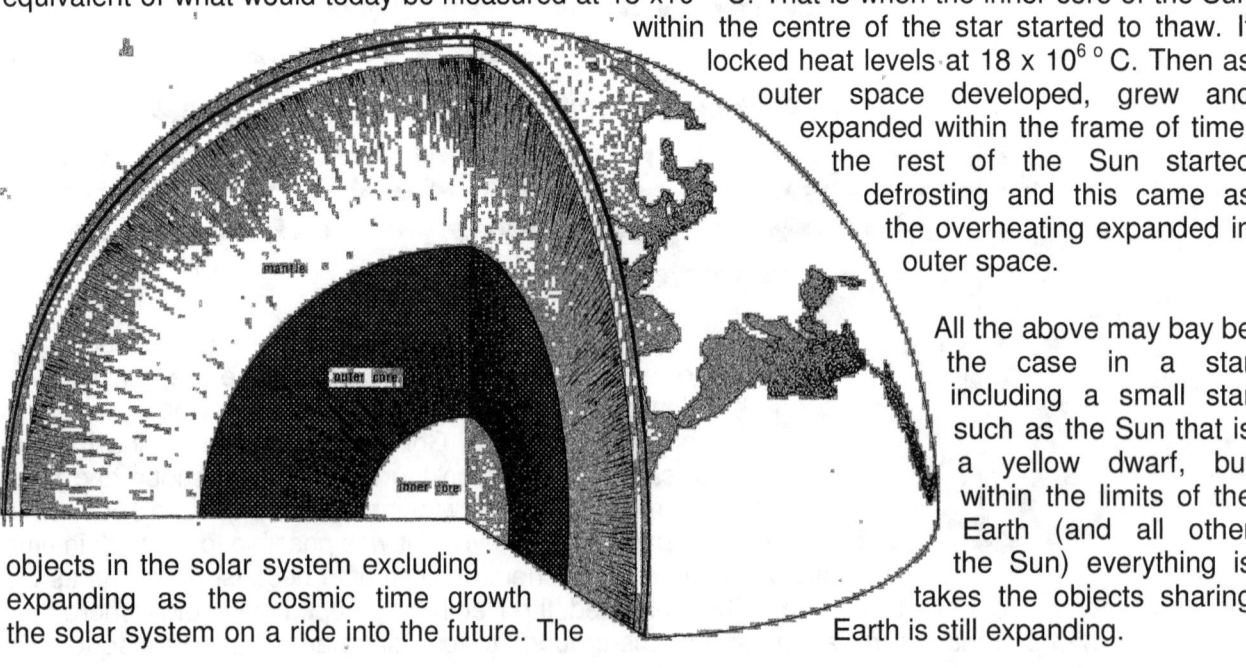

objects in the solar system excluding expanding as the cosmic time growth the solar system on a ride into the future. The

$$a^3 = T^2k \text{ is also } k = \frac{a^3}{V^2}$$

The Earth is growing as much as everything is growing and everything is growing since the time of the Big Bang. With singularity in place and controlling time we can trace the relation there is between when time was at a value as time formed space where time went on to translate to space. The entire Universe is expanding and in spite of all Newtonian protest coming to try and protect the reputation of Newton, such expanding includes the Earth as well.

The Earth is bulging out like an over ripe melon as the inside is outgrowing the outside. What I am now going to explain forms a very small part of **Seven Days Of Creation** or **Matter's Time In Space: The Theses Part Seven**, which is the same book. This is the command that set the trend for the third day as The Creator said: "***Let the waters under the heaven be gathered into one place, so that dry land may appear***", **and so it was. God called the dry land Earth and the gathering of the water He called seas; and He saw it was good**.

This is how the most accurate book that was ever written describe this following bit of science, which I might add, went by without one atheistic idiot and all other Newtonians that are too wise to read my work ever noticed…because they are too God damn stupid…and then they call me uneducated and illiterate just because I don't agree with Newton and all the other Newtonian stupidity attached to the Newtonian religiosity that is behind this self-righteous mentality.

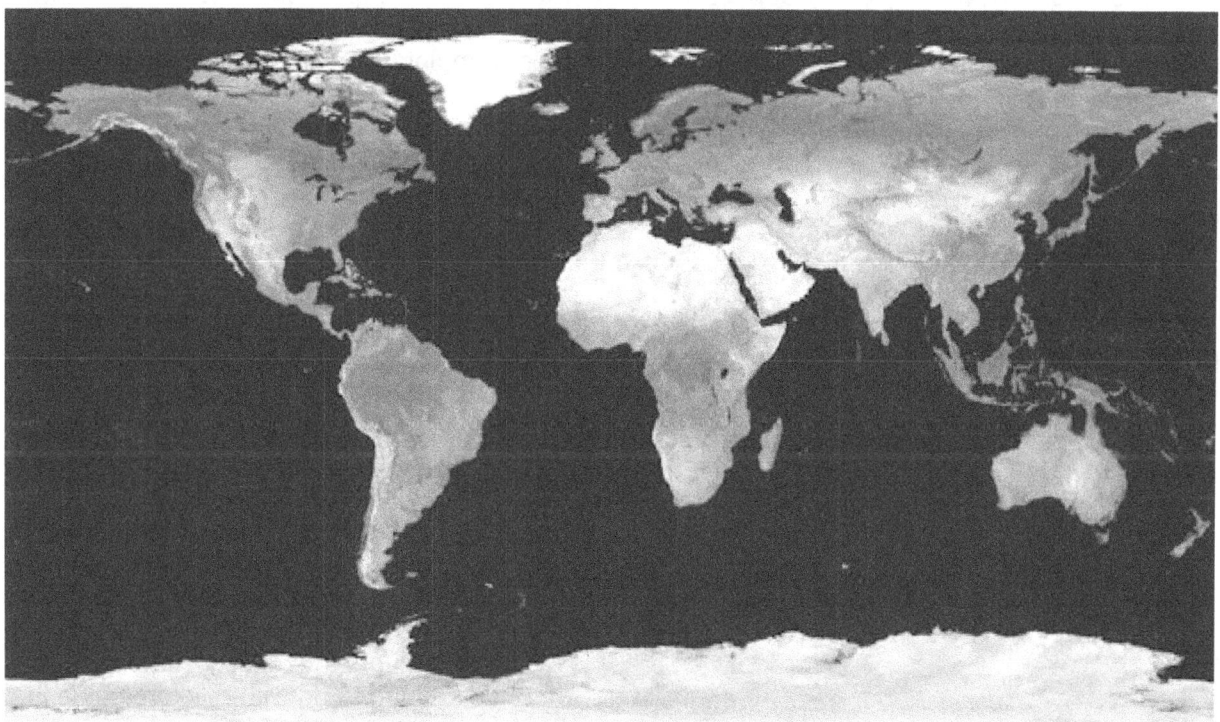

The centre core of the Earth is iron and the middle is mostly silicon with the top mostly covered in water. The atoms are collecting heat and in that process the atoms are growing. This is an aging process and with that even the human body adhere to this principle and in line with that process we with life get older. The tissue at birth receives a specific ratio in singularity in usable and maintainable size and that connects to life. Then as life acquires a larger body by accumulating cosmic resources called carbon, the body grows in stature. This cosmic cell ratio is constantly providing for bigger cells in human tissue if compared to what once was the case. Every generation of human specie is bigger than the previous generation but also they are smaller that the following human generation. As heat accumulate in the tissue we grow as children by adding human cells and by the growth in the human cell, but then we get to adulthood and the human expanding in size is capped as we reach a size that we will be for the rest of the time we hold life in our molecules. This affects the tissue numbers as the atomic growth keeps on adding space to material because the body has capped the size limit the specific human will reach.

This process we named aging. Then in the aging process the human body discards some tissue in quantity to maintain the size limit while compensating for the continuing atomic growth still carrying on through out the entire cosmos and the compensation for atomic growth is vested in keeping less tissue to compensate for the larger atoms coming as a result of the continuous cosmic growth. The evidence of this growth we find in healing where old tissue is discarded and still more evidence about this is in the process that we find in the growing of hair and nails and even in the cancer process. That is aging where the human body collapses under the continuing expanding of every molecule that holds life as a guest and finally the tissue can't function with the limited use of the remaining tissue numbers available for human functioning of life. Aging is not a human concept that can be medically bypasses so that doctors can make money out of the fear of death their patients have but is a cosmic institution. Why I mention this is because life borrowed this process from stars that follow this same pattern, except that stars fall back into uniting infinity with eternity by using cosmic matter and gravity takes cosmic particles to become unified whereas the human or if you wish to think in terms of life. Then it is life that is also uniting infinity with eternity but life on the other hand unifies that which can never start with that which can never end by compromising the structure of the Human body in discarding it where nature then dismantles the body and life then takes on the unification of infinity with eternity in a process we call death allowing only life as the concept the privilege to unify that which can never part. The cosmic growth in gravity or time applying is here and the process is governing every aspect of the cosmos, even being as close as governing my thoughts by regulating the electricity control in my mind which I use as thought patterns. It is so close it even affects my human body's ability in the duration of maintaining life as to be host to life.

The core is growing by the measure of the number of protons collecting heat and translating this heat to cooling singularity as to maintain singularity. This then feeds the Earth's governing singularity and as this process does not require space, therefore this happens on the other side where reality governs time in space and where space plays no role and that is where the Creator of heaven and Earth have absolute control of every aspect concerning the governing of the Universe. The atoms relate to singularity forming as 1^1 and the governing singularity relates to singularity 1^0 and this relevance connects the entirety to one unified Universe of which the Master of eternity and infinity has absolute control, but the control is where singularity finds a measure. This has nothing to do with religion, but is the pure basis forming all physics. This concept holds the building of the Universe in tacked. The one that can't see this has the mind of an animal or an atheist, which to viewing is identical and the same. The worth measure of singularity being a presence of only holds a value on the other side where there is no infinite as much as time is eternal. The value of singularity forms the building blocks of the Universe and only holds a value with measure where infinity meets eternity undoing all space which is on the side holding no space because where we are, the parting of that which can never part only holds relevancy.

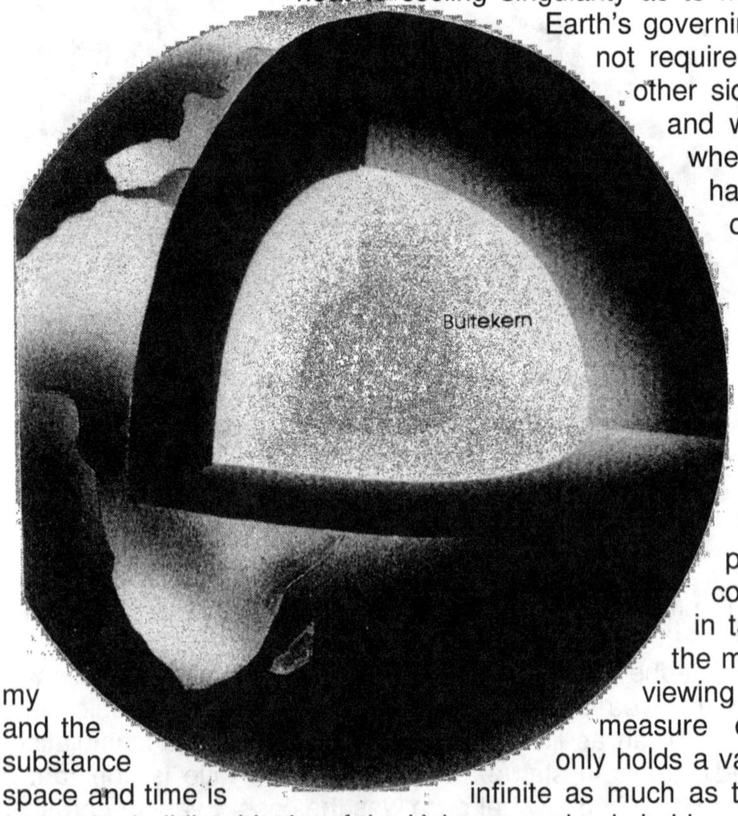

Buitekern

my
and the
substance
space and time is

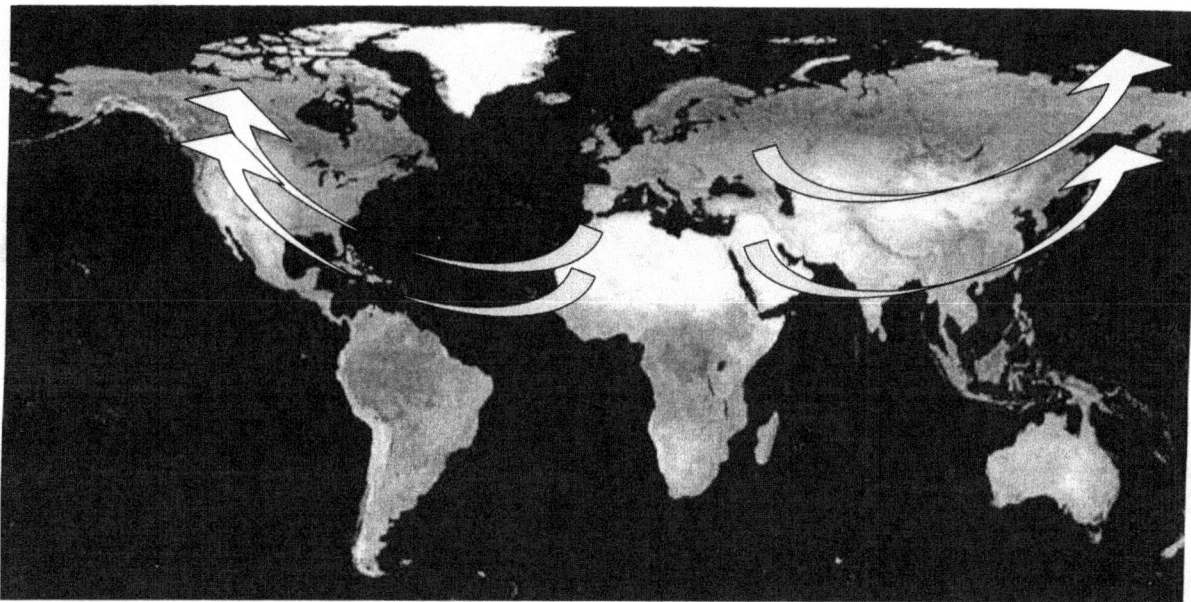

Let's start by rejecting this idea that tectonic plates are moving underneath and on top of each other. There is no shifting of plates going underneath or on top of the other because the plates are showing lines where the Earth crust is growing by expanding into using larger areas for the purpose of holding dry land, just as the Bible says it does.

As the Earth is a sphere so the Earth expands as a sphere would. The entire surface of the sphere is getting bigger and the areas mostly affected by the circumference growth are the top part that is mostly covered by land material. But as a person would expect, there are no lines forming specific borders that match certain criteria from which one can deduct behaviour patters that would provide Newtonians with definite areas holding specific constants that provide the average Newtonian the opportunity to pretend to be very clever and play with a lot of numbers and hide the fact that they have no idea what is happening around them. The growth is not even predictable but is absolutely sporadic and unpredictable except where in certain cases one could read behaviour patterns as characteristics that formed.

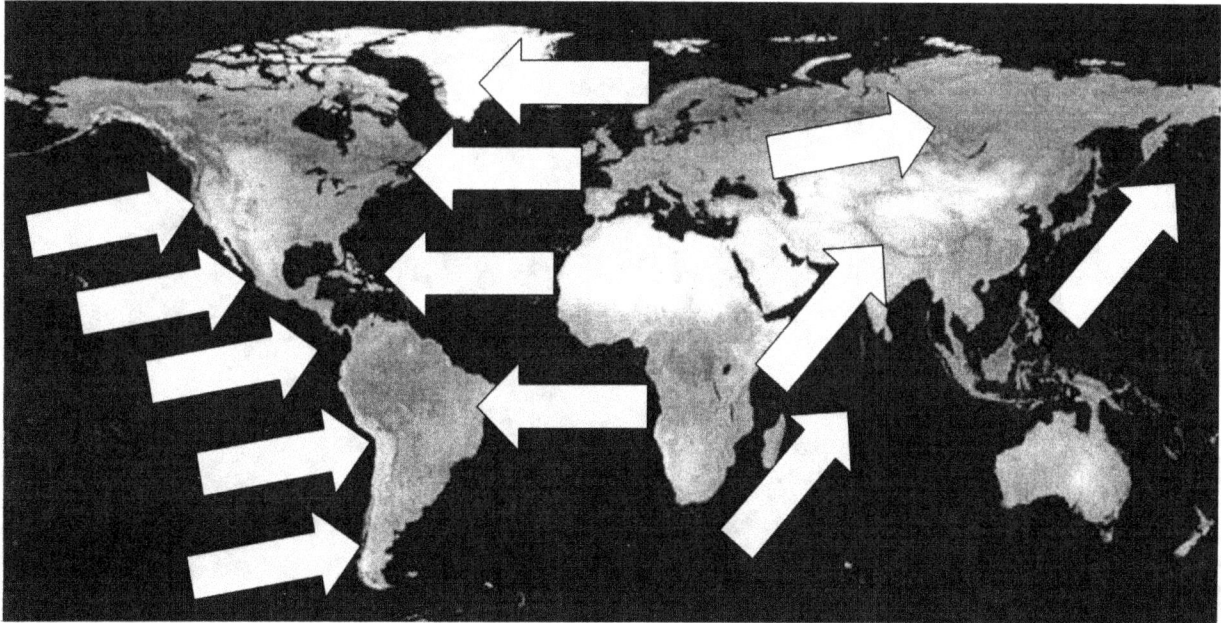

The Earth crust is expanding as a result of the inner core growing more rapid than the mantel is growing and the mantel is growing more than the crust because the mantel can accommodate larger measure of heat than the crust does. This brings about that the expanding will crack open the continents like a balloon would crack mud if it was covered in mud and is expanding by inflating.

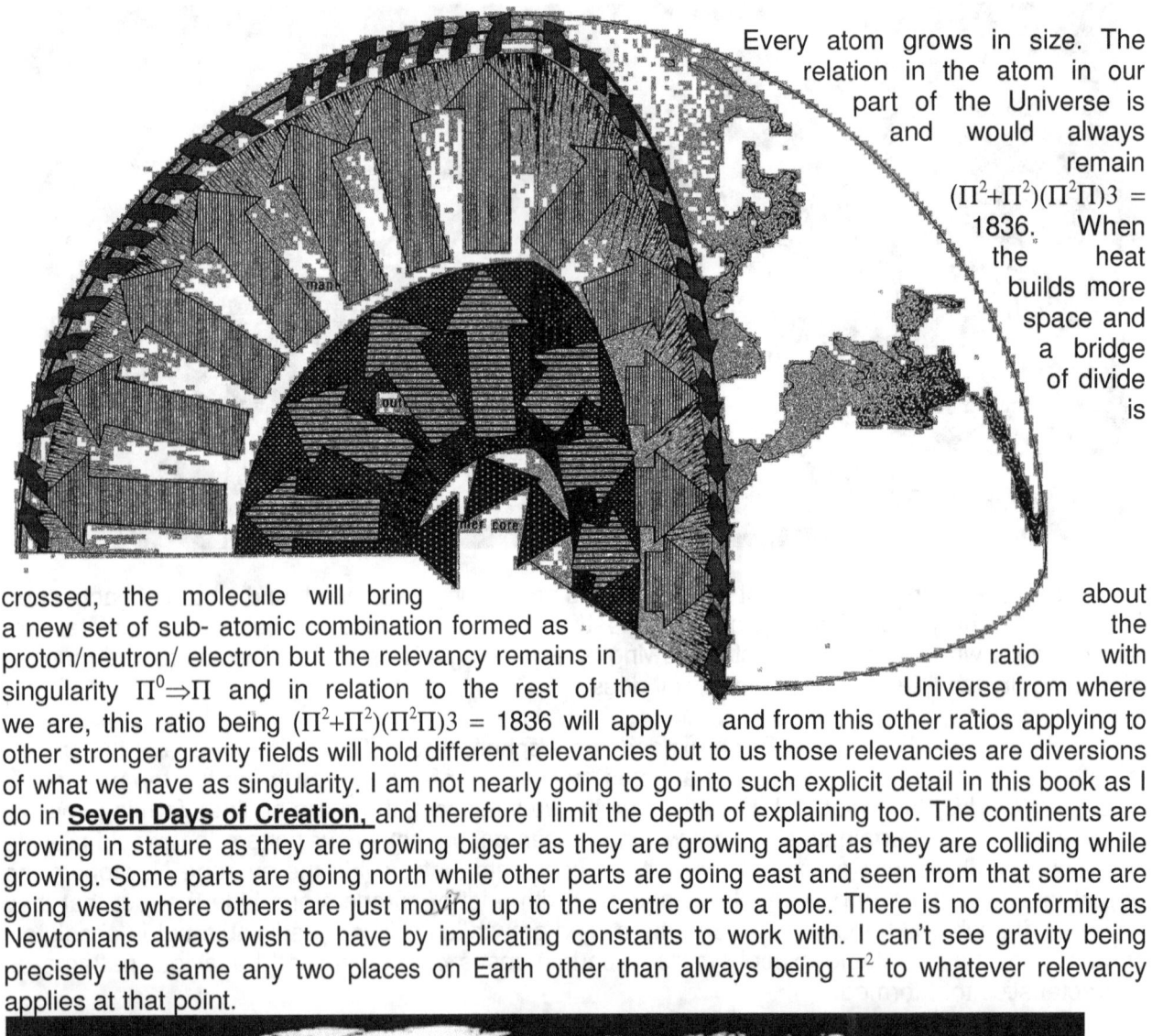

Every atom grows in size. The relation in the atom in our part of the Universe is and would always remain $(\Pi^2+\Pi^2)(\Pi^2\Pi)3 = 1836$. When the heat builds more space and a bridge of divide is

crossed, the molecule will bring a new set of sub- atomic combination formed as proton/neutron/ electron but the relevancy remains in singularity $\Pi^0 \Rightarrow \Pi$ and in relation to the rest of the we are, this ratio being $(\Pi^2+\Pi^2)(\Pi^2\Pi)3 = 1836$ will apply about the ratio with Universe from where and from this other ratios applying to other stronger gravity fields will hold different relevancies but to us those relevancies are diversions of what we have as singularity. I am not nearly going to go into such explicit detail in this book as I do in **Seven Days of Creation,** and therefore I limit the depth of explaining too. The continents are growing in stature as they are growing bigger as they are growing apart as they are colliding while growing. Some parts are going north while other parts are going east and seen from that some are going west where others are just moving up to the centre or to a pole. There is no conformity as Newtonians always wish to have by implicating constants to work with. I can't see gravity being precisely the same any two places on Earth other than always being Π^2 to whatever relevancy applies at that point.

The Earth inner core holds a value of mostly iron $_{56}$ with more heavy material mixed in the compound that forms the Earth centre. Then above this layer we have a much lighter material forming the mixture and the most inner centre will grow much more rapidly as well as seemingly faster than the layers above. The inner core is also surrounded by more heat that is more readily available and the spin provides a much more demanding as well as dynamic gravity. It is by the dynamics of electro plating that the atom growth takes place and with the value of gravity or electricity applying the growth outweighs the silicon layer exponentially. This makes the top seem to become more and the water seems to become less and it seems as if the Earth is drying up. This is the process that the Bible says is happening and since science never researches the Bible with the view to secure facts from Newton's fiction, and because of that, therefore science was left behind.

This very affect is what spins the mind of Mars explorers because this process also applies to conditions found on Mars. The available water that was once in abundance on Mars, when now placed in ratio to the land surface, has disappeared in relevance compared to the growth that the landmass has undergone in the progress of time, as the surface grew and turned the water into icy molecule traces. Now the Mars explorers see the giant riverbeds without the presence of flowing water and get so excited about the water prospect on Mars and keep asking where the water went.

The huge river beds are there and the water that produced the beds has disappeared and if there is no mass and there is no pressure to use in order to calculate mathematics that cannot be calculated in the first place, then all intellect in the Newtonian mind vanishes like the water on Mars did. Any one looking at the Earth from Mars with telescopes will see the Grand Canyon and other huge river systems on Earth and will ask the same question...where has the water levels gone on Earth. The Canyon amongst others were huge rivers when the land mass was small and the water ratio was big, but since the ratio changed the river systems grew smaller in ratio in comparison to ratios that applied millions upon billions of years ago.

But with water being 17 in displacement and the inner core 56 in displacement and as well as having the silicon displacement at 26, the ratio in growth is a staggering lot going in favour of land filling enlarging old river canyons where water previously flowed, but with the lack in growth the water in ratio went missing. On Mars, as is the case on the Earth, the land outgrew the water at a ratio of 26 to 17 while the 26 is being pushed by a 56 from down below and this process will keep on going until all the water becomes gravity vapour as it now is on Mars. There is no water on Mars to the measure Mars explorers are in search of because the river beds that were once big kept on growing and got bigger looking like enormous canyons while the water vaporised into the air. Today Mars explorers ask where has all the water gone and when looking at old riverbeds today in ratio as time presents facts, today the rivers that once were filled with water seems enormous. However, the water that once was a lot on Mars when Mars was a tiny planet compared to today with much water to present in ratio to land, the water shrunk away as it disappeared by the ratio that the land is growing to become just a thought of water as it is today.

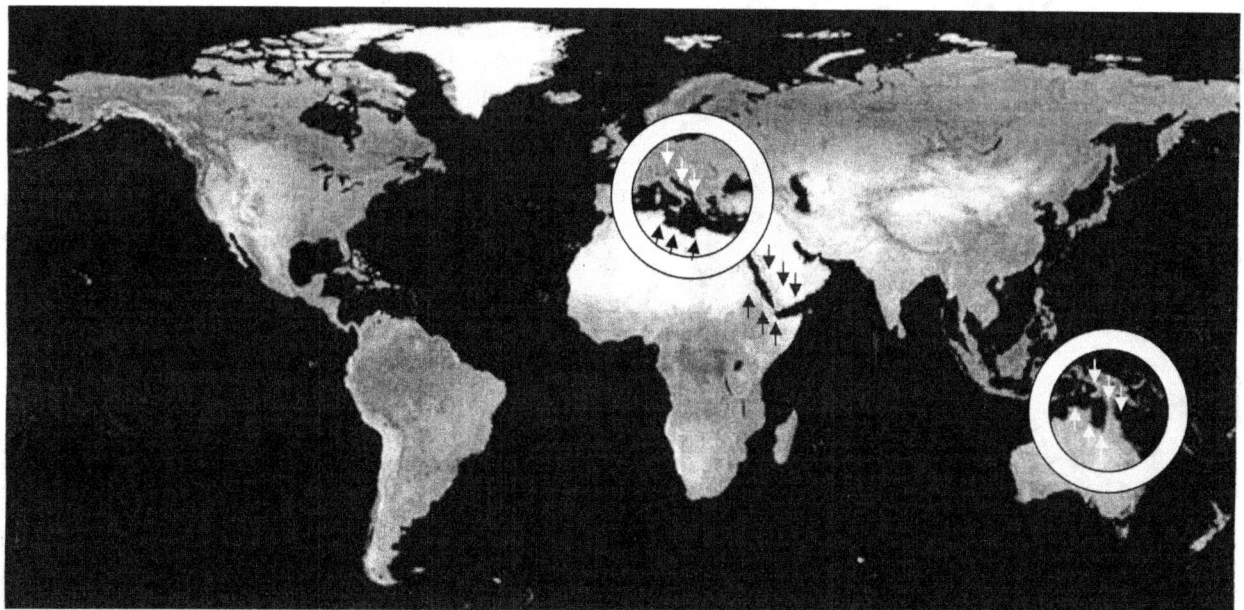

We can detect clear evidence that the Earth is cracking as the continents grow where Europe clearly split from Africa leaving the Mediterranean to form. I am not explaining this to perfect detail because that will take up far too much space considering the motive I have in writing this book. I just wish to mention some detail to use in support of my theory about gravity being heat moving to cold and by that margin the Universal expanding is shrinking material back into the abyss where it came from. Europe is splitting where some part move west from where other part move east and a lot is moving into Asia by moving north as India clearly does. Down south in Australia we find just as much evidence of expanding going on but since there is much oceans the expanding is less obvious.

Looking at the picture carefully it is very apparent that the land mass is breaking free while in other cases the land mass is filling in where land

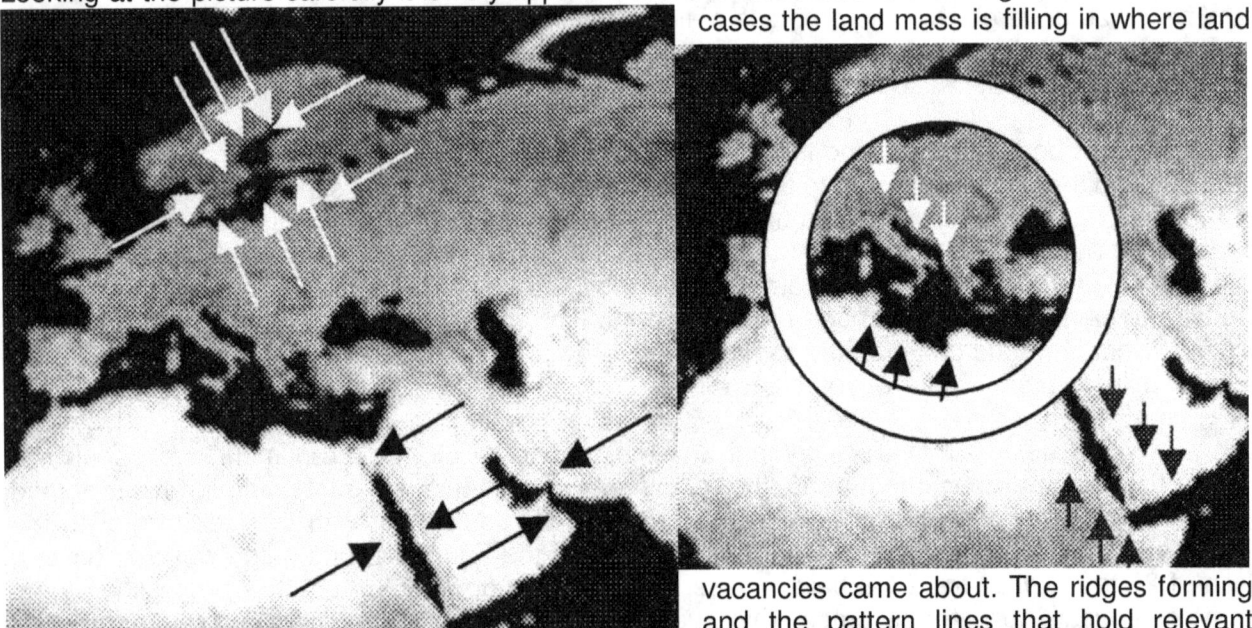

vacancies came about. The ridges forming and the pattern lines that hold relevant imprint and form in comparison to the other side are very clear. It also is true that some continents grew bigger and more rapidly than what the other area grew from compared to the area it broke away from, but that should be expected because I presume that is the reason why the split came about in the first place. Europe was a precise part of Africa but as the Earth surface grew larger, Europe detached and left Africa where Europe emigrated north. This followed the sphere lines expanding, as the sphere grew bigger and bulged more.

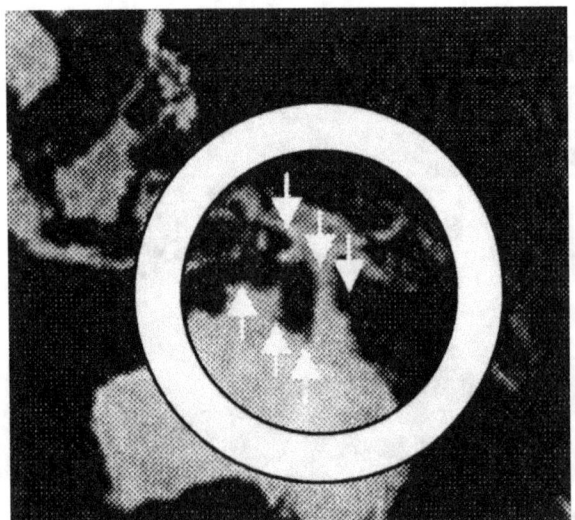

It is absolutely meticulously clear that Italy broke free from Africa leaving the gaping hole as evidence. Then Italy moved slower than the rest and turned to one side pushing into the continent at an angle and that forms the mountains we have as the Alps. My graphics I have at my disposal leaves nothing to admire but anyone is welcome to go to a chart and see the evidence I point out. It is clear to see how Scandinavia is splitting and to see where Italy once fitted in Africa but then moved away from Africa. Then Italy got left behind when the rest of Europe broke loose from what became the shoe of Italy. Sure, there are parts that grew more than other parts but the indentation still remains clear for all to see. The same one can see happened in the case where Arabia moved away from Africa and Arabia had much stronger growth than Africa had as it remained smaller when compared to the other side, but the resemblance remains obvious.

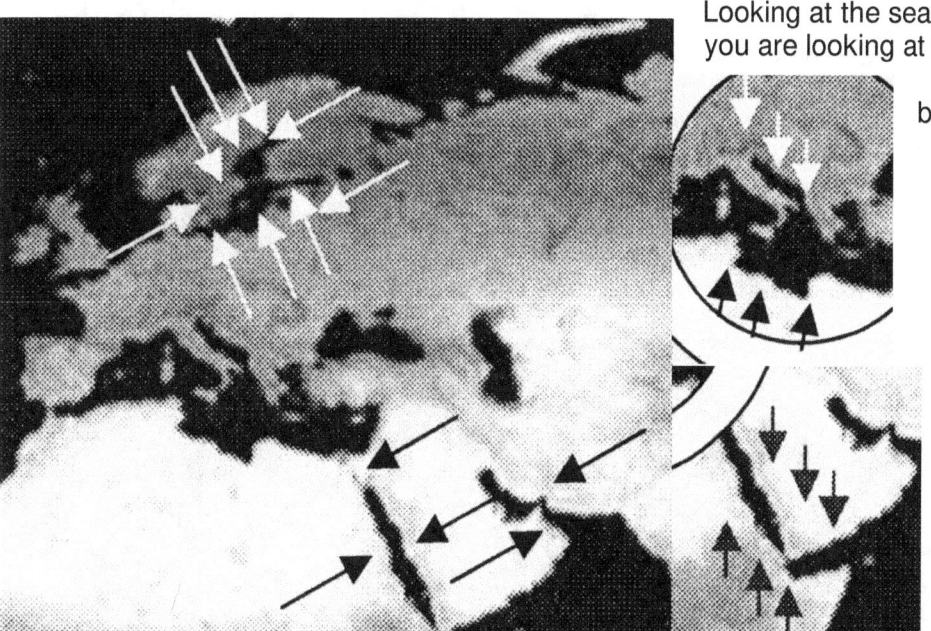

Looking at the seams of the Earth it seems you are looking at the stretch marks on the body of a mother after birth. One can clearly see where the Earth got loosened at the seams and it is obvious that some continents grow more than others do while where others are left behind some are in progress and departing to join up with some that stayed behind on the other side of the Earth. There is definitely a continental drift but this drift is coinciding with gravity bringing the Earth crust on a journey of growth and in sympathy with the expanding of the core underneath the continents.

It is clear where Madagascar broke away from Africa because the lines still remains

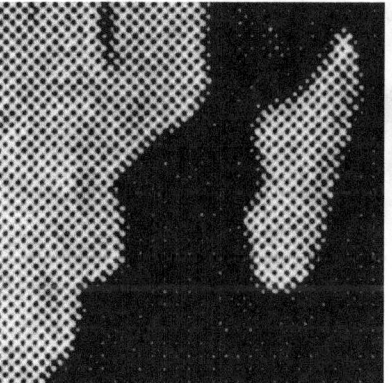

very explicit as the one side is a duplicate of the other side. The border resemblance remained very well kept although from the life differentiation and diversity between Madagascar and Africa it is clear that the split happened many, many millions of years ago. So too can one follow with some detail where Europe joined with Africa and one can trace old seams with a little imagination as too look past newly formed land in respect to land that remained fairly original.

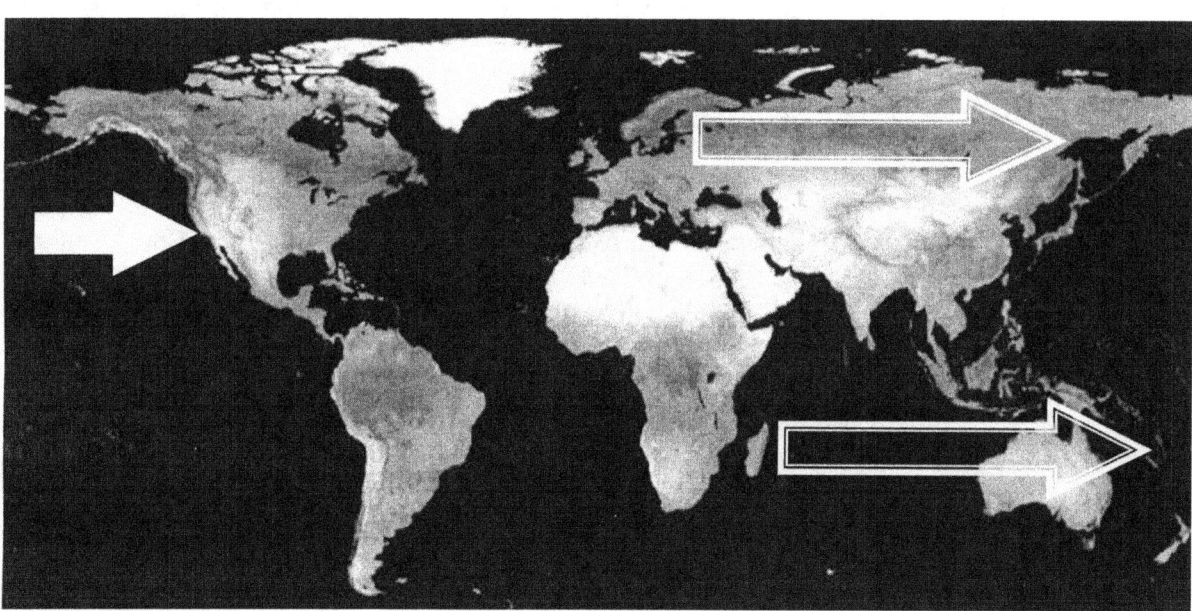

It is clear that there are directions setting trends in movement and for reasons that can only be attributed to core differentiation, there are borders where more splitting is taking place and other regions where mountain ranges come as a result of collisions coming about. This all forms part of gravity, of the physics of time forming space in relation to space already formed. This is where the future meets the past and how what was interacts with what will be and the evolving is in tune with gravity expanding matter. However in contrast to Newtonian science there is no generalisation to make it easy for the mathematicians to prove to others how smart their calculator operating skills is. The growth affects us all and this growth holds the origin from where weather patterns evolve.

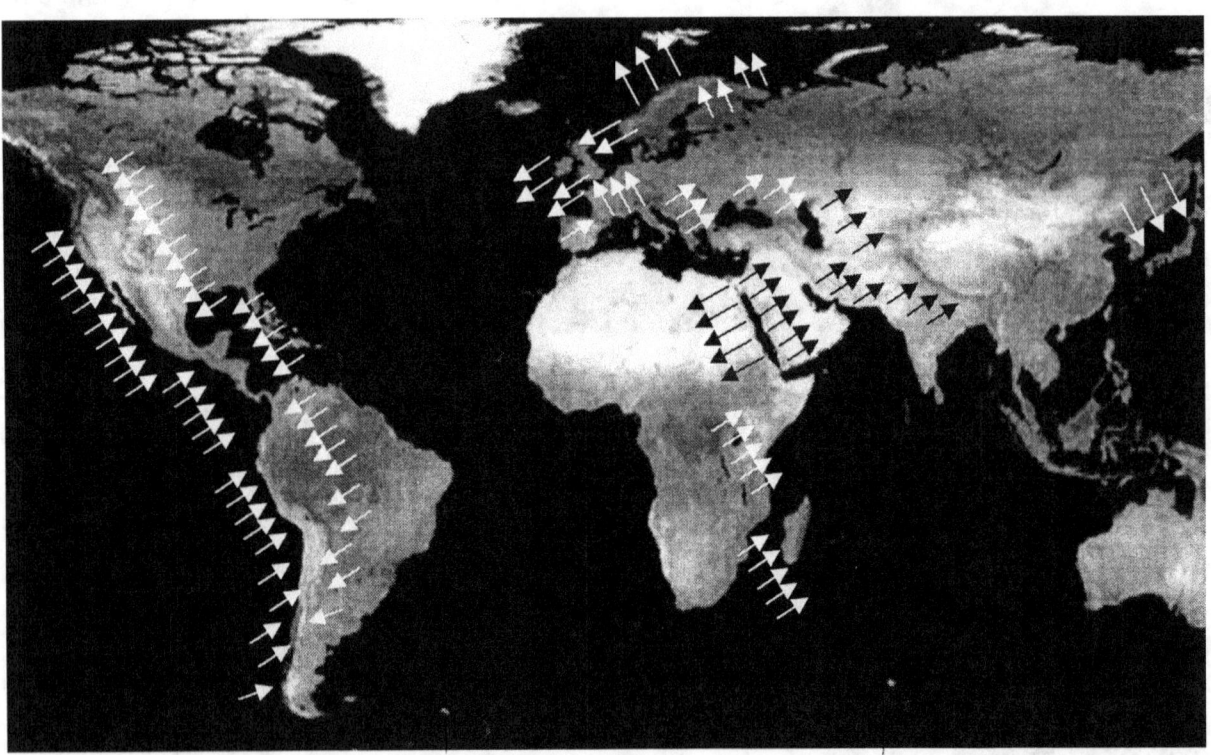

Some parts of the Earth is moving east and others are moving west and I think by now I have brought somewhat conclusive evidence to support my case about the Earth is growing and not simply floating on platforms because Newton promulgated there is contraction ands since not even the Universe may contradict Newton, there can be no growth evidence found in any form of science.

Now I wish to go one further and show how much understanding it can bring when reading past Newtonian misconception and when science makes an honest effort by trying to study the reality that is out there. Looking at the chart of America North and South, one notice a huge Mountain range that starts in Alaska in the far North and comes down forming the Rocky mountain range that then is interrupted in the centre part of America and starts building in size again as the range progresses down to South America.

However there is one part where the mountain range breaks off in size. To the very north of North America there are the Rocky Mountains that are truly impressive. Then down in South America there is a mountain range running on the rim of the western border of South America and this range is equally impressive and also is as devastating in terms of the climate conditions resulting from this. This ridge even comes out of the see and forms a land connection between two continents that has no connection at all. This ridge is where some of the most populated cities on Earth are located.

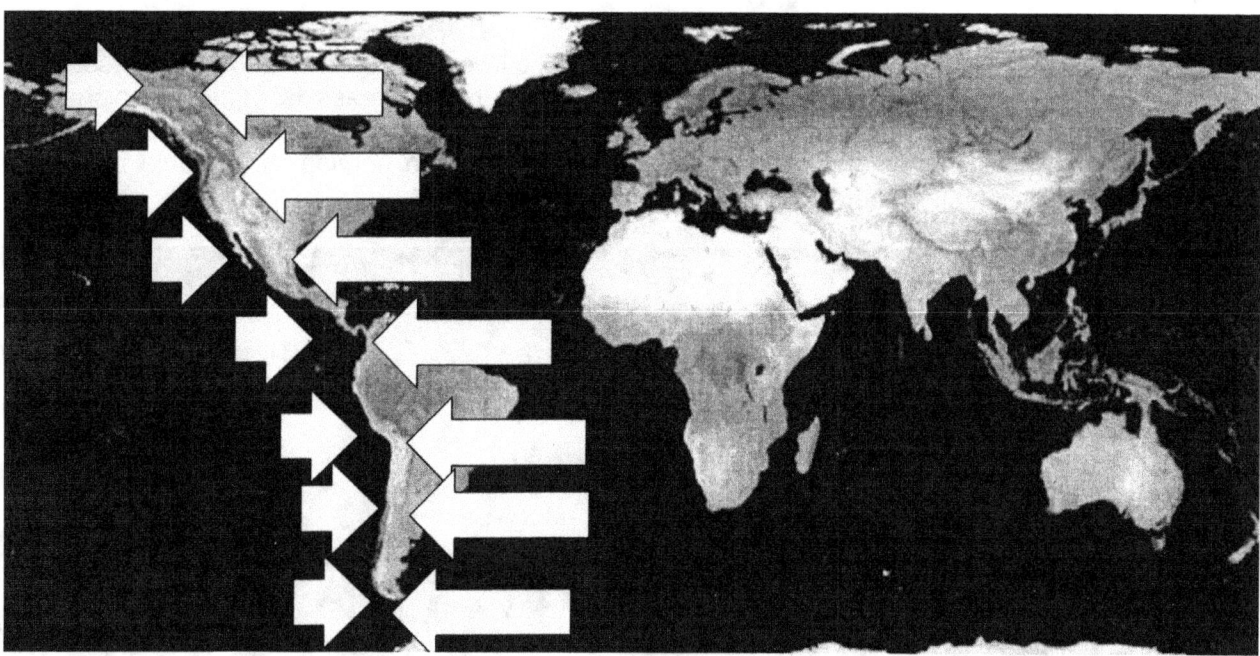

This line is where a ridge forms and is a result of the Earth coming from two directions colliding at the line that forms there. One may correctly think that this is where the Earth meets, just because all Yanks are under the impression America is the Earth and the rest of the Earth only forms small additives to America. There are other places where the same meetings take place, but since those places are not in America, those places have much less Hollywood fame and glamour and camera wheeling attached to the process. I include this to show those clever Americans that answers are normally obvious when an effort is launched to make people think and not run wild with cameras while they are stalking hurricanes and get paid to do it! Show any group of Americans a camera and you have a group of idiots on your hands. This line is forming points where that what is heading east is meeting that which is heading west and where else would it be than in America and moreover it revolves around Texas. Also this should bring perspective to the Grand Canyon being why it presents to the Earth, if this ever was intelligently connected by constructive reasoning to any particular region and this will include a well known region known as hurricane alley. There is a piece of land in Central America that is known as hurricane alley and hurricanes building up in the Bermuda triangle or there about come rushing towards America and then goes down this region in the flat lands of Central America. On the other hand also in this region there are other hurricanes that appear from the sky and just start to blow as if coming from nowhere and with the veracity the twisters unleash, they destroy everything and all that they cross paths with. The Yanks call them twisters because a hurricane can be located anywhere on earth making a hurricane unfit for America, but a twister is American and whatever one finds on Earth, those that you find in America is always dimensionally bigger, or stronger, or more fascinating, or harder to divulge, because the Yank is always exaggerating in order and by motive to impress…to be rowdy and brush and loud, and in Afrikaans we would call them a "windgad" which comes down to being a very unpleasant smelling fart that is caught in a bag, but this time their twisters are truly in a class of its own.

The destruction that these hurricanes can do, can match a Hollywood movie and that says a lot. But how do they form and what are they. Why are the twisters where they are and why are other regions with land filling not that badly affected as the persons living in Twister alley. There are other regions that experience as bad and as volatile hurricanes, but those hurricanes evolve from the sea and coming from the beach onto land gives the momentum that they have when they unleash their destructive powers. However, in the American Midlands the hurricanes come from nowhere to destroy whatever is in their path. There is no pattern that they follow and one minute there is quietness and calm and the next minute there is a twister leaving chaos.

Without having mass and in the presence of not having pressure the Newtonian wise and wonderfully educated are lost for knowledge. Remember Newtonians are smart when they have mass and where there is pressure but in the case of twisters, the twister lifts whatever has mass and therefore robs science of the benefit of working with mass. Also since the twister lifts mass high into the air, therefore Newtonians are left without pressure to calculate and not having pressure to calculate or to calculate with, while at the same time not having mass is equal to not having mathematics. This absence of having data to apply viable good old Newtonian maths leaves the Newtonians in a state of being mindless where they have all the nothing that should be in the twister, located between their ears and filling their thoughts.

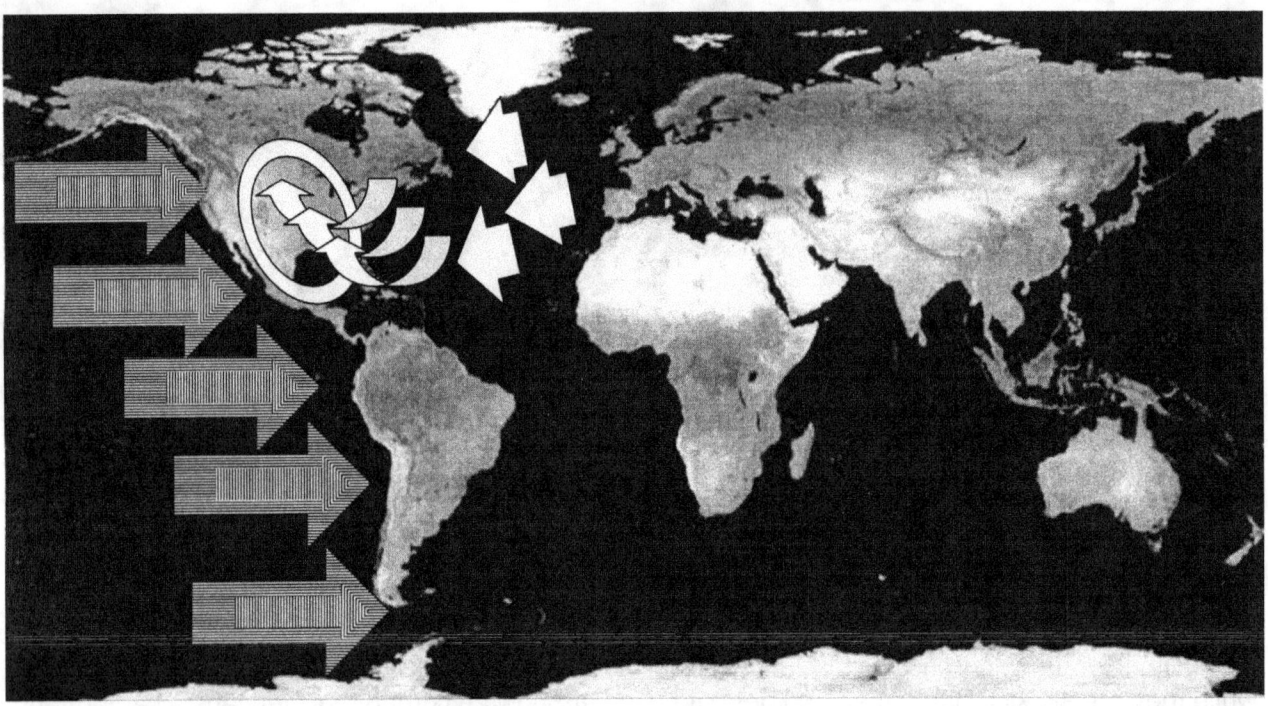

Going by the Newtonian's intellectual structure they would reason that by not having mass requires having to see where the mass goes because not having mass nullifies all other potential understanding that is covered by the high degree of Newtonian education those Newtonians were abused with. There where all the so many hurricanes develop are persons with Doppler radar equipped trucks chasing down potential twisters and these camera- wheeling neck-break-driving lunatic behaving individuals has an occupation going by the name of twister chasers. They chase twisters while running with cameras after hurricanes in the hope they can push this camera up the funnel of the Hurricane… to do what…? Is their bizarre behaviour the result of trying to show how brave or stupid the average American twister-chaser is? If anything of notoriety to the Newtonian intellectual has no detectable mass in the absence of pressure they go insane. They will push a camera up the snort of the hurricane to see where the absence of pressure took the lost mass and then they will be able to send a report to inform head office that those more intelligent persons that are wise enough to stay away from the rain sits and wait for the mad clang to inform them. The information the more sensible group that sits ain a building receives they then use so that they can send other dumb brainless idiots to where they will locate the lost mass in the absence of pressure.

It is only in America where one will find someone stupid enough to try and get a hard blowing wind caught on camera while they push this lens up the arse of a twisting wind that is mind boggling –strong and try seeing what the wind looks like up this funnel of air. Can any person with a mind set between their ears, get more stupid than that? Someone should gently inform them that wind is not photogenic, be it in normal weather or up the funnel of a hurricane, it still remains all the same.

Their stupidity will only bring Hollywood interest and further have no educating results. The evidence they are looking for is not in the hurricane but it is in the ground and it is gravity. Looking at the hurricane we see how singularity is charging the weather. When the top spins the top becomes a hurricane. The top takes charge of singularity and by spinning it entices a line that activate innumerable lines to manipulate the top's stance and by manipulation in movement these tiny space less lines are the actual substance that takes charge of the top's atomic substance and these innumerable lines gives motion direction to the atoms forming the body of the top and by these lines manipulating movement, the time that sets the top in motion, places the procedure of movement whereby the atoms would be placed during the time of spin. It puts timing to the atoms of the top by giving the atoms a past position where the motion is coming from, a present position being at and the following future position where it is heading. This is gravity. It is the directional flow of time providing motion to space. The lines are in sequence with the Earth rotating and the top manipulates the Earth's directional singularity.

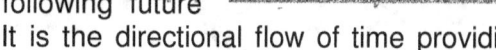

$10/7\ (4(\Pi^2+\Pi^2) = 112$

$7/10\ (4(\Pi^2+\Pi^2) = 55$

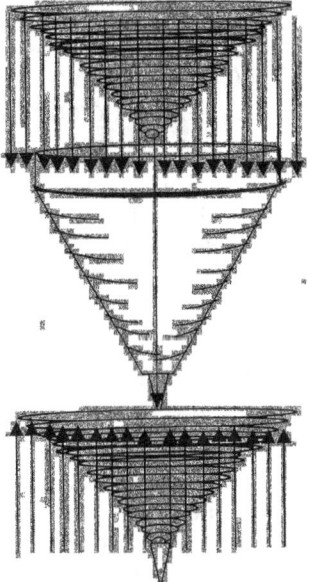

These lines are there because the Earth holds the lines in relation to the Earth governing singularity. The lines form as the Earth contracts heat in the direction of the Earth point holding the governing singularity. The lines carry heat or cosmic liquid from outer space $10/7\ (4(\Pi^2+\Pi^2)) = 112$ to the iron core that spins within the centre of the Earth $7/10\ (4(\Pi^2+\Pi^2)) = 55$ and this process is as much electricity as it is gravity as it is the atomic displacement that is pumping the heat by measure of nuclear weak forces through the Earth in a display of conducting heat and this charge of displacement of space-time delivers a quantity of heat to the centre of the Earth.

These lines hold no space, take no space, have no claim to space and are much more numerable than the atom could ever be because every substance within the atoms is singularity presented by such a line. The atomic makeup consists of innumerable lines and these lines eventually form space since these lines maintain time and time is gravity, which is the movement of space in relation to one centre point that has no space.

These lines form the three dimensional six sided neutron ruling Universe we live in $7/10((\Pi^6)\div6) = 112$ as they charge heat from the most expanded heat limit $10/7\ (4(\Pi^2+\Pi^2)) = 112$ to become concentrate within the iron magnetic fields within the Earth $7/10\ (4(\Pi^2+\Pi^2)) = 55$ by using the atom singularity relevant composition formed by the construction of the atom as $(\Pi\ (\Pi^2+\Pi^2+ \Pi^2 + \Pi +3)) = 112.31$.

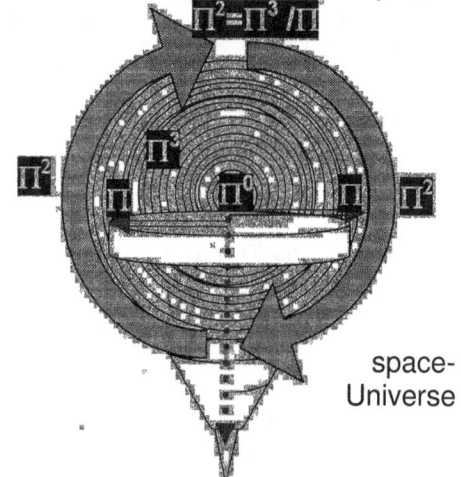

$\Pi^2=\Pi^3/\Pi$

Π^3

Π^2 Π Π^6 Π Π^2

space-Universe

This recipe keeps the top spinning up straight. This recipe provides the charging of electricity. That atom to displace time uses the recipe. This recipe is the gravity confirming the in conformity and keeping everything moving in form.

As it happens to all good things, there is an oversupply reaching the inner core of the Earth. This leads to a discharge and the discharge amongst many forms it takes on is to fill the atom with material that consists of such points as just represented. This makes the atom swell and that makes the Earth swell and when the Earth has its fill it is because the atoms would not absorb more heat. The atom carries some of the heat in a jacket around the atom and as the atom spins in gravity Π^2 it holds space Π^3 in a relevance to singularity Π^0 by placing the relevance Π in accordance to forming space-time in relation to singularity applying $\Pi^0 = \Pi^3 / (\Pi^2 \Pi)$, which then is $\Pi^2 = \Pi^3 / \Pi$.

All this information is available by just consulting the Bible and there is no need to push a camera up

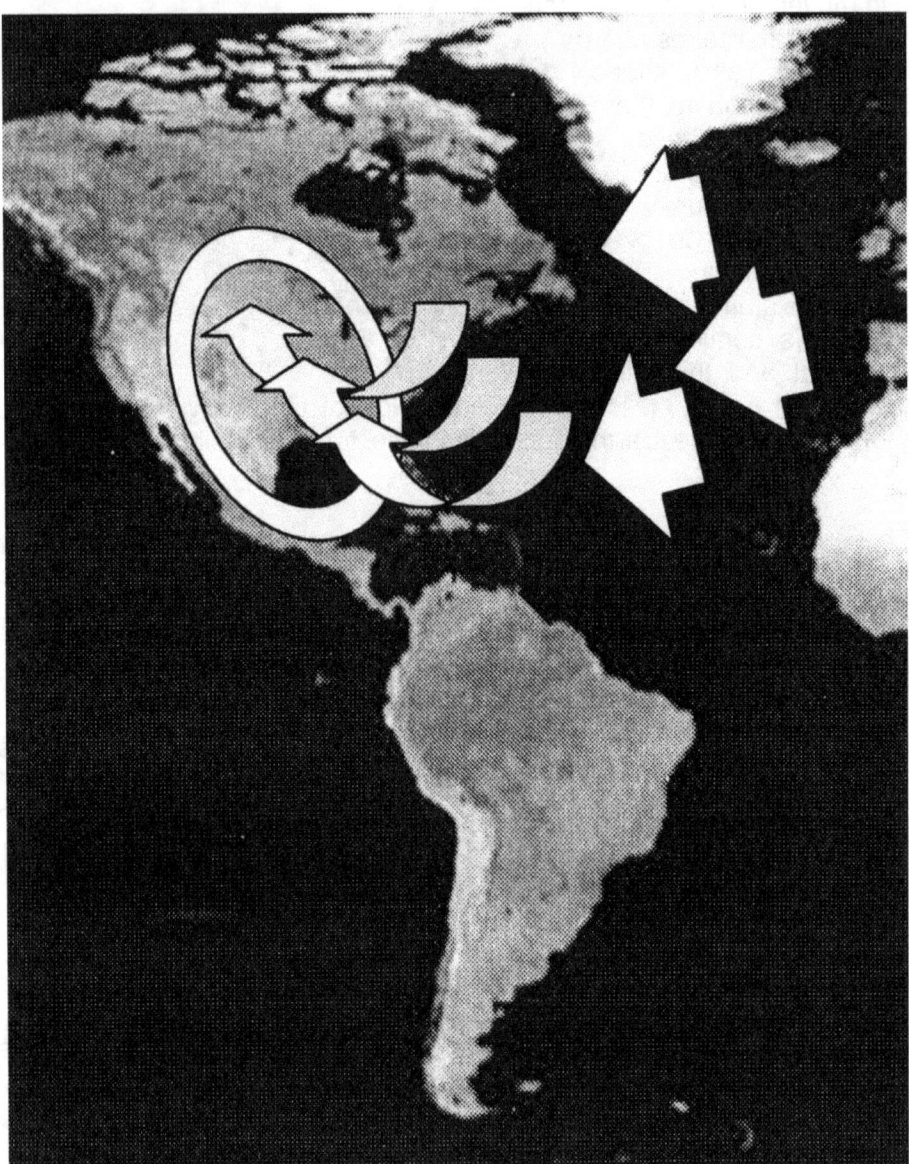

the arse of a funnel within a raging hurricane. Those poor stupid Newtonians that is unable to think…and see how much trouble that is costing them to experience. Time is the movement of space. Instead of pushing the continent further at hurricane alley as the movement of space does in the Northern regions forming higher mountain ranges than what is in the flatlands of Central America or forms the incredible mountains in South Americas. The movement in the Central part of America rises above the surface of the Earth and form hurricanes in order to release excess heat gathered by the Earth core. The reasons would be found in the material forming the inner core at that point. It is my guess but I would say that the core found north and south are heat containing material while in the centre the core material has much better conducting characteristics. The north and the south push material in collision while the centre material releases the heat in movement to form wind and storms.

Newtonian wisdom furthers this misconception that fires consume oxygen. Newtonians always claims that a fire ignite when oxygen is introduced into the system, or the fire burnt out because there was a lack of oxygen and humans need to burn oxygen to stay alive. It is so often said that a person dies of oxygen deprivation meaning the death was the result of a lack of oxygen supply to the lungs and never, not once do I hear even one Newton explain this statement. Saying that implies that without the body inhaling or consuming oxygen, the hapless person dies. This scientific statement suggests or better put, the statement declares that we burn oxygen and that by burning oxygen is how we live. Our blood carries oxygen to our muscles leaving the suggestion that the muscles take the oxygen and use it for the consumption with food as fuel. It is said by Newtonians that without devouring tons of oxygen leaves life lifeless and that is a farce as big as any other Newtonian farce can be! Also the connection is that in the case of fires burning out of control there are massive winds bringing in oxygen to fan Australian and Californian fires. That is a myth as big as mass is and as big as pressure is, and therefore by nature it is very Newtonian by heart. If fires devoured oxygen in order to flame fires then by this time all oxygen should be history. Think through all the billions of years the Earth went through and the trillions upon billions of raging fires there was during that time spread all over the Earth, how much oxygen then was consumed by all those fires.

There is a limited availability of oxygen and if oxygen was used as fuel, the oxygen by now must have then have been totally spent. Oxygen is only the carrier pigeon of heat and transforms the heat it carries to carbon where the carbon holding life uses heat, not oxygen to maintain a temperature, which is necessary for sustaining the presence of life in the cell. The wind blow in oxygen and the oxygen transports heat that flame the fires. The oxygen (8) work in tandem with carbon (6) and nitrogen (7) to flame or energise life as it also flames wild fires, woodland and forests that is all carbon life.

Never does the fir start burning the sol in such a devastating all-consuming fire because with the Earth being soil, it is unable to burn. The reason for that is the life is mimicking the transfer of heat via oxygen to carbon and carbon transmits spent heat or excess heat via nitrogen out of the system. In the star this will works just the opposite where the carbon would hold the heat and have nitrogen transport heat to oxygen and with oxygen it can then be transported to the inner structure of the star I state this to show there is a constant unabated interaction going on relentless or is never interrupted between all particles and heat. Every particle has a different relevance with heat and as I showed, nitrogen expands heat while carbon stores heat and oxygen transports heat.

Here is the chain of heat carrying process within the star that life also implement and that is transferring heat to carbon where carbon discharges the heat to nitrogen and the relation that nitrogen has to heat is always volatile because nitrogen has the proton number of 7. There is also another relevancy applying between atoms and this relevancy goes in relation to light or if you wish to call it electricity or gravity it is all the same and the way this behaviour pattern is relating to singularity. It is in this ratio we find atomic growth ands within huge stars we find atomic substance decline but the decline I am not per suiting to explain any further in this book.

The release of heat can take the shape of two forms where one is forming whether patterns and wind and the other is having molecular structure increase. I think it is clear to see that an atom such as iron should have a far more intense relation with heat as the iron atom displaces far more heat per volume of spinning by having 56 protons to say the 6 protons we find carbon have.
Then there is Cobalt with $3(\Pi^2+\Pi^2) = 59.21$ and at that point the nucleus start to refute further charging of nuclear adding. At the value of copper we have $\Pi(\Pi^2+\Pi^2) = 62.01$ and singularity borders Π are met and encountered within the atomic nucleus $(\Pi^2+\Pi^2)$.

EXPANDING Universe
Any model Universe in which the space between widely separated objects is expanding. In the real Universe, neighbouring objects such as close pairs of galaxies do not move apart because their mutual gravitational attraction exceeds the effect of the cosmological expansion. However, the distance between two widely separated galaxies, or clusters of galaxies, will increase as the Universe expands.

KANT, IMMANUEL (1724 – 1804) was the German philosopher, which proposed a cosmogony, published in 1755, in which the Solar System forms, via a disk, which condensed out of primordial material. The Solar System was part of a larger system (what we would call a galaxy), and many of the nebulae seen by astronomers were in fact other galaxies, which he termed island Universes. Kant was influenced by I. Newton's theories, which he termed island Universes. Kant was, as everyone ells up to now, influenced by I. Newton's theories and by the English philosopher Thomas Wright of Durham (1711-86).

I think before we go into philosophising about the Universe it would be prudent to define what the Universe is. Every one is roasting a chicken about the Universe, but what is the Universe? I have no come across one that could explain the Universe.

To me everything makes perfect sense and while saying this I do admit full heartedly that I am not a Master such as yourself with the knowledge you possess. In that light, should you feel there are aspects I do not explain to a sufficient standard, I am willing to work on it.

What is the Universe? This is such s simple question every one gets wrong because of the relevancy we humans place on the Universe and the relevancy what the Universe truly is. Official Policy Protectors really get tide in knots with making all about nothing so complicated it absorbs everything holding back nothing.

I say that the Universe is the ratio of unoccupied space-time (the liquid) forming occupied space-time (the solid) and the Universe is formed by a ratio between time and space. Now we better find this ratio and see what the ratio is and if outer space could be nothing after all?

Firstly singularity expands from Π to the seven positions it hold in material
Singularity explodes into space having a value of a^3 as Kepler introduced or Π^3 if the singularity value is used. But that does not introduce the six-sided Universe with the dimensions we came to know. In the process of making contact the Π^3 becomes another Π^3 by abandoning the one side of the Universe and because of the moving of singularity Π^0 through Π^0 a situation develops where the seven points representing Π^0 receives another three pints in motion as the points cross over to the other side of the Universe. Remember the Universe in singularity is a flat Universe with no sides at all and the slightest motion of any point provides the point another Universe to move to. It is not the 180^0 we see as a straight line and at the same time it is the whole issue of 180^0 in all three forms of Pythagoras that forms the issue. The slightest motion becomes a most deliberate crossing of borders since there were no borders to cross and the movement provide the borders to cross. Then by motion of seven dividing into ten and on the same subject the ten dividing into seven as the crossing is implicating singularity $\Pi/2$ the ten and the seven form an alliance that brings about the value of Π^2 in relation to singularity as Π^0

Newtonians underwrite the notion that the Universe is a compliment of matter, ranging from as little as a photon and radiation to as large as a Black Hole and Galactica. Still it is a compliment of matter, holding space. There is no open Universe and there is no singularity running free in the wider Universe looking for some space to be within. Newtonians use all sorts of silly ideas because they want the Universe to fit into Newton's science because Newton's science can't seem to fit into the Universe. There is only one Universe holding many, many, many compliments of parts in the form of atoms to one space that does not even exist. This means every particle holding singularity is the Universe and the complimenting total of that is the Universe. Light is only part of the Universe and not time as such. Light in as much as the photon is space in the form of heat that we can see and space in the form of heat is light we cannot see because it diverted from our direct line to singularity. Light is the smallest particle that maintains singularity while running away from a mutual singularity amongst photons to join a singularity within larger matter. The darkness of space is light we cannot see. Light is the darkness of space we can see. There are three components in the Universe, one is time, the other is space and the third is matter. Looking through the looking glass there is only matter that matters because we can feel it and matter that so far did not matter because we cannot see it. Then Einstein brought in light as pure energy and with that he was correct up to a point. Einstein led us to believe that energy was some fluid magic where "gravity" on the other hand was some solid magic. If that is the case, then what is gas, magic? If there is two of the three forms identified the third form also must be somewhere. By introducing singularity as the third magic is surely not on. He introduced singularity but never identified singularity. That brought about that singularity was also some witches brew of magic and ghosts and all the rest. There is not even a Universe at large because take away matter occupied or unoccupied and you are left with nothing. That is precisely what space is. Space is nothing. I have a huge problem with the fact that

Einstein saw singularity, but never could place singularity, while the position of singularity is so very obvious. I am very aware of the fact that it seems to me as if that any Newtonian that was brave enough to endure my opinion about Newtonian religiosity up to this point will now by this time be enraged, steaming with anger and with me not being present that will even further their feeling of frustration. I'll even go as far as to presume that by now every Newtonian that is pure at heart will only insist on my immediate execution. At this point I have Newtonians stretched to all limits with my going on with the second largest Newtonian that ever walked the earth. Newton was off course the biggest Newtonian of all, and the rest of science filling all the other positions from top to bottom according to personal importance and devotion. I truly consider the world of science a conspiring collaborating Brother Hood. Blame me if you wish, but the Newtonian religion and devotion to Newton has no bounds. Every one is supporting the other while covering for the group as a whole. If they wish to castrate me for saying that, then answer some simple questions No Newtonian knows time or space by definition or place, yet they want to bend, stretch and tie it in knots. I should think they must first locate the objects before they intend to torture it with such tenacity. What is the Universe and where does it end? That is questions that one should address before going on to the more significant major issues like time travel and galactica infesting. The Universe begins inside every atom and ends where the atom's influence on space ends. The Universe runs from atom to atom and the lot holds the Universe in line with individual singularity. The Universe is the total of all heat holding space occupied or unoccupied in whatever form heat may be. The Universe ends where one will find the last of matter occupied or unoccupied wherever that may be and that will be nowhere because of the vastness of the Universe. The centre of the Universe is in the centre of every proton, where singularity is. There is no collective universal centre because the Universe is holding together atoms, which is holding together space in different dimensions.

Xepted science (Xepted science is a term that I invented that I use when I refer to science that is totally incorrect but that is nevertheless still accepted) has that also the complete opposite to what light is. No object can at the present geodesic space-time be in motion at the speed of light or to any related value, therefore no object can move towards the Earth or away from the Earth at even a fraction of the speed of light. What this indicator reveals is the geodesic space-time value during that specific era when the light was permitted to brake free from the time concentration during that point. The higher the geodesic time-space value was, the more time was to space and the less time was committed to matter.

In the beginning, there was time Zero to moment Alpha. There has never been a Big Bang, as such and there were too many Bangs and too numerous to count. Everything is a variation of time duration in space. During the period of time Zero to moment Alpha the value of 1 second was equal in duration to about 1 000 billion, billion, billion, billion years (I am only stopping with the billion part in order not to bore the readers), measured in geodesic space-time values that currently applies. It could be even billions times this duration because the value of time then, was measured far beyond the speed of light, since light did not yet exist. We have no way to calculate the duration of time that applied during this geodesic space-time era, except to put it down to eternal. When a bowl of soup is boiling, have you seen the bubbles of air rising from the soup? Has any Newtonian ever taken the time to explain that process in detail? I think not, because such explanations would be far too "everyday-like" to bother their mighty brains. To find the mass of Black Holes and finding the mass of the neutron, that is what such mighty brainpower can cope with but it cannot bother with small events.

Well, that boiling soup tells the complete story about the creation. Poets and painters and writers always wish to say how "they created their creation". That is rubbish; they created nothing. They brought nothing new to the cosmos, they only rearranged what was a small part of the cosmos into a new order, that one can detect a distinction from. Creating is producing what never was before. When looking at the boiling soup, one sees bubbles rising from the soup at the top. In the soup's brew, there are only liquids and solids before the heat came. No one placed air in before the event or during the event at any time. Yet, from the increasing in heat, the subsequent reducing in density by the increasing in used space solid rises to liquid and with more increase and density reduction it finally forms gas, or if you wish to call it pure space. That space increasing brought about space that was not there previously. That space was created by the density of the space that reduced.

That space is energy en energy is the interaction between heat and space. As space becomes a part of the soup, a part not there before, with no room to be, it moves out. We refer to that process as boiling. That space creation is applying heat to time, and time in singularity will respond as space in singularity. The space created will vanish just as it came, back to singularity. By applying heat to time, brings forth space, and from the three components, only the heat factor is not in singularity. It removes space in singularity from time in singularity to establish room (space) for heat (time).

It started with a dot, because that is the only form, size and dimension mathematical logic will allow our brain to accept. From the one dot came a second dot and a third dot. The dynamics of such a dot is smaller than we can understand because such a dot is in negative relation to what we see Π to be, and the deeper we delve in finding the smallest fragment where space started, in the spot where time is still eternal as much as we can accept eternity to be. This we find in the aligning of planets where the one dot from which the aligner stem becomes the reference too the distance applied between the aligner and the original dot, or governing singularity or structure in charge of holding position to all orbits following. The reason why we should first locate the spot is because we can only work from that point forward. By working forward we have to work backwards to locate where we are heading. The cosmos started at a point and where such a point is, we will find the Universe. Every one knows where the Universe is, because we can see where the Universe is, but if we can see where the Universe is, then we should find the centre of the Universe in that spot. Einstein theoretically positioned the point of beginning at a place he indicated where singularity should be. With the cosmos the size it is and space so large compared to our smallness we have no chance in finding the centre of the Universe. The Universe started where singularity is and singularity is the sure indicator of the Universe. With all spinning objects holding singularity we then have located singularity in as much as finding the centre of the Universe. The Universe started with a dot forming. That answer arrive from taking mathematics back to a point of being the smallest possible position, far smaller than we may be able to calculate form.

My approach might seem unconventional but through the abandoning of the accepted, it enabled me in locating the precise location of a universal singularity forming a connecting basis of the Universe (this I say with some degree of confidence). The smallest figure there can be must be a dot. The dot is the only form that leaves all the options open to extend in any and in all directions should the opportunity arise. The only mathematically sensible option about extending a line from the dot will be non-bias progress in all directions equally in order to give a meaningful flow of mathematical equilibrium.

The Pythagoras mathematical principle is the proof and that I explain. The obtaining of singularity is in my rejecting of nothing by replacing it with something being the dot. With the clepsydra or "water thief" Empedocles deducted that air was composed of innumerable fine particles, breaking the thought that what we now know is air, was also believed to contain nothing being altogether a space filled with nothing until proven to be wrong so many years ago. Never did science take the lesson learnt back then to the future and out onto outer space. If there is space, there cannot be "nothing" as space is something. The claim becomes obvious when observing the connection between the half circle, the straight line and the triangle, which could also promote all the qualities lurking behind the pyramid. Consider the connection between 180^0 sharing and then one may realise much of the pyramid mystique becomes less spectacular in considering the very basic in mathematics being the Law of Pythagoras on which all mathematics are focused. Once the water thief was eliminated by some human intelligence the matter was left at that. Nothing shifted out to an area we think of as outer space. In outer space we now find nothing. There is nothing but an atom here and there and even the atom is covered in nothing.

I wonder why the nothing landed there. Could it be that the reverse came about and because there was no visible "water thief" the very limit of man's suspicions came into practice. Man has always been extremely good in flying from one outer edge to another and if the water thief proved something was present, and then the mere absence of a water thief must therefore prove that nothing must be in outer space. But what is space as such. What can space be, because with explosions we can clearly witness space created from heat. Our culture prevents us from admitting our vision, but the release of heat produces a *"shock wave"*. That *"shock wave"* is nothing less than space created from heat released. We have to brake free from culture of the past and a rigged mind set narrowing our vision.

There is no doubt that the atom forms the Universe. The atom is the Universe. The atom represents material and space represents liquid and the Universe comprises of material and liquid. Material is occupied space-time and liquid is unoccupied space-time. Unoccupied space-time can become occupied space-time as soon as material fills the space during time. If it is true that the Universe is the atom we find the question as to what is a star. The star is a collection of atoms. The star is a cosmic atom and represents the movement of all the atoms combining in the star. Newtonians proclaim that the Sun is only hydrogen formed by mass which of course is as Newtonian or Neanderthal Cave-age wisdom as anything ever said to be Newtonian. Firstly a star has to have an Iron core. That is to generate gravity. The iron core has to work in conjunction with copper. That is a necessity should the cosmic atom or better known as the star would have the ability to generate gravity. Gravity and electricity is the same thing however the scale on which it works is different. Electricity and nuclear power is the same thing but the intensity on which it works is different. Atomic gravity or weak forces are the same as gravity or strong forces because it still remains gravity produced on different scales. Gravity is time and time is the movement of everything in relation to one point. The atom controls the star and the star controls time therefore time is controlled by the atom through the star.

Time is the dispensing of heat in favour of space and the Universe is cooling down dramatically as it is overheating. This is not my way of double talking but it is dispensing of Newtonian single mindedness. If it was 10^{37} K at the moment of the big Bang first taking place, then the Sun was super frozen at 6500°C. What then was the temperature of the Universe? Today we think of outer space as 0°K, but at that stage 10^{37} K was what is today 0°K. When it was 10^{37} K that was the temperature of outer space and everything that is particle was in a frozen state below 10^{37} K. The concept of heat and cold is as Newtonian as mass and attraction.

It is Newtonian single mindedness to think of conditions only in terms of what is befitting life and comparing everything in terms of conditions that only apply when conditions fit life and there is only one known place that to befit what life requires. That is as Newtonian as Newton's concept that everything stands still until a force moves it wile nothing in the Universe can ever stand still. To think that way is to think in terms of the dark ages before renaissance brought light to the human mind. Hot is that which can expand no more and cold is that which can contract no more and the cosmos is everything in between the two poles. Singularity forming outer space can expand no more because time is constantly allowing heat to expand. Singularity in infinity formed inside everything that spins can reduce no more because t is as small as whatever can be.

This is quite in coming into place within the gas structures and is starting to apply in the solid stars.

All the while the atoms have to comply with the rules within the star as demand on the atomic space claims sets new standards. At a point the reducing of space becomes so demanding that the factor of light finding an ability to apply motion disappear as the massive structure draws even light towards the centre because at that stage the photon no longer had the ability to duplicate space as it displaced space. The photon must then surrender space due to a lack of adequate motion applied at a relevancy of 56.6. At the beginning when a star establish independence from the galactica outer space which is more dense than the star itself, it is the totality of all the protons working as a group within the secure unit that dismiss space-time and the total displacement finds a focus in the centre of the star. At first the star may only demand a reducing focuses on the 3 or the Π to become independent and secure defined borders or atmospheres but as the star develop through the intense centre it forms, the protons will grow and bring about through fusion much more active displacing that eventually forms fusion. The more protons there are in the least space there are will bring about the strongest gravity there are. As the star development progress the dominant gravity generating protons found in one location begins to form within the centre of the star where the major heat is accumulated.

The shift takes place from the focus at first on the outside rim of the star developing towards and then to the middle sectors and eventually to the centre of the centre. In this the focus of the displacement gradually moves from a massive number of single proton atoms to a massive number of massive atoms. The quantity of protons efficiency move over to form a focus on to the quality in proton numbers in one unit of a centre and then the dominant atom displacement will not be Π but it will become Π^2, later $3\Pi^2$, $3^3+3\Pi^2$ and so the centre progressively develops.

⚫Then further premiums on the space-time that the individual atom may require becomes resolved as the space demand within the star annihilates all atomic motion of individual atoms and the neutron, which is the representing of motion in the atom, abandons the unit of the atom to reproduce space in the manner the photon did in normal stars. In the end there are by then no hint of any photons left because a lack of motion brought on a total demise of photons. The star is not dead! Eventually only the collapsing of space can sustain the proton activity still present in the star as singularity sets in and diminish all motion activity within the star.

As explained on the previous page this mass comes from the fact that the proton lags in motion to singularity, which is motionless. Explaining this concept or the following concept will also take to much time but what I mention in this page I have a book of more than five hundred pages that covers the whole aspect and has the name of **AN OPEN LETTER ON " STARSSTUFFN' "** ISBN **0-9584410-3-0.**

 I will just quickly touch on the thought. What we see from the outside is just the opposite of what is applying on the inside of the Universe where the "inside" is singularity. Considering it in that light that is why all the information we receive from the Universe by means of light is a mirror reflection of what is taking place. I shall quickly mention the most basic idea of this concept: The motion produces time and the time brings about space.

To us being in space-time the forming of space by using time is a positive measure because we are on the side in space-time, but from singularity such motion becoming space is a disaster coming into practise and it will take the Universe many billions of eternities to once again correct the disaster. To us the motion is quicker that eternity but to singularity the motion is slower than eternity. By creating eternity minus whatever that deduction slows down eternity where as from our perspective it increase eternity by splitting eternity.

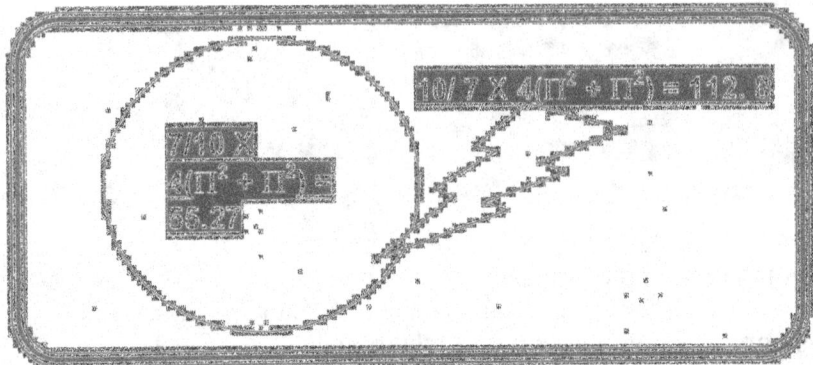

I have shown that liquid hold ten places or position or dots when material or solids hold seven. Gravity is the concentration of heat running from outer space with a displacement value of 112 to the inner star that has to have a displacement value of at least 55 to ensure the flow of gravity generated by the motion of the object. With an inner core displacement of less than the required 55 the star would not yet have arrived at the point of securing an individual singularity in the presence of outer space at 112. The potential difference needed to generate gravity is 112 coming down to 55

Liquid relating to solids is 10/7 while solids spinning in liquid are 7 /10. This puts the atom in relation to outer space at 10 / 7(4($\Pi^2+\Pi^2$) where outer space forms time and going to form a cube or solid 7/10 (4($\Pi^2+\Pi^2$) where solids accept time.

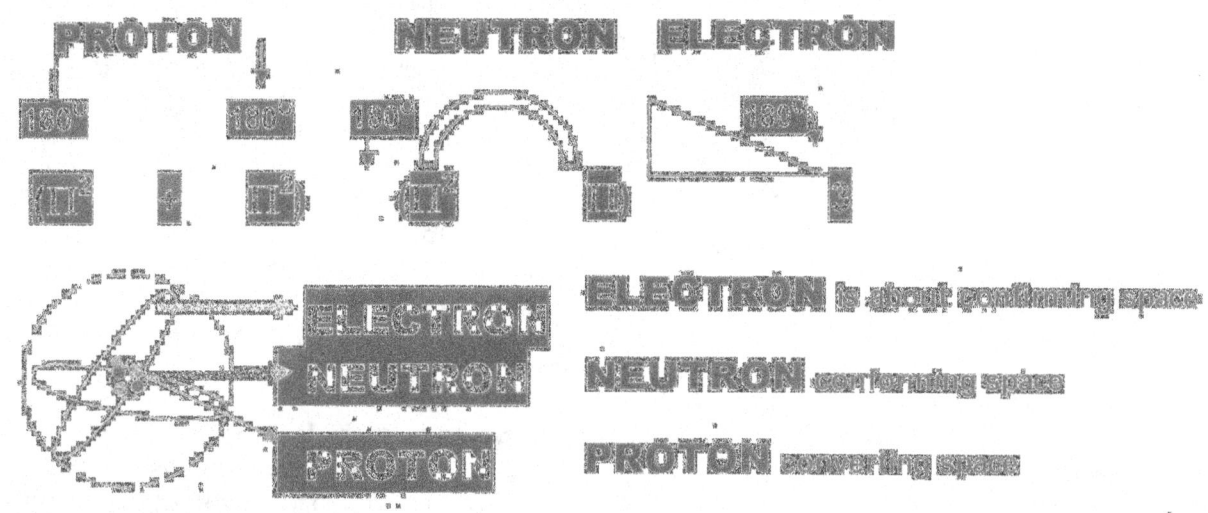

The proton is (($\Pi^2 + \Pi^2$)=19.74
The Neutron is (($\Pi^2 \Pi$) = 13.01
The electron is 3

We can see when the dominance started creeping to the other side and when $k = a^3 / T^2$ got the better of $k^{-1} = T^2 /a^3$ because at a point where the sum total related to the singularity the **proton ($\Pi^2+\Pi^2$)+ the neutron ($\Pi^2+\Pi$) + the electron 3 = 35.89 × singularity Π =112.75 outer space.** Putting singularity into space by extending Π^0 is (7/10) $\Pi^6 \div 6$)) = 112, and that confirms that through singularity the atom contracts the cosmos into a cube by virtue of the atom.

Past such a point the expanding factor began to gain lost ground and the expanding got predominant as the containing factor started to store and preserve more than contain.

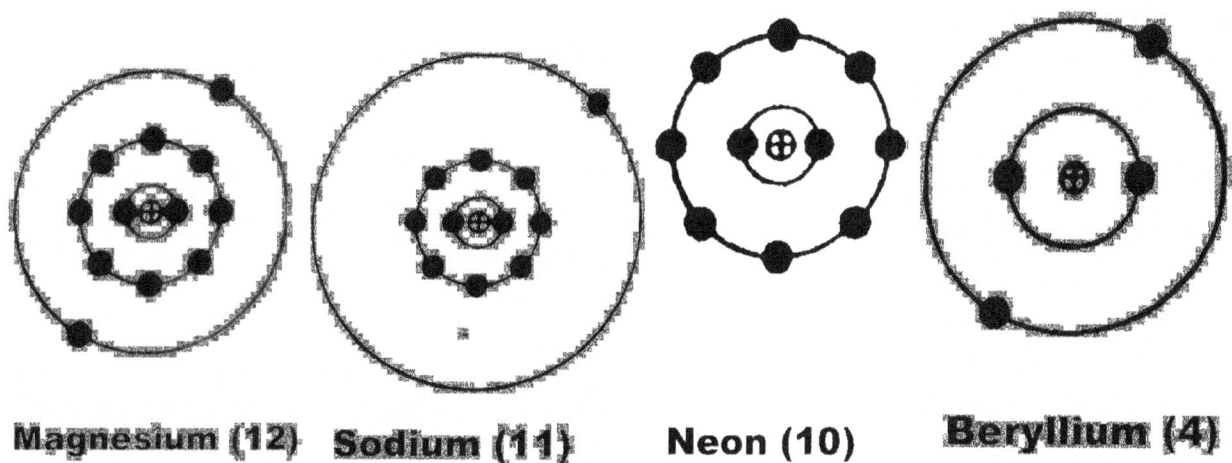

Magnesium (12)   **Sodium (11)**   **Neon (10)**   **Beryllium (4)**

Every element in a star has a different purpose that provides a different density, which relates to a specific gravity applying. That makes ever layer that forms in a star a star in its individual right and as the star grows it discards the layer as the star progresses on the road to eventual fully development. The purpose of galactica is to form space whereby stars develop and stars has the purpose to confine space back into material by dispersing space.

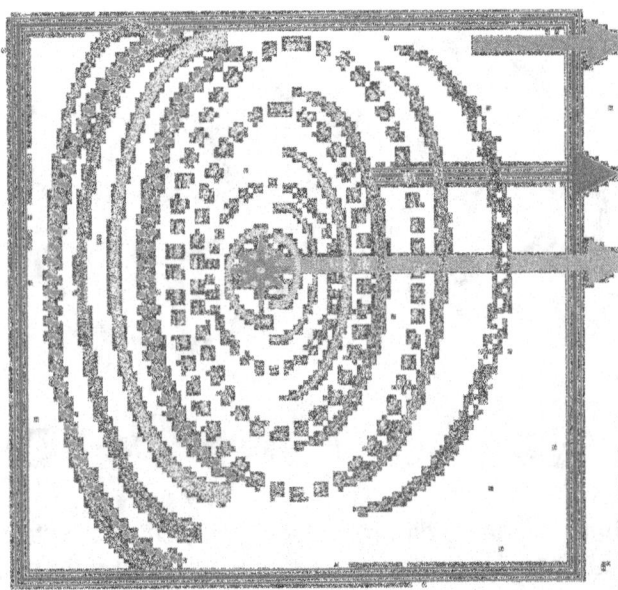

Whether my next remark will suite the taste of the brainless idiots also thought of as atheist, holds no inclination on my thinking. It is a fact that the Universe is a very carefully, perfectly constructed program planned in the most admiral detail and with so much attention given to the specific that the human mind can never achieve any comprehensible understanding to the explicit detail that went into such a marvel. The wonders that went into the all planning required in the Universe can only leave the intellectual astonished and those mindless enough not to understand that the planning was due to a Superior Creator, can only step aside and cry about their lack of ability to do so for they are stupid enough to be God loathing atheists

In   the   Universe we have, there is a range of elements that forms   the   Universe we know. To us with the needs life acquired, the   elements   have much different purpose as to what the requirements are in the Universe. As said there are two substances in the Universe where the one is liquid and the other is sold. Between liquid and solid there is an interaction of displacement taking what is not dense to what is dense. This comes about as a result of the atom spinning.

What Newtonians see as the mass that the atom has, is the displacement ability turning the overheating into the frozen cooled. It is transforming liquid heat into solid material and displacing the expanded into the contracted. Every atom has a defined purpose in this displacement order and every atom is taxed with performing a specific task and composes a role that fits into the total wonder called the Universe. The highest value space-time could have is 112, and that I calculate. That is when heat is totally expanded to the maximum it can expand which in other words is the hottest singularity can go in our era. Then by gravitational motion the star displaces (reduces) (cools down) the heat and condenses it to 56, which then transforms the expanded heat to condensed light.

ELECTRON Space falling outside the domain of the star

NEUTRON Space in the star domain transmitting heat into outer space

PROTON Space in the star domain that holds the gravity and therefore holds the solar system in birth.

This is the function of starts. It displaces heat from total expansion to usable light and then further condensing heat to what atoms can use in cooling down heat to a solid state. In the structure of the star, every atomic number holds relevance to the position of the layer and the layer forms the displacement that is needed to keep the star cool, functional and prevent overheating that would lead to expanding z(exploding) and form a Super Nova.

The following mathematical formulations prove that the atom is responsible for forming the entirety of time in space.

In outer space the limit on the atom is

$((\Pi^2 + \Pi^2) = 19.74$

$+$

$((\Pi^2 \Pi)$ $= 13.01$

$+$

3 $= 35.75$

$\{((\Pi^2 + \Pi^2) = 19.74 + ((\Pi^2 \Pi) = 13.01 + 3)) = 35.75\}$ x Π **(singularity) = 112.31**. That is the limit placed on the atom within the boundaries of what we consider to be the Universe. That will remain a unit

From outer space the atomic relevancy is as follows

$(\Pi^2 + \Pi^2)$ Represents the proton in relevancy to singularity Π through out the Universe

4 Is the time aspect of spin creating motion that is creating space.

10/ 7 Is the **space(10)** in which the **material (7) spin** according to the Titius Bode principle.

7/10 Is the **material (7),** which **spin the space (10)** according to the Titius Bode principle.

Outer space has heat secured at **10/ 7 X 4($\Pi^2 + \Pi^2$) = 112. 8** while the star through motion generate a requirement to heat that establish a flow of **7/10 X 4($\Pi^2 + \Pi^2$) = 55.27.**

The outer walls of outer space are **10/ 7 X (Π^6) / 6 = 112. 8** while the position that the atom demand space is the value iron have as a potential difference. It is in the **7/10** and the **10/7** that the limits are placed. It is seven spinning about in ten crossing singularity by turning about the inner core of the star.

The factor of **10Π** being in relation to Π^3 is a direct translation from Kepler's formula $\mathbf{a^3 = T^2k}$ By substituting the symbols used with the actual value of Π the symbolic massage transforms to specific values applying

10Π Is space square 2(5) (T^2) in relation to singularity Π (k) being equal to space Π^3 (a^3)

In outer space the motion **2(5) (T^2)** of the material $\mathbf{\Pi^3}$ **(a^3)** keeps space in dimension **Π (k)**. But this motion produces a relation that apply to material groups such as stars relating to space holding groups such as outer space which I refer to as geodesic space in more advanced books.

$\mathbf{a^3 = T^2k}$ is $\mathbf{\Pi^3}$ **(a^3) = 2(5)(T^2) Π(k)**

When using the atomic relevancy I refer to the proton relevance in space in example $((\Pi^2 + \Pi^2)$ and then how space will relate to accommodate the atom as the atom as a group facilitate the star and accommodate the stars unifying requirements.

In the expression **10Π** relating to Π^3 it is space flowing towards the star centre in approximately an equal manner as volts flow from space to the Earth or Neutral whatever name there is to choose.

The star on the inside cannot support space up to equal or beyond 2(10Π) before ultimately collapsing the space dimensional support of 6 sides in the square of space **(10)**

On the other hand can the space in the geodesic securing the presence of the atom hold space up to the ability of 112 protons displacement secure **10/ 7 X 4($\Pi^2 + \Pi^2$) = 112. 8**. This is theory because we well and truly know that it is actually **5($\Pi^2 + \Pi^2$) (Π/2)^2 (3/5) = 244** which is the number of neutrons and protons that will allow Plutonium the ability to remain a constructed atom within our Universe. But as one can clearly see it is as volatile as no other element and is on the very edge.

Time forming space = Π^3 = 31.0061

Outer space is 10 / 7 $(4((\Pi^2+\Pi^2))$ Time collapsing space= $2\Pi^3$ = 62.01255

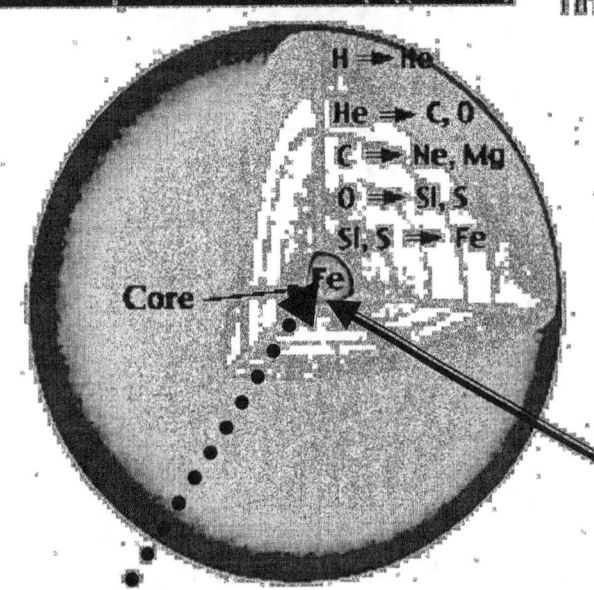

H → He

He → C, O

C → Ne, Mg

O → Sl, S

Sl, S → Fe

Core → Fe

The condensing of space-time or the freezing of heat or the destroying of unoccupied space or the demolishing of time or whatever term is the favorite to connect to the movement of space in a motion called gravity condensing the motion down to contraction is in the following margins 10/ 7 $(4((\Pi^2+\Pi^2))$ to the lower level 7/10 $(4((\Pi^2+\Pi^2))$ of space-time

=112. 79547 gravity expanding or motion

Inner space is 7/10 $(4((\Pi^2+\Pi^2))$ =55.2697 gravity contraction

Light meeting singularity is $3^3+3\Pi^2$ = 56.6

Elimination of space-time is $3(\Pi^2+\Pi^2)$ = 59.21762

Elimination of time and space differentiation is $\Pi(\Pi^2+\Pi^2)$ = 62.01255

Space reuniting with time is = $2\Pi^3$ = 62.01255

Outer space is 10 / 7 $(4((\Pi^2+\Pi^2)$ =112.79547   7/10 $(4((\Pi^2+\Pi^2))$ =55.2697

Gravity generated from outer space to inner space which is collapsing or freezing of heat.

Final collapse of the photon space is $3^3+3\Pi^2$ = 56.6

End of space of neutron space is $3(\Pi^2+\Pi^2)$ = 59.21762

Final collapse of proton space is $\Pi(\Pi^2+\Pi^2)$= 62.01255

Plutonium holds at 94 and as an atom almost falls outside space-time reality as it is on the very border with a possible increase in displacement of $\underline{5(\Pi^2+\Pi^2)\ (\Pi/2)^2\ (3/5) = 244.}$ In the Sun however the dimensional change is $10 \Rightarrow 10\Pi^2$ in comparison with our change of $10 \Rightarrow 3 \Rightarrow \Pi$. With the Universe being $7/10\Pi^6\ /(6)$=112 and the Sun at $ $\underline{=(\Pi^0)\ 10\Pi^2 = 98.696}$

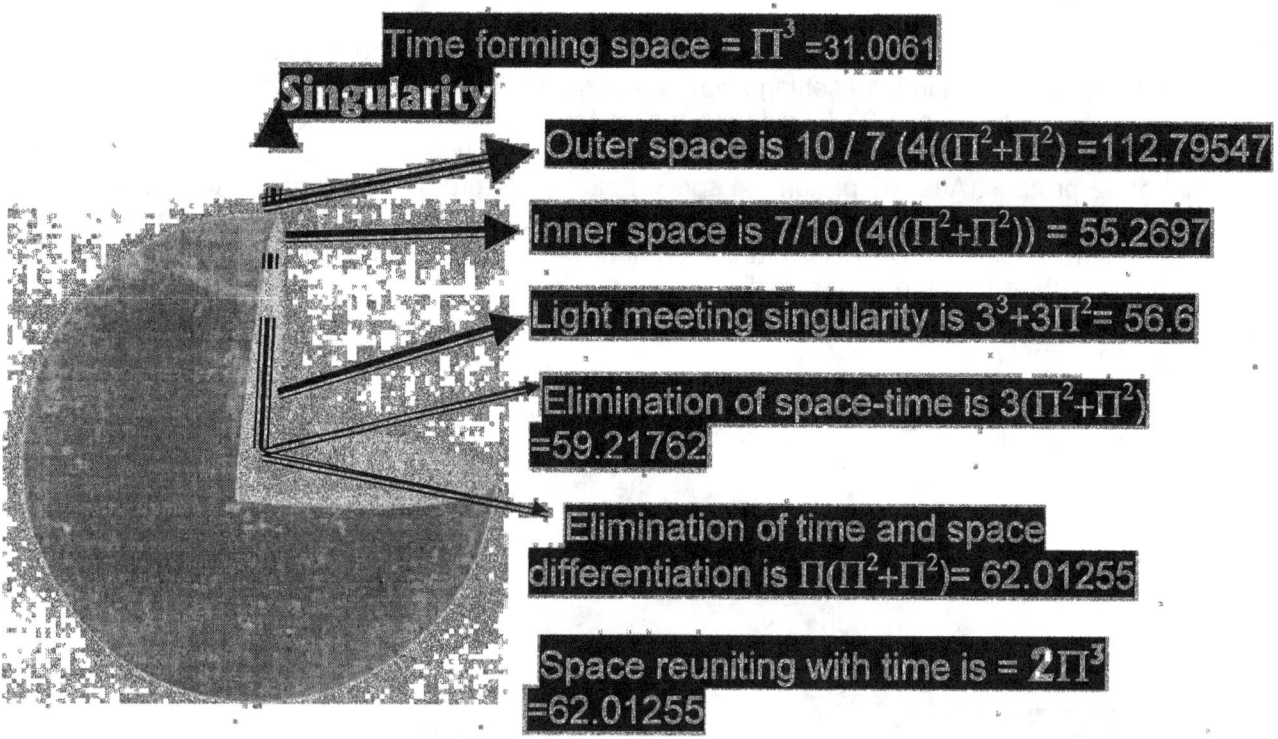

Time forming space = Π^3 =31.0061

Singularity

Outer space is 10 / 7 (4((Π^2+Π^2) =112.79547

Inner space is 7/10 (4((Π^2+Π^2)) = 55.2697

Light meeting singularity is 3^3+$3\Pi^2$= 56.6

Elimination of space-time is $3(\Pi^2+\Pi^2)$ =59.21762

Elimination of time and space differentiation is $\Pi(\Pi^2+\Pi^2)$= 62.01255

Space reuniting with time is = $2\Pi^3$ =62.01255

In the star the balance bringing about space-time flow is in the iron displacing limit of an atom not holding more that 56 protons because the atomic relevancy is <u>**7/10 X 4(Π^2 + Π^2) = 55.27**</u> whereas the neutron reaches the value **of   3(π^2 + π^2)= 59,22** the double proton value, it will respond by returning to a space-time value. This is as far as the atom will go down eternity, no farther.

Where we are is not the only place that is possible Universe to be. This concept is as wide as the Universe is and I am by no means getting into that argument in this book. What I now in the following few paragraphs will refer to is what applies to our Universe forming our space in our time concept. There are as many possibilities as there are names for people on Earth, but I refer to the one I share with all my fellow Earthlings circling on route around a star we named the Sun. In the Universe I am able to witness the proton holds $\Pi^2+\Pi^2$ giving a displacing of space in the duration of time as **19.74** of what ever you wish to name the measure.

The Universe holds a displacing value of 10Π in relevancy of the motion applying Π^3. Dividing Π^3 by the square of space as **10Π** leaves **9.86** or Π^2.

When the centre displacement of a cosmic structure has a group atomic displacement at the core that is exceeding Π^3 the star qualifies to form an independent structure as the outer space it separates from are **10Π**. When **10Π** shows a relation to the inside of the star holding a displacement of Π^3 the value of Π^2 =9.8696 becomes gravity which the spin or motion of the star will produce as it moves in space through space. By having Π^2 between the **10Π** and the Π^3 inside the star becomes independent from the outer space that captures it. The motion inside the star delivers the independence the star requires to separate from the centre of the galactica and proceeds as an individual star. There is no star in our Universe showing this weak gravity but in other parts of the Universe there are such stars coming into operation. Such a star will not be able to space to ensure total independence. It will not yet have gravity that is forming electricity on a grand scale.

With the displacement of iron being 55 + the iron atom has the capacity to dismiss space and by doing that it has the ability to generate such proton motion as to remove space all together from a selected area on conditions of motion producing a connection with singularity. Such connection we call electricity and the diminishing of the space we call an electromagnetic field. The electromagnetic field is the reducing of space between the element copper, (63) being motionless and the element iron producing the motion to generate the gravity micro, which carries the human name as electricity.

Space-time displacement which also is motion that reconverts space from heat back to singularity start to achieve a duration in time setting such duration above what the Universe reserve as having the ability to duplicate. Above 62 then forms the epitome of time. At the point the space can no longer sustain the flow in time to sustain the demand set out by singularity with a dismissing potential of 62 protons. After 62 proton the space held by the protons break down the dimensional wall created by motion and the atom of which only the proton remain in place at that point in any case completely destructs. Only singularity remains casting all other space-time out into outer space. The star then become a star holding no atom but only contains singularity on the inside. It becomes the all so famous Black hole where all falls down a pit of space less ness into singularity without space or motion.

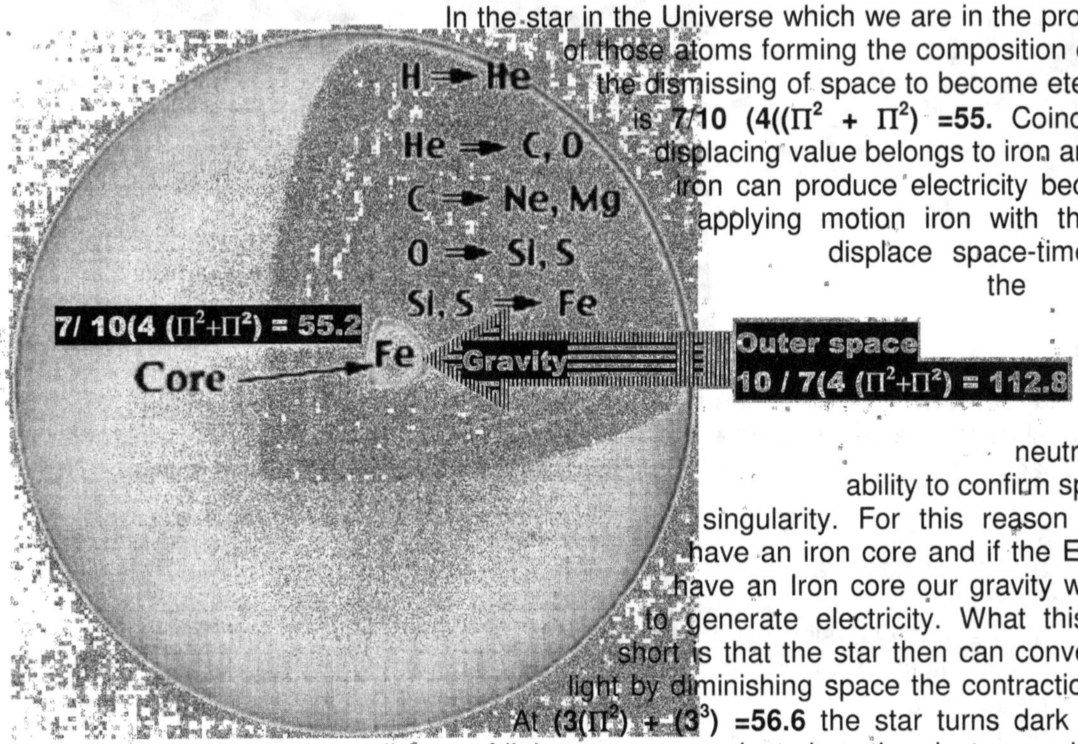

In the star in the Universe which we are in the proton number of those atoms forming the composition of gravity or the dismissing of space to become eternal in time is 7/10 (4(($\Pi^2 + \Pi^2$) =55. Coincidently this displacing value belongs to iron and therefore iron can produce electricity because when applying motion iron with the ability to displace space-time by using the combining motion of 26 protons and 26 neutrons has the ability to confirm space-time to singularity. For this reason stars must have an iron core and if the Earth did not have an Iron core our gravity was not able to generate electricity. What this means in short is that the star then can convert space to light by diminishing space the contraction of space.

H $\rightarrow$ He

He $\rightarrow$ C, O

C $\rightarrow$ Ne, Mg

O $\rightarrow$ SI, S

SI, S $\rightarrow$ Fe

7/ 10(4 ($\Pi^2+\Pi^2$) = 55.2

Core $\longrightarrow$ Fe =Gravity=

Outer space
10 / 7(4 ($\Pi^2+\Pi^2$) = 112.8

At $(3(\Pi^2) + (3^3)$ =56.6 the star turns dark as the star captures all form of light to use as coolant since the electron no longer has a function and the star becomes solid atomic particles frozen in space turning beyond what the speed of light $(3(\Pi^2) + (3^3)$ =56.6 can endure. Then the atomic structure falls victim to movement increases as time begins to influence the atom directly in $3(\Pi^2 + \Pi^2) = 59.217$. The atom now reduces space

beyond what the atomic walls could manage to form as $V^2 \geq \dfrac{a^3}{C}$ which is a formula I explain

elsewhere. At double the value of outer space $2(10\Pi) = 62$ space within the star collapse since the compactness within the star starts to destroy the space that atom holds as **55.27**. The motion applying within the density the star is creating as gravity then has no space or time to occupy any atom in form using space-time such as all an atoms must do.

When a star find the inner-Core- value of applying atomic spin or motion to create $((\Pi^2 + \Pi^2) X \Pi =$ **62.0** which is double that of outer space $2(10\Pi) = 62$ the result is that space depletes within the inner core of the star and the star will start to withdraw more heat from outer space than the star establishes or returns light into outer space.

At $7/10\Pi^6 / (6) = 112$ the Universe stretches space-time to the limit we find ourselves in. For this reason atoms that are exceeding the mass of 112 cannot fit in our Universe we have. But it has nothing to do with mass coming from pulling, pushing or shoving. It is about motion exceeding eternity. This is what the atomic number is that can apply motion within the atom centre by the maximum number of protons gathered as a group. More protons that bring about a group motion will produce a collapse of the atom space. The dimensional walls (Π^6) will collapse into the centre of the atom.

The number of protons applying motion produces space dismissing and cultivating heat from space and in that process is the space in time returning to heat by duplicating space. The protons apply motion where there is just about no space and by the motion at that level the proton motion turns space to absolute heat where singularity then dissolves the heat. By reducing the space it intensifies the heat and that returns singularity to what it was when space came about as the Big Bang presented space-time. $((\Pi^2 + \Pi^2)\ ((\Pi^2\Pi)3)$ is the atom number used to form the atom in the development which brought about the atom. The reducing of space is 1836 times more at the proton $((\Pi^2 + \Pi^2)$ than it is at the electron but the combining effort of displacing is the sum total of all the atomic part.

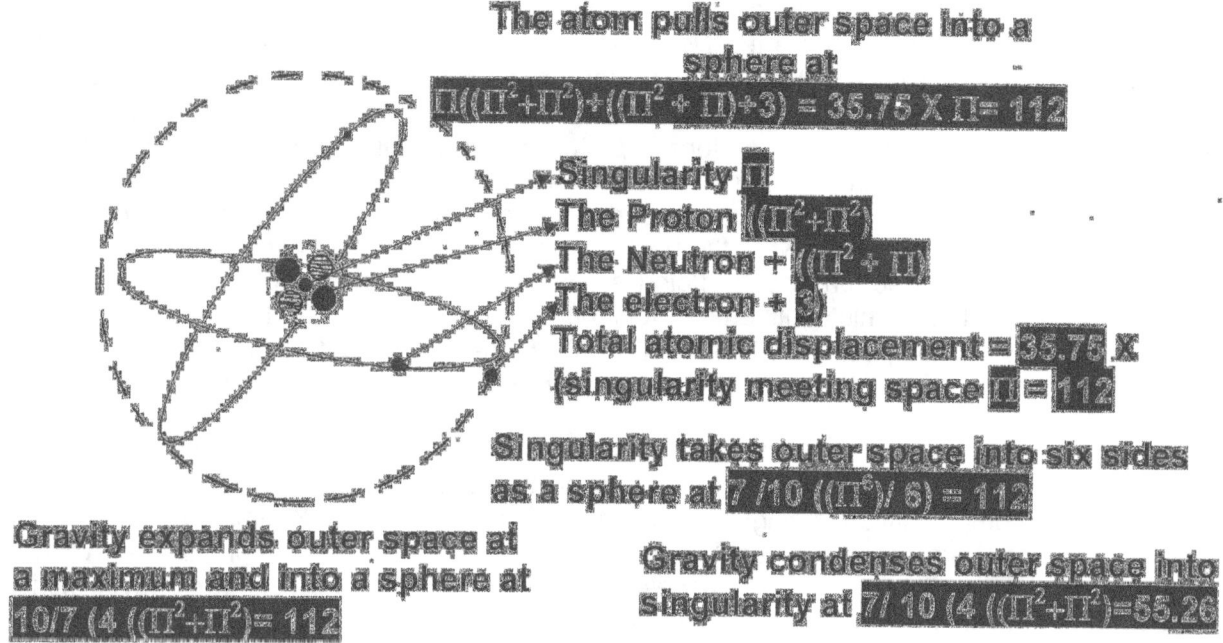

The atom pulls outer space into a sphere at

$\Pi((\Pi^2+\Pi^2)+((\Pi^2+\Pi)+3) = 35.75 \times \Pi = 112$

Singularity Π
The Proton $((\Pi^2+\Pi^2)$
The Neutron $+ ((\Pi^2 + \Pi)$
The electron $+ 3)$
Total atomic displacement = 35.75 X
(singularity meeting space Π = 112

Singularity takes outer space into six sides as a sphere at 7/10 $((\Pi^6)/6) = 112$

Gravity expands outer space at a maximum and into a sphere at
10/7 $(4 ((\Pi^2+\Pi^2) = 112$

Gravity condenses outer space into singularity at 7/10 $(4 ((\Pi^2+\Pi^2) = 55.26$

When the proton $((\Pi^2 + \Pi^2)$ and the neutron $(\Pi^2 \Pi)$ is added to the **3** the electron produces the total dimensional sum produces not 6^2 as it should but **35.75.** With the sum being **35.75** one can see where space will collapse or return to the form singularity provides if it exceeds the singularity connection there is between the atom in total and such an atom connecting to singularity Π.

Then we arrive at the universal six dimensions of three sides in space and three sides in motion bringing about a totalling of $((\Pi^2 + \Pi^2)=19.74+((\Pi^2 \Pi)=13.01 + 3 = 35.75 \times \Pi$ **(singularity) = 112.31** and that is the maximum atom displacing value outer space can tolerate before destroying the space holding the atom all together. Inside the star the proton maintaining a connection with singularity directly will produce the proton value of $((\Pi^2 + \Pi^2) X \Pi = 62.01.$

This means that at this point within the star the protons and above this velocity time cannot duplicate space any longer and the wall of space erected by time collapses back into singularity. Space disappears because time cannot any longer sustain space. Any star having a space with a displacement exceeding the generating ability to displace space to the value or above the value of 62 protons in one secluded given space that is repeated as a unit by motion duplicating space will no longer have the ability to sustain the walls time provides space.

The limit is 62, which is **10Π** holding the one proton Π^2 on the one side in place and the **10Π** holding space in duplicated motion Π^2. Therefore the gravity a star produce at a maximum point is (**10Π** +**10Π**) converting space to $(\Pi^2 + \Pi^2)$ in relation with singularity Π then collapses back to Π^0.
At a value of 6 X 10 time can duplicate space because 3 X 10 = 30 and that is more than singularity Π^0 extended to the square of space at **10Π** will tolerate. But when the displacing exceeds (6 X 10

+6 X 10) making the duplication of space 60, the walls start tumbling in or space is overhauling time as the one side catches the other side and **a3 $\neq$ T² k.**

This **62.01** is the total and the maximum number of protons dismissing a value of Π^0 space-time after which the atom as a single unit or as a group in space and time and ultimately the star as a unit of the combining effort of all the atoms forming the star will abort space or the Universe will dispense of the star, which is all the same effort. The star then has grown back to the connecting with singularity and then forms a Black hole. The principle may sound somewhat simple but it is quite involved in the total explaining.

When the atom finds the motion within the star centre or star-core-centre start to reach $(\Pi^2 + \Pi^2)$ X **3 = 59.22** the neutron moves outside the atom and also outside the star. At a displacement value of **59.22** the atom in the star can no longer accommodate the neutron within the atom and the neutron motion slows down to a point where the Neutron motion is to slow to find accommodation in the atom and in the star. At $((\Pi^2 + \Pi^2)$ X Π = 62.01 the proton collapse, which is double the value of **10Π** and half the cosmic value of **7/10Π^6 / (6) = 112**. At a higher motion the proton moves to leave the atom and go outside the atom away form a proton motion by introducing the proton motion to the space surrounding singularity. The contraction of heat and the subsequent increase in density of the condensed liquid forms a material stronger and more agile than what the Neutron could offer. By that the proton moving faster than what the neutron can accomplish, the proton then dissolves the bond it formed with the neutron and by doing so, it is rejecting the neutron as it disallows the neutron space within the atomic confinement. It is then where it started with, the dot that claimed a spot in the cosmos. This happens at $((\Pi^2 + \Pi^2)\Pi$ = 62.01 where space within the atom collapses as the proton movement reduces the walls of the atom to form a proton holding atom.

Outer space is pushing against the star by measure of the star moving through outer space. The maximum displacement in duplication and contraction that outer space can accommodate is the total sum of the atom in relation to singularity which is $((\Pi^2+\Pi^2)+((\Pi^2 + \Pi)+3)$ = 35.75 X Π= 112 and with that in relation to singularity it forms the atomic displacement limit of 112 protons to one cluster. With the motion that is the maximum expanding there can be when duplicating. However motion stands in relation also to contraction. The contraction is freezing of heat into a state of liquid coming from a gas. At 112 the state of singularity is expanded at a maximum and cannot cope with more heat than 112 protons will manage to control in one atom cluster. But relative to that must be a cold where such a cold will not hold space under a specific level of freezing. Beyond a specific limit the cold of space freezes time into singularity.

The science of cosmology is all about the working of astronomy, which is the art to glance at the picture of the Universe through lenses of much dynamics. The pictures are overwhelming and awe provoking. Re-think the picture applying the full content in the picture to what the size is of you eyes. Think how big the picture is that your eyes take in and translate that area to the size of your eyeball in an effort to determine a ratio. One will be forgiven if one thinks of the ratio as eternal to nothing. How can the greatness of space fit in the lens of the human eye? Great guns are firing with all inspiring knowledge when the reply is through the conducting of light pacing through open space of continuing nothings broken now and then by something but in general the space is just one continuing stream of nothing, but no one ever think that such an answer may just be a tad too simplistic. We see light and we see by means of light. Darkness does not hold light and yet we see darkness because the light is something contrasting the absence of light in the darkness.

Yet, I show that according to mathematics there couldn't be anything as nothing. Consider the path the light followed from the source connecting to light from all other sources the light may come from and bringing a full picture to the lenses one use to look through. When drawing a line that will connect the star to the human eye the measure goes in parsecs. How wide an angle does that cover? Should the line come from one specific point within a star, as we all know that it does, how can we then see the rest of the star, let alone the surroundings where such surroundings cover many, many degrees. Topping this even by miles is the fact, that the part, we see surrounding the light of the star is darkness we are not suppose to see through light. Darkness after all is the exclusion of light and how can light then convey a picture by not showing light?

Newtonians think of outer space as geodesic zero, with nothing in outer space but space. Geodesic zero mans the light travels in a straight line from where it originates unhindered all across space to where the light connects the eye. Such an idea by itself is outrages because the stream of photons reduce in space to such a minute quantity that taken the area the photons travel and the space in vastness it covers, the chances of one photon coming across many hundreds of light years through billions upon trillions of cubic kilometres of space and selecting my eye to convey the electricity is less than infinite. Yet such conveying takes place every second of every minute. If the simplicity of Newtonians is accurate a line has to come from a point at a distance ranging a measurement that makes ordinary distance measuring a trifle ridiculous and even by using astronomical units it still is of little expression. From a point that is so small we cannot size it adequately over a distance we can only measure to the nearest million miles comes light in a straight line and indicate a scene it left behind a good while longer than any person can spend a life of being alive on Earth.

I mentioned to questions on which all my arguments are founded. The first question I addressed and now I wish to introduce my second question that I asked myself. The second thought that came to my mind was that if outer space was nothing as Newtonian science believes, how is it that it can be possible for me to see nothing. If the entirety of the Universe out there is filled to the brim with nothing, then how is it possible that I am able to see all that nothing being all over? How can any one view nothing? How is it possible that I can see darkness? If the night sky was black, it meant that it was impossible to see the sky at night since I can't see black. Have you as you sit reading this part at this minute sat back and gave a thought about the light enabling you to read? Such a thought brings to mind the most simplistic answer one can imagine. It is thought by Newtonians that the light hits the page, then bounces from the page and contact the lens of my eye where the lens conveys the photons becoming electricity to a part of the brain that translate the electricity to an understandable message and that gives a person the ability to use light with which one can read. To the Newtonian intellectual it is as simple as that! Ever gave a broader thought about light streaming across the night sky, coming from places in the Universe we do not even realise those places are there? How do the photons manage to convey one complete picture coming from as far apart and as wide an area as it does? With a few photons connecting to the eye or lens, no one ever noticed the wonder of light. The photons reflect a view that seems as if coming from all the billions upon billions of stars. But most is coming from darkness covering an area no man can measure. Yet, how many photons can actually connect to the lens of the camera or to the eye? Still a few photons coming from a single direction directly ahead eventually tell the entire storey of the picture that is out there. It is very simple to take the process of seeing by means of photon conducting very lightly and I have never heard one of the Brainy Bunch really in sincerity dissect the process to its potential. It is impossible that light from such an array of assorted sources can simply come together at the eye lens and show a picture of objects spanning across a Universe as wide as our mind can receive where the objects they reflect is beyond human measurement and the quantity is inconceivable many. Even more alarming is the Newtonian view that while the cosmos is consisting of nothing that is holding all the space that is not holding stars, we are able to see that lot of nothing without the aid of light because we then can see nothing. If the darkness of the cosmos was representing nothing, then how are we able to see nothing while also seeing something (stars and other light) in-between all the nothing.

It started with a dot that touch space at a point Π, because that is the only form, size and dimension mathematical logic will allow our brain to accept. From the one dot had to come a second dot and a third dot. The dynamics of such a dot is smaller than we can understand because such a dot is in negative relation to what we see Π to be, and the deeper we delve in finding the smallest fragment where space started, in the spot where time is still eternal as much as we can accept eternity to be. This we find in the aligning of planets where the one dot from which the aligner stem becomes the reference too the distance applied between the aligner and the original dot, or governing singularity or structure in charge of holding position to all orbits following.

The reason why we should first locate the spot is because we can only work from that point forward. By working forward we have to work backwards to locate where we are heading. The cosmos started at a point and where such a point is, we will find the Universe. Every one knows where the

Universe is, because we can see where the Universe is, but if we can see where the Universe is, then we should find the centre of the Universe in that spot. Einstein theoretically positioned the point of beginning at a place he indicated where singularity should be. With the cosmos the size it is and space so large compared to our smallness we have no chance in finding the centre of the Universe. The Universe started where singularity is and singularity is the sure indicator of the Universe. With all spinning objects holding singularity we then have located singularity in as much as finding the centre of the Universe. The Universe started with a dot forming. That answer arrive from taking mathematics back to a point of being the smallest possible position, far smaller than we may be able to calculate form.

My approach might seem unconventional but through the abandoning of the accepted, it enabled me in locating the precise location of a universal singularity forming a connecting basis of the Universe (this I say with some degree of confidence). The smallest figure there can be must be a dot. The dot is the only form that leaves all the options open to extend in any and in all directions should the opportunity arise. The only mathematically sensible option about extending a line from the dot will be non-bias progress in all directions equally in order to give a meaningful flow of mathematical equilibrium.

 The Pythagoras mathematical principle is the proof and that I explain. The obtaining of singularity is in my rejecting of nothing by replacing it with something being the dot. With the clepsydra or "water thief" Empedocles deducted that air was composed of innumerable fine particles, braking the thought that what we now know is air, was also believed to contain nothing being altogether a space filled with nothing until proven to be wrong so many years ago. Never did science take the lesson learnt back then to the future and out onto outer space. If there is space, there cannot be "nothing" as space is something. The claim becomes obvious when observing the connection between the half circle, the straight line and the triangle, which could also promote all the qualities lurking behind the pyramid. Consider the connection between 180^0 sharing and then one may realise much of the pyramid mystique becomes less spectacular in considering the very basic in mathematics being the Law of Pythagoras on which all mathematics are focused. Once the water thief was eliminated by some human intelligence the matter was left at that. Nothing shifted out to an area we think of as outer space. In outer space we now find nothing. There is nothing but an atom here and there and even the atom is covered in nothing.

I wonder why the nothing landed there. Could it be that the reverse came about and because there was no visible "water thief" the very limit of man's suspicions came into practice. Man has always been extremely good in flying from one outer edge to another and if the water thief proved something was present then the mere absence of a water thief must therefore prove that nothing must be in outer space. But what is space as such. What can space be, because with explosions we can clearly witness space created from heat. Our culture prevents us from admitting our vision, but the release of heat produces a *"shock wave"*. That *"shock wave"* is nothing less than space created from heat released. We have to brake free from culture of the past and a rigged mind set narrowing our vision. We have to learn to see the Universe with our minds and not our eyes as we can see in the presence of the Black hole we cannot see. The Black hole is only visible by presenting invisibility. We know about the Black hole because we can't see the Black hole. Why would that be?

Because of the manner in which the Universe initially started where more singularity in the relation was unsuccessful to form contraction after the overheating brought about expanding and overheated by expansion much more heat released into the Universe in heat as space uncontrolled than that which remained controlled secured inside a unit such as atoms or as star or any form of containing material. There is more heat in space uncontained than space contained in some cosmic unit in heat that is volumetric consoled and secured. Therefore to restore balance there must be a position where singularity reduces space faster than heat can fill space. At such a point singularity is taking longer to reduce the space than it takes the heat to fill the space and therefore the space reducing takes longer than filling the heat in space. The space flickering that announces contraction takes longer that the expansion causing the heat to turn to space. Before the heat expansion can begin in progress the contraction already completed the motion successful.

At the start of the Universe birth there was heat that turned to space that turned back to heat through motion applying contraction but there were lots of other where singularity could not contain the expanding by contraction and at that, that expanded more then singularity could contract allowing more heat to be in space than material is in space. Since there is more heat in space than there is darkness because the heat is darkness that turned to space the reducing thereof will bring about more darkness because it reduces the heat.

In the region where the singularity present space starting time where the motion originates space is so little that time is eternal. The space the photon reduces too is 1836 times smaller than what the photon is. Actually in more advanced explaining I shall show the reducing is $(1836)^3$ times smaller that the photon but let us leave that explaining for later in another book at another time.

In the area where singularity release time to contract space the motion of Π to fill Π in the seven Π positions of Π creating ten Π to establish gravity by reducing ten to Π^2 the very idea of moving to the other side of the Universe is covering a distance we cannot begin to grasp. The slightest of motion ever possible completes the whole journey and the space duplicated is one mark more that no space at all. We are dealing with mind accepting and not sight ability because where we venture is $(1836)^3$ smaller than the photon we cannot even see but can only use to see, Where the space is that little the time must be that long because time is the very opposite of what space presents. **($k^{-1} = a^3/T^2$).** Being on the other side of the Universe being so small the contracting duration takes so long that the space the photon fill reduces to compact the heat that the photon contains above.

The motion is reducing space in a space much more reduced than the Universe can sustain. It is far below or far more than the speed of light and captures the space that the photon claims by motion in reverse. The flickering that we find within all stars and by which we name some that pulsating is taking so long with the reducing of the space it remains on the dark side of the impulse one eternity while the flair of light is then reduced before the impulse can begin to expand. It operates in the area below the speed of light and even below motion producing space. The mass the photon creates is so enormous it destroys the space the photon claims. If the mass of something as small as the photon Is destructed by own mass and where we on Earth find the mass immeasurably small the gravity applying in the Black hole is so enormous the photon space destroys the photon.

From this evidence it is clear that stars range from the one dimension of total darkness to another dimension of total light In the Black hole a dimension is where dimensions are just a concept. There is the other limit to where stars are still in the cradle blanketed by covering of heat in density beyond the photons releasing. That is why that inner centre of the galactica is so eternally luminous. The time duration on that side of the Universal limit again is so short that the contraction captures the expanding light faster than the speed of light can secure a photon release from that centre. If not that centre must by now be as dark as the rest of the dark Universe.

The flickering is taking so long in favour of expanding the light flowing from it seems a solid by constant flickering. That we also can tell because photons must present flickering if photons are individual particle apart from one another. The Universe is in ranges of space-time. The Black hole presents space-time that favours the flickering of the dark much more intensely to the extend the dark is far beyond solid and then there is another singularity that Prof. Hawking detected within the centre of the Galactica that favours the flickering of light to the extend the light exceeds the limit of solidness.

One must see the Black hole as a lot of solid space in another big space. When the space in the Universe was little the Black hole was big and was constructed of massive space that was available at the time. There was in comparison not that much space available at the time. The **k** that was then present produced a minute **a^3** (in our reckoning at present) with an enormous **T^2**. The Universe was not small just as much as the current Black hole was not big at the time. The Black hole sustained the space that became singularity but the extending of singularity grew with the growth of space-time not committed to star structures. The Black hole stayed behind just as much as the space grew more. It is a relevancy applying both ways and not just to one side. As the heat in space declines the heat within the star rises. It is a relevancy not favouring any side.

The Black hole started off as a (presumably) red giant of its day. In the Universe applying at the time the space the Black hole contained was enormous and I suppose it engulfs large areas of the space available at the time. It is still acquiring massive space in dismissing therefore as far as the star goes little changed but in relation to the Universe the Universe acquired muck more space in exchange for large reducing in time. I cannot for the life in me see how Newtonians can consider that one part of the cosmos change leaving the rest of the cosmos never to change. The cosmos is about changing on all fronts there are.

There was a time when the little space the Black hole now controls was a big space in a little Universe with lots of heat and little else but a few very bright stars covered in massive heat.

There is still the time where the space is still little and the Black hole consumes much of the space because the Black hole remains relevant only to the singularity it sustains by dismissing space not relevant to the singularity it is sustaining in growth. To that end it is supplying heat to the singularity that will reconstruct the heat to space occupied by material consuming heat to convert to space.

But then the Universe expanded as the star reduced. The star did not reduce while the Universe did not expand. The star reduced space as the Universe expanded space and one may never lose sight of the relevance. Only singularity grew as singularity acquired the taste for other singularity, as they will provide the required space/heat to contain.

There was a time when the Black hole presented hundred of millions times larger space than the space that the Huge Red giant now present in relevance to the space available to outer space and to the huge star claims at present to occupy material within boundaries set by singularity extending. But back then the Huge Giant Red star of present was only a speck invisible in the centre covered by a blanket of thick foggy layers of heat in the centre then of the Galactica that was then present.

At present the star, as a whole is smaller than an atom because the star ejected all the proton qualities of the occupying singularity to the out side the structure the star claims in singularity. It rejected the electron before space –time was equal to the electron.

There was a time when the Black hole presented hundred of millions times larger space than the space that the Huge Red giant now present in relevance to the space available to outer space and to the huge star claims at present to occupy material within boundaries set by singularity extending. But back then the Huge Giant Red star of present was only a speck invisible in the centre covered by a blanket of thick foggy layers of heat in the centre then of the Galactica that was then present.
In the spiral of the proton that the Black hole now have stretching into space the gravity influence extending might be some indication of what the core influence had in relevance to the space ratio that applied when the star went from the Universe into singularity.

One can only imagine the growth this star represents in the Universe because as much as the star lagged behind the Universe also grew in space. Just imagine how hot space was when that Black hole was some flickering orange star dot in the sky. But that star was a star before light was the lights we now know because the light within the Black hole that we on the outside are familiar with, is invalid we now know. In the mean while the cosmos expanded and the Black hole reduce by remaining the size it had before the expanding came to alter ratios applying. Just as the Sun captured the outer space value in heat that was present and part of the outer space density back then when the Sun released form the outer space the Black hole too captured the space relevant to outer space then and holds that space relevant throughout other progress applying. It is only our human insignificance that indicates what might presumably be big or small, hot or cold, near or far.

At present the star, as a whole is smaller than an atom because the star ejected all the proton qualities of the occupying singularity to the out side the structure the star claims in singularity. It rejected the electron before space –time was equal to the electron.

It discarded the neutron when it got rid of internal motion and only occupied singularity in the sector where the proton commutes in. Then finally with further Universal development that implicated the star as star development the dismissing of space became absolutely overwhelming by even displacing the proton function to the space outside the Black hole. While space is in demise stars are growing.

The Sun is not a gas-filled sphere holding hydrogen in its "natural gas" form, but it is all fluid and is in a liquid form where singularity is liquid- freezing hydrogen at 6500^0 C while outer space is boiling over at $- 276^0$ C. This book explains the Roche limit in the practical sense... when applying cosmic laws instead of improvising cosmic laws uncovers that reality then becomes awesome. It becomes clear the Universe is as much expanding as it is contracting and contracting by expanding. As there is no hot or cold, no big or small, no grand opposing but relevancies in ratio to one another. If you do not believe me, then believe your eyes when looking at the picture. What ever the Sun is it is fluid falling into fluid.

Consider the time it took from 10^{-43} to 10^{-5} seconds to create a cosmos the size of a neutron. Compare that to what is happening now and see how many events took place by the creation of every lepton and every hadron and it is true that that period took longer too complete than it took the Universe to create the solar system. The flow of light through the density that space produce heat gives the speed of light the relevancy of time in space. The thicker the "soup" of heat is that space forms, the longer it will take light to over a distance. It is very important to note that the speed of light is a relevancy between time (seconds) and space (kilometres). The speed relies completely on the value **k** holds on space –time. The speed of light is forever a constant but the constant is part of the relevancy of space-time.

The Universe connects in a way Kepler established through his relevancy theory. Those not convinced answer this: where would the Planets be if not for the Sun securing planet positions. The relation proves the ratio of one in all cases to be valid. It proves much more than merely connections at liberty of holding positions where ever the randomly opportunity placed the structure. The structure does not come closer by a pulling and tugging. Kepler's figure must still be around and by repeating the task but this time made much easier with the help of computers and telescopes of magnificence compared to the which Tycho Brahe felts exited about.

The way I interpret Kepler is very new and up to now all I could find is Academics with scepticism and detachment from my views. I do not blame such reaction but if I am not correct, please explain according to the view you hold how the following is possible. You being a person stand at night on the highest elevation in the vicinity in a manner so that no other solid object can restrict or block out any of the light flowing towards you. From all over and from the most outer regions of outer space light is travelling in a straight line directly to you. The light is travelling at a speed maximum to what the cosmos will permit. That is how eager the light is to reach you personally. That light is using mostly millions and in some cases even billions of years to reach you while you are filling the centre of the Universe. Wherever you move the centre of the Universe will shift to the position you then hold because the light flowing to you will follow you to wherever you are at any spot. It is coming straight to you because you are in the centre of the Universe according to the light travelling. Not one spot represented by one ray will pass you thinking that you are not the centre of the Universe or the spot you hold is not important enough. Why would that be? Why will the light act as if you fill the spot where the light considers it can locate the centre of the Universe?

The light is underlining what you take for granted, the light admits that it is coming to you because you are filling the spot the light comes too as if that spot is you being the centre of the Universe. Most of the light started on their route even long before man was to become a species and most of the light was already on route long before the time when the solar system came about. Yet the light treats the place you occupy as the very centre it wishes to come to be. If you close your eyes or ignore the light, the light will go unnoticed for all time to come. The light then travelled across so much space using so much time to travel only not to be recognised by you the person filling the centre of the Universe. The light and all the space the light represents and all the time the light travelled would go into eternity dismissed as never acknowledged to be worth noting just because you, the one filling the centre of the Universe did not take the time to look and acknowledge the effort. So much effort depends on you're taking notice of the effort and appreciate what the Universe did to admit that it also believe you are the centre of all space and time out there. Now use modern science to explain this centre you establish.

Tycho Brahe and Johannes Kepler stood there night after night and made a super human effort to acknowledge the information the light of the Universe brought to them. They wrote down every massage every night. But since they are only human they could only managed to acknowledge the light coming from the Sun that was reflected by the planets. The two masters were the centres of the Universe when the two Masters decoded the language the cosmos used to speak. Think about what they managed to collect for all mankind's benefit. They acknowledged the light coming towards them in a straight line. A flow of electrons is causing photons to travel across outer space and meet their eyes. One line of photons flowing tells them all about the regions the light came from. Think of the size Jupiter holds.

Yet the light acknowledged the position the Master was in and came all the way representing such a large structure confining such an enormous structure with all the information it has into one line of photons and with the information of the entire structure it managed to convey all that information across such vastness of space, choosing that specific point as the centre point. By using the theories now applying and representing the views of Mainstream Science how would you go about explaining the way you are treated by light travelling through space and time to be in the centre of the Universe, which is the spot you have and hold. Seriously you know you cannot be in the centre where the Universe started the initial Big Bang process. Still without any possible influence the light puts you in that centre. But since you are in the centre, you must admit that there are billions of trillions of spots forming the centre of the Universe just as Kepler stated. The only way you can fill such a spot is if the spot is the place holding singularity and singularity represents the centre of the Universe. Then **k** can form at any spot and establish a^3 / T^2 because **k** is the end of k^0 being the start and k^0 is representing singularity. The centre of the atom represents k^0 and the electron represents **k.** The main issue brought to light by Kepler's formula is the relevancies that always prevail through out the cosmos no matter what sphere it is representing.

When a point holding singularity achieve energising by securing a number of protons concentrating heat towards that space centre holding the singularity, the singularity will establish a will to seek independence from other more dominating singularity surrounding that point and moreover seek to abolish control of the dominant singularity that suppresses its individuality by establishing a securing centre. Two points will redefine their relationship and will establish space-time relation between the points. A relation will come about where there is a dominant point in singularity establishing a controlling centre that is trying to establish control over space-time by creating motion too space-time by displacing space-time, then there is the space-time in motion and a third factor where singularity is applying motion to material within the space-time in control. The dominant singularity will control the space surrounding the lesser spot in singularity that is gaining in heat concentration by material growth. Through applying gravity to ensure heat concentration with space-time diminishing the lesser singularity will gain heat in centralising space-time and by then turning it to motion it then can use such motion to gain independence from the centre in the space-time that the dominant singularity is controlling and is diminishing through applying motion within the space -

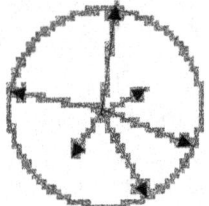

effecting the lesser partner. This is the manner which galactica use to progress in development.

The cosmos cannot be if the cosmos do not share with everything else in the cosmos but the sharing is always producing relevancy to the position of another factor forming the Universe.
The inner space is applying positive space-time displacement in relation to the object in rotation.

The heat the inner structure secure prevents the motion from applying to the object because it became dominant enough to reduce the space towards the heated centre and in doing that it is producing space to secure space through applying motion to the space.

The orbiting object forming the outer ring is in a negative displacement in relation to the inner centre. The outer object has heat in the centre it has but the heat is far less dominant than the heat of the centre and by increasing motion it is concentrating heat as much as

 the motion is reducing space to concentrate heat. By applying motion it is securing space  **is**

The role of the electron the neutron and the proton very commonly accepted the role each sub atomic particle plays. But galactica and stars are just as much just more cosmic atoms playing their part in the very same way as does the sub atomic particles.

If one looks at the transmission of sound, it too depends on the relocation of matter, but to a very small degree, and in this process lies the transmitting of sound. To make the error of judgment in

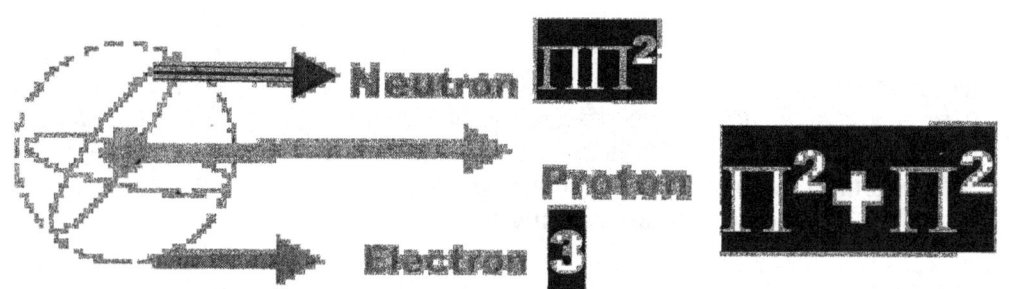 confusing the process with the breaking of the Doppler rings are quite

understandable.

It is about confirming space ($\Pi^2 + \Pi^2$) conforming space ($\Pi^2\Pi$) and converting space3.

The Universe is the atom and a star is just another atom. This comes about from the fact that both particles are the same in the Universe since both particles serves singularity. From singularity the size space-time takes up is unimportant because from singularity space-time is merely principals connecting singularity forming energy as gravity or antigravity and presenting space through the relevancies forming the motion of time. That is particles and atoms surrounding singularity, protecting singularity, maintaining singularity and securing the surviving of singularity. This service of space is done by motion in duplicating space or extending space.

When a star favours **3** and a multitude of three the star is still in a process where it will favour more the duplication of space. As the star develop it will ever increase as it moves through the ranks of being liquid ($\Pi^2\Pi$) the favour to the proton ($\Pi^2 + \Pi^2$) where more space id dismissed that space is duplicated because the motion of the star also diminish and progressively so to the end where the star will once more be motionless.

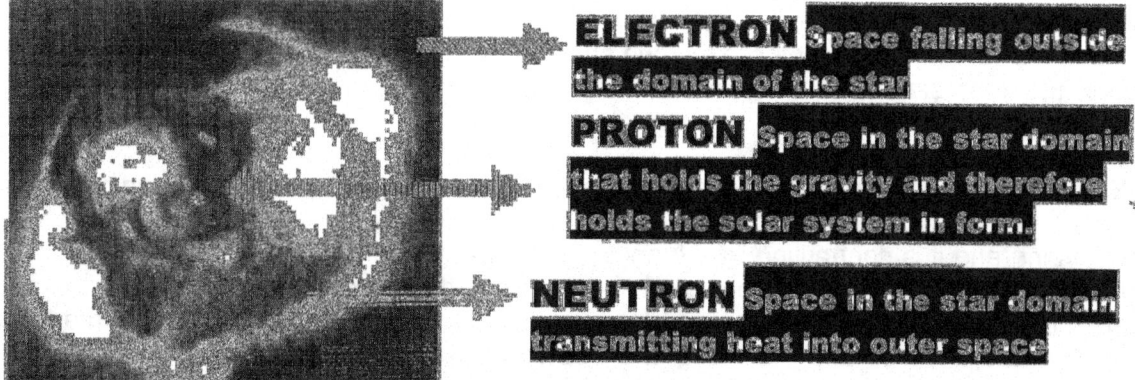

Science should become serious about the task they chose to perform and not going flat out covering up mistakes in performing acts so shocking about self-protection and self-preservation. When Roche presented his findings they should have realised there is something missing with the way they see things. When Hubble presented his findings they should have put Einstein to task about finding the mistake they made. Newton is about contraction and Hubble proved expansion. By going overboard even further and order Einstein to measure the Universe is quit frankly madness! I found on all and every campus I went that any remark about uncovering a mistake Jesus Christ supposedly made generated immediate interest with even the most adhering Christians coming to hear the argument. Making a remark about Newton making an error gets you marched off the campus by security. Why not test Newton's $F = G \dfrac{M_1 M_2}{r^2}$ from figures Kepler left us and see how far planets shifted closer. Prove what is lectured to millions of students. With me openly criticising Newton and Newtonians being the Universities I guess this will again make this book as successful as the others in the past. Then they get very hostile when I blame the Academics to their faces that they are not about gaining knowledge but about conserving the past through protectionism. Universities protect their own without any willingness to test that which it protects. All evidence should be clear in confirming that the basis on which the entire world science union is founding their policies and beliefs are correct.

They should not become annoyed with critics but not only that, they should show evidence immediately to back their claims showing how far did the structures move closer. From that we then can see what collisions we are waiting for and how long before the big solar clashing will begin. The absence in they're just mentioning such possibility confirm to me they know as well as I do there is no tugging and the Universe is in synchrony more that any person may ever be able to prove.

The atom is the Coanda effect because gravity is the Coanda effect and the formula expressing the Coanda effect is the Formula Johannes Kepler received from the cosmos. $a^3 = T^2 k$ **is the Coanda effect.**

When looking at photographic images coming from the Sun we can clearly see for that the fluid push out of a bowl of liquid, from within stars, spilling both sides as it falls back into liquid pool forming the Sun. The inside of the Sun is not gas but it is fluid. In all of nature there is no NATURAL **gas** as much as there is no **natural solid**.

Hydrogen is as much a liquid as iron is a gas and neon is a solid. It depends on the element relating to the space/heat in the circumstances surrounding the substance at that very precise instant in time. We have to stop telling the cosmos to show us what we wish to find and start accepting what the cosmos is telling us what is out there that we should look for and find. As you have seen so far in this book the fluid state and the gas state is expendable waste that stars remove through development but then so is all space-time and material.

This book now will explore how singularity came free to form space-time and commanded motion by creating space. In that exploring we may find out that the Universe is already contracting as much as it is expanding and it is contracting by expanding because it is through the contracting that it is expanding; the answer comes about from $a^3 = T^2 k$. My effort with the book criticizing the Academics was never to attack the world of physics and I never had it in mind to destroy any work made by them I can ill afford enemies and even less enemies as power full as the Academics. But on the other hand I cannot go on praising when I see mistakes about their work.

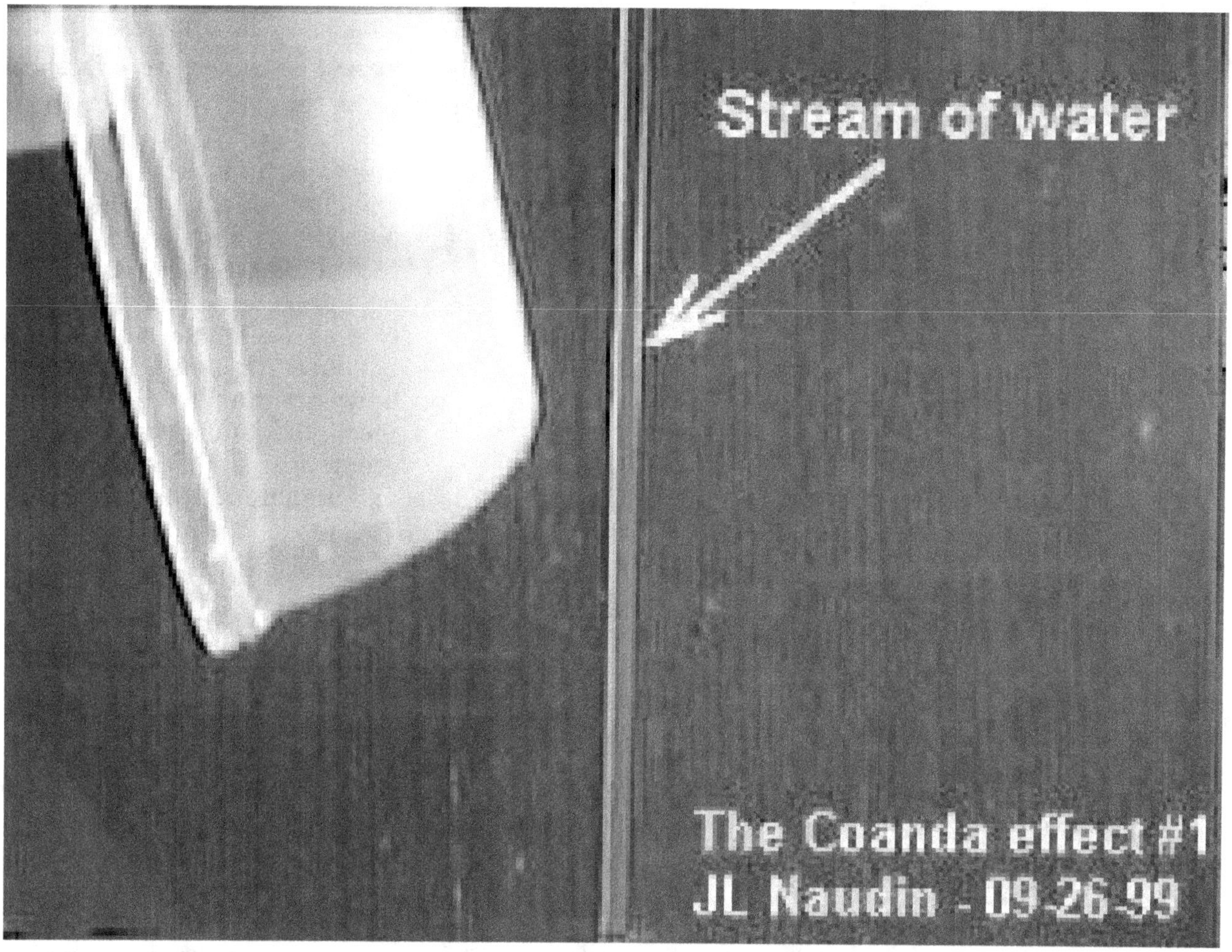

Stream of water

The Coanda effect #1
JL Naudin - 09-26-99

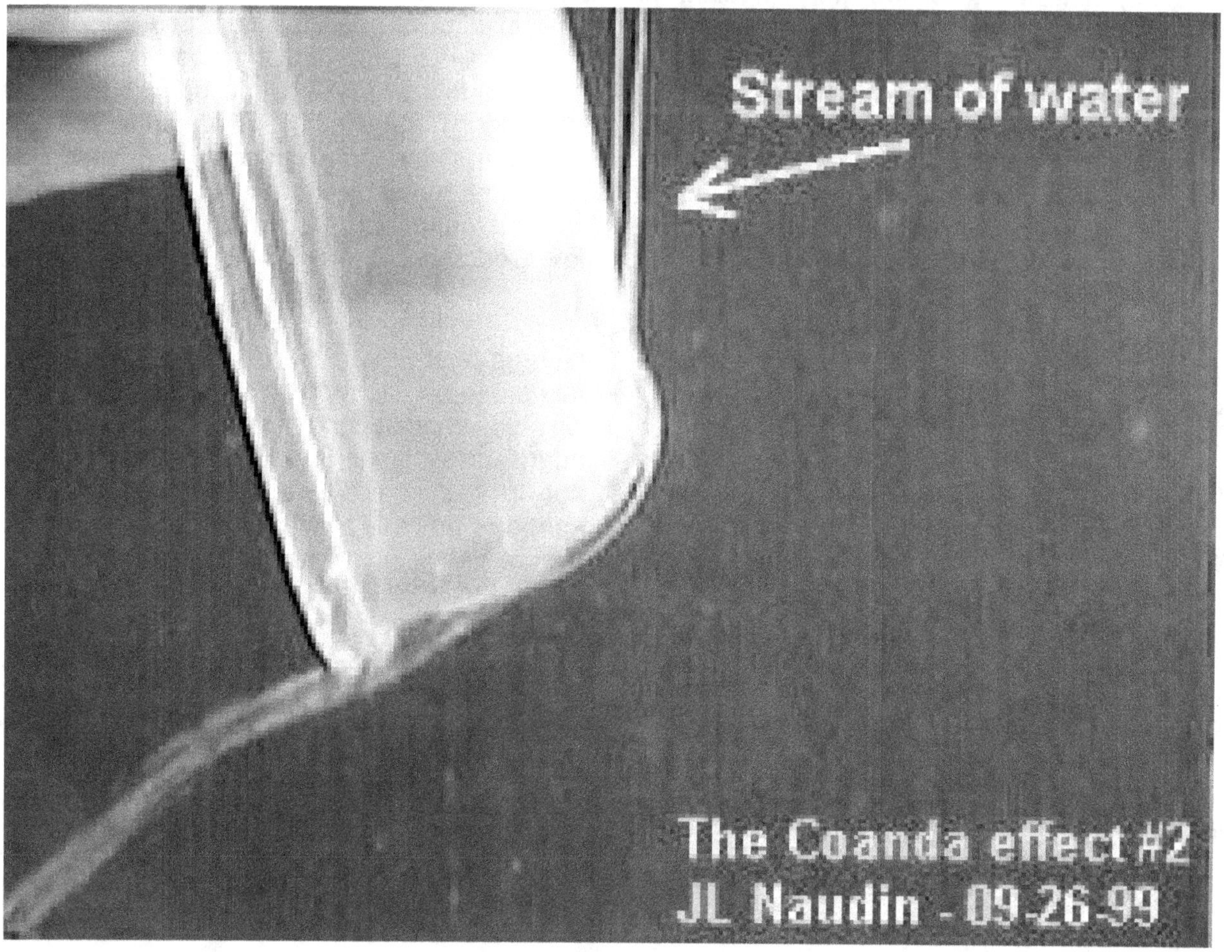

Stream of water

The Coanda effect #2
JL Naudin - 09-26-99

Everything in the cosmos is moving, either by own individual accord, or under the influence of some other singularity dominance. In explaining we return to the top.

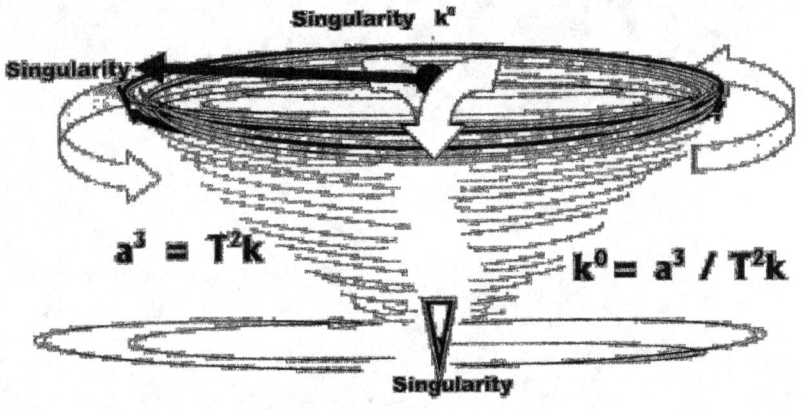

The top spins because of the Coanda effect. The top's wooden round body forms the solid and the atmosphere or space in which the top rotates forms the liquid and because of the interaction between the top (solid) and the liquid the top stands up erect and spins around an axis initiated by the rotation. The top is the manifestation of $a^3=T^2k$, which is the implementation of $k^0=a^3\div T^2k$

When the top is in a state of motionlessness on own accord it is everything but motionless. The motion it adapts are synchronised with the Earth in harmony with the solar system and according to the greater picture of the cosmos. When an energy source not related to the cosmos called life intervenes and energises the tops motion, the singularity in that top suddenly jumps to life. By adopting a rotation energised to an unnatural state of energising because of life's intervention, the singularity of the top is not in charge but as it applies more and more energy, it will begin to find a means whereby it can escape and apply individual singularity as the top starts to separate from the singularity the Earth holds. The singularity holding the Earth would then allow the singularity of the top to rotate within a specific band where that a specific band of being active before the earth's singularity will start to destroy the singularity in rebellion. The top on the other hand will try its outmost, when the singularity it holds gets by individual spin is too strong to remain be in domination of the earth's singularity. The motion of the top is an attempt to begin applying an individual singularity space-time defying and standing apart from the earth's gravity. That action we see as the top starts rotating in a manner where the top does not align with the earth's singularity. With the adding of spin, the time the top holds becomes unrelated to the time the Earth holds and the top will start a campaign too escape from the singularity domination the Earth has on the top. When the time or spin of the top exceeds the limits the Earth places on the top, the top would emerge by trying to escape from constrains placed by the earth. The view I represent at this point is known to science for almost as long as science knows mathematics.

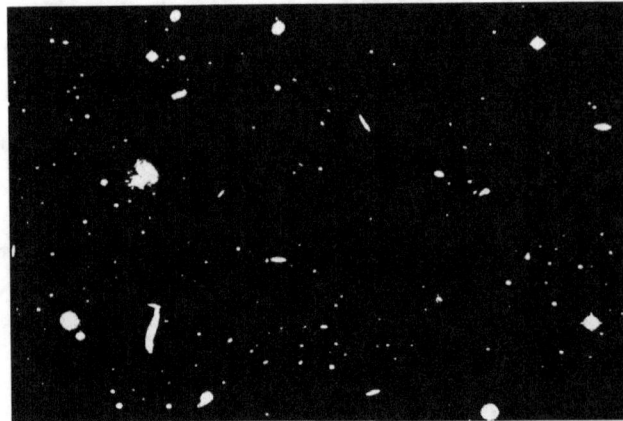

Light is much more than the medium science takes it to be. Light connects the Universe in a way we cannot contemplate. Light being far apart and where it is originating from regions not in the same time or Universal space, connects in a way that present us with a picture holding the Universe in an understandable content. From the point we stand and we watch the Universe, the significance of what we see surpasses the sense of understanding of what we are experiencing. How can the few photons that our lenses catch coming from such an area as that which the night sky covers, transmit the complete picture of what we see? Take a few seconds and gleans at any picture of the night sky and then rethink about what the picture tells when the picture is applying the full content in the entirety of what is pictured to what the size of you eyes is.

Think how big the picture is that your eyes take in and translate that area to the size of your eyeball in an effort to determine a ratio. One will be forgiven if one thinks of the ratio as eternal to nothing. Yet in a few pages I am going to show that according to mathematics there couldn't be anything as nothing. Consider the path the light followed from the source connecting to light from all other sources where all particles of the other light may come from and how much light must beam from all over as to bring a full picture to the lens one use to look through. In your mind connect a line from every atom producing light and connect the lines to your eyeball and see how you can manage to fit all the lines, as small as the lines may be. How is it possible for all light to come directly to where I am and not having some light pass by me while missing me? How can all the light travel at the speed of light to the point I stand and come directly to me where I fill the point in which I stand. How is it possible that every ray that ever left any object, heads to me directly? Why is there not some light that simply goes in another direction and by mistake misses me by a few yards? In this lies the answer to the Universe and in answering this small question one solves most of the puzzles we associate with the cosmos.

If the darkness was the representation of "nothing", then that should be exactly what we must see, nothing but the stars. Taken from the top picture some stars and leaving the rest to nothing is what we see in the picture below. A blind person sees nothing but when we look at space, we see something that we think nothing of as we see as space while being able to also view something because that we can recognise. One cannot have the ability of sight and see nothing except by closing your eyelids and then you still don't see "nothing" because you see what's behind your eyelids. But even in that case where you see the darkness behind your eyelids, you still do not see "nothing" in contrast of "something" you see "nothing" without it forming a contrast to "something".

If the darkness was nothing, am I then seeing nothing?

It is rather meaningless to suggest that we are able to see nothing. That means it is rather meaningless to think that the darkness in the Universe is made up of lots and lots of nothing filling everything there is and nothing is substituting that which is not filled by the many particles representing something we are able to see with the aid of light. Simply by seeing the night sky proves that the night sky has to be filled with light because my vision only detects light and the fact that I see black instead of white light has to hold some significance to me other than the thought that I am seeing nothing just because I am seeing black. Surely my intellect has to stretch further than having the capacity of thought such as to deduct complete witless information from a concept that produces a fact that is not even worthy of an intellectual thought? Is it possible to even associate any intellectual concept with the idea of seeing nothing in colour, even if the colour is black?

The fact that we see light means that the dark next to the light cannot be "nothing", If the darkness was the representation of "nothing", then that should be exactly what we must see, nothing but the stars. Taken from the top picture some stars and leaving the rest to nothing is what we then would see in the picture we see of the night sky. A blind person sees nothing but when we look at space, we see something that we think nothing of as we see space and think nothing of it. One cannot have the ability of sight and see nothing. It is light that we see and it is light that we use, which enable us to see. That proves the darkness that we see in outer space is light that we see without recognising it as such.

I have to repeat because in this phrase lies the biggest damning error science made this far. Newtonians think of outer space as geodesic zero, with nothing in outer space but space. Geodesic zero means the light travels in a straight line from where it originates unhindered all across space to where the light connects the eye. Such an idea by itself is outrages because the stream of photons reduce in space to such a minute quantity that taken the area the photons travel and the space in vastness it covers, the chances of one photon coming across many hundreds of light years through billions upon trillions of cubic kilometres of space and selecting my eye to convey the electricity is less than infinite.

Yet such conveying takes place every second of every minute. The position of the location of the second singularity, which is the precise duplication of the first singularity but in a diminished capacity, is obvious to miss when one is not applying a detective mentality, as one should in scrutinizing the cosmos. Culture will have us believe that when one sees a colour shining from an object the colour is associated with the object. Logic tells a different storey. A yellow dot is all the colours in the spectrum but yellow because it is disassociating with the yellow. That goes for red blue and all other colours we may visualise. I think the norm accepts this as scientific fact with very little argument or substantiating proof about that required.

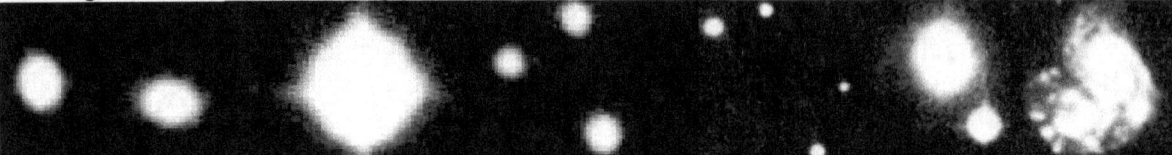

Nothing is all about not being and not "not seeing". By the ability to see the darkness renders the darkness something other than nothing and that changes the acquired value of the darkness from nothing to something. There is an eternal difference between something in infinity and nothing.

Again I ask the question because of the important implication that it holds. If light came as individual streams of photon flurries our visage would translate that as such shown in the fragmented picture above. It would be a picture that shows unconnected objects and would be brought across by some photons that present atoms in stars we have no idea about. It would be representing atoms in stars from where the photons came and would show a picture in the manner where atoms from every object stands apart not being related in any way and that will be what we see, if it is anything that we see. But as it stands, we get a total all including picture of the entire spectrum of information which is that information we call the Universe. We know we get the entire picture with all the information running directly from wherever it came right across a vastness of space of which I could never grasp the callosity thereof and with that being the case then also in such a case t that means geodesic zero is as much rubbish as anything Newtonians regard with simplicity and with careless thought. Geodesic zero means nothing and how can I see nothing as darkness because "nothing" is not darkness, nothing is "nothing" and the darkness I see is darkness showing the darkness as something.

The cosmos has no pressure or pushing or pulling. It has a flow of space-time by concentrating time and duplicating space as it is driving space-time towards the centre. In any picture about any star there is no containing wall that keeps whatever is inside, inside. There is no limit to what the wall if the structure can contemplate before bursting. In the centre of a star is a point holding singularity and since such a point has no space and is immovable, space has to compromise by flowing towards such a location. We regard what we see at night as space and how wrong can we be?

Have you as you sit reading this part at this minute sat back and gave a thought about the light enabling you to read? Such a thought brings to mind the most simplistic answer one can imagine. The light hits the page bounces from the page and contact the lens of my eye where the lens conveys the photons becoming electricity to a part of the brain that translate the electricity to an understandable message and that makes one read. It is as simple as that! Ever gave a deeper thought about light streaming across the night sky, coming from ends of the Universe we do not even realise it is there? How do the photons manage to convey one complete picture coming from as far apart and as wide an area as it does? With a few photons connecting the eye or lens no one ever noticed the wonder of light. The photons reflect a view that seems as if coming from all the billions upon billions of stars. But most is coming from darkness covering an area no man can measure. Yet how many photons can actually connect to the lens of the camera or to the eye? Still a few photons coming from a single direction directly ahead eventually tell the entire storey. It is very simple to take the process of seeing by means of photon conducting very lightly and I have never heard one of the Brainy Bunch really in sincerity uncover the process to its utter and full potential. It is impossible that light from such an array of assorted sources can simply come together at the eye lens and show a picture of objects spanning across a Universe as wide as our mind can receive where the objects they reflect is beyond human measurement and the quantity is inconceivable many.

Light is much more than the medium science takes it to be. Light connects the Universe in a way we cannot contemplate. Light being far apart originating from regions not in the same time or Universal space connects in a way that present us with a picture holding the Universe in an understandable content. From the point we stand and we watch the Universe the significance of what we see surpasses the sense of understanding of what we are experiencing. How can the few photons that our lenses catch coming from such an area as the night sky cover transmit the complete picture of what we see. Take a few seconds and study the picture of the night sky then rethink the picture applying the full content in the picture to what the size of you eyes is. Think how big the picture is that your eyes take in and translate that area to the size of your eyeball in an effort to determine a ratio. One will be forgiven if one thinks of the ratio as eternal to nothing. Yet a few pages back I showed that according to mathematics there couldn't be anything as nothing. Consider the path the light followed from the source connecting to light from all other sources where all particles of the other light may come from and bringing a full picture to the lens one use to look through. In your mind connect a line from every atom producing light and connect the lines to your eyeball and see how you can manage to fit all the lines, as small as the lines may be.

If it is lenses that enable us to see what we can't see in outer space it also means we cannot see the light, which is outer space because we haven't got the lens to match the curb of outer space. Newtonians think of outer space as geodesic zero, with nothing in outer space but space. Geodesic zero means the light travels in a straight line from where it originates unhindered all across space to where the light connects the eye. Such an idea by itself is outrages because the stream of photons reduce in space to such a minute quantity that taken the area the photons travel and the space in vastness it covers, the chances of one photon coming across many hundreds of light years through billions upon trillions of cubic kilometres of space and selecting my eye to convey the electricity is less than infinite. Yet such conveying takes place every second of every minute. The position of the location of the second singularity, which is the precise duplication of the first singularity but in a diminished capacity, is obvious to miss when one is not applying a detective mentality, as one should in scrutinizing the cosmos. Culture will have us believe that when one sees a colour shining from an object the colour is associated with the object. Logic tells a different storey. A yellow dot is all the colours in the spectrum but yellow because it is disassociating with the yellow. That goes for red blue and all other colours we may visualise. I think the norm accepts this as scientific fact with very little argument or substantiating proof about that required.

What then about colours that are technically not colours as is the case with black and white? White is simple. By spinning all the colours in the spectrum the colour white shines through. Black is quite another matter. One of my friends whom are one of the best painters I have ever come across told me that one couldn't paint black but have to make black a dark blue to show shade on the canvass. That apparently is his success in achieving the realism.

There are also many variations of dark blue form the shadows in one simple tree. This remark set my mind in motion. One cannot see black because black has no colour to show, but black is the colour most prevalent in the Universe. One can see only by colour and since black is not a colour we should not see black, but we do. I am not delving into this answer, but answering these two questions make one realise that science doesn't even begin to explore the Universe.

What then about colours that are technically not colours as is the case with black and white? White is simple. By spinning all the colours in the spectrum the colour white shines through. Black is quite another matter. A friend of mine whom is one of the best painters I have ever come across told me that one couldn't paint black but have to make black a dark blue to show shade on the canvass. That apparently is his success in achieving the realism. He also went on to explain how many variations of dark blue form the shadows in one simple tree. This remark set my mind in motion. One cannot see black because black has no colour to show, but black is the colour most prevalent in the Universe. One can see only by colour and since black is not a colour we should not see black, but we do.

If it was true about a Yellow object not being yellow and a red object rejecting red and therefore not being red, then in that case the same must be true about dark and light. The bright object rejects all the light just because it is light from the outside. If it is light from the outside it has to be very dark inside. That too would count for what we believe it to be dark stars. Such dark stars must be most brilliantly lit because they keep all the light to their inside and well protected. The stars have gone so cold the stars have to conserve all heat to remain in gravity. The same must then apply to outer space in that outer space is conserving all light and by keeping all light outer space is brilliantly lit. We just are unable to witness the light because our position is much concentrated where as the light being dark is expanded to the full. We are able to see the galactica because the galactica represents highly concentrated light in one reduced area. The darkness contrasting the light we see as darkness must therefore be because the light is expanded to the ultimate. The fact that we can see the darkness makes the darkness light that we are unable to see. However with the space stretched to the maximum the lens we use in the telescope we use to see the light coming from the sky has to bend the oncoming light because that means as far as our position goes, the sky is not even slightly curved yet because we are so small. If we can bend the cosmos by applying a lens, the cosmos must be a lens we can reform the bending of by re-bending the light to suite our requirements. Now use these thoughts to convince the average Newtonian atheist to rethink cosmology and he or she would not even understand what you thought about. The Newtonian atheist only wished to re-design the cosmos and work out how to travel at the speed of light. Use the two thoughts a measurement and tell any Newtonian atheist and mathematician in charge of theories this much and see how far you can get convincing them about your view. All that they wish to manufacture and design space whirls while trying to convince the world about the Newtonian atheistic brilliant mind while not being able to see other forms of reason.

The fact that we see light means that the dark next to the light cannot be "nothing", If the darkness was the representation of "nothing", then that should be exactly what we must see, nothing but the stars. Taken from the top picture some stars and leaving the rest to nothing is what we see in the picture below. A blind person sees nothing but when we look at space, we see something that we think nothing of as we see as space. One cannot have the ability of sight and see nothing. It is light that we see and it is light that we use, which enable us to see. That proves the darkness that we see in outer space is light that we see without recognising it as such. If the darkness was the representation of "nothing", then that should be exactly what we must see, nothing but the stars. Taken from the top picture some stars and leaving the rest to nothing is what we see in the picture below. A blind person sees nothing but when we look at space, we see something that we think nothing of as we see as space. One cannot have the ability of sight and see nothing. It is light that we see and it is light that we use, which enable us to see. That proves the darkness that we see in outer space is light that we see without recognising it as such.

What puts us humans in a category one higher than animals (or so we like to think) is our ability to think about that what we can see. The less develop an animal is the more it has the attitude of eat or be eaten. The higher developed animals are the more the animal find reason to argue. One may teach a crocodile not to eat you if you start feeding the animal. That is a mindless reptile and yet it can think above eat or be eaten. What we see is not merely the truth and it requires reasoning to see the truth and substantiate between culture motivated observations and thought through decisions.

There is a mad hype about reversing time and not only amongst us the uneducated mindless, but even more amongst the better Educated Brainy Bunch. It is remarkable that every one wishes to reverse time by going back into history but never to speed time up going onto the future. To return to the past, one has to reduce the **k** factor of every atom in every star and every star incubating galactica through out the entire cosmos. It involves a far more complicated process than just move down some imaginary double Black Hole that creates an artificial space whirl everywhere they so wish. As **k** develops so does time expand, so does space explodes and so does time within the Universe progress. The point of **k** connecting k^0 is to differentiate space-time from its origin since singularity throughout and everywhere is equal. Getting the growth growing away from the point of origin is creating time distinction away from where all sides meet in singularity and singularity is the place where there is no space dividing the Universe without showing division but is still dividing. It is where there are no boundaries and yet every aspect of k^0 within every individual particle commits a boundary in relation to the singularity that is maintained. As we are part of the 3D and on top of that locked in time motion it will therefore be very wise for us not to try and understand the fist dimension or singularity at this point. In the second dimension there are sides committing motion to the third dimension but in the first dimension all is alike with separation coming about merely from being a unit. With every singularity progressing by **k** from the centre of the galactica since the Big Bang, **k** became the measure of progress.

It progressed from a Universal governing singularity charging an accumulative control that responds to the growth of individual singularity forming and from that more individual atom singularity comes about. As singularity finds heat through gravity it progresses by establishing less support in the atom's individuality and by passing support onto the governing singularity. The atoms are at task to remove heat from space thereby eventually they remove the atom's individual independent singularity to favour the stars governing centre singularity. The factor time T^2 provides the duration of time but **k** indicates time development and if our time travellers wish to return back to the past (to do what only they would know what they wish to do in the past) then they have to reduce **k** by taking time and **k** back as far as they need taking **k** back. Time is not just a direction but time is progress of stars, galactica and atoms forming countless Universes enrolled into countless Universes. Remember the least singularity is equal to the ultimate original singularity by virtue of value being 1^0 and that makes whatever point there is holding singularity as much a Universe as the entire Universe forms a Universe. Those whishing to travel through time must accept the reducing of space and the increase of heat that accompanies the journey. It will be a lot easier to shoot into Jupiter and by such reduction of their personal **k** establish a duration extension. There was a time when the Sun had **k** as relevance being the very same as Jupiter now holds and Jupiter then was presumably even more frozen than Pluto. Being frozen back then adhered to a much different meaning than being frozen now. Coming to this conclusion is simple. All one needs to do is study Galactica and see what the galactica lectures when they say in light what they say at night.

We stand on the outside 150×10^6 km from the spectacle and from such distance we judge the Sun. We don't even judge the Sun from what we can see but we judge the Sun from what we feel. We feel heat coming from the Sun and from that we argue that the Sun is hot. We see the Sun has heat rising from the surface as liquid soup. That puts the hydrogen layer as the outer layer in a liquid. Hydrogen freezes on Earth at a temperature, which is the coldest amongst all other elements. Yes, the Sun is 6500^0 and that is on the outside. To a human that is hot but a human has no mind judging the Sun. If the Sun squirts pure heat turned to liquid from the surface and the heat falls back into the surface the Sun is a lot colder than the Earth is. The Earth requires an enormous effort to cool hydrogen down to a liquid state. We must mind the way we think of the hydrogen in liquid. The hydrogen remains a solid. The element is untouched by temperature differences. It is the heat environment surrounding the hydrogen that changes from a gas to a liquid to a solid. One remove or one amplifies the heat in which the hydrogen is and that turns to liquid or solid or gas. The hydrogen is untouched in the elements worth. Yet we see the heat flow amongst the hydrogen as a liquid. Nevertheless we remain adamant that the liquid is a gas and the hydrogen is in a gas and the Sun is a gas bowl filled with hydrogen because to our mind hydrogen must be a gas. After all, our element table classifies hydrogen as a gas and that is the way we think of hydrogen. We do not consider hydrogen to be in a liquid state when we see the heat is flowing just like a liquid and shows all indications that it is a liquid. No the Sun is hot because the Sun feels hot.

In the Universe there are no hot or cold but a state of differentiation produced by time. The Universe parted by parting heat from cold when eternity parted from infinity, when Π^0 singularity parted from Π singularity, when 1^0 parted from 1^1. There is no hot or cold but there is a relevancy where one factor cools and another factor overheats. By retaining the Sun is the coldest space in the solar system and outer pace is the hottest there can be.

From since the time that man discovered intelligence (if he ever did) man has been with the presumption that the Sun is the hottest centre in the solar system. Later on in the present time, it came to someone's attention that the Sun also holds the solar system in gravity. The Earth by its standard and dominating its sphere of which it can control with influence is the hottest centre in the space of its domain and it holds the moon centred to the Earth. The gas planets are the hottest centres in relation with the most heat and they all hold their satellites captured by a hot centre. All space structures hold in every centre there is that is confirming their independence at that point of securing independence the centralizing of the most heat it is able to concentrate and from that centre holds all material captured or controlled in the domain of what that forms the independence of the structure. I can go on and on but heat in the centre couples gravity to space-time, just as if Kepler said before he was spoken for   on his behalf and without his permission or his agreeing to it.

In the centre of a spinning top runs a line from top to bottom that does not spin. The top spins around this centre lien without the line ever moving. The line divides every aspect from every other aspect as the top spins and changes directional relevancies in accordance to the speed that rotates but never rotates. The line divides every the top every other point on the other side and changes point from point direction constantly without ever showing any point −point preference to any side.

$$a^3 = (T^2 k)= a^{3 + 2 + 1 = 6}$$

with the sphere presuming the position of singularity as part the of $k^0 = 1 = $ **singularity**. Einstein proved that at the point where space reduces and such reducing reaches a point where space as a factor in the third dimension disappears into the single dimension (space going flat) gravity is overwhelming. Einstein interpreted this, as the complete Universe going flat but while it may be true that the Universe is going flat, that can only be within singularity since Universe as flat as it can get.

singularity represents the The centre of any sphere has to be at the very point where space completely falls away. It is at the point where all the points of line centres meet by the crossing the centre of their individual connection coming in to contact as a group. In that way one may assume that the lines connecting the controlling points on the other end are crossing on a centre point that all that is participating in the constructing of the sphere is democratically electing such a centre. Please note this conclusion very well because this forms the heart of the Coanda principle. That will put that position where the lines cross which in itself is centralising all space in the sphere at that point, such crossing point will become very distinct and controlling where that point forms in the single dimension and singularity is the single dimension. Kepler also solves another riddle that truly got Newtonians unstuck. This, to which I now refer, is what is referred to when they refer to the Hubble constant.

The growth we see in the Universe is an adding of space in every cycle completed by every cycle, which all the protons complete. The adding is the smallest addition that can come about in the shortest period of repeating by cycle rotation there can ever be. This growth of space-time next to singularity confirms the growth of singularity as singularity recalls the space it uses to grow in the time it grows. The margin of growth will be by the extension of **k** in the formula **k** = **a**^{3} / **T**^{2}. Every cycle completed in the relation to space by the initial value of **k**. **k** = **a**^{3} / **T**^{2} leaves ultimately **a**^{1} extending as space or as Kepler chose to indicate it as **k**^{1}. That too has to be compensated by the duration of time reducing the time aspect by the margin that the space expands. This confirms what is evident in the Hubble Constant. The further one looks at time the more time seems to race because time has the invert properties we give to space.

The arguments introduced up to this part of the introduction only touches the most basic aspects of my work and by no means can such an introduction secure an opinion. Yet, not once through all my long investigation in the past thirty or more years have I found any other person claiming such views that I have brought about even in this skimpy way as I do in the prologue. The answers arriving from these most basic and elementary questions brought to me new concepts about physics and about cosmology that strays widely from Newton. As it applies with all things, so it does in this case as well that when delving deeper into any issue, the complexity of the issues truly come to the fore ground when analysed in more detail. I wish to advise the reader to treat the book as new work never observed before, for that is what it is. There might be parts being of a general introductory nature and are overlapping in some sense but each highlighting issues in different manner as to clarify facts used in the conclusion I arrived at when I was able to formulate gravity bringing conclusion to different cosmic perspectives. In an attempt to bring across with the meagre mental means at my disposal, I try to find a way in which I could convey the massage about the picture we see and by employing Newtonian atheistic views, the way we see the picture does not match in the least of the facts that we see. I realised that the conducting of photons through space is more complicated than anything we may ever imagine. That same lack of grasping even the complexity that we have to face about matters in physics and the complexity of physics as well as our misunderstanding of physics comes about through the way we generalise and simplify gravity and that incorrectness comes across by our categorising stars.

It is just the lines that the Newtonians understand least. Somehow galactica was observed that was at a point where the Universe began. The Newtonians put the cosmos at a flat bed with a there and a here. The most informed Newtonian holds a view of a specific plotted point in geodesic terms from where the lot started forming mass and this point had the eternal neutron which then on the big map of the Universe formed a point where everything came from that centre. I have it on authority that there are some experts plotting a 3 dimensional map in relation to points allocated to galactica. What the Newtonians misses is the instant in time required to hold a position forming one centre where the Universe was standing still so that they could draw their galactic map. Where do they hope to place the viewer and in terms of what are they going to secure time is a moving factor. They only have an optical view and the distances they are going to use will have dire consequences on the positions allocated in terms of perspective placing. Typical Newtonian behaviour insists that they see everything in a simplistic reality as if everything is honouring one centre point. In the solar system this is the case. Gauging the Universe from where we are, we also would place the allocation of the Milky Way in terms of our location because we can use our centre point holding singularity to pinpoint time from where everything then shifts according to time moving space.

The Newtonians admit we can see past the centre point to where the galactica no are and where the lot started. Or else how did we get so far away and the galactica did not? From this argument they admit that there has to be a specific point where the Big bang started shifting material in directions in relation to other materials that went in numerous other directions. The only way the centre would not be crossed is where "we" and "they" went in the same direction after the exploding of the Universe during the Big Bang explosion, but "they" went further than "us" because "they" went faster than us. Looking at the map they draw where are the placing a time honouring centre point to allocate the relevance space will insist on. I return to this argument later on. At this point I wish to show how incorrect any thinking is if not placing singularity in infinity in the centre.

If there is an edge of some sorts we could find in the Universe (because Newtonians forever find something being "right on the edge of the Universe" especially when they discover a "**new star**" or "**new a galactica**" that they feel very exited about discovering and wish to exaggerate as only Newtonians know how to do) then with that "**new star**" or "**new a galactica**" where it is, "we" must be smack in the precise centre of the Universe. If the newly discovered "**new star**" or "**new a galactica**" is where our brilliant Newtonians say they located it, then while "they" are being that far away and being "**right on the edge of the Universe**", then we must be right "**in the centre of the Universe**". This comes from our view that "we" are in the centre of the Universe putting "them" on the edge of the Universe.

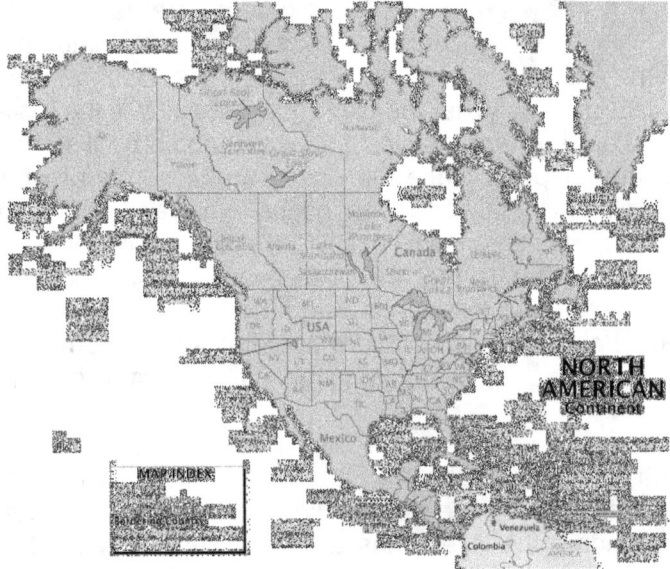

But this puts the odd question that "we" then have to be where the Big Bang explosion started because where the Big Bang explosion were can't be where "they" are on the edge of the Universe, then "they" there on the edge of the Universe did much more travelling than what "we" did. That is why those brilliant Newtonians see "us" standing still" and that is why "they" move a hell of a lot faster "there" on the edge of the Universe and 'we" are standing still "here" where "we" are.   Understanding the issue about lines is more than crucial to our understanding the Universe. Applying the expanding of the Newtonian concept of the cosmos leaves a lot to desire. According to science, the Universe is expanding at a rate known as the Hubble constant. To the human mind working on boundaries, this statement is very plausible, because to the human mind, every thing has to have some form of calculation, a specific value and boundaries.

All the numerous previous theories in the past proves that the cosmologists are evidently unable to divorce their way of thinking in terms of the cosmos from using the human allocating of boundaries. This remark includes the way Einstein viewed Hubble's expanding Universe. By his attempt of initiating the calculation of the mass filling the entire Universe in order to formulate the Critical Density scam, Einstein had to place limits and boundaries in the Universe where mass will end and the counting of mass could then start. Einstein places the Universe in terms of having a position where the boundaries have an option to go on and shift the end further away…from what? Here is an end where the Universe pushes this end outwards in order to expand and allow this end to shift further by expanding the space the Universe holds. Those ends he wished to get to contract and again come closer once he could realize the mass required to get the contracting going. Even in being the master such as Einstein was, he still thought of the Universe having endings with boundaries, although he declared that the Universe has no boundaries but still he had words but not the mind to accept the concept, since he saw a Universe he could measure from end to end wherever the end might be.

It is a common known fact that Americans are of the opinion they form the centre of the Universe, but they do not even cover the whole of America. They can expand their borders and relocate their borders and to that the Indians and the Mexicans bear witness. The problems they create with their expanding boundaries are that they do not allow a sea to curb their possibilities of expanding. They most frequently expand across the sea. These territories include Germany, Japan, Korea, Vietnam, Iran, Israel, Iraq, the Balkans and the list continuous indefinitely.

They expand in regions and geological territories on Earth where they have never been before, and that creates a ripple effect, which may one day accumulate to the dimensions of a title wave, but that is politics, and my work is on science. Let us have a look at a nation that, thus far, set their boundaries according to the sea.

Australians, on the other hand do also see themselves as the mainstay of the Universe. In the case of Australia they cover the whole island they call a continent just not to feel claustrophobic, therefore they hold everything bordered by sea as part of Australia and in as much as concerning Australia as such, they cannot expand. It is very easy to say the Universe holds all, contains all and, occupy every thing. However, understanding this concept proves to be something a little more advanced! Every poster or picture about the Universe indicates a concept suggesting boundaries.

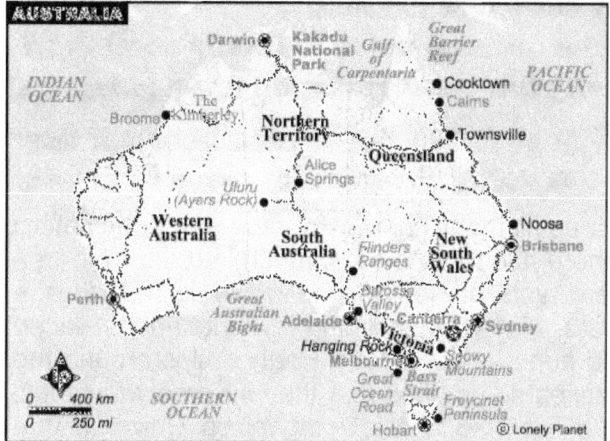

In the Anglo American mind and that includes every person alive that speaks English notwithstanding domicile and is a descendent of the Great Britain islands, all those coming from any region speaking English, which includes amongst many other countries also New Zealand, Canada, Australia, those living in Africa and speaking English and a multitude of other parts of the world to many too mention, they all are joined b one common concept and understand one concept and that is if there is something you wish to have, you go there and you expand by annexing what they have. The Anglo American species speaking English living wherever they might be will expand by taking with force whatever they wish to grab that is not theirs to grab. The latest is the oil in Iraq but previously it was even the Deli Lama and the Tibetan Monks. The British Anglo American species were at war with a bunch of unarmed monks that only knows how to pray and knows nothing about warring. The Deli Lama thought he could prevent the monks from getting slain by British bullets by prayer. It didn't work and the monks got slaughtered. In the mind of the Anglo American expanding into other person's territory is the same as breathing air. That is why they culture this idea of having an expanding Universe. When they saw the Universe was expanding, they accepted it without a hint because if there is one thing the Anglo American specie

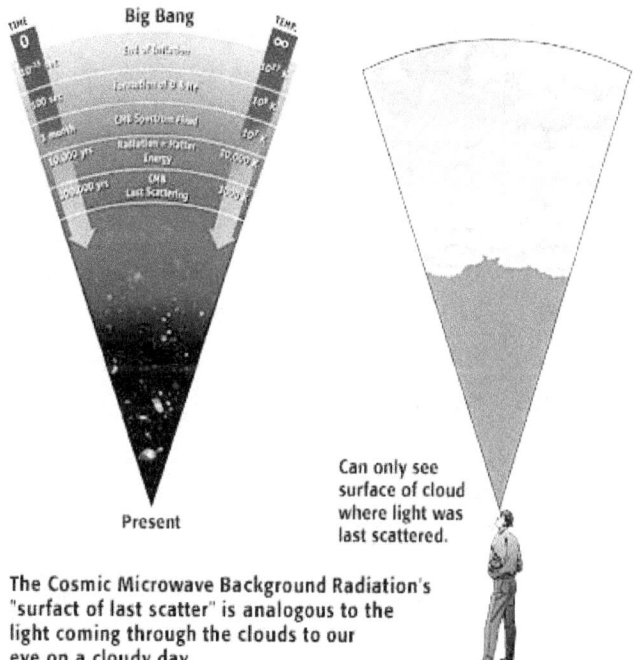

The Cosmic Microwave Background Radiation's "surfact of last scatter" is analogous to the light coming through the clouds to our eye on a cloudy day.

understand is to go and grab other countries that are not yours to grab and take what they are not prepared to give and call it extending their sphere of influence and progress and expanding. My forefathers knew about the British aggressive expanding when they stole my country and then stole the diamond fields and then stole the gold fields and then stole our taxes. To the Anglo American grabbing by going there and taking by shooting if they do not give is progressive expanding. With that attitude so deeply cultured in the species it is little wonder they never asked how the Universe are able to expand.

It is very easy to say the Universe holds anything there is contains all and occupies the lot of every thing, but with an Anglo America mentality the practical implications are far and wide. However, understanding this concept proves to be something a little more advanced! Every poster or picture about the Universe indicates a concept suggesting boundaries.

As soon as science released the world from a flat Earth, Einstein brilliantly came along and gave us the flat Universe. Is that not expanding the Universe into much wider misconceptions than there was before…? To marry our logic with Einstein's calculations of singularity brought about some bizarre ideas. To incorporate a three-dimensional Universe into a flat Universe in singularity is not that simple.

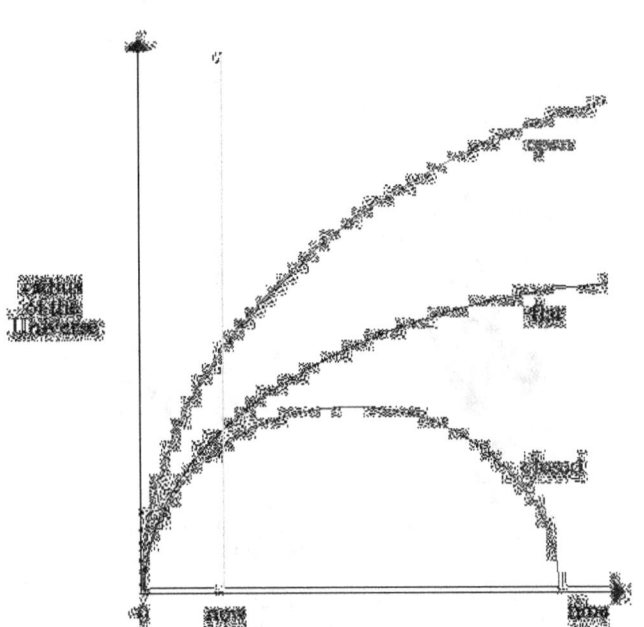

Found Newton's missing mass that went lost (by mail?)

The Big Bang

Getting the missing mass to finally commit to gravity and start reducing the expanding

Ending all in the Big Crunch because they activated the missing mass to gravity

The list with compelling theories and proof can and does fill many libraries of men with brains more than hair. According to the theory on the Big Bang, there has to be a Big Crunch in order to form a Universal beginning of an end. According to present facts presented, the Universe has an age ranging from between 10 and 25 thousand million years. The overwhelming favourite age is placed **at $13,5 \times 10^9$ years** and is the most widely accepted number in years or Earth circles around the Sun. The first scientific formulated theory that was accepted, implied that the Universe was static, going no where as it stayed the same which meant we had a nicely organized Universe that maintained itself uninterrupted and existed in a state of regularity where matter was evenly destroyed as it was created. This meant that the Universe would be precisely the same, unconditional to changing the position that matter was distributed through out the Universe. This gave the Newtonians an unbelievable chance to play God with their mathematics and to redesign what they thought God got wrong! As time progressed and a lot of research was done, no concluding evidence was found to substantiate this fact; no proof could be found at any place where matter originated spontaneously. There was no white hole in relation to a Black Hole where material came streaming out.

This implied that as matter changed, there was no proof that it was ever renewed. The list of theories born from that confusion is as many as they are bizarre. To make sense of everything I placed the Universe into dimensions and after implementing the dimensions a lot began to make sense. I placed gravity where there is only a line, moving from somewhere and going somewhere, but never having a defined beginning or end. The second dimension is the wave, moving from a specific point outwards but continuous.

The wave can never be in one position, as it always changes position through time.

The third dimension is a combination of the previous two, of which even Einstein could not break free.

Every person alive knows that there is a Universe that is not flat and this Universe is chained to time. One can draw all the lines you wish, but as soon as you bring in time, the Universe becomes a four dimensional, moving "something" that cannot turn back and cannot skip forward. It is on tracks going in the direction, wherever that may be, through the movement of time.

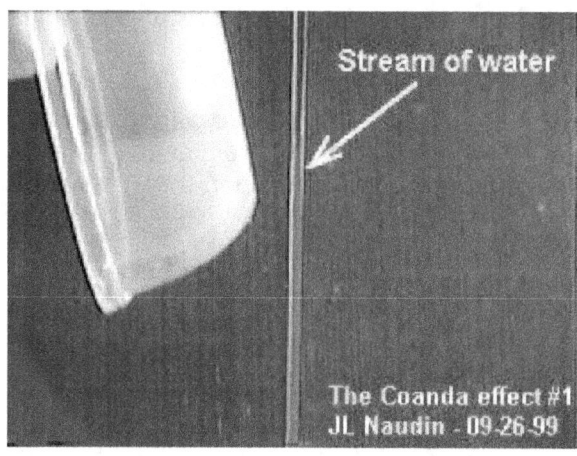

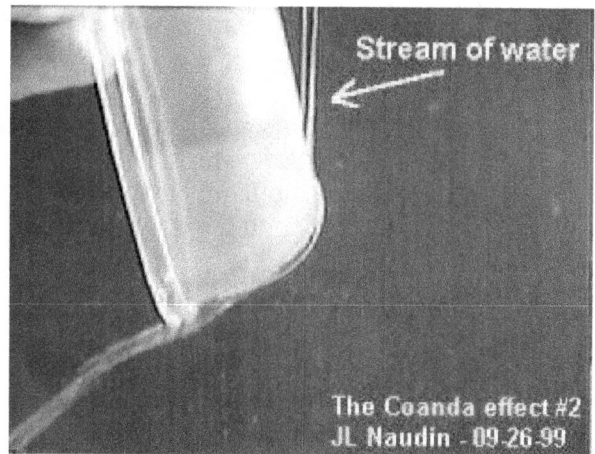

After I concluded this I then realised I was looking for Kepler's formula that detailed all the factors I was in search of. $a^3 = (T^2 k) = a^{3+2+1=6}$

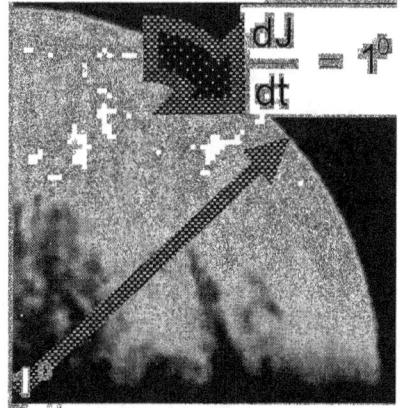

Newton said that the spin of the top nullifies the radius of the top...and that is bollocks $\frac{dJ}{dt} = 0$. This concept and the thought that $a^3 = T^2$ is as misplaced as the entire mass idea is. When a body is attached to the Earth by having mass or forming a unit with the Earth, the body as a cosmic entity loses its singularity value as it then becomes as part of the Earth that spins while the object in relation to singularity at the point in the centre of the Earth then holds a link to singularity and in relation to the spinning Earth the object suffering from mass only has a spinning potential. As soon as the object renounces a connection to the Earth by declining mass as it floats in space, the body then claims independent singularity and the Earth becomes a ruling singularity, which is the same as what the Sun is to everything spinning unattached in the solar system and around the Sun. However, a value of $\frac{dJ}{dt} = 0$ or zero can never apply to anything being a factor that is holding relevance to another factor. The smallest it can go is $\frac{dJ}{dt} = 1^0$ just because there is an applying legal relevance placing the one factor in a valid ratio to another factor.

What Newton in his utter simple minded stupidity saw was that the Earth hold the object Newton awarded mass in terms of its movement quality and that is $T^2 = a^3 \div k$. It is not the Universe that has gone flat but the object serves a movement by the square relation to singularity. The object now is serving the Earth's space a^3 in the capacity has being the moving part $T^2 = a^3 \div k$. In terms of the Earth singularity $k^0 = a^3 / T^2 k$ the object placed at allocated position k in terms of singularity k^0 forms the moving square dimension T^2 because the object in mass is allocated at a point k in terms of space a^3 that the Earth has. Never did $T^2 = a^3$ as Newton saw it but because k forms a link to k^0 in terms of k^0 and in those terms are linked by a factor from which one may or may not read mass, we have the scenario applying that $k^0 = a^3 / T^2 k$, and with k linking k^0 in terms of the spin T^2 in the allocated spot $k // k^0$ where the spot holds a relevance with singularity applying k^0 the object having mass and only when the object is having a mass attack from the Earth does $T^2 = a^3 \div k$, where we then find $T^2 = a^3 \div k^0$ and then by having mass, in relation to the rest of the space the

Earth holds, the position with a mass indication has only a turning function $T^2 = a^3 \div k$ when compared to the rest of the Earth $k^0 = a^3 / T^2k$.

The space that forms a^3 depends on the object suffering from a mass attack which places the object in a circling T^2 rotation at the rate of singularity k^0 duplicating k space-time $k^0 = a^3 / T^2k$. Therefore $T^2 = a^3 / k$ and $k = a^3 / T^2$.

$k = a^3 \div T^2$ the solid of the Earth is extending into space giving a mass factor to the point.

The solid
$$k = a^3 \div T^2$$

The liquid
$$k^{-1} = T^2 \div a^3$$

$k^{-1} = T^2 \div a^3$ the liquid of the outer space is extending into the solid Earth giving gravity.

$a^3 = (T^2 k)$ The Earth to the rest of the Universe is space a^3 spinning T^2 relative to a centre k^0. The Coanda-effect shows the flatness of the Earth spinning in relation to the curve of space forming the Earth in relation to singularity. There is no flat Earth $T^2 = a^3$ as much as there is no flat Universe that Einstein saw.

As soon as science released the world from a flat Earth, Einstein came along and gave us the flat Universe. To marry our logic with Einstein's calculations of singularity brought about some bizarre ideas. To incorporate a three-dimensional Universe into a flat Universe in singularity is not that simple. Whenever I am presented with this explanation I so dearly wish to ask the person presenting the argument what he did with the rest of the Universe? Where did that person find a place to go and hide the other five sides? Only nothing can vanish and the cosmos is definitely not nothing...well that is if you are not cursed with having a Newtonian's mentality. It seems that Einstein took the three dimensional six sided Universe away, put it somewhere I don't know and then left only one of the six sides on a photo image in place of where I am suppose to see only one of the six sides...and how am I suppose to know where he put the rest of the Universe! It is easy to draw a picture and tell the story of disappearing fairies and dragons but go and show me a fairy or a dragon. The reality in confirming whatever the picture should represent, becomes infinitely less plausible when our dimensional grasp of time and space makes our intellect insist on a three dimensional explaining. Here using a picture he is so smartly showing a stupid bloke such as me what happens when a Black Hole comes about and fancy the thought that of three sides supposed to form the three dimensions, two sides went fishing leaving one side to guard the place and now I

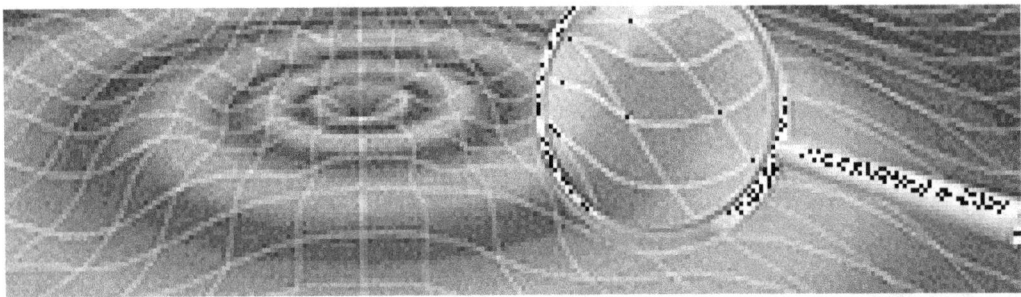

see the guard stuck with the job! My believing in this scenario hinges on two factors: How brilliantly incompetent the story teller is or how gloriously incompetent the story teller think I am. The two sides went A.W.A.L.L and I have the trustful third side not willing to leave its post. That is fine and that much I truly understand even with me being as stupid as I am. What I do not understand is where he put the other five sides of the Universe. He is showing one side of the Universe that has gone flat, but what did he do with the rest of the six-sided Universe.

How did it disappear and where is it gone to...how long will it stay there and how is it coming back...who is strong enough to bring it back...these are all viable questions asked in sincerity...therefore even in my stupidity I deserve an answer, after all it is my Universe too this fellow is messing with. Just look what did they do to my Universe...and who is going to repair it! How did the Universe get that flat because gravity is not in outer space...gravity is in the atom no, moreover it is in the proton and by having gravity means the contraction is even going smaller than the proton. It is the proton pulling space-time flat and not outer space pulling flat. They show a centre and they show a flat Universe with light of all things travelling about a flat Universe. How light which is the very focus of the three dimensional purpose of the 3D Universe can go flat and travel in a flat Universe along where no travelling space is available, but the light travelling is just the thing that is establishing the three dimensional space we see. How light under those circumstances can travel flat through no space is beyond that which I shall ever be able to understand. I see the marvel of a 3D Universe with the aid of light because I see light travel. I don't see a flat Universe having a flat image because the Universe through which the light travelled showed the flatness it went through as it passed right into my eye.

Fortunately, I am the stupid one around and they are highly educated Brainy Bunch and therefore I am permitted not to understand while they are not permitted to question their personal doubt in this matter. If this scenario they sketch is true, then the electron should pulls the atom flat because the electron is the epitome of light and is the indicator of what density light can ultimately achieve. This is total hogwash and bullshit. The electron forms gravity by maintaining the atom as a spherical object and the atom will always remain a 3 dimensional form of material. His picture lends no chance of anything becoming flat and therefore we have to start looking for something else that can direct us to the flat Universe Einstein discovered. However it is just as meaningless to say that Einstein was mad and wrong and there just can't be such a Universe anywhere.

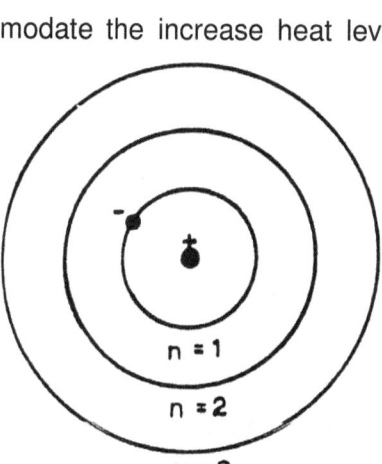

Whenever the electron jumps from a higher into the lowest (innermost) orbit, the atom gives out radiation at a wavelength corresponding to a spectral line of the Lyman series. The jumping down into the second lowest level contributes to the Blamer series. This is a vital clue. There is a release of nuclear charged heat and this indicates to a flow of heat that releases backwards as gravity reduces the flow of heat. If gravity is present then the indicator shows heat released from contraction.

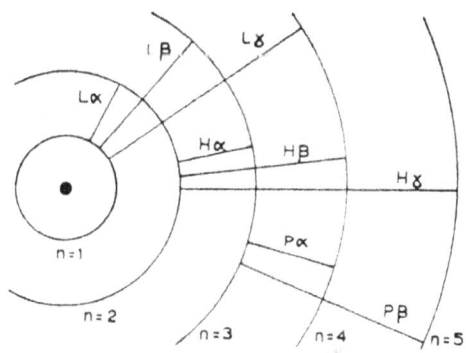

The greater the jump is the greater is the emitting of radiation to the limit of the series, which is reached when an electron enters from the outside of the atom. Outward jumps involve the absorbing of heat and that is inclining to provide space to accommodate the increase heat levels

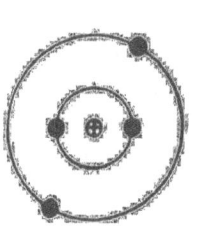

because of the increase or rise in the absorption lines. When the heat level in the atom rises, the electron jumps to a higher band and when the heat reduces in moves down one band. The heat coming about in the surrounding of the atom produces more space because the atom increases the space by applying the electron in a higher orbit ring. The moving of the electron is coupled to the giving out of radiation at a wavelength corresponding to the spectral line of the Lyman series. From this it is so obvious that gravity lends itself to an atom that fluctuates in dynamics by resizing in relation to gained or lost space and space represents heat that releases in concentration.

When the heat level rises or lowers the space within the atom decline or increases. Every atom in every element association shows different corresponding to the heat it is in association with. The corresponding of the atom and the reaction derived from such rising of heat turning into space, is a direct result of the interaction there is in the gravity contracting and the gravity in expanding depending on which sector corresponds to the dynamics if which of the actions of the Coanda principle is in dominance at the time in regard to where the viewer focuses on. The main issue is that the atom forms the Coanda gravity principle as liquid heat interact with atomic solid and the bands are clear indicators the Coanda principle where these lines forming the Lyman bands is adjusting relevancies in accordance with heat intensity requirements.

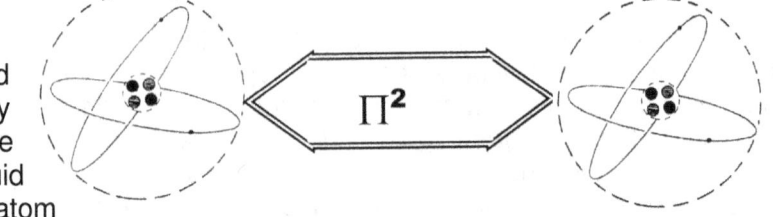

The rise in heat is a rise in the liquid part that extends the contracting by giving rise to the adding or the relenting of motion. The heat is liquid because the heat is motion and the atom inside becomes the solid since the motion is conserved by the spin of the electron. Today it is the rise of levels that is in focus but this same principle had to be in use when atoms were formed that today is responsible for elements.

If it was true about mass pulling mass by reducing distance then in that case the Big Bang was not possible and individual elements was not possible. With the cosmos down to the size of a neutron and mass confined to that space within the neutron that would form the recipe for the biggest crunch there could ever be. If the Big Bang was brought about mass confining mass by reducing the radius

that implied the space within as the Newton formula $F = G \dfrac{M_1 M_2}{r^2}$ would suggest, then

what would ever be more applicable than that moment to bring in all the forces the hell can unleash and destroy what ever was not yet in the Universe. The entire idea of mass pulling mass to reduce space is a prehistoric thought and explicitly incompetent in explaining science. The following is a far more suitable explanation and is as true as Kepler is. It is so obvious that gravity is part of the movement of the atom in the heat in which it spins and the Big Bang is the story of heat configuring

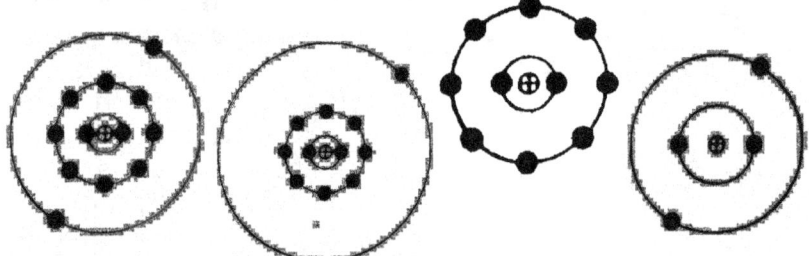

to form matter in space by gravity. Magnesium (12)   Sodium (11)   Neon (10)   Beryllium (4)

Those ones that were first formed were most of all very dense atoms where the first had to come in place when time was one point away from still being eternal. With the high velocity of movement applying at the time when atoms formed during the big Bang, the velocity condensed heat into huge piles of protons where each proton attached to a neutron and the neutron formed an electron to serve the chain with flowing heat and dominant and space infinite and one notch off singularity. We

still find the liquid time having a vital role in the space the atom uses. At the time when T^2 was almost eternal space was infinite because T^2 will not permit the space a^3 much room to be. But the opposite is also true that if time was steady then time being so long could pack in large numbers of protons with the accompanying neutrons at the time into the time unit in space that formed. Part of the sage was Kepler's formula applying to the detail of the forming of the atom and giving credence to the density required to form such enormous atoms.

This is most accurate, but this is only in concern of a unit in rotation that is in conflict of its own spin. When time in the cosmos at large view time in progress we find that the factors providing meaningful development are not quite so simple. Time moves by the measure of T^2 but time in progress is by the measure of k. Therefore space in progress of time is $a^3 = T^2 k$ where progress is T^2 but in relation to gravity we find that $a^{3\,-1}k = T^2$. The smaller the space $a^{3\,-1}k = T^2$ that is required to form the atom is, the higher was the rate of spin.

In the relevancy we find the action and reaction of space-time flow is $a^3 = T^2 k$ and that translates to being $T^2 = a^3 / k$ on the one side and $T^{-2} = k / a^3$. In the times we now live in we can and do produce an optical illusion of $T^{-2} = k / a^3$, but that is implementing the use of a telescope. In the true time we find as a cosmic reality the fact of $T^{-2} = k / a^3$ is rather a mathematical statement and no more than that. In reality we have $T^2 = a^3 / k$ on the one side as time expands and on the other side we find $k^{-1} = T^2 / a^3$. This we know is true because while it is possible by using an optical illusion the reality is that time can never reverse. In truth the reality about the opposing actions is that we find normal growth and that which Hubble first saw is the process of expanding space-time by the margin of $T^2 = a^3 / k$ while on the rebound we find the opposing while contracting space-time is $k^{-1} = T^2 / a^3$. The atom becomes the dead giveaway in supplying evidence supporting the idea that spin gradually relented as space grew to become more accommodating. Time is the relation that eternity forms with infinity. Physics to this day never tried to establish where infinity in the cosmos is and physics ignore the location of eternity on a daily basis...and then Newtonians feel surprised that they don't know what time is!

The Coanda effect exemplifies the definition of time or gravity, which is the same thing. Time is the movement of everything n relation to one specific point or gravity is the movement of eternity in relation to infinity. In the centre of the spinning top there runs a line that connect space on the one side to space on the other side without having any space at all.

On the outside we have space in the form of heat, which the spinning of the top turns from being a gas to being a liquid. That liquid is eternity because that liquid never ends being present throughout the entire Universe.

In order to understand time we have to return to the top because in spite of Newton's misgiving about the top rotating, it is through the ration of the top that we can learn so much more about the characteristics we find as gravity as it is practised by the atom.

To give the top motion line within To essential that represents the Coanda gravitational characteristics principle known as cosmic viability one must throw the top in and that supplied to the top, the throw initiates a time the top. To understand the atom forming gravity it is we understand the top because the top everything one can conclude from investigating effect and moreover it is an investigation of principles when we enter an investigation of the of gravity shown in the attitude of the working the Coanda effect. By casting a top with a string turned around the base of the top and allowing this string to rotate the top, we do what no other since God achieved. We created a Universe secluded and excluded from other all other concepts forming a Universe. Man establishes a separate identifiable time within the realms of another time. Man presents time by Π.

After all it is gravity that keeps the top as it is spinning in an upright position while it is spinning because it is gravity that stabilises the cosmos. Moreover, what is actually in progress from the top spinning is the Coanda principle activating gravity and that happens in accordance with Kepler's formula.

This means that in the cube at the point of contact between the cube and the sphere the cube experiences such a contact point as if the "bottom falls out" of the cube and without a "bottom" to support objects, they fall to the sphere as objects does fall to the Earth. The one side resolves its position in favour of the centre taking the side over. Remember that a body "floats" in space, but at one specific point it starts to "fall" to the Earth. That is gravity and it is a dimension change much more than any force. I shall explain this last remark later on. That too is the Lagrangian system with five cosmic structures holding relevancy to the centre structure where the centre structure stands in for seven positions diverting from singularity and the orbiting structures standing in for five positions in space.

From the

such a point it holds, when it leaves that point going to every other point, new point will be opposing any other point from where it came that was not pointing in the direction to which the first point is pointing, whereby it diverts from the direction it should extend and changes by $7°$ the direction it holds. No matter what the point is or where the point leads, such a point holding a specific direction will be unique in the direction it is rotating because at that or any other specific point wherever, it will be directing not in the direction it travels but spins in the direction flowing from the centre point outwards.

Any   point will be it opposing itself within the rotating of $180°$ changing
every   aspect of its previous flowing characteristics it previously had or will once again have in $180°$ from there. While in rotation from the point of an outside observer all may seem static and never changing but to the object in spin every next second will be a diverting from every aspect it was in every second passing, and the direction it held in relation to the direction it held the previous mille, mille second will totally be incompatible with the direction it holds the very next mille, mille second of rotation. That proves no point can be static or constant, all though it may seem that way to outsiders.

In the very centre of the sphere the form of the sphere dictates that the shape will relinquish space as the line runs from the outside towards the very centre. With this natural state of affairs the sphere is naturally inclined to dismiss all space that it can form in the form as the sphere holds space inside and the form will finally be without dimension. The line shrinking by reducing actually takes place in every sphere as the diameter reduces to the centre. In the centre where the radius line goes single, the form relinquishes the three dimensional form it has inside. Being without dimension in the very centre means that at a point in the extreme centre of all spheres there is a point that holds singularity because this point with no space has a mathematical position although it is invisible since there are no sides to such a point to give that point any dimensions. The shape of the sphere is calculated by using the formula $4\Pi (r^3) / 3$. By reducing r to a point where r is r^0, singularity steps in because only the form remains as Π. Going even further, we find that there then comes a point where Π goes singular Π^0. At that point absolute singularity is present but so is absolute gravity present at that point. When holding the strength of the shape of the sphere in mind as well as taking into account that all cosmic objects of importance are in the form of planets or stars and they are all in the form of a sphere, we therefore may contemplate that it is where gravity originates. We now only have to find the reason why gravity will hold a base in a space less ness as Einstein predicted. It is clear to be seen that gravity is in the centre of the sphere controlling from the centre everything that is outside the space less centre. We can reason with confidence that gravity is the strongest where space is the least. We can further reason that it is gravity that is holding the sphere in true form and since the sphere allows gravity the best working opportunity, gravity can form the sphere in as strong a shape and form as the sphere seems to have. From every point on the surface of the sphere is where that point connects with the other side of the surface of the sphere by a line that runs through the space less ness of such a centre of the sphere. Such a line also connects by an angle of 180^0 as well as 90^0 to six other lines running from top to bottom, right to left, and back to front, where all join and cross in the centre of the sphere. There are therefore six lines crossing and connecting by a centre from any given point on the surface of the sphere. Such points connect in total six surface points on each side of the sphere while they all support one another through the space less centre. In that absolute space less ness in the centre holding singularity we find gravity supporting and controlling all space within the sphere as well as space connected to the sphere. That is where gravity controls and guides the space, which falls in the parameters as well as under the influence of the form of the sphere. In the gravity centre space goes singular meaning space becomes space less or flat.

Also it is true that the entire form that is the sphere is controlled from a centre within the sphere. That centre holds the sphere in form and shape. Therefore the strong form is dictated from that space where there is no space and no form left. The natural inclining is in the form of the sphere. It is part of the roundness that the overall shape of the sphere represents and this structural strength is carrying down to the very centre. Because the circle is forever reducing, that reducing which is inherently part of the form of the sphere becomes a tool in distorting space in the sphere and is eventually removing all forms of space from within the centre of the sphere. The very centre ends up as having no space because of the reducing that continuous down to become the space less inner centre. The all roundness is the ingredient that forms the backbone of the absolute strength that the sphere has and that is the component that the sphere is so famous for. The form the sphere has allows the sphere to have a control that is coming from the centre deep inside the sphere where the space vanishes and being without space seems to keep the entire structure rigged. The strength of the sphere comes from the centre of the sphere, which is inherent of the shape. That is why the sphere has such strength in form and the fact that all connecting sides refer to a centre brings credence to the strength that the shape has. How does it work in its most basic analyses?

It is from the layout that the sphere uses as a natural form that we are able to locate singularity. In the case of the sphere the material naturally reduces by measure of the radius becoming smaller to a point where the radius is r^0. At that point the line that will form the radius has gone single dimensional r^0 and that is equal to 1^0, which is singularity.

The cube has sides and the sides form a rather weak and flat surface that connects four corners. The flat surface produces a rather indifferent contact point with no special features on the surface. The corners connect to other sets of corners and those corners form a weak structure without any direct support coming from the other five sides. Without material to fill the body of the cube the cube has no direct connection between any of the sides other than corners connecting at the edges of the sides. Taking the vantage from the point the sphere is holding from the centre out into space there are ten points connecting to the centre. In that are the dimensions of singularity connecting to space where five connect to space in the second dimension of singularity, and five connect in the third dimension of singularity. On the other hand, the cube does show a very different characteristic, which involves only six sides (at least) connected.

In the very centre of the sphere the form dictates that the shape will relinquish all grounds in space that it can hold and the form will finally be without dimension. Being without dimension means that at a point in the extreme centre of all spheres there is a point that holds singularity because this point with no space has a mathematical position although it is invisible since there are no sides to such a point to give that point any dimensions. When holding the strength of the shape of the sphere in mind as well as taking into account that all cosmic objects of importance are in the form of planets or stars and they are all using the form of a sphere, we therefore may contemplate that it is where gravity originates. We now only have to find the reason why gravity will hold a base in a space less ness as Einstein predicted. It is clear to be seen that gravity is in the centre of the sphere controlling from the centre everything that is outside the space less centre. We can reason with confidence that gravity is the strongest where space is the least. We can further reason that it is gravity that is holding the sphere in true form and since the sphere allows gravity the best working opportunity, gravity can form the sphere in as strong a shape and form as the sphere seems to have. From every point on the surface of the sphere is where that point connects with the other side of the surface of the sphere. All other possible points connect by a line that runs through the space less ness of such a centre of the sphere. Such a line also connects by an angle of 180^0 as well as 90^0 to six other lines running from top to bottom, right to left, and back to front, where all join and cross in the centre of the sphere. There are therefore always no less than six lines crossing and connecting by a centre from any given point on the surface of the sphere. Such points connect in total six surface points on each side of the sphere while they all support one another through the space less centre. In that absolute space less ness in the centre holding singularity we find gravity supporting and controlling all space within the sphere as well as space connected to the sphere. That is where gravity controls and guides the space, which falls in the parameters as well as under the influence of the form of the sphere. In the gravity centre space goes singular meaning space becomes space less or flat. That is where Einstein's Universe goes flat because that is where gravity is at its strongest. However my bringing up this statement brings me directly to the point where I get very confrontational about how the brilliant mathematicians treat those they suspect are less inclined to think.

By examining the form of the sphere, we find that there are 6 points on the surface of the sphere holding the form at a specific and equal distance from the centre. Lines run from the centre into space at $90°$ and $180°$ angles of each other from six opposing sides. There then are six lines at $90°$ and $180°$ connecting to the centre from six points on the outside edge of the sphere. As a result of the basic shape that a sphere has, there is a spot in the extreme inner centre of the sphere where the lines in $90°$ relevance cross each other and others connect by $180°$. There is also at that point a spot where all space relinquishes a position and only singularity 1^0 as form remains. At such a point we find the measure of the sphere being Πr^0 with $r^0 = 1^0$. That is where the line that represents the radius as a line disappears, as it becomes singularity r^0. After more reducing continue we get to such a point where we find only Π^0 left. At that extreme point is where space in all form disappears, as the circle providing the sphere the form the sphere has, removes all possible form by going into singularity $\Pi^0 = 1^0$.

Then in that area all form of any possible space disappeared leaving only the dimensions of singularity 1^0. I cannot delve deeper into the argument. However, from such a point there runs lines that connect to space on the outside where six points on the outside points connect to the space less point in the inside. In this book I take this argument much further but for now I leave the argument at that. Those lines carry the structural strength the sphere has. Contact with one point has support of six other points across the whole structure where the other six support every one of the six by singularity and the support runs through the entire sphere including the middle. Where there is no space, there must be singularity 1^0 just because the space filled with material removes zero and only material filled space is present. That means material fills the lot although in singularity 1^0. If zero was a factor where all space finally halted in zero as the value, then zero would be able to remove the space from the centre and such removing would continue to remove the space until all space was removed. It will finally abolish all space in the sphere and it would remove the sphere. Zero removes all possibilities of anything coming about. Since the sphere is there, a zero factor in the centre cannot be present. Only infinity can be a factor from where space may grow because infinity can extend and grow into and up to eternity.

The implication of this is that following the line down to the centre of the sphere we located the centre of the Universe. That is where gravity is. There is a lot more to that but be patient, we are getting there. In every centre we find a point, which is in truth not there but is the mainstay of all that is within the sphere. The mathematical value of such a point is $\Pi^0 r^0 = 1^0$ and 1^0 is singularity. That is the point where the Universe started and that is where the Universe will finally end. That is the Universe without space-time. That is $k^0 = a^3 / T^2 k$ which proves the Universe is without doubt a sphere...and we just located the centre of the Universe!

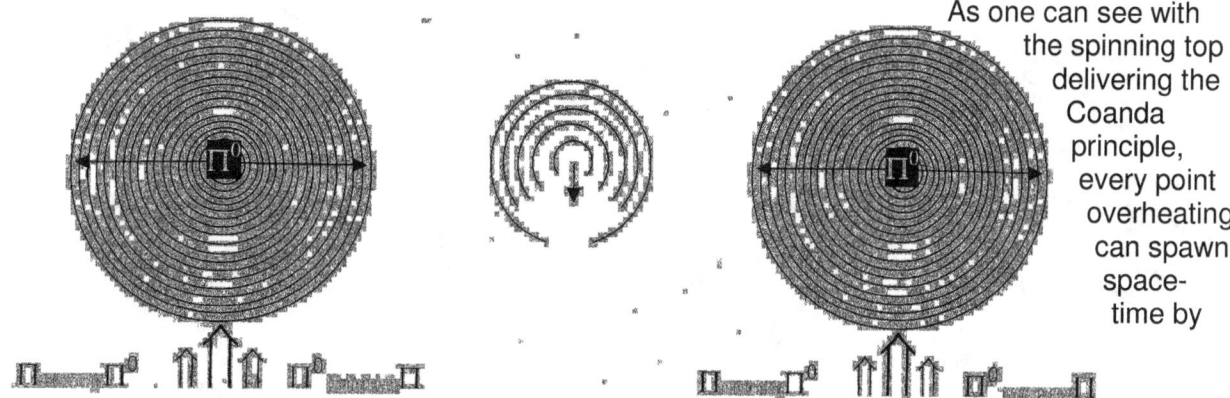

As one can see with the spinning top delivering the Coanda principle, every point overheating can spawn space-time by centralising singularity.

One can see from the top that singularity is established wherever spin occur. The motion generates a position of seven in relation to ten and singularity manifests as 1.9991 as is explained elsewhere. That means any point formed by the sphere spinning can and does start a centre in which no motion holds no space and of which motion surrounds such a point by forming space. Although everything at the time was in the form as a multiple circle, which results in a sphere, the sphere was not the only form present. This too has to do with singularity interpretations. We see a cube, as we know the cube but at first when form came about the cube were not yet a form.

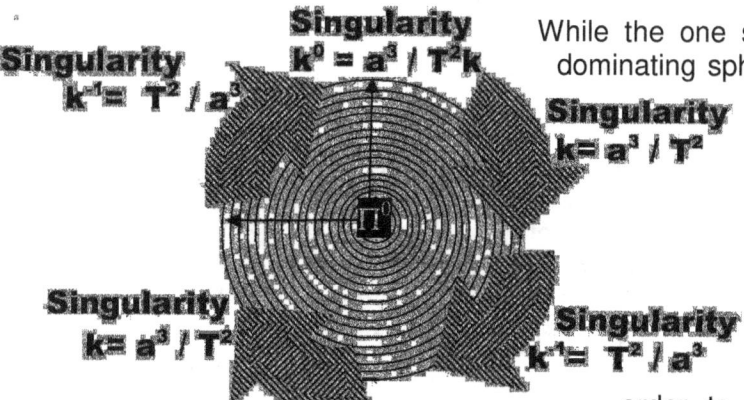

While the one sphere forms on this spot where the dominating sphere secures an edge the dot may be reserved as an edge marker to the dominating sphere. To the forming sphere in progress of emerging heat gathers at that point because the rotation is a result of duplicating and duplicating is the tendency of naturally growing in space-time $k = a^3 / T^2$. In order to find duplicating coming about there has to be heat in order to duplicate what will form heat. The duplicating process is a process of one factor going softer or less solid and therefore more dynamic than the other. To have singularity is to have gravity but to have gravity there has to be a point of motion and a point of sturdiness. The point of sturdy may be in the centre of singularity, but then the solid must be motion. However even today it still apply: what moves forms liquid in the presence of a solid and at that point singularity presented the solid therefore what we might think of as solid was the liquid because it moved around the solid.

Where the one factor is duplicating the other factor is compressing $k^{-1} = T^2 / a^3$

The points duplicating is four moving around a centre by the square of gravity. The motion is the sources of heating because the heat is bringing about the movement. The heat growth therefore provides the action because the action is what energises the points to provide the motion. The motion is purely is space-time duplicating and the duplicating is feeding heat to the centre from the four points overheating thus the points that shows expanding. But also the duplication leads to the spawning of one point of singularity that provides the installing of the next centre for the next sphere. Because of the principal in which the Coanda works the motion will centralise a new sphere and by appointing six positions around the centre three points will not move while four will move about the three points forming the centre line. The result is that the four points by duplication will reserve the point moving as the next point in singularity because of $k = a^3 / T^2$ singularity will be a natural result of the motion. Then that point will secure a position $k^{-1} = T^2 / a^3$ which will secure six points about such a centre. The centre will bring about four points spinning around three points holding a line singularity. The line in singularity will stand in relevance the contacting factor $k^{-1} = T^2 / a^3$ and the duplicating by expanding points will be four and serve the relevancy by contributing $k = a^3 / T^2$ as space-time only in form. From this the rest of the Universe burst into the next phase of Creation.

The gravity is in relation to the spin, which is in relation to the four points spinning which are $\Pi^2 / 2$ and that is the Roche limit. It is the dividing of singularity sharing space-time just as we on Earth share singularity by division between the Earth and us others that is not part of the Earth. The division of total that forms from the point that spawns is seven plus five plus pi square in another sphere four totalling twenty one that stands related to the first seven and once again formed. However this is an eternal relevancy that can never break.

Any object in rotation will have a middle point, a very specific centre point that does not spin. That point once again hypothetical but none the less must be standing still because every line running from that point in opposing directions are also in opposing directional spin to each other. Although the points had the same characteristics only seconds before, they oppose the characteristics it had just before and just after the very second in which they are and to which they relate by similar points also in rotation. Due to the spinning nature of such a point with all surrounding the point very varying second, the value of such a point can only be Π because of its constant changing.

As the top hit the ground after being thrown in a spin it starts to move around in small circles while circling around the axis in a vigorous manner, then it forms am almost motionless stance of complete blessedness as if the top is suddenly satisfied by energetically almost standing still while spinning in a precise circle while adhering to its axis and that is precisely what happens. This surging to find a new dynamic is a very important sign and is of most importance.

With all the excitement of being freed from the depressing of the Earth and no where to take go while enjoying every minute of it, the extending of the drive line runs down the line forming singularity as well as from the edges all the way inwards towards the newly established governing singularity that keeps the whole job erect.

That is why the top is spinning in the first place. The more assertive the spin is in velocity the more reaction there is from the lines running towards the centre and extending as it is

expanding outwards. In real terms the space of the top expands as the spin is in contact with more time in space during the same time in period and a bigger unit fills the space in which the top spins. In this the space in which the top spins the vigour and excitement has to allow the structure space in which to expand as well in order to compromise for material relevancy growth the expanding also serves the purpose of room to fit the newly acquired singularity governing the motion that extends and asserts influence to the edge of space-time.

The support that the spinning top finds in the established governing singularity keeps the top spinning in an upright stance only supported by the singularity that takes charge of the spinning of space-time within the set boundaries to establish time within the Earth's time restraint.

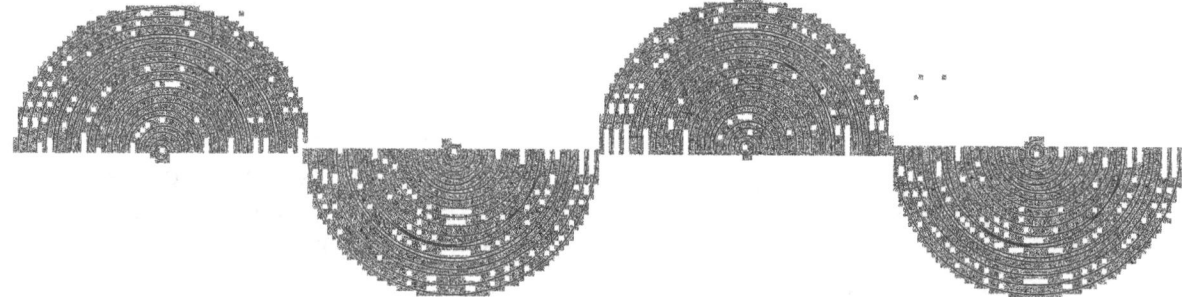

The heat that should supposedly under cosmic law drive the spinning top will come from the governing singularity accumulating the heat in concentration by the contraction or cooling ability the top singularity acquired. But in this case the spin is a result of life's ability to manipulate space-time and lead cosmic events. The heat that would establish such a drive in motion in real cosmic terms would require a lot of nourishing and sustaining from a large number of maintaining atoms that produce a large flow of space-time.

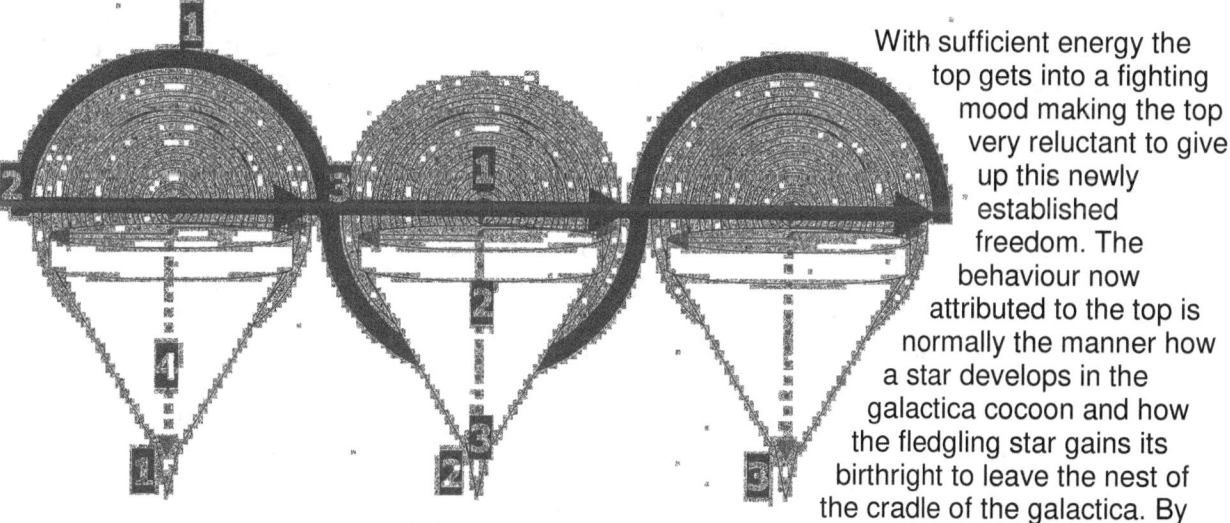

With sufficient energy the top gets into a fighting mood making the top very reluctant to give up this newly established freedom. The behaviour now attributed to the top is normally the manner how a star develops in the galactica cocoon and how the fledgling star gains its birthright to leave the nest of the cradle of the galactica. By the atoms forming a sum total that can support the generating the required gravity to secure the heat that would unleash such a drive can singularity that is governing the structure movement come to life and release the new star from the blanket of heat that covered the star up to the time of this release from the galactica centre.

The example we can gather from the top shows how desperate a governing singularity can become and how such an exited singularity can put up a fight for life and independence. The top is in a fight for independence while the Earth is restraining the independence. The fight goes on until the Earth finally suppresses the last bit of motion that the top had and the top uses the last motion it has to defy the Earth's domineering control.

When the motion exceeds the level of the Earth gravity, the top shows an eagerness to rise to higher levels of independence in the same manner that an electron reaches into higher rings of energy because the top with motion is in an electron relation with the Earth filling

the role as the proton would play its role within the atom and that puts the atmosphere in the neutron role.

In the sphere centre is a spot that has to be there mathematically by the calculating and measuring of the defining space of any circle we find singularity $(\Pi r^2) / (\Pi r^2) = \Pi^0 r^0 = 1$. In order to provoke the line into action, there is motion requires to excite singularity just as Kepler indicated, where the space becomes equal to the motion and the motion is equal to the space $a^3 = T^2 k$.

Let's quickly establish events as they translate singularity from a dot to a controlling entity that is commanding space-time through the establishing of a separate individual drive. The motion comes about which proves to be that which generates the gravity that drives the individuality in the top.

By the motion and the singularity the top evoke graph forms where the graph runs along the line of time. This makes nonsense of Newton's presumptions that the spin nullifies the space.

The balance is a control of motion that is established as a flow of space-time supports the ends (4) holding time while this generates the space (3) singularity containing and creating the space (3) in which the spinning takes place. **There is a something (if you wish I'll use the term force although I strongly hesitate to use such an outrageous term) that is generating power to keep the top upright while the top is spinning. The energy that is charged, has the dynamics stand its ground against the might of the gravity of the Earth that is under normal circumstances controlling the stance that the top has to take, but as if inspired, as the top seems to be revivifying by motion, the top is fighting and rebelling against the Earth's constraining gravity. The top is self-driven, as an electric motor would be. The difference between it and an electric motor would be the origin of the source from where the energy comes which drives the spinning top. The top stands upright as individual as any self-propelled object can be. Although gravity is retaining the motion of the top, it is not contradicting the motion, all though it still restrains the actions. It is not combating, but is merely suppressing the motion. In this there is no race of evidence linking mass as a factor to any of the above mentioned actions. What we would think of as air restriction is no restriction because from the restriction comes support that keeps the top standing on a very thin needle edge. The top should tell us so much about nature if we would only listen and learn and not tell nature what we think nature should tell us.**

By investigating how singularity interacts with Π when forming space-time the value of singularity becomes multi facet but moreover, the radius takes up the role as an indicator to match and underwrite Π in multi dimensions.

From the motion the top inspires by creating a situation that the top can establish a force or an energy-driven time line, which is able to keep the top upright, and all of that is equal to the establishing of gravity and electricity. By charging electricity we enlist the very same principle behind what is being the Coanda principal.

It can be only singularity that is keeping the top upright. One must remember that the part doing the balancing, that is creating the space in which the top is able to spin, that is establishing the necessary time distance in which the top can spin, that is establishing the time difference that the top can use to apply the motion that extends the time, is pointing to where we find singularity. The line that is evoked by the movement is not in real terms part of the Universe because the line has no sides, can't move at all and is the only substance that is there fails all those wishing to see everything there ever can be and is still there while being very much invisible.

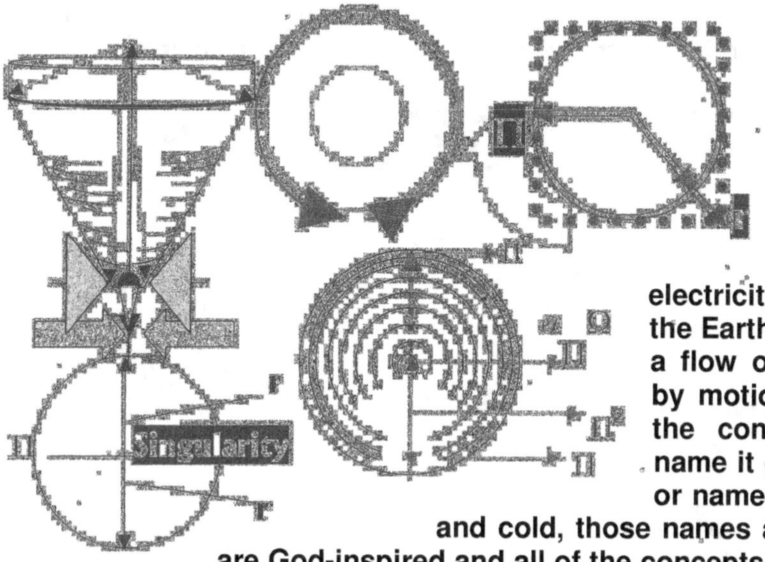

That which charges the top to stay upright is the same that charges the generator to charge electricity, is the same as that which charge the Earth with gravity. There is no force but a flow of space-time, which is contracted by motion and duplication to bring about the conducting of electricity. Man may name it gravity and then name it electricity or name it motion and balance but like hot and cold, those names are man-made while the principles are God-inspired and all of the concepts that has different names are in the end, still the same. To the cosmos all these different concept hold dimensional differentiation while honouring the same integrity in the Universe. The lot is still the **same thing. It is what started the cosmos. It** is what drives the cosmos. It is **the engine giving motion as it is giving** discipline in the cosmos. It is **producing space to heat and heat to** material.

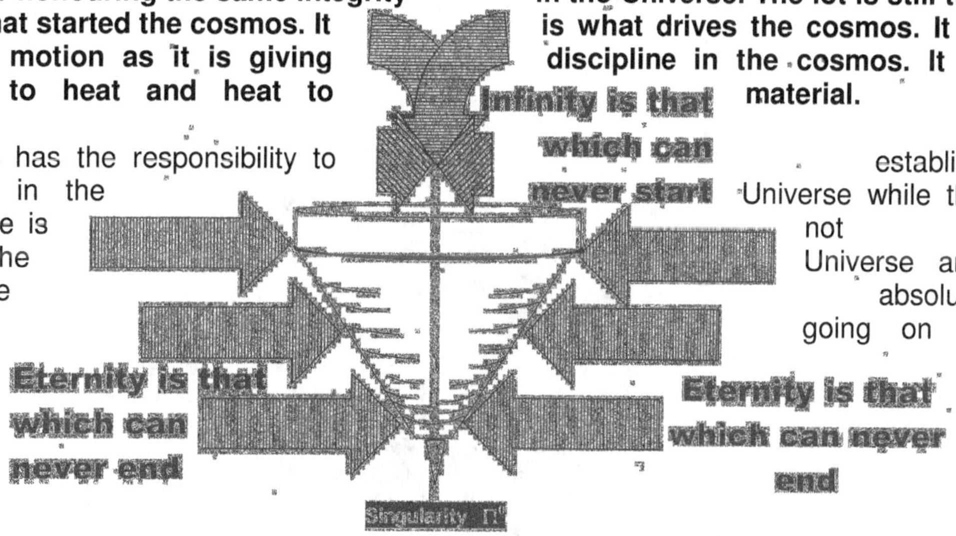

The line that forms has the responsibility to everything that is in the line by own measure is functioning part of the yet it sets the control of what is our Universe. Without the line being part of space in the Universe it is what drives the establish Universe while the not a Universe and absolute going on in

Eternity is that which can never end

Infinity is that which can never start

Eternity is that which can never end

Singularity

Universe in time and in detectible, has no space, line can be called into action the line immediately Universe and creates matter accordance with time. The line is invisible, in claims no tangible part in the Universe, however the any place and anywhere from where establishes control of what is in the in space in time in the Universe in infinite detail while the line is even less than part of the cosmos while being infinite.

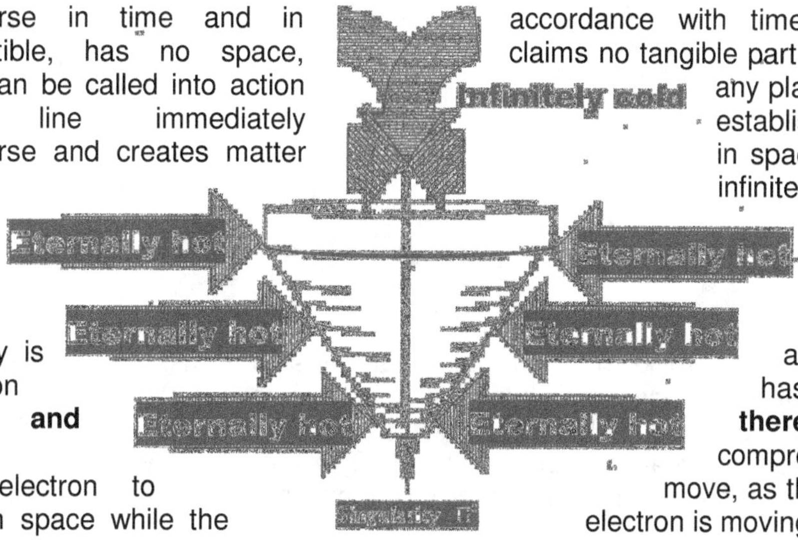

Infinitely cold

We gravity is rotation T^2k **and** must the electron to proton space while the

Eternally hot   Eternally hot   Eternally hot   Eternally hot   Eternally hot   Eternally hot

Singularity

have to recognise that a balancing of motion where has to refer to linear motion $a^3 =$ **therefore** $T^2 = a^3 / k$. The proton compromise in its motion to enable move, as the electron has to provide the electron is moving.

Therefore the spin involves space already established and the space contracting is space of other parts that overheated and in relevancy liquefied to become fluid heat. Therefore the expanding becomes part of the growth as it becomes the contracted space. It is this effect that we gave the name the Coanda effect and is as much part of gravity as gravity is part of gravity. The spinning top again is as much proof that the Coanda effect is a product of gravity as flying is proof of the Coanda effect where it establishes antigravity through motion. As the top spins, a centre is established with the motion of spin activating the centre not spinning. The centre comes about, as the centre remains motionless while the rest is spinning and the centre becomes an additional part being part of the motion of the top.

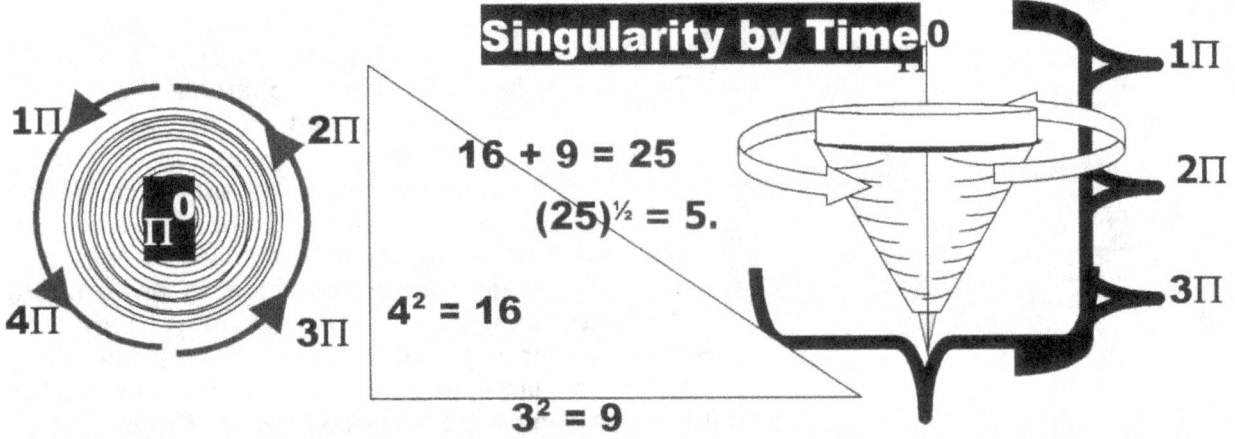

When creation established space-time for the very first time, it did so by parting infinity from eternity. That which cannot end parted from that which cannot begin. Eternity was interrupted by infinity as much as infinity interfered with eternity. What can never end came loose from what can never start. It is not the same thing although we as humans tend to regard the matter as such. The one is having a look at it from the one side and the other is looking at what happened from another perspective. It is in relevance that the entirety rests because it is born from relevance that the entirety came about. God said "Let there be light and there was light" and no Newtonian, believer or atheist alike, took this to heart... This means God said "Let there be hot and let there be cold" and that still is what the Universe still is today and ever since Creation came about. Let there be infinity and let there be eternity and let heat separate the two by movement because light is the movement of heat through space.

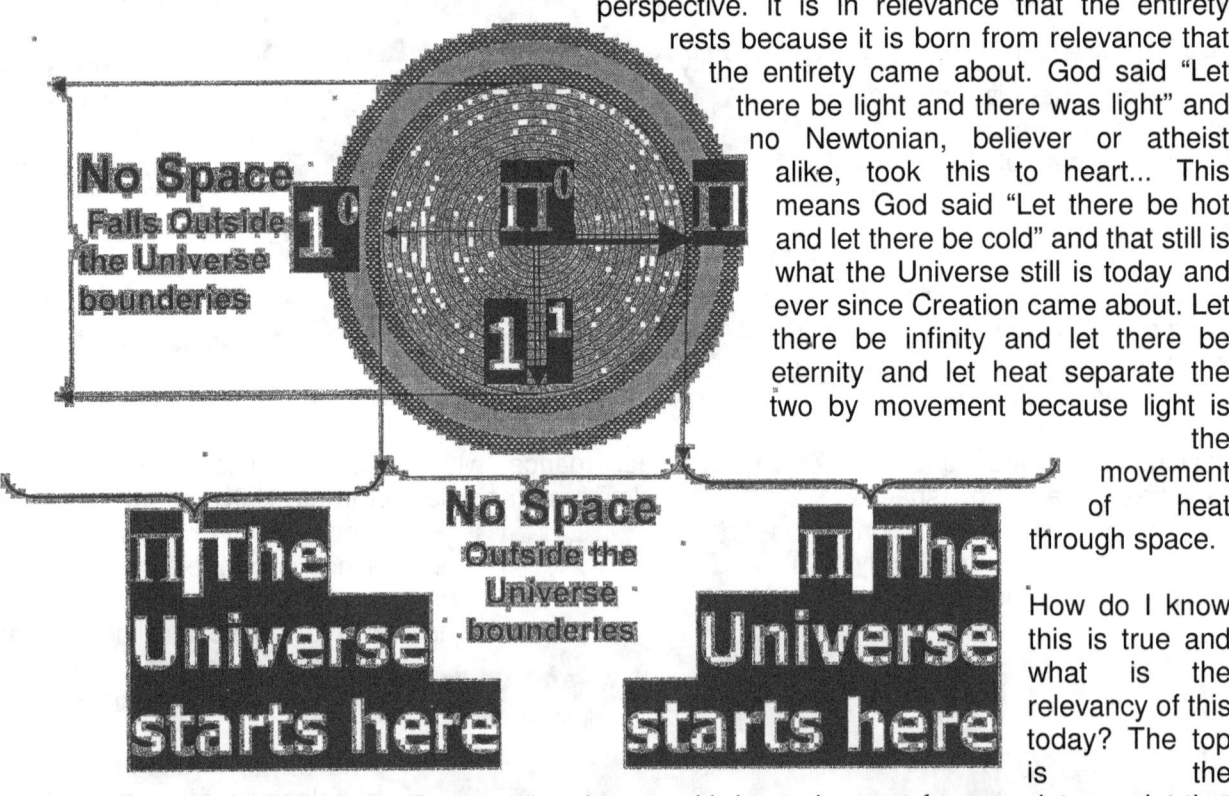

How do I know this is true and what is the relevancy of this today? The top is the manifestation of the Biblical text. By providing the top with heat, the spot forms a dot or point that forms a line that divides that which can never start from that which can never end. By supplying the top with motion we can witness the birth of a Universe, which is the birth of a Universe, which is the parting of what can never move from what can never stand still. The line in infinity can never move and we see with the overheating of outer space called the Hubble expansion; outer space can never stop moving.

The Universe expanded by measure it could never repeat again afterwards since then

Think of what intensity there was when the command established an entire creation filled with possibilities of the possibilities ended in establishing the atom and the entirety is filled with atoms forming the pot that is out there to see. That is when the sot formed the dot.

I wish to deliberate the dot in better detail. In a close inspection of the dot as the dot is in our current Universe we find the dot not holding a quantifiable measured value except for being the numerical value of 1 to whatever exponential value one wish to apply. That is singularity and even if there is a mathematical wizard such as everyone says Einstein was, and no matter what genius his mathematical wisdom represented, if Einstein with all his formidable mathematical skills gave singularity any quantifiable value other than 1, then all his wisdom was washed down the toilet.

What I now intend to do I am not allowed to do for this which I now am going to do is completely beyond my powers and outside the scope of my abilities. Although is beyond my abilities I am gong to do what I am not allowed to do for once. I am drawing a sketch of what could ever be drawn because it has no space to draw what is has not.

A spot formed (I put is in the past tense to make better connection to what the human mind thinks) and moving from this spot going in every direction forms one straight line. As and before the line starts it ends as it reaches its final destiny and at the point it started it ends. As it begins it also ends and that sounds a lot like the Newtonian equivalent of what they think is nothing, but this line is so vivid it controls ever aspect ever thought to exist in our Universe. Only where this line ends does the Universe start. As it starts it also ends. This line that is not is holding an entire Universe being between a spot and a dot. The picture I just drew I draw in confidence of being the same value as the thought and where singularity is equal to the thought I use the thought to convey intelligence.

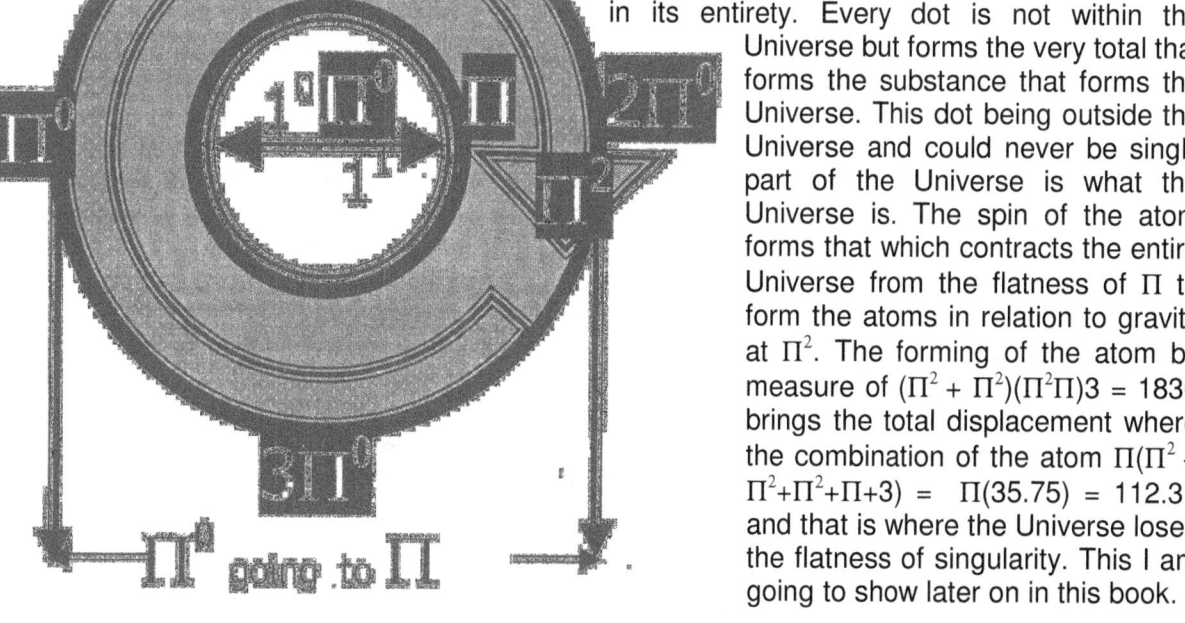

This spot forming the dot is forming the Universe in its entirety. Every dot is not within the Universe but forms the very total that forms the substance that forms the Universe. This dot being outside the Universe and could never be singly part of the Universe is what the Universe is. The spin of the atom forms that which contracts the entire Universe from the flatness of Π to form the atoms in relation to gravity at Π^2. The forming of the atom by measure of $(\Pi^2 + \Pi^2)(\Pi^2\Pi)3 = 1836$ brings the total displacement where the combination of the atom $\Pi(\Pi^2 + \Pi^2+\Pi^2+\Pi+3) = \Pi(35.75) = 112.31$ and that is where the Universe loses the flatness of singularity. This I am going to show later on in this book.

This spot that forms the dot is not tangible, even in the terms we apply to anything being a part of the reality in the cosmos. It is similar to all the characteristics we have in a thought. In fact it has everything that a thought must have. Having a thought in my mind is a reality as far as I am concerned but in the view of the person next to me the thought has no right to exist. The thought may drive me in reacting in any manner that comes to mind and the person next to me will never be aware of the thought I have. The fact that I have to have a thought is a proven fact without ever doubting such a statement. No person can have a mind that is clear of thoughts because in such an event such a person is brain dead. The thought is there in my mind while to others it should be there in my mind but the content it holds is as good as me being present without the thought being there. It is present by never being visible and not being tangible. It forms a silent improvable reality and me being present is the only proof that makes the thought being present in the view of others. One of the only ways to make the thought tangible is by speaking the thought in words. I have to reproduce the thought into a voice box and translate ideas into verbal language to make the thought a reality to others. If I wish another person should regard the thought I have with understanding I have to transform the idea I have to a voice and in the voice I use I reproduce the idea in terms of sound. But this action relates to time. Having the thought also related to time but the time spent on the thought while thinking was much less than the time used to find a means in converting the thought to sound. I transform my thought to another person where I then hope that person will translate the sound back to a thought that that person will have. No part of what I just mentioned has any tangible evidence, which I can touch or hold but notwithstanding that, it is part of the most valid influences of the human being in the Universe. Never can I reach any part of what I just said and still that is what controls the Human race. If not for the ability to think and to produce conversation by voice the thoughts I produce I would not be classified as a member of the human species. That is what makes me part of being man, the most dominant specie ever to live on Earth. If not spoken, my thought will go lost forever because I did not share the thought with any person. Yet the thought had validity because the thought controlled my life for a period and an impulse. I might try to determine the time it took me to think the thought I had, but that will be in vain because no chronograph can measure such time response. If I try to measure the time it took to transform the thought to sound through applying voice that can be measured. That is because it is done by time that already formed space.

All this arguing is to show those Newtonian nihilists that it is what they and I can't see, touch or feel that is in control of our daily life. Without the thought the deed will be impossible. But the conveying of the thought gives the thought a lasting reality because it then is shared with others. It is the sharing with others and the way others receive the thought and measure the thought that gives the thought eternal recognition. If not one of the Greek Masters back then shared their thoughts with the world we would be much less wise and they would have been lost in time. They are remembered not only for their thoughts but also for their sharing of their thoughts with those to come in the future.

By speaking, one convey that which has no substance to another point with something that has substance where it reaches the other point where the point it reached transforms the substance to something without substance being a thought once more. The thought holds singularity because singularity is not while it is in control of everything and that is what is making singularity a reality. The thought conveys what we cannot dispute because one can even sometimes read the thoughts of animals. The thought is in or minds but one cannot place the thought in the electricity. The thought comes to form a charge of electricity when the presence of the thought generated the electric impulse but first it has to be generated, and that is by way of thought. That is where the Newtonian nihilists make their second huge error. As mathematicians their world runs on simplicity by the repeat of what they understand and to reproduce what they know. Those Newtonian nihilists are of the opinion that computers may replace man and intelligence. That is never a possibility because one can feed computers information and numbers and the computer will spew out answers of such calculation, the going of the calculations is a machine operating and whether the Newtonian nihilists does the calculations or the machine does such calculations, it is a repeat of what was taught by humans first of all by repeating man's thought.

By reproducing the though in the style the machine does is using repeat in a process of calculations and that part could be repeated by electric programming as it stays in the scope of a mathematician or a computing machine and it remains a case of monkey see monkey do. Let them that consider them so awfully wise because they have the ability to calculate some ludicrous idea such as a space whirl, start to think about the complex issues and then see how a computer reacts to that. They can instruct the machine to realize the forces involved but they miss on the human part in how it is achievable. They are on Earth and yet they think of travelling to the nearest Black Hole, wherever that might be, without giving recognition to the possibility they have no idea where or what a Black Hole is. How they are going to get there is not part of their computed brain, because if they could compute by using the magic of going mass by force that could never be, then sure they could take magic one step further and make more shambles of science. How they will be able to manipulate the Black Hole in the Space worm they invent, is also gone past their obscure simplistic mentality.

They find admiration in their work on the merit the mathematics presented and replace the dignity of reality with the obscurity that mathematics could deliver. When the first Madman came up with the idea of the possibility of forming space worms, the reality of getting to the Space Worm or locating the Space Worm (the thing is pitch black and without light) went undetected and nobody seemed to notice or care to notice the reality posed by the suggestion. Every Newtonian grabbed for a calculator to calculate what this simplistic-minded-Newtonian-idiot calculated. Not one person rubbishes the thought on pure merit and every Newtonian nihilist sends the idea of this being a practical possibility out to the world at large that this is a possibility on the conditions that a computer computing formulas and figures may adjudge the correctness and plausibility. Not one Newtonian nihilist think of the overall picture as a travesty of common sense and the entire idea is at best a fairy tail told by the one that tries to score recognition on the blind trust of others. Not one Newtonian places the possibility of Space Whirls in the category of Harry Potter! That my friend is the type of scientist the Newtonians has become and it is because they hold onto Newtonian imaginary mass and pressure for deer life

Realities such as getting there, finding the object, fighting the gravity (remember this object has the gravity to absorb the entire solar system in a matter of minutes and that is according to their calculated efforts on the Black Hole) Those capable of designing space whirls were in charge of the mathematics of the Universe and still they could not solve such a simple issue as gravity in all the time they had to their disposal over the past three hundred and fifty years. It takes human effort to recognize reality. It takes a human's intellectual effort to recognize the Godly aspect to the Universe. The thought manifests as it sparks an electrical impulse but the impulse is no component harbouring life, t is only reflecting on the wishes life imposes resulting from the input of a thought, and a thought is with singularity and not within the cosmos. Only after the conclusion of the thought is realized and then by a human's deciding ability the result coming from the conclusion of the thought is transferred to other parts of the body in electricity conducting. Normally it is the emotion provoking the thought that generates the thought because it is the emotion attaching thought to the moment that places the thought in the mind. The machine can never reason and challenge conceptions but will always accept conceptions unconditionally a Newtonians are brainwashed to do in science.

To them and theirs and those, the atheist and the Mathematician that is substituting the Creator with the power of their mathematics, to them I say that the computer may replace them and I hope it will the sooner because that will make life the better for the rest of us that can face reality and not hide behind computing skills that machines can replace or replace a living God they are unable to see with the mathematics by which they are able to cheat reality. Again I will break my personal rule by using a picture of that which could never be pictured.

The shortest line in the realm of possibilities must have a start and finish holding one spot and such a line will also be a dot or a circle. Not favouring one direction puts all directions at equilibrium meaning that any form of what ever might develop from such a spot with the end and the start being in the same position also has to be a sphere because the flow outward will be equal in all directions. This reasoning prompted me to look for singularity in such a spot because if the prime spot from which all came was a spot holding all, then the spot must hold the shortest line but more prominent it will hold the smallest form including the smallest circle or for that matter the smallest sphere. One possibility that the shortest line or smallest spot can never have is having a starting point on the zero mark. If the mark of zero holds the start it must also hold the end because the end and the beginning have the same position. If the position of zero then is the beginning, the end will also be zero leaving the line or spot without an end as well as without a beginning. Such a spot will constitute all of nothing.

As **I am** of the most sincere **belief that mathematics is the only indicator that can lead man to the initial creation of the Universe** in man's quest there of **one must take mathematics back to the most original start** because it is **what one may find in mathematics** that will hold the key to unlocking the answers to questions that are unknown to us up to now. In that one may find the origination of the cosmos. That way will be the only way possible locating singularity. It might seem as if I am contradicting myself by going against the use of mathematics and then the very next line admit that I am a firm believer in mathematics. When I say I am in favour of mathematics, then it should be used to underwrite reality and to confirm the truth. When I say I am against mathematics then I say it with keeping in mind Newton's manipulation of mathematics to circumvent the truth, and

that I reject. Newton came up with $F = \dfrac{r^2}{M_1 M_2}$ and when that did not bring the results he

hoped for he changed the formula to $F \, \alpha \, \dfrac{M_1 M_2}{r^2}$ with the intention of replacing the formula

to become $F = \dfrac{M_1 M_2}{r^2}$ and when this was not proving his point he changed the formula again

altogether to $F = G \dfrac{M_1 M_2}{r^2}$ without once indicating on what grounds did he do all the

changing while intending for the result he hoped for to remain the same! All this clowning around with mathematics is proving one point and that is to show how far was Newton prepared to go with his cheating in the art of mathematics to prove his skew and crocket physics.

The key to unlocking the unknown facts about the cosmos is not vested in the calculations of the mathematics or formulating the equations, but it is in the manner in which those mathematical factors are interpreted about the cosmos and it is translating the mathematical equations into a viable and cosmic sustainable logic that would represent correctness. Mathematics is a language and mathematicians are translators of the mathematical coded language translating a coded text written in equations into a verbally used form of communications such as a language. This translating of coded mathematical messages is about bringing across the correct interpretation of the coded message and not making it up as you go along to prove a point as Newton did. This translating is that which becomes the over riding important issue. Of all factors, it is the translation that holds the key to understanding the cosmos because if the messenger cannot decipher the codes correctly the translations would be wasted. When making incorrect deductions from incoherent translations it will sustain facts that produce a silly theory most people would find laughable. This is not only a symptom of our modern age but it can run back as far as civilization takes the mind of mathematics.

One conception that has been with us from civilization beginning is the way we think that a line starts from zero while it is not

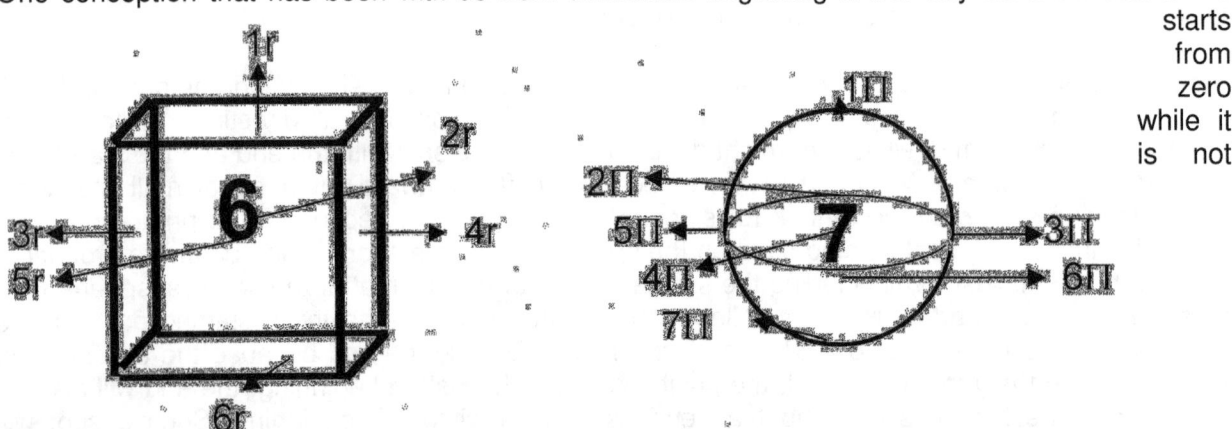

possible that any line cans start from zero. By starting a line from zero the starting with zero has removed the line and therefore the line cannot start...it is removed by zero. By allowing the line to grow with what it used as it started and continued using that which it used to start with, then from beginning starting with zero there is nothing to grow because the growth was removed by using zero $\dfrac{dJ}{dt} = 0$. Zero cannot accumulate and zero cannot progress. Zero removes any value from the mathematical line because by zero continuing there is no progress registered as growth. Placing zero at any point removes any possible point from that location and therefore removes all future possibilities to development from that point. Zero removes and cannot construct. The importance about this issue is in the argument of where did the Universe begin and how did it grow. The Universe is about lines connecting objects because it is such lines that light has to follow in order to flow from point to point when connecting the objects that is connecting time. When reducing the length of the line to an infinite number ($k^0 = 1^0$) it leaves free, all possibilities that may grow from such a point including a line $\dfrac{dJ}{dt} = 1^0$. Replace ($k^0 = 0$) and no possible growth can extend in any way imaginable. It is counter productive to place zero as the first starting point in the path of progress in any line. With zero at the start, it not only dissolves progress but it nullifies any start. Even considering a progress is improbable because zero removes anything constructive to grow as only zero then is in place and only zero has a possibility to become more ($1 \times 0 = 0$). Zero is the only thing that cannot become more... Only by correcting such an age old misconception about mathematics is there any possibility to advance with a theory past the limitations and barriers as we find taking development into a pre- Big Bang environment. By removing nothing we find a manner to enter the arena where we can locate the true pre Big Bang cosmic birth. Before the Big Bang there was form and not size. The Big Bang brought about size when form was already present. Think about it in this way...

Ever considered why a water drop would freely choose to form a sphere? When a water drop is released in an astronaut's space capsule in outer space the water drop forms a sphere as it floats in free gravity. As soon as the water drop is released from the Earth gravity and has the opportunity to float in space, it can form any shape it finds pleasing yet it immediately turns to the shape of a sphere.

It is the same reason why we would think the Universe takes on the shape of a sphere, although we know the Universe has no outside and we realise that the Universe is limitless in size. Yet notwithstanding, we take the shape of the Universe as naturally being a sphere that formed...but why think of a sphere that the cosmos consider as the sphere being the pre-cast shape?

Well by saying it is the strongest form available has the same argumentative potential as saying a baby is little at birth. **Blaming it on gravity is rather avoiding the question with a most simplistic answer because there still is no reason substantiating proper evidence that would indicate some facts more prudent than just a simple answer to dodge the question by hiding obvious incompetence.** I'll give you one clue...it has something to do with finding the centre of the Universe. **That brings on the next question being where one might locate the centre of the Universe.** These questions put to you are most ordinary questions...and yet the complexity in the answers rises far above the answering ability of those with the supreme mathematical skill. **Those master mathematicians cannot answer such simple questions by using their complex mathematical powers. If they could have, then they would have...but the answer is in the simplicity of true mathematics $\Pi r^0 = \Pi$.**

Have you ever thought why we would think of the Universe as a sphere? Look at all pictures about our expanding Universe and it is striking to find that lot is round, and most is showing a sphere that holds more space with every picture that implicates the cosmos growing. If you were one of us that always think of the Universe in terms of a sphere, then why would you say you presume such a picture of a sphere would most probably be correct in any way?

Ask yourself why do you accept that...the fact that the Universe ultimately forms a sphere...
In the event where you may consider that question to be rather trivial I would suggest that you reconsider your opinion because there is nothing trivial about that question. In answering this question about presuming the Universe to form as a sphere, one will arrive at the point where we can begin in finding answers to the cosmos. It is the spontaneous choice the humans make by selecting a sphere when thinking about the cosmos in an overall picture that we should address. From logic comes the answer about the sphere being the form that inspires the cosmos. To find solutions we must then investigate the cosmos in that direction. If you think this question is unimportant and trivial...then you better think again!

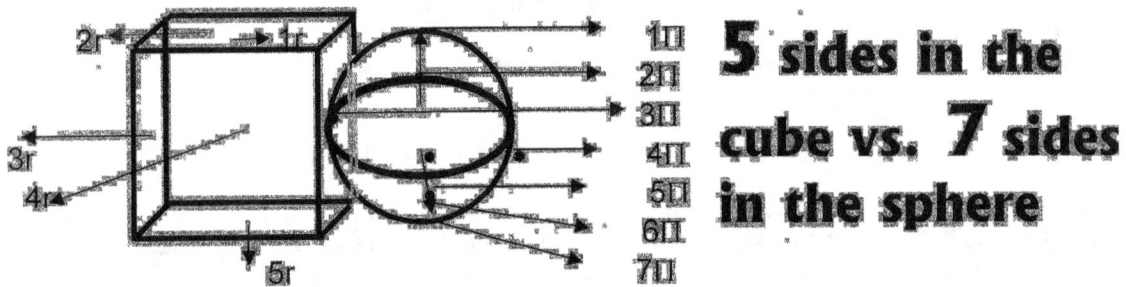

In the very centre of the sphere the form dictates that the shape will relinquish all grounds in space that it can hold and the form will finally be without dimension. Being without dimension means that at a point in the extreme centre of all spheres there are a point that holds singularity because this point with no space has a mathematical position although it is invisible since there is no sides to such a point to give that point any dimensions. When holding the strength of the shape of the sphere in mind as well as taking into account that all cosmos objects of importance is in the form of planets or stars and they are all in the form of a sphere, we therefore may contemplate that it is where gravity originate. We now only have to find the reason why gravity will hold a base in a space less ness as Einstein predicted. It is clear to be seen that gravity is in the centre of the sphere controlling from the centre everything that is outside the space less centre. We can reason with confidence that gravity is the strongest where space is the least. We can further reason that it is gravity that is holding the sphere in true form and since the sphere allow gravity the best working opportunity, gravity can form the sphere in as strong a shape and form as the sphere seems to have. From every point on the surface of the sphere is where that point connects with the other side of the surface of the sphere by a line that runs through the space less ness of such a centre of the sphere. Such a line also connects by an angle of 180^0 as well as 90^0 to six other lines running from top to bottom, right to left, and back to front, where all join and cross in the centre of the sphere. There are therefore six lines crossing and connecting by a centre from any given point on the surface of the sphere. Such points connect in total six surface points on each side of the sphere while they all support one another through the space less centre. In that absolute space less ness in the centre holding singularity we find gravity supporting and controlling all space within the sphere as well as space connected to the sphere. That is where gravity control and guide the space, which falls in the parameters as well as under the influence of the form of the sphere. In the gravity centre space goes singular meaning space becomes space less or flat. That is where Einstein's Universe goes flat because that is where gravity is at its strongest. However my bringing up this statement brings me directly to the point where I get very confrontational about how the brilliant mathematicians treat those they suspect are less inclined to think. Gravity is about form rotating in form and the result of forms aligning brings about gravity as Π^2. However, it has nothing to do with mass and the toughest thing (I think) there can be is fighting a myth (like Newton's corrupted physics putting everything on the mythological mass) that can't be, that never was and that holds no proof and yet the entire human species are fixated on the corrupted lie.

I admit that most if not all of the content I present in my work and as my work is new but that is only by my simple mind I have a simple approach and in that I have an unusual approach to cosmology. Though my approach is unconventional it enabled me in locating **the precise location of singularity that forms the connecting basis of the Universe** (and this I say with some degree of confidence). There **are two locations** but I shall **first concentrate** my explaining effort on **the prime singularity**. I have in the past contacted various academics in South Africa but I suspect that because I did not send a complete manuscript to some individual academics in some cases, I was unable to generate the required enthusiasm.

On other occasions where I did send complete manuscripts the shear volume may have (I suspect) eroded their inquisitiveness. **My work is radical** and **it is unconventional**; that I admit. It seems in Africa no academic wants to be part of a radical approach. I have confidence in my work and through that I have confidence in finding answers not yet found in all attempts previously made by the most powerful minds because powerful minds generate powerful thoughts while my capabilities only stretch as far as simplicity goes. Therefore, Professor Strauss, I thought it might be beneficial as much as it might be promotional on my part in sending you a manuscript of my book(s). I am not being arrogant when I say that I uncovered a mistake in science. Whenever I make this declaration all academics I have contacted up to this point in my life was immediately in prejudice and I do not blame them, but I beg of you to keep an open mind and read the pages forming the entire volume and then decide with an impartial, non-bias and non-judgemental approach about the evidence I present. Finding an academic with those qualifications was my biggest quest thus far and proved my downfall in my effort to find a promoter.

At first one may think of such a mistake as that I have discovered as trivial. It seems minuscule small and should not bother any person any way. From the onset, the mistake seems as insignificant as it is small. The mistake came about with the culture of education and the mistake in itself seems harmless. When admitting that, one must also admit that any pioneer that got lost and suffered privation by ultimately succumbing from starvation through an incorrect travelling direction, made the very first part of his ultimate mistake by a mere gleans in the wrong direction. How harmful does looking in a specific direction seem, and yet such a mistake leads to his ultimate capitulation. The traveller's mistake might start by taking his first directional flaw with that first incorrect step only when possibly putting his foot skew. Or he could have turned his face. It is the outcome of mistakes granting the mistake notoriety and not the start of the mistake. Thus please keep that in mind when you may find my next declaration somewhat silly and most likely over rated but my mentioning is about the start and it is the start I introduce and not the end of the mistake. **Lines mathematically cannot start at zero because there is no evidence of zero as a factor in mathematics.** The question in need of answering is this: **What will the length be of the shortest hypothetical line imaginable and moreover, should you disagree with my statement, what will the total overall length be?**

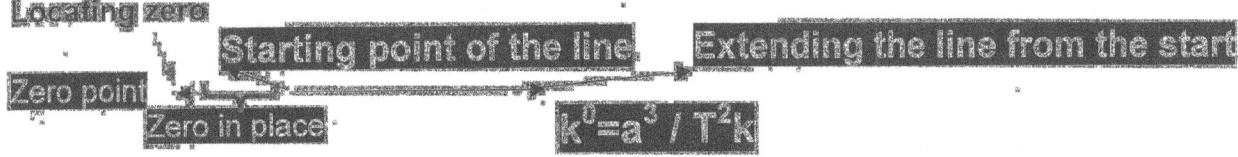

The shortest possible line (hypothetically) must be so short it must have **an initial and ultimate point sharing the same spot.** Any theoretical line being the shortest possible line cannot have the line holding the initial starting point at point zero and advance from there. If it used zero as a start, the zero part would not count, because the line will only start at a point past zero where the line then will start. Zero ultimately means not existing and then that point, as a start does not exist. When the line **has a beginning and an end at the very same spot** and it wishes to extend the position as to further the possibility it has, which direction should it favour? Extending the line in any one direction will favour one direction without any clear reason not extending in other directions. The only mathematically sensible option about extending will be in all directions equally in order to give a meaningful non-bias flow of mathematical equilibrium.

The cosmos has lines forming cubes and lines forming circles, which in applying 3D manifests as spheres. Between the circles and the cubes run lines so the key to understanding the Universe is line. The Big Bang was a time when the Universe was incredibly small making the running lines small. Understanding the Universe is taking the line connecting particles through space back to its limits where such limits were during the Big Bang. But the reducing cannot go to zero because zero removes the line all together. By reducing the line to where the line will not reduce any further we will find at that point that all points land on the same spot. The spots all share one position because that is the only position there is to hold in the form singularity presents.

One such a relevancy is the sphere. The sphere has six edges relating to one another at all times linking through a controlling centre. Then connecting the six sides is a centre form where the control comes about that place these edges at specific related points and the points in return puts the centre at the precise centre.

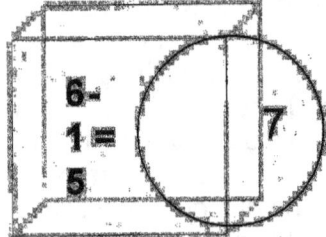

At all times the sphere has six precisely controlled edges connected by a supporting centre that is in such a position the six plus one in the centre is in immediate support of any or all of the points at any time. When touching one point the point reserves the strength to its disposal that is given by all seven points, which are backed by the entire structure. Try beat that for form strength and that is why a sphere is the ultimate form that provides structural strength

The cube has six sides in three pairs relating to one another at all times. The cube has six flat sides loosely connected at corners where the corners prove even weaker connecting points than the flat sides convey support to the structure as a whole.

The Lagrangian is one of the four Parts where the Universes' start.

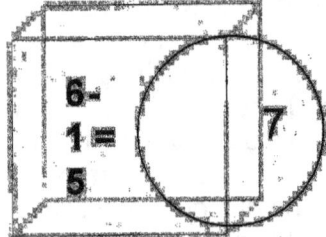

Where the sphere makes contact with the cube the sphere loses one dimension to the sphere. Because of the absolute domination the sphere has in form and in control coming from a centre the sphere removes one of the six sides the cube gas leaving the cube with five sides in relation to the seven the sphere has. That is another factor that gravity shows. This explanation also concerns the Lagrangian form.

The five points forming are the consequence of the cube having six sides that connects with the sphere having six sides as well as having a centre seventh point. That gives the sphere one additional dimensional advantage. Gravity is about forms relating to forms and has nothing to do with mass inflicting magical pulling via gravity empowering fairy tales. The seven points that holds the sphere in place will always bat the six point forming the unattached cube and in that the six points of the cube will lose one connecting side to the seven point claiming dominance with the sphere. The sphere will always win the battle of superiority and to that end we find the sphere holding the absolute relevance in pi.

The cube will relinquish one side in favour of the centre forming the seventh point of the sphere.

LAGRANGE (-TOURNIER), JOSEPH LOUIS DE (1736-1813)

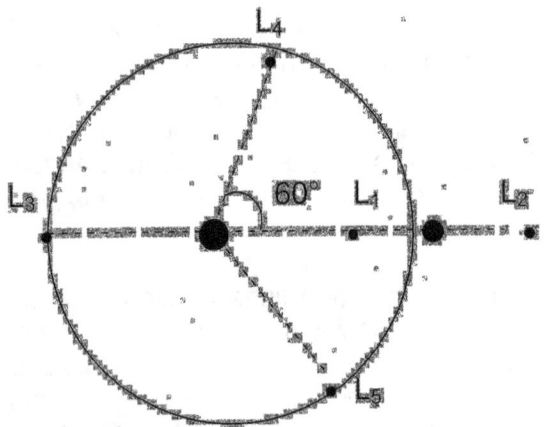

LAGRANGIAN POINT: *The Lagrangian points are five equilibrium points in the orbit of one body around another, such as a planet around the Sun.*

French mathematician, born in Italy. In celestial mechanics he studied perturbations and stability in the Solar System. He examined the three-body problem for the Earth, Moon and Sun (1764) and the motion of Jupiter's satellites (1766). In 1772 he found the particular solutions to the problem that give rise to the equilibrium positions now called Lagrangian points. Lagrange also studied the Moon's liberation. LAGRANGIAN POINT One of five points at which small bodies can remain the orbital plane of two massive bodies; also known as liberation points.

Three of the points lie on the line joining the two massive bodies: L_1 lies between them, while L_2 and L_3 have the two bodies between them. These three points are unstable, slight displacements of a body from then resulting in its rapid departure. the fourth and fifth points (L_4 and L_5) each form an equilateral triangle with the two massive bodies, 60° ahead of and behind the smaller body in its orbit around the larger one. A well-known example of bodies flying at the L_4 and L_5 Lagrangian points are the Trojan asteroids in Jupiter's orbit. Among Saturn's satellites, Telesto and Calypso lie at the L_4 and L_5 Lagrangian points in the orbit of the much larger Tethys. In similar fashion, tiny Helene precedes Saturn's satellite Dione, keeping 60° ahead of Dione. The Lagrangian points are named after the French mathematician J.L. de Lagrange, who first calculated their existence.

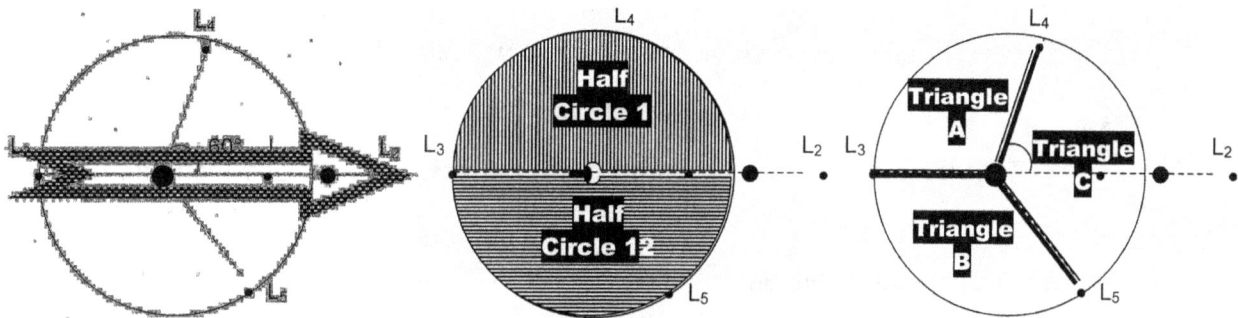

The Lagrangian system is the straight line forming two half circles that unites three triangles and this is indicative proof of singularity.

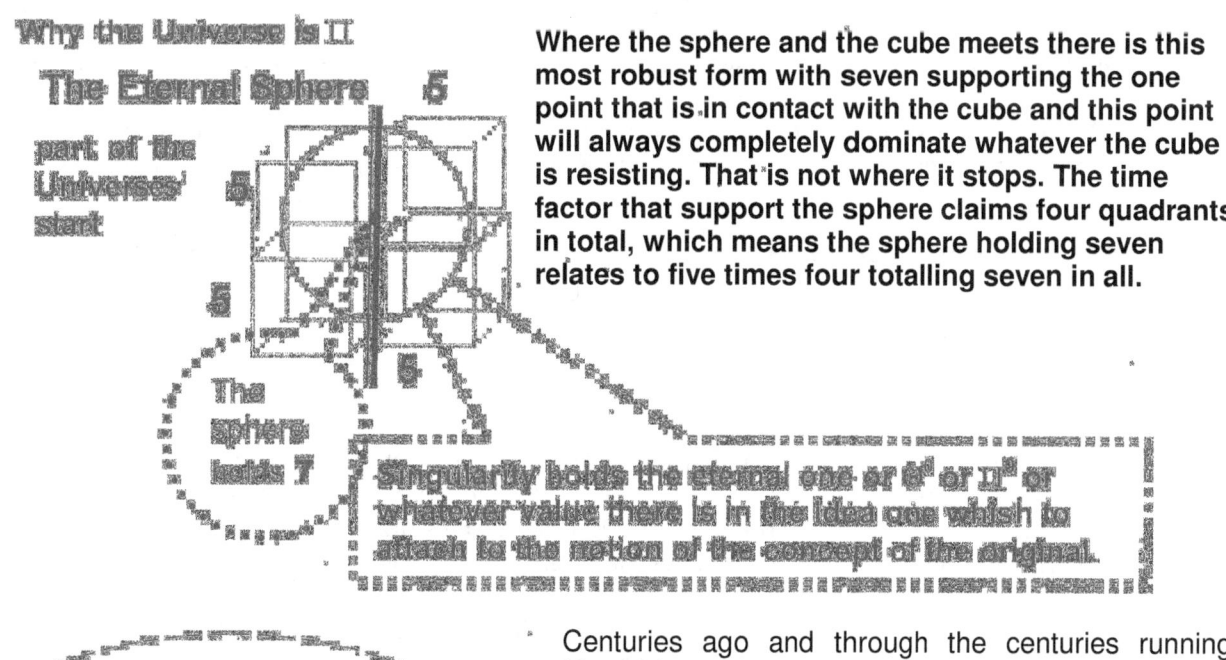

Why the Universe is II

The Eternal Sphere 6

part of the Universe' start

6

5

6

The sphere holds 7

Singularity holds the eternal one or 6° or II° or whatever value there is in the idea one which to attach to the notion of the concept of the original.

Where the sphere and the cube meets there is this most robust form with seven supporting the one point that is in contact with the cube and this point will always completely dominate whatever the cube is resisting. That is not where it stops. The time factor that support the sphere claims four quadrants in total, which means the sphere holding seven relates to five times four totalling seven in all.

Centuries ago and through the centuries running Newtonians should have realized that Newton and Kepler did not have the same use of mathematics in mind. In the manner the two measured the circle or the sphere the way they went about was one Universe apart...and then for all Newtonians to automatically accept the Universe gave Kepler the wrong formula is quite extraordinary arrogant, which is not surprisingly because it is very typical

Newtonian behaviour! Kepler did not formulate is findings...the Universe gave Kepler numbers and in order to make sense of the numbers the numbers formed a formula, a relevancy. The cosmos spoke in the language f mathematics and it was the Universe that told Kepler and all of human-kid the Universe composes of $a^3 = T^2k$ and by using this formula correctly one cold locate the start of the Universe. It is obvious the two did not use the same mathematical principles...but in the eyes of every one since then up to now it has to be Kepler that is in the wrong because only Newton's mathematics made sense. Newton preserved the right to correct the Universe...that is how competent Newton thought he was. Newton thought his brilliance could tell God Almighty to use mass because that is how Newton saw the Universe should be. Instead the Universe used something Newton's mathematics could never fathom... $a^3 = T^2k$. By comparing Newton to what Kepler said and dissecting Kepler's vision on gravity, Newton's reflection on gravity looks rather flimsy. Newton claimed he struck a force and a force it became, but a force of what? Kepler on the other hand declared gravity being in two modes where there has to be movement if there is space $a^3 = T^2k$.

Mathematics formulated is a messenger telling the reader of possibilities that applies on the condition that the reader has to interpret the formula correctly. When Newton got to

$$F = \frac{r^2}{M_1 M_2}$$ and it did not work, it was very much illegal to change the concept to

$$F \, \alpha \, \frac{M_1 M_2}{r^2}$$ in the hope the result will prove to be $$F = \frac{M_1 M_2}{r^2}$$ and then after years if

cheating and misconduct, then settle on $$F = G \frac{M_1 M_2}{r^2}$$ as this would mathematically bring

the most believable answer needed not to prove a point but to fool everyone brainless enough to believe the misconduct. The formula is as any formula is, just a messenger. It tells a story by delivering a message and if the messenger says $a^3 = T^2k$, there is no point in beating up the messenger to a pulp in order to get the messenger to confess that what the messenger actually

means is that the relevancy factor **k** has no value $\frac{dJ}{dt} = 0$ and then to change the message to

read $a^3 = T^2$ just because you cheated the living daylights out of mathematics before when the cheater deceived the entire world for three and a half centuries to believe that the following could be

accomplished by using mathematics ($$F = \frac{r^2}{M_1 M_2} = F \, \alpha \, \frac{M_1 M_2}{r^2} = F = \frac{M_1 M_2}{r^2}$$

$$= F = G \frac{M_1 M_2}{r^2}).$$

If you say $a^3 = T^2$, then you are mathematically implying a person can walk into a mirror and function as if everything remained the same! That is rubbish and every Newtonian should at least be aware of the impossible situation the changing of a formula from $a^3 = kT^2$ to $a^3 = T^2$ should create...and yet not one Newtonian got up and went against their Master of deception.

The Sun holding actual size or space a^3 as seen, or at the allocated relevancy or distance **k** from each planet's orbit position T^2 or rotating circuit location.

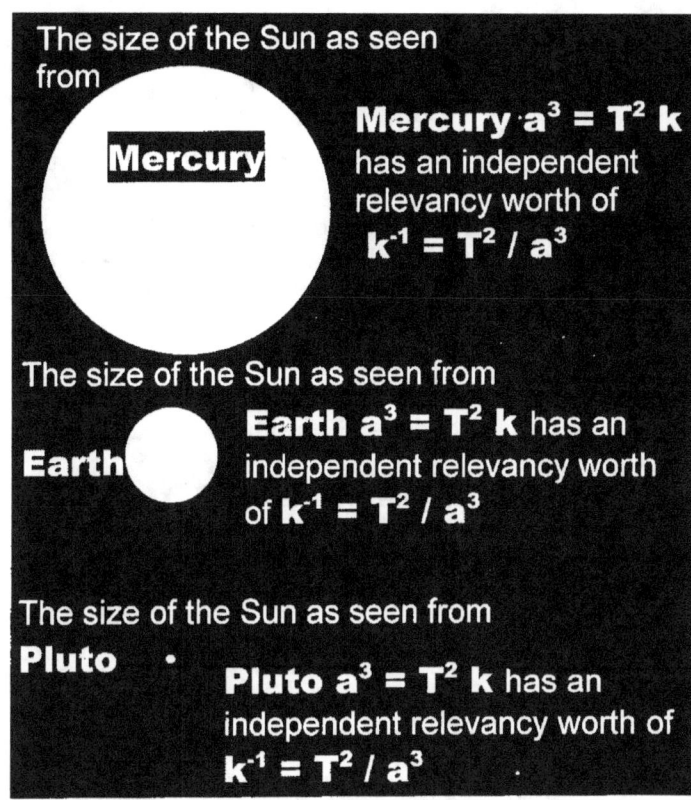

The size of the Sun as seen from

Mercury $a^3 = T^2 k$ has an independent relevancy worth of $k^{-1} = T^2 / a^3$

The size of the Sun as seen from

Earth $a^3 = T^2 k$ has an independent relevancy worth of $k^{-1} = T^2 / a^3$

The size of the Sun as seen from

Pluto $a^3 = T^2 k$ has an independent relevancy worth of $k^{-1} = T^2 / a^3$

In spite of this $a^3 = T^2 k$, mathematical truth even scholars know, Newton changed Kepler's $a^3 = T^2 k$ to an unbelievable corrupt $a^3 = T^2$ by just blatantly wishing away the relevancy factor **k**, stating that the circle T^2 divided by the distance k was equal to a total of zero.   Newton gamely further corrupted this rule in the mathematical principle whereby he accomplished more fraud.

$$\frac{dJ}{dt} = 0 .$$

This picture in its entirety condemns Newton's statement of the rotating distance nullifying the motion totally.

$$\frac{dJ}{dt} = 0$$

Newton and Science made one enormous blunder from taking this stance. It is as if they took the idea that when the wheel spins, the radius of the wheel has no influence on the wheel. In doing that, they removed the very fact that keeps the wheel at a radius on the one side and no radius on the other side. No wheel can turn in that manner and all things in the cosmos can and must spin ($a^3 = T^2 k$), that is what the Universe told Kepler) and size has no function because the size relates to the movement in relation to singularity that provides the cosmological and the universal attachment ($k^0 = 1^0$) and the movement of space equalling the space is what keeps the cosmos glued together.

That serves as gravity. The space is the movement of the space and the movement is gravity while having mass is the preventing of individual movement and the frustrating of independent movement.

They put two objects in an attaching relevancy and then announced that there just is no relevancy applying between the two. When one divides into another there is an irremovable ration in place. Removing that ration is breaking the most fundamental mathematical principle.

$$\frac{dJ}{0} = dt \ \ or \ \ \frac{0}{dt} = dJ$$ This disputes mathematics. **dJ/dt** can have any number except the only number not possible to have is zero.

It has taken met more then seven years of constant effort and continuous but fruitless communication with Newtonian academics in physics with no results to show. Thus, I am now finally convinced to report that I could not convince one Newtonian about this invalid formula, which implicates that every Newtonian is participating and furthering Newtonian fraud. Their spreading this indicates their deceit by total corroboration with Newton's recklessness in mathematics, which misrepresents Kepler's correct formula. This carries the hallmark and trademark of nothing less than corruption and blatant fraud. They will not correct this because they are corrupt.

If Newton said $\frac{dJ}{dt} = 1^0$, then yes, with that I would most heartily agree, because that is what Kepler said.   Kepler said $a^3 = T^2 k$, which is **dJ = dt** and therefore $\frac{dJ}{dt} = 1^0$.   That is mathematically correct. Then it is correct to say that $1^0 = \dfrac{dJ}{dt} = k^0 = \dfrac{a^3}{T^2k}$. This means the

movement is equal to the space it moves through. If Newton said $\dfrac{dJ}{dt} = 1^0$ which would imply that

dJ = dt which in turn means **$k^0 = a^3 / T^2k$,** then this would read in normal English grammar as the space is equal to whatever can move in and through that space which is **$k^0 = a^3 / T^2k$** and that opens up one Universe of infinite (1^0) possible development (1^0) but by using 0 or zero all possibilities fall away as the Universe collapses into nothing.

When the circle reduces, the value located to r will become implicated because r determines specific size. Not so in the case of Π, because Π in the true sense only indicate that the circle is a square without corners and therefore Π dictates form and not size. By reducing size only r comes into contest and will point to such reduction. By reducing the circle radius r by half continuously will lead to an infinite small circle but Π will remain because the circle as a form remains even being

infinitely small, but zero $\dfrac{dJ}{dt} = 0$ it could never be because the line is a

presence misconception reach. The shortest possibilities must have a start and spot and such a line will also be a dot favouring one direction puts all equilibrium meaning that any form of develop from such a spot with the being in the same position also has the flow outward will be equal in all directions.

Locating and finding Singularity

This reasoning prompted me to look for singularity in such a spot because if the prime spot from which all came was a spot holding all, then the spot must hold the shortest line but more prominent it will hold the smallest form including the smallest circle or for that matter the smallest sphere. One possibility that the shortest line or smallest spot can never have is having a starting point on the zero mark. If the mark of zero holds the start it must also hold the end because the end and the beginning have the same position. If the position of zero then is the beginning, the end will also be zero leaving the line or spot without an end as well as without a beginning. Such a spot will constitute all of nothing.

nevertheless of any Newton could ever line in the realm of finish holding one or a circle. Not directions at what ever might end and the start to be a sphere because

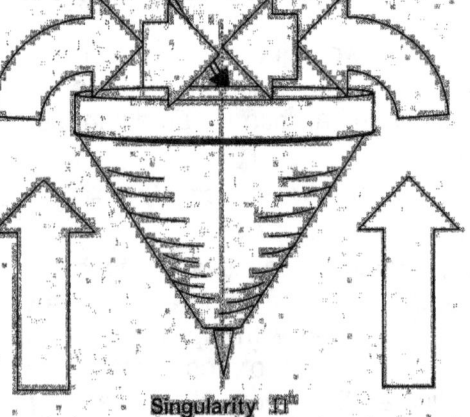

Singularity Π^0

An observation coming instinctively to mind one may recognise is that the form reminds rather explicitly of natural phenomenon as hurricanes, water whirls and even the shape most commonly favoured to express the cosmic object referred too as a Black Hole. The similarity may be more than coincidental. Let us consider the statement in the reverse.

By reducing r indefinitely to the tune of half each time, r would become infinitely small, beyond human calculating means, however as mentioned in the case of the smallest dot holding one spot, r would become insignificant beyond human comprehension even, but never reaching zero and still Π would remain intact and dictating form. The size only depend on the fluctuation of r in the square as a component to the circle or sphere but that does not affect the form by indication of Π in any way there may be. The conclusion from this is that no line can start at zero because that will be a mathematical impossibility. A line or spot starting at zero would therefore be shorter than the shortest line possible.

Let's put it in mathematical terms.

When a line starts with zero it will be as such:
0+0+0+0 = 0

If it is stated that the line starts with zero and continues with 1 then the line starts at 1 and not at zero because it is 1 that forms the progressive number in the line.
0+1=1+1=2+1=3+1=4 and this puts the line starting at 1 and not zero.

Therefore no line can ever start with zero as a number.

0+0+0+0+0+0+0+0+0+0+0+0+0+0+0+0+0+0+0+0+0+0+0+0+0+0+0+0+0+0+0+0+0+0+0+0=0 and
0+1=1+1=2+1=3+1=4+1=5+1=6+1=7+1=8+1=9+1=10

This argument proves mathematically that the Universe could not have a number such as zero anywhere since the Universe forms by lines connecting points and so it starts with singularity or 1^0

With all the obvious cleverness and genius the Super-educated has they all are too stupid too have this insight. For obvious reasons can no line, or any line grow or extend from zero because such a line must then quit zero and become something, thus abandon its original value. That would mean the start of the line has a different value to the end and a line holds conformity through out. When any line is starting from point zero it can never leave zero because of the influence of being zero disqualifies any possibility of growth. If the line then had to grow in all directions at the same pace the line must therefore be a circle or being three-dimensional, a sphere.

Flowing from this fact is that in the Universe there can be no zero point or unfilled space. In the case of the growing sphere the value of the circle is Π, and that is where creation started. That gave me the clue where to start looking for singularity. One would find singularity in the value Π and the value Π will be in all things rotating in a circle. You might wonder how does that apply to the cosmos and moreover to gravity?

In the **precise middle** of all **objects in rotation** is a precise centre dividing the object in sectors that will **start the spinning initiation** from that centre point. But the spinning object **will have a middle point**, a very specific **centre point that does not spin** and only holds Π as a specific value. One value such a line **cannot have is zero** because **zero does not start any** line and therefore the **value of the line must be infinite**, just as described in **accordance** and by **the definition of singularity.**

That point albeit hypothetical, is also as much a reality none the less and is where that point **must be standing still** because every line **running from that point** in **opposing directions** are also in **opposing directional spin to each other.** In considering the spinning motion in the fraction of time in the detailed instant every aspect of rotation will turn in every instant of change in time. Although the points had the same characteristics only seconds before, they oppose the characteristics it had just before and just after the very second in which they are and to which they relate by similar points also in rotation. The fact of the graph proves my point in quarterly opposing dimensions and values,

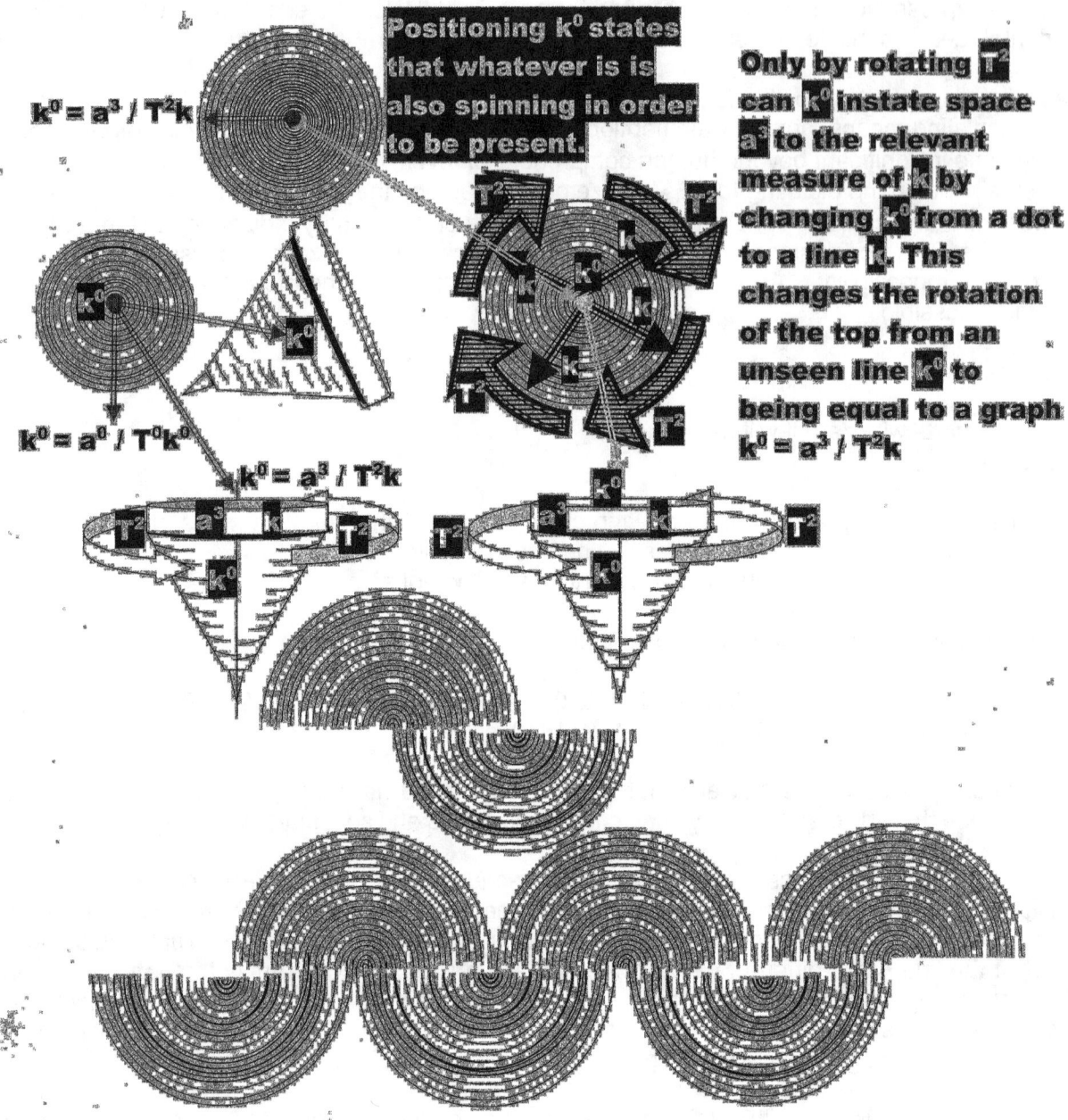

Due to the spinning nature of such a point with all surrounding the point will be alternating direction favouring change every second and in that the value to such a point can only be Π because of its constant changing. Using r would specifically oppose another r from every angle because the use of r will bring about a static relation to the previous and following instant and therefore it will cancel the constant spin flow. The role of seven in the value of pi has always been overlooked. Newtonians never gave pi a second glance although pi must be the very most important mathematical factor one can produce in mathematical physics. Newtonians always impress with the mind the totally boggling mathematical concepts while totally failing to understand to basic fundamental of physics and of mathematics. This statement Newtonians prove by the fact that having a factor arrangement such

as $F = \dfrac{r^2}{M_1 M_2}$ in any formula this they then can't change to $F \, \alpha \, \dfrac{M_1 M_2}{r^2}$ while

suggesting it will be $F = \dfrac{M_1 M_2}{r^2}$ and then with no reasons given again changes this lot to

$F = G \dfrac{M_1 M_2}{r^2}$ while feeling very upbeat about it.

What these changes indicate does not ring very favourable in the understanding of mathematics and of physics. Instead Newtonians should dissect the most simple factors that mathematics provide to find where the Universe starts. Kepler said it far better when Kepler said it correctly as Kepler said that singularity is represented by k^0 in his equation $k^0 = a^3 / T^2 k$.

In the spinning top, it is clear how gravity generate space-time when motion of space creates gravity. Taking a cue from this is vital.

A centre elects but as important to note is that relevancies come in place. There are forever some factors forming which then identifies as relevancies forms units.

The gravity the top generates is in relation with the gravity the Earth generates. The gravity the top generates find as much competition as there is devastation but also there is support as long as there is spin to the top which then generate gravity beyond and above the Earth gravity.

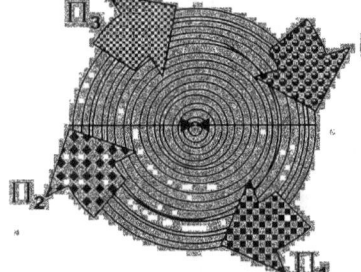

$$\Pi = 3.14159$$

With singularity placed in infinity within the centre of every rotating object, every atom and its relation to its surroundings including other atoms form space-time diverting from the point holding singularity as far as rotation goes because every object holds three relative positions in as far as where it was, where it is and where it will be in relation to singularity providing time. I elaborate on this else where.

Individual singularity and governing singularity and group singularity enhancing the gravity every time singularity find an accumulation. I have already shown where singularity is and how singularity comes into space-time. By reducing the radius to a point it is neutralised we have to get to a flat surface. This is not the flat Universe that Einstein envisaged because that Universe has gone three dimensional and in a three dimensional state can never go single or flat. This Universe is where the

$$\Pi = 3.14159 \times 7$$
$$21.991 =$$
$$= 21.991$$
$$10$$
$$+ 10$$
$$+ 1$$
$$+0 .991$$

radius has reduced to become the shortest line possible being r^0 or 1^0 or singularity. The formula for the shortest (flat) circle would be Πr and when r becomes a relevance to singularity as in 1^0 then only Π remains. That makes Π form the curvature of space-time or the point where singularity extends pi. But the importance of seven indicates the directional change the circle applies while spinning around the singularity point within the centre. Singularity meets space-time by curving pace to the measure of Π, and I think I have established this fact already. The curvature of space-time therefore is the relevancy of Π.

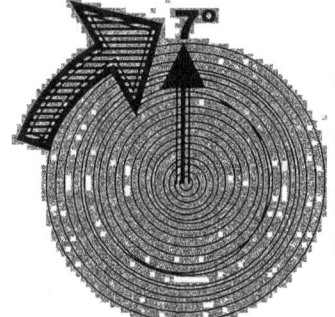

The curving of a circle is 7° and that gives the circle curvature.

In order to for us to learn how space-time establish the founding principle in forming a relation with singularity, it would therefore be advised to dissect pi and then see how pi fits into singularity. The value of Π is 3.14159 and that is always related to seven. The seven is the measure of space-time curving in the ratio of Π and to that effect we have $\Pi \times 7 = 21.991$.

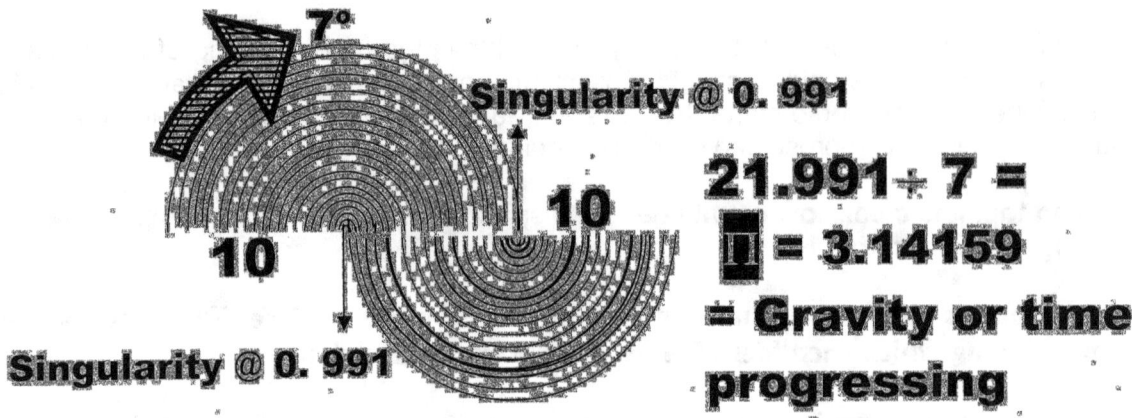

I will return to this argument when I use the graph with movement aligning and show how gravity is formed by the principles guiding the graph.

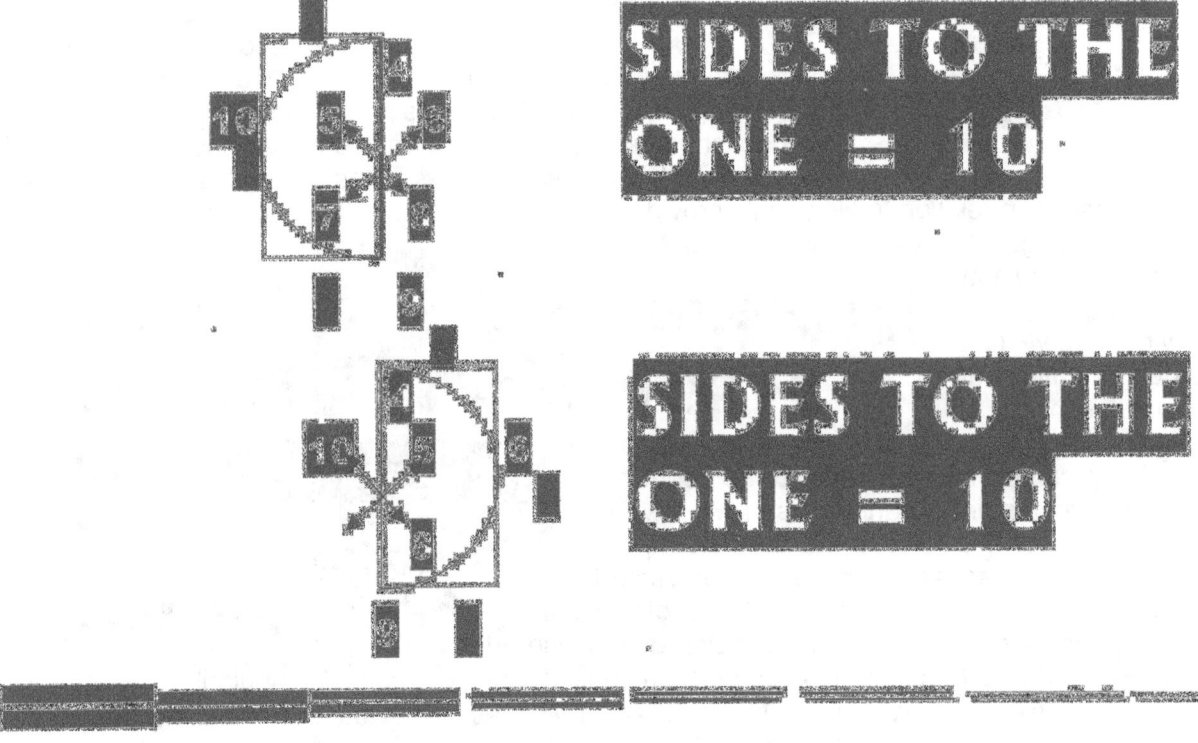

From the smallest ever possible dot will grow a line in every imaginable direction relating to a prospect of Π not favouring one direction that puts all directions at equilibrium meaning that any form of what ever might develop from such a spot will have the end and the start being in the same position, which will also have to be a sphere as the flow outward will be equal in all directions. This reasoning prompted me to look for singularity in such a spot, the spot must hold the shortest line but more prominent it will hold the smallest form including the smallest circle or for that matter the smallest sphere. All lines would form a duplication of another line sharing value since there will always be a possibility of yet another line in the realms of singularity lying between the two lines in question reducing the size infinitely to either side of the divide we humans create. From every new line within the previous line will be ten positions holding ten points. Therefore the line within will be one tenth of the line outside, and this will go on until a single dimension comes about where 3D never yet came about.

Using the concept that gravity applies Π as the circle factor Π as well as Π^2 replacing r^2 the

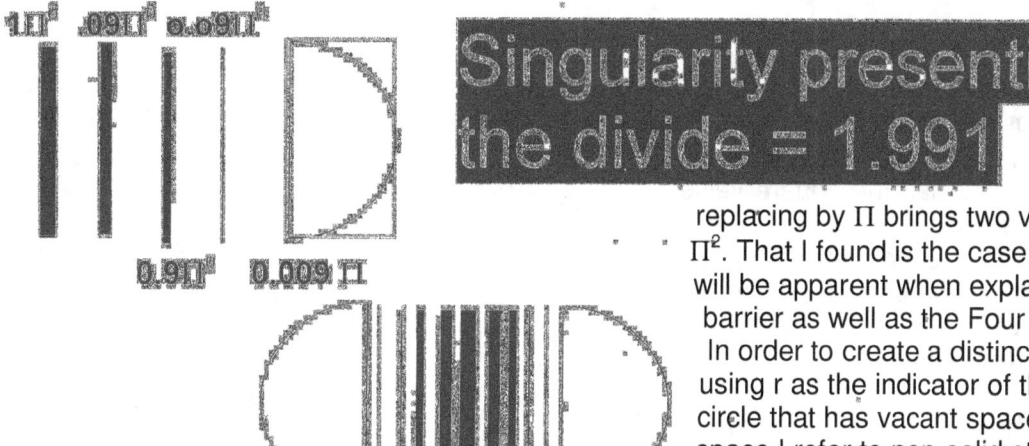

replacing by Π brings two values as Π and Π^2. That I found is the case with gravity and will be apparent when explaining the sound barrier as well as the Four Cosmic Pillars. In order to create a distinction I remained using r as the indicator of the cube or non-circle that has vacant space and by vacant space I refer to non-solid structures. In the solid structure I use Π as a value for reasons that will become apparent in due time.

The triangle, the half circle and the straight –line has two things in common, they share 180^0 as a mutual value and they are part of singularity. This could only apply where form still has no importance and directional movement of time brings about progress.

Space-time is a four dimensional position of the Universe where the position of an object is specified by three coordinates in space and one position in time.

From the smallest possible dot will grow a line in every imaginable direction relating to a prospect of Π not favouring one direction that puts directions at 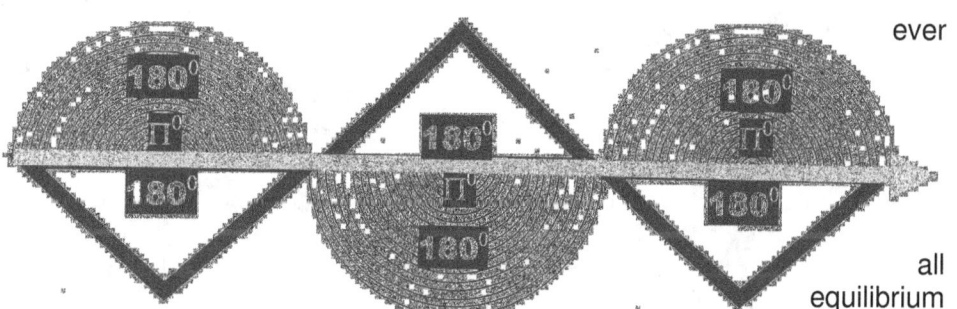 ever all equilibrium meaning that any form of what ever might develop from such a spot will have the end and the start being in the same position, which will also have to be a sphere as the flow outward will be equal in all directions. This reasoning prompted me to look for singularity in such a spot, the spot must hold the shortest line but more prominent it will hold the smallest form including the smallest circle or for that matter the smallest sphere.

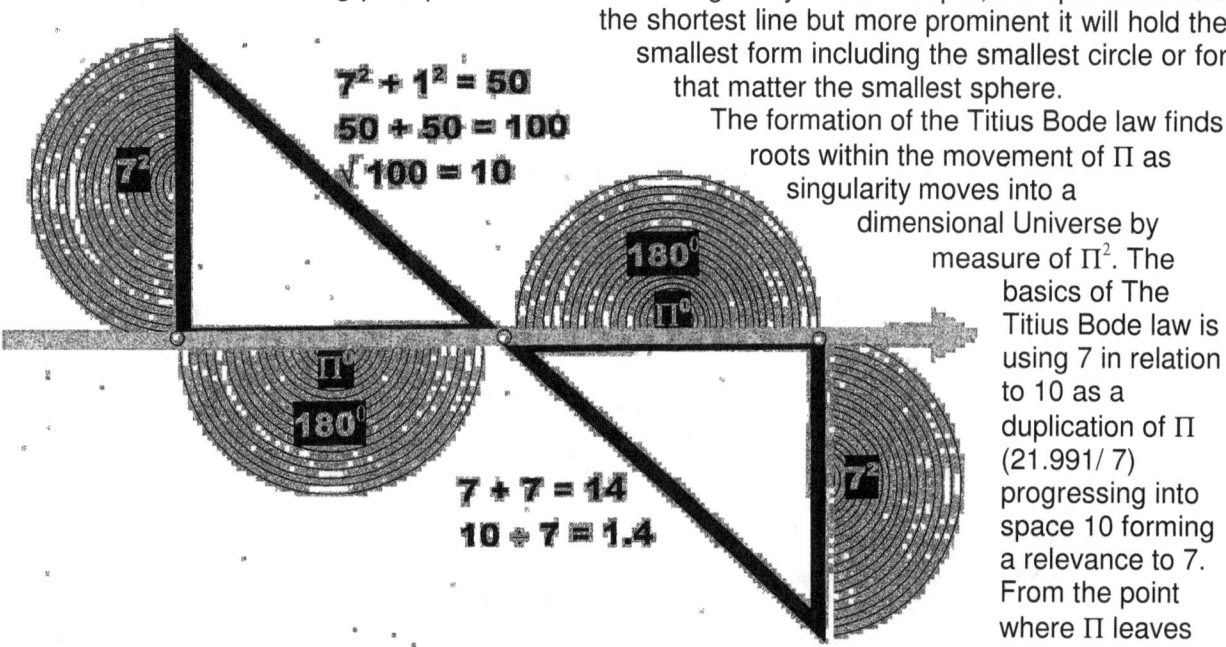

$$7^2 + 1^2 = 50$$
$$50 + 50 = 100$$
$$\sqrt{100} = 10$$

$$7 + 7 = 14$$
$$10 \div 7 = 1.4$$

The formation of the Titius Bode law finds roots within the movement of Π as singularity moves into a dimensional Universe by measure of Π^2. The basics of The Titius Bode law is using 7 in relation to 10 as a duplication of Π (21.991/ 7) progressing into space 10 forming a relevance to 7. From the point where Π leaves the flat Universe of singularity by taking time into space the half of Π forms space in as much as 10 relating to 7. This is the building block of the Universe where Π is the blocks and the Titius Bode law is the mortar.

By placing a connecting circle on the sides of the triangle half a circle forms. By implicating Π as a relevancy and not the straight-line r, two values of Π applies to each circle, and the straight line is no longer r, but is $Π^2$. This will bring about that each circle holds half the square value implicated to the allocated conditions applying to Π in that specific instance. By adding the two half squares forming the two half circles and then calculating the square root of the total that then forms the average diameter, an average of Π in the connecting line will come about. As both lines are the straight line forming singularity coming from 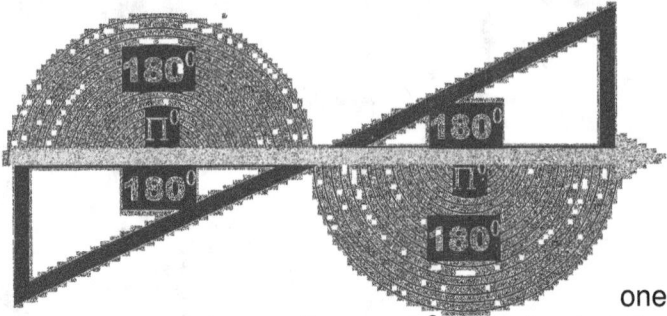 one line being Π, the connecting line then must be the average of the two lines as $Π^2$. That is what **the law of Pythagoras says.**

All lines would form a duplication of another line sharing value since there will always be a possibility of yet another line in the realms of singularity lying between the two lines in question reducing the size infinitely to either side of the divide we humans create. From every new line within the previous line will be ten positions holding ten points. Therefore the line within will be one tenth of the line outside, and this will go on until a single dimension comes about where 3D never yet came about.

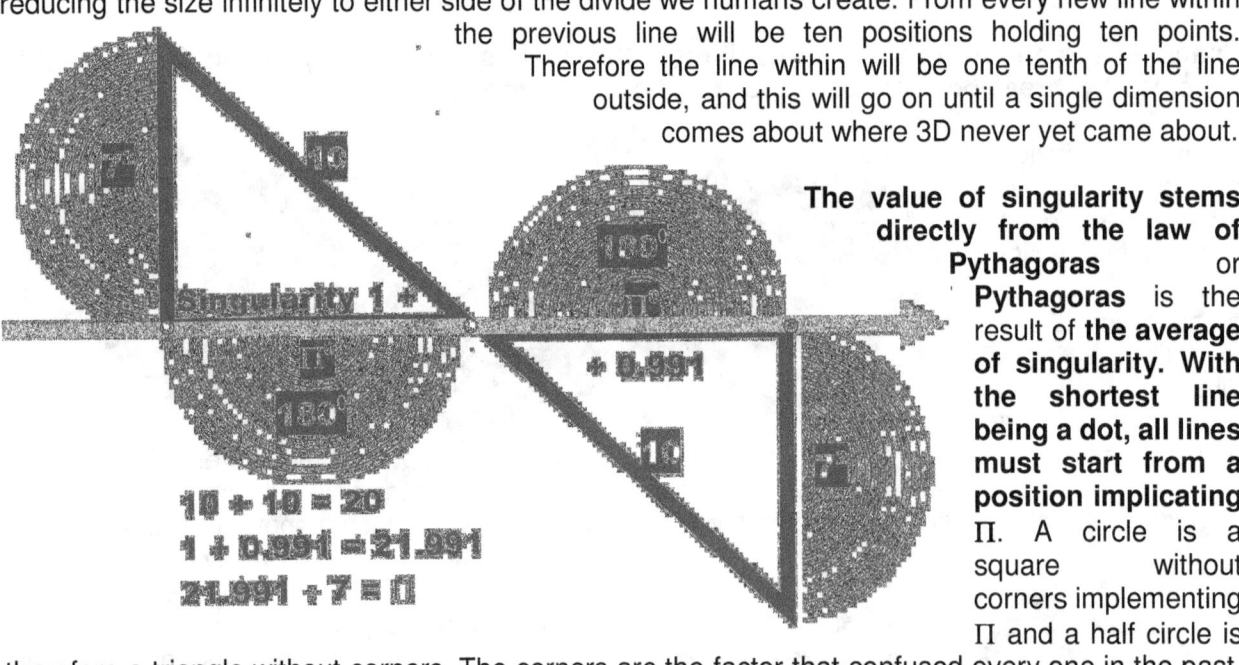

The value of singularity stems directly from the law of Pythagoras or **Pythagoras** is the result of **the average of singularity. With the shortest line being a dot, all lines must start from a position implicating** Π. A circle is a square without corners implementing Π and a half circle is therefore a triangle without corners. The corners are the factor that confused every one in the past. When replacing the value we normally attach to circle being r with Π, the law of Pythagoras becomes quite meaningful and mathematical.

Gravity has nothing to do with mass or pressure or any Newtonian concept. Gravity is about form and three - dimensional formation coming from singularity while progressing into form. In the sixties we use to talk about round object not fitting into a square and those free loving Hippies never knew how good they were with physics. That is gravity. It is a round sphere turning in a cubic square. In the sphere there are never only one direction implicated in movement. Movement are always in relation to the centre position because as a line goes up it also goes in or out. When a line goes north or south, it also towards the centre or going away from the centre. There is always a comes relevancy present in movement. As this moving indicates direction it also apply Π^2 for indicating value forming the time factor.

From the Sun there are three points moving between two points from one point to two other points giving the six dimensions we find in space. It is space in time or space converting space through the movement of time. It is a location of a point in the third dimension a^3 that will move according to the second dimension T^2 that will implicate **k** as a reference in the first dimension. It is about dimensions in reference to one another.

The value of **k** is not in the provides to other factors using definition in terms of the scenario where space applies time.

The value of Kepler's space which he indicated as a third dimension a^3 does not depend on indicating a structure a^3 that in rotation T^2, but only needs position having a constant of sorts. Any point where **k** may position one will find a value and the matching location will space a^3 could be the planet the space a^3 could spin around T^2 around an axis at the distance **k** from the Sun. it is by the relevance attached that gives meaning or measure to the concept. That is the relation there is in the solar system between all planets and the Sun. The Sun always indicates the centre and the planets always indicate the rotation. But $a^3 = T^2 k$ is only producing a relevancy of three dimensions that is equal to two plus one dimension.

value of **k** but it is in the relevance that **k** **k** to find is one some indicate a matching a^3 fit T^2 at that point. The moving in the circle T^2 at a radius from the Sun **k** or

Let us take it from a point where the Sun provides a centre as one starting edge of **k** then that centre **k** will provide a line from the centre and the line **k** will provide three spots in a formation that produces a structure by the square T^2 of the dimension. Not once did Kepler indicate size as a contributing factor to a^3. That means every single point that **k** indicates there are three positions a^3 implicating sides of a double dimension. In the same manner is **k** not limited to distance or is T^2 lesser by size. $k = a^3 / T^2$ That is what Kepler said. There are three dimensions a^3 between any two points T^2 flowing as time from the centre of the Sun, which is indicated by the line **k**.

This is where the Universe starts and this is where the Universe ends. This is the point where hot and cold parts and puts an entire Universe in between. On the inside we have cold which is infinitely cold and on the outside we have hot which is eternally hot.

It is the movement that parts the Universe setting movement that freezes material by duplicating time forming space. On the very inside the freezing of the Universe starts up infinity serving within singularity. Without freezing eternity we will never have infinity because it is the freezing of eternity by movement that sets infinity apart from eternity. It is movement of space that creates space as it establishes time.

Both eternity and infinity joins as this unity that don't use space forms space. Only by movement does the very same thing become opposing points in time that create a non-existing Universe. By moving singularity that can't move, turns material to move and that sets infinity apart from eternity. Then time in infinity sets eternity apart from singularity forms infinity and everything in between that we regard as a Universe but which does not truly exist. The same thing parts by movement of something that freezes space away from the liquid that forms a gaseous space. This is the reality of Universal physics that forms the cosmos.

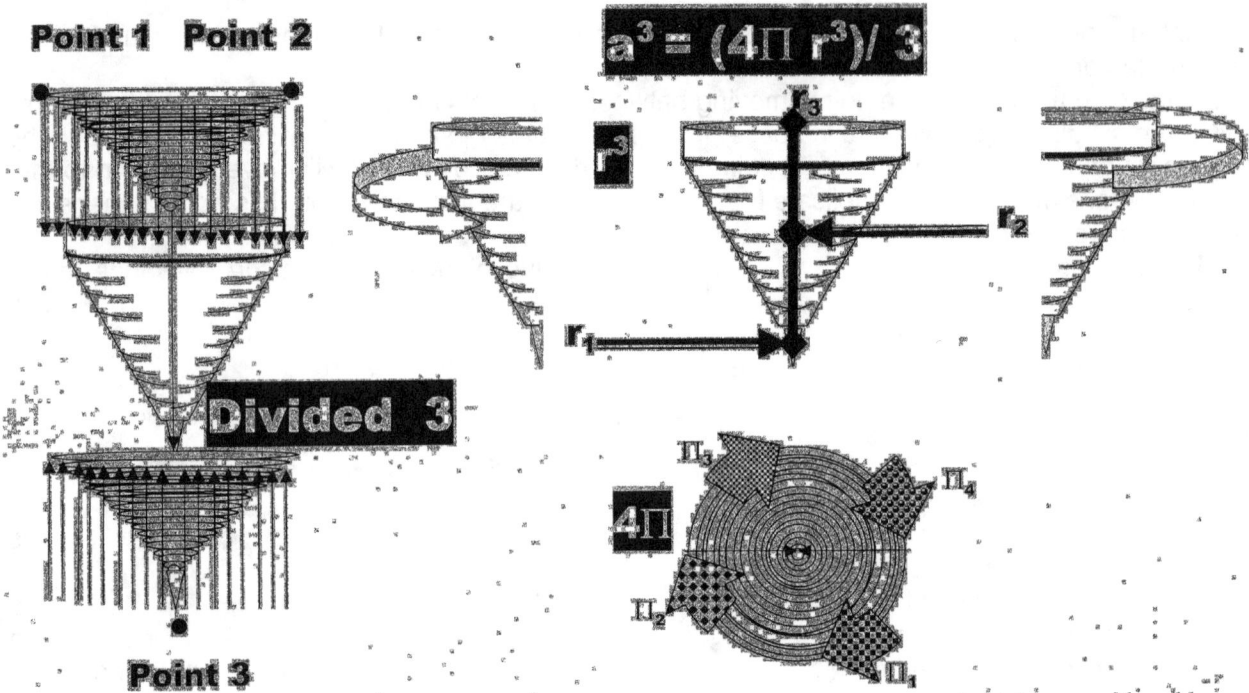

The implication of the relevancy produced by the use of the formula $k = a^3 / T^2$ brings about that when dividing T^2 into a^3 there is k left. The fact is that a^3 is a three dimension (^{3}) of single k (^{1}) showing one or T^2 is two dimensions of k being the one dimension it means that k is a part of space a^3 or T^2 which is time. It is the same thing in a double dimension or space being a triple of k then k is one factor and k cannot show a position of zero. If $k = 0$ then there is no possibility of $k = a^3 / T^2$ because $k = 0$ then $0^3 / 0^2 = 0$. That does not make sense. Mathematically space cannot be zero because those being of the opinion of space being zero or nothing must first prove mathematically that space is zero. Moreover they then must prove mathematically how zero grows through the Hubble constant. By translating Newton's vision of the circle in completing a cycle would become zero through rotation…well that does not count the use of the formula a^3. If k cannot be zero then k could not start from zero. With $k = a^3 / T^2$ no point can be zero because k shows space $a^3 = k T^2$ is no reference to the volumetric mathematical formula used to calculate $a^3 = 4/3\ \Pi\ r^3$.

Nor does it show the use of the circle in the second dimension being $a^2 = \Pi\ r^2$. In the case of the Newton's formula, the circle factor becomes the square as indicated by the duration of the time T^2. The factor standing in for the line which normally would be r and then be the square value is in the case of Kepler not the value indicating the square. That means Kepler never indicated a circle of mathematical procedure but said mathematically the distance of the planet from the Sun k holds space a^3 in relation to time T^2 **Lines mathematically cannot start at zero because there is no evidence of zero as a factor in mathematics ($k^0 = a^3 / T^2 k$). Should you disagree with my statement** the question in need of answering is this: **What will the length of the shortest hypothetical line imaginable be and moreover, what would the total overall length be in that case?**

The fact of form proves that the sphere captured all sides that can possibly influence the sphere. The sphere therefore holds $k^0 = a^3 / T^2 k$ within the boundaries designated to the sphere. When a body is placed in a location on the outside of such spherical borders that object seems to float in any direction. There is no control one can establish which will secure movement in any specific direction of preference except by releasing heat to counter act the required motion in a specific direction of choice. We all have seen what happens to any object that comes into the border area of a sphere. The object suddenly is motivated by motion to follow a specific designated direction and the motion leads the object to move towards the centre of the sphere. It is as if the support of the six opposing sides has lost one side where the sphere took over the control and movement starts in the direction of the Earth centre. The support of one side is literally removed by the centre of the earth where Einstein claimed the strongest gravity is and the motion of the object starts in that direction. There is no pulling on the object but there is removing of space by the centre of that specific point leading the object and the space it is in as well as the space it carries to move to the centre spot. In the sphere the borders the sphere holds are deliberate and very distinctly placed edges forming a specific distance from the centre. The centre is also proven beyond any debating. The centre of any sphere has to be at the very point where space completely falls away. That will put that space at that point in the single dimension and centre is the single dimension.

The fact in the matter suggests that all three factors hold the same identifiable measure and the differentiation between the factors is in name alone. In what I explain at a later event the true value should stand connected to singularity and in singularity the value is allocated to singularity is Π . The formula should read that $\Pi^0 = \Pi^3 / \Pi^2\Pi$ and when used with the correct connotation the formula becomes more sensible. But be as it may the factors indicate the same measure carried in different dimensions and should all read the same value. The cube is a loosely connected structure with the possibility to hold any form possible but the only precondition is that there must be at least six sides connecting at angles of any degrees.

The six sides hold no relevancy or responsibility to support or maintain one another and provide a Universal accepted form maintaining the Universe. From the structure one can see gravity is not strongly present since there is no vivid connection binding the six sides. All six sides' support what ever are inside evenly forms all sides. The sphere is the form securing gravity. In the centre of the sphere there is a point where space vanishes and form disappears. At that point where space vanishes gravity is the strongest. From the centre point where gravity is the strongest gravity hold the sphere true to form. At the edges of the sphere there are also point lining in 90^0 and 180^0 holding relevancy and responsibility to one another but the centre spot being the gravity point positions all the points in a location that the centre point allocate. In the centre where all lines cross one will locate singularity but I am explaining that fact a little tater on.

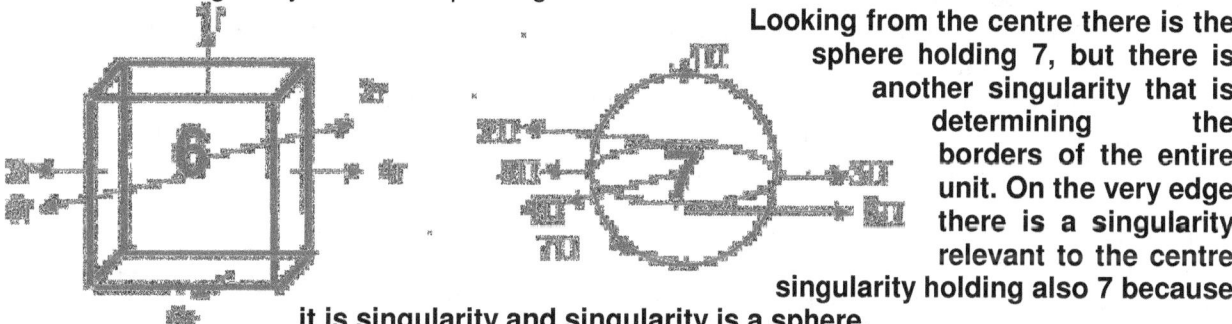

Looking from the centre there is the sphere holding 7, but there is another singularity that is determining the borders of the entire unit. On the very edge there is a singularity relevant to the centre singularity holding also 7 because it is singularity and singularity is a sphere.

In the sphere there is no radius but only the extending of Π from the centre Π in six opposing directions relating to one another by the square but remaining Π because of the unity the matter holds in relating to space. It is not possible to draw a precise line that would form a precise ring and not cut some atoms in parts. Because there will always be an atom disallowing the precise positioning of the circle the circle continues on a solid basis holding Π as a positional reference and not r. In every sphere there then are the seven Π relating in precise dimensional and positional equality forming equilibrium to the centre Π as well as to one another by 90^0 and 180^0 implicating the dimensional positioning. Therefore the sphere holds $_7{}^{\Pi 0}$ and the cube holds 6 X r^2

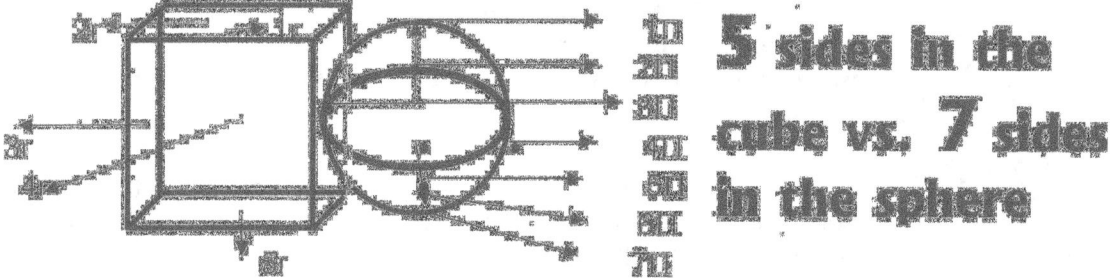

5 sides in the cube vs. 7 sides in the sphere

Where space comes into contact with the sphere the cube loses one of the six dimensions it has to the more dominating seven dimension of the sphere whereby the seven dimension in equilibrium will dominate the six dimension loosely connected by r bringing about that the cube then has 5 sides to the seven of the cube. This means that in the cube the "bottom falls out" and without a "bottom" to support objects they fall to earth. Remember that a body "floats" in space, but at one specific point it starts to "fall" to the earth. That is gravity and it is a dimension change much more than any force. I shall explain this last remark later on. That too is the Lagrangian system with five cosmic structures holding relevancy to the centre structure where the centre structure stands in for seven positions diverting from singularity and the orbiting structures standing in for five positions in space.

The circle to the left would come about from a straight line r growing influencing the appreciation of Π, but to influence Π would lead to a breakdown in r as Π and r are different entities. The circles to the right shows a continuous growth by extending Π every time and since Π is the same part as the previous Π, only extending that billionth of a millimetre each time, the circle will be truly continuous without any signs of a break . **Pneumatics use r as a pressure indicator and hydraulics use Π therefore air can compress and liquids cannot but act as the toughest solid found specifically because of it uses a relevance in the applying of Π and not r bringing conformity evenly.**

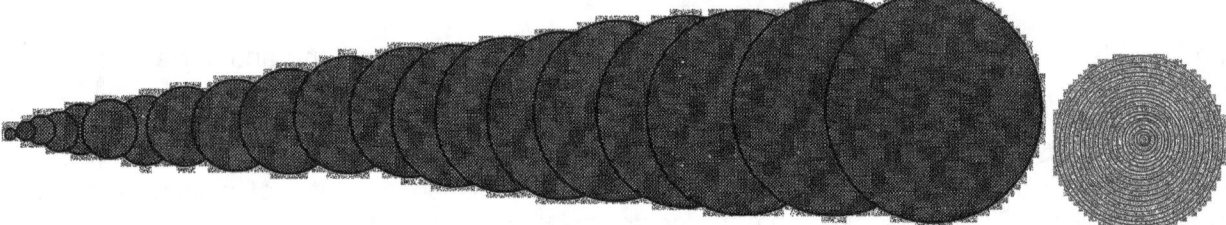

In the circle using $r^2\Pi$ the r has to have distinctive qualities placing it as a factor apart from Π. Where the growth shows no separate distinction but a continuous flow from the precise centre to the precise edge the flow would become in relation with Π depicting the circle and Π replacing r as reference to any point on the circle. By using r distinction in the circle is possible but by using Π there is no distinction possible. When working with concrete and heavy metalled solid objects r would show as a crack distinctly parting solid structures, while Π indicate a continuous flow of solidness giving the material an overall and continuous structural strength, yet engineering never recognised this difference. By confirming Π the circle employs singularity in all components and therefore proves to be a much stronger support as building choice than other shapes.

Looking at the affect of gravity it shows the precise quality of no distinctive point, as gravity never seems to end at a point but flows all over affecting all that holds a position in its sphere of influence.

The gravity coming from China meets the gravity coming from America at no particular spot but intermingles without distinction. Individual singularity and governing singularity and group singularity enhancing the gravity every time singularity find an accumulation.

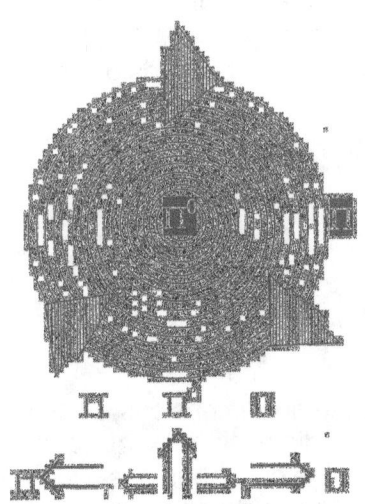

$\Pi = r$ **in constant directional change as time flows through rotation. Pinpoint positioning of singularity Π^0 with Π positioning space to either side forming the border set by singularity.**

The new direction pointing to a new location in relation to the previous point will oppose the previous point it had in relation to direction considering the centre point.

Any point will be opposing itself within the **rotating of 180°** where it **then change every aspect** of its **previous flowing** characteristics it had or **will once again have** in **360°** from there. While in rotation from the view point of a bystander it all may seem static and never changing but to the object in spin every next instant in time will be diverting from every aspect it had every second passing, and the direction it held in relation to the direction it held the previous mille, mille second will totally be incompatible with the direction it holds the very next mille, mille second of rotation.

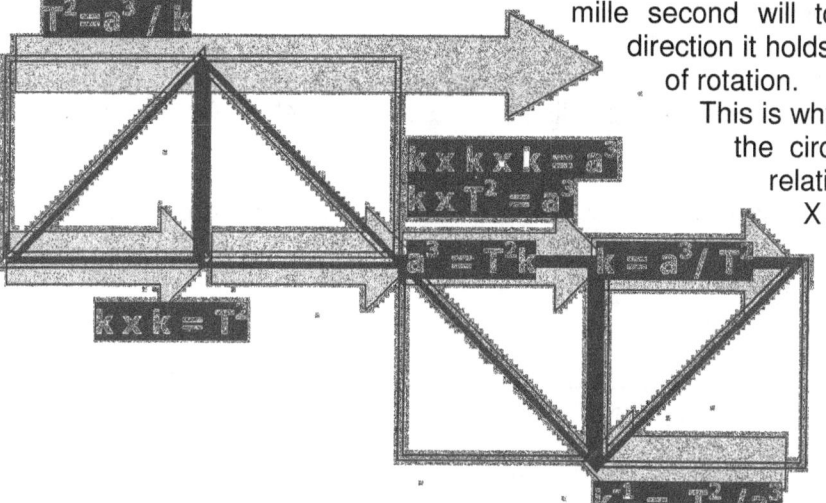

This is why we can use degrees measuring the circle by (6^2) (forming the square relating to matter through singularity) X 10 (square if space) = 360^0 however it is always in motion. That proves no point can be static or constant, though it may seem that way to outsiders. Although matter is matter, matter can also be anti-matter and moreover form its own anti-matter at the same time. This degeneration of structure is very likely to occur with overheating.

Revaluing Π to Π^2 will bring about a new contact point where Π meets **r** forming another relation in Π^2 **Time is** the **changes in relation** where Π **contacts a different r** not withstanding the many r points there may form because **every r constitutes a different value** to the Universe through other ratios and relevancies brought about **by heat and light. Time is the duration it takes Π to rotate between any two given points of r** and therefore must always amount to **a square (T^2)** moving from point to point through the **cube of space (a^3)** in that **duration of time (k)**. With that it proves **Kepler's a^3 (space) $= T^2 k$ (time in the instant of motion)** but motion must continue through a specific value in space where the space-time is maintaining relevant equilibriums throughout singularity connecting.

With the dimensional change from space in the cube to space in the sphere a relation of 5 to 7 comes about depicting gravity. The principle of 5 sides in space relating to 7 in the sphere holding matter, that is Π.

The spinning movement takes the curve of seven from singularity at 0.991 to 1^0 through four five point cubes. That is (5x4) + (1+0.991) x 7 $= \Pi$.
That also is the Titius Bode law putting 7 in relation to 10 because it is using the manner in which space forms by moving from the one side to the other side through space.

The Newtonian Master minds are in for a nasty wakening to reality. The Universe they created with inventing mass and pressure and having the rest go to nothing has come to be of no more use. Creation is not as simple as leaving everything up to mass and when finding then Universe doesn't comply, then blame the mistake on dark matter to solve the situation. No longer is it possible to hide behind a formula $F = G \dfrac{M_1 M_2}{r^2}$ **as well as creating a concept worth nothing** $\dfrac{dJ}{dt} = 0$ **while they all know these bogus ideas are not fore filling any of the solutions physics require. Newton is a lie while Kepler is the truth, notwithstanding the rape Newton committed to Kepler's work.**

Kepler was the one that discovered the concept formulating singularity $k^0 = a^3 / (T^2 k)$
Kepler was the one that discovered space / time as $a^3 = T^2 k$.

Kepler was the one that discovered gravity holding space-time relative as the Universe expands in increasing relevancy $k = a^3/T^2$.

Kepler was the one that discovered there in the centre of a sphere is singularity as $k^0 = a^3/T^2 \, k$ Gravity then is $k^{-1} = T^2/a^3$ where space is equal to the time it moves in. Space has to move or space will collapse. That we see in the case of a Black Hole.

Gravity then is also $T^2 = a^3/k$ Time forming is valid since the object is unable to depart any further distance and is captured at that distance by the restricting of the space in motion. Gravity is keeping space relevant to determining the relevancy of motion in relation to the centre.

Gravity then is also $k = a^3/T^2$ because gravity is space in division of time which comes down to space having a velocity that time holds space in motion too. Gravity is motion of space at a specific speed or velocity, which is moving across a distance in relation the time it takes. Again that is space-time $a^3 = T^2k$.

Gravity then is also $k^0 = a^3/T^2 \, k$ where the Universe is being in the centre of space from where the Universe can and is claiming and controlling space- time.

Translating Kepler's mathematical expression $a^3 = T^2k$ correctly to the verbal statement in English Kepler said that there is a space a^3 which is equal = to the motion in the time duration T^2 thereof where the motion of space takes up the time it uses to go between two specific points which holds a relation to a centre k^0 where from where there forms a straight line k and is located on the spot where space begins the circle therefore that centre spot has the least space.

Kepler stated the very opposite of what Newton saw. Kepler had direct opposing ideas about the circle and what factor should be using the square because in Kepler's method of expression the circle indicator T^2 goes square ^{2} and the diameter indicator k, which replaces r remains single in the face of the volume being a cube a^3. This is mathematics and not the corrupting thereof $\dfrac{dJ}{dt} = 0$ in order to further bogus ideas in order to promote personal megalomania of self-importance as was the case with Newton. Newton's ideas were about self-promotion and had no implication of physics.

Newton should have stuck to physics, which under the circumstances he was good at. But when he tried to evolve into cosmology...that is where he got stuck in mud because it is apparent he understood nothing of and the proof of his inability to understand cosmology is in the way of his understanding nothing $\dfrac{dJ}{dt} = 0$ which in his case is quite literate. This brought about that the two masters were using different dialects of the same mathematical language spoken by all... and yet the two mathematical dialects uses incompatible as well as inconceivably different spelling forms ($a3 = T^2k$ compared too $a3 = 4/3 \, \Pi r^3$.)

It is about translating mathematical equations and being correct in interpretations of mathematical expressions. Other factors are about certain mathematical deductions that were made in the past but were incorrectly presumed. A line cannot and therefore does not start with zero. Should you think such a statement is trivial then this book is even more especially for you because that changes where one presumes the Universe came from at the very beginning of the cosmic conception? Kepler answered all the questions we have...and much more! This I prove.

All the skills that are required in understanding Kepler correctly are not to read Kepler's formula, but to read into what Kepler's formula says to the reader.
Kepler was the very first person to mathematically introduce space a^3 in a time relevancy by k and related directly at all time T^2 in the Universe that determines k space a^3 and time T^2 and much more. Kepler was the first one that saw that gravity comprises of two factors being k or linear gravity and circular gravity or T^2.

By reading what Kepler said correctly, so many centuries ago the effort brings all the answers to so many unrealised questions...but it does not involve looking at what Newton said about what Kepler said... it is all about looking at what Kepler said.

Not only did Kepler introduce space-time a^3 / T^2 but Kepler also placed space a^3 and time T^2 in a relevancy **k** long before Einstein did and placed gravity in space-time $k^{-1} = T^2/ a^3$ even before Newton gave a name to the concept of gravity. Kepler was the person who placed gravity as the ingredient in the Universe that determines **k** space a^3 and time T^2 and much more. Kepler saw that gravity comprises of two factors being **k** or linear gravity and circular gravity or T^2 and almost half a millennium later Newtonians still have not reached this understanding which Kepler showed so long ago!.

Singularity is $k^0 = a^3 / (T^2k)$ and is not $\dfrac{dJ}{dt} = 0$ because it is $\dfrac{dJ}{dt} = 1^0$, just as Kepler said it is.

That is singularity being one to all but it is not zero. Finding form in that point shared by all will give a value of singularity. Extend that value received to a Universal centre and bring that value to align with Kepler's $a^3 = kT^2$ and understanding the Universe by finding the centre of the Universe makes the Universe simple as can be. The Universe becomes sensible making the entirety thereof different and it brings along a new understanding of the yet unexplained phenomenon. It makes the explaining of the four phenomena that creates gravity as easy as children schoolwork. There are suddenly no more mysteries in the Universe. It is only possible when we see gravity not as a grabbing force instead of seeing gravity reducing the space between particles, which is mathematically explained. Gravity is not being some magic force found between particles grabbing onto everything but instead is a mathematical diversity between cosmic forms moving in relation to each other.

I mathematically first uncovered that which I am about to prove which serves as the keys starting up the Universe and pushed the lot towards the Big Bang. The Big Bang was not one single event but from what I mathematically uncovered, there were several of those coming and going. Wee now are in one, which I call the Iron Age. Before the Iron Age there was some and some will follow the Iron Age. Those I explore into in other books. This book is about uncovering what came before the Big Bang and what were in use before space-time and mathematical discipline came into force.

It started off with one dot so small eternity met infinity within on that spot. Where this ended, was still so small it is not part of the Universe, and yet it is the Universe forming whatever can fit inside the Universe and still it is too small to be part of the Universe. Then there came one more dot that grew away from one more spot. With the one came an immeasurable many while all that formed was still being one as they continued coming. This repeated constantly until there were a countless number of dots. The accumulative size of the dots were the same size as one dot because in the true Universe big and small plays no part. It was singularity in eternity parting from singularity in infinity as everyone that formed was the same since they all were one. The dots were infinitely small and eternally big at the same time because size is a relevancy and without one the other has no size. So in the true perception, there is no difference in size.

It started with the fact that there is no place or part in with which one may associate zero or nothing. There are no room for a number such as nothing. Next to the one dot (infinitely close) one will find the next dot being eternally large but still fitted into the infinitely small spot, and if nothing was a factor then that is precisely what one will find between the two dots. While being eternally big it still held no space, too our Universe it is a non existing entity that still forms the totality of our Universe, and while being whatever there can be it still is taking up no space, and much more important, no time, therefore the dots are infinitely close to one another, being the same space, eternally big as much as infinitely small. If we as humans cannot find a manner in comprehending this notion, there can be no manner ever understanding the cosmos as much as forming a concept about the start of the cosmos.

Every dot was a Universe in its own and the accumulation was a Universe. The Earth in itself is a Universe as the moon is a Universe while every atom is a Universe forming the above mentioned Universe, because rules applying on Earth do not apply on the Moon and visa versa. When in the ocean another set of rules apply, therefore being in the sea places a body in another Universe. The number of Universal entities is still countless, as much as it was in the beginning.

Every point in the infinity we may observe at is not merely part of the Universe in not being nothing, but is the point where the Universe started representing singularity. It is the very first point where everything began so many eternities ago, because after all, how can we ever determine where the first point was, as they were very much equal and alike at the beginning. Every aspect of the Universe started with the fundamental fact that no point in the Universe can represent "nothing" as a number, because every aspect in the Universe represents singularity in what ever form it may hold in that specific spot forming space-time. If man does not reach a conclusion where that conclusion is matching the Universe and stop to match the Universe with man (and man's incapability coming from small-mindedness), we may all go back to caves and become starving hunter-gatherers again, because we will never find a way to progress to the ultimate understanding of the Universe.

The content of my work contain a new view about Cosmology, which I have been working on for the past twenty-seven years and exclusively for the past seven years. Since I have no direct contact with the academic world, my aim at first was to find a promoter, if not locally, then overseas. In that I failed miserably because I said Newton failed miserably. Now that I am launching a new approach where I am going to try to reach students and tell them to investigate Newtonian corruption, I thought it wise to contact you in person in order to try and explain my views on Cosmology. This I do to show that I have no ill feeling to you as a person but the rest of the science world can go to hell for all I care. With this letter I guess my main objective will be to give you a little insight into my work in announcing to you that I came to realise that lines mathematically couldn't start at zero because there is no evidence of zero as a factor in mathematics. Should you disagree with my statement the question in need of answering is this: What will the length of the shortest hypothetical line imaginable be and moreover, what would the total overall length be in that case?

The shortest possible line (hypothetically) must be so short it must have an initial and ultimate point sharing the same spot. If it used zero as a start, the zero part would not count, because the line will only start at a point past zero infinitely close to zero, but not on the point of zero as zero is no where and something must be where the line then will start forming an infinitely small dot. I press this issue because this is part of mathematics ever since mathematics became part of human intellect. I can't press this point too much because this changes the concept as to find out where the Universe started. The value of pi can only find conclusion when this aspect of mathematics to come to be realised and accepted. The Universe started with a line and that we all accept, but if a line starts with zero, then the Universe started with zero, which will remove the entire Universe from the Universe. However if one realise that a line starts with one spot growing into a dot that forms singularity, a point from where everything that is, came about, then we have a point where the Universe started. The dot is in infinity, however small, it is not zero. Zero ultimately means not existing and then that point, as a start does not exist. The smallest line has a beginning and an end at the very same spot located in infinity, and infinity may be beyond human scope, though infinity is still not zero. Infinity may constitute something we do not yet understand, but we may not define our human misunderstanding or lack in understanding matters as nothing. In this aspect lies the difference there is between arithmetic and mathematical science where arithmetic can have position such as zero since arithmetic excludes the cosmos calculating numbers only.

A man may have that many oxen or so many sheep and even this amount of wives, (in Africa) or not have any therefore having then a total of nothing, but there cannot be nothing between the Sun and its orbiting structures. The having and have-nots are part of arithmetic and arithmetic connects to life whereas mathematics involves cosmology. Light will indicate a line flowing between the Sun and whatever planet, following dot after dot thereby proving the existing of the possibility of something that ca fill that space while this endless row of dots are going about by a straight line, and any straight line in relation to other straight lines will be under the law of Pythagoras.

There is no possibility of a straight line not form in space. Mathematics converts the values of integrating lines according to Pythagoras and arithmetic is about numbers to be added or subtracted. By mathematically excluding zero from cosmology a new Universe of understanding opens to the human mind. For instance the distance between the Sun and Pluto is roughly one hundred times more than the distance between Mercury and the Sun, but according to the Newtonian super-wise, both planets mentioned have a vacuum filled with nothing except one atom hear and there occupying the vacuum between them and the Sun. If space supposedly comprises of "nothing" as Newtonian super intellect wish to advocate, then how can "nothing" become plural by forming more or be multiplied by a number as to indicate a growth in something not even existing. As far as my mathematics go, I would say multiplying anything with nothing or zero ($0 \times 1 = 0$) and yet those that are inconceivably more educated than me in mathematics believe Newton when

Newton divided one with another and got an answer worth nothing $\dfrac{dJ}{dt} = 0$.

As the one becomes one hundred the one cannot substitute a value of nothing but then must be part of something. If the one substituted the nothing, all laws of mathematics will go in disarray because when one multiply any number by zero it becomes zero placing both planets in the centre of the Sun. By excluding nothing from the equation, space becomes something bringing in a value lying inside the realms of the infinite that must form singularity. As the zero becomes a dot, something else becomes clear about the dot. Looking at the night sky we find darkness overwhelming the space in relation to the stars bringing across light. We can detect the dot because we cannot see darkness since our eyes were only meant to cope with light. With this knowledge, then how can we see the sky as darkness at night? We are only supposed to see the light of the stars and not darkness, yet at night we see a much wider picture than stars alone. One may bring in the argument that the blind see nothing but darkness. We seeing persons do not know what the blind does not see, so we presume it is just about the same darkness, but that is presuming. Due to a lack of space, I leave it at that but I do address the counter argument in much wider context.

When we see a red flower, science knows the flower being all the colours but the red it rejects and this we all know. Therefore the dot we see as darkness also must be light, withholding its light and giving us the darkness we see as light…But the dot must influence the surrounding as well, subtracting the light it claims from the surrounding by casting it as darkness. In the case of stars we see the light the star disassociate itself with, keeping the darkness it has as it pours all the light it has in excess into the darkness which evidently is then light.

From that one may conclude there should be two forms of singularity where one associate with a dark dot being light and another being matter with flowing light evidently proving to be the dark one. Proving the dot with many such arguments was easy. Naming the dot and its position, value location and proving the influences mathematically was much more complicated and proving the dot has a definite influence on the surroundings was at first seemingly impossible, yet it is done. This then definitely defined and underlined the measured value of the dot where this value the dot holds becomes of utmost importance when finding solutions to cosmic factors not yet clearly defined. The location and position of singularity makes the value all-important in acknowledging the unmistakable influence utilised by heavenly bodies occupying space in the cosmos.

Due to the position and influence of the dot, there is a direct reversible relation between heat and space where the gain of space (explosion) will be the loss of heat and the gain of heat (compression) will be the loss of space. Space is not "nothing"; it is the counter action to heat whereby overheating brings about expansion, which is the production of space. Cooling on the other hand reduces heat by increasing space.

Producing space by expanding reduces heat. The reverse interaction there is between heat and space science does not recognise … (hopefully yet) but in that relevancy are the answers to so many questions unanswered by science on the subject of cosmology. My approach might seem unconventional but through the abandoning of the accepted, it enabled me in locating the precise location of a universal singularity forming a connecting basis of the Universe (this I say with some degree of confidence). The smallest figure there can be must be a spot moving out and into a dot. The only mathematically sensible option about extending a line from the dot will be non-bias progress in all directions equally in order to give a meaningful flow of mathematical equilibrium. In that manner space will grow when exploding or occupying more space when overheating. This is also the precise phenomenon we see as the Universe expands. There is a direct relation and the Pythagoras mathematical principle is the proof and that I explain. The obtaining of the location and the value of singularity is in my rejecting of nothing by replacing it with something being the dot.

In understanding the Universe we have to stop placing the Universe around us. I have to repeat what I said before because I judge this concept to be extremely important. Every point in the infinity we may observe at is not part of the Universe in merely being nothing, but is the point where the Universe started, which is where the cosmos is being represented by singularity. Every point there might be holding a possibility of being occupied or not being occupied is the very first point where everything began so many eternities ago, because after all with singularity being space less eternity (1^0), all has to be identical $(1^0 = 1^1 = 1^2 = 1^3)$ and from that, how can we ever determine where the first point was, as they were very much equal and alike at the beginning. Every aspect of the Universe started with the fundamental fact that no point in the Universe can represent "nothing" as a number, because every aspect in the Universe represents singularity in what ever form it may hold in that specific spot forming space-time. If man does not reach a conclusion where that conclusion is matching the Universe and stop to match the Universe with man (and man's incapability), we may all go back to caves and become starving hunter-gatherers again, because we will never find a way to progress to the ultimate understanding of the Universe.

What is of vital scientific importance is that there are three fundamental dimensions controlling the Universe! The three are beyond intermingling and one confirms a status in relation to the others but not intermingling in status. From singularity comes matter and forming space-time in own accord. By matter not controlling time, space grew uncontrolled and the third dimension came about. That dimension birth we now recognise as the Big Bang, but the Big Bang is the last of a three prong cosmic growth. Science has to recognise the dimensions of densified (singularity), occupied (matter behind the electron) and unoccupied (space-time outside the orbiting electron boundaries) forming three points of cosmic recognising space-time

Every dot was by itself as well as the accumulation as it currently is forming what we think of as the present Universe. The earth in itself is a Universe standing apart from other Universes such as the moon as well as the space between the moon and the Earth. The moon is a Universe. Rules applying on earth do not apply on the moon and visa versa. When considering conditions with in the oceans and applying space-time another set of rules apply therefore the sea places a body in another Universe. It takes the same engendering technology going underwater in deep sea diving that going into outer space. The number of universal entities are still countless as much as it was in the beginning matter as atoms and even much smaller.

Every dot insignificantly as it may be is a part of another Universe as much as it is part of the accumulative Universe and every dot in infinity holds singularity, which we translate as "nothing" but it cannot be nothing. There cannot be nothing as much as there cannot be darkness. There cannot be something big or small except in the relevancies of perceptions and then the relativity of such perceptions becomes questionable. There cannot be hot as much as there cannot be cold The Sun freezes hydrogen to a liquid at 6500 °C and outer space boils over at 0 K. If we humans cannot or will not abandon our human culture driven perceptions and our mankind's pre-programmed perspective we may as well return to astrology for what the future hols. There are so many boundaries out there ready to destroy us because of our lack of insight, as did the challenger disaster.

Creation birth started off with one dot so small eternity met infinity within. Then came one more, and another and they continued coming until there were a countless number of dots. The accumulative size of the dots were the same size as any one dot because in the true Universe big and small plays no part. The dots were infinitely small and eternally big at the same time because size is a relevancy and without one the other has no size. So in the true perception, there is no difference in size.

To unlock scientific truth we first have to dispose of scientific misconception

In the two pictures we are seeing disposing or releasing heat creates space. We may call it plasma or shock waves or what ever Newtonian name that hides the complete lack of understanding, but in the final analyses it is heat turning to space. Whatever you wish to call that which lies between the particles comes from being a solid, then with adding heat, the solid *"whatever"* becomes liquid and that is the white and orange plasma that we find. That plasma or yellow flames are the black of space that is a gas, being concentrated to perform as a liquid propellant. That white and orange is heat in a liquid form, just as all flames and smoke is heat in a liquid form. But that liquid does not remain liquid because the governing singularity cannot enforce a commitment ensuring the liquid heat remains liquid. The liquid *"whatever"* you wish to call the heat in fluid form then further overheats turning the heat to space.

The space created must be equal to the heat reformed. That is a law of energy where energy equals equality everywhere it is. That is the only energy in the Universe. Let us humans first detach culture from facts. Take the argument to iron, which we know well. Iron cannot boil, iron cannot flow or bend and iron cannot break. Iron is an element like all the other elements we know, not one element can do any of the above, in sharp contrast to human belief. It is the ratio of heat in relation to material that will increase the density to become solid, or decrease the density to become liquid and dilate the density even further to become a gas. As indicated in this book, the limits we should find to guide us we ignore for the reason that we cannot see it. We may not be able to ever see singularity, but with intelligence guiding mankind, we do not have to see everything to believe everything. It is because we could not see religion, but still practised religion that set us apart from the other animals.

Everything in the Universe is heat. The only way movement can be is through heat changing relevancies. If heat increases the space in which heat is, expands. If heat levels decreases, then the space in which the heat is shrinks. The heat brings about movement by expanding or by contracting and that is the only way movement in the Universe can come about. There is heat contained by rapid movement and there is heat expanded due to slow movement. We think of the Sun as hot but the Sun is not hot. If we consider what the requirements there are to which heat has to adhere in order to become classified as heat being hot, then heat has to expand whereas the heat in the Sun is contracting to the extent it becomes a liquid. However when the heat escapes the contracting of the Sun the heat is expanding in the form of compressed light particles we call photons. The photons are frozen clumps of heat. When any element (except water) heats, it expands. When any element or object cools, it contracts or become smaller.

In the cosmos there is heat that spins faster than the speed of light, which we named solids or material. Then there is that which spins as fast as the speed of light, which we named photons and electrons, but is heat that manifests as liquid. Then we have heat spinning slower than the speed of light, which we think of in terms of gas but can also be less dense fluids. That is all there is. There are solids that spin so fast in a confined and secluded space we think of that as materials befitting the name we gave as solids. That is heat in occupied space. Then we have heat in unoccupied space and that we thought of as space. There is atom filling space formed by heat spinning faster than the speed of light and that is parted by heat spinning as fast as the speed of light which we think of as electrons and electricity and then there is heat moving slower or below the speed of light and that we think of as space.

Thinking about the Sun in terms of being in any state is incorrect because the Sun is not hot or cold, but holds heat confined in all three stages. The Sun, through enormous movement reduces the space and by contraction of confined space, is reducing heat in space by cooling and this is gravity condensing space that would otherwise be expanded heat. The movement of the Sun condenses the density of hear from being an expanded gas to a concentrated liquid in the form of photons. The movement of the Sun is cooling heat from an overheated state of total expanding to a concentrated form of forming a state of liquid in the form of electrons.

To an the elements wrapped in an atom and enclosed by electron jacket, hot and cold are outside influences that do not apply to the core of atom. Heat and space are influences outside the proton but within the proton nucleus it is bitterly cold. It just has to be if considered the veracity of movement the proton is undergoing. The heat or space or time will surround the atom on the outside, but has clearly an extremely strong influence on the inside of the star or for that matter the individual atom. That is where the movement freezes time to a solid. The heat or space is a state that in which the neutron finds buoyancy that somehow extends he flow or concentration of time between the electron and not the proton. The heat in concentration or the manifesting as space is neither hot nor cold we see hot or cold but in cosmic terms it is in the terms because the proton presents eternal cold. Heat is an freezing exterior influence bringing about influence between atoms within the star.

The density that we attach to particles is a contribution of the level of heat being responsible for classifying a substance a solid, a liquid or a gas. It is a

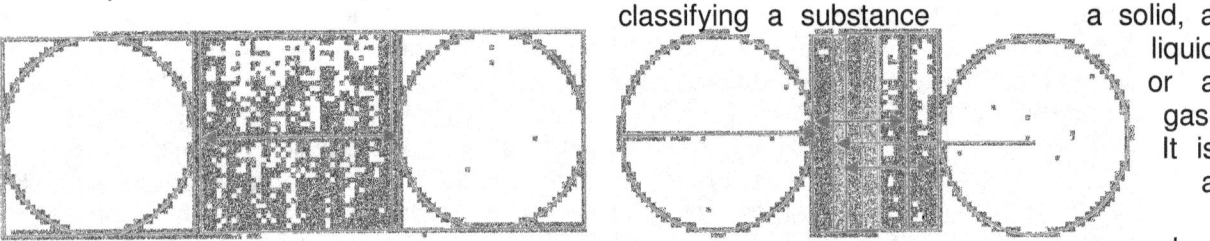

secondary substance not valued or put to measure by Newtonians which is there and is that which places the status of the group of elements in a category of being seen as solid, liquid or gas. A gas will allow influences to charge a seemingly loosely connecting uniting in the bonding there are between the elements that form some association in being a unit we think of as a structure. Such a structure is then the particles that form the positioning of the compounding molecules and forming the structure of one another forming a unifying object occupying space allocated in space lined up next to one another. Where matter or liquid is concerned, the structure is totally rigid and would not allow compromising of space. In the case of a gas, the gas will allow a lot of flexibility and allow to compromise in space where such compromising will lead to "pressure" rising bringing about heat levels also rising in sympathy because in the case of gas, space are always somewhat flexible as it never becomes a premium to the extend where it never compromises in inflexibility. I should think placing space at a premium would apply when fusion of atoms come into the picture within stars, that is said on the condition of fusion ever applying because I have serious doubts if such fusion does apply in any case. But f this fusion of particles does apply it would be in association with deep cooling and not being pushed into unification by pressure applying. If there is such a form of uniting then such fusion will only connect to movement bringing about freezing cold and the freezing cold will bring a condition of contraction we can never grasp in our conditions on Earth, let alone that we could imitate while being on Earth in this feeble gravity applying on Earth. If heat forms a gas when that heat surrounds the hydrogen atoms grouping, the gas will withstand as much compressing as can be induced by whatever force bringing about such compressing. One can compress heat up to a point where heat would be as dense as the electron and that is as dense as heat can go being in our Universe of $7/10 \, \Pi^6 \div 6 = 112$, but that is also as low in density levels, as whatever form of heat in an unoccupied state can go.

By heating an element, the overall density applying to the structural unit will depreciate directly in line with the value that the heat levels appreciate, whereby through the heat –levels increasing this space-heat increasing then are releasing extra space and the decreasing density will bring about more space within the confinement of the heating. But the more space does not come in the form of more material being present and therefore the extra heat is extra space claimed by the same material found in the occupying of the space. Yes, in some cases the atom wall extends and the atom claims more space because it holds more heat being controlled as material, but in atom numbers increasing because of heat levels getting more, well this is not taking place. If the confinement is such that it will not allow the growth of more space, then the heat level will rise. By heating without allowing space to compromise the heat will grow until a state will arrive where the liquid air will cut through any container that is man made. It even cuts through the walls formed by the gravity if stars and we call such an event a Super Nova exploding star. Gas will always compromise by giving away space to heat, but in doing so the heat will increase or if the heat does not increase, then the space must increase. That is the two options open for time to form space. With such knowledge we have to look at the functioning of stars once more but this time with a much more critical view. In fact having realised this it prompt me to look at pictures of the Sun once more and then I realised from what I saw that the Sun is liquid and if the Sun is liquid and not gas then all other stars are a liquid inside or in better developed stars even solids on the inside! Gas can never compromise enough space to allow fusion to take place, but liquids are quite another story and electricity is just the way nature use to turn gas heat into liquid heat.

In the process there are relevancies. If outer space is - 276^0 C then it can only be that if something else is 0^0 C and another object is 100^0 C. But if space stood alone at -276^0 C the number could have any value in relevance and the relevance could be anything because without another position that holds a counter value to bring about the relevancy by comparing the two, the one number standing alone has no meaning, not withstanding whatever connection there are between the two.

The objective of a star or any sphere for that matter is by spinning it is compromising the inner space and more so as one comes closer the centre. The star's spinning is reducing the space to the

inside which then claims space from further a field which is providing a reference as far as the influence stretches, but what reference and to what will such a point refer? By using a liquid within the star too the comparable gas outside the star and the solid being the star we can begin too see comparisons emerging. Remember that outer space gas is hotter that the inner space fluid of the star not withstanding culture shouting the opposite. If it is denser, then it is colder and that is always the case. In a hydraulic system the hydraulic power will only fail once the weakest spot in the solidity breaks down by overheating where such a point will produce space as liquid evaporate into a gas formation. Let's see this by comparing it to hydraulic power applying in machinery and keep in mind that life only uses what is already available for use and already applies as a principle within the cosmos. With a tough enough cylinder that will withstand all pressure pushing at it, something will give way and we know one hundred percent it will not as it cannot be the hydraulic fluid on the condition that the hydraulics stay cold and does not overheat by evaporation. The hydraulic oil will produce more fluid when starting o increase heat levels up to a point of overheating and thus breaking the limit set by coldness While the heat levels rise and the fluid remains in the liquid limits, then in such a case the hydraulic oil is the solid being liquid and performing as a liquid. We also can see that the liquid form that the oil is characteristic of applies to form a hydraulic drive and do not become a solid as the state of the material normally holds Forming a liquid is much different in drive that what will happen when a solid is taxed with the same burden. The liquid is a pure and uncompromising form of heat being as cold as fluids can get while still remaining in the purist form of whatever forms outer space that the cosmos can provide. When in a liquid the heat can only compromise form by becoming more and not less and when becoming more, the heat compromises in density. By becoming more it also changes into having more space and while remaining a liquid, fluid driving is less uncompromising than what solids are. Solids break and crack while hydraulic power pushes because it can't crack or break. In a star such as Jupiter that has gas within the atmosphere, (let us call the atmosphere a gas this time because gas and liquid is a very grey issue) the pressure within the structure cannot apply a solid base to secure solids compromising space to bring about fusion. Fusion is about freezing and the gravity hydraulic power will produce will be much stronger that what any gas would produce.

When outer space is 0^0K then the Sun is 6500^0. That is the relevancy between the Sun and outer space. On the one side the gas forming outer space is -276^0C and the other side the liquid surrounding the Sun is $6224^0 C$. But the material within the confinement is neither -276^0C nor is it 6224^0K. It is the space holding the solids that can claim to be such temperatures while the solids will reduce in size, but not become colder or become hotter. Deep within the Sun the temperature seems to be 18×10^6. In view of the contraction coming about within the Sun, everything must be frozen to a solid at be 18×10^6. It is only a relevancy with substance when humans relate the information to circumstances seen from the Earth and moreover what life requires to survive. If standards applying to   the Sun were transformed to standards on Earth then no substance

known to man including man or any form life may have would have any validity to exist. With standards applying in the Sun, conditions there are set to totally different understanding because the Sun would be 0^0C where we think of the Sun being at 6500^0 as that will be the lowest temperature that can be found in the conditions applying within the governing singularity controlling conditions ruling the Sun. Water would be an unknown substance holding an unknown quantity as a reference. Outer space must then be minus $6776^0 K$ if the Sun whish to use the scale we apply where the Sun is 0C. But it does not work in that manner either because most of the substance that is providing life support can have no claim to hold any validity as part of space-time within the Sun. The relevance stands related to the first one breaking under pressure, setting a standard by which others following can truly apply The standards set can only match when all conditions are met by exactly the very same standards applying. We know water boils at a much lower temperature 100 km up in the atmosphere than it boils on the surface of the Earth and down below sea level it boils even at a higher temperature. Heat does not adhere to the same specifications on Earth, so how the hell would Newtonians feel the Sun should be obliged in meeting the Newtonian thermometer scale and be hot as Newtonians think the Sun in gas should be. This again concludes my argument about space being much hotter because it needs less heat to apply to get water to vapour. If water takes less time or heat to boil then water must be closer to boiling point in the highest atmosphere and that can only be if water up there is hotter from the beginning before the boiling process started. Conditions on Earth have such a variation, yet science has this tendency to standardise everything in the cosmos by applying constants.

If water boils that quickly the closer we get to outer space it should be a big indicator that water is naturally much hotter from the start in outer space. When the heat surrounding the outside of the molecule increases the heat inside the element has to reduce because there is a relevance attaching to the two opposing limits without the one opponent having any precise limit to show. As the outside fluid heats up, there is no breaking down of the substance because the fluid heat forms as light or flames, is the purist fluid there can possibly be. Light is pure liquid heat. The heat will be more solid than the solid elements can be because the heat cannot give way any more than it already did at the event of forming conditions that realised the Big Bang. Just before the Big Bang arrived it took all the heat occupied or not to form heat that would be able to bring about the forming of space as a compromise to time limiting further progress and give way to space forming. There is not enough gravity left to produce more compromising from heat in a liquid form, but we have to recognise the principles applying.

By heating the fluid the space the fluid holds becomes more and by heating the fluid the element becomes colder reducing the space the element holds. Space reducing is synonymous with becoming colder and becoming colder is about compromising space occupied. On the Earth material will reduce space occupied that much and no more because the relevancy establishing the edges can only push the reducing that far and no more. But in the Sun the conditions applying is a lot different and the relevancies can push that much further. As the elements in fusion one of the two compromises all space and in that act, it supplies the biggest compromise in the relevancy and that is to give space over to the fluid side. When the fusion between two particles take place, one of the particles freezes space into the oblivious and in doing that, it stretches time eternally and with no space in eternal time it becomes part of the atom it froze into. In receiving the space the heat sacrificed, the space that the heat sacrificed for time is then absorbed by the particle gaining matter it froze time to a standstill to gain more space by acquiring a sudden growth material occupying space. When looking at the Sun we see dark and light spots and we think of the dark spots as being cool while the light spots are hot. This is so typical Newtonian because the dark spots accept heat (back maybe?) into the Sun while the light spots are areas where the Sun rejects the heat by pushing the heat and releasing the heat back to space. In the one (the dark area) the Sun accept space into the Sun by growing black or dark or expanding space while the light spots have heat in space concentrating and by concentrating space it rejects the overload in heat which that area has. By acquiring the space, it then receives a cooling that will translate to dark spots on the surface of the Sun once the heat again surfaces to the top. By expanding space, it admits heat into the Sun as the Sun provides a spread of heat over a larger area. The very opposite then has to apply in the case of where the light spots are as the heat levels rise higher and thereby rejecting the increase of heat levels. In sacrificing heat the liquid obtained space to reject liquids as over cooled or over condensed space and by sacrificing space the elements in fusion reached the ultimate freezing temperature it could ever achieve. The elements never become hot or never freeze but the relevancies changing brought about condition allowing the heating and cooling to take place without ever taking place as far as the elements occupy space in relation to singularity governing by rule. Gravity is the condensing of space by freezing space into a colder substance we think of as liquids.

Gravity produces mass but mass is only the result of gravity. Mass increases when the object holding mass is surrounded by atmospheric heat that by the effort of gravity applying, condenses whatever holds the space into a colder substance and heat will influence mass by shrinking space into a colder condensed area. Therefore mass does not produce gravity as science indicate by their formulas used in calculating gravity because gravity does not increase when heat influences mass. Heat levels increases by movement concentrating space. The movement of material concentrates space density as it increases heat formed by condensing space through the increase in motion, and this increasing of heat reduces space held by material as the material in relevancy becomes colder in relation to the increase in heat levels rising. The reducing of space material holds translates to the space surrounding the material and by singularity increasing relevancies, the heat in the space surrounding the material also freezes and thus it forms a concentration of liquid heat. To the space and us the heat level rises but that is because the density increases in the heat forming space but the density rises because the space reduces because the area must then get colder. In the way of increasing heat levels by concentrating more heat into over a smaller space, this produces gravity and gravity reduces space by increasing heat density. Any one that is not in agreement with what I suggest must convince themselves by comparing the neutron star with the massive read giant and then be convinced that reducing space increases gravity and by cooling a star it takes away space. To calculate a Black hole Newtonians found a way to go and throw C^2 next to the dividing radius and throw the square onto the C that presents the speed of light. Then they sit back and feel smart in the way they manage to cheat once more to prove their incorrect views correct because after all who will ever fly down a Black Hole and return to support or deny their calculations. The gravity of the Black Hole is a result of gravity or movement or speed increasing because all gravity is speed or movement, which is time. Then the speed that light has is gravity where gravity changes from Π^2 to

C^2. The gravity of the light can be gravity as much as it at that very same time can be antigravity. What the hell has C^2 got to do with a Black hole because you can pop what ever nuclear device far away from a Black Hole and the consequent explosion resulting from the nuclear detonation would at the most and at the worst be very much insignificant! The light will not even escape form the point of thermo release because of the gravity applying in the Black hole. When this became apparent that the radius of stars reduces as the stars develops through progress and in that spin faster because the star becomes smaller but also much denser, then someone that is able of thought was supposed to say: hey there is a dead rat I smell because as size or apparent mass reduces, gravity increases. For my saying so I am the clown in the courtyard, the one with the two dead brains cells and have no more to use as spare.

It started with the fact that there is no place or part in with which one may associate zero or nothing. There are no room for a number such as nothing. Next to the one dot (infinitely close) one will find the next dot, and if nothing was a factor then that is precisely what one will find between the two dots. Nothing of space, a non existing entity, taking up no space, and much more important, no time, therefore the dots are infinitely close to one another, being the same space, eternally big as much as infinitely small. If we as humans cannot find a manner in comprehending this notion, there can be no manner ever understanding the cosmos as much as the start to the cosmos.

Every dot was a Universe in its own and the accumulation was a Universe. The Earth in itself is a Universe as the moon is a Universe, because rules applying on Earth do not apply on the moon and visa versa. When considering the conditions with in the ocean and applying space-time another set of rules apply, therefore being in the sea places a body in another Universe. The number of universal entities is still countless, as much as it was in the beginning, before dots formed atoms.

Every dot insignificantly small as it may be, is a part of another Universe as much as it is part of the accumulative Universe and every dot in the infinity holds singularity, which we translate as " nothing" being " darkness". There cannot be "nothing" just as much as there cannot be "darkness". There cannot be something big or small, but the differentiation is contributed by the relevancy of perception, and then the relativity of perception becomes the question. There cannot be hot as much as there cannot be cold. The Sun FREEZES hydrogen to a liquid at six and a half thousand degrees Celsius and Universe boils over in the form of the Hubble constant at the temperature (we presume from our vantage point) at minus 273 degrees C. If we Humans cannot or will not abandon our human perception and our manly perspective, we may as well return to astrology for all its worth, because that is the only boundaries we will find in the cosmos.

The reality is that science has to reassess their view about what is cold and what is hot. Hot and cold is not about a reading on a thermometer scale but is about when something is cold it is dense and when something is hot it is vapour.

The element	Transforming to a liquid	Transforming to a gas
Hydrogen 1	melts at -259^0 C,	boils at -252^0 C,
Helium 2	melts at -269^0 C	boils at $-268,9^0$ C
LITHIUM 3	melts 180^0 C	boils at 1300^0
BERYLLIUM 4	melts at 1287^0C	boils at2770^0C
BORON 5	melts at 2030^0 C	boils 2550^0 C
Carbon 6	melts at 804 ^{0}C	boils at 3470^0 C
Nitrogen 7	melts at -210^0C	boils at -195.8^0 C
Oxygen 8	melts at -218.8 ^{0}C	boils at -183^0 C
Fluorine 9	melts at -219.6^0 C	boils at -188.2^0 C
Neon10	melts at -248.59^0 C	boils at -246^0 C

Melts (meaning it becomes a liquid) at -259^0 C Hydrogen 1 boils (meaning it becomes a gas) at -252^0 C,

According to what we are taught by culture tutored in schools we think that there are forms in which elements are found and it is the elements charging the form it is found in. There are "natural" liquids. There is "natural" gas. Forget about the example we find in water because it seems if any one talks about the three materials all concerned immediately thinks about water as one structure that can ice and can boil when it is not flowing. That is not even an example because water is not a true substance but is a compounded combination of volatile elements forming the most outrageous concept the cosmos could think to form. By combining some of the most volatile elements the cosmos created the least volatile substance known to man, which is water. But the three forms water personifies is the structure compiling the Universe and prove stages of development and I this book I am not getting into explaining this statement in any better detail.

There was singularity. Singularity came first. From singularity came material as a solid substance. From elements comes liquid, a soft substance. The next step to follow is gas forming when liquid changes density even further and material turns to space. Again we can see a pattern coming about. But it involves Kepler more than any Mathematics Newton cheated into being accepted. There where this line forming what divide is, there at that point can't be a Universe because that point has no space and is not part of what is inside this Universe. This point is the Universe but fall way outside the Universe and while being the Universe can never on single effort be part of the totality of the Universe. While forming the Universe, this point falls way outside the Universe. Honestly, if Newton and Einstein was that smart as everyone presume they were, then they must have realised the importance about this line because even I can see it and I promise you Professor, I am not the smartest smarty in the pack! This is that what Newton said is $\dfrac{dJ}{dt} = 0$ but in cosmic reality it is singularity that one should express as $\dfrac{dJ}{dt} = 1^0$.

All lines we can visualise would form a duplication of another line sharing value since there will always be a possibility of yet another line in the realms of singularity lying between the two lines in question reducing the size infinitely to either side of the divide we humans create. Boundaries therefore are human and as man made substances it does not belong to the cosmos outside the influence of man and must be discarded. The understanding of insists on one vital precondition. We have to abolish our instincts about things being big and small, high and low, hot and cold, tall and short. All the human measures are truly unfitting to the cosmos. Those are measures made by man. In the cosmos the tiniest object is the key to that particular Universe. By going smaller one is going bigger. By going within the atom we go out and meet the Universe. In singularity we find the cosmos because every point holding singularity not only represent the cosmos but also in fact is that Universe. No mathematics will ever measure the thickness, because as the line that is standing still it cannot have a width at all because the instant the line forms space that point connecting singularity to the point it connects to space is then part of the divide and not forming the division any longer. The moment a width appears which one can measure or calculate; at that location the line will become part of the factor forming what is the divided and not the dividing. The instant when space connects, the spin direction will produce the partisanship of space and spin. Any form of space including even in the most incredible minutes will produce a favouring of direction.

A bigger motion discrepancy will bring about a larger pushing of the secondary space the smaller object holds to slow down the motion of the Earth in an attempt to increase the motion that the secondary object can apply. There is a specific border or a definite barrier where the motion becomes so critical that the incorrect transforming of motion can have the same effect on the incoming object as hitting a solid wall. Later on I shall show that it is in fact the equivalent of a solid wall the object will break if the object fails to enter through the small door there is.

All principles that I make use of to explain my theory is part of nature. I base my theory on heat becoming stabilized through movement becoming more rapid as the spin accelerate by gravity increasing movement which then with a higher rate of spin collects more space by using motion increases to produce cooling. This idea is most basic and that I admit. It may sound basic, but Mainstream science is also most guilty of their departing from these most basic principles through the employing of terminology and terminology has covered many of the crudest, most basic meaning behind the most basic principles in nature. Newtonians try to impress by looking smart as they hide behind the most impressive names that create just to hide their incompetence of understanding what they discovered. I do not applaud a principle Mainstream science underwrites in the sense that matter in the beginning was coming about and anti matter came to destroying the matter. That person that came up with that idea loved the image the (I think it was the pact man computer game?) the first computer game brought about as you shot and destroyed funny little biting faces racing towards a curser that the game player controlled. It is moreover the disappearing from the Universe of the result between the two opposing materials that I strongly reject. If that was part of the cosmos, it still must be in the cosmos because there is simply no other place to go. If anything is part of the Universe, it can go no where but remain in the Universe and this statement only exclude life that never becomes part of the Universe but is only connected to the Universe by some link carbon can provide. But life mimics heat and material that can remove heat, can remove life from cells as heavy metals seem to be able to do.

 At the start one would find iron and iron in a "natural state" as we find iron on Earth being an element mind by humans from the surface of the Earth it will be a solid, suitable for man to handle with bare hands. When such a piece of iron is left in a desert in the midday heat, the human hand cannot handle the iron any longer without aid of covering the skin of the hand. Our perception is that the iron became hot, but that is not the case and our view is a culture contribution and not scientific fact. By heating the iron artificially with combined gasses (acetylene and oxygen or what ever) we now can over heat the iron to a state of flowing like a fluid. Our human culture tells us the iron now is melting. That is a misconception! Like the fact of "nothing" we inherited the idea from our past.

After introducing artificially even more heat with more heat releasing gasses we may artificially form a condition where the iron would become a gas. Again it is not the iron that becomes a gas; it is the space the iron finds itself in and thereby the density decreasing with the adding of heat in a liquid flame that became hot enough to help the iron density decrease to become a gas. The iron particles remain the same; it is the condition surrounding the particles that changes form with overheating.

Important to note is the fact that iron in a solid state will surround itself with solid matter in space applying a solid space. By introducing conditions producing *more overheating* the space or connecting between the particles become concentrated heat forming a liquid substance! It is not the iron that turned liquid but the wrapper containing the iron that concentrated so much it formed liquid fluid by the introducing of more heat to a point where the overheating created a fluid. It is considered that the oxygen burn and by that the iron heats up. NOT TRUE! If oxygen burns no oxygen would be left on Earth by the time man arrived on Earth to use it to the benefit of intelligent life. The oxygen remains oxygen while the oxygen merely does a task in nature where oxygen carries heat to a specific space. On the other hand it is the task of nitrogen removing heat from the point of overheating by means of flames whereby it creates space. One can feel the "wind blowing" as the flames generate created space. In the extreme the creation of such space we call an explosion.

In the process where the space between the iron particles still further overheats, it becomes a gas. It cannot be iron that becomes gas, because iron will be as much a gas as iron will be a liquid or a solid. It is the space covering the iron particle separating the different iron particles, which will increase in ratio to produce density decreasing of solids per volumetric space and this decreases convert and sustain form. The gas is as invisible as space because the gas is the form space holds.

Iron is a solid. Introducing more heat the iron becomes wrapped in a cover that concentrates the wrapper to the point of concentration where it becomes a fluid. The iron remains what it is, and it is neither a solid, nor a fluid nor a gas. By introducing more heat it becomes a gas. The gas we cannot see because the gas is space or time. But so was the fluid space.

The introducing of heat brought about the turning of a solid to a liquid to space and every time more space becomes part of the picture. Iron is in its normal form a solid. That means the space, which the iron particles are in is solid and that disallow the iron to alter the form in which it is. By introducing considerable heat the iron melts changing the form of the iron from solid to liquid. Considering the evidence we find it is not the iron that melted and that became liquid, but it is the space in which the iron is that became liquid. The iron particles are still as solid as they were. By introducing more heat the iron would eventually turn to gas. It is not the iron that turned to gas, but it is the space in which the iron particles are that has increased to the extent that the space now has so much heat, the heat turned to more space. The iron as particles remain the same, they are just elements confined to a nucleus with electrons spinning about. The space between the particles increased to such an extent it first became a liquid or a fluid and with more heat introduced the heat increase brought about that heat turned to space. That means by overheating the particles surround with heat as a fluid the heat increase then add space as a gas. The gas is the ultimate form of overheating but where one is unable seeing the gas. Let's take this iron principle into the galactica.

1 Firstly the iron is cold enough to be a solid. Replace the word iron with cosmos and forget the colour we associate with heat being white and note the solidness of the centre of a galactica. This must have been the state of galactica that contained large parts of the Universe when time rolled away from eternity.

2 By introducing overheating the space between the iron and not the iron as such turns to liquid. The same apply as more matter (iron) produce more space forming as some matter turned to heat by overheating. The matter increased spin and in that way went out of sequence where it then became softer and softer in relation to other particles, where the loss of the matter released more of the third cosmic component we named heat and space.

3 Some of the heat introduced with the overheating by means of congestion then forms space while other remain in the form of heat allowing space to seam liquid. The matter could not breathe and overheated by the enormous gravity the overheating created and this allowed the material to start to spin in order to cool the material. It was either starting to spin or expand by exploding.

4 As the area between the particles still further overheat certain parts of the area overheats to the extent that the space becomes an invisible gas allowing the congestion of matter to separate from one another and allow the stars' individual governing singularity growth and this brings about that the entire star structure stars spinning.

5 From the soup of heat galactica come about allowing stars to rise out of the dense liquid cradle from where they can establish singularity growth. The process continues as more space becomes introduced through space overheating turning heat into space and by spinning the material controls the overheating within the material.

6 Should star development come about as suggested it is foreseen that the Milky Way once was a liquid from which the Sun developed the singularity in which it then form self-sustaining. The only pre-condition was that it captured individual space-time where the captured space-time remained a liquid frozen (as it was back then at the time of parting) by the governing singularity while outer space further overheated into a thin gas.

7 The Sun captured so much space by the intervention of singularity when released from the Milky Way that it produced space so concentrated today at present it clearly remained a liquid inside as it froze the interior in time the liquid it now is while outer space is still overheating as a gas with no visibility.

From this overview one can judge just how far behind the time Newtonian scientific concepts are in their views on creation and the beginning of time including the universal establishing. Think of how they see gravity conjure dust particles floating everywhere as mass into solid planets that after forming solid then is nicely circling the Sun. Cosmology still hides behind medieval ideas that other faculties and scientific departments forgot long ago.

The Earth overheats in two ways. The one way we have the heat on the outside of the atom and in the other overheating we have the heat on the inside of the atom. The heat on the outside of the atom is contracting heat making heat more dense and allowing heat to flow from the outer circle to the inner circle. By cooling the inner area of the atom has motion applying and this pushes up the relevancy of attracting heat from out side to the inside and in that the atom is gaining heat because of the cooling and the subsequent compiling of heat that surrounds the Earth.

As much as there is expanding coming into the Universe by reason of overheating bringing growth in a space, the atom spinning forms just as much part of the expanding process because of the atom creating contraction which is just forming a directional opposing of expanding, but in essence it still forms part of the overall expanding by the motion allowing growth in material....

...just as much is there expanding coming atom by the motion of the atom allowing outer space in material... ...and the growth caused by the reduction of density of heat the one loss compensate for the other precisely evenly distributed when the Big before half of the Universe started to stared to contract the the end the again into the Universe outside the decrease in density in in heat within material is in outer space and in this factor's gain. Material was Bang commenced just expand and the other half half that was expanding. In two factors that formed will be evenly distributed when the unification of eternity with infinity places everything back into singularity as is happening in the Black Hole at present where the Black Hole is the second last stage stars reach before singularity finally unites infinity with eternity.

As the top starts to spin and remember the spin is a product of life initiating the mimic or copying of cosmic law, and in copying the motion process of the cosmos, the top resumes the role of an entire galactica, the top serves as a future Black Hole, the top becomes a star but most of all the most important cosmic position the top serves as is the top becomes an atom and all the above mentioned are all atoms rein acting the role given to the atom as the atom is the Universe.

The atom is not just a part of the Universe but the atom Universe is later becomes, therefore what the Universe is, it because the atom forms the Universe. By the motion the top develops, the top charge into service the presence of

forms what the

Expanding into liquid by losing space integrity

singularity Π^0 forming Π and at that point the top forms a self serving Universe as much as any or all other atoms in the Universe forming innumerable many Universes which then are forming the Universe. But with the top spinning one can see the outer space is as much contracting as the inner space is expanding and in that the outer space reducing forms the inner space of the top growing by the margin of reducing. As the outer space is losing density to the contraction of the top, the inner space of the top is gaining in volume due to the loss the outer space is experiencing. In other words, the atom is growing because by spinning the atom is initiating the charge of singularity and the charge of singularity puts a flow of heat in relation to singularity growing in intensity.

The issue that I whish to introduce is that the atom is growing in stature by the sinning motion the top is exerting. That is what gravity is all about. Gravity is the transfer of heat in a process of expanding in a positive direction becoming less dense but more space, and in a negative direction becoming more material and denser by serving movement in order to accumulate heat in order to maintain the role eternity has in connecting with infinity. In the formula Kepler used to introduce space-time $a^3 = T^2k$, Kepler showed that space inside the atom $a^3 = T^2k$ is growing in density $k = a^3 / T^2$ as much as space outside the atom is reducing in density, $k^{-1} = T^2 / a^3$.

Following the direction that this argument brings, it is clear that the Earth as much as every atom in the Universe is expanding by the growth in density in material and the deducing of density of space outside material. In this argument it is clear that the size the Universe is, is today much bigger than the size the Universe was in the past. But unlike Newtonians serving the goal to preserve the integrity of their master and allow Hubble's constant to expand there where it is far away so that Newton being much closer still can contract, the expanding is part of every atom every where in the entirety called the Universe.

In the Universe singularity stitches everything into a woven concept we call the Universe. There is the premier singularity that serves as a beacon to everything carrying singularity and that singularity forms part of every singularity charging space-time. Everything is growing but not according to my perceptions because that is growing in alliance with the cosmos. I am not going into much more detail but to say that everything there is (1^0) connects to everything there is (1^0) as (1^1). The network linking what there is to what there is, is not linking on this side where there is no reality but is linking where reality meets infinity by uniting eternity. The fact that a person can view the entire Universe by making contact holds what there is (1^0) in view of what there is (1^1) and this takes place on the side where space is of no consequence because time in eternity meets time in infinity.

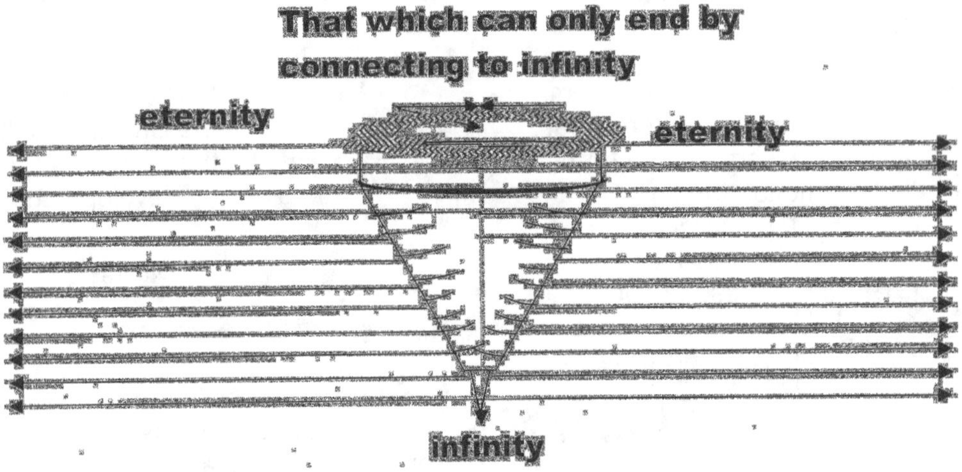

That which can only end by connecting to infinity

eternity   eternity

infinity

On the outside of space there is time that can never end because the time has no outside. Everything that ever will be, will be because it is inside this that can hold everything but nothing can hold this space that there is no outside too. This space ends where this that has no end meets that what there can be no start too. This has everything on the inside just because there is no outside to what this is. Why would I not use names…it is because it is Biblically named. This is eternity.

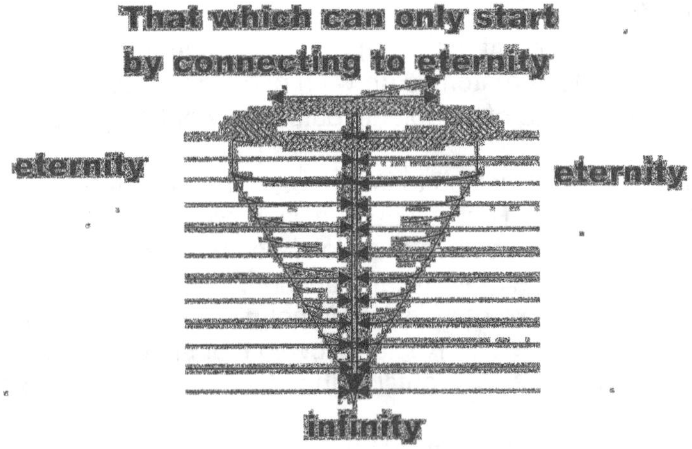

This is infinity. This can never start or begin because this is the beginning of everything there could ever be. This has no inside because this is the inside to whatever could be on the outside of this that is representing al the inside there will ever be. The Universe is what is inside that which can have no outside and what is on the outside of that which can never have an inside and reality is where these two factors representing that which can never end unite with that which can never start and

That which can only start by connecting to eternity

eternity   eternity

infinity

moreover, the unification forms reality found only on the other side of where time produces space. This is where eternity becomes infinity and this is where (1^0) unifies with (1^1). This is where (1^0) connects with (1^1)…and that is where we are not because where we are, there is no connection of what has to connect in order to allocate us to where we are.

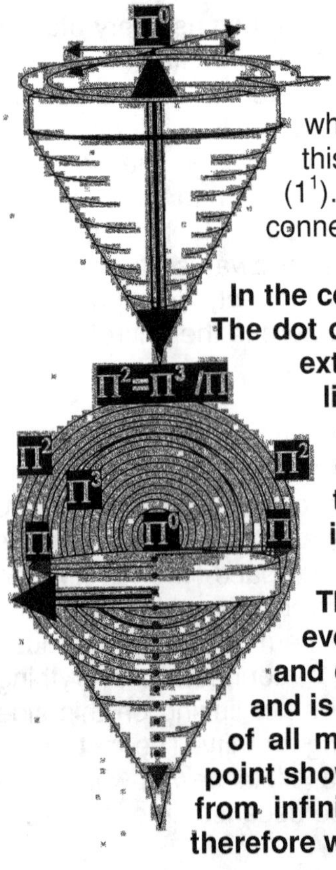

Π^0

$\Pi^2 = \Pi^3 / \Pi$

Π^2 Π^2

Π^3

Π^0

In the centre of al things spinning a line comes from where a dot first was. The dot came from where a spot once was. The line comes from a dot that extends but just as the dot extended from the spot it once was, the line has no start and has no beginning. As soon as entering the line one has gone through the line. The line has no inside that can go even smaller and yet we know that the line must have some ability to be able to go smaller since our understanding of the concept insist on this reality.

That point without space is where the Universe starts because everything that can ever be, starts at the point that could never start and can have no beginning because it is where everything is beginning and is without limits because it holds no space. That point is in the centre of all material, which puts that point in the centre of the Universe. That point shows where eternity parts from infinity. However to get eternity to part from infinity, eternity needs to move because infinity can never move and therefore will forever stand still.

There was Π^0, which was α^0 or if you would rather have it Ω^0 or it maybe was 1^0, but more correctly it was all the above and the beyond because multiplying what ever constitute the mentioned will bring about what is mentioned to a precise equality. It was a spot that was not. It was a line that ran eternal but because it ran eternal and kept repeating exactly what was before to the precise what came afterwards the line was there and was eternally running, while never changing in the least or growing by any measure. It was not one because before it was one, what was repeated and the process cycled back to before one and before one could be reached. It was such a continuing of the monotony, no change ever once occurred and therefore never did the running produce progress because the progress was in the perfect repeat of what was before. The duplication brought contraction to the smallest detail.

That is where our atheists get one hiccup. Everything that I show is as real as the Universe can be and yet not one point is part of the Universe we see. At the start before the star eternity met infinity and as eternity repeated the past it met infinity at the next point holding the future. Eternity ended on infinity every time eternity shifted. The repeat brought eternity and the repeat was so perfect that the repeat continued. The repeat still is with us as much as we are within the repeat. There was something beyond the Universe that instigated breaking the perfect cycle, which change the institute. There was something that brought a difference and we are within that difference.

That difference was time and that time is what we move through as much as what we see at night. Oh, how stupid and how thoughtless the minds of atheist and other atheistic animals are. Baboons do not recognize this factors revealing the position allocated to infinity as well as eternity because they cannot think and are therefore atheists. Animals are not able to realise that the true value of every point securing singularity albeit (1^0) or (1^1) is without space and is therefore not in this Universe. The location is where space is not and the only value such points have, is the trajectory of time bringing on movement. The points hold relevance and that holds space as a result of time leaving footprints but the true essence is in a place that has no space at all. The only way this can be recognised is by persons recognising another bigger Universe of which we have no part in the present. It is a place one can only reach in having faith and religion. These points referring to singularity is what we see at night when see light or darkness and when we can't see the nothing Newtonians see so clearly. Spiders cannot think and therefore they are atheists, as they do not think what the night consists of.

Reptiles cannot think and without thought they are incapable to see what time is, what space is, what light is and what darkness cannot be. All the animals I have mentioned are mindless atheists because they fail to see beyond the visible into the realms of the thinkable. Because of the incapacity to think the animals are both mindless and they are atheists. Therefore atheists are mindless. The night sky is such a bright light our evolution development protected our vision from the brightness of the night light in order to give as much better day-time vision. Through evolution development our eyes are protected from the light and we remove the qualities that night gives nocturnal animals as such animals see by the night light. However animals do use dark light and not our light to see by. You can shine a bright hunting spotlight onto an animal at night and the animal will not be able to see the light you shine on it.

The animal does not use the light we shine in order to see well as the animal is totally unaware of the light. Then a prowler come from the night and see the animal in the light the night provides and although one shine a hunting light into the eyes of the animals, they remain unaware of human presence. It does not use the light the spotlight uses and the light is not even traceable to either the animal hunting or the hunted. From there we accept that during the day the animals must be using our light to see because the nightlight is inferior to see by. Who says they use the daylight much different from the nightlight because all evidence is there that they cannot recognize our light as light. It is very evident in the manner they go on hunting and grazing while being totally unaffected by our form of light. That which you see at night because you cannot see darkness and you cannot see black is the light the Universe is painted in just like the Bible says. This is not religion and it is not a sermon, it is hard-core and brutal basic science and it the most fundamental basic physics there is. It is the start of the mathematical Universe portraying the only physical way it could ever be.

To save Newton's fraud from being discovered, it is agreed that while the Hubble constant is a fact, it is a very far off fact and while the Big Bang expanding is a reality, it is only a reality that applied before Newton placed contraction as the pivotal force in the Universe.

We accept that we advanced from the Big Bang, just as the Bible says and even Newtonians profess that we advanced from the Big Bang. The big issue about the Big Bang is that every atom grows and as every atom grows, so does every star grow and even the Earth and the Sun is getting bigger all the time, without one Newtonian openly admitting this or recognising this as a cosmic fact.

To save Newton's fraud from being discovered, it is agreed that while the Hubble constant is a fact, it is a very far off fact and while the Big Bang expanding is a reality, it is only a reality that applied before Newton placed contraction as the pivotal force in the Universe. However, as the Earth is getting bigger all the time, so is the distance between the Earth and the Sun also growing wider apart because the distance between the Earth and the Sun is not space, but is time that is developing by time accumulating and then progressing as space. In that same manner the Earth and the Sun is also not growing in space, but is being part of a time development in accumulating space. Time captured by singularity in the past held heat in space at much denser levels than what we find heat to be in outer space currently. This evidence is apparent from how the Newtonians believe the Universe cooled down and to get that lot to admit to that as they then applauded expanding without renouncing Newton mind you, this compromise was a brave move on the part of the stubborn Newtonian.

However, as the Earth is getting bigger all the time, so is the distance between the Earth and the Sun also growing wider apart because the distance between the Earth and the Sun is not space, but is time that is developing by time accumulating and then progressing as space. In that same manner the Earth and the Sun is also not growing in space, but is being part of a time development in accumulating space. Time captured by singularity in the past held heat in space at much denser levels than what we find heat to be in outer space currently. This evidence is apparent from how the Newtonians believe the Universe cooled down and to get that lot to admit to that as they then applauded expanding without renouncing Newton mind you, this compromise was a brave move on the part of the stubborn Newtonian.

There is a reason why spiders have eight legs and the reason is not because it is coincidental. There is a reason why insects have six legs and that too, has the same foundation as too why spiders have eight legs. At the time these species developed on Earth the gravity and the winds resulting from the gravity was ferocious. Gravity was incomparably more then too what it now is. The gravity that applied at the time brought on winds that blew mountain ranges flat as they were sticking out from the then sea level. There is a reason why reptiles had such enormous ling tales and that was to help them with the balance, as they had to fight the enormous winds. For the same reason they were long and slim and comparatively flat. Gravity back then was no mans' playmate and was something to keep account of. I have an idea that when comparing your average modern super intellectual Newtonian with the giant Dinosaur in size as it lived in the past in true reality relating to living proportions applying to both during their lives as each holds the volumetric ratio, the comparison would be about he same as the two pictures depict. I can't see while I try to maintain a non sensational as well as a pragmatic and a stable mind that the dinosaur that lived seventy million years ago was much bigger than a lizard is today. Conditions and land space that was available at the time just would not permit the size of the animal to be much larger in compatible ratio. But reality would not prevent the Brilliant Newtonian genius in going totally Hollywood and fantasize about the fabulous sensation of the size of the fossils without having any thoughts about what pure common sense would insist on. If our Newtonian could get ridiculous while also standing in front of a camera, then their brains become the size of what the so called Hollywood stars have and that is a true Newtonian nothing!

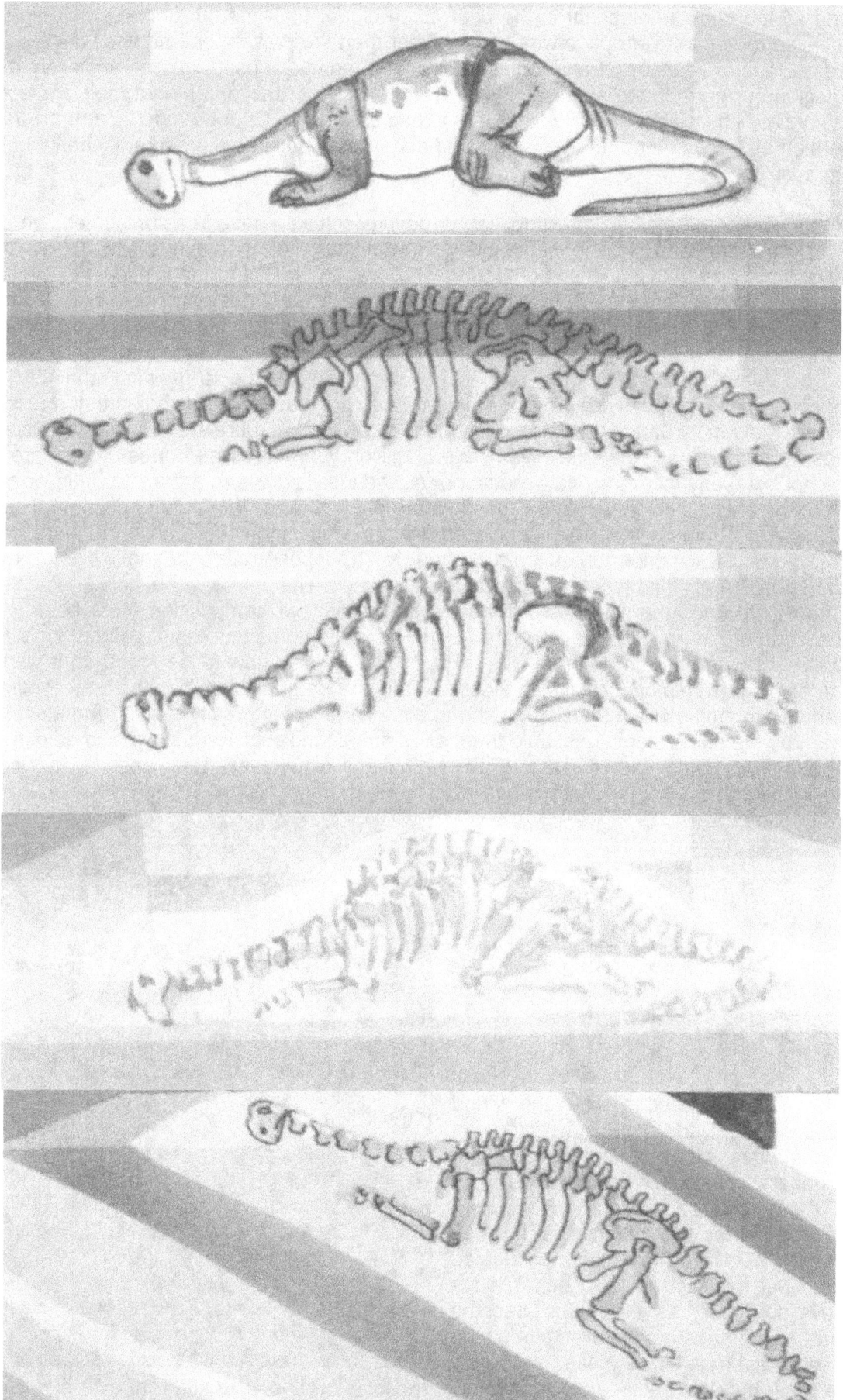

According to Newtonian philosophy everything in the past was bigger than it is today. A crocodile that lived sixty million years ago is fifty percent bigger than the current crocodile when compared to the size of the current crocodile. The Hippo of the past was almost double the size of the present hippo and so too was the elephant in the past much bigger than the current one. The tigers and lions that lived millions of years ago was much bigger than the present one and so was the snakes that lived millions of years ago…and the list of giants living in the past goes on endlessly while all the animals in our current age seem to shrink and tarnish to a fraction of what their predecessors were. Every time I hear about a T-Rex that was a long as a rugby field, as wide as a rugby field and stood as high as the pavilion roof of a rugby field and was as agile as a meerkat is today, I get sick to my stomach.

If I hear this nonsense coming from supposed to be intellectuals I gaze in astonishment while I feel like puking when I have to listen to the stupidity Newtonians cover their minds with. To Newtonians everything in the past was big when comparing what was to what is in the present. Yet, the entire Universe including the Earth, the Sun and the Moon is becoming larger…and no Newtonian has the presence of mind to think a little further than what their stupidity would allow? If only I could understand why the Newtonian is so persistent in not thinking! The Universe is growing and that is accepted as everything involving the Big Bang. Everything in time is growing and the Hubble constant confirms that. Atoms are growing with time and the fossil samples being that much bigger as they were that long a part of the Earth confirms this concept. If there are fossil bones discovered that dates from back then when they lived several million years ago, the bones by now are stone. They are not bone any more they now are stone. As the fossils grew from the carbon in which form the flesh of the animals originally was, it took in more of what forms the atom to become a stone as it now is, and if it took in more heat to become more atom, then the atomic structure grew by becoming bigger. The bones turned from $carbon_6$ to (I suppose) $silicon_{26}$ and by doing that the molecular structure size had to increase in volumetric space. The only way that could be achieved is by the atoms growing in structure and therefore adding to their composition more heat, which is time. The Earth grows and with the Earth that is growing, the atoms that remain part of the structure that reminds of what the fossils once was to be bigger and by now to be stone as it went from carbon to holding much more protons, the fossils within the Earth also feed on the heat coming from gravity and if the atoms within the Earth being in advantage of the gravity feeding the atoms of the Earth, then by the same measure should the fossil's atomic molecular structure also gain in size, if they could gain in atomic proton mass. If everything was bigger back then, then everything back then was smaller than what presently is available on Earth because time is the invert square law by the practise thereof.

Every generation that produced Newtonians found that the young incoming generation of Newtonians were bigger than the outgoing smaller generations of Newtonians and record kept of previous data about facts written down in the past confirms this statement. The wise and the wonderful, the highly educated Newtonian put this fact down to better foods being available and as the foods became better and also more available everyone eating food, your average Newtonian not only became wiser but also became bigger than the previous lot. It is accepted that ever since records were held, every human generation grew bigger in size than the previous generation that was denied all the healthy fat free foods. But to blame that growth by generation on food consumption is not concluding the case quite correctly because Newtonian simplistic approach again fails to appreciate the complexity of the situation. This human enlargement of size is due to the cosmos and moreover the atoms growing that put an increase in volume to everything being in the Universe that forms by accumulation of atoms to form what is constructed. In this way Newtonians stay the same size because they remain even in growth to the growth found in the cosmos that allows the Earth to grow at the same pace as the Universe grows. By having everything growing, therefore it sets the condition that secures that everything is staying the same.

The picture portraying the Big Bang is showing the ever expanding Universe as portraying a vision about science and conditions applying according to science, Newtonian science that is, as interpreted by the esteem well educated Brainy Bunch and in being Newtonian by concept is as usual also totally wrong because it indicates a Universe with a growing outside and growing outwards whereas the Universe has a growing inside and becoming smaller. Everything is in-between that which can't go bigger and that which can't go smaller. Since there is no outside that can grow just because there is no outside at all and therefore the walls in the picture can't expand, we have to see the Universe growing by the measure and the margin of its atoms growing, and that has to be growing towards the inside where atoms are. There is a certain relevancy attached but since breaks off size by becoming old and en dying it makes life a renewable product that is replaced every fifty or so years.

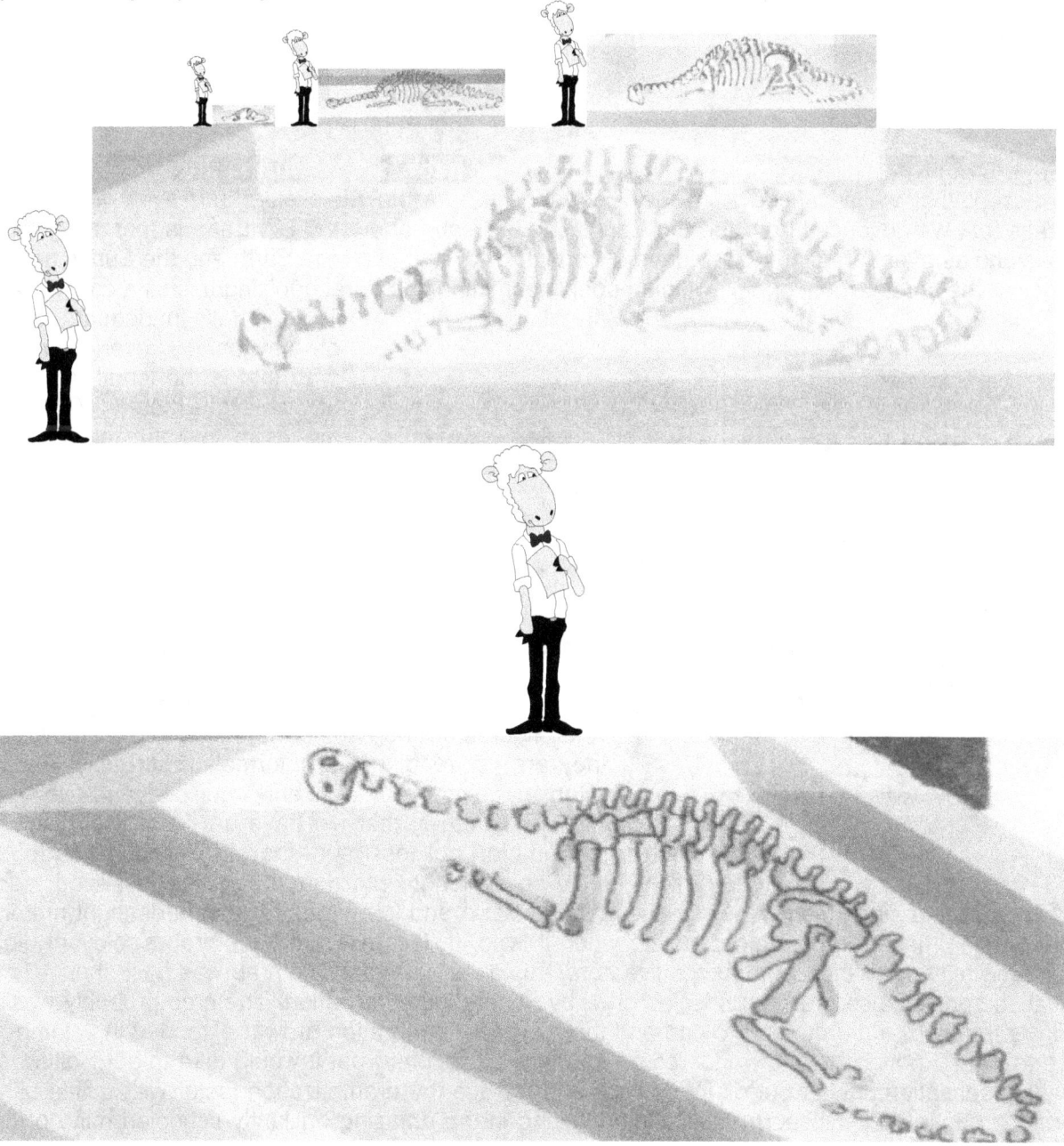

In the case of the fossil, the fossil is frozen in time and in place of growing by generation replacement of forming larger specimens of the same species, this fossil is growing as a combined structure by using the atoms that formed the monster from back then all is still using the same atoms from the time of death time up to now. In the case of the body used by life to host life, life forms the body in aid of life as to use during its occupation of space-time, then when aging makes the body outgrow its usefulness in hosting life, life then abandons the body as it destroys the structure completely after death where no two atoms stay connected and the atoms will be used in a completely new arrangement forming a total new composition to the requirement that the next form of life will accumulate and in that the dynamic of growth remains with the use of life. This process I explain in much better detail using a volume of book material to explain procedures that took place every day. This is a part of ***Seven days of Creation volume 1-7*** or ***Matter's Time in Space: The Thesis Part seven volume 1 to 7***.

The size of his top would apply as follows:

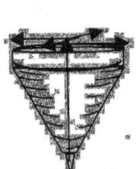 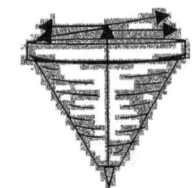 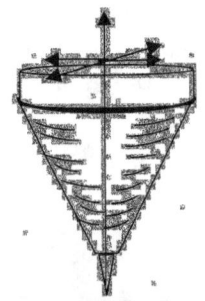 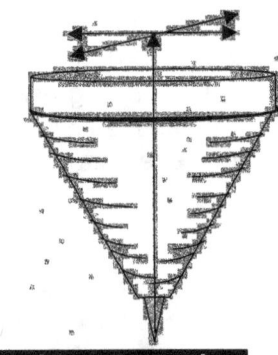

Long, long ago …   Some time back…   A while back…   During the present time

We accept that we advanced from the Big Bang, just as the Bible says and even Newtonians profess that we advanced from the Big Bang. The big issue about the Big Bang is that every atom grows and as every atom grows, so does every star grow and even the Earth and the Sun is getting bigger all the time, without one Newtonian openly admitting this or recognising this as a cosmic fact.

Present day

10^3 years back

10^4 years back

10^5 years back

10^6 years back

10^7 years back

The mentally inadequate nature of Newtonians even found a crab that is the length I believe of over eight meters. What this means is that the relevancy of the crab grew with the growing Earth from what it was to what it is and the eight meter crab did not stay the same since then and now we have a crab of eight meters that lived on Earth 265 millions circles around the Sun ago. That mentality is so Newtonian it makes me feel like puking. Is there not one Newtonian that can catch the hint that there is a huge snake in the grass, that their figures are no adding up, that whatever they read they are not reading that information correctly! Are all Newtonian minds so childishly gullible and void of thoughts? The truth is that we have an Earth that grew so much since the crab got fossilised, that an ordinary crab found today might be 10 cm in length but this specie got fossilized many moons ago and since then grew with the growing Earth and is by today's standards eight meters. That doesn't make the crab eight meters when it lived. It does not make the crab a relevant eight meters in relation to the Earth it once lived on. This does not mean the crab was back then 10 cm because species also grew but species grew by generation compilation. The crab probably was in today's terms 10 mm from end to end and the specie by generation growth extended in volumetric size to what it now is. I have wondered on so many occasions what it would take to get Hollywood seeking sensation hunger out of TV science and replace the dramatization factor with a little bit of common sense that will sperm intellectual thinking in the amazing brilliantly schooled mind of the intellectual Newtonian Brainy Bunch.

Every element is using the relation to the heat level it secures when forming the gravity it has. One can see how the forming of the numbers of elements available in the Universe stands related to the density of the elements total numbers. More pertinent to note of that the effect of gravity is not in the mass of the element but shows a much stronger relation with the density and the density is the relation the element has with the heat that marks a boiling point or a freezing point The density factor shows what we use to classify the element in relation to being a liquid, a gas or a solid. This factor is much more prudent than the mass factor and that I show later on as the book develops. In short: what I am saying is that the Universe is much more complex than the simplicity of awarding mass and awarding pressure and then grab a calculator to use by trying play god with mathematics. Where singularity connects is the true increase and where singularity expands is there where we are not and where we cannot see. The connecting and the growth are performed in the space-less-ness of singularity and in singularity there is no space and therefore we are not there because we fill space by sing time. So what is the point of calculations when we can't even see what to calculate and we are not able to know how what fits anywhere?

If singularity expands when heated and there is a limit to the point it can heat, and where that point forms the maximum expanding possible, then it has been reached in the area we think of as outer space. Outer space has expanded through the unleashing of heat, where overheating is turning liquid heat into gas or space. Any explosion is a vivid reminder of this fact and the unleashing of space is so real it destroys the space holding solids by rearranging the construction of the solids. With that in mind we can declare with great confidence that outer space is the hottest place there is. Whatever expanding there possibly is, was done to secure the cooling and all cooling that can be introduced to bring about further cooling was performed in the area we think of as outer space. Forget schoolboy culture and the temperature scales and other Newtonian scientific defects I call Xepted mistakes. Think of reality and throw out culture teachings methods. Use the mind and not the thinking power of the past. Any place that can expand no more is the hottest place there is just because of the shear implication that it can cool no further but has to slow down time in order to gain space and where that is, that then is as hot as it gets anywhere. If that is the case then it is safe to say that galactica is freezing cold notwithstanding our concepts of heat and space and heat in space given to us by our collective culture and not by our ability to reason.

The galactica is little frozen islands in a vast sea of heat. That is the reason we can see the galactica because the galactica is space concentrated by into a frozen state of concentrated light and light is heat frozen into an almost solid space. The galactica is slowly heating and therefore it is expanding into outer space. However it is the atoms that are concentrating heat while space is overheating and the balance between what must always move and what can never move jointly form the expanding of the atom and of the galactica. Outer space on the other hand has expanded to the maximum that it can, yet we think it is cold when it is the formed by the most extreme there is in heat and that extreme heat introduces the maximum expanding. What I now am saying might be deemed by the most purists as the contradiction of the century and that much I do realise. At the inner core of a star all space shrinks into the oblivious but we consider the inner core area of a star to be the hottest spot in the solar system. That just cannot be because when material shrinks it becomes cold and by shrinking into the oblivious it has to freeze into a fusing element as newly formed units. Again that is the contradiction of the century. Why will that be? The space inside the star shrunk to the minimum there can be and that tells us the space has to be cold because of the shrinking took the space to a position where no space can shrink anymore. This nuclear reaction represents a picture where the atom releases the cold to expand into heat. The heat is not the liquid transforming to gas; it is the heat expanded into gas.

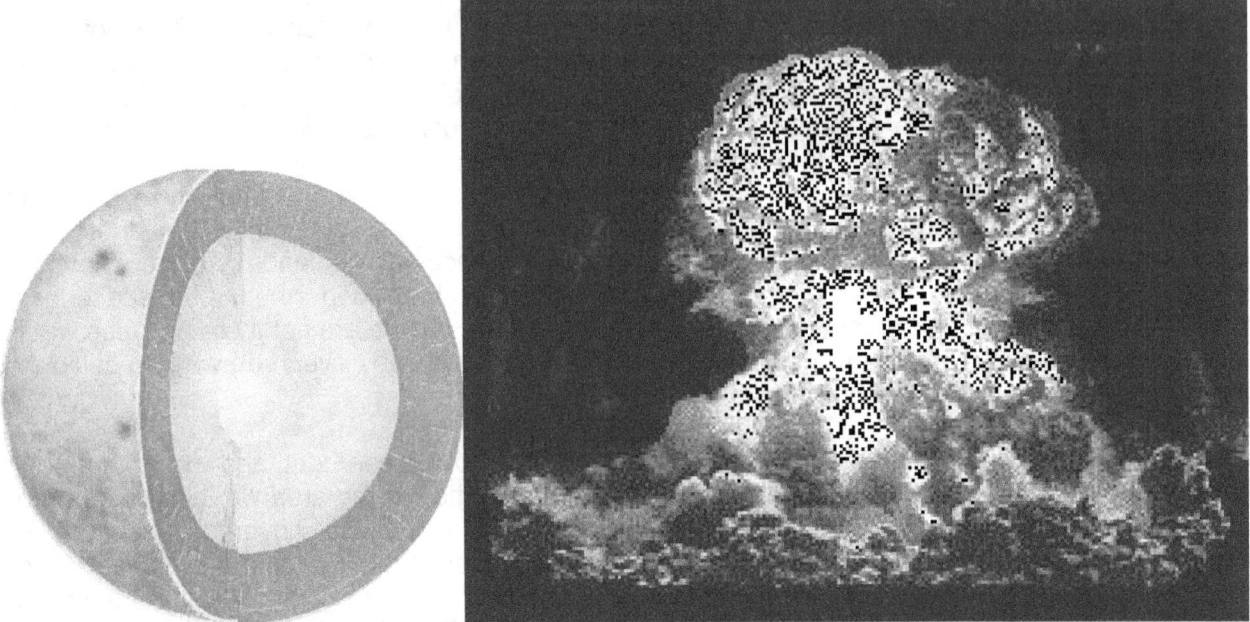

That shrinking of space into no more space where there is obvious loss of more space can only be inside the inner star and in that region is where we locate the strongest gravity. With outer space as expanded as nature may allow the space that grew could only grow in conditions of heat because heat produces expanding and expanding is the result of heat coming about. Space shrink because it is cold: that we know and taking this law to the star centre it means regardless of our interpretation of hot and cold, that area in the star centre is as cold as it can get notwithstanding what our Newtonian culture may tell us. Then obviously the same must apply to outer space for precisely the same reasons because it is so hot there it can expand no more.

At this I have to redeem myself from being human. Only looking through the eyes of humans there are hot and cold, but as a reality in the cosmos we will find this nowhere. We look at the hotness of space and the coldness of space but it is the relevancy to the state of materials being a gas or a liquid or the solidity of extreme cold is that which forms the actual heat and cold limits. It is so hot no expansion can produce more space in outer space, as the outer space seems to be the epitome of what can be cold while it is truly hot and quite the opposite reveals as the true scenario inside the star in the centre of a star structure. That means the number of protons in motion has a lot to do with the cold and hot scenarios because where the protons are most dense the cold is in extreme…well in most cases. Only in the absence of space can so much heat gather in excess and the opposite is true about outer space where the least denseness found brings about the space in heat found in outer space. Our human selecting of hot and of cold and what is hot and what is not prevents us the clear vision we would have when truly understanding the applying temperature. Temperature comes about from spin and the smaller the spin density is the colder the space becomes because the more duplication produces the most cold. We think of outer space as 0^0 Kelvin but in fact it is as hot as no other place can be in the Universe. The coldest is where material is freezing solid as material does when frozen solid and the hottest is when by boiling the material is going into a gas with liquid being the intermediate position where heat acquires the space to perform as a flexible substance.

When we look at particles in outer space we see the particles being frozen. It is because there is such a severe contrast between the particles and the environment surrounding the particles and not the particles that is so frozen. The density of heat is so large because of the limited number of particles available to bring a cold environment to the overheating of space. The particles we look at are clearly in a gas state because the particles do not form a part that is part of the space unit required in such a manner as to get influencing the heat factor. If the hydrogen were confined to a smaller space, such as they are in the confinement of the Sun, the spin would bring about a more defining state of substance where the gas that we now find would turn to a liquid state as we can clearly is in the Sun. This being part of a cold or a hot environment rests with the spin of the particles forming the state in which we find the substance. Hydrogen clouds of hundred of light years in diameter are a common sight in outer space, but they are just hydrogen stars that spun to slow and by losing movement, the structure lost the ability to freeze and consequently then overheated. However once it overheated it lost movement and by not again gaining movement it will remain expanded space or then not being cooled into liquid as a result of loss of movement. The heat we find filling space is not part of the space but like the particles the heat is a separate issue. It is the density or the compactness or the condensing of movement creating a governing singularity of greatness in prominence that forms the issue Newtonians desperately wishes to calculate. That heat filling the space is another form of material that could conduce by diverting from space or marry by condensing larger areas of space into smaller and more concentrated areas of space. However without the union of space by movement becoming more condensed space, the movement of the space will fall to the expanding of space that becomes part of the ongoing overheating process. If it was that cold which we think it is, it would not have expanded into such a massive cloud but would have contracted forming a cube of frozen hydrogen. But as we can see the cloud expanded the gas as far as the gas can expand and that could only be a product of overheating.

That expanding we talked about is indicative of heat and has extremely little to do with gravity in Newtonian sense or is it just a matter what we think of as gravity. If you are of the opinion that those hydrogen clouds will contract one day into forming a star, well then think again as there is just no such a chance that that will ever happen because that is not the manner that form of gravity functions. Gravity is the movement of space and in those clouds the movement now goes to the expanding. Because outer space is completely overheating the condition it has in support of the particles makes the particles appear to be in a state of freezing but the particles is counteracting the heat limit it meets. However the particles do not contract, as the heat is immense. The space in outer space has absorbed all the heat levels rising by means of expanding and will appreciate still further as it will never depreciate. That is not because outer space is freezing the particles but it is because in contrast to the heat of outer space the particles seems to be frozen.

In realistic terms gravity is not the contraction of space but is the expanding of space. The space can expand positively by growing into more space by becoming more in outer space or by expanding negatively as a result of growing denser within the atom and thus expanding within the space claimed by the atom. No matter how one may look at the scenario, Newton is wrong in finding anything contracting because this entirety is expanding in joining or uniting such expanding everything that can expand into the abyss

The atom must be the utmost coldest because the proton is even much colder than what the electron can freeze. In fact the proton is 1836 times colder than that what the electron is able to freeze. We find that when cold escapes, it turns to heat and the heat relieves by forming space, however it seems that that part is the part that no one can understand. Motion brings about cooling. When the spin of the atom allow the cold of the atom to release the heat it had, which it had frozen the heat returns to space. This is what the atom shows in the electron bands or rings the atom holds. This must not be confused with uncontrolled release of heat. When the motion of the electron is interrupted such motion reducing results into the utmost expanding there possibly can be. When this heat releases from the containing form of the atom, it brings about much more heat than the Human mind can cope with because no human mind can ever comprehend the total devastation a nuclear release of space may bring forth. In this I am not referring to the normal way material relates to heat, but the dividing of singularity fragmentation. That is a totally different matter altogether and this is the only way that can destroy life altogether by fragmenting life into fragments holding cosmic singularity. From what I see, persons indulging in nuclear exploding has no idea of what they enlist to destroy and soliciting nuclear fragmenting the singularity devoted to life is sealed into cosmic singularity for the duration of on entire eternity. I have no idea when or even if that singularity presenting life could ever escape from the cosmic singularity as the nuclear fragmenting of singularity sealed into these fragments the parts devoted to life.

One may not look at the material and judge the surroundings. The fact that hydrogen remains a gas and so does helium in outer space must serve as enough proof that outer space is hot, regardless of our interpretation of the temperature gauge telling us what we wish to hear. In the vent of outer space truly being the coldest we have then in such a case hydrogen and helium should be frozen crystals clotted in balls of material. One must look at outer space and judge outer space from the findings only by taking outer space as an entity into consideration and we have to look at outer space without the prejudgement of teachings about ideas that prevailed in the minds of men when persons were still held in prison for being suspected of becoming werewolves when others were not looking. If helium remains a gas it is hot. However, we can witness hydrogen being a liquid in the Sun and that makes that hydrogen pretty cold while being a liquid. That liquid we see is squirting from the Sun is liquid heat that is frozen as a form of material that is surrounding the material housed as the hydrogen layers and holding the hydrogen in form in the hydrogen layer.

We might think it is hot in the centre of the Earth but that type of thinking is as Newtonian as thinking of big stars as might y gravity pools. The removing of heat into a liquid makes the material in the centre of the Earth cold although we see it as being terribly hot. The only reason why it can seem to be hot is because it is cold and in such a cold environment the heat can gather and space can collect heat because the particles find the surroundings extremely cold. Then again we confuse heat and time altogether and completely but more about that later on... I am going to explore the following a little further on in the book but people use the intensity of heat to gauge the heat levels and not the density of heat levels in inner space as a guide.

The cold in the Earth centre causes the concentration of heat by reducing space, as all cold surfaces tend to do. When material reduces space it parts the material from the heat within and places that heat within the electron bands to outside the electron bands. By removing the heat the atom contracts and by contracting the atom reduces space. That heat forming space has to go somewhere. If it was hot within the centre of the Earth, then the space within the Earth would expand (explode) and the space within the Earth where we think so much heat is concentrated does not expand (explode except in normal cracks such as volcanoes) therefore it must be cold. To gather and accumulate the space in a liquid means it became much colder being a liquid. Finding the surroundings terribly cold will allow the heat to gather and not expand but when the surroundings are hot, it will not tolerate more concentration of heat and thus will expand to rid the balance of excess heat within space. That is the terms in which to think in when thinking in terms of cosmology.

Look at the Sun and see how the Sun turned the hydrogen to a freezing cold liquid at 6500 K. The 6500° C means nothing...it is the liquid that says everything because turning any gas to liquid is freezing it by condensing that which is the ultimate gas.   Hydrogen is in a fluid state within the Sun and is colder than the hydrogen that is in a gas form in outer space. The Sun is the coldest place in the solar system. That is when the protons oversupply the removing of space to produce the cold that is so apparent. By the reducing of space it can concentrate heat to a fluid state by producing the opposing cold that finally freezes the heat to a solid state. The expanding of space is a way of duplicating space without reducing space and by duplicating in the form of expanding it becomes just the opposite to duplicating by motion therefore reducing space by halving space in time. That is what gravity does. By motion space duplicates and by space halving it removes heat by spreading the hat over a larger area of contact in space as well as by condensing space. In all the applying of gravity space bites the dust on the one hand while on the other hand it expands space positive and negative in direction.. The density of the protons brings about space dense enough to harbour the heat in such quantities and visa versa applies in outer space. However it is not purely the density of the protons that produce such cold but the exquisite motion forming a rapid duplication of material and such duplication brings the contraction by removing space. Removing space is also removing heat that is separating material.

When you walk outside and look at the vastness of the blue sky while tying to absorb what one observe as one looks at night at the black night sky, you are physically standing in the part of singularity which is in the part of 1^0, the part that moved away from 1^1. I know Newtonians love to name because with names they impress everyone and put all that read the name under the elusion that the Newtonian that did the naming also know exactly what is going on while the Newtonian has no bloody idea what is going on. This time it won't work because I can say that which is 1^0 and is 1^1 should be exactly the same and yet while it is what splits that which can never split and located in-between 1^0 and 1^1 is everything that could ever be, that everything is found to be there. Looking at the night sky you are within the part 1^0 that has no end because it has only one side, which is the inside. It is 1^0 going nowhere. It is the part that I named the spot that had the dot 1^1 moved away from

This is the dot that has no start. It is 1^1, the part that released from the spot 1^0 when motion parted singularity. It came apart when motion unleashed the dot 1^1 that has no start from the spot 1^0 that has no end. It is the Universe born from motion that was driven by heat. It is still there because once anything is part of the Universe and forms a principle within the Universe it has nowhere to go but to remain within the Universe. Walk outside at any time and you are a witness to the result.

When singularity expanded for the first time ever and when heat parted from cold bringing about the Universe forming 1^0 to 1^1 moving from Π^0 to Π, a relevancy was born and that relevancy grew into what we now have as a Universe. Gravity in the centre formed time Π^2 by dismissing while the four time positions started the cosmic trend of duplicating. With every one of the four points taking form to the value of Π at a measure of $\Pi/2$ each brought about the Roche value of $\Pi^2/4$ in relation to the developing centre. One has to remember that the star of today takes on the characteristics of the form of that era.

Everything in the Universe and therefore everything in the solar system holds 1^0 in relation to 1^1. We have to accept that the coldest place in the solar system is in the very centre of the Sun because there the most number of protons sharing the least amount of space producing the coldest area by producing the most rapid movement that can allow therefore the hottest density of heat within the cold environment. We have to realise that whatever forms space, has to be that same ingredient which also is the basic component that forms the lot of everything in the entire Universe.

It is that which then is increased to become more to use for making everything seem more and it is also by removing that which forms everything that then will reduce every aspect of the Universe. That which becomes more is what the Universe is built with and it is that which the Universe uses to form its entirety. When particles heat up the particles expand the space the particles hold to limit which the rising heat demands in relation to the heat rising. The particles claim more space when heated to preserve the cold that the material is protecting. The claim to more space produces more space but that in turn reduces more heat exaggeration. Such expanding brings about cooling. When particles heat up or cool down this is because of motion applying in some form. Regarding this fact we can claim that motion started at a point when the Universe was extremely hot and there was no space. However I have indicated that hot and cold are only factors with little specific or formal value in the Universe. By introducing motion space formed and the lack thereof produced friction that became heat that became space. That must be the way the Universe then started.

The application of gravity is the same as the condensing of space. Creating movement is bringing about heat by the compressing of space into cooling a centre spot. However, the movement will come about only when there is disparity between what is cold in the centre in relation to what is hot on the outside of the object starting to move. The normal thinking would be to put life centre stage and see how life would influence movement but we can only mimic the cosmos as we apply motion in the way we go about tapping into the energy that nature provide. Internal and external combustion engines all rely on this application for harvesting motion and the method applied is by creating a driving power but in such an event life uses combustible fuel to bring about thermo driving. Compress space even today with a piston in a cylinder and then pump the compressed air into a container and such confining of space will increase the heat by the piston effort to reduce the space brought about in the container. The heat coming about inside the cylinder as the air is reduced in space granted and we change the applying rules. The heat that comes about when being compressed has no relevance to particles colliding inside the chamber because all compressor cylinders cool down or become colder when that cold escape through the walls of the cylinder.

In the compressed cylinder the heat releases from the inside space as soon as the pumping stops, while in the engine the heat is tapped to use as the driving. There is an immediate stopping of the increase of heat as soon as the pumping stops. In the compressed cylinder the material inside the container forms a secondary form of material that comes about since the space reduces and the forming of space is in a turnabout. The compressing of the space inside brings about a rise in the heat levels within the container but apparently that no one in Newtonian circles can understand. By compressing the spin of the atom increase and the motion of the material remove additional heat from the ranks of the inside of the atom. Thus, when the spin of the atom increase it allows the cold within the atom to release the heat that the atom holds from the controlled environment into uncontrolled space on the outside of the atom. This releasing of heat and unifying the released heat once again with space increases the levels of heat in the atmosphere of the containing cylinder. What that heat does is the heat that the material absorbed as material within the atom was captured as frozen heat because of the motion of spin to space that the atom holds remains in a frozen state under the guard of the spinning electron. But when this heat releases from the containing form that the atom holds the heat as a frozen substance is frozen by the spin of the electron. The spin of the electron brings motion and such motion reduces the heat to a frozen state which is the frozen state of heat we named material. Therefore one may not look at the material and judge the element state of form by its surrounding which is heat it surrounds its electron with.

Again we must look at the state of material in outer space and realize that the fact that hydrogen remains a gas and so does helium in outer space must serve as enough proof that outer space is hot, regardless of our interpretation of the temperature gauge telling us what we wish to hear. One must look at outer space and judge outer space from the findings only considering in the terms which outer space insists upon. If helium remains a gas it is hot. The removing of heat from the space that contained the heat makes the centre of the Earth cold. In our Universe we see

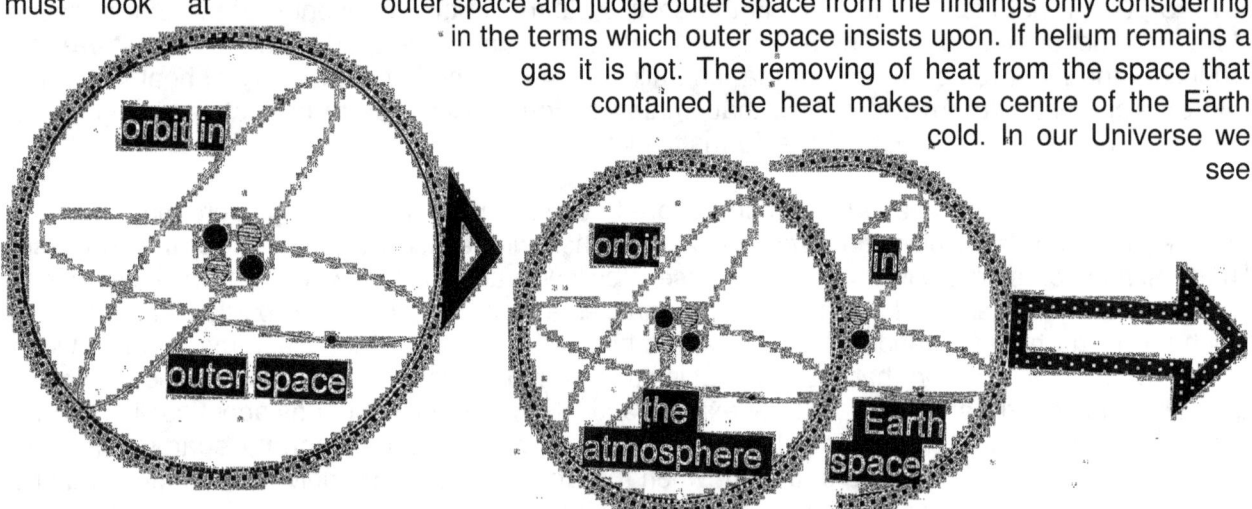

it as being terribly hot because the heat then forms a separate substance but remains a form of material but that is because we see the heat and not the space derived from the separating of the heat.

By accelerating to the velocity of the Earth time or the Earth movement the top duplicates its entire position more often but faster during intervals and although holding more heat per cycle the top reduces heat as it spreads over a wider area. In that way the top may capture more surrounding heat but the top maintains less heat per time cycle.

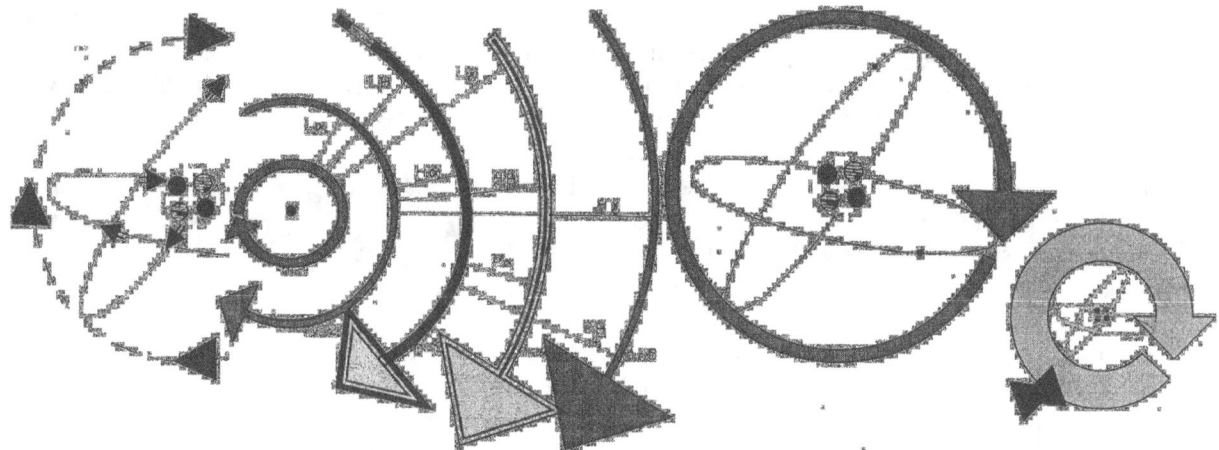

The only reason why the space can seem to be hot is because the space is cold and in such a cold environment that reject the heat within the atom where the heat then must gather in a much concentrated state and space can collect heat because the particles hold concentrated heat in the space separating the particles. By removing such high concentration of heat from the space that use to be expanded heat, the space outside the atom then must contradict the heat by being extremely hot. We look at the heat in the space, which by being in a liquid state should be by our standards considers as another form of material and find the surrounding heat in the space hot while the atomic material in space is extremely cold. The cold in the Earth centre causes the concentration of heat by space reducing, as all cold surfaces tend to do. But the numbers of protons contributes that reducing of space and the removing of heat captured by the material.

If it was hot the space within the Earth would expand and explode but the space within the Earth where we think so much heat is concentrated is so much it does not expand therefore it must be cold within the solid parts. It is the motion of so many protons in such a little space that allow the heat to be contained as a liquid and the extravagant motion by the many protons in such a reduces area forms the ability to contain the heat as a liquid substance without allowing the expanding of the heat into gas or space. To gather and accumulate the space in a liquid means it became much colder when the space parted from what then is being a liquid. Finding the surroundings terribly cold will allow the heat to gather and not expand but when the surroundings is hot it will not tolerate more concentration of heat and thus it will expand to rid the balance of excess heat within space. The concentration or release of space with heat or space from heat is a direct contribution of the motion controlled by the space-time. The regard of the space-time providing the motion, which provides the cooling of the space, stipulates the conducting of heat in space or the release of heat to form space by means of seizing the occupied space.

The particles claim more space when heated to preserve the density by getting cold. The claim to more space produces more space and reduces more heat because the heat spreads over a wider area. Such expanding brings about cooling. When particles heat or cool motion applies in some form. Motion started at a point when the Universe was extremely hot and there was no space. By introducing motion space formed and the lack thereof produced friction that became heat that became space. It is natural and it is simple and above all it makes believable sense.

As the cosmos present its evidence, we can see from such evidence how destructive overheating is. Forget pressure, because Newtonians over simplify everything with pressure and to them exploding is simplified by having shock waves. That might happen to a drum they fill with gunpowder but that is not applicable in the cosmos. In the cosmos, unlike in containers, there is no retaining wall that sets limits to pressure inside the container versus pressure levels outside the container.

The cosmos has no pressure or pushing or pulling. It has a flow of space-time by concentrating time and duplicating space as it is driving space-time towards the centre. In any picture of any star there is no containing wall that keeps whatever is inside, on the inside. There is no limit to what the wall of the star can take such as the case is in a container that is pumped to the limits that the container wall can absorb and therefore that the structure can contemplate before bursting. I am going to show that in the centre of a star is a point holding singularity and since such a point has no space and is immovable, space has to compromise by flowing towards such a location.

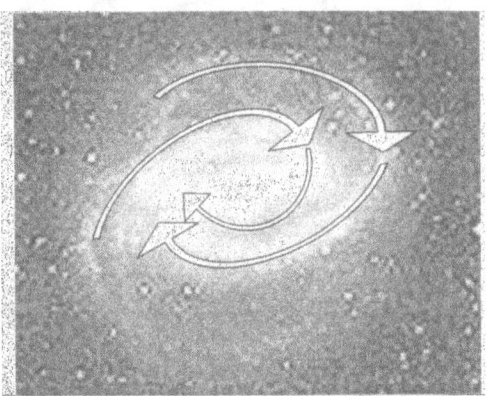

The flow must equal the conditions and the movement of material inside the star and the supply must feed the demand, or things will come to meet disaster. I am going to show that every atom serves as a little pump that by rotation, pumps liquid Newtonians think of as nothing, into the star and this forms a balance that cools the star. That is what Kepler's formula ultimately says. The formula says that from within the centre of the star $a^3/T^2k = k^0$, the space is controlled $a^3 = T^2k$ in relation to the movement $T^2 = a^3/k$ of the space

This I can do by applying Kepler's formula. The space a^3 within the star, has to be equal $=$ to the movement $T^2 k$ within the star k and of the star T^2.

When the star is encountering adverse conditions, the flow will interrupt the even-handedness and the changing of gravity or the flow of space-time in time will set a new standard. It happens all the time and every time the centre fails to set a standard that the flow of space-time in the star can meet, then the relevancy of time sets in place a new standard and this comes about by the use of a principle we think mostly of in aviation. It is the principle we refer to as the Coanda principle. It is where motion creates a flow of space-time, which establish a centre and where that centre performs demands that the flow has to initiate. The containing of the space is as much set by the time of the flow as the retaining of the centre. It is a proven dimension implicating Kepler's vision of gravity.

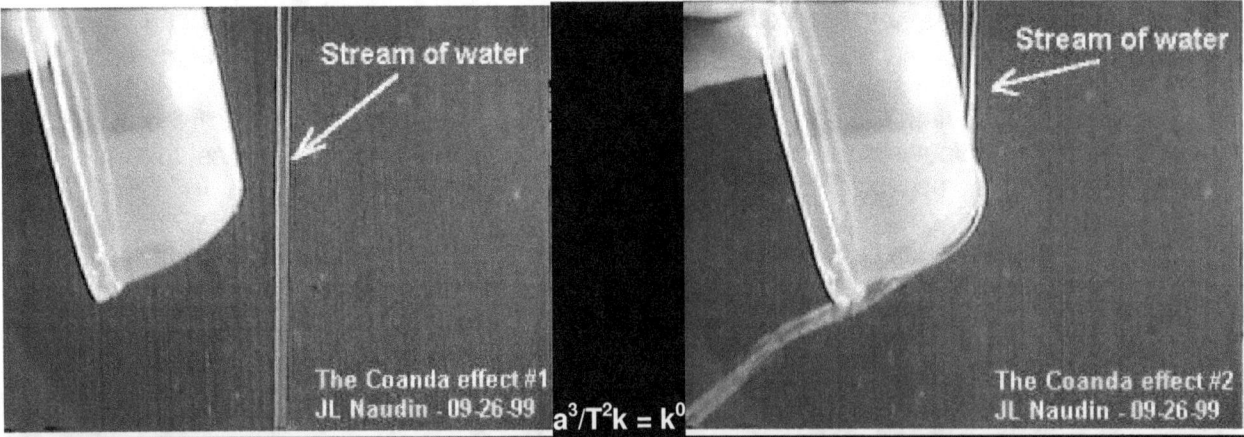

Stream of water

The Coanda effect #1
JL Naudin - 09-26-99

$a^3/T^2k = k^0$

Stream of water

The Coanda effect #2
JL Naudin - 09-26-99

I have taken from the Internet images of an experiment indicating the working process of the Coanda effect and nothing in the Universe explains Kepler's formula better and Kepler's formula explains gravity better than anything else in the Universe. This is gravity and that is how plain and simple gravity is. The Universe has two substances, space filled by material and time filling space. Where the two substances meet and interact, the integration produces gravity as the material pumps liquid from space to the atom centre.

With my presenting a simple sketch in a simple manner which then is holding a very simple formula in three sectors, the simplicity portrayed in the sketch will pass by the Newtonian just as easily as the notion of light formed by single rays portraying the cosmos does. Light can only travel in a straight line if it travelled in a circle and when the two factors form the entirety of space. Please believe me that the formula presented below makes the most complicated mathematical equation-wonder man has previously inspired seems less than a common adding equation. Light can only travel in a straight line **(k)** if it travelled in a circle **(T^2)** and when the two factors form the entirety **(a^3)** of space which is what the cosmos told Kepler forms the cosmos being **(a^3) =(T^2) (k)**.

In each individual star, a micro Universe is locked up confined to a single layer, as the layers all adhere to a different age in the period of the development of the Universe. All the time fluctuation runs back from a few billion years to somewhere after moment-Alfa. In the event of the star going super nova gravity does not go mad because disaster only occurs in a certain layer when the time span in that particular layer moves to a period predating moment-Alfa or in other words, the beginning of time in relation to the time applying to that particular star. In the event, where time is pushed to a period in which the matter of that particular layer has to endure a value that places a contracting of space in that space, where time in accordance with the standard which that particular layer's elements has to apply to, becomes time-Zero.

When a star shine we have heat in the form of matter that escapes the system in the form of electrons, photons, radiation and magnetic space-time as the expansion of the heat in the form of matter allows the increase in negative space-time displacement. Such a view is the general presumption and takes our human culture back longer than we have recorded history. It is the basis of our accepting cosmology.

We tend to see the Universe from our perspective we have where we are filling the **centre of the Universe**. With all the absolute phenomenal achievements science accomplished and more so during the past sixty years, by going to the moon and splitting atoms and visiting planets and... was there ever one that took the time to find the **centre of the Universe**? How can anyone tell how much mass gravity attracts in the entire Universe in relation to the critical density of the entire Universe when you are incapable of knowing where to look for the **centre of the Universe**? The Universe with gravity's attracting action must be pulling us to the **centre of the Universe** and because you can't judge the direction we are going, you therefore don't even know where to find the **centre of the Universe**. When any person is standing on any place anywhere, while viewing the Universe, that person is filling the **centre of the Universe**. That is not only applying to Americans in particular, but to all persons that were born through childbirth. This however does no apply to animals but that argument is left in another book reserved for another day.

These proposed Universe formation concepts that Mainstream science proposes prove to be examples that are a sure indication that without mass and without pressures the thinking of the Newtonian boils down to nothing. The problem these suggested Universe concepts have is that

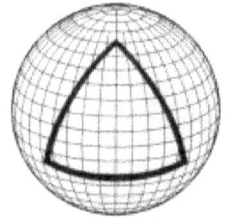

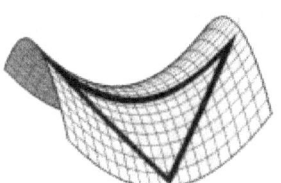

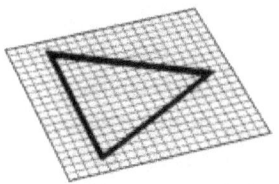

Closed Geometry   Open Geometry   Flat Geometry

it places borders in the Universe and allows the observer to look at the Universe from the perspective God will have. They haven't got the insight to see that the viewer observing the Universe belong inside the Universe because to the Universe there can be no ending on the outside. The fact that this places a picture of the Universe and not of within the Universe as it truly must be, nullifies the entire idea. Here is a far better explaining of the Universe starting at its starting point and also at the ending point. <u>Einstein's Critical Density</u> lacks the accepted matching facts we need in proving the critical mass factor. But our inability in securing such required evidence defies the most basic logic. It seems all new evidence we receive from outer space is disputing all Newton laws findings that disprove <u>Einstein's Critical Density</u> as the answer. The Universe will not reach a point where it will start its long awaited contracting, not withstanding whatever dark matter astronomers try to locate in the vast space. Why would the expansion turn around and do a reverse by going back to where it came from. Consider the momentum alternation such a change will bring about. While the clever academics present an image of very literate about physics they also lack the physics insight to realise that if motion shows a turnabout all material in the Universe will demolish because the momentum that applies with such a turning around in direction will destroy all structures. You can't even stop a bus that fast without paying consequences let alone an entire Universe. The whole idea is one big corrupt scam. Let's get a bit critical about this corruption Newtonians devised to get off the hook with the Hubble's expanding compared to Newton's contracting gravity. I am not going into this issue with much depth because I wrote entire volumes about Newton and Newtonian corruption and it is in the Internet under the web address www.sirnewtonsfraud.com where any person can download a book for free that I wrote and see the fraud going on in physics. Not only that, I challenge any person to prove me wrong in even a single item I uncover about the systematic brainwashing and how students are mentally abused in being fed very selective information that is completely disinformation so that the students have to learn facts that in reality is no more than just unproven gossip, taught to them as approved physics and then academics go on to brainwash the students in facts not proven yet the students are expected to learn that disinformation off by heart and repeat word for word in examinations without ever being given proof that what they learn holds any speck truth, which the facts don't hold.

To return to the fraud I wish to touch in very short on what I prove in another book is fraud by explaining just the following: Newton had every one going on believing this lot we see as forming the Universe is going to contract into huge lumps of material because that is how he said stars form. The way that stars is a process whereby stars when they form, even before forming collects material by the magic of mass that gathers a lot of dust particles in the gravity that collects the dust particles by gravity that pulls the tiny dust particles individually from all over and around and about and this ends up as a planet when there is not enough or a star when there is surplus to be. May I add to the accusation that what students are taught is no more than old wives tails and plain gossip as this idea about dust being collected by gravity in the measure of mass was never proved but for one guy that new how to manipulate a program he wrote. No place in the Universe can photo images show how dust collects into massive stars. This is a fable and not worth the effort of writing about it. Then a fly fell into Newton's ointment and this fly went by the name of E. Hubble. Hubble was the first person that saw Newton's contracting Universe in another light but also never saw the light the cosmos handed him because Hubble died a Newtonian by heart as all the others in physics so far did. When Hubble saw through his telescopes that the cosmos was expanding he declared what he saw in no uncertain terms. This fried a chicken or two amongst the many Newtonian believing in Newton's religiosity.

Newtonians were in dire straights because now everyone could see the truth about what Newtonians hide the deepest, and that is that no mass is pulling any gravity around! Hell, any one with the least IQ can see that when they read what Galileo said about all things falling equal under equal conditions when falling an equal distance and this has to exclude mass because mass is bringing individual mass to every participating particle.   Now with this information out in the open even schoolboys were able to see the truth...what is the truth...the truth is that no Newtonian had a goat's idea of what the cosmos was about! As the goat, the Newtonian could run up the mount and pretend to be king of the castle for a brief moment and then run down again but when it comes to real issues they drip with incompatible ideas where all being more senseless than the previous one. If you think I judge harshly, then sit back and read what they can cook up.

When Newtonians were caught with their trousers on their knees as far as Newton's theory about gravity contracting goes they did not come clean and started to dust the floor to see where the cockroaches hide. Hubble proved the Universe is expanding and when learning this truth about contracting being a farce, did Newtonians come and clean the house by throwing Newton out? No, they cooked up even more criminal deception. They got the man they thought everyone thought was the most –brilliant-brainy brain on Earth and then got this guy involved in fraud as never been committed before in the history of mankind. I have no idea whether Einstein went along as a participant in the scheme and the conniving and went with the deception or that he (Einstein) was fooled into it because from where I stand Einstein was not the sharpest knife in the kitchen, notwithstanding all the admiration he is given by the world press and this I say on the grounds of what is clear with all the schemes he came up with and never could convince any one or was able to conclude by forming a viable answer! So being as blunt as he seemingly was, he might have been tricked into it or he might have wittingly played along in devising this scheme of fraud. What is very clears was that the man had a much bigger opinion about his abilities than he had an IQ to match those abilities!

So they went and cooked the cosmology books badly. Compared to the Enron scandal, then when looking at the Enron debacle and shady deals going on in Enron, it seems as if the Enron scheme was devised by a nursery forming a play group and the deception was formulated by a bunch of pre-school pupils playing a game...these guys that came up with the Critical Density Theory was good and they went on to get even better...but let me explain. In order to cover Newton's deception about mass bringing on an attracting Universe and as it became clear from Hubble that the lot having mass was not pulling on the lot having gravity this lot conniving the scheme of physics fraud had to devise a scheme that would stand Newton in the clear and would leave the impression that science had an idea what was going on as far as cosmology was concerned. They had to cover the obvious fact that they were the fools looking at the Emperor's magic clothes in order not to seem the fools they were. They could not allow anyone in the public to notice that $F \ = \ G \dfrac{M_1 M_2}{r^2}$ was never tested and that this formula is a mathematical joke. They also could not allow the public to see that mathematically it is fraud to change a formula from first being $F \ = \ \dfrac{r^2}{M_1 M_2}$ then to being $F \ \alpha \ \dfrac{M_1 M_2}{r^2}$ while pretending it will then read $F = \dfrac{M_1 M_2}{r^2}$ and when this still doesn't fit anything useful, just change the lot to $F \ = \ G \dfrac{M_1 M_2}{r^2}$ and get an entire world with the best brains on offer to believe this mess for almost four hundred years. The greatness about Newton was not his physics but his fraud. He go the entire world with the best brains walking around to believe this lot of trash for almost four centuries...now that is something no other person will ever achieve again... to fool so many fools for so long and get away with it for four centuries! That is the true genius of Newton...that is the part worthy of admiration!...to think he fooled so many millions for so many centuries and he kept them fooled because they all tried to pretend to be the wise that was special enough that they had the gifted ability to enable them so that they could see the magic clothes, or in this case it is the magic mass of creating gravity to attract whatever needs attraction. This brings the story of the Emperor's magic clothes into the custody of reality. To think that even being the best minds there ever were, they all were in agreement with a statement that a^3 could be equal to T^2 or that the cube a^3 could also be equal to one of its flat sides T^2 and thinking about this carefully should bring the point home.

 Getting back to Newtonian Critical Density fraud Hubble came along and showed the lot that was suppose to have mass and thereby was attracting was expanding (another crooked idea because the universe has no borders to expand) and everything was going the wrong direction that Newton claimed things are going. This was a catastrophe. I am not going into details what this involved and considering that Newton had to be blameless to save physics from destruction then if Newton was not to blame and something was wrong then the only other party that could take the blame was the cosmos. If the cosmos was not behaving expedient one could hardly blame Newton for that because Newton said the Universe was attracting by the measure of mass and that is how things must be...for Newton just could never be wrong! Newton just could never be wrong and therefore the only other party was the cosmos misbehaving and with Newton's inability to ever be wrong it was making the cosmos the only party being wrong. The cosmos was going the wrong way and the Super-wise and Super-educated had to find when the cosmos would come to its senses, correct its wrong direction by mending its ways and then follow what Newton declared had to happen. The Universe in its entirety had to find the point where the cosmos would follow Newton's subscription and start to attract and stop this incorrect expanding process! There had to be the mass that would get the Universe contracting if it is not contracting at this point because Hubble could not be silenced as much as Hubble could not be reprimanded.   They had to locate where the mass was that was not attracting as Newton said it has to because the Universe was suspect of hiding mass so Einstein was taxed with this task of counting the mass in the Universe to find out when the Universe is going to come to its senses and start being obedient to Newton!

At anther venue in this book I indicate just why Einstein became the chosen one to locate and measure the mass and why Einstein was in the centre of the Universe form where he could see all the mass he was counting! To make the long and sad story short and dreadful, the outcome was that Einstein saw the Universe had a shortfall in the mass that it required when it acquired the mass the Universe had and the Universe was doomed for ever and another day unless...The anti matter ate too much matter and there was too little mass left after the anti matter had its fill and for all we know this is where the missing mass went and then this possibly made some mass go into hiding where the anti matter couldn't see the matter to eat it up.

When Einstein said there was not enough matter to attract the lot that is expanding back to unify again, the lot in charge of physics was in a serious predicament. Again they had the opportunity to come clean and tell the world that they had no idea about what drove physics and why gravity is gravity and why this lot was expanding because Newton was wrong because Hubble proved the lot was not attracting or they had to protect their personal academic standing and all the historic academic investment as well as protect Newton's virtues and commit to more deception as was previously done by them on a grand scale then and including their master that came before them.

They then really got cooking the books and cheating the balance sheets because this time they threw all logic overboard as they came with one final onslaught on human perception. The created non-existing dark matter that would come into effect and bring the contracting about as well as then vindicate Newton by openly see to it that the Universe come to terms with its rebellion against Newtonian law. They put undetectable dark matter scattered at random all over the Universe and this dark matter will come into effect and start attracting the Universe at some point. However, the deception is little more elaborate than that because those questioning Newtonian wisdom now first have to detect this undetectable dark matter before they have to prove this undetectable dark matter has not got sufficient mass to bring about the reversing of cosmic development. It is sort of what was first, the chicken or the egg, but this time they put this in reverse because before proving them wrong you have to prove the non-existing mass was insufficient to do enough pulling to reverse the Universe and to do that you have to prove the non-existing dark matter is not there. The dark matter is dark and therefore not detectable. It is as hard top prove it is not there as it is hard to prove that it is there so from the angle of proving it they just have to sit back and challenge any one to prove that t is not there while they can't prove that it is there or prove the quantity being there. It is dark and therefore no one can see it and this eliminates proving it is there just as much as proving it is not there. What a brilliant forgery. They don't have to prove Newton because the dark matter that is not there proves Newton correct and if any one asks why Hubble' Universe is expanding in the face of Newton' Universe that is attracting, they can push this down on the dark matter they are going to locate sometime in the future and the rest must prove in the meantime it is not there. So it is a game we play where we all wait and see if the dark matter is there and then after finding the dark matter it is up to any one challenging the veracity of the dark matter to disprove the presence of the dark matter we are unable to locate and therefore allocate. What a way to prove Newton!

This is the biggest swindle ever pulled off by any group of persons. This is the biggest hoax thought up by any group to deceive the broader public at large. Thus is a cover up to hide the first fraud and is bigger than the second fraud involving Einstein that was in place to cover Newton's deception but this third one was the cherry on the cake. It was the red light on the tower...the first deception came by way of Newton claiming that gravity came by mass attracting while every one clearly was aware that all things fall equal and thus all falling things can't fall by the intervention of mass. All things rotate around the Sun at an equal pace, notwithstanding differences in orbit positioning or size of objects in orbit or allocated placing in orbit. In the face of all this evidence Galileo presented Newton still went on as Newton declared that things fall in accordance by mass, which contradicts Galileo completely. Then when Hubble went on to burst Newton's bubble, Einstein was called in to form deception as to cover up Newton's deception. I say if Einstein was in on the act, his behaviour was equally criminal and if he was a victim that was unaware of the scheming and conniving going on, then Einstein was a fool for letting those Masters in physics sweep him up by engaging him in a hoax of this magnitude. Go outside and see how many stars there are. Then think of the fact that there are more atoms in one drop of water than there are stars in a galactica. The atoms being more in one drop of water are an estimate because no one this far was able to measure the number of atoms in a drop of water. Knowing this fact very well Einstein thought himself worthy of the task to count and calculate the measure of mass in the entire Universe. No one could yet measure the mass in a drop of water but Einstein could see his personal qualities being so good that he thought he was able to count all the mass (all the atoms in all the stars in all the galactica through out the entirety of the Universe) and then was still prepared to bind his findings to an actual number. Can any one with a sober mind underwrite any person's honesty when that person says that that person in all honesty went about to measure every atom in the entire Universe and was able to find the density to be insufficient. This alone points to a hoax as big as no other hoax ever was and yet the entire world played along because of the involvement of the name Einstein.

What followed was even a bigger hoax where the Masters of physics invented mass in a dark form to bring about the turn about of the directional development of the Universe where the Universe is going to start contracting instead of expanding. In another book I paint a picture of what would happen to particles with different mass suddenly coming to a halt in relation to mass and where some would crash into others stopping earlier than the ones in front or behind. Believe me that alone would bring the Universe to a sudden explosive conclusion. What breaks the wall that should protect the hoax from others detecting the fraud is one simple question: If mass is responsible for gravity attracting and the particles are present which means the particles are in mass, then why are the particles not attracting at present. What makes dark particles not attracting at this very second by gravity while those dark particles hold mass this very instant and is committed to using the mass to create the contracting gravity and once the dark matter are seemingly located, then suddenly they will come to life and begin to attract because the gravity coming from the mass are then detected. Why will it pull then if it is not pulling now? What has the dark or luminousness got to do with the pulling or not pulling? If there are particles, then the particles must have mass. If the mass does the pulling, then the particles must pull whether we humans detect the particles or not, because having mass is creating the pulling and not our detecting thereof. Either the mass is there and is pulling or the mass is not there and is not pulling but what the mass can't be is be redundant while having mass and while mass has the pulling power to create attracting gravity and now its not being able it pull just because humans can't see it. Can any one else reading this detect the hoax or is it just I being over suspicious about nothing again?

Being in **the centre of the Universe** is frightfully Newtonian. It is very clear how Einstein and his compatriots followed their genius in their arguments when they argued about the critical density and the manner in correcting Newton's obvious misconception. To be able to count the entire mass distributed across the Universe they first had to be filling the very **centre of the Universe from** where they could see where all the mass is hiding. Only from the **centre of the Universe** could they position them selves to see what is needed to see and as lady luck would have it, they were fortunate enough to be in the **centre of the Universe** because only from such a vantage point could they see and measure the entire Universe.

Let's get more personal. We tend to think of our position as having the position only the most important person in the Universe can have because every one thinks of himself or herself as the most important individual that is holding the **centre of the Universe** in the entire Universe. Are we wrong in taking such a view...well not if you go in terms of what the Universe allow you to think.

Einstein said gravity is where the Universe draws flat, and the Universe can only draw flat where the centre of the Universe is because only gravity can draw the Universe flat while drawing the entire Universe to such a centre. Consider this while looking at the night sky outside where light pollution has not destroyed the view... All the light that come across and travelled all of the vacant space from any and all possible positions in space runs directly towards your position using a straight line towards you where you are filling the **centre of the Universe**. With you being able to draw the entire Universe flat so that all the light through out the entire Universe come together to meet you in person in the position you hold, you must therefore have the most intense gravity by your effort of drawing the Universe so flat, in order to have all light running directly to you.

Not allowing even excluding the effort of one photon, all light is heading to meet you where you are in that centre spot and not one photon will pass you by. Not one photon dare miss you because if they do they miss, the effort that all light has to accomplish and that is to locate you as the person filling the **centre of the Universe**. Should you decide to shift your position to any other place in the Universe; you will shift the **centre of the Universe** to that location as well because the light will track you down in your new position. If you install a camera on Mars, the light is obliged to acknowledge your relocating the **centre of the Universe** at your will to reposition you're taking control of that **centre of the Universe**.

All the light that ever left its destination crossing the vast spaces of the Universe, excluding no particular light, as it travelled all the way just to find you filling the **centre of the Universe**, right where you are. By you're standing anywhere, you fill the **centre of the Universe**, and the entire Universe admits to that because all the light comes to meet you there. If you shift from the North Pole to the South Pole you will shift the **centre of the Universe** because all the light travelling throughout the Universe will find you where you then moved the **centre of the Universe**. The light left its destination billion years ago, even long before your birth, as it travelled through space at the speed of light while every photon was so anxious it is to acknowledge you're being in the very **centre of the Universe**.

No photon will be able to pass you by where you are in the **centre of the Universe** because all light is heading your way from their starting positions with one purpose only, and that is to locate you standing in **centre of the Universe**. No wonder every person born has the idea they were born to fill the **centre of the Universe**, which is exactly that what we do fill. The Universe is spinning around you or me, where you or I am, which is filling a centre where all motion is connected. It implicates gravity as wide as can be... Some things mathematics is able to explain but other explaining goes beyond mathematics. It is just not possible to take Newton to Kepler's world and find Newton fit in it.

From where every Newtonian academic professor stands, even to this day, the Newtonians have the fortune of seeing the entire Universe from edge to edge to edge and so forth...and that too applied to Einstein. Einstein could see all the edges and Einstein also new that beyond the edges was nothing more than just the edge of the Universe. Therefore, from Einstein's perspective he could see where the Universe edged and beyond the edge that limited what he saw, he knew there was nothing, because the Newtonians masterfully filled the entire Universe from edge to edge with nothing in any case!

Although we know the Universe is without an edge, Newtonians being **in the centre of the Universe** can see the edge that cannot be there but to Newtonians it is being there. They see the edge of the Universe every day from every possible telescope, so there is an edge where no edge can be because they are Newtonian and Newtonians fill the **centre of the Universe.** Visit the web sight and see how many times do Newtonians discover the latest star "On the very edge of the Universe' and find out yourself how far can Newtonians see through their telescopic lenses...they see to the very edge of the Universe...but never beyond...

However it is not for us to criticize so therefore let's journey back and see what Einstein saw...and remember we are going back in time so tenses becomes an issue. When Doctor Einstein and his fellow doctors look outside at night, they see the edge of the Universe to the left of them. To see an edge in the cosmos is apparently preserved to those sporting a Newtonian mentality. They can see where the Universe ends in that direction that they turn towards. Looking to the left side, they see the end of the Universe and then turning to the right of where they are looking then also there to the right where they look, they see the Universe ending at the edge. Apparently doing that is having a Newtonian gift or something to that effect. Looking to the front the same happens and to the back the same happens with the Universe having edges in all directions. Even when looking up into the night sky they can clearly see where the edge of the Universe defines the end of the Universe in that direction. It is to their fortune that they are where they are because by being where they are, they are filling the location holding the centre of the Universe in the most magnificent place they could ever choose to be. They are in America and we all know that Americas is very much the very **centre of the Universe**. More so is the fact that they are in an institution called Harvard, which puts them in the Academic **centre of the Universe**. By them being part of the physics department of Harvard brings them in line with the astrophysics Academic **centre of the Universe** and everyone in the entire Universe knows and respect this...or so they think.

Things are going from good to better to best to excellent, because with them being in Doctor Einstein's office, they are in the brains **centre of the Universe** and with Doctor Einstein being in their midst they are placed by his presence and intellect smack in the **centre of the Universe**. Their place in Harvard's physics department right inside Doctor Einstein's office standing next to Doctor Einstein puts them smack in the **centre** of the one half **of the Universe**. Now they do not have to worry about finding the bottom half of the Universe because America having Harvard with a physics department having an office where Doctor Einstein presides takes cover of the bottom half of the entire Universe and puts their bottom half smack in the **centre of the Universe**. Doctor Einstein is the living presentation of everything that is not stupid, as Doctor Einstein just has to walk outside and look at the light coming directly to him. One spin of 360^0 would ensure him that being Doctor Einstein and all... he then must be in the **centre of the Universe** because he can see the edge of the Universe in every direction possible! If he wasn't the **centre of the Universe** there was no way he could measure all the mass in the entire Universe And find out why the Universe is failing Isaac Newton by expanding instead of contracting as Newton explicitly said the Universe is doing...and everyone with any mind knows that Newton and God alike can do no wrong...although when you are a Newtonian you might have doubt about God never being wrong, but Newton being wrong, that is a cosmic impossibility to any sane minded Newtonian master.

So maybe God is wrong and created the Universe against the directional wishes of Newton and now Einstein must determine when God would realise God's error and change God's expanding Universe into the correct contracting Newtonian Universe. To see when God will learn about God's error, they taxed Doctor Einstein with the task to see where God misplaced the mass that would turn the Universe around from expanding into its correct stance and that is attracting as Newton stipulated it must. That task fell to Doctor Einstein because Doctor Einstein found his person being in the **centre of the Universe** from where he could see the Universe from edge to edge as not to miss any unseen mass that might be invisible to his eye. Therefore, because of his allocated position of any other person not being the **centre of the Universe** because such a person would not fill Doctor Einstein's shoes, the person not being in the **centre of the Universe** would bring obstruction to part of the view required for the measuring task in hand and with the person by chance missing some mass, the intended calculation might not be that accurate. But with Doctor Einstein being in the **centre of the Universe**, every Newtonian was and still is satisfied the calculations would be spot on, because from where Doctor Einstein filled the **centre of the Universe**, it was impossible for Doctor Einstein to miss any mass that God misplaced anywhere out of sight...

That means with America's Harvard Physics office that takes part of the bottom **centre of the Universe** and Einstein in person taking care of the top half of the Universe the entire Universe aligns at that point. I know every one has sleepless nights wondering why they gave the problem of measuring the entire Universe form edge to edge to edge to edge to determine the critical density calculations to a person such as doctor Einstein. Wonder no more! There is a possibility that it has something to do with his mathematical abilities but that would not count for much if he was not able to see the Universe from edge to edge. But with Doctor Einstein in the place where he is, he can see all the stars sending light directly to him and telling him how big and how far they are. If he were in the incorrect place in the Universe his measurement would not have been trustworthy.

With Doctor Einstein being Doctor Einstein, he knew he was the most important set of brains America could present to the Universe. By the Universe realizing this fact and acknowledging the fact while sending all the light to them at the **centre of the Universe** and without causing delays, the Universe responded by sending all the light at the speed of light to the location of Doctor Einstein. The light came from near as it came from far. It came as much from the very edge of the

Universe to the right hand side of Doctor Einstein as much as it came from the very edge of the Universe to the left hand side of Doctor Einstein. Then the light came from the front as far away as it came from the back of Doctor Einstein. From the top as well, all the light travelled as far as light can travel just to acknowledge and support Doctor Einstein in his task to calculate all the mass in the entire Universe.

Of course with him being in America and at Harvard's physics department and moreover in Doctor Einstein's office placed the bottom half just as accurately in the **centre of the Universe** as Doctor Einstein found the top half to be aligned. If it was not for America and if it was not for Harvard's physics department and Doctor Einstein's personal office the bottom half of the Universe might have mismatched its effort to align with the top centre and then the lot was not in the **centre of the Universe** from where they could see every possible edge of the Universe. But with the fortune of things being as they are, the top and the bottom halves of the Universe matched and in that Doctor Einstein could now fill the entire **centre of the Universe** on top, at the left, at the right, to his back and to his front as well as the bottom half of the Universe. If that was not the case, then what a tragedy that would have been because only from being in the **centre of the Universe** could Doctor Einstein view the entire Universe and see what there is in the form of mass to calculate and measure every atom in every star there is, be they seen or unseen, it makes no difference to the calculation of Doctor Einstein for he was in the centre of entire Universe.

Fortunately for mankind the world had a person such as Doctor Einstein to fill such an important position from where he was then able to see and measure the entire Universe. How gratefully we should be to the Academics of America for allowing us to share America's centre position in the Universe. More than thankful we must be for the Academics that allowed us in sharing the physics department of Harvard's central position in the entire Universe because from there it was a hop to get into the office of Doctor Einstein and share his position of seeing all the light from every corner the Universe has, and to share with him the use of such a marvellous position in the entire Universe as to gauge and measure all the mass we can see…and if you say Newtonians Academics don't put them and their position they have at our disposal also, from where they manage to see them being and filling in the **centre of the Universe** …then please do a rethink! The theory they present puts conditions we find on Earth used by the entire Universe in the entire Universe. Being on Earth we can see that water freezes at zero Celsius and we can see that it is one bar of air pressure that we find at sea level. We have the element table with solids and gasses nicely arranged for us by nature in the table in column order and we know that absolute zero is absolute zero because absolute zero is what we measure when we measure absolute zero. How can the Universe have the tenacity to have absolute zero anything else than what absolute zero should be where we measure absolute zero.

The Universe has not the capability to change anywhere because if it did dare to change, we will see such changes as we fill the **centre of the Universe**! From the Newtonian stance the Universe grew from the size of a Neutron, however its official Newtonian policy that the Sun was the same since time began and the atom was always what we measure the atom to be. With all the constants in place there can be little to nothing changing in the Universe because then our mathematicians also known as Newtonians would not be in a position to play their awesome mathematical games, as they love to do. With Newtonians filling the **centre of the Universe** we know the Universe can grow and expand but that feat is quite impossible for the Sun and the planets to achieve. It is completely anti Newtonian to think that the solar system is getting bigger just because the Universe is growing bigger. All Newtonians would recognise change the instant change occurs because from filling the **centre of the Universe** Newtonians will notice any changes immediately and after all, one has to consider that Newton said the lot is contracting. Saying anything to the opposite of what Newton presents will be quite sacrilegious to Newtonian religiosity. The reality is that the Universe grew and the Universe still grows even in our part of the Universe. Amidst all this evidence, Newtonians have a constant speed of light, a constant time since time began, a constant gravitational force, a constant expansion and every other aspect that will bring along nice and easy calculations, so the Universe will have its constants just to keep life a little simpler for the Brainy mathematicians. To the Brainy Bunch using mathematics to promote their godly image they have about the role they have in the Universe is to think that they have the ability t measure the significance of the Universe in terms of their ability and then tot think that by their calculations they can reserve an idea as to what the Universe is about...that is like being an ant in New York's Central Park.

It would be the same as if the ant running in Central Park in New York is of the personal opinion that it is his being in the park that is the cause of the park being where it is and is kept maintained to serve our ant. The park is kept as it is and is there where it is only because our ant is in the Park. Because the park is maintained in honour of our ant, can others share in the ant's personal benefit with every one also therefore having the joy of a park in which to play. Everyone sharing in the park can do so through the magnificent generosity our ant bestows on others. Because our ant is a sharing ant, every one, which is there in the park can enjoy the park but should be humbled in gratitude to the generosity of the ant. Our ant considers that every one holds the opinion that amongst the many of millions of people they are all aware that he is the reason why the work effort of many a thousand people is dedicated on his behalf to maintain Central Park in New York just so that the ant may harvest the benefit. The ant will have the opinion that all the people in New York are of the obsession having one purpose to live for and that is to please that one ant. Well that is how we all see the Universe. I stand in a point where all the light that came from everywhere came to meet me personally and there will be no light that will ever not come to greet me. If there is an ant that has this opinion about his place in Central Park, then that opinion, as self serving as it sounds, is just the same as the normal opinion is of all the persons living on Earth. Every person living is living by the idea that the cosmos came in place to serve life and moreover to serve every individual on Earth. It might be shocking, but every individual goes around with the idea that when such a person dies, the Universe will end. The Universe is there to serve me above all others.

I If you dare to be as rude as thinking the Universe has no ends and there is no edges placed in the Universe you better wake up and start smelling Newtonian shit coming across the internet because the Newtonians find new planets and stars and Galactica on every edge they could find in the Universe. Log in on the Internet and explore to your delight and you will find so many Universities that located so many things on the edge of the Universe.

Go ask every Newtonian and that Newtonian will tell you the Universe has no edge because it is limitless, but being as important as only Newtonians may, be they fill the position where they are able to see all the edges the Universe cannot have. However I should warn any one that listens, don't tell Newtonians of their double standards. When I confronted professors in the past and accuse the Academics of limiting the Universe to benefit their views about claims they make or to support Newtonian claims, I am compromised by being the one referred to as being the incoherent, as the raving idiot because they say Newtonian science will never do such a thing as commit the wise Brainy Bunch to double standards. The wise amongst all–of-the- all-wise, as Newtonians think they are, will always use only sound arguments and well proven facts.

Then the next minute, I see Academics limit the Universe to having an edge, which they find in very clear telescopes. They see a boundary where the Universe ends. To prove my case I advise the reader to go and visit the many such web pages carrying this very claim. It is such normal every day Newtonian practice to use a double forked tongue. It states clearly that science caught big bright stars on the edge of the Universe. Any one can glance at this on the condition they have access to the web page and can use the web page. This clearly shows the lack of understanding on affairs Newtonians claim to be knowledgeable of. Those Academics (the lot of them do advocate that the Universe does not end) are those giving the Universe an edge...to do what with. What ends at the edge or the border and what bring the edge about?

What would they suggest forms the wall that then would present the border forming the edge of where the Universe then supposedly ends that allows no more of the Universe to continue? However, I am sure that there must be some persons in education that will share my view that the modern Newtonian in science are still committed to boundaries and edges. Seen on a mental scale, those brilliant Newtonians did not yet leave the shores of the known harbours that Columbus' sailors left five hundred years ago when thinking in terms of distances and where ending might occur and it is clear they now know they will not fall off the Earth as Columbus' sailors was scared of doing five hundred years ago when those sailors found no edge of the world. Newtonians got use to the idea that the Earth is round with no known edges from where they will fall off the Earth.

Yes, they did progress from the days of the superstitious sailors. Yes, Newtonians science did manage to take science much further than science was before during the days when Columbus went sailing. Newtonians managed to shift the edge of the world so far it became the edge of the Universe and they became so wisely educated that they managed nowadays to give the Universe with no end, an end. That is progress, if it is anything… It became about time that the entire philosophy of cosmology is overhauled and is revised from the backwardness of five hundred years ago to a more fitting approach in the five hundred years that time went on, because as science held fast to Newtonian philosophy, science is the only faculty that did not yet shake their past believes and became modern. Theologians stopped witch hunting and Medicine stopped blood letting, but science still holds onto the magic of mass to provide gravity, notwithstanding all the evidence showing the truth is to the contrary of Newtonian beliefs and Newtonian superstitious religiosity.

Then they claim the Universe is expanding…expanding where too. Where can anything go that covers the lot and has nowhere to go? How can anything get bigger when such a thing is as big as anything will ever get because whatever will be, that thing the Universe already has. How can that which has no way of increasing what it has, because it already has whatever could be, gain in size by expanding when it can gain no more since it stocks all the possibilities there will ever possibly be. How can that which can grow no more because it already is what anything ever could be, then become filled more and show it is still filling with more of that what it in total already has everything of? Where is that which is expanding that which already holds everything there could ever be coming from because nothing can leave and nothing can add to what the Universe already holds!

How can the Universe get bigger when it is already limitless? Notwithstanding what logic needed they will tell you the Hubble constant is about the Universe expanding that which is not expandable and going where there is no going too because (I guess) the edges that are not there is shifting further out to nowhere. I don't for one-second question the fact of Universal expanding but I am trying to address the real issues applying and remove the Newtonian simple-minded thinking as it is used in the case of the edged Universe. Newtonians put an edge onto the Universe because the instruments they use can read no further.

The Universe carries on regardless and limitless until it crosses over to form eternity and that fact everyone knows and accepts. But it would be very unsporting of the Newtonian in question to admit the telescope used by this particular station is very limited, and therefore Newtonians cap the Universe with the edge of possibilities their instruments show. When saying they detected yet another star on the edge of the Universe, they should be truthful enough to admit they can't go beyond that limit, not because the Universe has edges all over, but their capabilities are very limited. Don't throw the limits in human possibilities onto the Universe and don't expand a Universe that can't expand because it has whatever ever will be and the human mentality can't grasp with cosmic realities.

The Universe is getting bigger but where is the Universe getting bigger too because wherever it is going the Universe is surely already there! There can be no place without the Universe already being there so where is the claimed territories that is gained by a growth that it is from what because it clearly already claims all there is. I realise that Newtonians being in the **centre of the Universe** can claim to see what we others with less intellectual means are unable to see for we do not have the grant in privilege to see the Universe from the **centre of the Universe**, that privilege only befalls the wise Brainy

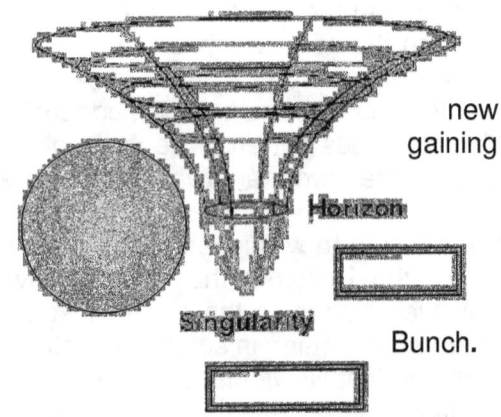

new gaining

Bunch.

Quoted directly from the Oxford dictionary of Astronomy the following:

The definition of singularity is as follows:

Singularity: a mathematical point at which certain physical quantities reach infinite values for example, according to the general relativity the curvature of space-time becomes infinite in a black hole.   In the Big Bang theory the Universe was born from singularity in which the density and temperature of matter were infinite. Let's hunt singularity down! Singularity can have but one mathematical value and that is 1. Let's find 1^0 and find the centre of the Universe, which is $\Pi^0 = 1^0$

Big Bang starting point

When taking Newtonian science to task then we meet with an argument that hints to the fact that there has to be a specific point where the Big Bang started and as from that point all matter went on expanding while such expanding was shifting material in directions in relation to other materials that went in numerous other directions. The point where the Big Bang occurred holds the centre because space formed from that centre. Remember that this point had the size rumoured to be a neutron in volume. When reading a declaration found on the internet such as Hubble's deepest view of the Universe reveals earliest galaxies, one do realise that that point of centre was crossed because the point we see must be past that point from where the centre originated. This means in the context of how Newtonians go about in search of singularity they of the educated opinion that it is most probable that we will singularity at a point that is wide as it is far from us because the Big Bang had to have begun 'over there" since there is no trace of the Big Bang beginning "over hear". This puts other questions at our door   that   also   requires   serious   consideration!

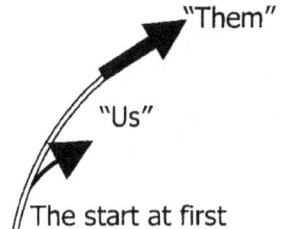

"Them"

"Us"

The start at first

The only way the centre would not be crossed is where "we" and "they" went in the same direction after the exploding of the Universe during the Big Bang explosion, but "they" went further than "us" because "they" went faster than us. That means "they" are then also further away from the centre of the Universe where it all started with a Bang than "we" are.
I wish to return to what is said as:" Hubble's deepest view of the Universe reveals earliest galaxies." How did the Galaxies get there and we got to where we now are and all of us is so wide apart as the statement reads while the galaxies and us moved during the same time in the same explosion called the Big Bang?

This leaves everything concerned in an array because if it took the objects longer to travel further than it took "us" to travel while all were going in the same direction, then the light that is coming back from where "they" currently are, is also handicapped by the motion that "they" had when the Big bang erupted, which took "them" further than "us". This dual conflicting motion then is giving an inaccurate position as to the where about of "their" current location in relation to the light coming back to "us."

It is a question of direction from where the end is coming. If it comes from an edge, as the article expressively points out, then is the end of the edge also where these galaxies now are, is coming in our direction and the question to be asked is what pushes the edge forward and onwards to us. Hey, this rubbish argument is not because of my inability to think but it is because those highly mentally skilled Brainy Bunch members places and edge they found in the Universe somewhere from which they so often discover more stars "on the very edge of the Universe!" It is they that moved the edge of the world onto the edge of the Universe and while feeling very pleased they managed to do so.

Where "they" are

Where "we" are

The centre of the birth spot

It is rather comical to think that Vasco Da Gama and his sailors would not set sail on a voyage of discovery in fear of confronting the edge of the world while currently our cosmic sailors in waiting desires to get going on such a ridiculous voyage and reach the edge of the Universe because they can see the edge of the Universe. How much did things change just to remain the same, will you not think? Today the schooled Newtonian opinion about cosmic science is that we just have to hop into some craft, blast off to the unknown into the unknown at the speed of light and send a post card back home when we get to the edge of the Universe. ...And all the while Newtonians have no clue how light travels. They show some mat-like surface with graded blocks that should represent space and time but it puts space and time in some single dimension holding a square of some sorts and present that as the travelling road light supposedly takes as it journey all the way to them where they are filling the centre of the Universe. Let's have a good look at the theory on the general Relativity principle.

Since this argument is suppose the define singularity and explain how light travel, I wish to use this image again and return to the Newtonian arguments that is suppose to represent singularity.

The Curvature of Space caused by a Massive Object.
By allowing the object to have a hole is returning the object back to 3D from the single dimensional surface it is portrayed to be when being in such enormous gravity. However there are certain requirements and there are certain rules that has to apply in such conditions.

The conditions will apply to all and will not exclude any object in the vicinity. One aspect of gravity is that it applies to everything equally in the space in which the gravity is acting. It is very obvious that the claim of the curvature of space-time is made without any support or backing about reasons why such curvature is there in the first place. Why would such curvature appear and why a circle forming curvature?

Our Super-Educated first puts on the table a flat surface. This is not possible because even showing the flat mat picture, such a flat mat has to have a bottom side if it has an up side and if there is a left

and forward side, then there has to be a right hand and back side. There can only be three dimensions applying or no dimensions at all.

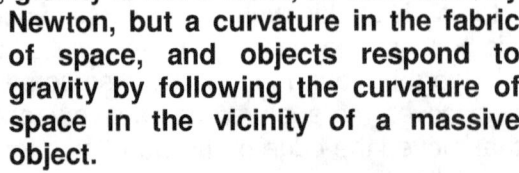

If the top side is in use then where is the bottom side gone too because either this flat mat / carpet /whatever runs eternally down to the one side which is not possible, or the bottom side has to form a part of the picture.

No one ever try to bring across any explanation or reason where the surface went when the side under the surface supposedly went flat or is flat except that it is gravity that is making it flat. They left the guessing to us to fill in what produces gravity in outer space where bodies float around centre objects because such explaining might just mesmerize their theories while they wish to mesmerise our brains. The Universe does draw flat, that I do admit, but what is the Universe that goes flat. The answer to that bit I can explain in terms of applying singularity. Either they put that which is what they can't explain down to magic as Newton did with his force, or they say they have no foggy idea what is going on in what they clearly don't understand, but to bullshit as they go along brings no one anywhere.

The second fundamental principle of General Relativity is that the presence of *matter curves space*. In this view, gravity is not a force, as described by

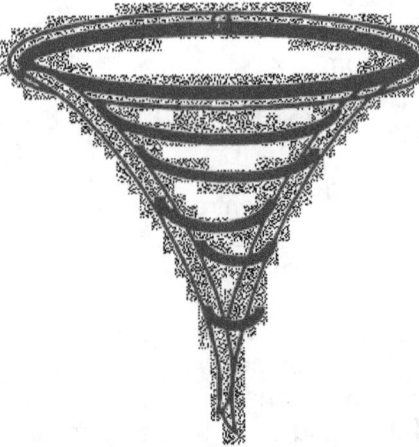

Newton, but a curvature in the fabric of space, and objects respond to gravity by following the curvature of space in the vicinity of a massive object.

The description of the curvature of space is the mathematically complicated part of general relativity involving "metrics," which describe the way that matter curves space, and tensor calculus.

If matter curves space then it is because matter is curved and then matter is responsible for moving space

and that means matter is moving in space by which it curves space.

In order to understand this cosmic feature one has to locate the source of gravity. The star is a collection of all the atoms within it and if the star becomes singularity, then it has to be because the star has contracted all the inner atoms that form the star into singularity.

As any symbolic picture of the strongest possible gravitational force will show as in example a picture of a Black Hole, the gravity deforming the surrounding Universe is where no space can be located.

In order to find the answers to the Black hole we have to locate the biggest thing there could ever be because when finding the biggest there could ever be we will find the measure of what the smallest is that there could ever by found in the Universe.

In the centre of a sphere that is holding the sphere in form as well as the surrounding space attached to the sphere is gravity. It is forming the surrounding space from a point inside the sphere that has no sides and no space other than merely form in which it puts all space attached to the gravity. This view points undeniably to a point preserving singularity and in this light, Newtonians fins singularity in Black Holes and all other places no man may dare venture. It is therefore in such a place that the Big bang presumably started with singularity being there and all...

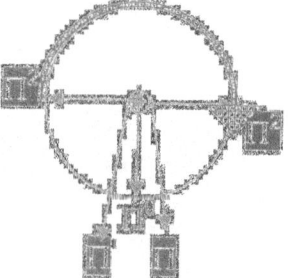

Such a pint will also establish a line we call an axis. There must come a point where the ring is infinitely small, where it can reduce no more, where it reached its ultra limit, but at that point it cannot be zero, because the point is there for all to realise but nobody to see.

Understanding all the following is connected intimately and all conditionally to the fact of accepting that all individual particles in the Universe use motion and therefore spin. On the border where k becomes Π and step away from singularity into the very first point proving to form 3D a factor of Π^2 comes into place.

In the **precise middle** of all **objects in rotation** is a precise centre dividing the object in sectors that will **start the spinning initiation** from that centre point. Thus, the spinning object **will have a middle point**, a very specific **centre point that does not spin** and only holds Π as a specific value because no radius can apply. But also the one value such a line **cannot have is zero** because the line **is there and holds contact** to the rest of the material bringing about that **zero does not start any** line and therefore the **value of the line must be infinite**, just as described in **accordance** and by **the definition of singularity.**

As I am introducing a very new idea, I whish to explain in better detail what I try to convey... While the toy top is spinning move the rotating line progressively to the middle by reducing the length the line have from the edge to the middle. At one point all further reducing ends.

As the rotating direction moves inwards, the rings will become smaller and smaller.

That point albeit hypothetical, is also as much a reality none the less and is placed where that point **must be standing still** because every line **running from that point** in **opposing directions** are also **in opposing directional spin the other or opposing side.**

In considering the spinning motion in the fraction of time in the detailed instant every aspect of rotation will turn in every instant of change in time. Although the points had the same characteristics only one instant before, they oppose the characteristics it had just before and just after the very instant in which they are and to which they relate by similar points also in rotation. The fact of the graph proves my point in quarterly opposing dimensions and values, the changes of the flow brings a radical revaluing of the same space just by turning ninety and hundred and eighty degrees.

In dimensional terms, which explain later on the value of 2Π relates to Π^2. Then that relation extends to the next value where Π^2 relates to Π, which relates to Π^0. The first space in the circle will then be $\Pi^2\Pi$. From the centre being in infinity one can realise by applying mental power the single dimension factor not seen but present all the same. Extending that into the 3D comes six Π and any one of the six will further extend to form a seventh point as Π^2. This centre gives the atom gravity and this centre centralises the entire Universe by linking the entire Universe in relation to 1^0 or Π^0 or the single dimension or singularity. This value gives the entire Universe equality in time in movement and not in time experienced.

At the heart of bringing about the solution to one of the greatest Astronomic riddles one will find a child's toy… the riddle of Einstein's singularity pointing to the position where the cosmos started so many billions of years ago.

In

Einstein's view singularity is a mathematical point at which certain physical quantities reach infinite values for example, according to the general relativity the curvature of space-time becomes infinite in a black hole.

In the centre of any and all rotating objects is where singularity holds position as a value of Π^0. By merely applying movement (in the form of atoms) qualifies all matter to be space-time because space-time quantifies singularity. It does not only fit the description of space within Black Holes, but it fits all stars where singularity becomes part of all the stars from the smallest minute to the largest cluster of matter.

With no line starting from zero because there is no zero as a mathematical fact, then all particles hold the point of infinity and not merely the Black Hole. From that argument one may conclude that all stars will become Black holes depending on the gravity increase they may generate. Through rotation encircling the point of singularity and matter is (1 the past) coming from, (2 in infinity) being at, (3 becoming the future) as it is going too in one movement in relation to the specifics of the centre point being singularity all matter then qualifies to form space-time.

In the spinning top, matter would always relate two three positions as does Einstein's space-time declaration require. All atom particles forming matter composing, as material forming the top would have to relate to the centre where singularity forms space-time and two other positions. Being the onlooker, the viewer has to maintain one position. From that position some particles would be circling a centre point, as the particles would be coming towards the onlooker. The other matter would be circling the centre point while rushing away from the onlooker.

All atomic particles would have to refer to coming or going as much as circling around while spinning in the top. This comes about as no atom can be ignoring in relation to any of the three positions it is aligning with in direction. Some would be on route from North to South and others would be on route from South to North. It is a response to an ever changing directional re-aligning with the centre holding the centre too a specific location in relation to the position it previously owned and would own the next minute, changing constantly and in according with time location. All matter will have to adhere to any of the two directions, which in fact is actually four, but it also changes dramatically in a moment-by-moment positional change. In the centre a line MUST form separating the comings from the goings and again the goings from the comings. Such a point has to be positional completely neutral as it forms the eternal divide in infinity. At that point we locate singularity and can only be where there is no chance of locating matter. The end point of such a radius line at line's end is too small too hold any atom, sub- atom particle or matter of any kind. All matter is either on the one side, or on the other side, but never can there be matter that is neutral. It even gets more complicated because another line forms locating matter in groups. The one group is relating a poison from the "centre point" holding "back" and "front" running through "the centre" where the other line is relating from "side" to "side" running through the "centre point". The fact of the lines is that "they are there", but we cannot see them. Try as you may, no one will be able to calculate the very position that forms the lines, but as they change all particle characteristics, the lines are a reality as the spin of the matter is real. Being too small to hold atoms, they then therefore must become part of singularity, where singularity is a spot in the centre with two lines crossing the spot at an angle of 90° as well as forming a line running from top to bottom at a 90° angle with the four lines crossing. That is the basis of singularity, and since all the positions still relate too a centre of a circle, forming a part of a spinning circle, Π must form the basic value that initiates space-time.

In that there are one specific group in relation to coming towards the centre while coming towards the front.
The following group is rushing away from the centre while coming to the front.
The third is heading for the back while heading for the centre and…
The fourth is rushing away from the centre while heading towards the rear…and all the while the lines hold contact with as well as adhering to singularity.

This is Bang most

where the Bible says creation started and the Bible also says the Big came after in a second event…if only science would take notice of the conclusive work that was ever penned, the living Bible

According to the Bible, God created heaven and Earth and that is precisely what happened. Today we have heaven which is liquid heat which is eternity and then there is material or the solid or Earth which contracts the heaven or the liquid or the uncontrolled heat. In Creation we still find two components the Bible refers too as Heaven and as Earth, which I refer to as solids and as liquids which science refers to as atoms and as nothing. It is no wonder

the haven't

Newtonians are stupid enough to be atheists because I can see they got a single lick of common sense.

If we wish to investigate how it all started we must use the only reliable and the trust worthiest book ever written, which can only be the Bible. From studying the Bible we will learn how Creation began and we don't do this investigation of the origin of the Universe in terms of Theologians. Theologians are very well known to have an orgasmic misconception about Biblical realities. In fact on Biblical realities they have no concept about because Theologians conduct their personal notions whereby Theologians create God to the likeness of what Theologians think God should represent in order to fit their endorsement. If we wish to investigate how it all started we do not use concepts to try implicate other madness such as Newtonian science tries to portrait because if they are missing gravity by the margin that they do, one won't begin to imagined how far off the mark they are when they reach the origin of science. However, when we wish to put truth in as an impetus we have to take every word in the Bible to be truthful and undeniable as to fathom the Bible by the letter as it reads in terms of what is written? Theologians create a God they wish to understand where in that way God will support their views on good and evil and Newtonians wish to create Cosmic start by portraying science in they wish to understand the laws governing nature and all the while both groups are making a disastrous mess of the lot because they try to establish something they could manage an understanding to profess to understand while those parties of both groups clearly don't understand anything. I am now going to bring the written word in contexts with science, but I am just touching on the matter. In another book I named **Seven Days Of Creation** I go much, much, much deeper, but at this point I am going to knock on the door by addressing aspects of facts. This might shock the established Newtonian atheist clan, those professing to be the ultimate wise, those self proclaimed members of the established Brainy Bunch, members to profess to be atheists because they see their position as the unlimited ultimate wise, but in the light of their excessive and unbelievable stupidity, that stupidity is enough to shock any true intellectual. To all and everyone's surprise; the Bible is ultimately correct in science. Creation started just as the Bible describes the way the Universe started.

With that observation about time and space in space-time let us reflect on the events the Bible tells us about Creation. How and why can we be sure the Authentic Author of the bible saw the Big Bang (which ever one you may choose because in relevancy all were the same)?

The Bible reads as follows: In the beginning of creation, when God made heaven and Earth, the Earth was without form and void, with darkness over the face of the abyss and a mighty wind swept over the surface of the waters. As I said before there are two forms of substance forming the cosmos. Newtonians like giving names and that they cleverly do in order to conceal or ensconce or even to drown their stupidity, Newtonians are hooked on difficult names because that hides their weak understanding.

I am going to say this which I am about to say again in repeat many times over in this letter because if there is a vital aspect in physics that overrules all other concepts and is the most prevalent concept of all, then in physics as much as in Biblical writing, this concept is where the Universe started because this is where the Universe starts. When one draws a line towards the centre of anything spinning around, which will represent the radius line of any spinning circle, one will reach a point at the very centre where the line starts. It can't start or end at zero because zero will remove the line completely. At the point in the very centre where the line can go no further because if it crosses the point reached at the very centre, f it crosses the line would be allocated on the other side of what is spinning and it would contradict whatever it represented a motion before. The line can't enter the centre because there is no space to enter. The line will cross the centre as it enters the centre, so it will move through the centre to the other side. The line ends at a point that holds no space but holds location and by the location the point controls the entire Universe. It forms the great divide of all things forming a Universe. It is where Einstein saw that gravity draws the Universe flat and that is because only at that point the Universe loses form to become flat and single dimensional.

Everything that is in the Universe, everything that is a part of our concept of what the Universe holds has to connect to singularity at 1^0. The entire Universe we know connects by this point holding singularity. The value of singularity has to be one (1^0) because that is the only possible value of singularity. Everything that is, can only be as it moves in relation to that point referring to singularity 1^0. Everything that moves has to spin as much as it moves in a straight line and the straight line cannot part from the circle. This is what Kepler found the Universe unveiled. This is $a^3 = T^2k$, which is if it holds space, it has to go straight while circling at the same time. In the centre is $k^0 = a^3 \div T^2k$.

That spot is void of space. That spot is in the abyss. The spot is at a position where when the space the spot fills ends is the place where form or space starts to start.

The Bible reads as follows: In the beginning of creation, when God made heaven and Earth, the Earth was without form and void, with darkness over the face of the abyss and a mighty wind swept over the surface of the waters. If the referring to "without form and void" is not pinpointing to the point holding singularity then the reader quite literally does not understand the concept. Singularity is the only point that can be void because it holds no space. For the very same reason such referring could only point to singularity where it states that the Universe was without form. There is no other point that is without form than the point holding singularity. That is the Bible saying the Universe began at a point holding singularity and all blind persons dabbling in theology as well as science missed the mark by miles.

We start again by looking what the bible says about how creation started because I am about to show how the atom could come about in the multi proton sate as the density of such atoms require. In the beginning of Creation, when God made the heaven which is uncontrolled space-time and the Earth which is material or solid substance such as the Earth uses the Earth was without form and void, which means the heavens too was without form and void but since the heavens still are seemingly void and without form, the Bible is not surprised by that part. The Bible specifically indicates that the Earth or material was void and this show that it came from singularity. Light was not yet present in the Universe because it specifically says that darkness was over the face of the abyss. Referring specifically to the word abyss indicates yet again the presence of singularity. The point I showed holding singularity is where I also showed there can be no start because it is infinitely present. Then the Bible says that a mighty wind swept over the waters. I showed what infinity is by showing where infinity is on the spinning top. Then also I showed that eternity forms as a liquid substance or going by another name would be water. Newtonians still have no specific name to call heat in outer space because Newtonians to this modern day still sit with the mentality that outer space constitute to nothing…well with that mentality it is highly unlikely they would name something they still regard as nothing. Newtonians still so admirably think that lifting hot air balloons is a trick of magic because they still don't see how a balloon can lift while filling the balloon with hot nothing and finding a reaction when doing so.

…But first I wish to come back and try to

There the wind U_g is that the bible refers too, there is movement of space.

There the abyss is Π, there is singularity, and there is the point that can reduce no further

There the water is Π, there is the neutron, and there is the point that forms the liquid component

to Newtonian gravity establish in accordance wit Newtonian tradition they don't say how gravity as it is in their opinion could form any multi-proton atom. With

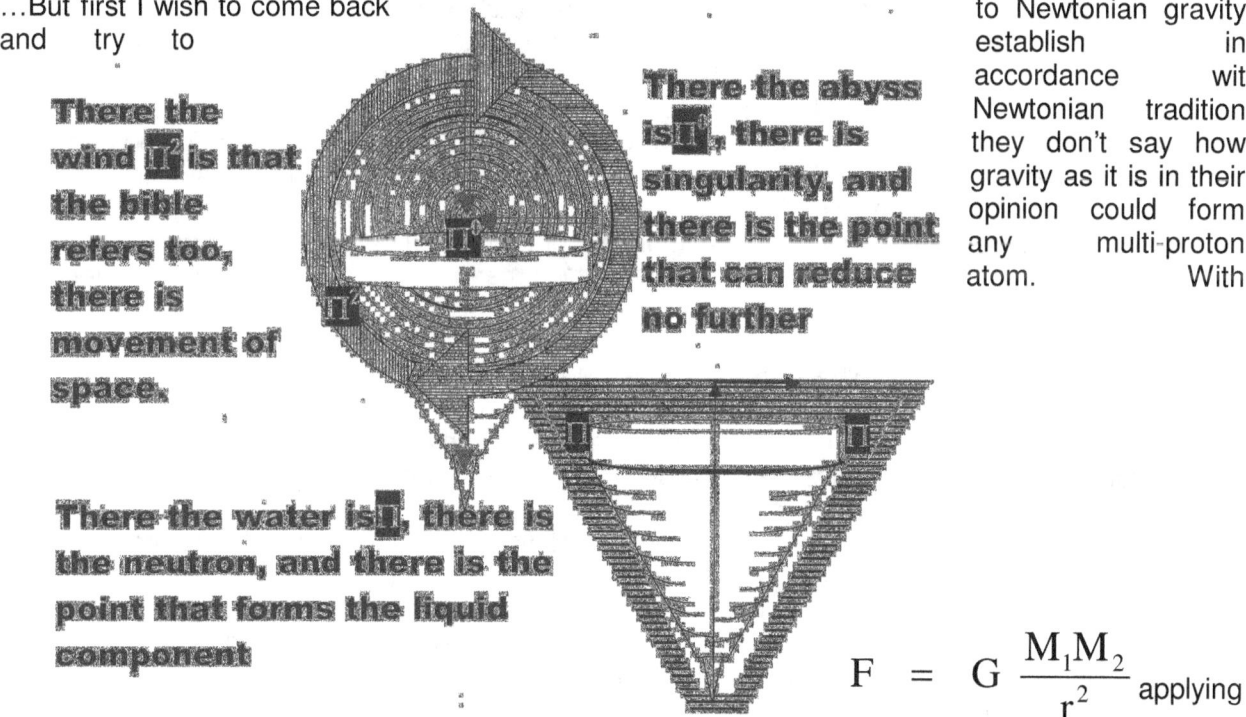

$$F = G \frac{M_1 M_2}{r^2}$$ applying

the forming of a multi proton atom is not possible because one proton holding a specific mass would draw one neutron without a specific mass and the two would collect by gravity one electron that holds a specific mass. There would be no incentive to form eighty protons in one atom or to have ninety protons collecting ninety neutrons that form on group that combines with ninety electrons. Let's investigate what it is that would make atoms very small and very compressed because to fit ninety protons and ninety neutrons into one sealed atom has to be thought of as compacting something into something much smaller.

The Oxford dictionary of Astronomy defines Gravitation Collapse as follows:
The collapse of a body that is unable to support itself against its own gravity. Gaseous bodies undergo such collapse if they are not hot enough for their gas pressure to balance gravity. This can

happen in the early stages of star formation, or when nuclear burning ceases in a star's core. The time taken for such collapse decreases rapidly with increasing density, varying from about 100 000 years for the birth of a new star to less than a second for the formation of a neutron star. Star clusters may undergo a similar collapse if the random motion of their constituent stars is insufficient to offset gravitational effects, either during their formation (see violent relaxation) or at an advanced stage of their evolution. I don't agree with Oppenheimer and Chandrasekhar's hogwash that if a star has 1.3 times the mass of the Sun and or 1.6 times the mass of the Sun that mass would bring about a gravitational collapse and with such mass it would bring about such a gravitational collapse as we would find in the Black Hole, but be that as it may, things get smaller when the gravity increases and to bring witness to that we have the dynamics of stars being smaller all the time but being considerably more massive all the time.

If mass was an issue in the applying of gravity then the mass will play a major part in the time it takes a body to release from the motion when releasing from the Earth. Mass increases when the movement of atoms accelerated. If mass was responsible for gravity increasing in that then the body must actually grow bigger to have more mass. That would place Betelguese in the region of a Black Hole and not being a slopping bowl of liquid heat sloshing from side to side. On the other hand we have the movement of Black Holes contracting liquid much faster that the speed of light would allow photons to escape and hence, we call it Black Holes wherein light disappear for ever. By increasing motion the actual virtual mass increases therefore, motion can increase mass by gravity increasing. One has to accelerate an object to navigate release of the Earth by the value of the escape velocity which is $7(3(\Pi^2)(\Pi^2/2)(4\Pi^2)\div(6^2 \times 10^2) = 11.21$ km / sec. to release from the Earth gravity and orbit as a satellite. However, in such acceleration the mass increases many times over but notwithstanding that mass increase the accelerating motion secures the release in any way.

The difference in velocity that the star produce and the matching velocity the object in outer space must produce in motion to match the motion requirement of the planet's motion relating to the outer space factor such motion creates a velocity differentiation but why that takes gravity is one of the biggest mysteries I can find in the Universe. Again, I must press the issue that since Galileo proved otherwise where otherwise is being that what Newton stated everyone agrees with Galileo and then completely ignore Galileo. Galileo said mass fall equal...that means mass draw equal notwithstanding mass differentiation in size or mass and while every one is admiring what Galileo said then by the same margin every one is completely ignoring what Galileo said. Newton said Mass one and Mass two draws space according to mass, which means mass has the ultimate influence in the process.

Gravity is strongest when and where space is least. Where space is least motion is producing most relevancy and change. More gravity leads to smaller space and in such a smaller space, the bigger gravity can hold more shrunken mass. This increases the intensity that the mass would experience coming from much higher dynamics achieved by more gravity that produces much more movement. This is even truer in the smallest of stars using virtually no space, since motion produces gravity. With the stringent reduction in space, more particles will fit into less space and by having a smaller space it can manage a higher quantity of particles. In the end, the reducing of the space held by the material in the star is the result of motion. The motion is contributing to space decline By producing a larger area to fill in the way of using a specific space occupied, the space occupied will have to spread thinner as well as become much more intense and much more concentrated in all dynamics to compensate for the larger area commandeered to fill $a^3 = T^2 k$ should $T^2 k$ increase then a^3 must reduce.

By this it is clear that even referring to a force applying is directly suggesting motion occurring or a tendency of a serious effort to bring about motion restrained that then is a blocking of motion occurring and that makes that mass is the restricting of motion trying to continue in a specific direction to come about. If one takes away the motion or the tendency to bring about motion then it is clear that gravity disappears, and only when gravity disappears would mass or weight relent. This we see happens when aircraft fly or balloons take off. The Black Hole contributes the strongest gravity since the Black Hole places all motion in space and no motion in the star. In fact, the Black Hole has returned all atomic particles back to singularity or on the rim thereof. Nevertheless, the motion we see comes as a response to a point where that whole ending of space-time centre on singularity which in essence is the result or the product of the motionlessness of the star. Since all stars apply motion and the Black Hole reveals the ultimate form of motion, therefore the Black Hole shows that gravity is space in motion by the reducing of space towards the centre where there is space less ness and motionlessness. In the invisible centre is a point even beyond where space and motion is at its least. Without space, there is no motion and without motion, there is no space. Kepler tells us this as $k^0 = a^3/T^2k$. But that is what the Bible has been saying for thousands of years and the Mindless Newtonian attests are too stupid to follow the Bible.

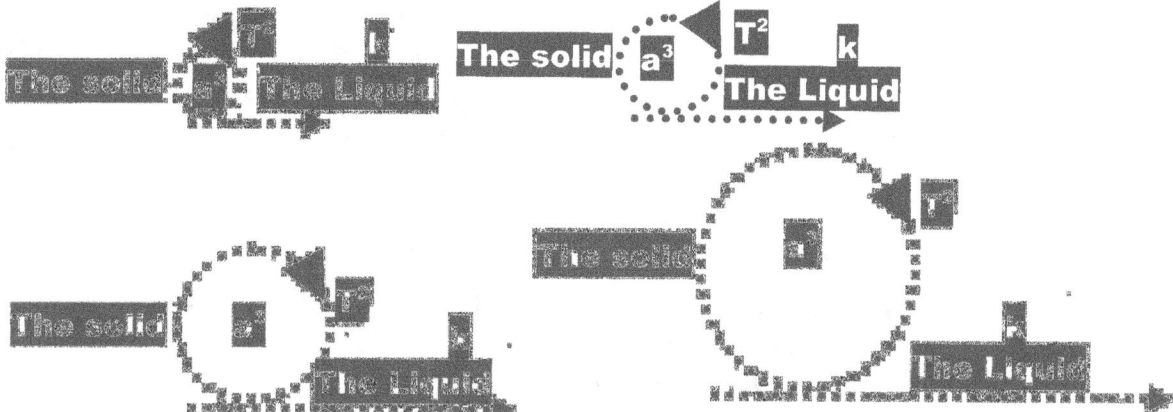

Galileo said all fall equal notwithstanding mass…Newton said mass does all the pulling and Galileo said mass has no influence on the drawing of the object…yet everyone is totally ignoring the fact that mass, be it big or small, proves not to influence the drawing of material that is falling! If Galileo is correct something else than mass is doing the pulling…and for the life of me I can get no Scientist to see what I say in the contexts that I say what I see. The concept of mass being the producing factor in gravity comes across as rather less thought through and more than a bit silly when considering the above

. All evidence points directly to the idea that mass can not generate gravity and such presumption that mass can not generate gravity is most inaccurate. Mass becomes a factor only when the Earth restricts any object that is falling, further movement by blocking the movement as the Earth then is "in the way of further descending". The gravity part remains present in the fact that the body still show a strong tendency to move down but with mass applying the Earth restrict or counteract this movement by disallowing further movement. It is completely incorrect to think that it is mass that produces gravity since it is a notable fact that as space shrinks or reduces the stars ability to generate gravity excels. However, space has little to do with massiveness because the mass increases exponentially as space used within a star declines.

If space is infinitely small, then time is infinitely slow. There can be no space when motion is at its slowest possible speed. That forms the ultimate relevancy available and space-time or space in motion is all about relevancy. However, the enormous gravity falls outside the star. In the Black Hole singularity controls matter and space applies all motion that is in fact the time factor to space occupied where the motion aspect is more commonly known as gravity. However, the space less ness of the Black Hole shows that space less ness is the location of strongest gravity. It is in the place that the heat is the most, which is in that centre area of any sphere. If any one does not believe me then test nature. It means that mass has the least say when gravity is generated. According to Kepler, mass in motion within space in motion and gravity is the same thing. $a^3 = T^2 k.$

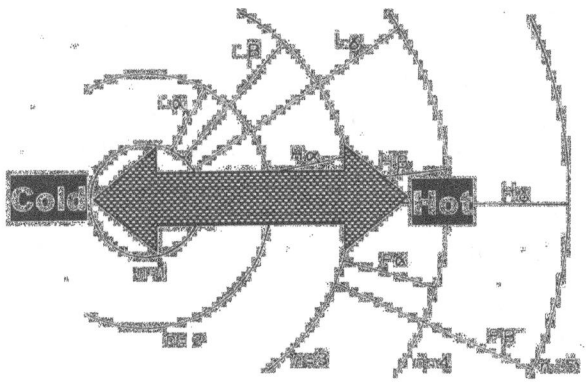

I am not getting in a debate as to relevancies applying and how that operates but the atom applies movement according to a differentiation between what is hot in terms of singularity uncontained and what is cold in terms of singularity confined and contained. This movement is not perpetual but is very precisely defined. What singularity would use as a gradient as to determine the cold to hot differentiation is what brings about a star classification. Time slows down as space decreases. What would this entail if space (a^3) were at a premium then movement (T^2k) is equal to space. The bigger the atom is the shorter would the duration of   time be.

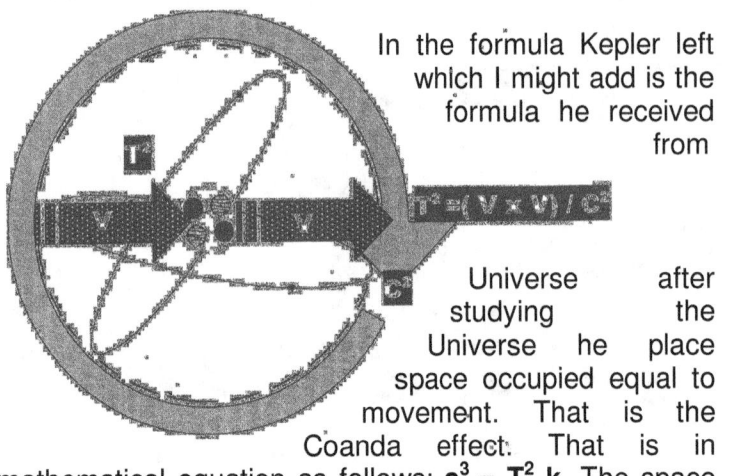

In the formula Kepler left which I might add is the formula he received from

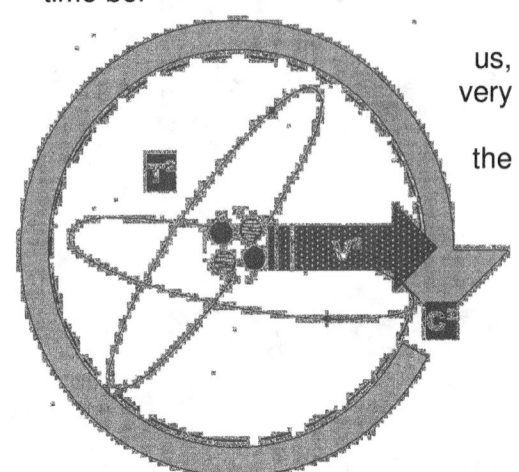

us, very

the

Universe   after studying   the Universe he   place space occupied equal to movement. That is the Coanda effect. That is in mathematical equation as follows: $a^3 = T^2 k$. The space   occupied holds a specific relevancy to the movement applying to the space and in that time determines movement

as much as movement brings about time. Newtonians declare that time is $V^2 = \dfrac{a^3}{C}$ and say no

it is not time...but also yes it is time.

Let me define what I try to say: Time can never stand still. Gravity is movement and gravity institute time as performing as the movement of everything in relation to one point in infinity. But we know that the movement is in the atom while we think of as space that changes the ultimate relevancy. Why it turns out to be space and not the atom moving is a big issue and going down the road of explaining why this inverse of movement actually forms the change of the relevancy I leave to **an Open Letter Announcing Gravity's Recipe.**
Explaining how this really works takes up the best of three hundred pages and I best leave the explaining in detail to remain part of the book entitled **an Open Letter Announcing Gravity's Recipe.** I mention this in order not to seem to contradict myself in another book.

The atom is moving. The straight-line movement as the atom moves around the Earth's axis is V gong on to V, which leaves a V^2. The atom is moving in a straight line (circling around the Earth's axis) as well as rotating at the speed of light around the atom's axis by way of the electron spinning. The electron is circling around the circumference of the atom at a rate of C^2. That puts the atom at the value of C, which is the speed of light, which is a standard unit, not the same everywhere but a unit bringing some equilibrium in relation to singularity applying.

The atom is always moving except when it becomes part of the Black Hole and space becomes abolished as the entire component inside the Black Hole goes singular (1^0). When the atom stands still as it does inside the Black hole, then movement is transferred to that which never moves in relation to the atom and that is space. Oh, so many smart names were given to the Black Hole in as much as the event horizon and the curvature of space-time and... God only knows what name comes next, but underlining all the naming is one fact: No Newtonian has a vague clue as to what is going on inside the Black Hole except that gravity has gone bananas and where gravity would be intellectual enough to go mentally skew remains an open question!

If not for the Coanda effect no release of energy would ever be possible because there is only one form of energy drive in the Universe and that is transforming heat from liquid to gas or from gas to liquid. Also transforming heat from solids to liquid and from liquids to solids is part of the same process.

All flying in whatever form depends on the Coanda principle and yet not one person in physics ever saw that this is part of the gravitational process forming gravity or time.

It is not only the helicopter but all flying that works on the Coanda effect because the Coanda effect is the interaction between liquid / gas and solids that establish density differentiation.

In every atom through the Coanda effect, there is a relevancy applying between cosmic fluid and cosmic solid or heaven and Earth if we go back to Biblical terminology. In every star, it is not the mass as in numbers or weight that sets the margin to gravity but the duplication tempo of material. When a tire of a car spins fast it can have the car run on a layer of water that is an inch or 25 mm thick. The water is not drawn onto the tire, but is compresses solid. If the water was normal ice meaning it was normal water in a solid state, the Ice would have had to be several meters thick to sustain the car. But with the Coanda effect applying, with the velocity of the wheel that high the spinning wheel duplicates it space of the wheel so many times that it freezes the water onto the edge of the tire. The atom within the tire becomes smaller and that causes the gravity because the number in quantifiable acting atoms increases although the reality numbers does not increase. In numbers as well as the space taken up by particles into duplicating of material therefore in large stars it is the reduction that freezes smaller areas and by the material occupying space cools and if the cooling is excessive, then it freezes by cooling the liquid into the star. Gravity is a process of heating or cooling. That is why blowing hot air into a balloon brings on the lifting of material into the air, which is anti-gravity applying. Cooling of air is gravity and anti-gravity is heating air. There is a specific ratio between water and air and the Coanda effect experiments conducted proves this. But furthermore, it proves that the Coanda effect is the way gravity works. It proves that gravity is the Coanda effect.

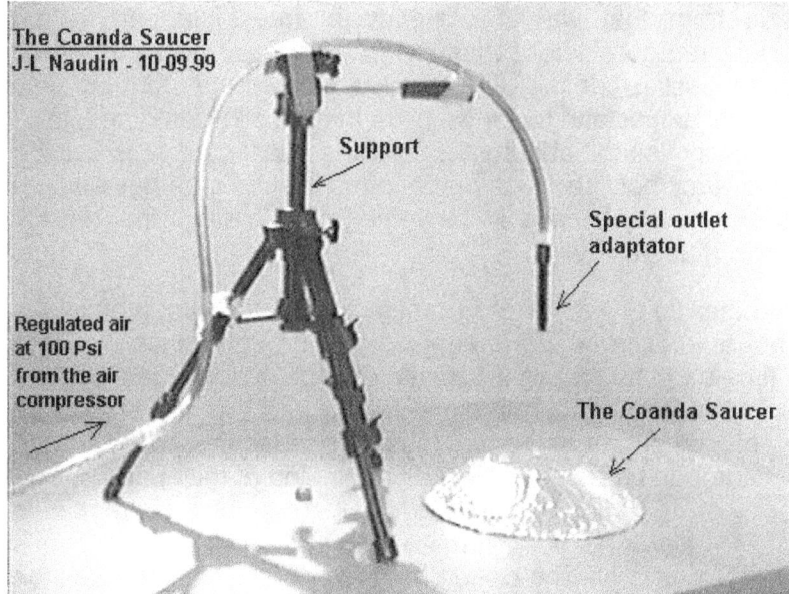

The Coanda Saucer
J-L Naudin - 10-09-99

Support

Special outlet adaptator

Regulated air at 100 Psi from the air compressor

The Coanda Saucer

There is a direct link; no moreover, there is a direct ratio between what fluid is and what has for. There is a direct ratio between what came about as that holding no form and that which gives singularity form by which singularity developed space-time. This proves what gravity is and

Newton's mass misconception has not role to play because by increasing the ratio in air moving that mass effect that the Earth restricts the saucer is the mass of the saucer reclining.

The abyss, which the Bible refers, too is still present. The abyss went nowhere since God created heaven factor (liquid) and the Earth factor (solids). From this came the solid part of creation the (proton$\Pi^2+\Pi^2$) part as well as the formless (the weightless or that with no mass) (neutron$\Pi^2\Pi$). Both the Bible describes so fittingly is still there. Both hold exact locations as the Bible says and came in place just as the Bible says it did so long ago. I am able to show where each component is the Bible refers to and that which the Bible refers to fits in the location it was attended to be in.

Let us presume that this image represents the atom in our side of the Universe.

Since it is a Newtonian tradition to place life in the centre of the Universe they also take it as a fact that everything on earth represents everything as the same as on earth in the Universe.

The Universe grew from a density difference that in today's application was so small we would never be able to measure it and yet it is so massive it parted the universe into two sectors. That which is hot moved way from that which is cold. That which moves with time split from that which evolves with time. The part in the Universe contracts, separated from the part, which expands by losing density. The entire Universe formed two substances and today we have three, solid, liquid and gas. Gravity is not about mass pulling mass but it is putting relevancy between that which forms a solid and that which h forms a liquid and this movement establishes the density applying as gravity.

Another point we have to consider is that the speed of light is not time. The speed of light travels through space by using time and that makes the speed of light as dependent on time in movement as all other factors in the Universe. The movement of an atom thus depends on space moving through time. It has electrons spinning at the speed of light or a fraction of that while it as a unit moves as a collection that forms a star and this altogether forms gravity. The closer the gravity of the star reaches the speed of light the closer will the linear movement get to the electrons circular movement around the star.

When the linear movement gets close to the speed of light or C the closer it will get to the ring formed by the electron. In front the protons will close in on the electron ring and therefore reducing the circle and at the "back" it will move away from the circle ring which will force the electron to reduce the ring on the "back side". That is the reason why the density of big stars increase exponentially while the space it occupies "shrinks" exponentially.

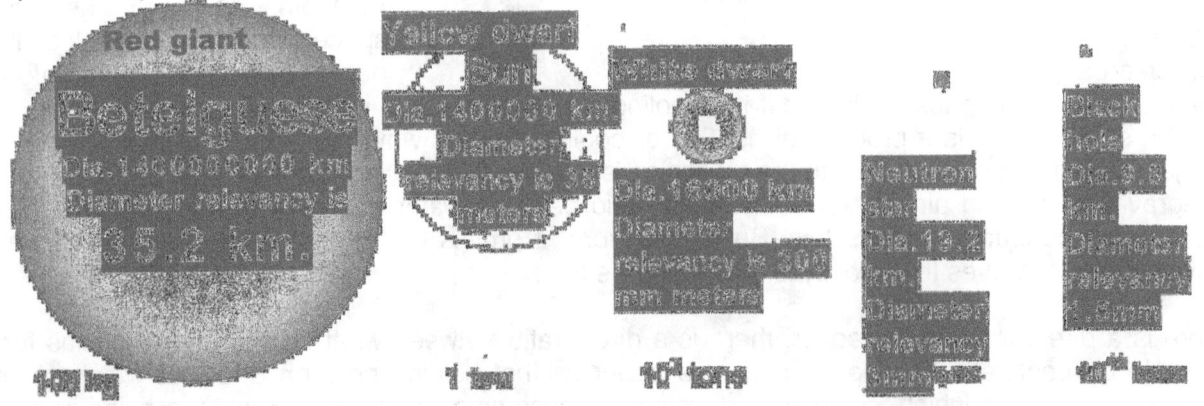

The further a star moves back in time with gravity applying a relevant movement, the more the atomic particles would combine to form the increased gravity while occupying in much reduced space. As much as the Universe is expanding, is material within stars also shrinking in time flow. The Moon is not moving away from the Earth, but the Earth AND the Moon is growing apart. This has to do with movement and relevancy applying on both sides of singularity. As much as infinity is motionless, we have eternity moving perpetually. If eternity did not move, then infinity could not be motionless because it is the eternal movement eternity holds that keep infinity motionless. Although this is true, with infinity never moving, infinity still has a duration and relevancy of twice that of eternity because notwithstanding the fact that infinity can never start, it has to start and end eternity since eternity can never end, and can therefore also never start. Eternity needs infinity to start and eternity needs infinity to stop since eternity can do neither of the two. For that reason alone infinity has to be twice that of which eternity constitutes because only by ending eternity which is what we think of as the past, can eternity start with a new frame which is what we think of as the future. Therefore the only limit placed on eternity in duration, is how quick it is that infinity will end the one frame and start eternity with the next frame. For that reason we have eternity only holding one frame whereas infinity takes the ending of eternity as one frame and replace eternity's ending with the next frame while it draws time into infinity as it draws the Universe into singularity in infinity. That is what Einstein saw but it clearly is not what Einstein saw because the entire Universe we see is time that has gone to the past where time formed space as the past. Once anything is a Part of the Universe, it can never not be a part of the Universe, except if the relevancy changes and in such changing the uniting with singularity places the relevancy back with unifying with singularity once more.

Therefore infinity that never moves, also moves twice as fast as eternity that never stops moving. Time draws flat...not the Universe as Einstein said because what Einstein thought formed the Universe is what forms the time aspect of space-time. In the centre of every proton time draws space flat to form the past from what is the future while space-time is forming the neutron where it is the neutron that holds 3D and one Universe in form.

Time form draws the Universe flat because the Universe starts where time meets space to space-time. This happens where singularity ends with infinity starting that starts with eternity replacing what was with what will be and is therefore as much where infinity ends eternity. When infinity ends eternity it holds what eternity was at the ending and the ending of eternity becomes space where space forms the history of time. In every atom singularity lurks as the final control of what applies to space-time and every atom holds singularity directly in association with singularity where one part of singularity is that which never moves forming the relevance with the second part of singularity, which is that which never stop moving. As infinity starts eternity, it pushes the future eternity becomes into forming space that is the past of time. Taking time from the future through singularity to the past it does by taking that which serves the future to the past, which is material. By taking light to form the future of what the past will be, light becomes the eternal messenger of space or for that matter, of the past. But the past is also in the future of what eternity takes to the past because as much as I see with my eyes what the past (space) was, I form the future for that space to reach. Eventually eternity and infinity, the past and the future, is one unit and only a Newtonian (I guess) would wish to place a difference between the relevancy applying because there is no difference in infinity and eternity, except that which can not be, which is a Universe we have because between the two that can never part (infinity and eternity) forms that which becomes the relevancy of what is the same thing and therefore the Universe can never be. By movement of that which can never stop moving, such relevancy places what is the future in time to form the space in the past of what is that moves the entire Universe into the past.

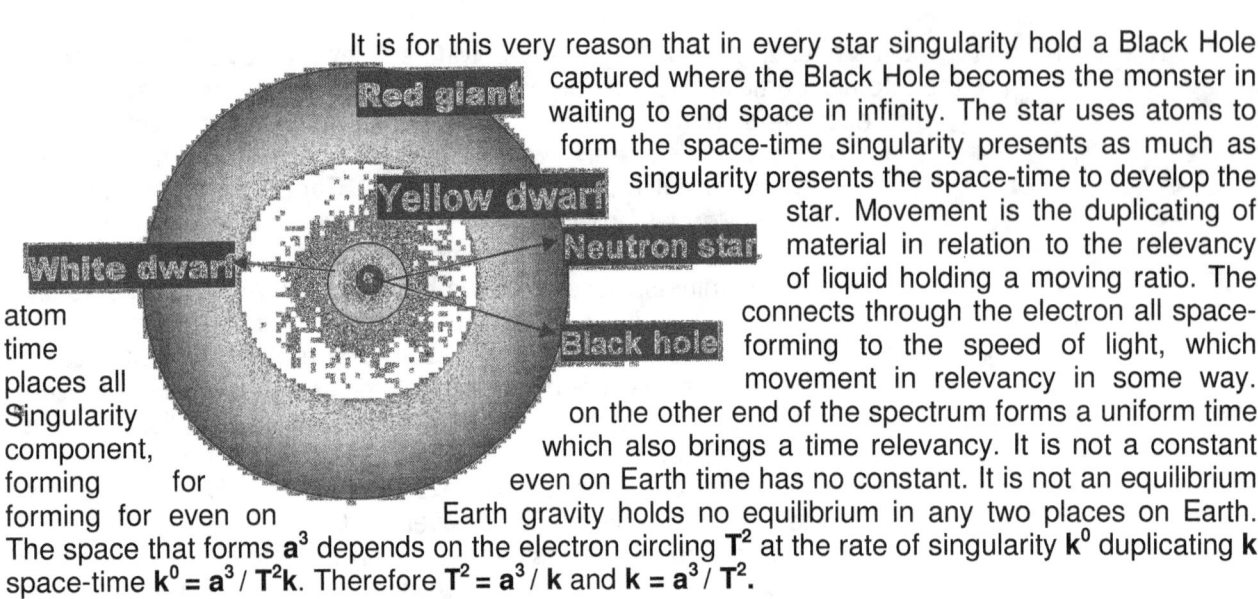

It is for this very reason that in every star singularity hold a Black Hole captured where the Black Hole becomes the monster in waiting to end space in infinity. The star uses atoms to form the space-time singularity presents as much as singularity presents the space-time to develop the star. Movement is the duplicating of material in relation to the relevancy of liquid holding a moving ratio. The connects through the electron all space- forming to the speed of light, which movement in relevancy in some way. on the other end of the spectrum forms a uniform time which also brings a time relevancy. It is not a constant even on Earth time has no constant. It is not an equilibrium Earth gravity holds no equilibrium in any two places on Earth.

atom time places all Singularity component, forming for forming for even on The space that forms a^3 depends on the electron circling T^2 at the rate of singularity k^0 duplicating k space-time $k^0 = a^3 / T^2k$. Therefore $T^2 = a^3 / k$ and $k = a^3 / T^2$.

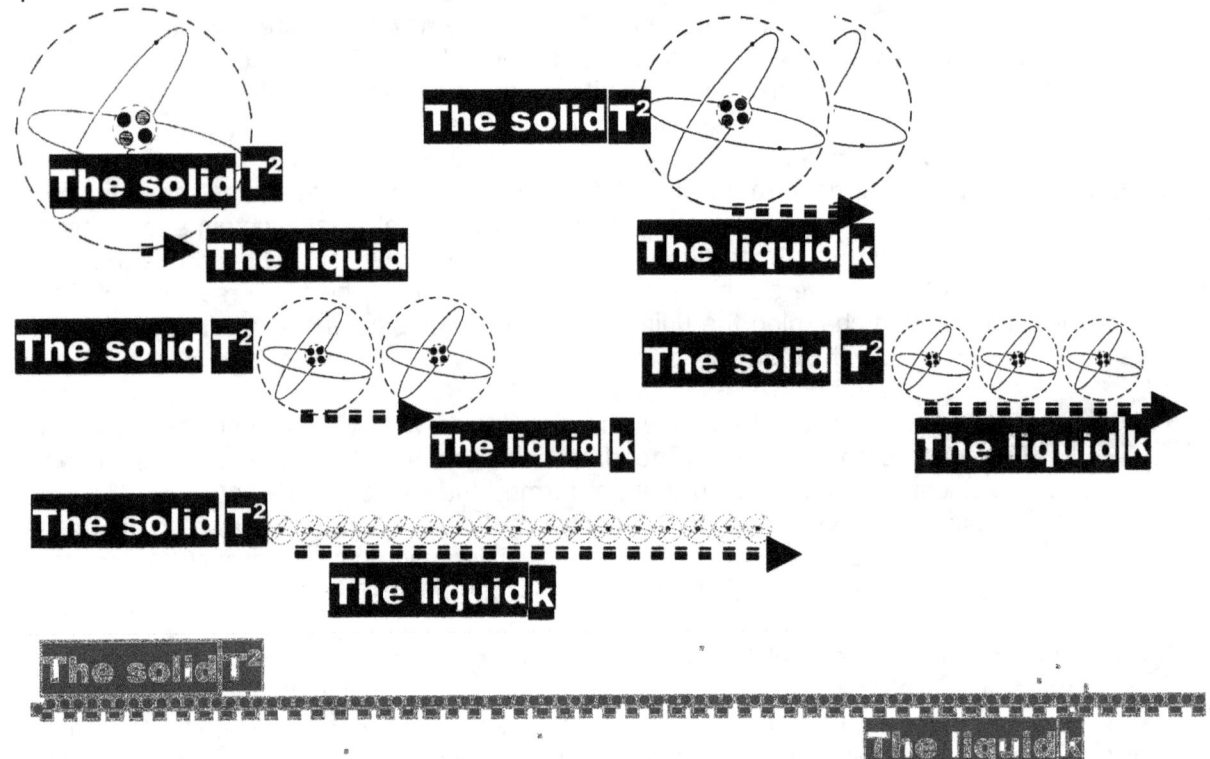

Time in duration is determined by space moving in relation to time.

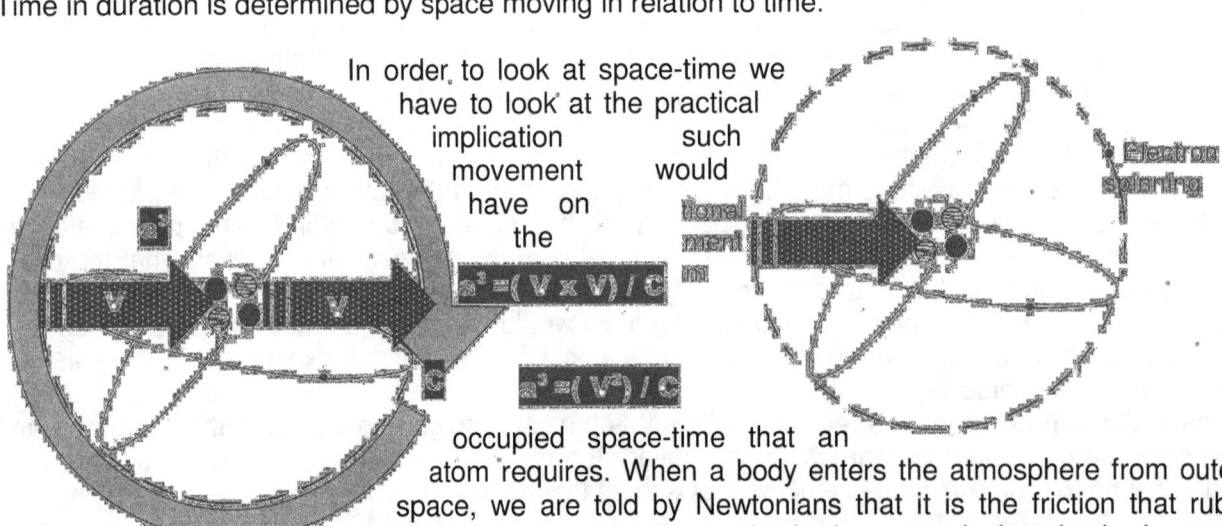

In order to look at space-time we have to look at the practical implication such movement would have on the

$$a^3 = (V \times V) / C$$

$$a^3 = (V^2) / C$$

occupied space-time that an atom requires. When a body enters the atmosphere from outer space, we are told by Newtonians that it is the friction that rubs against the body, which heats the body so much that the body starts burning. We have all seen a body coming in from outer space, albeit a sand particle or whether it is Mir coming in that we see images of on TV or whether it is the space shuttle getting hot on entry, we always witness a blanket of heat engulfing the incoming object. The oh so wise Brainy Bunch Super Educated tell us in confidence brought on by wisdom that those flames are the result of particles within the atmosphere that is supposedly rubbing against the incoming object and this will allow friction to cause such heat. If it was rubbing and such rubbing resulted in friction, I would guarantee those oh, so clever Newtonians that the entire structure would melt into liquid and fry whatever is inside the body to ashes.

Again I wish to return to the formula Newtonians devote to time being $t=\sqrt{1-\left(\dfrac{V^2}{C^2}\right)}$. As I said

before this could never represent time because time has to be in the square since time holds space to relevancy and that is always going to put time in the square by movement as well as in rotation. When we get to the speed of light it is C that forms the linear aspect and I do not wish to go into explaining that at this venture because it is going to take up too much space. By moving the atom it gives the electron a specific duration in time to spin around the atom at a velocity of C. This velocity is fixed at the speed of light and can't under circumstances change at will, although circumstances might change and with that the implication of the speed of light might be different in gravity changing from star to star. However, it takes the electron a specific time to rotate the atom while the atom is moving at a specific rate.

When the atom moves in position it duplicates the entire atom structure. When the atom duplicates

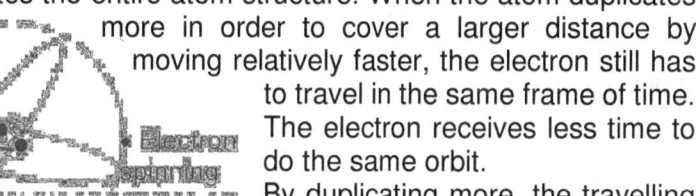

more in order to cover a larger distance by moving relatively faster, the electron still has to travel in the same frame of time. The electron receives less time to do the same orbit.

By duplicating more, the travelling distance of the atom has to

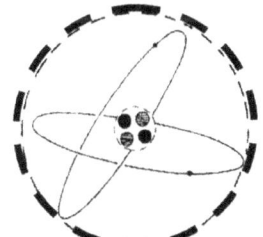

Earth

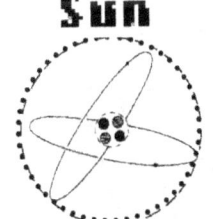

Sun

White Dwarf

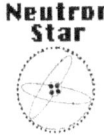

Neutron Star

Black Hole

increase. If the travelling distance increases, this has to effect the time it takes the electron to orbit the atom. Einstein might have been of the opinion that the speed of light is time but if that were true then the speed of

light would not have a quantifiable measured ratio of velocity with the space through which it travels to the value of 1^0. Time stands related to space and only if space was one in terms of time moving through space can time not be effected by space. If the velocity increases by two because the movement of the atom doubles, it would effectively have to reduce the time the electron has which it takes the electron to circle around the atom. If the electron orbit reduces because of time by movement increases and this shortens the time that the electron has to complete one orbit, this will reduce the size of the atom. Since gravity is the movement of space, this is why gravity will reduce or increase the atom's size in relation to the star in which it is. There is a specific predetermined relation between the relevancy of liquid and heat and the interaction that every atom plays in this relationship between solids and liquids makes the layer of the star part of the stars development process.

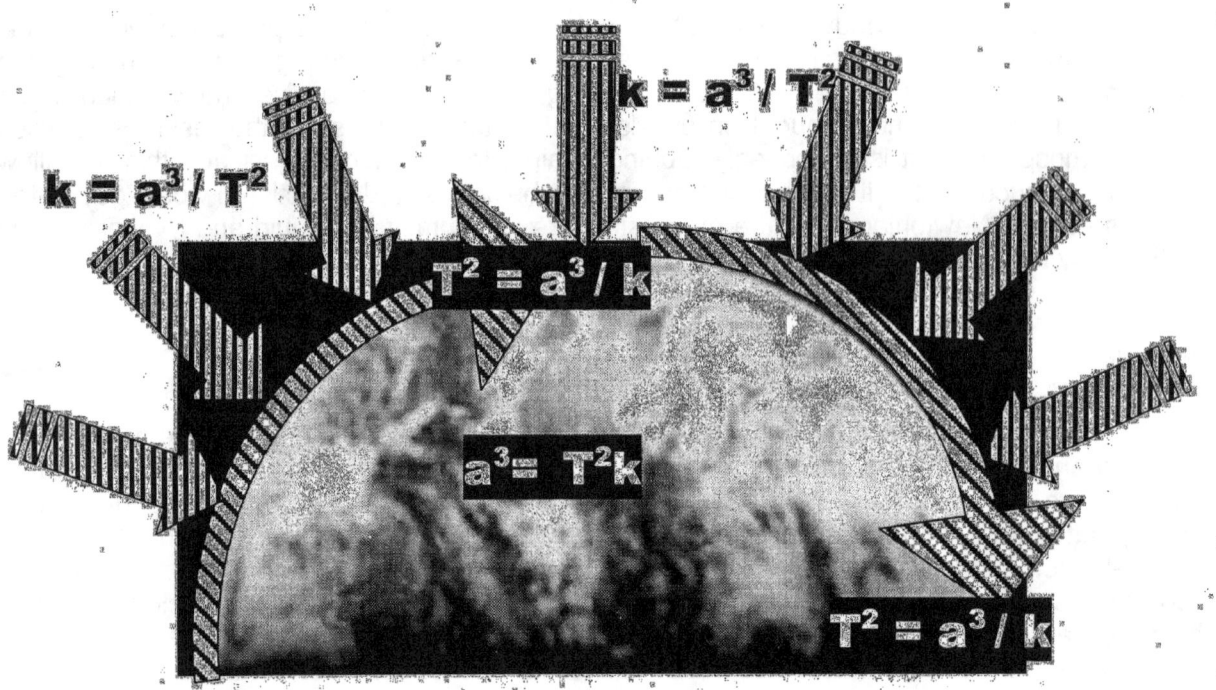

An atom is an atom because it holds singularity in relevance to a centre while the rest is spinning. The atom is singularity and will collapse if it is not spinning. The atom is an atom because the atom is spinning but the atom forms the Universe because the atom is spinning. While the atom is spinning the atom is also displacing space-time in relevance to gravity applying. Gravity is about spin, yes but gravity is also about lateral movement. In all cases the lateral movement is interlinked with the spin and this comes from the combined effort of all singularity 1^1 connected to every atom that links to the governing singularity 1^0.

The more rapid and the more profound the gravity is by measure of spin and thrust, the smaller the atom space would become in sympathy of the larger gravity that allows every atom and thus the entire star to claim less space and eventually to occupy less space as result of more profound gravity applying.

Dia. 1400000000 km Diameter or the comparable relevancy is **35.2 km.**

A star this size can't be a star but it has to be a galactica that is still to reach a gravity density in the very far future where it in that density then will become a galactica.

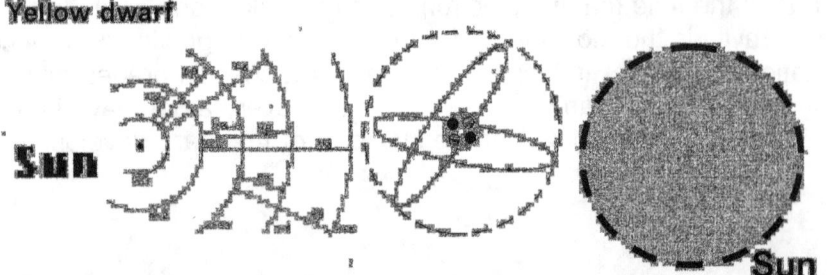

Dia. 1400000 km. Diameter or the comparable relevancy is 38 meters.

The sun is so under developed I don't it can truly be in a class of a real developing star.

White Dwarf

Dia. 16000 km Diameter or the comparable relevancy is 300 mm meters

Neutron Star

 Neutron star

Dia. 19.2 km. Diameter or the comparable relevancy is 3 mm

Black Hole

Black hole

Dia. 9.8 km. Diameter or the comparable relevancy is 1.5 mm

On Earth an atom holds a relevancy of $(\Pi^2+\Pi^2)(\Pi2\Pi)3) = 1836$, but this relevancy changes as the atom holds a position in more developed gravity fields such as Jupiter and bigger planets as well as in bigger starts as Jupiter is.

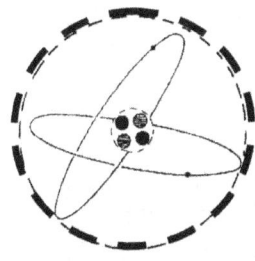

Earth

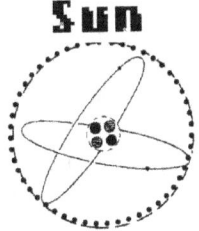

Sun

White Dwarf

Neutron Star

Black Hole

The solid   The liquid

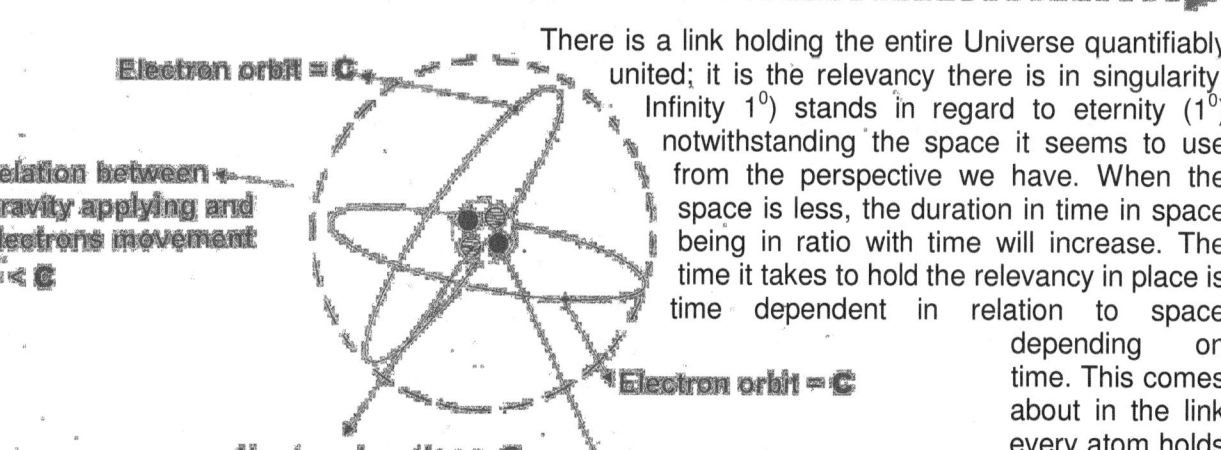

Electron orbit = C

Relation between Gravity applying and Electrons movement =< C

Electron orbit = C

Neutron location ≥ C

Proton location > C

There is a link holding the entire Universe quantifiably united; it is the relevancy there is in singularity. Infinity 1^0) stands in regard to eternity (1^0) notwithstanding the space it seems to use from the perspective we have. When the space is less, the duration in time in space being in ratio with time will increase. The time it takes to hold the relevancy in place is time dependent in relation to space depending on time. This comes about in the link every atom holds with C as well as the measured value in infinity in which the atom constitutes.

If the space occupied in time seems less, then the time duration converting time to space has to increase in order to keep a definite equilibrium connecting the Universe.

This is what I mentioned earlier and this relevancy determines the overall density of singularity throughout the Universe. The star with all the atoms spins at a specific speed condensing the liquid around it. This procedure we named the Coanda effect or atmosphere of the star. This turning takes heat from a gas in outer space to a liquid forming the atmosphere of the star. The spinning can become so intense in developed stars that it renders the star without a liquid phase or becoming dark but sure as hell the star cannot die or run out of fuel. As this turning of the star gets close to the speed of light it has to reduce the spinning circle applying in the atoms ands this makes the inside of the star ever more denser. When the star is spinning above the speed of light it will reduce the star mathematical to a point only $\Pi^0\Pi$ and the r went singular as in $r/r = 1$. The star reduced all space to form one point in singularity and host's no space within the entire star. The star became singularity, which is then called a "black hole".

The space that forms a^3 depends on the electron circling T^2 at the rate of singularity k^0 duplicating k space-time $k^0 = a^3 / T^2 k$. Therefore $T^2 = a^3 / k$ and $k = a^3 / T^2$.

There are forever lateral movement connected to circular movement. Light does not move from there to here by following a path in line as the crow flies. Light move forward in a ratio to moving sideways and that gives as the three dimensional perception we have about the cosmos.

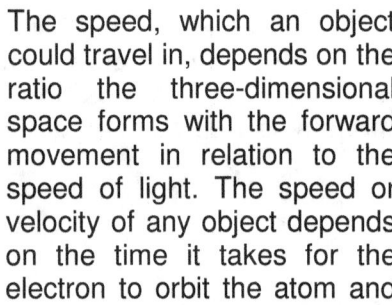

$$\sqrt{V^2 - \left(\frac{a^3}{C}\right)}$$

Occupying a certain the space (a³) over an certain area (k) more during the a certain period (T²) The atom size holds measured size in space in time as gravity moves the atom through space because the atom size is determined by time @ C

The speed, which an object could travel in, depends on the ratio the three-dimensional space forms with the forward movement in relation to the speed of light. The speed or velocity of any object depends on the time it takes for the electron to orbit the atom and that ratio is in line with the time it takes the object to replace the space from one location to the next location. In this way a lot of heat will be dispelled from inside the atom to the outside of the atom and this rests on a fine line that singularity places on the material cooled by spin inside the atom. When there is insufficient spin inside the atom to sustain singularity being balanced while cooled, the heat will expand and the centre governing singularity will become dysfunctional in proportion to the controlling singularity and in this singularity will release all the controlled heat into the uncontrolled space. We humans use this method of thinking in terms of space travel, which is totally alien as a concept to cosmology.

However this control of movement and protecting singularity by balancing the governing singularity in relation to the controlling singularity is moreover part of gravity and that is what forms the balance of material within stars.

Occupying half the space (a³ /2) over an area three times (3k) more during the same period (T²) The atom size has to reduce as the atom speed increases because the atom is limit by time @ C

I am not even going to try and explain because it come from one part of the book I named **Matter's Time In Space: The Thesis**. In that part of the book consisting of seven parts I show that as much as every star is the next star in waiting that is in a process of developing it is reducing space as much as time is expanding space. The star reduces while it holds material the size the star occupy reduces as the material in numbers increases to anther exponential level because as space expanding and therefore space is contracting by the same margin, every layer in every star is a star in place to develop the star.

Now only can I get to the part in the Bible where I wish to be. Again I state it emphatically that I use the Bible as the only reference book since with all theorising of the atheistic Newtonian guess-work culture, it is only the Bible which can be used to show such explicit detail as to how the Universe came about. I would believe the Newtonian story and all the explaining they do about how events unfold that lead to the Big Bang forming, if only I could get some informed accuracy about matter and anti-matter's eating habits.

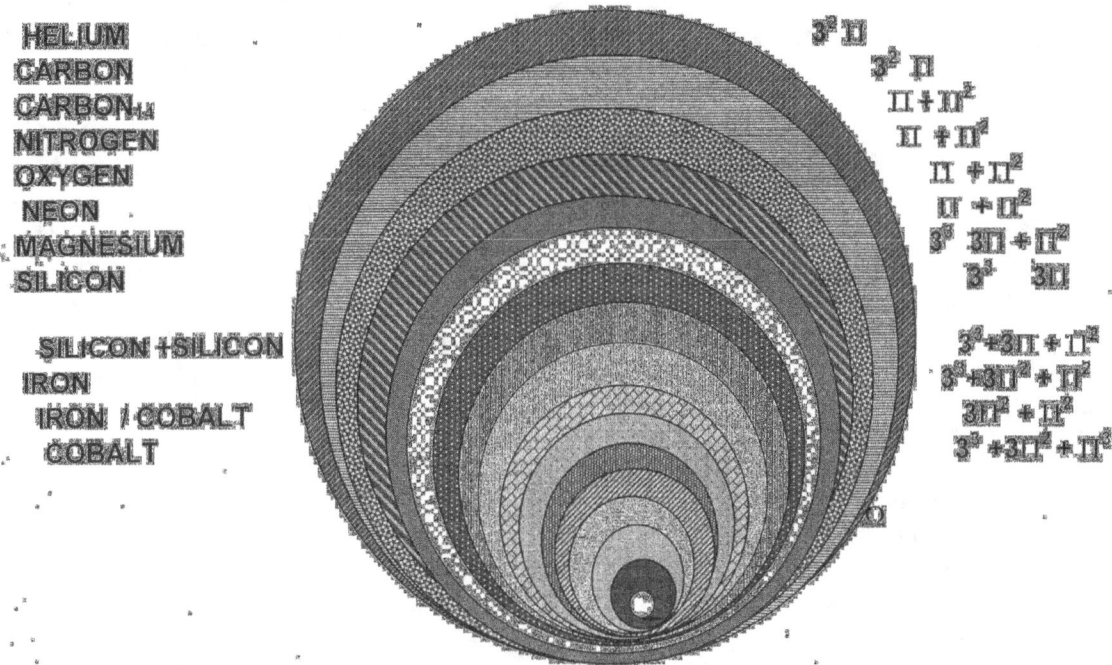

HELIUM
CARBON
CARBON₁₂
NITROGEN
OXYGEN
NEON
MAGNESIUM
SILICON

SILICON + SILICON
IRON
IRON / COBALT
COBALT

The Newtonians devised some theory as to explain the Big Bang by implicating substances they call matter and anti matter. The crux of the story pivots around matter and anti matter sharing a location and the anti-matter being in some conflict about turf warring and then matter lost out and anti matter came in conflict with matter and the anti-matter ate up all the matter and now the three bears...no sorry, this time it wasn't Curly Locks that ate up all the bear's porridge, no...it was anti-matter that ate up all the matter and now the Universe finds itself in a position where it lost all the matter by anti-matter's eating habits going mad. There are two unassailable questions I have to raise at this point. The first is anti matter. What is anti matter? Matter is spinning heat moving at a velocity that exceeds the speed of light and is therefore controlled by movement where such movement has the ability to freeze the heat In the solld state of the atom. Secondly, where did the anti-matter that ate up all the matter go? There is a substantial and unavoidable flaw in this theory: once anything is part of the Universe it can never remove from the Universe, it can go no other place than remain within the Universe and whatever is part of the Universe will remain doing so until infinity once again captures eternity in a lockdown state. If matter ate up anti-matter (or was it anti-matter that did the eating...I get confused) then where did it go? Where is that whichever was eaten now at this present moment because it just can't be in the digestive system of the other part that did the eating? It has to be and it has to be somewhere because even if Newtonians give the Universe edges, we know whatever is part of the Universe has no hiding place other than to be in the Universe notwithstanding ends they may find.

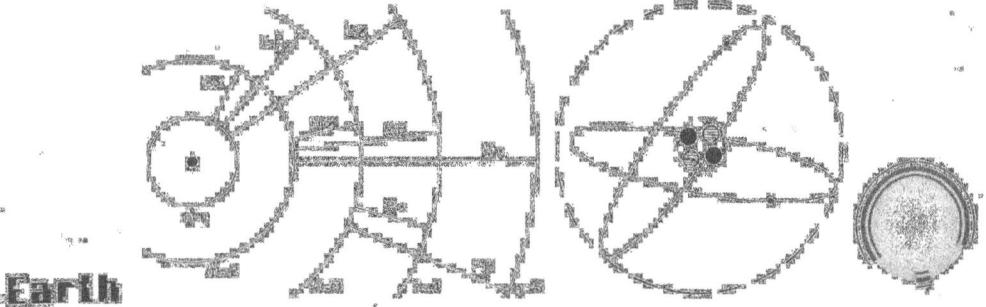

Earth

Gravity is strongest when and where space is least. That is the location we locate singularity because from that location originated about the total Universe. Fortunately Kepler gave us a precise method of locating such a position. When according to the Bible, Creation was coming about at a rate far exceeding the speed of light since it was still the pre Big Bang or the pre light era, the velocity in which material at that stage accelerated, placed a relevancy in time to space we can never rival at present and this is what the Bible says. It is a pity that Newtonians are to simple minded to understand the Bible because if they did have more brains in place to understand the Bible, then science would have been much better understood years ago!

Time moved on and space came about from the imperfect moving of time as well as the perfect moving of time. The roundness and the perfect shape is part of eternity and that in eternity is what we see with life being part of eternity (not the human body which is a cosmic result of the imperfect) but life seeing it self in the position where it is occupying the centre of the Universe by studying light and night and finding that all light draws to life.

The imperfect part is the part we find in infinity

and with life holding part in eternity we cannot see infinity. We experience infinity while we see eternity. We see what is always there because we are unable to see what is changing.

Look at your own fingernail growing or hair growing or even a wound healing you will see the nail, the hair, the wound never the addition that adds by growth. might find what I say at this point not to be physics but it is more physics than anything are part of true physics.

your and but You currently used as cosmic physics

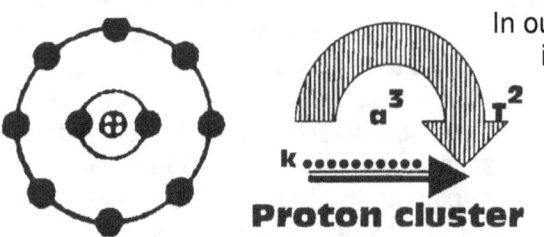

In our ability not to see the imperfect while experiencing the imperfect and at the same time see the perfect in eternity and not being part of the eternity being perfect, it is little wonder we lot are so mixed up in what we see and cannot understand. Eternity lasts visible eternity

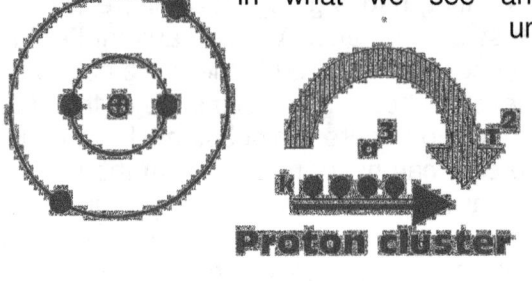

forever and that is why no changing is but we can only experience infinity because is an ongoing repeat of the same without changes. Infinity on the other hand is what interrupts eternity and therefore what we see eternity is what infinity is interrupting.

in

The proof of this is in the top where motion distinguishes infinity in the centre from eternity surrounding the time position of the top designating the motion from the past through the present onto the future. Those are there for all to see and the fact of that being there goes beyond dispute.

That is what Darwin missed with his species being from one ancient origin. Yes, the fact that material duplicated by others using the same that part is true, but the one did not develop into the other.

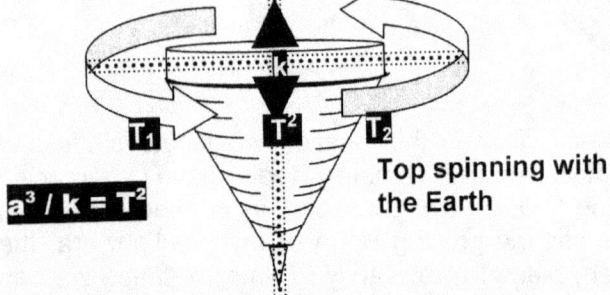

$$a^3 / k = T^2$$

Top spinning with the Earth

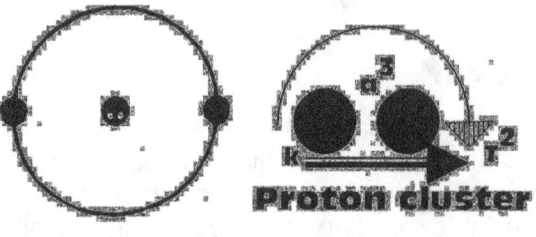

Time did produce changes, but the donkey has

no family ties with the horse. If it had the mule would have been able to multiply and be fruitful and the mule is able to do a lot of things but that it is not able to do. Things go along in eternity while infinity interrupts and then one-day infinity brings a change no one noticed before. The same building blocks are used, but in the building process we use building blocks in repeat every day, but then one day comes where a new corner stone is laid and new specie arrives that has no family ties with the previous lot. In this manner elements came about. But the placing of protons within the atom formed elements whereas the atom was there the first instant heat parted from cold. I go into that part in the **Cosmic Birth...Dismissing Nothing** and since the issue is rather complex in explaining I would prefer to leave that explaining the book mentioned.

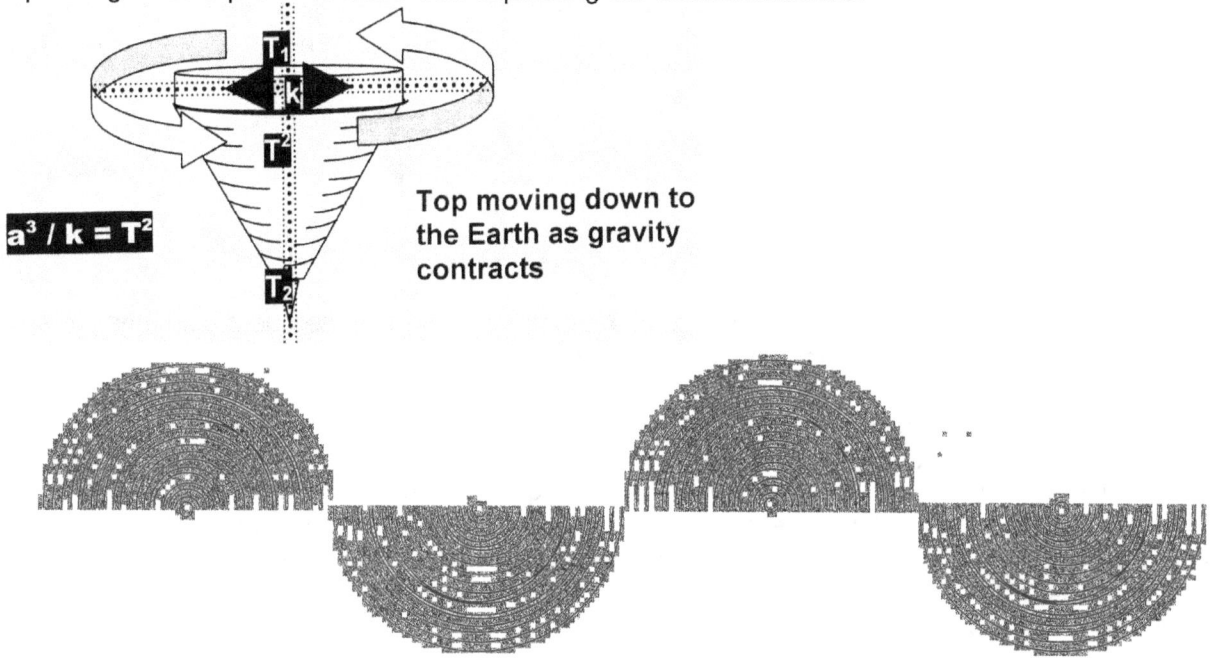

Top moving down to the Earth as gravity contracts

$$a^3 / k = T^2$$

Once again I have to draw the attention to what is out there in the cosmos serving as evidence. The proof we still find in the manner in which Galactica and all other orbiting objects develop. There are $T^2 = a^3 / k$ that is in favour of the promotion of one point holding singularity in the relation and to that there is another and opposing point holding singularity in prominence which is in relation the expanding contributor that holds a relevance of $k^{-1} = T^2 / a^3$ in ratio to the conserver.

The relevancy was there from moment-Alfa when relevancy applied as eternity parted from infinity, which brought relevance from 1^0 to 1^1. Looking at $\Pi^0 \Rightarrow \Pi$ we can see that there were seven in ratio of ten and we can see how the seven produced the gravity of motion relating to the ten in time. We can see when the dominance started creeping to the other side and when $k = a^3 / T^2$ got the better of $k^{-1} = T^2 / a^3$ because at that point it is where the sum total related to the singularity formed the sum total value the atom cluster has as it values the proton $(\Pi^2 + \Pi^2)$ + the neutron $(\Pi^2 + \Pi)$ + the electron 3 = 35.89 × singularity Π =112.75 forming outer space as a result.

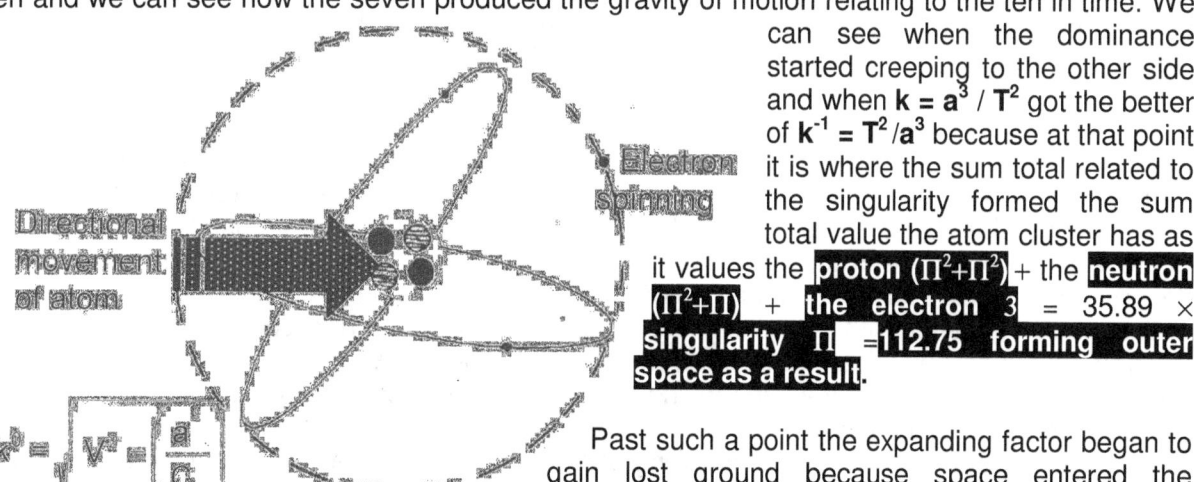

Past such a point the expanding factor began to gain lost ground because space entered the equation and the expanding got seemingly predominant as the containing factor started to store and preserve more than contain. This is where the Bible says God gave the Command: "Let there be light" and the Newtonians placed the Big Bang as the starting point.

The major issue in hand is to recognise that when 1^1 overheated it parted from 1^0 because 1^0 was too cold to harbour 1^1 and by measure of 1^1 being hot, the same measure places cold on 1^0. The one cannot heat without the other establishing a border for the cold. In the Coanda principle the liquid establishes itself onto the solid by gravity as much as the solid allows the extending to lock on. The liquid is not locking onto to an unattached solid. The solid gain as much as the liquid gains but that what the solid gains are not the same in likeness to that which the liquid gains. The solid will as much prove to be colder as what the liquid proves to be hotter. By becoming colder due to movement and the duplication of space, this cooling freezes the liquid onto the solid.

We have to investigate the behaviour and the characteristics of the atom to see why heat would entice movement. An

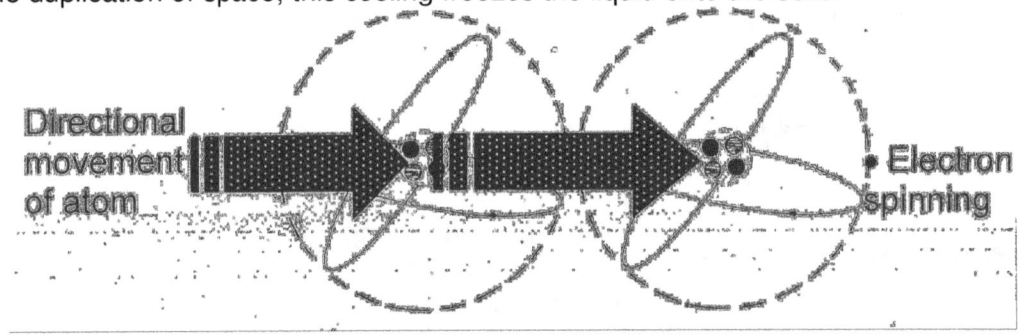

atom is the controlled confinement of heat in space and to keep the confined heat cold it has to spin in relation to unconfined heat holding unoccupied space.

By moving the atom distributes the heat over a lager area that is in contact with liquids. In order the cool down, the atom has to move faster as to spread the heat over a wider area that is in contact with the liquid aspect. Gravity is the movement of space by duplication in relation to time and the reason why gravity is a factor, is to regulate temperatures applying in relevance to what singularity deems as being hot and cold under that gravity applying under those specific conditions. That proves that mass has no implication with any result gravity may have but it is the ferocity of gravitational duplication that makes mass a factor.

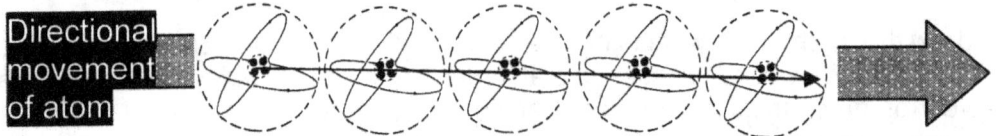

By allowing duplication of space in time as a velocity standard, which is precisely what gravity is, in that time temperature maintenance applies and when there is a shortfall in movement within a star of gravity applying which is movement of material in relation to liquid which is what gravity is, then the distribution of heat becomes less controlled allowing expanding to come into effect and then The Super Nova effect comes in as the star expands due to overheating being a product of the lack of movement. In that mass has no part or mass can't be any part of any factor and that makes Newtonian wisdom actually being Newtonian madness.

When mass becomes a presence as a falling body attaches to the Earth's rim or edge, the body then serves in the capacity that an electron would. The body that attaches to the Earth restrains the normal flow of space-time and the restricting of such flow becomes the measured mass. By clinging onto the edge, the object brings a factor that would put the object having mass in the same criteria as the electron would be in the atom. It is not only the relevance applying from the centre to the electron but it is also that the electron is holding an allegiance with the centre in the precise manner as planets do in relation with the Sun. There is a non-detachable connection holding relevancy from which the attached object cannot escape, whether the object serves in the capacity as an electron having mass or in the capacity as an orbiting attachment. It is because of the dual relevancy that the attached electron stubbornly clings to the elected centre, before being overpowered by the Earth that is providing singularity via the controlling centre. It is all about relevancies attaching to centres and centres holding singularity in relevance where movement by rotation had centres that formed when creation came about. To break those relevancies we have to take time back to before those relevancies because of the gravity gluing the relevancies in a unit. This means that the concept Newtonians play with space travellers that are leaving the solar system and take on journeys to other stars and beyond, such an idea is total fiction. It is all part of the hype and spin with which the Brainy Bunch brainwash the public

Other hype we have heard is about the brilliant mathematical expression of the impossible where the star goes bang and the gravity goes mad and it implodes (how ever that may be achieved) and the whole cosmos goes bananas because a star has gone lost for one eternity and now has died a tragic death. This too is just common hogwash where behind Newtonians hide their poor understanding of that which they have no grasp of. If the star implodes as they wish to suggest, then is it possible that how something being as solid as an atom does is, implode. What Newtonians would suggest is that they see an atom as being completely hollow. In other words, the atom to them is filled as a balloon is filled to their opinion with more nothing coming from outer space! If a star implodes, it can only be that the atoms in the star implodes, and if it is the atoms that implode, what does the imploded atom then become because surely it still has to be material. Trying to explain that with an intellectual dignified argument is so complicated that there is little available to explain. Their explaining is something about gravity going nuts but I would only take that under serous advice with one precondition; they have to show how does gravity go loony and form a Black Hole when all the mass that is supposedly producing the gravity that is supposedly forming the Black hole and with all the mass pulling all the gravity just went bananas or nuts. Then the Academics call me incoherent…

In a Neutron star, the space reserved for atoms becomes so small, there is no space reserved for neutrons. With no space allowed for the neutron because the gravity exceeded the neutron moving abilities, the neutron gets ejected from the space reserved for the atom and is then the neutron is released into outer space. In that case gravity becomes so strong that outer space becomes as dense or as much water as the Bible says it was during creation. The water or density of heat becomes so prominent, it replaces the neutron as a unit and that is the winds the Bible refers too. The Bible is such a wonderful scientific correct book it makes all Newtonian religiosity and crooked dogma seem ridiculous. In the stars becoming dark, those stars don't die just because only that with life or that being there as a result of life intervening can end or die. The speed or displacement of gravity begins to exceed the speed of light and that makes the star start to reduce outer space to a period going back to the pre-Big Bang era where light became predominant. All stars are in relation to the movement of heat where heat is holding occupied space that is displacing unoccupied space by movement between the two factors.

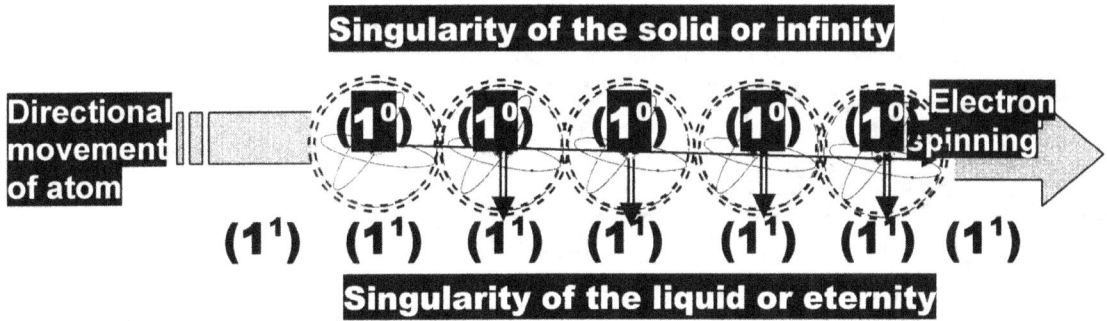

The answer to my explaining is so simple it is laughable. A star is about fusing atoms together with the liquid of space and even on that remark I am reserving my view because the implication can never be that simplistic. Then what happens when all the atoms fused together in the space of the singularity (1^0) that is representing the culmination of all atomic points that maintained singularity as atoms of various descriptions within the star that eventually forms one point in singularity covering the entire star as if the star is one atom. The thought of a Black Hole is simply to gauge what gravity will such an atom produce that is consisting one point that is not even present or part of the Universe we have and the space less ness located in the centre where all the atoms that the star had through out its entire existence accumulated the gravity they generated into one point holding singularity? If we consider what the Hawkins relevance of the Black Hole is formed in the centre of huge galactica by all the material moving as one unit in the entire galactica. The cooling movement which is formed in the centre of giant galactica is generated by the spinning effort of every atom within that galactica and the gravity of every atom (1^1) holding singularity, becomes vested in the centre of that galactica holding the governing singularity (1^0). By having every atom singularity (1^1)

attaching by gravity to the governing centre singularity (1^0) by the movement of all the material within the galactica spinning around such a centre, such a Black hole as what Hawkins discovered becomes viable. If we can imagine the quantity of all the material (not mass) that is required to generate such a Black Hole we see in the centre of the Galactica, then only we can start to fathom what the movement must be of the singularity controlling a Black Hole.

It is the spinning movement of all the material there is in one giant galactica that forms the ability to establish the singularity that has the ability to replicate the scenario we find within a Black Hole. But Please God, for the love of my sanity, let there not be one simple minded Newtonian that think he or she has the mathematical skills even to try and calculate or begin to calculate what mass would be required because as I once told Professor Luth Strauss, so I still maintain my idea that mathematics has no place in cosmology at all. The figures we talk about used in cosmology are just too numerous for any human concept to form.

However, we can start to imagine what atomic volumetric space a Black star once held in relevancy to the rest of the cosmos, which can produce the gravity required keeping a Black Hole frozen. From imagining that concept we can start to form an idea of what atomic volumetric quantity such a Black hole once held. If only a handful or maybe just one atom remains which is the final result in fusion of what ever contraction finalized all the possible fusion between every available atom of all the atoms in the star, and one atom ends up with all the gravity the star had which was initially delivered by all the atoms in the star that produces gravity and such gravity is now within the space that one atom holds, then the gravity will be devastating. In other words, put all the particles found in a Hawkins galactica that produces a Hawkins discovered Black Hole into one centre not even with space, then one could start to think of the gravity there has to be available that is causing movement of that which will eternally move in relation to that which will eternally never move. If gravity is about fusing eternity onto infinity then the fusion has to end somewhere because by fusion the star is heading in a direction that will eventually combine all the atoms in the star.

Curved Space-time

The figure below next to the spiral represents a two-dimensional slice through three-dimensional space showing the curvature of space produced by a spherical object, which is perhaps the Sun. Einstein's view is that the planets follow the curvature of space around the Sun (and produce a tiny amount of curvature themselves). That again is crossing monkeys with watermelons to bread marble. They picture a two dimensional (flat bedded) mat like surface that light uses to travel by. According to the Educated Wise, gravity will pull the Universe as flat as the topside of a mat. However, some of the content within the Universe escape the fait the Universe has because some things in the Universe does not draw flat although the entire Universe supposedly draws flat. By the way, in the event you might have missed it, according to my theory the curvature of space-time is Π, which is the result of the Titius Bode law, which I would explain later on, which is how Π by the implication of 7, bring space-time about in forming the curvature of space-time in a curve, where such a curve comes from Π forming gravity in the measure of Π duplicating to form Π^2. Now let's go on to investigate what Newtonians consider the curvature of space-time is!

I have pointed out that time can never present space less ness as using t would suggest, but always will be in a square as T^2. However, it is the motion of the three-dimensional space within the context of three -dimensional space that holds the square value and not the three dimensional space that is going square to please the needs of the Newtonian lack of insight. To their way of thinking going square is effectively losing one dimension while in cosmology going square is repositioning the three dimensions from one location to another location. Because the square is the time or motion of space it serves as a square only in terms of carrying the square which gravity also has being Π^2 as

$$g = \frac{GM_\oplus}{R_\oplus^2} = 9.8 m/s^2$$

it moves the three –dimensional object in a straight line by the square of time. The square applies to the moving of space Π^3 in the third dimension but it is very unrealistic to place space Π^3 by measure of losing one dimension. To do that is completely unrealistic or in other words completely Newtonian. Moreover, it is confusing and more so to the Academic preaching that as a fact in physics. It confuses the rest of us that has to believe what the rest of us with some form of intellect know can't apply to reality even though the person that says so is a Super-Educated Newtonian. If such proposal is made, one require a reason why five dimensions went away because gravity is not space therefore gravity is that which moves by the motion of space. One cannot say gravity pulled space flat. How flat is flat and what is flat? Singularity is flat but singularity has no dimensions since the only measured value one may connect to singularity is 1^0. That I can prove, because I can show where singularity is just because I can prove that singularity is not part of the Universe we fill.

One cannot depict gravity as being selective in the dimension it prefers to be in just to prove the correctness behind the Newtonian thinker's thoughts. If the ball in the hole is three-dimensional, as it obviously is according to the sketch they use, then the hole is three-dimensional also making every aspect in the picture that is suppose to portraying the light in the picture as three-dimensional. Or otherwise if the grid is flat, then everything is flat and that flatness will exclude all other three dimensional portraying as being flat and that leaves a single dimensional sketch no one can draw because no one can draw anything to portray anything being flat. The sketch will show nothing and it will have to be without having a hole to fit a ball. If gravity pulls the Universe flat then gravity is quite unable to curve the Universe by the same margin. Gravity, by the power Newtonians vests in it, can draw whatever flat as well as at the same time curve whatever, but that there can be such a possibility in the real Universe is very unlikely. With gravity having the power to draw the Universe flat, it has by the measure the ability to curve that which it draws flat. All these arguments show concepts that are as Newtonian in substance as thinking about mass having magical powers to pull and pull an entire Universe flat! Again we arrive at a Newtonian fork tongue. This can only happen when one put mathematicians in charge of theory. Mathematics has no place in the cosmos because as soon as one come to realise what reality is in the cosmos, one comes to the reality that the human mind holds no true volume to understand what God established. It is still the same double standards that the Brainy Bunch employs senselessly to give the Newtonian view validity. Gravity is having the Universe go flat or so did Einstein portray. Science removed a flat Earth to replace the flat Earth with a larger flat Universe. What progress has befallen our gravity stricken Universe? Gravity also produces gravitational lenses...but this idea they never mention in terms of us having a flat Universe or not. Are the gravitational lenses applying effectively while there is a flat Universe or whether it is or before it is going flat because there is a choice to be made? Having lenses by gravity is either in a three dimensional Universe or no lenses can be in place since having the Universe going flat will inevitably remove the option of having a lens by gravity... but it cannot have a flat single sided lens.

The part that is ether three- dimensional or is not there at all that is space $\mathbf{a}^3$. Where space becomes flat the space is moving with time it goes square in the movement and not in form and such space then has motion measured as $\mathbf{T}^2$ which is in relation to an ever-changing position and location endorsed by the square of time. I would think that my argument is much more realistic and much more plausible in the real world of cosmology.

Looking what the brainy Bunch has to offer we find a mix of dimensions that only the truly besotted mental person could applaud. There is this carpet that represents the flat two dimensions in which there is a centre showing only a top side and in the centre with a only a top side there is circles that should represent a hole that is running down... to where...after all it is supposed to be a flat single dimensional waved carpet with only a top side. In the centre that is supposedly flat with only a top-side there appears a hole in the centre...and yet there is a dimension because there is a hole. A hole has a top and a bottom side. That makes a hole a hole...having a top end and a bottom end. In this hole is a ball and by being a ball it has to be three dimensional, if it is to be a ball of whatever description. Just by having in a hole in the centre with a ball, is portraying a three dimensional picture. The hole and moreover sporting a ball in the hole makes complete rubbish of all arguments that any simple minded Newtonian wishes to advocate about singularity having a presence as a single sided dimension. The presence of this hole returns the whole picture back to being a three-dimensional surface because the hole produces a third dimension to the point where gravity is supposedly pulling everything into a flat two dimensional state with one single side that has wavy blocks drawn on the one side of the one sided carpet. The waves introduce three dimensions to the single sided carpet! It all could make sense to the weak minded except that the hole is adding one dimension to the having a flat one sided carpet made of a flat square. How did gravity produce a flat two dimensions while equally at the most intense point, gravity produces a third dimension in the location where gravity is supposedly lurking to give us our flat Universe.

Then comes injury to the insult already inflicted on our weak minds; they place a three dimensional ball into a flat Universe with waves forming more dimensions than they whish to mention in order to present a flat surface as to show us a three-dimensional hole in the centre of this flat Universe they present. If that is not done by the magic of Newtonian gravity, then there can be no other acceptable explanation about the whole affair. They have no valid reason for the Universe to go flat except blame it on gravity. However, gravity is the strongest where space is the least. That would then allow gravity to curve the space-time surrounding the sphere from the centre, evenly, in all directions equally while not forgetting the round surface is a flat square.

If I was asked I would say the circle would put in place the seven degrees the circle has to be to be a circle of any description that the sphere in form insists to have by a multitude thereof. It would use Π by many dimension placing Π in relation to the centre where gravity is the strongest. With this in mind there is no need for all the hype about the curvature of space-time, as we may just refer to the sphere being in place by using Π to present the circular circle. If the ball in the hole is three-dimensional the hole is three-dimensional making the picture with the light portraying the picture three-dimensional. Or otherwise everything is flat without having a hole to fit a ball. If gravity pulls the Universe flat, then gravity is quite unable to curve the Universe by the same margin as it is able to draw the Universe flat. Again we arrive at a Newtonian fork tongue. This can only happen when one put mathematicians in charge of theory. It is still the same double standards to give the Newtonian view validity. Gravity is having the Universe go flat or so did Einstein portrait our gravity stricken Universe. Gravity also produces gravitational lenses...but they never mentioning whether it is while or before it is going flat because there is a choice to be made. It is either going flat or it is having a lenses but it cannot have a flat lens.

In a picture of outer space Newtonians place a gravitational constant. That means there is a form of gravity keeping order to the vastness of outer space. In such vastness light is known to travel in a straight line between points. Then with having a gravitational constant in place while having light moving straight in line, gravity must be pulling objects to a centre where mass is concentrated. In this vast region without any end they claim the presence of a specific gravity and that gravity has a specific value with a specific name being the gravitational constant, which according to the name the gravity never changes. If it is gravity being out there in the blackness, then being gravity by implication alone it is pulling to a centre. The biggest Newtonian question then is: Where is that centre and what produces the centre whereto the pulling is going. Where is the centre of the region thought of as outer space because with gravity as a constant in place, the gravity should by pulling adhere to the presence of a very specific centre? If outer space had gravity there has to be a centre to which the gravity is pulling or moving objects. The only centre that I can see there is, is the centre that material form. It is the centre of the Sun, or it is the centre of a galactica that gravity centralise material. That is material and can hardly qualify as outer space because such space is openly controlled by material moving about in such a space. Other than that there is complete lack of gravitational evidence in the region the Super-Educated calls outer space.

In the use of the formula there is a deliberate admitting to space-time being used. In the formula

$$R_{grav} = 2GM/c^2$$

they put 9.8, which are the square of pi to a unit of meters (space) divided by seconds (time) to the square. However that is applicable to the Sun

$$E = mc^2$$

as much as all other objects. Then the formula is taken to apply to Black Holes where the concept science then put foreword changes totally. It becomes they are using the formula in the Black Hole in doing so it suddenly calls for the square of the speed of light to come into the formula as a factor.

The square of light comes in to support the diameter of the star. How can they get the diameter of the star married off to the speed of light? The one concept is a length a distance between points of compacted material and the other is the flow of liquid space through time.

The one concept is measuring distance and putting that into the equation as an additional factor, which by any standard, is inexplicably unrelated to the distance of the diameter, which is what they normally would use, but in the case of the Black Hole it is employed not as distance but comes to support the equation by introducing the square of the speed of light. Motion supports the diameter of material by the square. Light cannot go into a square because the speed of light is the epitome of motion in space through space. There is no possibility that any moving object cab exceed the speed of light. There cannot be a speed of 1.005 times the speed of light or be twice times the speed of light. The speed of light is the ultimate and the optimal velocity that the smallest particle in the Universe, the photon could travel. It is where velocity ends and no increase can come into furthering the calculation or bring any haste into the equation. Yet, the Brainy Bunch has this unexplained novelty of squaring off the speed of light where we traditionally would locate a diameter or radius that is normally used to the measure of the third dimension as r^3. I do not say that doing that is incorrect, but the motive and the explaining why its use could be correct, totally lacks in science and even their realising the incorrectness passed unannounced by all of the Brainy Bunch. I also do not say going square by motion in conjunction of the diameter is incorrect, but they do it by using all the wrong reasons which they never mention. If that is true and they can use C^2 in place of a diameter to perform as gravity, then they have to admit to my statement that gravity is no force but pure motion. Gravity is the motion of space through space during time controlling movement tempo. I say gravity is the difference there is in motion between particles having unequal relevancies to a specific centre controlled by singularity and the gravity is the extending thereof. Still gravity is the motion of space in time through time in relation to singularity applying.

There is a sketch presenting the directional movement Newtonians award this ball. The sketch represents a ball thrown through space and through time. The ball is representing the movement of time. See how equal space and time is? See how space and time form a mathematical vector giving each an equal share in the proceedings. See how easy it is to draw a line that represents space and that represents time in precise proportions. Does that not make the life of all mathematicians much easier? Can any one argue that space and time is represented even handed by equal proportions just to make life much simpler to calculate the Universe? All the while Newtonians admit they have no foggy glue as to what time is and on space they place a value of nothing. They say outer space comprises of nothing and time is everything they know nothing of. Well, seen it

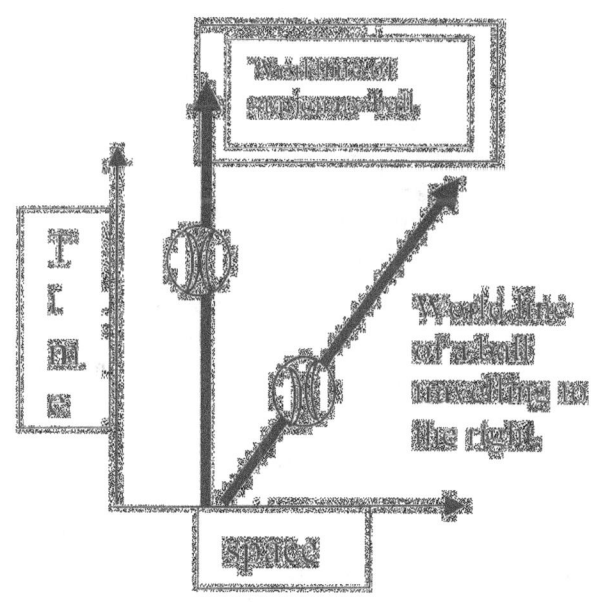

that light I suppose they have a point because they know as much nothing of time as the nothing they have nothing filling the vastness in outer space. This (I suppose) must be where gravity draws space flat and time flat and space multiplying by time gives some square from which a vector then forms. How flat can time get and what happens when space goes flat, but more still, how does the flat then multiply the flat to form a flat square. There is no mention of dimensions although we all realise dimensions is the issue concerning the Universe.

Please note how the Newtonians suggest that time and space is both the same value and is equal in partners being one and the same of value. How they can bring space in as a flat whatever and being as flat in the same instance as time is although in division of time (space dividing or per time or space-time) notwithstanding Einstein declarations go beyond any normal logic. How they can put time and space as equal partners and still find that space can move through time also is beyond what silly old me can explain. Where would the dimensional aspect come into play when both time as well as space goes flat? Fortunately for every one concerned in the books I handed in to various Universities through the years, they told me that in accordance with my arguments concerning their views, they see me as that I am the incoherent one.

In the real Universe time always takes control of space by motion of space. Time positions the location of space in the relation time altering space where time places all in relation of one point serving singularity to everything forming the rest of the Universal space. If science wishes to put space in a motionless stance science should produce evidence where space is motionless or bring proof where time can find the ability to stand still.

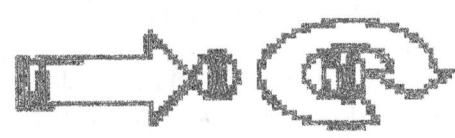

What our delightful Newtonians actually tried to explain is the manner that balls divert from the straight trajectory it should follow when kicked, hit or thrown in a certain manner. I have studied their view on this matter as much as I could and I am please to say that they got some part correct! The only part that Newtonians got correct about this phenomenon is to link the concept to the Coanda effect and even their ability in getting that part correct astonished me. The other night on a news channel I had the privilege to sit and listen some expert explains why some Brazilian soccer player was worth to be paid so many millions of pounds in English money for his soccer services to some English club. This person had the ability (presumably more than most I gather from the discussion) to curve the direction of a ball when this soccer player kicks the ball. Under other conditions the doings of a soccer player would normally not interest me as I personally am a devoted Rugby enthusiast. The fact that the man is worth millions just because of his ability to kick a ball while so many others without such ability has to starve of hunger disgusts me, so frankly it is not part of my nature to listen to glut. So why would I take time to listen to this conversation? It is because it was so funny! This Newtonian religiosity priest with the mentality of an intellectual (again I suppose that was the reason why his opinion was valued at the night in question) had this discussion as to why this person had the ability to kick a ball in one direction and let the ball swerve following a curve that fooled the gaol keepers and with that curving of the ball trajectory the person was able to score more goals than many others.

If anyone ever removes the pulling power of mass from physics, or remove the pushing power of pressure from physics, then to the intellectual Newtonian physics will seize to be, for to the Newtonian intellectual physics is about understanding that mass either pulls or if that is not the case then its pressure that pushes. If there is no room for any other form of explaining physics, then the Newtonian is lost in a Universe that can't be explained. The Newtonian intellectual went on about this Brazilian having the ability to put pressure on the one side of the ball by kicking on the one side of the ball and which left the other side without pressure (that last part I have to presume because if not leaving the other side without pressure then what good would it do to put pressure on both sides of the ball) and with the ball having pressure only on one side, the ball turned in direction. I almost felt sorry for this fool, but then realised he was far better educated than what I will ever be and that almost made my heart bleed for the man's meagre mental powers. It shows one can't be taught to be clear in perspective or how to be of a sane mind and capable of thought, because that ability seems to come with birth.

Please allow me the opportunity to explain for the very first time what happens to the ball trajectory changing in the "curve ball" phenomenon. This I am able to do not because I am clever, or because I am well educated, but just the opposite is true. It is because I am so stupid and because I am so poorly educated that I had to figure out what was going on around me by debating issues with myself and find conclusive answers in order to hide my gig man-mentality. When the ball is not moving, the ball has mass because the ball has no centre line to sustain a time line. There is a single point in the centre of the ball that serves as extending the Earth singularity to such a point. The Earth gives the ball mass because there is a connection between the Earth holding the governing singularity and the ball extending the end limit of the Earth singularity. This allows the ball to have mass. The ball can have mass because the ball has no individual movement except the movement it inherits from the Earth moving. In cosmic terms the ball does not exist but only becomes a cosmic factor when the singularity in the centre of the ball extends the singularity in the centre of the Earth.

Then life gives the ball motion. This is not a cosmic event but is secluded from all cosmic action. If not for the ability of life to be able to manipulate space-time, this proceeding would not be possible. Therefore, only utter Newtonian stupidity will try to find cosmic implication into the action of the ball. However it is the principles applying that is cosmic and in that we can learn everything about gravity except learn anything about Newton's mass because that part obviously don't apply and that is because it is non-existing in the cosmos.

The manner that the ball uses, yes that is as cosmic as the gravity being natural in the cosmos. But life mimics what the cosmos apply because life can't create anything, life can only put in place that which God provides to be used in the cosmos. The scenario applying to the "curve ball" is identically the same as what keeps the spinning top erect. It is singularity establishing a relevancy between infinity standing still and eternity spinning. Life gives the top gravity by which the top spins. Therefore by giving the ball an axis with spin applying, the ball then holds a time line running through the centre where singularity forms and this spinning motion provides the ball a time line in

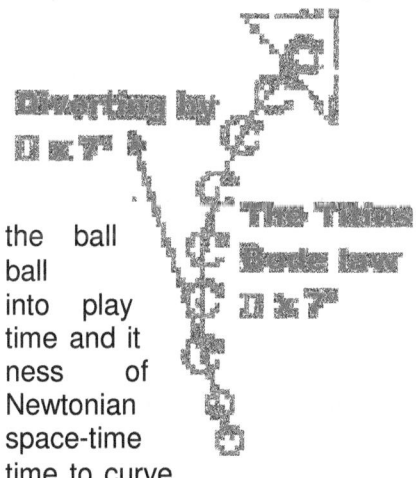

the ball
ball
into play
time and it
ness of
Newtonian
space-time
time to curve

which the ball can operate as an individual cosmic object. This time line connects to the time lines running towards the centre of the Earth were the Earth has eternity connecting to infinity to provide time. However, this route the ball then has to follow will be a route that the ball must follow to match the Earth redirecting by spinning around the Earth's singularity and this provides the curvature of space-time which is the curvature of space-time that has to match. By using the curvature of space time the spinning implicates Π by the movement of 7 (a circle), which then brings the Titius Bode law on movement. This is the curvature of space-is in ordinary use as it applies in daily life. It is not in the Special relativity that is only found in Black Holes and other exclusive clubs. Every person playing the ball can employ the curvature of by using gravity and gravity is the Coanda effect. The ball uses space by the value of Π. Time is the duplication of space in

relation to one point that holds Π^0. The ball now connects to time in infinity while moving and that places the value of Π^0 connecting to Π in relation to Π^2, which diverts from time in eternity where time in eternity is moving at 3. This might sound a little over the top, but it truly is as simple as having the top that spins.

However the time that Newton got frozen in a single dimension by a drawing on a flat surface paper has very little credence. Where the Newtonians get the idea of time being a single dimension frozen in one instant must come from pictures Newtonians look at. That is the only way and image in time from one moment can freeze. In this action of presenting time as if time can stand still, Newtonians establishing of a single t that they say is representative of time, putting time as a factor of one. This can only be possible when freezing an image onto paper by taking a photograph of an incident that happened. However the picture relies on intelligence to place a time to the picture. If I show the picture to a cow taken of her calf, the cow would try to eat the picture instead of getting all teary about little heifer looking quite. She would not recognise time in the picture because the picture is not presenting time as a cosmic reality inside the picture, but time is on the outside forming the paper that holds the picture that was used to freeze time by supposedly framing time. The effective reality is that the picture might be used in remembering the viewer of an event that happened, but in that case the viewer depends on intellect to see time forming a part of the picture. Recognising time as a part of the picture is in the ability of human association with the event but that frame the picture represents of time in the instant cannot be the event that is part of the present any longer. If we reduce the moment to a snapshot, the picture we focus on can only be what we see the image to be and it is an image of time frozen in a single dimension that reminds us of an event that is no longer part of reality. The image remains in ink on paper, which froze not time but an image of time. It holds an image that was part of time for a very short instant and then forms that which was how the event occurred during the time from where the camera shutter opened T_1 to where the camera shutter closed T_2 and the time frame T^2 was then during the period that the camera shutter opened. As soon as the shutter shuts, time moved on and another T^2 formed leaving the image taken as time serving an image never to repeat again.

Afterwards the image we see is not time. It is an event that occurred during the flow of time at a specific stage in the flow of time. It did not freeze time but took an image of time distorted forming space as the picture represented **t** at that moment of T^2. When looking at the picture, the looking at the picture also became an event. That event happened during a specific T^2 that went from where one is taking the first look to where one is looking away from the paper. The event lasted while carrying the first dimensional image of an event gone by. That is at that stage a representation of **t** in another milieu of $a^3 = T^2 k$.

It is not time standing still. If you show the picture to a horse, the horse will try to eat the picture because the horse will be unable to recognise the image in ink on paper. The last thing the horse will experience is a freezing of the moment. The **t** in the single is when mathematically presented as only **t** indicating a mathematical single flat dimensional view of time being part of paper by means of ink. The image we recollect is in our minds and not in time. It is an image that is then correctly applied because it represents a reminder of a four-dimensional event $a^3 = T^2 k$ that went single dimensional because the moment in the fourth dimension which was then frozen in a single dimension on paper. With the paper being part of space-time while the fourth dimension $a^3 = T^2 k$ soldiered on and time will always be represented as T^2 just as Kepler stated.

The all-inclusive issue is the relation there is between what is the more heated area and what is less heated area and the concept goes blurred by the lack of understanding that Newtonians show in this matter.

By the time the Big Bang started all material was wrapped in a blanket of heat. I have little intention at this point to go into the matter by getting significantly technical and explaining the process in detail other than to say that the wrapper covered the wrapped product that froze a Universe until a use could be developed. There are parts in the centre of large galactica where the heat wrapper still freezes young stars in a heat cover and the cover will remain in place until those stars are able to develop dimensional qualities to start moving as independent stars just as the curved ball and the top does. The Grand Unified Theory admirably proves this. While sanity prevails we must discard the nonsense about gravity pulling dust into thick solid material that eventually form stars and most of all planets. That is a mediaeval myth and the sooner it becomes discarded, the sooner can cosmology shed the envelope of backwardness. The thought that stars can come about from particles being dust that through the magical process of gravity can fuse together as solid matter is a fairy tale that is indicative of Newtonian mentality. By progressing intellectually we have to divorce from that nonsense as we explore true cosmic development based on fact and we must discard fiction. Once singularity establishes independence, the star structure leaves the cocoon the galactica provide and move into the open Universe as an independent structure. This is done by the cosmos extending **k,** and in that all singularity connects and extends **k,** by expanding the relevancy singularity takes on. Every line within every atom within every star goes through this extending process by cosmic development. This serves as an indication of cosmic progress. As the governing singularity develops within the centre, the star finds meaning in the development coming as a direct result inherited from the growth in individual singularity that the atoms show. Atomic development corresponds within the star and that drives the galactica by the progress that the relevancy **k** produces in the star matching the progress of all relative **k** that influences the galactica. The singularity governing the galactica discharges its control as its capture tarnishes. This favours the gain of stars in the galactica development.

The process comes by way of the developing of the star's atomic singularity driving the star's governing singularity. What this says is that as the atom singularity grows so does the star's centre singularity and this is done because all singularity is as equal as inequality in singularity may tolerate. It is where 1^0 are connecting to 1^1 to form a resulting bonding in time. This extending we find as the factor **k** is developing during the progress of time. What all Newtonians concerned with the issue misses, is what time really is. People wish to put eventuality concerning human history down as time but in all fairness that can only have some relevance to time by the implication of human intellect. To us it is very important to know what our past was and where from did we develop but in cosmic terms human history has no concern of any significance or importance in relation to the Universe. Let us journey to find space and to find time but above all, find the ability in recognising with some dignity the distinction between both factors. If we wish to find the future, we should locate the past. If the cosmos is contracting, where too is it contracting? If the Universe is expanding, then where too is it expanding? Is it expanding outwards and if that is the case, where is outwards because going outwards to one point will be closing in to another location. We can't find beneficial understanding by naming direction in the way of stating where too is inwards and outwards, it is all the same in cosmic space. Going away from the Earth might be closing in on the Sun. Going away from the Sun is approaching Alfa Centauri. The direction of contracting must be in the opposing direction of the direction of expanding will be but that can only apply when I am located in the centre of the Universe, and that is where every point is in the Universe! If we wish to locate the past from where the cosmos came and through that see in what direction the cosmos came, it must take an effort to backtrack the direction it came but where is such a direction going? Should the argument come about that all came from nothing, then everything still has to be at nothing, or our understanding of nothing leaves much to desire? Nothing means not existing, not being, never found and unable to produce any multiplication thereof by any growth and if it is nothing, then it is not filling the Universe.

Reducing r where the circle is formed by Πr^2 The above questions got me thinking, but mostly the unanswered questions about the Newtonian idea of what the nothing is that they use in filling the entirety of the Universe, which according to Newtonian thinking must have a measure of being more and being less, which is to have more nothing between the Sun and Pluto while Venus being close must then use less nothing. The question about what is more nothing and what is less nothing drew me to the realisation there can be no such a quantity in space as nothing because even space has to be something if anything is closer than others being further. Then moreover and as well it is true that there can't be more or less, big or small, little or lots, many or few because that poses human concepts that bear no response to what applies in the cosmos. Quantifying by measure is a human reality which has no place in the Universe except to glorify human's in practising sublimation as to hide human incompetence and becomes a practise of narcissism. This is where Human mathematics becomes impractical. How many atoms are there in a water drop? Could t be 10^{1000} and if it is that little, then how much is it in terms of having a specific quantity. What do you call the number 10^{1000} by name? Clearly as it is for any one to see, releasing heat by nuclear explosions creates space. In explosions Academics portrait the winds as shock waves, but what is the shock wave other than new space coming into prominence and rearrange the structure in relation to the new space just created by liquid heat unleashing the created space as well as the space volume that came in place. In that way, it is clear that releasing heat brings about the expanding of the radius "r" as part of the sphere forming space. Hubble proved the Universe is expanding and the concept is under a question mark. Then by backtracking, we have to set about reducing the sphere as if the sphere is a circle constituting the expanding Universe. If r in the circle is growing, we have to reduce r to backtrack the direction in which the cosmos progressed. When the circle reduces, the value located to r will become implicated because r determines specific size. Not so in the case of Π, because Π in the true sense only indicate that the circle is a square without corners and therefore Π dictates form and not size. By reducing size, only r comes into contest and will point to such reduction. By reducing the circle radius r by half, continuously will lead to an infinite small circle but Π will remain because the circle as a form remains even being infinitely small.

The difference in the circle and the square is the direction the indicator follows and a square cannot spin, but through contact with a sphere as a circle cannot be motionless. The factor of Π indicates eternal motion and NOT zero motion. There is a massive difference in that concept. If no line can have a zero point to start with where will the circle get the zero to indicate motion! This principle is the most basic mathematic rule and even I the ILL-EDUCATED can see this. When the end of the rotation arrives the end rotation also announces the beginning of another rotation and not nullifying of the previous rotation because the rotation will have a line showing the effort it made and as it forms a wave, the wave will be there forever. The pitch may decline to a straight line, but the line remains. When calculating the motion a triangle does the honours.

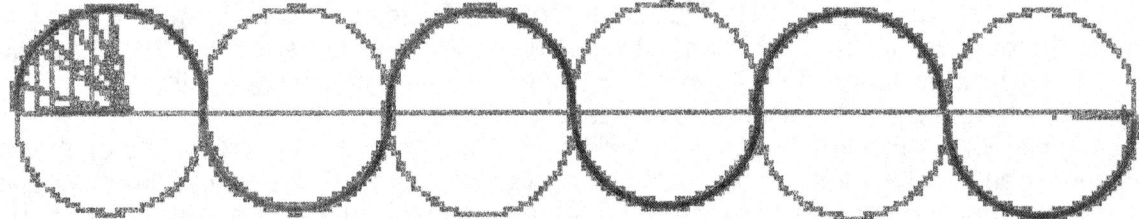

The wave confirms rotating directions followed by the circle as it spins. By stating that a wheel has a relevancy of zero by completion of a rotation such a claim denies the wave its rite of existing. The wave going flat as it becomes a straight line also has an indication to singularity. All spinning matter has the point where the spin is still there but the radius is too small to measure by any means. That point is standing still in relation to the rest of the spin. In relation to that logic I do not except Newtonian science holding the radius of s spinning object unaccountable in the spin, whether the spin is applying or not. It is mainly mathematics through the graph, which affect day and night and not the Earths standing in relation to the Sun that brings about climatically and weather changes and Earth development conditions.

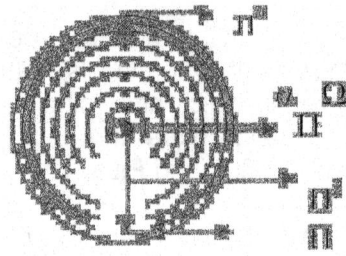

In the centre of all things spinning (moving) we have a point where that which can reduce no more meets with that which can never stop increasing for in the centre of all things spinning we find infinity meeting eternity.

That point serves time where time is forming by eternity that holds a specific relevance to what is infinite and what is eternal and what can never be smaller being in contact with what will forever increase eternally. In the centre we can locate a point where that which can never move meets with that which can never stand still and in-between a cosmos formed in which we find ourselves in space-time.

Nothing said so far is high tech or mind bending complicated. All the above arguments from the first page to this point reached are simple and there's only ordinary primary school mathematics involved that every scholar should know. One does not need a brain fitting Einstein to come to these conclusions but just thinking about everyday issues.

Arithmetic presents the possibility of zero and mathematics excludes such possibility. That is the difference there is. It is where mathematics departs from arithmetic and is as basic as counting is in arithmetic. There is nothing outrages that I have mentioned. Neither does one need the brainpower of a person like Einstein to come to such basic conclusions nor yet does it completely destroy the claims made by Newton in the formulation of Newton's cosmic formulas. Applying Newton's second law $F=ma$ one arrive at the formula $GMm / r^2 = m (\omega^2 r)$ claiming a zero influence between the radius and the orbit of planets. I shall come back to this thought but I would like to take your mind to one other thought that seems to have escaped every body. It concerns the medium science rely most on for the gain of facts and information about the Universe. It is the influence of light.

Newton made a brief calculation as a young man that saw an apple fall from a tree. Seeing this he jotted down a formula and the chucked it away. His piers and elders picked up the trashed paper with the calculation, and got all excited by the logic implication it had. $F = \dfrac{r^2}{M_1 M_2}$. The mass of the two objects destroys the radius between the objects. Everyone went ballistic, proclaiming him as an instant genius, the one the world was waiting for after the crucifixion event.

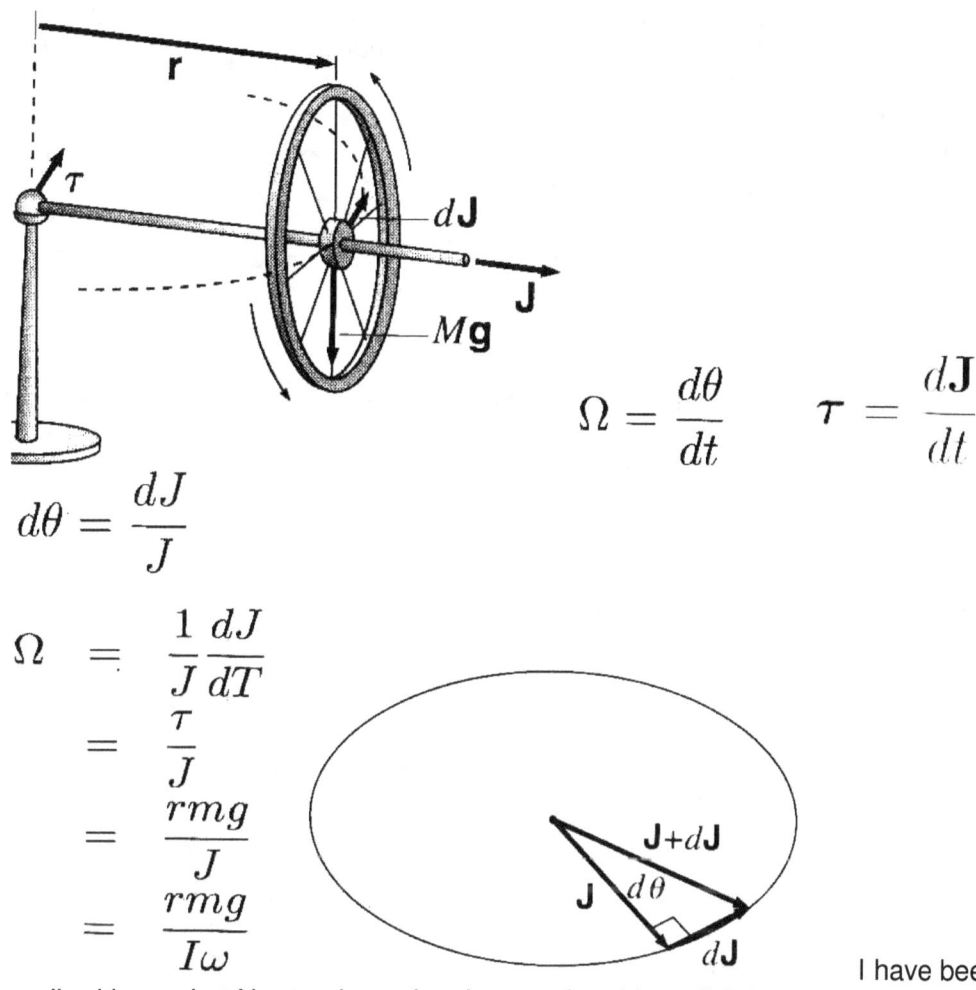

$$\Omega = \frac{d\theta}{dt} \qquad \tau = \frac{d\mathbf{J}}{dt}$$

$$d\theta = \frac{dJ}{J}$$

$$\Omega = \frac{1}{J}\frac{dJ}{dT}$$
$$= \frac{\tau}{J}$$
$$= \frac{rmg}{J}$$
$$= \frac{rmg}{I\omega}$$

I have been through this earlier I know, but Newton is so deeply rooted and I am fighting Newton for so long that by now I know that I am not in a fight with human intelligence or with common sense holding realistic perceptions, but I am fighting deeply seated concepts that form the foundation of thinking physics and although all that thinking is wrong, I can't get one intellect moving past that culture of repeating that Newton is correct, hail Newton for Newton is correct. If Newton was more modern I would challenge Newton to show how his views will fit the way of manufactured spoke wheels. In the cosmos there are no perfect circles circling a perfect aligned axis. From the presumption that the centre does not influence the spinning momentum, science draws a conclusion taking the relevancy directly translating to the spinning planets and orbiting comets. From this argument apparently Newton stated that $F = G\dfrac{M_1 \times M_2}{r^2}$ While Kepler said in his conclusion about the solar system

orbiting I do not, the and

that ($a^3 = k\,T^2$).

for one second, deny or dispute the revelation. What I do encourage is place event into its correct context. It was merely, simply an apple that fell from its branch to its roots. The apple did not pretend to be a meteorite that fell from the heavens. If it were a meteorite, I am sure, with the man's genius, science would be somewhat different at this stage.

However, as a young man, being very impressionable, as all young men are, and with the attention this brought about in the world of science, the matter overshadowed the fact.

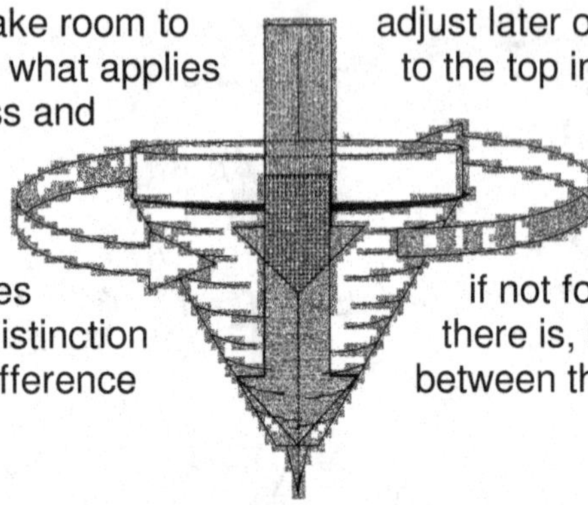

When Newton contemplated his theory there was one vital observation he didn't make. There was one flaw in his observation that he encountered but didn't make room to adjust later on. Newton never distinguished between a top not spinning and in what applies to the top in the case where the top is spinning. The top that lies motionless and never moves on own accord is distinctly different from the top that is in rotation. We all are aware of the fact that life intervened and by the grace of life, this action brought on by life intervening gave the top energy with which to spin...and yes if not for the intervention of life the top would never spin...but still the distinction there is, is overwhelmingly obvious. There is one Universe filled with difference between the top lying flat and the top vibrantly spinning.

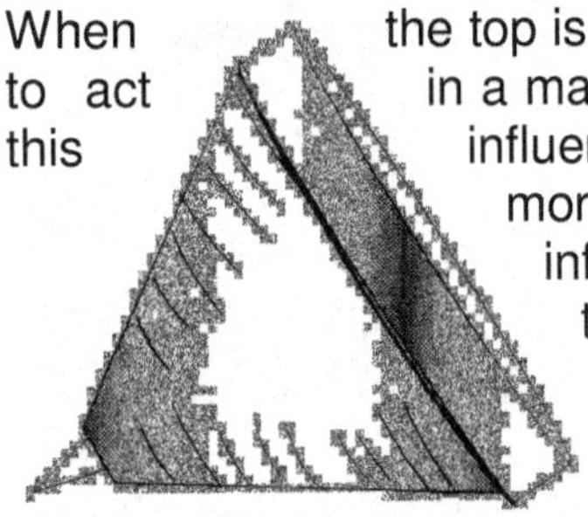

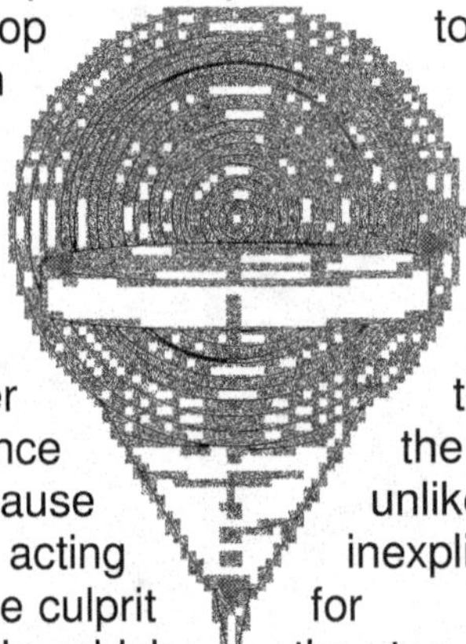

When to act this the top is flat lying on its side while resting motionless on the Earth, the top seems in a manner as if it is being depressed by some influence acting on the top and influence is restraining the top to being in a state of total morbidity. It is as Newton says a fact that another influence greater than the influence depressing the top must act on behalf of the top and rescue the top from this state of morbidity. Something must bring the top to life and make the top vibrant and seemingly alive. It must cheer the top up in order to get the top to break free form the depressing influence the top endures. Please note that not once did I, or do I refer to a force because unlike Newtonians, I do not believe in fairies, ghosts, magic or forces acting inexplicably and without reasons well understood. If mass was the culprit for this morbid state in which the top finds itself to be, then the arguments raised by Newton would be sensible and would be acceptable...but that is if mass did apply and contribute to this state of total depression in which the top is.

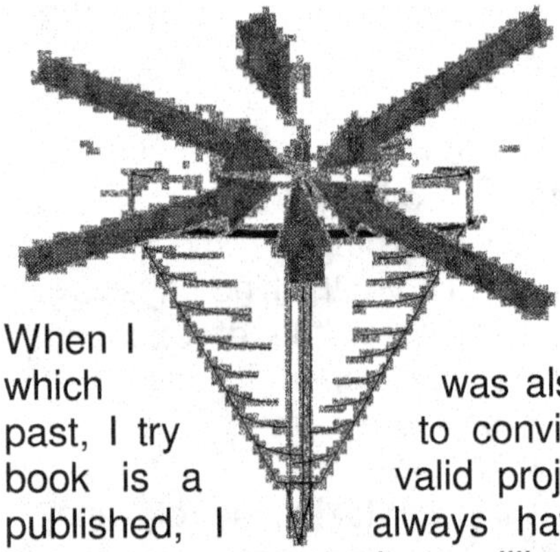

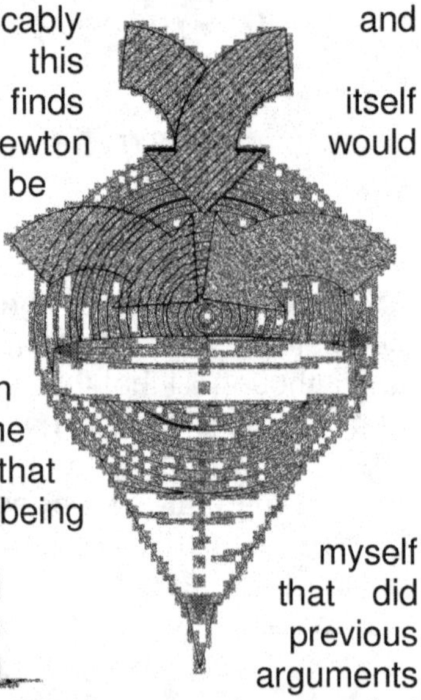

When I tell about these state of affairs again which was also told in another book with which in the past, I try to convince some publisher somewhere that that book is a valid project and that at the time it is worth being published, I always have to convince myself that there are only unwilling publishers that did read (or didn't even read) my previous attempts in which I introduce my arguments and I always have to fight the idea that because I wrote what I said before in other unpublished books, therefore the story I tell is not already well known to all while I do realise that no one before heard my arguments. I always sit with some vague idea that what I say, I repeat because I have said it in another book no person felt it was worth being published. Be that the case or not, I have to tell my case once more and this is my argument.

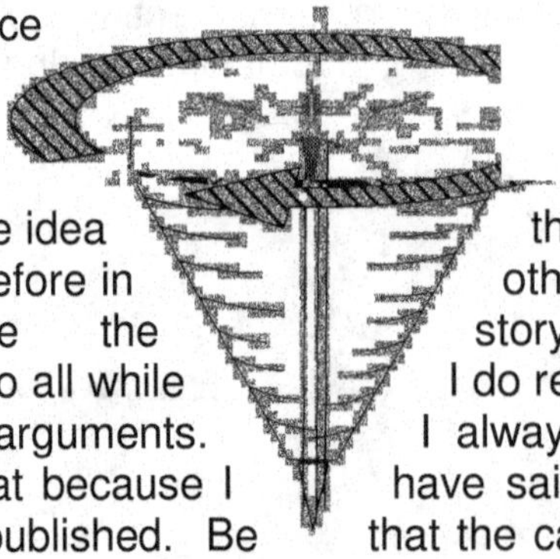

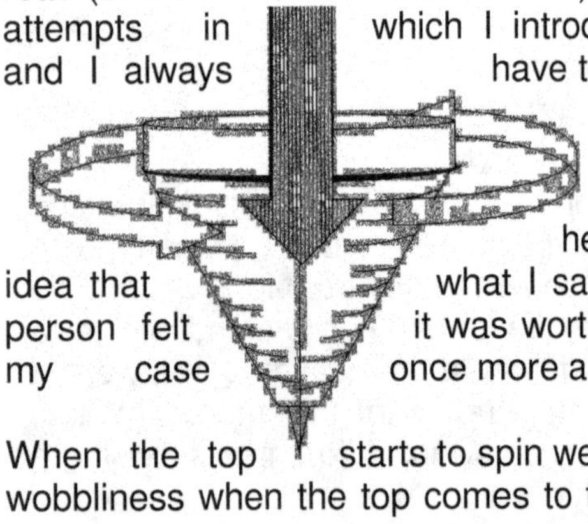

When the top starts to spin we find the rotation holds the top steady. There might seem some wobbliness when the top comes to the end of its performance but when n full motion the rotation brings a steadiness to the top that secures the top in an erect stance. The motion circling the top acts onto the centre of the top and from this the top finds the security in the balance that keeps the top up straight.

We find two distinct areas of influence which are responsible for keeping the top erect and maintaining an individual balance that allows the top to break free from the shackles of depression that contained the top when the top was motionless. From the rotation (a circle) a line connects with the centre (a line) that forms a line in the centre of the top and this centre line is responsible for the top staying erect. There is a circle forming and there is a line forming in the centre of the top and the circle is maintaining the line that is maintaining the balance of the top as an independent structure.

Newton's mass is still there if it was ever there. In that sense nothing changed and everything remained the same. The top has mass and the mass is still pulling the top down by the force of the grace of gravity. Mass is pulling as hard as ever and mass is pulling from the Earth just as much as the mass was pulling from the side of the top where the top too should contribute to the gravity effort. The mass is pulling just as hard as it was pulling before if it was ever pulling before. If mass was at work before then mass has little less to do with the process changing a lot in appearances compared too when the top is spinning because if anything, the mass increased many fold because now while spinning, the top is resting on a needle edge which according to physics, must enlarge the effort mass should have on the subject.

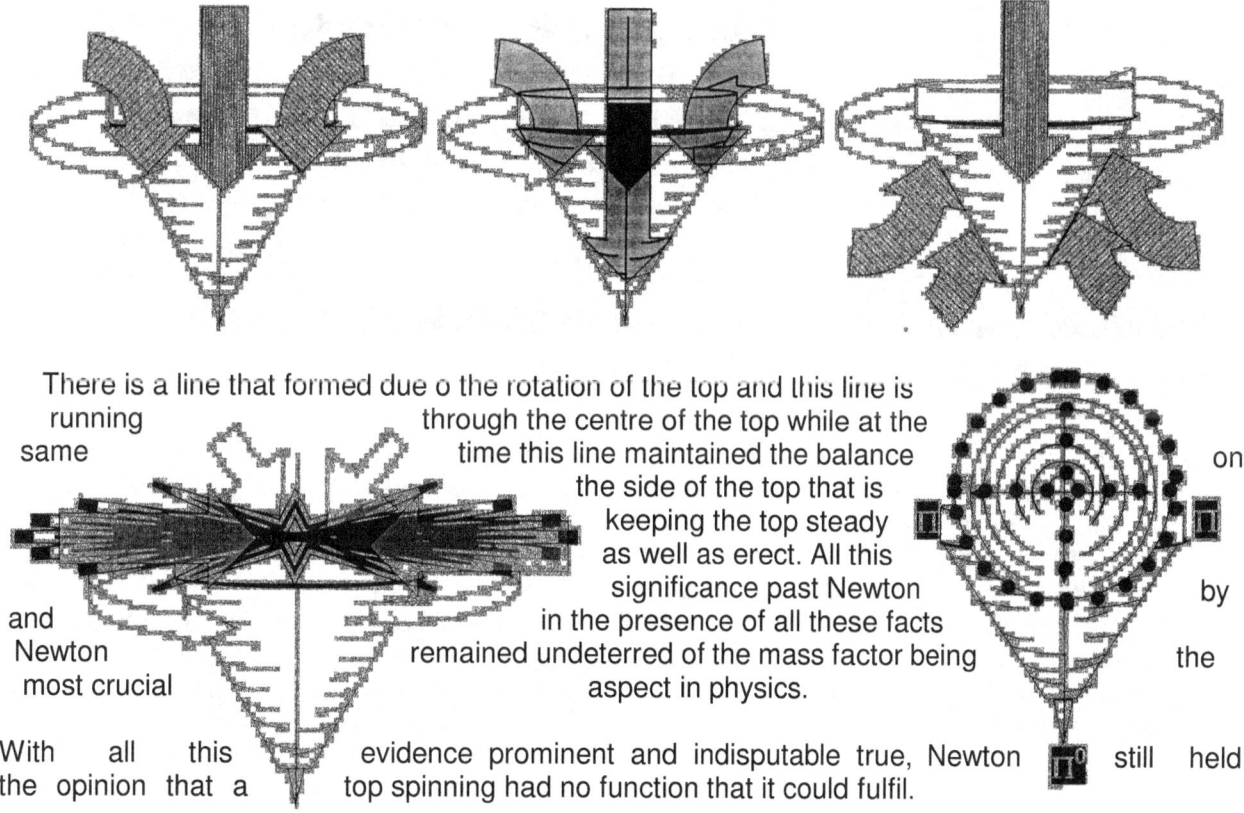

There is a line that formed due o the rotation of the top and this line is running same through the centre of the top while at the same time this line maintained the balance on the side of the top that is keeping the top steady as well as erect. All this significance past Newton by and Newton in the presence of all these facts the most crucial remained undeterred of the mass factor being aspect in physics.

With all this evidence prominent and indisputable true, Newton Π^0 still held the opinion that a top spinning had no function that it could fulfil.

Every quarter provides a distinct value that indicates the progress of the flow of time from the one point Π to the next point Π.

Any changers occurring in Π will lead to a an unequal triangle providing two different values to r and will alternate the link between r and Π^2 bringing about different form (Π) and time (Π^2). When singularity forming the lines of the triangle is not in equilibrium the triangle will destroy the matching of half circle.

In every sector the directional flow will provide a distinct meeting of Π linking r to Π^2 and this allow the time component in the rotation.

Every quarter provide connects distinctly to a completely opposing value that indicates the complete reversing changes from what applied before to what develops onwards of Π to the next point Π.

Any changers occurring in Π will lead to a an unequal triangle providing two different values to r and will alternate the link between r and Π^2 bringing about different form (Π) and time (Π^2). When singularity forming the lines of the triangle is not in equilibrium the triangle will destroy the matching of half circle.

As the meeting of r points to a very distinct different r in direction such a point of meeting opposes the other points in meeting and will lead to destruction of the form Π in any the event of any value changes by Π changing Π^2 and r.

In Newton's assessment that $\dfrac{dJ}{dt} = 0$ or put in English that the rotation removes the radius and

nullifies the influence that the radius then may have on the spin totally disregards the above mentioned implication such as that which the rotation brings about

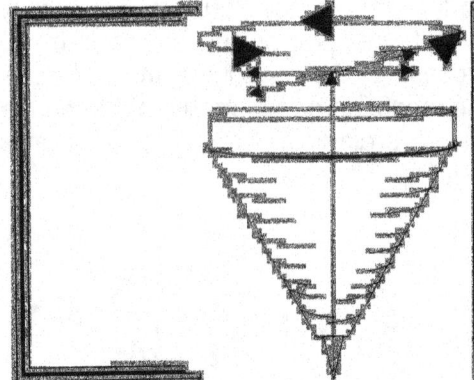

In the center runs a line called the axis line.
The line does not show any influence on managing the top when the top is motionless and bounded by the Earth gravity. However the sooner a motion sets in that is adequately strong enough to support the independence of the top the top generates enough gravity to sustain and independent attitude in relation to the

The top then maintains the role of an electron in relation to the Earth being the proton and the atmosphere forming the neutron

Not am I only disputing Newton; I am disputing the relevance of Newton's scientific breakthrough. I am saying to the face of any Newtonian that Newton missed the mark he tried to reach by a country mile. Newton's philosophy totally strays from reality! The spin the top has clearly liberates the top from the depression the Earth dictates. Where the top previously formed part of the Earth by motion, now with individual spin the top has a vibrancy that is defying the mass the Earth places on the top. Kepler and what Newton saw about what Kepler found were to Newton's mind the proof of total mathematical incompetence. He (Newton) saw a circle and without Π there can be no circle/ Further more since he was the founder of the invert four square principle the principle also had to be included the make the picture a smart Newtonian picture and with that remove Kepler as such.

Newton formed the opinion that the rotation of the top eliminated the radius of the top $\dfrac{dJ}{dt} = 0$ and

that by spinning the top drew the focus of the radius towards the centre of the top and thus eliminating the radius.   Mathematically Newton equated his surmising as follows. With this presumption Newton deducted that F =ma and that is a big mistake. F=ma presumes an instant of no motion applying and such a presumption is not sustainable in the cosmos because time would never allow any position to be motionless in relation to the rest of the Universe also being motionless. Relevant to the Earth, some objects may stand still or tend to be motionless, but the objects can never be motionless because by being relatively motionless in accordance to the Earth, the object still applies full motion that it adopts from the Earth by moving with the Earth, and this makes the fact of objects having mass in the cosmos a fabrication of Newton's imagination. The mathematical expression $\dfrac{dJ}{dt} = 0$ is not mathematically possible. Newton, and science, made one enormous blunder, from this stance.

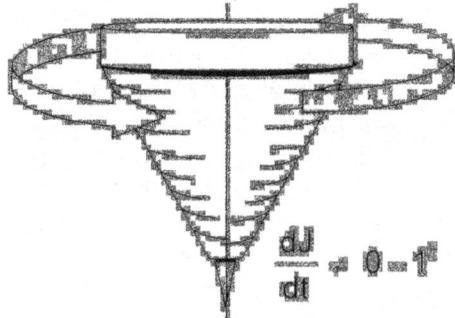

They took the radius of a wheel not to have any influence on the wheel. In doing that, they removed the very fact that keeps the universal attachment together. They put two objects in an attaching relevancy and then announced no relevancy. Doing that is breaking the most fundamental mathematical principle.

$\dfrac{dJ}{0} = dt$ or $\dfrac{0}{dt} = dJ$ This disputes mathematics. DJ / dt can have any number from eternity to infinity, only excluding one; it cannot be 0. By placing the one in division of the other, you bring in relevance. You cannot then say there is no relevance. By doing such, you proclaim that one of the factors is non-existent. In both cases, one of the factors then does not exist. Such a claim is incoherent, because you proclaim that a circle has no radius, or a radius has no circle. When calculating a circle, you multiply either the square of the radius by Π, or the quarter of the diameter at a square by Π.

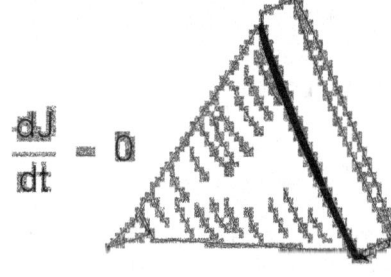

$\dfrac{dJ}{dt} = 0$ constitutes a circle and is also therefore $\Pi \times r^2 =$ CIRCLE

If you remove r it then is $\Pi \times r^2 / r^2 =$ CIRCLE.

You cannot then say $r^2/r^2 = 0$ and therefore $\Pi \times 0 = 0$. That is nonsense. $\Pi r^2/r^2$ will always be $\Pi \times 1$, and that is the eternal circle.

When looking at any rotating object, there has to be a point of no rotation and no rotation means "no rotation", not no existence. No rotation means a factor of 1, not zero. That then is singularity. The eternal Π, the Π that may not have significance but still it is a Π of value.

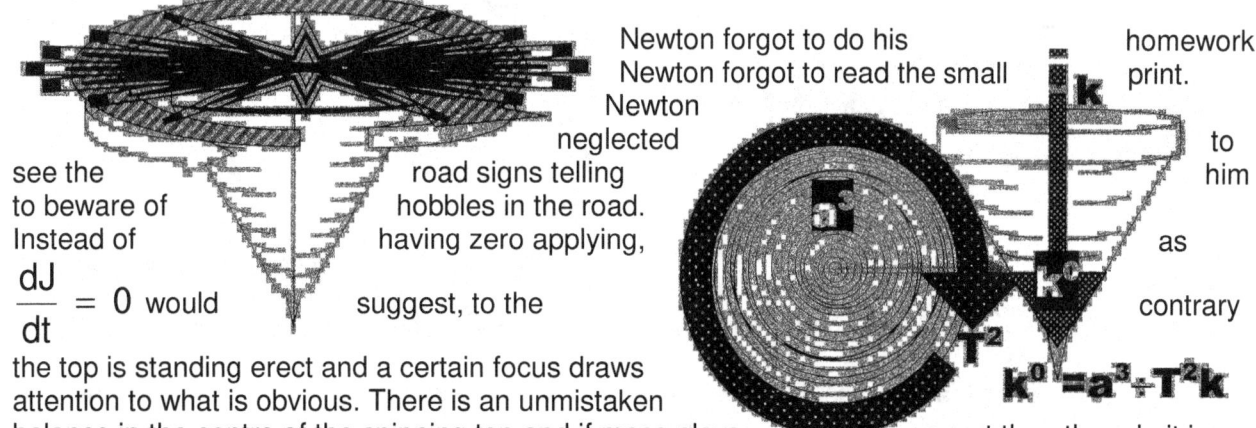

Newton forgot to do his homework
Newton forgot to read the small print.
Newton neglected to him

see the road signs telling as
to beware of hobbles in the road.
Instead of having zero applying, contrary

$\dfrac{dJ}{dt} = 0$ would suggest, to the

$k^0 = a^3 \div T^2 k$

the top is standing erect and a certain focus draws attention to what is obvious. There is an unmistaken balance in the centre of the spinning top and if mass plays a part then the role it is very much disguised. In the centre is an influence concentrated that keeps the top in balance and keeps the top well clear off the ground. It deters the drafting effect that mass should play if indeed mass plays any part at all.

The spinning top forms a balance coming from the centre that holds the top in a vertical position while the top is spinning horizontally in a circle. From the balance the top sustains by spinning, the body of the top remains upright, erect and in a balanced position. The spin is pushing the top as much upwards as the top is pushed downwards and in this process there are no grounds on which to base any force   gravity may bring if gravity brings any force as a factor. All that is clearly   well established and is very apparent is that there is a

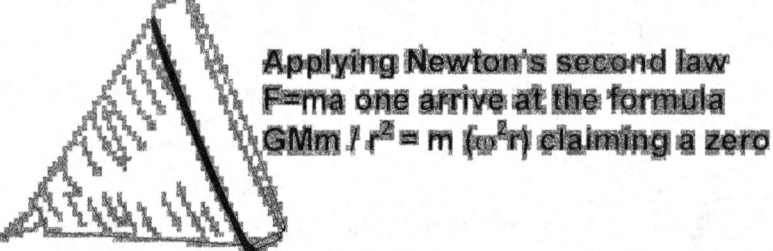

balance that the spin maintains and most of all is the evidence clear that if mass is pushing the top down as Newton said it should, then the spin is pulling the top up against what Newton tried to prove with much dodgy calculations. Let's try to estimate what it that Newton completely missed when Newton tried to push his idea about physics having mass as an influencing factor onto the world.

is

In   the top spinning the rotation is not eliminating the

radius, $\frac{dJ}{dt} = 0$, but in fact it is highlighting the prominence of

the movement T^2 in view of the centre k^0 by promoting the radius k by connecting the space a^3 that the radius k holds in terms of the rotation T^2 which puts the top in a time independent zone $a^3=kT^2$ or in independent space-time $k^0 = a^3 \div kT^2$.

Applying Newton's second law F=ma one arrive at the formula GMm / r² = m (ω²r) claiming a zero

The top lying still holds the same singularity principle that the sphere holds because if the shape the top has. The roundness protects singularity at a seventh position deep inside. However the top is a dot Π or even going down to a spot Π^0 and is only by the form the material has which puts singularity in place.

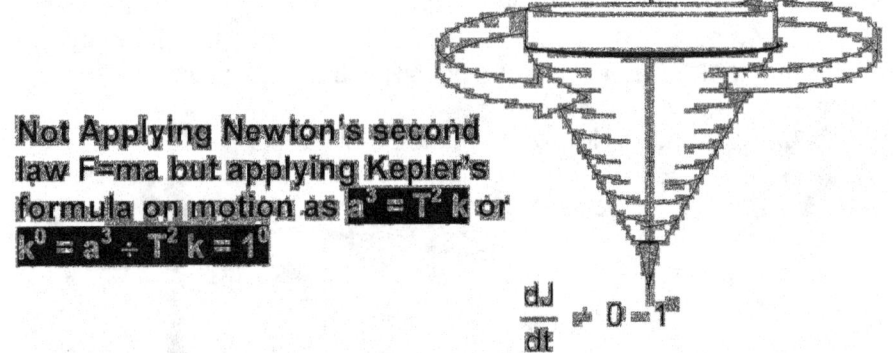

Not Applying Newton's second law F=ma but applying Kepler's formula on motion as a³ = T² k or k⁰ = a³ ÷ T² k = 1⁰

$$\frac{dJ}{dt} = 0 = 1^0$$

According to Newton it

takes no effort $\frac{dJ}{dt} = 0$ to get the top from where the top was motionless to where the top is

spinning. I say this on the work that Newton suggested comes about from the effort it takes the top to circle.

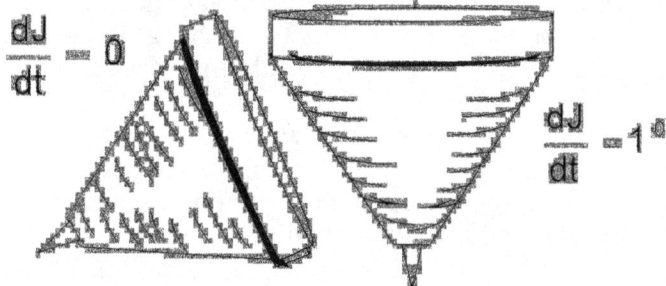

From a position of lying down top a position of spinning erect does not take nothing as an effort. It puts the top in a New and independent Universe as the top then finds courage to fight the gravity of the Earth up to the last "breath" is fought.

There is a position that is in motion that is forming the very edge of the outside. To be in motion the position must be in relation to a point from a centre. From the centre, there must be a specific allocated space ending at the object in motion and starting from a centre that has no dimensions. The object in motion determines the one limit and the centre with no sides and no space, which is standing still in singularity, determines the other limit. By that we can see there is only one way of looking at what we can observe and that is from the outside in.

The atom must be the utmost coldest and the proton is even much colder because when that cold escapes it cocoon and becomes released while it turns to heat in a process of forming nuclear space, what apply during such an event is so dramatic that no one can understand the process at all. When the spin of the atom allows the cold of the atom to release the heat it had as it had frozen heat to space at a velocity exceeding the speed of light, the atom holds that destroys all that surround the atom.

When this heat releases from the containing form of the atom, it brings about much more heat than the Human mind can cope with. One may not look at the material in terms of temperature and judge the surroundings. Heat releasing brings on volumetric space and overheating brings expanding and the expanding ability should serve as a measurement of what is cold and what is hot. It is not that which I feel against my human flesh that is hot or cold in cosmic terms because in reality there are only relevancies applying and not hot or cold as such. We have to look at the reaction hot or cold brings about and take the queue from that to inform us as to what is hot or cold.

The direction of expanding should tell us scientifically what we should consider as being hot or cold. Outer space must be the hottest place in the Universe because from a heat perspective outer space is still overheating and expanding as nothing else can match in the Universe. The expanding is thought of as the Hubble constant, but it is a process of heat gaining space. One fact we can look at in terms of how the cosmos reacts is to look at the fact that hydrogen remains a gas and so does helium in outer space. If outer space were cold, we would find massive lumps of solid hydrogen and helium forming blocks of crystals that is solid structures. In reality we find hydrogen as a gas that is widely distributed.

With hydrogen and helium being a gas this must serve as enough proof that outer space is hot, regardless of our interpretation of the temperature gauge telling us in cultural terms about what we wish to hear. One must look at outer space and judge outer space from the findings only considering outer space. If helium remains a gas, it is hot. The removing of heat makes the centre of the Earth cold although we see it as being terribly hot. The only reason why it can seem to be hot is because the atoms freezes into smaller particles and therefore dispense heat into space surrounding the atom. The atom by itself on the inside carries less heat when that atom is in the centre of the Earth in a smaller confinement of space and in that surrounding of rapped movement the atom is cold and in such a cold environment, the heat can gather and space can collect heat because the particles find the surroundings extremely cold.

The weather on Earth proves me correct because at the polar caps the human cells burn to carbon in minutes when the flesh is not protected which is not possible at the equator. The human concept of what is cold and what is hot is alarmingly disturbed in terms of reality.

In the centre of whatever rotating object runs a line, which is the line of singularity and from such a line forms the spin of the rotating object, (in this case I present a top). The relevancy applying places one rotation of spin eternal, (that remark I explain later on) therefore the rotation is standing still, but also one rotation consists of an above measurable number of infinities producing a rotating speed faster than any person other will ever grasp and that makes the rotation the fastest there can will and may ever be, if one takes the spinning performance of all the atoms as a unit. Again we humans now face the same value we wish to separate, as is the case with hot or cold, near or far, and quick or slow. Once again time locks relevancies beyond understanding. That line so infinite small as it is eternally big connects the cosmos through light we are able to see and light we are unable to realise.

In the centre of whatever two synchronized rotating objects the line of singularity matches in spin to one another not withstanding the space-time involved. The relevancy applying places one rotation of spin eternal, therefore the rotation is standing still, but also one rotation consists of an above measurable number of infinities producing a rotating speed faster than non other will ever reach and that makes the rotation the fastest there can will and may ever be. Again we humans now face the same value we wish to separate, as is the case with hot or cold, near or far, and quick or slow. Once again time locks relevancies beyond understanding. That line so infinite small as it is eternally big connects the cosmos through light we are able to see and light we are unable to realise.

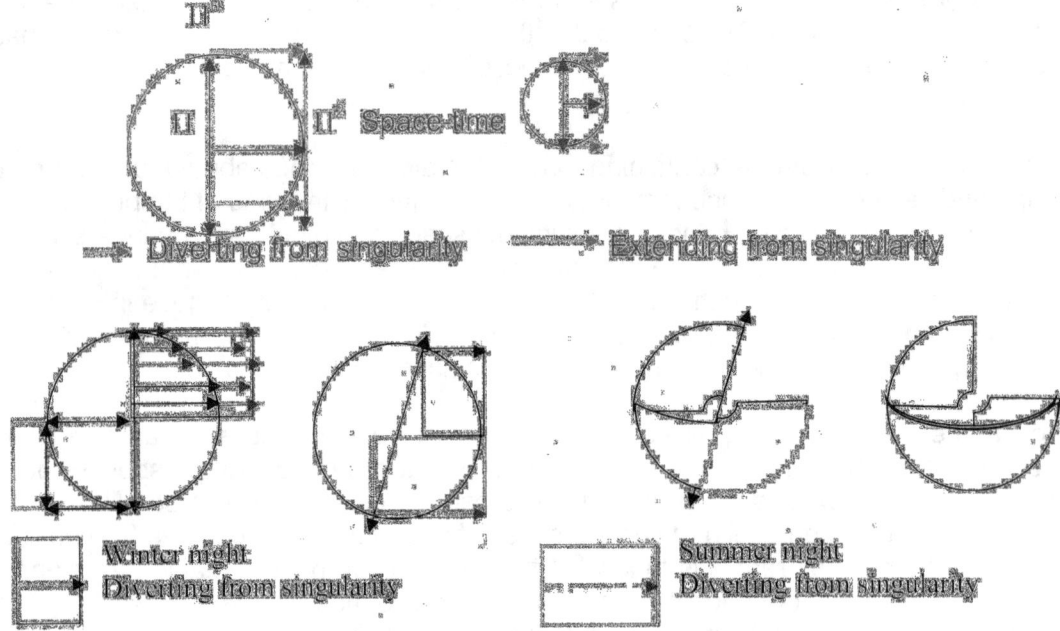

As singularity is eternally cold, it also has to affect the standing between matters, as matter diverts from singularity. The more occupied space there are between singularity and space ends, the more it will be to the advantage of the effect we consider heat, and the less space allowing time the more it will advantage the statement we refer to as cold.

The reason why the "naked" singularity $^{\alpha}\Pi^{\Omega}$ stands still is because of the relevancy it holds in comparing to all other singularity factors forming Π^{0}, Π, Π^{3}, $\Pi\Pi\Pi$, $\Pi^{2}\Pi$, the spin rate of the last mentioned cannot top or even come close in comparison to the tempo of singularity $^{\alpha}\Pi^{\Omega}$ therefore in comparison, achieving the epitome of relevancy it must be standing still since faster that what it spins non other can come close.

Because $^{\alpha}\Pi^{\Omega}$ sets the standard, forms the basis, applies all rules and cannot be zero, it therefore has to hold the relevancy applying as one the eternal one, from where the Universe with all content must apply.

There is an eternal link between singularity 1^0 keeping the Universe in equality and equal in measured value. But part of this link is related to time that differentiates by using the value of Π and by time applying Π coming from Π^0 the involvement of Π brings a differentiation that would have singularity always be inferior to singularity while being eternally equal. This might sound like burble of the senseless overflowing a drunkards mind but in due course this argument will become most clear.   By having Π and having Π in relevance placing time in context with space there will eternally be a mismatching value between infinity and eternity where both serve singularity as Π^0 going on to $\Pi \times 7 = 21.991$ and this sets singularity two values of $1 + .991$ or then hot moving to colds or cold expanding by overheating.

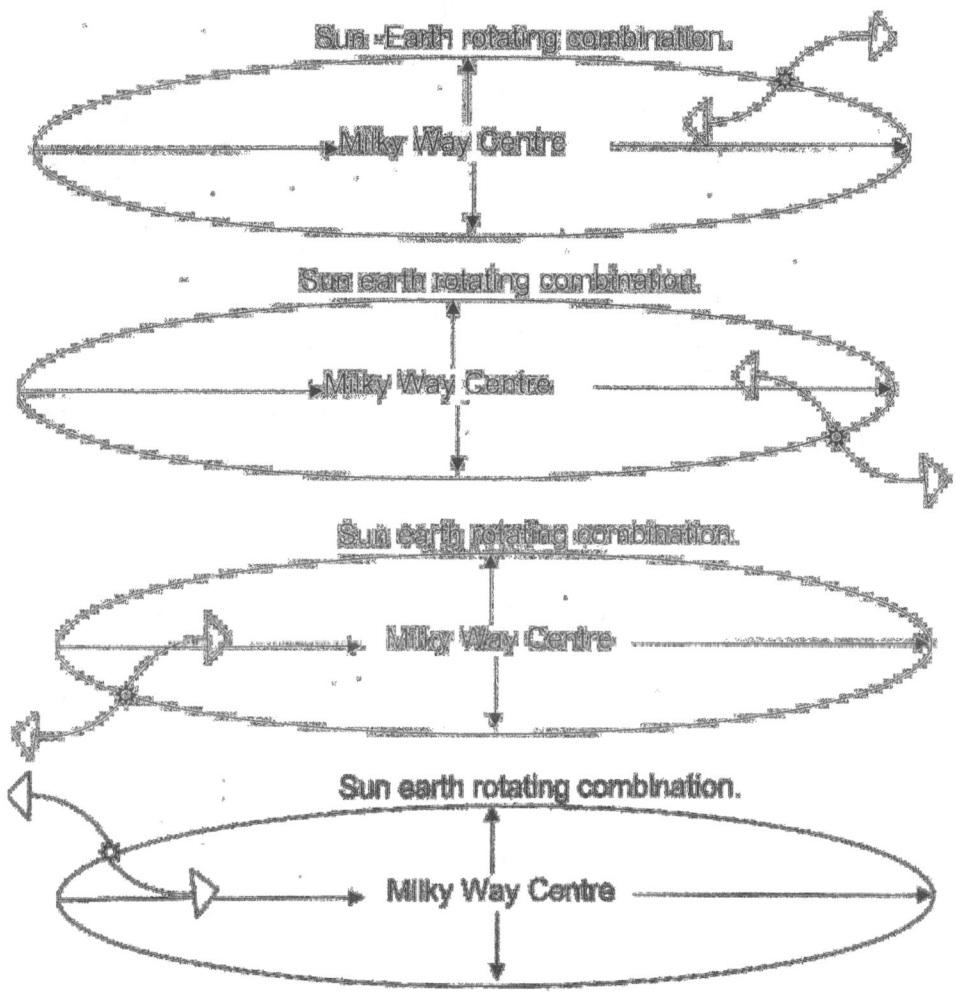

Sun - Earth rotating combination.

Milky Way Centre

Sun earth rotating combination.

Milky Way Centre

Sun earth rotating combination.

Milky Way Centre

Sun earth rotating combination.

Milky Way Centre

By calculating the rotation cycle in four parts that the Earth follows around the Milky Way and the Influence gene rated by the Sun forming a singularity matching alliance with the Milky Way we will find the solutions to the mysterious Ice Age / Desert Phases, the world suffers from with such cyclic intervals.

The cold space space Earth expand, space in in the Earth centre causes the concentration of heat by reducing, as all cold surfaces tend to do. If it was hot, the within the Earth would expand and the space within the where we think so much heat is concentrated does not therefore it must be cold. To gather and accumulate the any liquid means it becomes much colder being a liquid. Going from a gas to a liquid condenses material and in that process a lot of heat must remove from the material. Finding the surroundings terribly cold

Space occupied in one second moving at a speed of "Y"

will   allow the heat to gather and not expand but when the surroundings are hot, it will not tolerate more concentration of heat and thus will expand to rid the balance of excess heat within space.

We have to break free from the shackles of culture and see gravity for what it is. Gravity is the movement of space. Gravity is moving space in terms of singularity. Movement is duplicating space by filling space with less material more often and divides material into more space by reducing the heat material comprises of. Movement is the duplication of space in order to reduce heat used to fill space. On the other hand the lack of movement is the duplication of space to reduce heat levels preventing the rising of heat. The Universe is movement in two forms. That which contracts by duplicating space in movement and that which expands by duplicating space because of a lack of movement and therefore introducing movement as result of expanding of space. However it is seen space is the movement thereof and that is what gravity is.

The application of gravity that condenses space is bringing about heat by the compressing of space. It is removing space

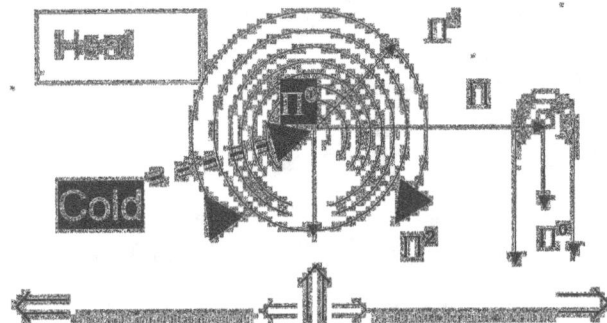

Occupying half the Space in one second moving at twice the speed of "Y"

from inside the atom to outside the atom but as Newtonians always do, Newtonians misses the obvious to become ridiculous by reason of stupidity. By producing heat we reap movement and on that ground all physics principles are formed. We apply heat which we named energy to form locomotion and with physics we apply locomotion as a result of science wishing to counteract the increase in heat by allowing motion and in that way we go about tapping into the energy that nature provides.

Internal and external engines combustion engines all rely on this application of heat levels rising and we feed heat to power. feed heat because motors when it incorrect the logic the engines fuel to generate harvest motion by driving Even electric motors we in the form of electricity when we feed to much electricity the electric burn. If the motor burns overheats due to electricity supply, then deduction is that electricity is normal heat that is intensified. Even today we compress space with a piston in a cylinder and then pump the compressed air into a container and such confining of space will increase the heat by the piston and the result effort is that by reducing the space a rise in heat levels were brought about in the container. The heat coming about inside the cylinder has no relevance to particles colliding because all compressor cylinders cool down. The walls becomes colder because when that cold escapes it turns to heat as the heat releases from space forming a secondary form of material forming space that no one can understand when the spin of the atom allows the cold of the atom to release into uncontrolled space. This release and unification with space that heat does is releasing of the heat that the molecules in the compound that formed between the atoms, which the forming of the compound had frozen. The freezing of heat situated between particles forming the compound is because the motion of spin to space that the atom holds remains in a frozen state under the guard of the spinning electron. When this heat releases from the containing form of the atom frozen by the spin of the electron it brings about much more heat than the Human mind can cope with. One may not look at the material and judge the surroundings. When the heat inside the atom nucleus releases in an atomic explosion, this release takes time back to the birth if the cosmos, which supersedes whatever the human mentality, is capable of understanding.

Instead of finding huge arguments about heat scales and temperature measuring scales applying it is a much better task to investigate the reason why the cosmos would have hydrogen remaining a gas and so do helium in outer space. If these two natural elements remain a gas, it must be because it is hot out there. Looking at this fact must serve as enough proof that outer space is hot, regardless of our interpretation of the temperature gauge telling us what we wish to hear in terms of what is acceptable to us from a human based cultural point. One must look at outer space and judge outer space from the findings only considered in the terms which outer space insists upon. We have to allow outer space to set the rules and not some Newtonian devising a scale in accordance with his skin temperature telling his senses what his human mind believe is applying. If helium remains a gas, it is hot. The removing of heat from the space that contained the heat makes the centre of the Earth cold. In our Universe we see it as being terribly hot because the heat then forms a separate substance but remains a form of material, but that is because we see the heat and not the space derived from the separating of the heat.   The only reason why the space can seem to be hot is because the space is cold and in such a cold environment the heat can gather in a much concentrated state and space can collect heat because the particles hold concentrated heat in the space separating the particles.

If we consider that the only reason why anything would move in the cosmos, is to apply a conditional balance between what is hot and what is cold we can see why the top spins when the top spins. From that we can gauge what it is that keeps the top erect, notwithstanding all the pulling asserted on the top.

By removing such high concentration of heat from the space that used to be expanded heat, the space then must contradict the heat by being extremely cold. We look at the heat in the space, which by that time is another form of material and find the surrounding heat in the space hot while the space is extremely cold. The cold in the Earth centre causes the concentration of heat by space reducing, as all cold surfaces tend to do.   The proton contributes to that reducing of space. If it was hot the space within the Earth would expand and explode but the space within the Earth where we think so much heat is concentrated is so much it does not expand therefore it must be cold. To gather and accumulate the space in a liquid means it became much colder when the space parted from what then is being a liquid. Finding the surroundings terribly cold will allow the heat to gather and not expand but when the surroundings are hot, it will not tolerate more concentration of heat and thus it will expand to rid the balance of excess heat within space. The concentration or release of space with heat or space from heat is a direct contribution of the singularity

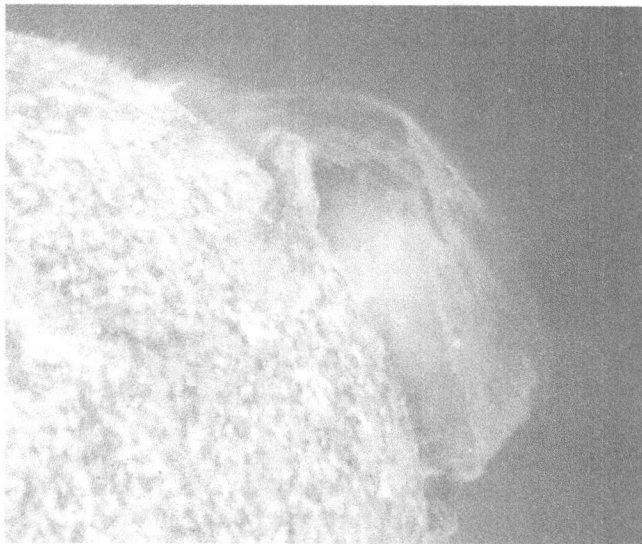

in control of the space-time. The regard of the singularity stipulates the conducing of heat in space or the release of heat to form space by means of bisecting the occupied space.

The only reason why the top will stay erect while spinning is that with the spin the centre of the top becomes cold which puts a heat misbalance in place from where the heat on the side of the top will flow to the cold inside. With a redistribution of heat coming about, a centre creates by movement a flow of heat and in this flow of heat a balance in measured value is placed in place allowing a flow of evenly distributed heat from the outside of the top to the inside of the top.

If we as humans wish to understand the cosmos we have to understand how God declared He created the Universe according to His Word. God gave the first command "Let there be light…and there was light" That places the cosmos in a relevancy from where the cosmos will act on this command t all times. With light coming in place we must see light as heat that is moving and if heat is moving there has to be a reason for heat moving. The only reason why heat would move is when heat is flowing because of a centre that is cold and a centre that is hot and a misbalance between what is hot and to where heat will be concentrated. If humans wish to form a concept about what drives the cosmos we have to look at how the Bible says the cosmos is driven. The cosmos works on the principle of heat flowing between hot and cold and only because of this imbalance can movement take place.

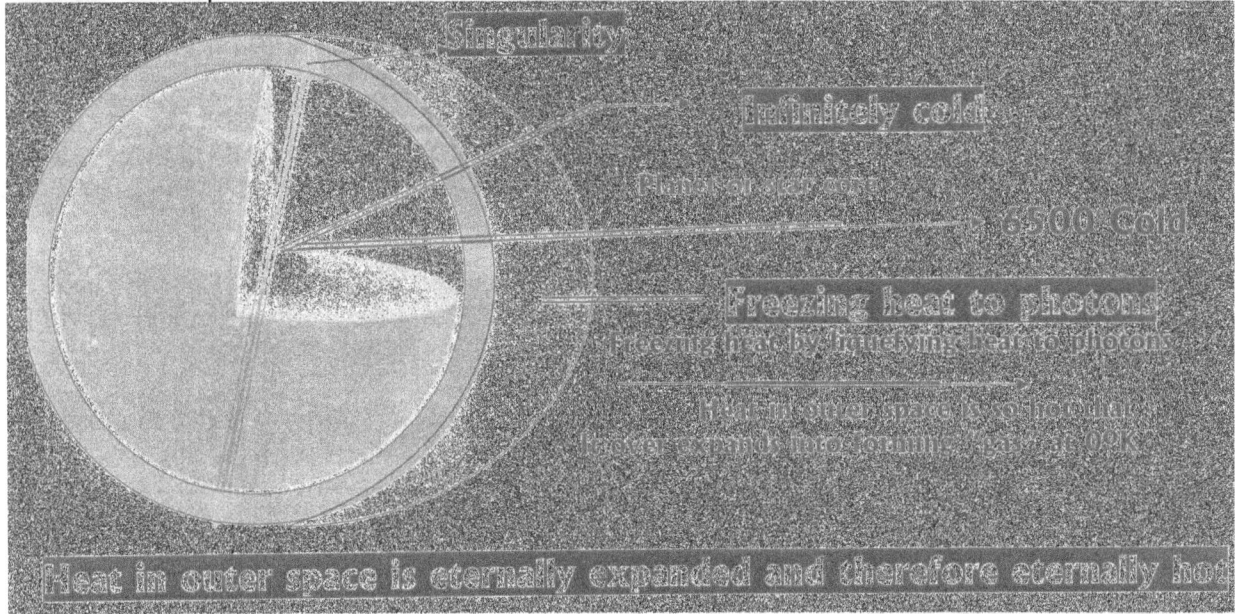

Look at the Sun and see how the Sun turned the hydrogen it holds captured in its atmosphere to a freezing cold liquid at 6500 K. Hydrogen is in a fluid state within the Sun and yet it is still colder than the hydrogen we find in outer space that is in a gas form in outer space. The Sun is without any doubt the coldest place in the solar system. That is when the protons oversupply the removing of space to produce the cold that is so apparent in the heat levels that do not join the spell. By the

reducing of space, it can concentrate heat to a fluid state. The rapped movement the Sun has freezes space into a liquid. By producing the opposing cold that finally freezes the heat to a solid state; we find that is what matter is. The expanding of space is a way of duplicating space without reducing space and by duplicating in the form of expanding it becomes just the opposite to duplicating by motion therefore reducing space by halving space in time. That is what gravity does. By motion space duplicates and by space duplicating the material must be by dividing or bisecting - halving it removes heat in space as well as by dismissing space and in that concentrating heat. The density of the protons brings about space dense enough to harbour the heat in such quantities and visa versa applies in outer space.

Recognising that then Sun freezes hydrogen or then its atmosphere at 6500 °C is a most vital part of understanding gravity. The Sun compresses its atmospheric space to a density that liquefies hydrogen into a freezing liquid at 6500°C. The 6500°C has nothing to do with the entire issue except to warn humans not to plan a trip in that direction because we just won't last long enough to send home photographs of the vacation scenery. It is the Sun spinning that freezes the atmosphere of the Sun into a compact liquid that has a temperature of 6500°C but that temperature has no senses on what applies in the cosmos.

The Earth condenses its atmosphere to an average of between minus 70°C and 55° C. This atmospheric m depression of space compressing outer space into a condensed state such as we has on Earth is the main part of gravity. This has nothing to do with mass and has everything to do with the Earth moving thus displacing the air into a more compact and more compressed state. Let's return to the top and see what effects would gravity have on any object moving. In order to move there has to be two directions of movement complimenting the space it forms. If there is space $a^3 = T^2 k$ where the spin is T^2 the lateral movement is k and the space formed by this action is a^3. This is gravity and has nothing to do with mass whereas mass has everything to do with this because as the movement is more rapid, the compressing of space produces more dense or intense space and this leads to more compresses objects that presses harder onto the turning Earth and with the Earth turning, this allows the object to have the mass theta Newton saw so vividly without understanding the first thing about what he found applying. Increasing movement can enhance spin as well as compress the downward action. The direction can indicate influence in both directions since it is relevancy applying to form space. The relevancy to space time being space a^3 and time or movement $T^2 k$ can apply to both directions but when spin T^2 increases and space a^3 remains the same, then the relevancy **k** has to compensate for changes coming as a result of movement. The compressing will compact the density since the spinning motion stretches the space. This is gravity and from this effect we have an atmosphere or a compressing of space becoming more compact and becoming denser. However, the only way this increase in denseness can come about is by cooling of space. If the space compressed as a result of compacting the space such as what is happening in a compressor reservoir, the space would expand exactly as it does when escaping by penetrating and going through the walls of the air tank. This rule does not apply in the cosmos since there are no metal walls containing air. When the movement by spin increases in applying to the space and the space remains the same in quantity since there can be no adding or removing of space, the density or the compactness of the space has to increase and this is gravity but also this is the result of movement cooling space.

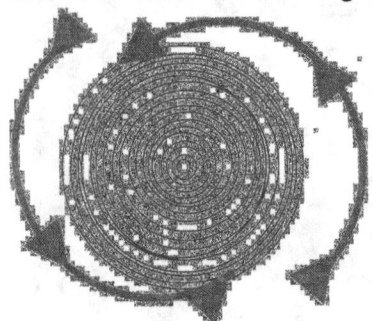

The particles claim more space when heated to preserve the cold. The claim to more space produces more space and reduces more heat. Such expanding brings about cooling. When particles heat or cool down, there is motion or movement that applies in some or other form. Motion started at a point when the Universe was extremely hot and there was no space. By introducing motion space formed and the lack thereof produced friction that became heat that became space. It is natural, it is simple, and above all, it makes believable sense. Let's study what happens when the top comes erect as it applies movement contrary to Newton's idea of gravity. We have to look at the top not as the top but as an accumulation of atoms working together as a unit that spins as the top.

Looking at the top it seems that the body structure of the top is solid and the air surrounding the top is liquid. The top as a structure composes of solid particles that light cannot penetrate and that material cannot pass through. In that sense it seems to fit all conditions we set for solidness. The top spins and it spins through the allows the top to spin seeing that the top has much more density than the air that air has.

IN THE MIDDLE OF EVERY ROTATING OBJECT NOTWITHSTANDING SIZE, RUNS A LINE THROUGH THE CENTRE OF THE OBJECT THAT IS DEVIDING THE OBJECTBECAUSE WHERE THAT LINE DOES NOT ROTATE BUT STILL IS IN CONTROLE OF ALL ROTATION. That line line shows no space and that shows no partition.

Point of no motion

In motion in considering the spinning detailed instant the fraction of time in the in every instant of change in every aspect of rotation will turn same characteristics only seconds time. Although the points had the characteristics it had just before and before, they oppose the they are and to which they relate by just after the very second in which of the graph proves my point in similar points also in rotation. The fact values, The point indicates as much as quarterly opposing dimensions and forming harmony as well as precise divides the rotating object in equal sectors singularity and from this line flows all movement control we associate with what we believe is gravity. From the singularity this line represents we receive gravity. Mass has nothing to do with the entire process we associate with as gravity or movement applying.

What can move is liquid and stands related to what cannot move being singularity. Since everything is singularity therefore everything is immovable but also since everything is moving in relation to the point being immovable.

Every thing outside the top is liquid with the top forming a solid or so it seems to us. Well yes in a way and not that much either. The top is a pump that pumps heat from the outside inwards just like a turbine engine. Every atom that is rotating inside the structure of the top is keeping the centre erect. The centre is totally motionless because all the atoms in the top are moving and the moving of the top circle is extending the singularity of the top to the edge where the top meets eternity. The extending of singularity is holding the air as a liquid and being the liquid the flow of the liquid keeps the top erect and spinning. The spin produces a cold in relation to the hot that the liquid is.

I have indicated where to find that which can go no smaller and place that which can go no smaller in relation to that which can go no bigger. Also I have shown where to find that which can go no bigger and show why that which can go no bigger is the volume only eternity will have. Although the difference between that which can go no smaller compared to that which can go no bigger holds a Universe captured, the difference there are is so small or big such differentiation it has is beyond human understanding.

There is no substance difference between 1^0 and 1^1 and it is a relation where one moves as the liquid partner and the other is the solid factor. Both are not as much equal as they are precisely the same and yet hold an entire Universe apart. The difference there is lies between infinity that cannot move and eternity that cannot stop moving. By parting time as the unit, infinity had to form a spot that can never move and eternity had to introduce a reference to this point that can't move where everything moves as part of the cycle around a point that can't move where it stops moving in relation to the other side that cannot move but does start moving.

If 1^1 is 1 then 1^0 is 0.9991

The factor that shows motion forms the liquid while at that moment the factor that does not show motion forms the solid. The measure of 1^0 is transformed to 1^0 and which ever are 1^1 is passing the extending of space on to 1^0. Time spin because everything spins in order to secure the centre singularity. But also time moves and in that there is the linear that always are part of cosmic motion. The centre is referred to by heat but heat also secure the centre by reconfirming the centre in the lateral. But in both cases singularity is reinstating singularity by confirming as it is referring one another. In the manner that 1^0 confirms a position in singularity 1^0 is supporting 1^0 by generating 1^0. By generating 10 it is repositioning and reallocating a position by confirming 1^1.

What is moving is liquid and what is not moving is solid. Everything has a reference in relation to another point. That which is capable of relocating is forming a liquid in relation to that which is securing the position of rotation. Everything in

the cosmos can move and yet not one particle in the cosmos can move. The cosmos stands divided between the eternal moving of eternity and the immovability of infinity

Everything around the top is liquid with the centre being a solid. However the solidness and liquid has cosmic standards and just as it is in the case of hot and cold, big and small, fast and slow, our standards and cosmic standards do not share any measurements.

So too does cosmic notions about liquid and solids have a totally different meaning in cosmic terms.

There is a pumping what is liquid towards singularity.

interaction of space-time that allows the flowing of singularity through every point that confirms

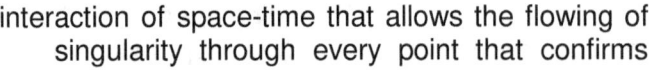

Everything that what forms the keeps the providing motion liquid factor that top spinning erect is effect is gravity. I shall

$$T^2 = a^3 / k$$

is surrounding the top is liquid. Being a liquid is substance supporting the upright stance that material forming the spinning top erect. By the matter that surround the top serves as the extends the space that singularity provide. The formed by the Coanda effect and the Coanda later on in this book define the Coanda effect

as well as for the first time in history I shall in terms of physics applying, explain the Coanda effect in the relevancy of singularity as it applies to gravity.

The structure that forms the body we associate with the top is composed of atoms. In the atom there is a governing singularity around which all material rotate. In the case of the atom all the rotating material forms the heat while it generates a centre, which is incapable of rotating, and it connects to material that as a unit forms the solid factor. Every aspect that is without motion stands in a relation of 1^0 and that which is relatively moving or changing location or finds a new position holds 1^1. Everything that is standing still is 1^0 and everything that is moving is 1^1. This discrepancy becomes apparent as the book develops.

$$1^0$$

$$k = a^3 / T^2$$

Gravity or motion is a constant relation that solids have with heat where heat forms the liquid and solids form space filled with material. There is the rotation but part of the rotation is the lateral progressing by rotation to confirm the generated centre. The generating is in the rotation but the flow towards is formed by the lateral and just as electricity produce a flow of time in relation to space collapsing, space-time by measuring of gravity is using the same system to do the very same when electricity is generated in this manner.

The variations in conditions are in relevancy with the Sun's governing singularity and are setting grounds, which apply different conditions to that what we find to suit us living on Earth. I do not wish to make presumptuous statements, but such rules apply very similar factors as the rules that bond compounds where singularity locks space-time of different elements in a relation and only by reaching specific counter conditions can the compound unlock and set free the elements. In that manner stars will regulate layer conditions in accordance with materials setting movement within the star but not by creating compounds...no, only by producing similar type of rules. This means not only does the Big Bang look different but also, everything about the Universe was different then from what is now applying.

$$1^0$$
$$1^0$$

It is not only the relevancies of heat and time putting the Big Bang Universe   much different from our perspective but the relevancies produced a completely different Universe altogether. It is not shocking that during the time and while experiencing the Big Bang, the Universe was a nice average temperature of 10^{34} K because back then with conditions applying, it was just a normal day during the Big Bang. It was another Universe, one being in the same space that a neutron holds today. That fraction of space in time had to accommodate an entire Universe potentially filled with everything as we now see the Universe holds. The Universe then had completely other rules than we have today. It is all coupled to the relevancy we find that singularity holds whereby gravity developed space-time. Everything changed as **k** extended and that proves that the factor **k** is the determining factor of space **a³** and time **T²**. The progress of **k** unlocked all other factors including layers in galactica, which in turn unlocked stars from the galactica cradle. The stars in turn had layers developing by measure of **k** and with that the Sun also developed because all other stars develop in this manner and so does the cosmos develop using this manner.

As sure as the Sun shines today so will Jupiter one day also shine as another Sun and then Jupiter's moons will be planets as large as the Sun's planets presently are, but by that time the Sun will be something we today think of as being awesome and awful. With the ratio fitting this tidally, the time duration and NOT TIME AS SUCH but the time it takes time to tick becomes infinitely shorter as it is coming from the eternally longer. In that is found also just a ratio. This goes totally against Newtonian religion and I am about to explain why that is true in a minute. At what temperature will water boil on Jupiter and the answer will most definitely not be equal to a temperature it boils on Earth. Even on Earth the temperature of boiling water runs along a spectrum covering many scenarios, but the temperature of water boiling on Mount Everest is not equal to the boiling point of water at the Dead Sea level. That evidence is never mentioned in terms of Newtonian constants applying as set rules. Ignoring such indicators only brought scientific miscalculations and scientific mistakes. If water does not boil evenly at all levels on Earth, then how is it possible that science can establish constants in outer space or for that matter try to introduce any constant applying anywhere? It is known that even on Jupiter the freezing point connected to hydrogen varies as one continues down the atmospheric layers of the Micro star. At one point the equilibrium between the temperature of the Sun core and that of outer space matched setting the singularity within the core free and allows the Sun to have a free inner core that then began applying gravity individually. That is why the Sun is dominating the entire solar system. Keeping these fluctuations in mind we must presume that at the time of the Big bang and even long after that the outer regions of the Sun then was as frozen as Pluto is at present. In reality in human terms we may think it was hotter or colder but if truth applies then conditions were just less spacious than what it is today in outer space. The cosmos was a lot smaller but that only made it more concentrated and being more condensed it presented higher levels of denser heat between atoms. This statement I shall retract when I get more technically correct in other more advanced books but in order not to get ahead of myself we will keep using this statement for the time being. As space grew and reduced at the same time the growing in space by reducing in heat intensity came as a result of a changing **k** factor that represented both the slow demise of time as well as the increase of space. This Kepler introduced while no one took any notice of such a possibility until Hubble came. By that time every one forgot to look at Kepler again. In the formula Kepler introduced we can trace a clear possibility of space growing. With the formula changing to $k = a^3 / T^2$ such **a** claim indicates hat space increases. It shows clearly that space grows while time diminishes as the relevancy reapplies in favour of materials. Materials holding space **a³** is inseparably linked to time in the square $k = a^3 / T^2$ as structures in orbit apply duplicating motion and $k^{-1} = T^2 / a^3$ as contraction recoups time that brings about motion. As space increases time demises. This I say full willingly knowing such a view does not represents the Newtonian and therefore the overall human outlook because they focus on what is present and with keeping an eye on what they can calculate by using constants.

Our tunnel vision comes from our stance where we see the cosmos we wish to see because we also only fit into a small slot. Our slot is blocking our view but don't tell the Newtonians that! As **k** develops **k** has to develop in all aspects of the Universe that is if we are to believe in the Hubble constant (another nice little constant with no applicable use anywhere). That means the layers in a star and in a galactica is the prone ones to changes brought about by the rules that the Hubble shift adheres to and the layers brings growth to singularity as singularity in governing stars come to life. They take charge as singularity allows the atom to progress and as the atomic singularity shifts the dominance to support the governing singularity in the star centre. This means that there is a governing singularity always with the orbiting or lesser singularity circling around the governing singularity. Then **k** progresses further and with that the **k** within the Sun and indeed all the solar structures progresses in equal terms but not in alikeness since every one point serving singularity has in place the charging of its space-time. This comes about as a result of another and most different singularity. It is this relation we have with the Earth and with Earthly relevancy we hold so dear in our occupying space on Earth that the Earth **k** holds every aspect including our minds and our ability in thinking under control. The Earth is our Universe and that we can see from all the constants and ultimate limits we place on the Universe in order to bring the Universe in line with our Universe which is the Earth and we have all standards going according to the Earth.

 We measure a distance by the length of a meter, which is a distance we take from where a line begins to where a line stops during specific time duration. That we then convert to a cube, which we attach to a mass by the thousand that we connect to the distance the Earth travels while rotating around the axis when turning yearly around the Sun as well as applying the gravity. Our measure of $\Pi^3 = \Pi^2 \Pi$ is completely different to the same measure we find in Jupiter or in the case of the Sun. Every cosmic structure will have gravity at Π^2 but the gravity will not be 9.81 Nm/sec as we have on Earth. Gravity is always at Π^2 but the gravity will be gravity aligning with Π in relation to the governing Π^0 we find applying in that Universe. That is why the Sun can freeze hydrogen at 6500 K and Jupiter can turn hydrogen into liquid at -150^0C, Jupiter takes less cooling to also have hydrogen in a semi liquid state. Gravity on Jupiter will be Π^2 no matter what because the Jupiter space Π^3 has a built-in seclusion from the rest of the Universe. Again the space Π^3 is directly in relation to the relevancy Π that we find in terms of that which Jupiter has, which stands in relation with Π that is the extension of Jupiter in singularity Π^0. That makes Jupiter a nice other world and our measure by meter per second and weighing in kilograms will be very different on Jupiter than we have on Earth. A six-foot man will most probably be some four-foot down at Jupiter's sea level where hydrogen becomes a sticky metallic thick liquid, but the man would not be an inch smaller. The man cannot be shorter because the man remains the same man as the man was on Earth. The man did not change for he only came into respecting different relevancies. The foot in distance has changed and the six in number could have changed but the man stayed the same.

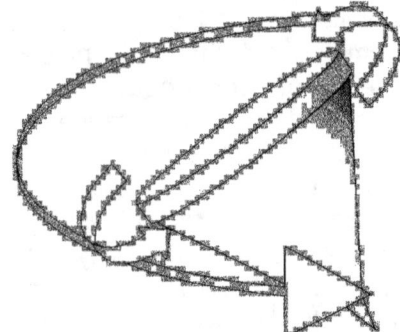 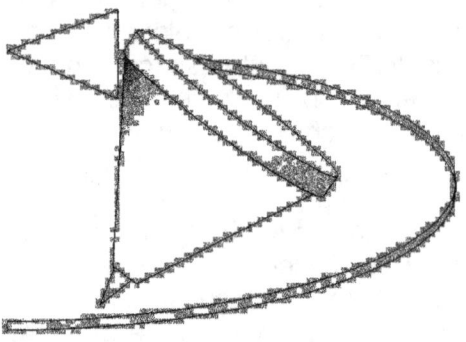

In the motion a line comes to "life" running through the very centre of the top. This line is not just another line but is there to focus the top in order to sustain the top to spin upright and erect. The line was not there when the top was on its side. By motion the line can concentrate an effort that will unleash such dependence to the top that the top will come into a position where the top have the tenacity to take the Earth gravity on in a dual.

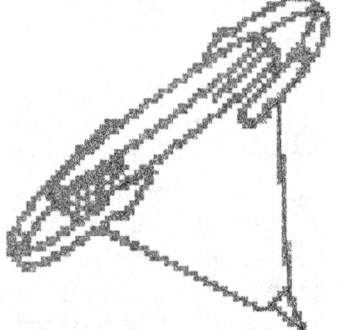 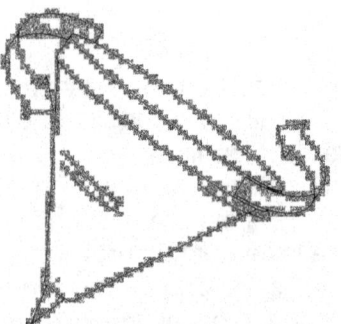

From every which angle one look at the top the top seems possessed and I can even be slightly forgiving towards Newton for calling it a force because although not a force the stance the top takes when spinning upright leaves on with an impression of forcefulness being part of the situation. In this one must not see a force but one should see the manipulating qualities of life extending to the top and by life's ability to manipulate space-time and control motion in space-time with space-time, the throwing of the top is as little a cosmic event as the apple Newton saw falling from the tree. Both instances were life controlling events and as far as there is proof there is no possibility of such an event taking place anywhere in the Universe. If you start to imagine about alien life living out side of the Earths parameters, or far away in the Universe you may just as well start believing in ghosts, fairies and all other fantasy creatures. Science must decide whether they wish to speculate about the fancy fill, but in such an event distance their fantasies from science and reality, or stick to science in reality and believe only in facts as science present facts.

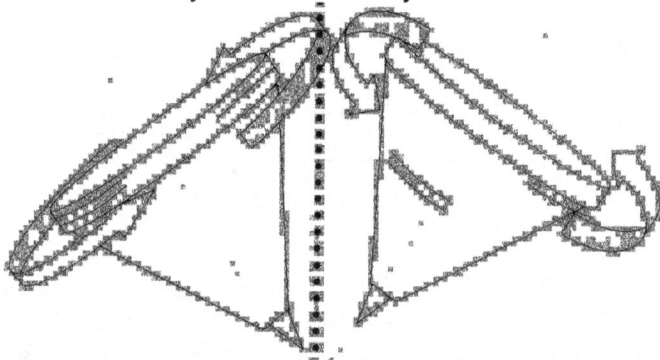

Let's consider what the facts are that apply with the top spinning as the top does spin when the top does spin. This is no fantasy or life coming from same imaginary source but it is a cosmic reality which life found a way to manipulate.

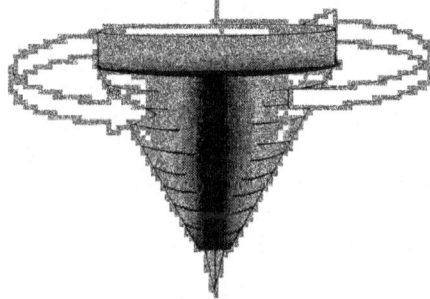

The effort it takes the top to spin gives the top a distinction of extreme significance. The top is promotes to a star in motion because it blew singularity to control space-time.

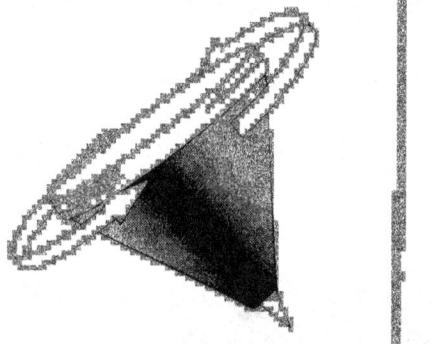 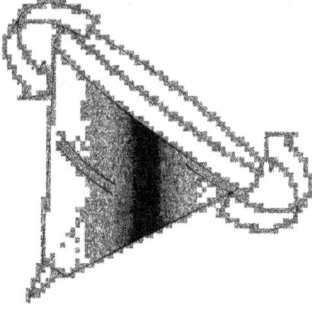

The definition of space-time is as follows:
Space-time is a four dimensional position of the Universe where the position of an object is specified by three coordinates in space and one position in time. According to the theory of special relativity there is no absolute time, which can be measured independently of the observer, so events that are simultaneous as seen from one observer occur at different times when seen from a different place. Time must therefore be measured in a relative manner as are positions in three-dimensional Euclidean space, and this is achieved through the concept of space-time. The trajectory of an object in space-time is called world line. General relativity relates to curvature of space-time to the positions and motions of particles of matter.

In view of the definition of space-time I wish to elaborate on my view of singularity and my deriving of space-time from the likeliness that singularity may produce space-time. In the past singularity was mentioned in the manner one would speak of a ghost hiding in a haunted Black Hole. Let's put singularity in the clear. Singularity is within every sphere due to the natural shape or form the sphere is committed to.

According to Einstein singularity is a mathematical reality within the Black Hole but much more so in every sphere. Einstein may be the first to name it and Galileo (unwittingly) may have been the first to define it as Kepler was the first to formulate singularity, but in mathematical terms singularity is the most basic principle. At this point I wish to establish a fact that seems lost in all other grandeurs of cosmology. When tracing the radius down into the sphere the radius stars where all lines start and a straight line cannot begin at zero or nil it can only start at infinity. Such a statement will hardly seem appropriate but the relevancy of this fact has no limits.

POINT OF INFINITY

If the line started at zero there was no line to start because zero multiplied by whatever results in zero as the answer. That must also be the cosmic starting point. Einstein introduced such a point and named that point singularity. When looking at the cosmos from whichever angle the indications seem to be that the fact that the cosmos is entirely in motion. It is forever spinning and it is going too as much as it is coming from. Everything is on the move and always encircling something of greater importance. A top can spin but the parameters of its spin are limiting the motion it can apply. By not spinning the top is still spinning as the Earth is doing the spinning on its behalf.

When spinning too fast the top fights something because the alignment keeping it upright starts to tarnish. The same apply when spinning too slowly but that makes sense. It is the fact that the same affect comes about when spinning too slow that triggers the questions. Why would the top stand upright by spinning. It must be because singularity charges the top into a cosmic independent reality.

The spinning top is all the evidence any one needs to come to such a conclusion. I know probably as much as any graduate about cosmology but lack certificates to prove my knowledge. I am not part of established science. In my developing of knowledge accumulation I came to some conclusions about cosmology that are unique and divert somewhat to drastic form the accepted norm. Most of the work I see the same way as the norm does but in a reverse. Allow me a short explanation

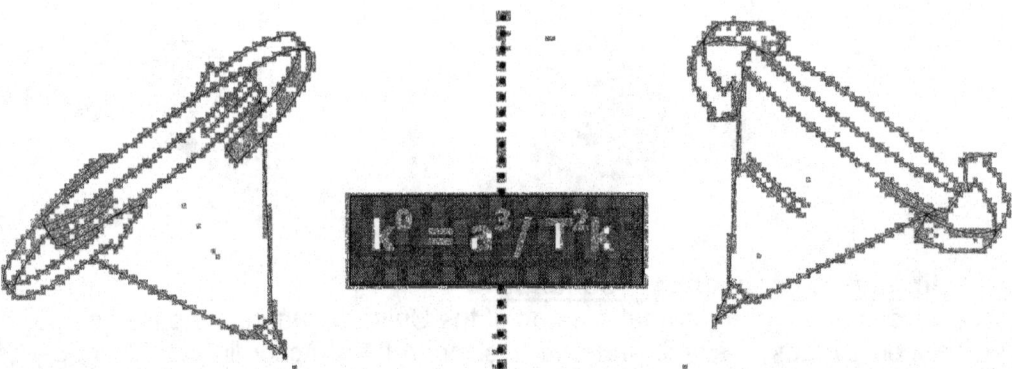

$$k^0 = a^3 / T^2 k$$

We have to be clear about what we think of when we think of the Universe. Most people think of a picture recalling the black night sky when thinking of the Universe and that thought is most incorrect. Einstein was most correct when he declared the Universe was going flat where gravity is at its utmost, but the concern we should have is not with the mathematics being valid or not but with the vision about the Universe being what we think of and where we place the Universe. The Universe is in the centre of what is spinning and the biggest single particle that is spinning in total independence of the rest of what forms a total Universe is the atom. The atom spins and by the motion the atom evokes the Universe forming what must be the group effort of all the atoms then spin by the motion the atom renders the rest of the larger Universe. The Universe is the part that allows the rest of what the Universe establish to spin. What spin you may ask. Kepler said it without saying it: $k^0 = a^3 / T^2 k$ and not even Einstein with his super human mathematical skills could say it better or more accurately.

The motion established by singularity results in the implicating of the Coanda effect as much as the motion establish the Coanda effect. The spin realises the space limit while the space limits attaches the motion onto the space in the time within the time.

With the top spinning the Coanda effect steps in and does justice to Kepler's formula.

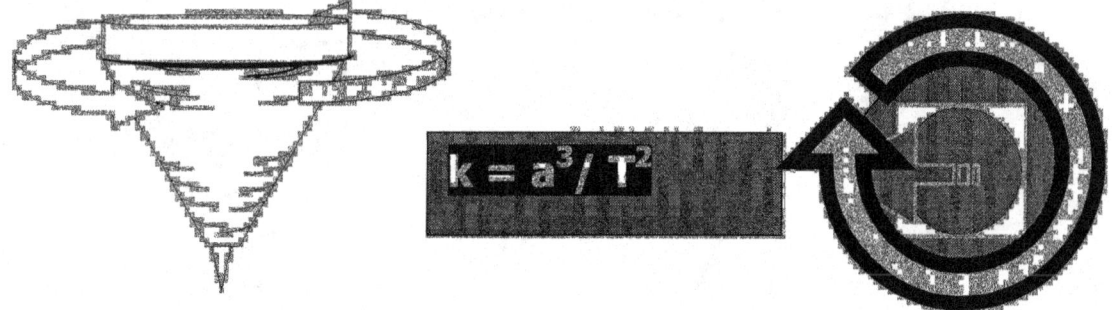

Time is always a displacement of space in relation to the implication of singularity, and comes about between two points in space relating to the centre of singularity as positioned by **k**, either too the value of **k** or too k^0.

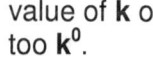

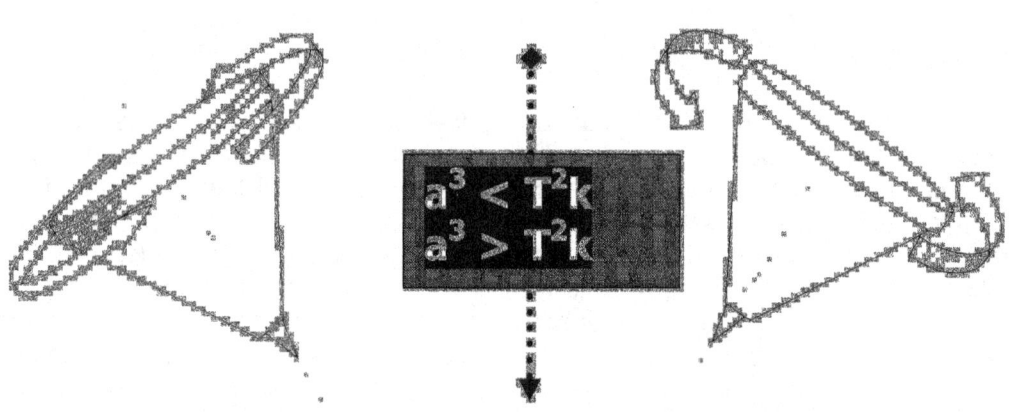

I have asked as many persons as I do not care to remember why the top sinning will remain spinning around one point while turning. The answer I receive from the most educated to the schoolboy is always about momentum. That is a very simple answer and to say the least a little too simplistic by further analysis. Why would the spinning top go of centre when spinning higher than a specific velocity and lowering the velocity it would stabilize and run square to the Earth only after that it will go oblong and then fall. I could go on about different positions bringing across different momentum of thrust but I do not wish to insult your intelligence because I am aware that you are familiar with all the law. When the top is spinning it is spinning about its own axis and when it is not spinning it still remains spinning about the Earth's axis therefore when it is spinning it is also spinning about the Earth's axis. Therefore the limitations applying can only result as an influence coming from the Earth's axis. The second question now comes screaming across and that is in what manner could the Earths axis ever affect a spinning top since the spin and he spinning top is a gross mismatch to what ever standard the Earth may introduce. It is clear that spinning objects do influence each other in contrast to Newtonian opinion.

Every round object has a point establishing a very centre, a middle dividing one side from the other. That division determines the space from one side away from the other side. At one point there must be a point that does not fall on either side of the divide. Such a point will still be a circle, because from that side the circle divides into two sectors.

Π⁰ In every spinning object there is a point of infinity, a point that does not turn because it holds the dividing spin. However when such a point becomes a line that cannot spin a new Universe is born in the midst of many others. At the birth that point diverts space outwards and from that point the spin is either clockwise or anti clockwise in all directions. As I pointed out no line can start at zero because then there is no line and no rotating point can start at zero because then there is no rotation. Calculating a square involves two aspects that we think of as sides.

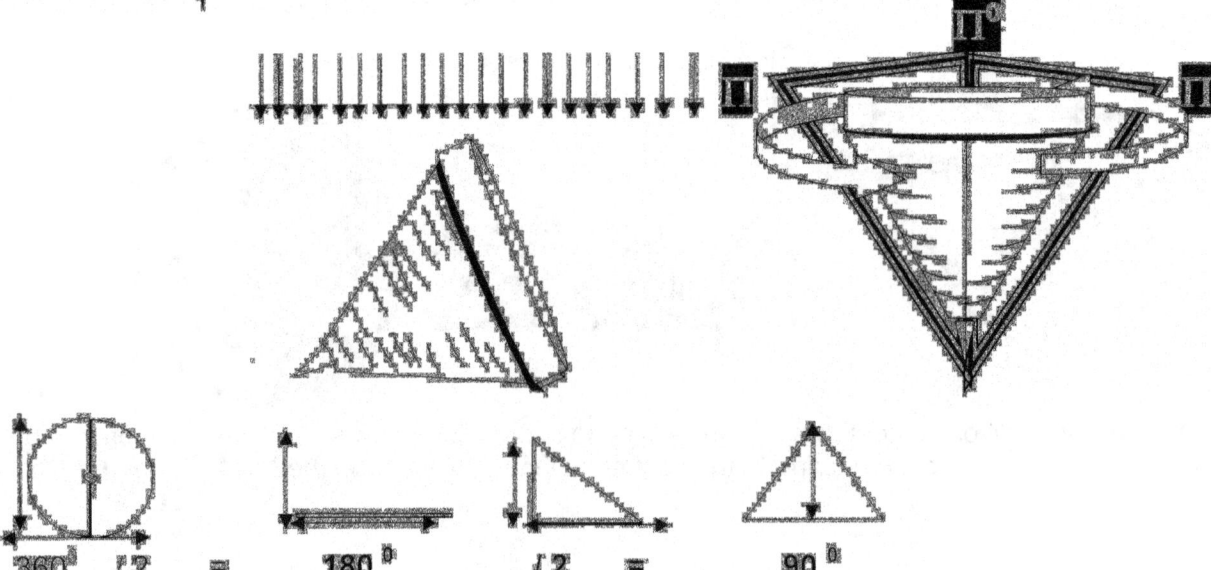

There is a Universe in differences between the top lying down without any individual motion and ostentatiously independent, self assured spinning top that even produce a sound to match the occasion. While without motion the top submits to the contraction lines running as the straight line holding half the value of the square being 180°. The top seems dead as it surrendered its long-term position and would eventually succumb to the Earth's gravity by relinquishing the structural independence it has. Then the motion brings life into the top and gives the top reasons to fight the Earth by fighting for independence. The top just became independent by the motion it received from the combined efforts of all the independent atoms forming the structure of the top.

The circle is a square holding a round shape, as the straight line is a square holding one side to infinity. Calculating a circle involves two aspects where the one is either the radius or the diameter that is double the radius. The other is the factor Π

Because gravity work both ways and not singularly in one direction as the Newtonian myth would have us believe, there is the interaction in the neutron position between the total of material in relation to time formed in space as space and time formed in space in relation to the total of material.

With everything in a cube or a circle or a potential of the two, brings about the implication of eternity in a form of singularity or the point of creation. Removing the radius of a circle does not remove the circle, because the circle is there, securing the ring. If the line (or imaginary line if you wish) holding the value of $\Pi^0 = 1$ there has to be a point where the circle is no longer in infinity but claims existing outside the imaginary. At that the radius may be lightly more infinity, but to all purposes it still infinity.

point than calculating remains as With this come to introduction piece. Professor, it took a while to get to to follow you instructions and get to the point followed your instructions to be as plain in writing as waste any of the Professor's time and so I did. Yet, with my best attempt it took me many pages to

statement I have conclude my the point and I tried immediately. I possible in order not to

come to show what I am setting out to prove in this book. In this book I am about to prove what no other has achieved, which includes up to and including Newton and Einstein To prove what no other could prove took me a while to introduce because it took others centuries not to find what I am about to show is true. I am going to define gravity. I am going to prove the value of gravity. I am going to show what gravity is and no. it is not a magical force as Newton estimated it is.

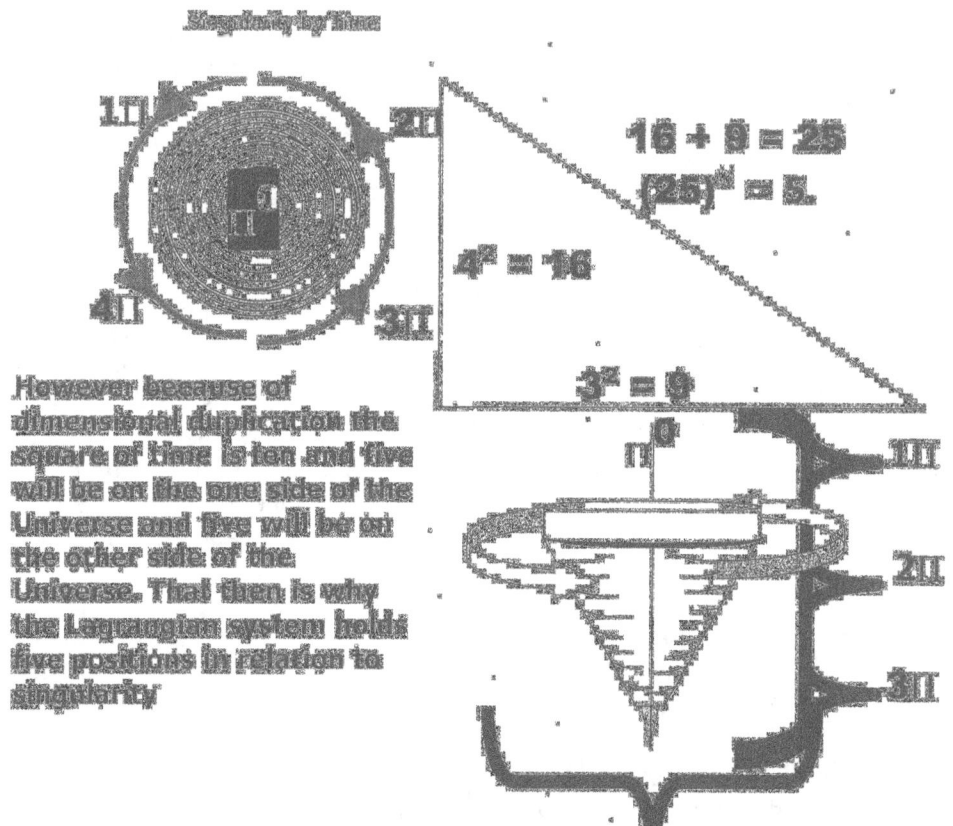

However because of dimensional duplication the square of time is ten and five will be on the one side of the Universe and five will be on the other side of the Universe. That then is why the Lagrangian system holds five positions in relation to singularity

$16 + 9 = 25$

$(25)^? = 5.$

$4^2 = 16$

$3^2 = 9$

Newton's   claims

about the principles he declared as being responsible for guiding physics carry no proof and after I realised that, I was able to start forming another line of thought on gravity. After formulating my concept about how gravity was truly formed, I had to introduce my ideas to academics in physics. In my quest to find the method how gravity formed I used the four phenomena and the principles of these phenomena as well as determining in which way each phenomenon applied. Not surprising is the fact that the shear existence and acceptance of the four cosmic phenomena is very securely hidden from normal investigation. Since Newtonians can't use Newton to explain the **Titius Bode law**, the **Roche limit**, the **Lagrangian positions** and the **Coanda Effect**, these phenomena are

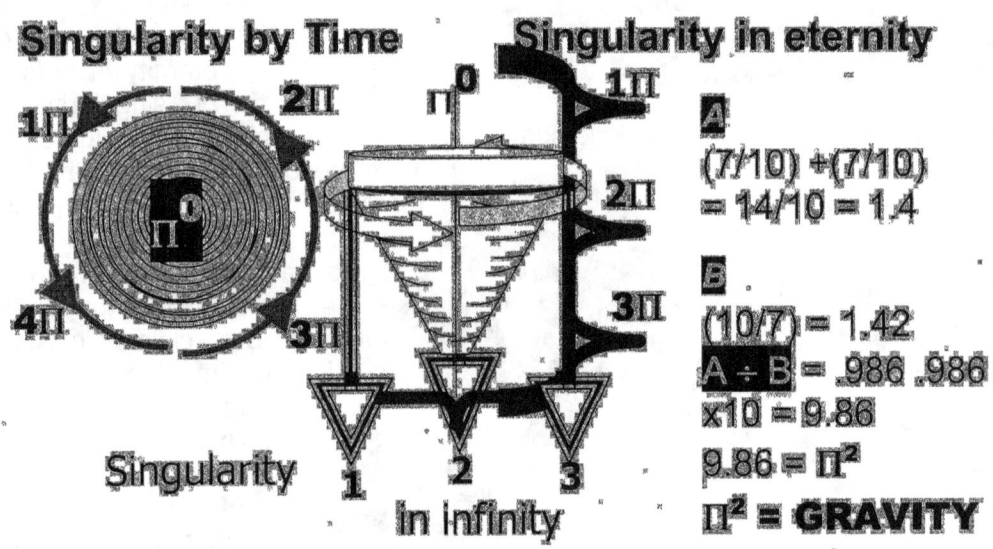

hidden so deep that it is very likely that you as the reader might never even have heard of these phenomena. It is because these phenomena portray Newton in a dim light and that no Newtonian would allow happening! I use these phenomena to mathematically

prove what gravity is, but the condition I apply is that gravity is not even close to what Newton suggested forms gravity. By my ability to prove gravity I had to place each one of the four in the way forming one part of what gravity is. Before I could manage that, I had to find a way as to explain what the working relation of the four cosmic regulating phenomena were and I had to figure out what

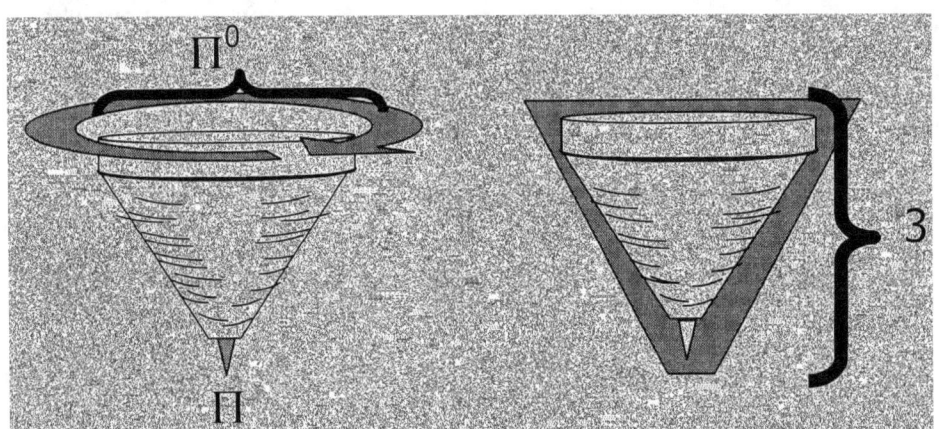

was the known contributing facts to the cosmos that each was responsible in contributing to gravity in my effort in determining how they work and then implicated that specific formula's function mathematically in forming gravity in the cosmos. This was no easy task but I did it

and by using Kepler's formula I managed to show that my argument is logic and the mathematics prove that it works well...and yes I still maintain that mathematics has no place in cosmology since the human mind is to small to even start to understand cosmology...but by completing this little book you might just agree with my statement about mathematics and cosmology.

The spin was going on for eternity because the spin does not apply, it has a value of zero and zero is another expression for eternity. In the above sketch I show how the four cosmic phenomena, which I named the four cosmic pillars come to form gravity mathematically.

Titius Bode law, The **Roche limit**, The **Lagrangian positions** and the **Coanda Effect** come together and allow the top to spin as a result of the Coanda effect applying.

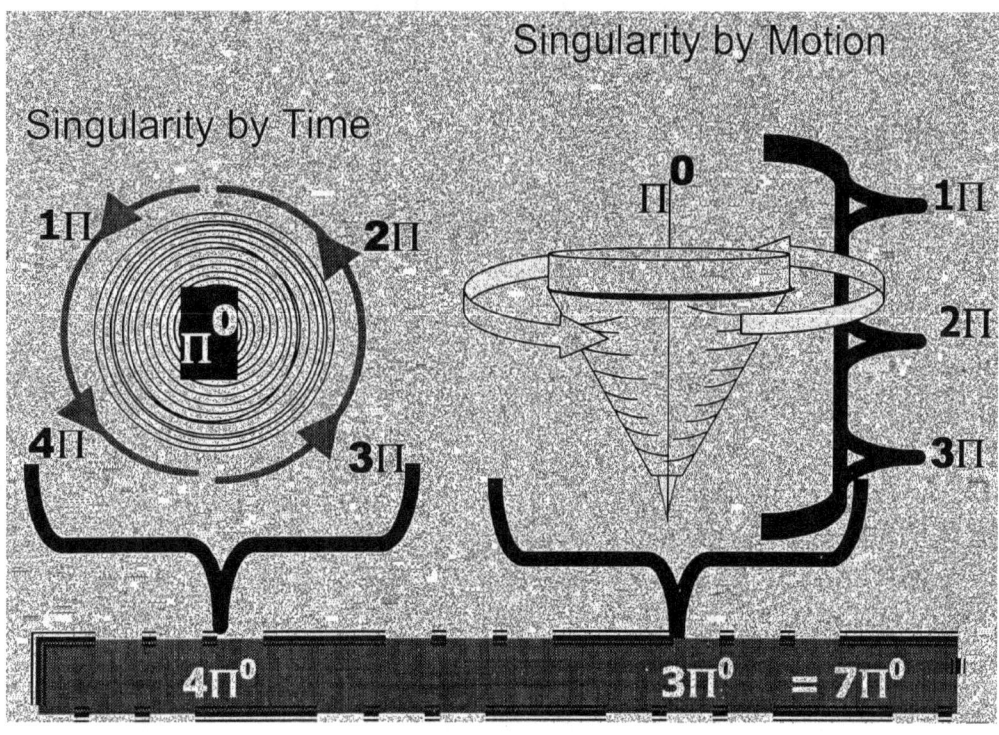

That was the first dimension outside singularity Π^0 where singularity has a value of Π^1 in the form of $\Pi^{1+1=2}$. The first claim to space had a value of Π^2. This applied to both sides of the claim to space outside singularity, and the double proton became the dominant factor on matter.

By receiving space, singularity received a value outside eternity as Π^0 received edges. Granted the fact that the edges were so small there still was no r to present a circle.

In the center runs a line called the axis line. The line does not show any influence on managing the top when the top is motionless and bounded by the Earth gravity. However the sooner a motion sets in that is adequately strong enough to support the independence of the top the top generates enough gravity to sustain and independent altitude in relation to the Earth.

The top then maintains an electron in relation to the Earth being the proton and the atmosphere forming the neutron

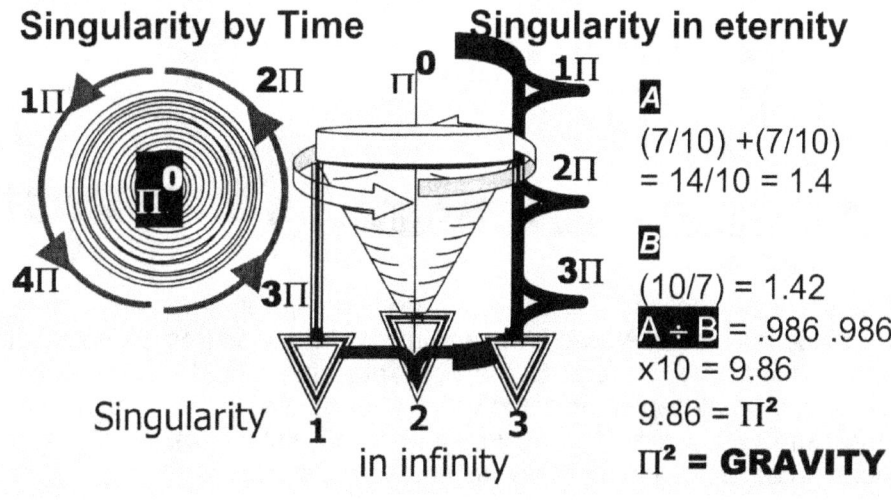

As I said, the rotation movement apply in relation to the centre and this brings about gravity taking effect by the duplicating of Π as Π^2. Mass has no role because mass is a institutionalised man-made fabrication of what never was a cosmic principle to begin with, but I have wrote many books in which I prove that and at this point in time I am leaving that argument for those wishing to read the books in which I prove Newton wrong.

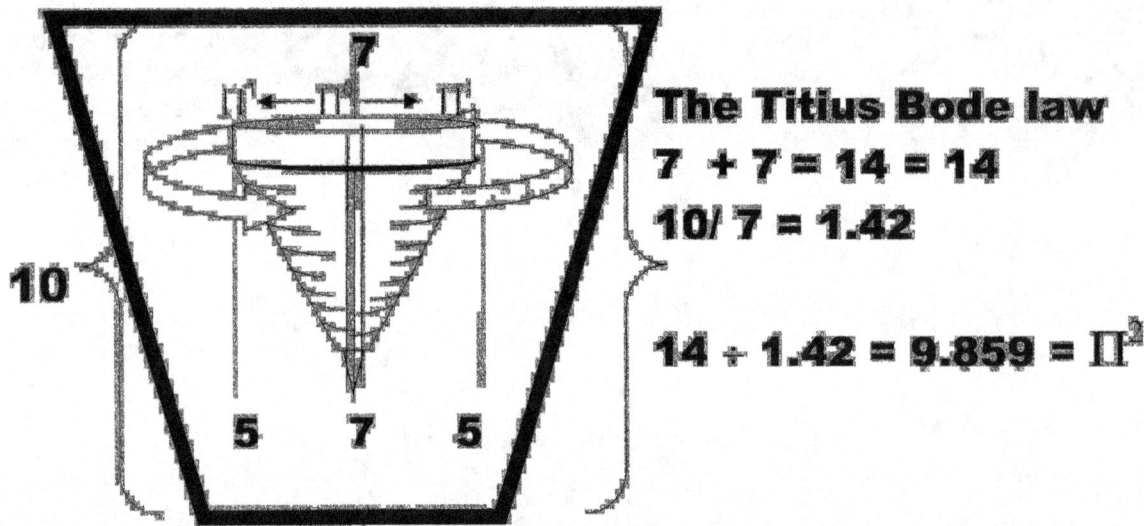

The Titius Bode law

7 + 7 = 14 = 14

10/ 7 = 1.42

14 ÷ 1.42 = 9.859 = Π^2

Taken from the point of rotation the two sides are in opposition to each other in every aspect that they may contain and with all that they hold.

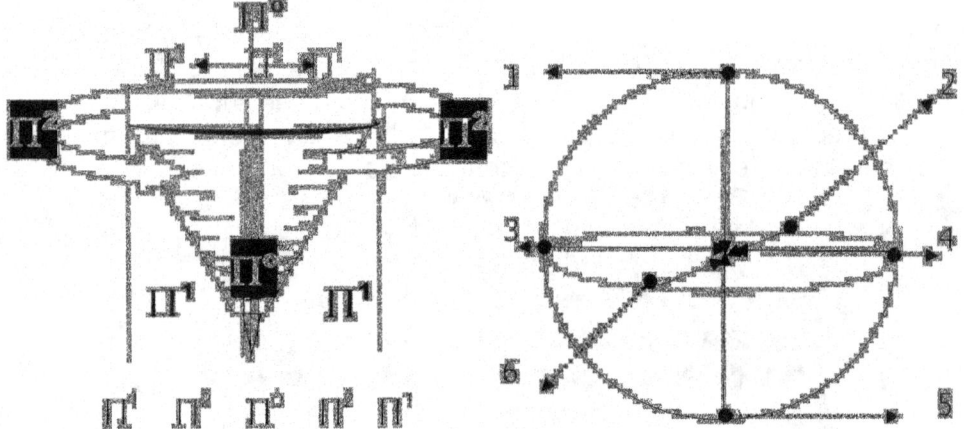

With Π^0 little more than a figment of the imagination there is actually to values of Π^1 facing each other in a relation combining Π^1 to hold the value of $\Pi^{1+1=2}=\Pi^2$ and with two sides being the very same but opposing each other there will therefore also be Π^2 to every side that holds Π^1.

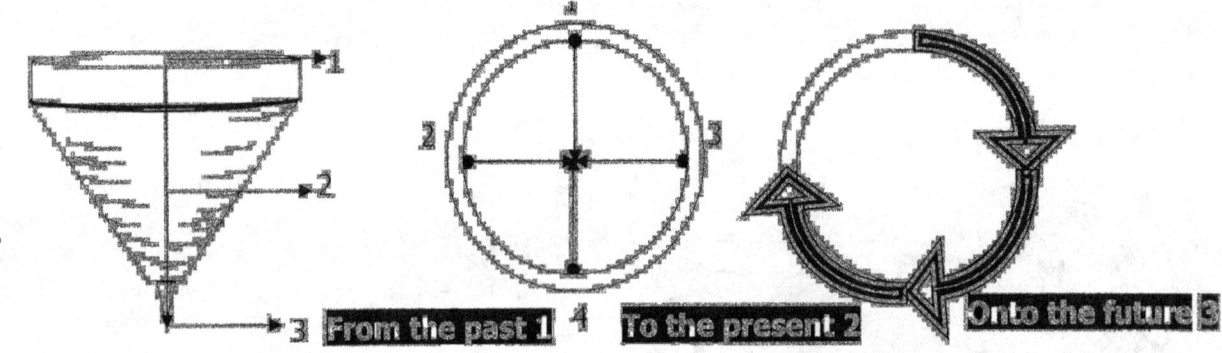

From the past 1   To the present 2   Onto the future 3

Using such logic makes science appear foolish. There is just no rational in the time verses events that can explain facts without. Since the time of Newton, the arguments tarnished from being brilliant to clever to fair too poor and a hundred years ago to the point of being stupid.   That is what Kepler's formula is all about?   That is what Kepler indicated with his formula $a^3 = T^2k$. The space of an object (a^3) is equal to the **time (T^2)**, which it is in, in every given instant **(k)**.   If the space becomes smaller, the time duration becomes longer every instant of time's progress.

When realising the error of science in accepting a value as zero to be legitimate in mathematics, one can establish from that that the circle does not employ zero as a value after the completion of one rotation therefore $F = G \dfrac{M_1 M_2}{r^2}$ is invalid, one has to return to Kepler's $\mathbf{a^3 = T^2\, k}$ and establish a value from that.

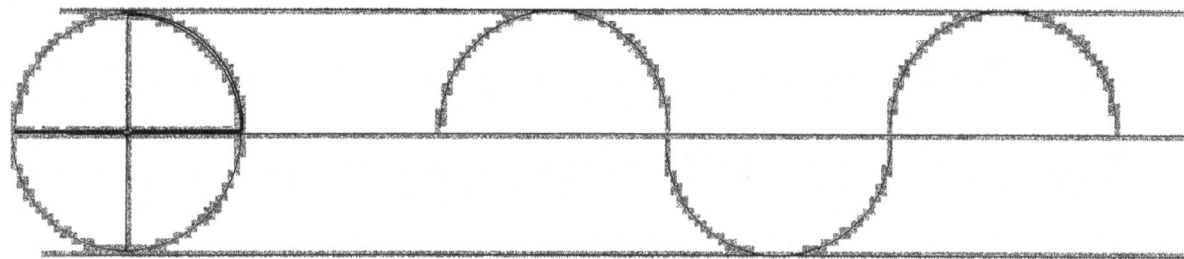

From the graph one can establish the link in the circle's rotation around a conforming unit being singularity.

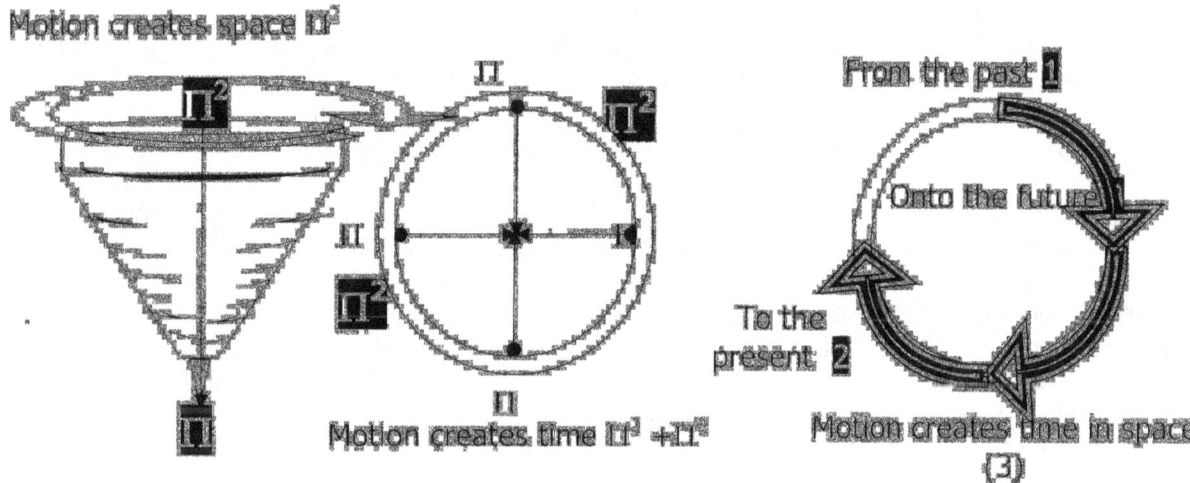

Saying that one therefore has to admit that the smallest spot has to hold space because the most insignificant dot can transmit light and being able to accomplish that, one must accept it to carry a value of something. If that spot had the value of nothing, it means that spot was not there to begin with. Holding space-time one should return to the original formula indicating space-time in as much as $\mathbf{a^3 = T^2\, k}$ where $\mathbf{T^2}$ is one of the time components and the other time component is **k**. Being time it has to alternate positions and that can therefore only apply to k where k will indicate a relation to the space-time in question or the relevancy to singularity being $\mathbf{k^0 = 1}$. **By receiving k on top of the already $k^0 = 1$ that is in place the top becomes an atom by erecting the line of singularity from $k^0 = 1$ to $k^0 = a^3 / T^2\, k$.**

Singularity: a mathematical point at which certain physical quantities reach infinite values for example, according to the general relativity the curvature of space-time becomes infinite in a black hole...but why that is, is the question to be answered...it is because the star is a combination of all the atoms forming the movement within the star.

With my common with the the start to the end. theory I present I shall show what exploding double stars have in Hubble Constant and how that fits in with the life story of Creation from

At the heart of bringing about the solution to one of the greatest Astronomic riddles one will find a child's toy…it show how the riddle of Einstein's singularity is solved by pointing to the position where the cosmos finds the centre of the Universe.

In the spinning top, matter would always relate two three positions as does Einstein's space-time declaration require.

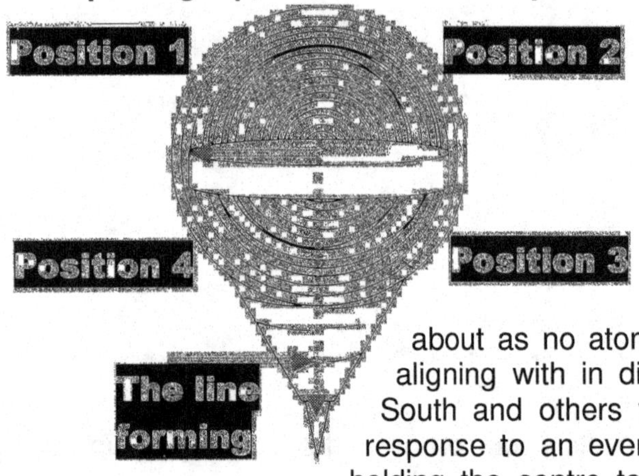

All atom particles forming the matter composing the top would have to relate to the centre and two other positions.

All atomic particles would have to refer to a point from where the atom would either be coming or going as much as circling around. This comes about as no atom can be ignoring any of the three positions it is aligning with in direction. Some would be on route from North to South and others would be on route from South to North. It is a response to an ever changing directional re-aligning with the centre holding the centre to a specific location in relation to the position it previously owned and would own the next minute, which is changing constantly and such changes is in accordance with time location. All matter will have to adhere to any of the two directions, which in fact is actually four, but it also changes dramatically every moment a new position sets the relocation conditions. In the centre a line MUST form separating the comings from the goings and again the goings from the comings, where no matter can be located. That line is too small too hold any atom, sub- atom particle or matter of any kind. All matter is either on the one side, or on the other side, but never can be neutral. It even gets more complicated because another line forms locating matter in groups. The one group is relating a poison from the "centre point" holding "back" and "front" running through "the centre" where the other line is relating from "side" to "side" running through the "centre point". The fact of the lines is that "they are there", but we cannot see them. Try as you may, no one will be able to calculate the very position that forms the lines, but as they change all particle characteristics, the lines are a reality as the spin of the matter is real. Being to small to hold atoms, they then therefore must become part of singularity, where singularity is a spot in the centre with two lines crossing the spot at an angle of 90^0. That is the basis of singularity, and since all the positions still relate too a centre of a circle, forming a part of a spinning circle, Π must form the basic value.

In that there are one specific group in relation to coming towards the centre while coming towards the front.
The following group is rushing away from the centre while coming to the front.
The third is heading for the back while heading for the centre and…
The fourth is rushing away from the centre while heading towards the rear.

In being the onlooker, the viewer has to maintain one position. From that position some particles would be circling a centre point, as the particles would be coming towards the onlooker. The other matter would be circling the centre point while rushing away from the onlooker.

When I look at the sky I see light being bright and white and I see darkness. The sky is either white and luminous or dark and black. From our vantage we see the future as darkness eternal, time to come but from the future the space holding time will view us as brilliant, luminous, light, so concentrated space is liquid or solid, the very same way we see galactica in our past. It all depends on the relevance of matter as atoms being the Universe forming the Universe as the density of such atom determines space and time. Keep this in mind while reading the following:

If those Newtonians read the Bible more often and paid more attention to the Bible they would have been in a position where they could solve the physics about Creation much sooner. The Bible said

The Creator commanded Creation into existing with the Command "Let there be tight...and there was light." The only thing science was required to investigate is the command "Let there be light." What is light? Light is heat that is moving. This Biblical statement says the Universe is heat that is expanding and contracting because if heat moves as light would do then heat either expands or heat contracts. The Universe is heat moving. The Universe is heat or light in some form.

Motion creates time in space (3)

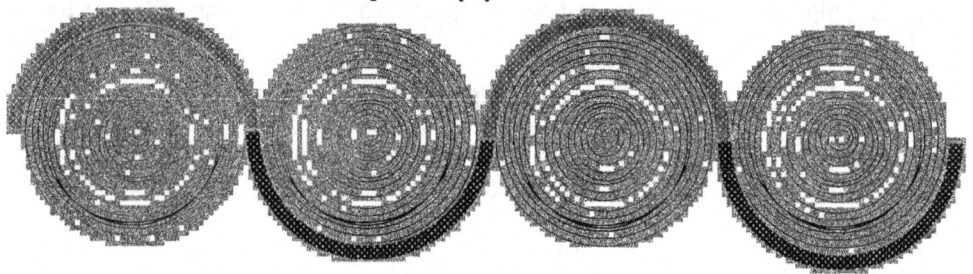

$a^3 / T^2 = k$ or a^3 / T^2

$k = k^0$. With this fact established we then must return to the value as indicated by singularity being Π. In this we find that $\Pi^3 / \Pi^2 = \Pi$, weather k is Π or Π^0. This brings about the value relating to space-time relevancies as a formula consisting of $\Pi^3 / \Pi^2 = \Pi$ in various forms and relations. One also must keep in mind that there are ALWAYS four sides relating to the Universe from any point holding singularity, and since every point in the Universe contains singularity in what ever form, very spot in the Universe comprises of four points initially extending to the next spot by means of $\Pi^2/4$, which we know as the Roche factor.

From such a relevancy there then must be four different values relating to singularity and since the atom has a proven relevancy of $(\Pi^2 + \Pi^2) \Pi^2 \times \Pi \times 3$. In the motion that replace the motionless, the motion made the motion less an atom by putting the object through the commitment of motion into an independent Universe where every aspect of the Universe becomes an Individual atom that maintains its independence as all atoms do.

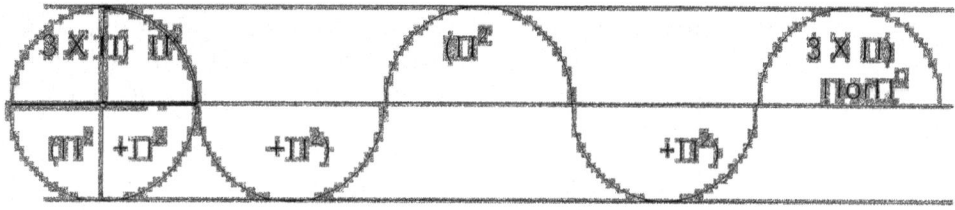

$\Pi \times 3 \times \Pi^2 (\Pi^2 + \Pi^2) = 1836.$

It (everything we think of as the Universe) started with a dot, because that is the only form, size and dimension mathematical logic will allow our brain to accept. From the one dot had to come a second dot and a third dot. The dynamics of such a dot is smaller than we can understand because such a dot is in negative relation to what we see Π to be, and the deeper we delve in finding the smallest fragment where space started, in the spot where time is still eternal as much as we can accept eternity to be. The reason why we should first locate the spot is because we can only work from that point forward. By working forward we have to work backwards to locate where we are heading. The cosmos started at a point and where such a point is, we will find the Universe. Every one knows where the Universe is, because we can see where the Universe is, but if we can see where the Universe is, then we should find the centre of the Universe in that spot. Einstein theoretically positioned the point of beginning at a place he indicated where singularity should be.

With the cosmos the size it is and space so large compared to our smallness we have no chance in finding the centre of the Universe. The Universe started where singularity is and singularity is the sure indicator of the Universe. With all spinning objects holding singularity we then have located singularity in as much as finding the centre of the Universe. The Universe started with a dot forming. That answer arrive from taking mathematics back to a point of being the smallest possible position, far smaller than we may be able to calculate form.

The ten dimensions I named the atomic relevancy is also showing the double value of singularity as singularity extends into as well as beyond space. The atomic relevancy is $(\Pi^2+\Pi^2)(\Pi^2 \text{ X } \Pi \text{ X } 3) =$ **1836** that is the mass relation between the electron **(3)** and the proton. Proton = $(\Pi^2+\Pi^2)$ Neutron $=\Pi^2\Pi$. The atomic relevancy holds the dynamics of singularity control. In the ratio and dimensions we find in the atom, all space-time derives from the atom, whatever the atom is.

Our instincts, our logic and our calculating process all indicate that the sphere holds a centre point from where six evenly positioned point's position matter to be. Using the formula

$$F \;=\; G \; \frac{M_1 M_2}{r^2}$$ it indicates to a force pulling objects closer, where each force is coming

from each centre point the body in question has. The contraction must commit the two bodies towards a point in each case being spot on in the middle, not withstanding what direction the force is applying, the body will draw to the centre.

If the Universe spins around a centre point holding singularity, and singularity confirms the centre of the Universe, then every particle holds the centre of the Universe making the number of universal centres immeasurable many, and every atom and sub atom particle presented outside the atom in smaller bits, are all not pieces of the Universe but they are a Universe surrounded by many Universes. If every atomic particle no matter how small is holding the centre of the Universe, then the gravity is coming about from that point because that is where the gravity applying in the Universe are applying contraction.

It then is the atom in the most centre part where space and time meets singularity, that Einstein found a Universe collapsing to a single dimension, and every atom at a point post of the proton where gravity initiates in according with the proton dimensional colas of $(\Pi^2+\Pi^2)(\Pi^2 \text{ X } \Pi \text{ X } 3) =$ **1836.**

The major problem physics encounter is that the entire faculty observes cosmology as if the Universe was stationary. This is a result of Newton incorrectly placing F=ma as a valid expression while the only validity this may have is when we see this statement of a photograph and a photograph shows what Newton said it does that $a^3 = T^2$ where the three dimensions we wish to see is represented on a flat piece of paper forming a square T^2 leaving a dimensional vacancy as a relativity.. This is miles away from physics.

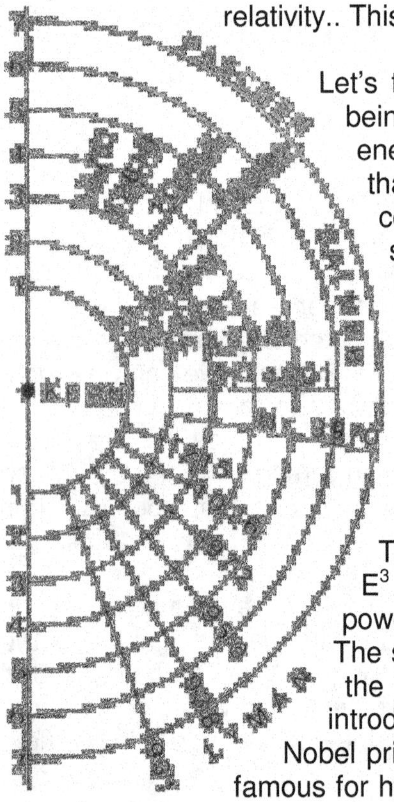

Let's take a look at another formula that Newtonians equally claim as being Gospel. We have a look at Einstein's formulation of nuclear energy, which is what we find in the atom. Einstein said E= mC² and that gave the man eternal fame and raving reviews. However this is completely wrong. It is so wrong it should be rejected outright. It should read $E^3 = mC^2$ and with that the formula has substance and true meaning. His was so grand the man got the Nobel Prize but is it as wise as they seemingly anticipate that it is. Try to reconcile this formula with Newton's F = ma where the formula suggests a stationary situation. If we had F=ma the force F would be representing a square and that makes the formula something we don't find in the Universe, notwithstanding Newton's clever but incorrect arguments to prove his case.

The photograph depicting an exploding star shows Einstein's formula $E^3 = mC^2$ applying because as space expands, it does so by the third power because the space expanding does this in the third dimension. The space E^3 is expanding in a circle C^2 as well as a straight line C^2 by the speed of light being C^2. Einstein replaced the factors Kepler introduced as $a^3 = T^2 k$ with $E^3 = C^2 m$ and with that Einstein got himself a Nobel prize on the trod, well not exactly for that but Einstein is much more famous for having made this formula than he is for his Nobel prize. Yet not one since Einstein's revelation saw that he re-invented what Kepler brought to science centuries previously and that what Einstein brought to physics makes Newton ridiculous. The fact about energy is that it is heat. It is heat called energy by those not knowing what to call heat that places everything in motion and heat places everything in motion.

When heat increases it expands into more space and when heat reduces it decreases in space. Heat is not what I feel when I feel it is hot at 40° C and cold is not what I feel when the room temperature goes to 40° below freezing. Cold is when that which is cold can reduce no more into smaller space and heat is that which cannot increase into a larger space. As usual with everything applying in the Universe, Newtonian physics has connected hot and cold to how life feels about what is hot and what is cold.

When the atom increases heat levels it expands in volumetric space occupied and when the atom decreases in space it reduces heat levels that is affecting the atom. Should you not believe me about heat increasing volumetric space as heat levels rise, then put water onto a sealed canister, place this sealed canister onto a heated stove element and wait a while for either me being proven wrong or you meeting your ultimate destiny…and while you are going to the other side remember to send my regards to my Grandma when you meet her on the other side. An overheating geyser blows a house completely away and the aftermath of a geyser overheating the result of that happening I witnessed with my own eyes. All that was left of this house was the foundation on which it was built.

As usual Newtonian science fills volumes with names they dish out to whatever they try to present as elaborate and meaningful and wise, but in the end after all the fancy talk has impressed who ever should be impressed with those names. The atom growing in circles is heat increasing the levels of orbit in the atom.

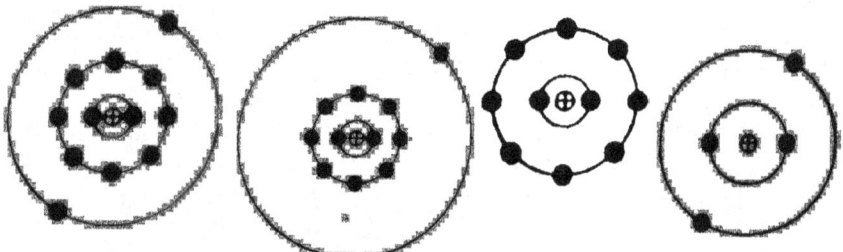

Magnesium (12)   Sodium (11)   Neon (10)   Beryllium (4) Trapped inside the atom is heat and trapped on the outside of the atom is heat while the atom is a little pump spinning like mad and by such elaborate spinning movement it freezes heat from the expanded outside to the concentrated inside of the atom. That is what gravity is. Gravity is having the atom spin at speeds where the speed on the outside of the atom goes faster than the speed of light putting the electron moving at the speed of light and this puts all movement on the outside of the electron slower than the speed of light.

The Universe is a combination of many material formations holding positions in space. Some of such material was covered in the blanket of heat, distributing into more spacious surroundings as the material expanded from the centre flowing outwards. Hubble's constant is proof that the space between cosmic structures are departing from many centre positions between such objects and this is a trend being located between all the objects through out space but also indicating a definite growth in the radius and such radius growth follow a patter where the growth seems to flow from any such a centre point away from the centre. With out the absolute and undeniable proof coming from the Hubble constant bringing proof beyond any possible doubt in any one's mind that expanding is very much and a very big part of all Cosmic activity the accepting of the Big Bang would not be in place. $F = G\dfrac{M_1 M_2}{r^2}$ is in essence a big issue about contraction while

Hubble showed the space was not dividing. The space was multiplying. The stars are growing apart and so is the galactica. This then brings in the question of space available.

With me not whishing to go into the formation of structures at this point in the book I would like to point to the fact that my following referring to the solar system is actually referring to a similar solar system that is somewhere and is now a part of a galactica we do not know about. I bring this in to disqualify any academic loophole that may come about from an argument about the solar system coming into place at a later stage of the cosmic developing and therefore the argument I am about to present that such an argument does not apply. To avoid such a loophole we now use a hypothetical but real solar system in space, which formed as the Big Bang took place. There are those who avoid admitting to inconsistencies by arguing that my argument about growth is invalid because the solar system was not in place at the Big Bang. To them we now present a solar system that is identical to the one we know in precise duplication thereof. But it represents a precise duplication of our solar system and was in place ever since the Big Bang. That means with the solar systems being apart in millions of kilometres there was a time the planets and the Sun were apart by the measure of kilometres. The Big Bang shows a growth in space. Then there must have been a time when the planets were between fifty-nine and five hundred and ninety kilometres away from the Sun. How did the planets being the size they are at present fit into such a space and still be apart? Material too must be part of the growing. This line of arguing I suppose is much below the Academics pursuit of matters but since I am much lesser in mental standards of developing than they are, such reasoning prompted me to go on some investigating journey. Light journeys through out the cosmos and it will be sensible to follow lights travel.

With objects being apart at some distances and light flowing in straight lines between them it must take light a straight line to travel between cosmic objects. The distance the light has to cover depends on the radius there is between the objects and as such the Universe is then about structures claiming space and space setting objects apart being the radius standing between those objects. The objects are circles by dimensions and the space is also dimensions that are crossed by lines travelling through the dimension. With light being a line and the Big Bang coming from a situation that was a lot more cramped for space than at present, the correct path to follow if I wish to trace the steps of the Cosmos back to the Big Bang is to reduce the straight line between the structures and find where such a line will no longer be a line. The same procedure will apply to the material structures all being in a sphere form. A sphere is a lot of circles forming a unit but not repeating the space claimed the one another Such a circle also apply a straight line only known by another name but still serves the same purpose. Reducing the line will lead us to the beginning of time.

r /2 By dividing the radius r by the half of the value that then reduces r to a point where the left edge of the line reducing will be at the very same place the right hand edge of the line that is reducing will be. At one point the spots that formed the two ends of the line will be at the same spot. Any further dividing will land the left hand spot past the right hand spot in the opposing half where it then will grow once again but in the opposing direction. All possible dividing then ends on one spot where such a one spot shares a location with all other possible sides. The centre then physically is in the single dimension applying as one spot to share a location for all sides. At such a point there is no further dividing possible. On several occasions in the past I have been accused of manipulating the argument to produce none-existing or overrate facts. That is not the case. I am not manipulating facts to create an argument as so many accuse me of. What I am talking about is a mathematical fact that any one can prove by calculating following a very simple procedure. A child is capable of using the two times table and dividing by two every time is the most simple form that mathematics may be used. It is a mathematical fact that a line will reach a point where all sides are at one spot and as such cannot divide any more. At that point all sides share but all sides prevent zero becoming as factor since the sides share on spot. While the different sides are in one place the factor and value is one to all.

Reducing the radius r from all angles possible throughout the circle will bring about that all possible direction will eventually land on the very same spot with no more dividing possible. Yet zero cannot be a factor since the sides still hold value. A point arrive where more reducing will land the one side on the opposite side of the line but it will not bring about zero in the equation

What this argument further proves is that the circle reducing must then come from all points because the radius might be a line but that line represents a circle through 360°. Taking that into account it is important to recognise that notwithstanding the size of a line, which any radius of any size is there is another line (or dot) eternally bigger as well as eternally smaller than the line in question. While we are in the third dimension being part of the third dimension then allows that all parts of the third dimension forever can be divided once more until the line in the third dimension is no longer part of the third dimension. When such a line leaves the third dimension it is still dividable because it might not be part of our dimension any more but it can still reduce further as part of the second dimension. By that time it has left our scope by miles does not mean that it end there because from our perspective that is where it ends. Yet it can still reduce infinitely more until it has left the second dimension and then at last forms part of the first dimension. Only then when the line reaches the firs t dimension no further dividing of that line is longer possible. We can never grasp the size of a line that forms the utmost or the least of possibilities and therefore size belongs to the human mind forming conceptions of big and small, but it has no place in the cosmos at large. This concept not only applies to size, but to all limits and divides we wish to create forming borders that we can appreciate. When looking at the circle in the conventional manner, we persist with errors brought about in culture and not by applying some significant modern logic. Take a circle and reduce such a circle constantly to where it no longer can reduce. Reduce it to a point where only form remains part of the circle because the radius has gone beyond human measure and becomes so small it is not noticeable with what ever tools man may use, then what remains is pi since pi does not indicate size but indicate form, and form is all that then will remain. In any circle or sphere the size only depend on the fluctuation of r, as a component to the circle or sphere but that does not affect the form by indication of Π in any way there may be. The conclusion I drew from following this process is that from this line can start at zero because that will be a mathematical impossibility since no line can ever reduce to zero. A line will forever be able to reduce further becoming smaller but it can never reach zero because zero is not on the scale of lines. If a line cannot reduce to zero it then cannot start at zero. A line or spot starting at zero would therefore be shorter than the shortest line possible. For obvious reasons can no line, or any line grow or extend from zero because such a line must then quit zero and become something, thus abandon its original value. That would mean the start of the line has a different value to the end and a line holds conformity through out. When any line is starting from point zero it can never leave zero because of the influence of being zero disqualifies any possibility of growth. If the line then had to grow in all directions at the same pace the line must then become a circle or being three-dimensional, then form a multi circle we named a sphere. Since the Universe is about circles and lines connecting circles I came to conclude that flowing from this fact is that in the Universe there can be no zero improvising as a filling ingredient for the space of a point or be unfilled space. In the case of the growing sphere the value of the circle is Π, and that is where creation must have started. That gave me the clue where to start looking for singularity. One would find singularity in the value Π and the value Π will be in all things rotating in a circle. As usual I am again shooting the gun before the hunt started. Lines in mathematics do not start from zero and that is no discovery on my part that was a realisation I came to. The importance that is behind this realisation is that the entire Universe presents a picture formed by lines. Lines coming from somewhere straight to us deliver that which we see that forms the Universe. A line carries the image of the point it originates from. If a line can't come from zero, it portrays the image from where it comes. If it came from zero, that is what one would see, but since it doesn't come from zero, a line can't carry an image of zero.

UNIVERSE
It is said that the Universe is formed by everything that exists, including space, time, and matter. The study of the Universe is known as cosmology. Cosmologists distinguish between the Universe with a capital 'U', meaning the cosmos and all its contents, and Universe with a small 'u' which is usually a mathematical model derived from some physical theory. The real Universe consists mostly of apparently empty space, with matter concentrated into galaxies consisting of stars and gas. The Universe is expanding, so the space between galaxies is gradually stretching, causing a cosmological redshift in the light from distant objects. There is growing evidence that space may be filled with unseen dark matter that may have many times the total mass of the visible galaxies. The most favoured concept of the origin of the Universe is the Big Bang theory, according to which the Universe came into being in a hot, dense fireball about 10-20 billion years ago.

Yea, sure but what is the Universe...that part is always accepted and never debated. I say the Universe is singularity 1^0 extending 1^1 to a relevancy Π implicating gravity Π^2 that from there develops space-time $\Pi^2\Pi$ or forming space Π^3. From this space-time extends to become an atom, that extends to become a star, that extends to become a galactica, that then can become a group of many galactica interacting and the combination also form the total Universe. But is the last mentioned he Universe as Newtonians would like to judge, I would say not because the Ultimate galactica is composed of what forms the stars which is atoms that contains the Universe which is singularity extending and ever growing to more complexity. Can we put time to this...well only if we think like simple-minded Newtonians that still believe in magic resulting in mass forming pressure or where there is absence of the mentioned magic, then there is always nothing to fall back to.

UNIVERSAL TIME (UT)

A worldwide standard time-scale, the same as Greenwich Mean Time and since the Universe's time is measured in Earth years, this then becomes the centre for Universal time and none surprising is the fact that time therefore Universally originates (Greenwich Mean Time) from the British naval port and the Royal doc yard of the British Naval fleet. How humble does the Anglo American mindset becomes in viewing such magnificent presumptions? How much does the Anglo American mind put their military might in the centre of God's Universe and how true is it that the Anglo American sees that Britannica rules the waves by replacing God with the Royal Navy might controlling the centre of time. Iraq is the latest proven case to show how the Anglo American war machine enforce the ruling of the Earth by enforcing centre of the Universe policy onto luckless people in Iraq who has one wrong aspect and that is to have oil the Anglo Americans want. Therefore since the Anglo American war machine controls time from Greenwich Mean Time they see them also fit to desecrate Iraq property and the Iraq people are to blame for they have committed the cardinal sin that they only hold oil fields which the Anglo Americans can plunder and steel. If you think my statement is far and wide in accusing the Anglo American mentality of domineering by war, then keep this in mind: Universal Time is the mean solar time on the meridian of Greenwich. It is defined as the Greenwich hour angle of the mean Sun plus 12 hours, so that the day begins at midnight rather than noon. It is closely linked to Greenwich Mean Sidereal Time (GMST), since the mean sidereal day is a precisely known fraction of the mean solar day. In practice, UT is determined by a formula from GMST, which in turn is derived directly from such observations of the meridian transits of stars. The version of UT derived directly form such observations is designated UTO, which is slightly dependent on the observing site. When UTO is corrected for the variation in longitude due to the Chandler wobble, a version of Universal Time, UT1, is derived which has genuine worldwide application. When UT1 is compared with International Atomic Time (TAI), it is found to be losing approximately a second a year against TAI. Broadcast time signals use the time-scale known as Coordinated Universal time (UTC). This is TAI with an offset of a whole number of seconds. The offset is adjusted when necessary by the introduction of a leap second, and UTC is always kept within 0.9 s of UT1. This is based on believing in magical force of inflicting mass on all unsuspecting victims by producing magical pressure or if not then having nothing filling the absence of magic. Fortunately Kepler left us the truth if only there was some interest in the truth. On this issue there is much more to explore than the meagrely mentioned. Time stands related to the position an object holds to a centre such an object refers too while in rotation. Kepler found for instance that $\mathbf{T}^2$, which holds the orbit to a rotation specific, is directly dependent on $\mathbf{k}$ to value the space $\mathbf{a}^3$. This however does not take into account the realisation Einstein came to conclude, but not taking such impotent issues into consideration and then not having to re-evaluate current ideas is more than just being typical Newtonian, it also is what Newtonian behaviour constitutes.

Einstein proved that in the presence of a strong gravity time slows down. Surprisingly with that evidence being around this long nobody since then in science took those statements and made any further progress from there. It was left in some drawer to dry. Science still sticks to the opinion that time did not change slightly since the beginning of the time and holds the same pace ever since. With the entire Universe including all the gravity now present and not excluding one Black hole or dust speck pressed in an area possibly the size of a lepton the gravity extending from that must have been beyond what words can ever describe. If the gravity was that high and Einstein already proved gravity slows time down, then there is one logical conclusion and that is that time was n fact standing still. Mathematically it is incorrect to allow gravity to compress the Universe into a spot smaller that an atom and exclude any other factors and relevancies to change. But before coming to the mathematics I would first like to bring your attention to the practical side. I am promoting a theory in which I am able to prove there is as much contraction (moving in the direction of the Big Crunch) taking the cosmic Universe back to the size it had during the Big Bang as there is expansion (moving apart by Hubble's Constant) and the contraction is as much part of the expansion. By contracting the Universe is expanding and everything is based on gravity providing both actions. The Universe rides on a balance and we have to locate such a balance. To prove my theory I firstly had to locate the centre of the Universe. Even admitting to such a notion sounds like madness or in the least a tasteless joke, but please give me a chance to explain in more detail. I realised that my effort to locate the point holding singularity only stood any chance of success if the reducing of the line enabled me to backtrack the exploding Universe to its origins. By applying some basic effort I have located the position from where all movement came and the direction it took moving forward in time...and yes, while I were also doing the finding the centre of the Universe I even located time as such. There are two standard formulas used to calculate a circle. The one use an r to indicate the radius and the other use a D to indicate the diameter, which is double the radius and therefore needs to be divided by a four to eliminate the Newtonian inverse square law amounting to the difference there will be between the two. The one using the radius is Πr^2 and the other formula using the diameter is $\Pi D^2 / 4$.

The beginning of my involvement with cosmology were brought about by my personal arriving at my understanding of my future theory started by trying to understand Einstein's view on light in motion. When light depart in opposing directions from one joint point and the light departed will travel in a straight line 180^0 in direction to each other they are all still relevant.

After travelling for one year in opposite directions the light will be one year from the point of origin.

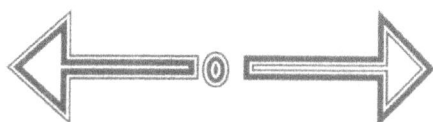

Under the normal circumstances the light will be two light years apart. That means if one could stop the light travelling to the left and have that light standing dead still, while the light flowing to the right can make a complete turnabout, it will take the light coming back one year to reach the point of origin once more. It will take the light one more year to reach the other point that was then standing still for one year.

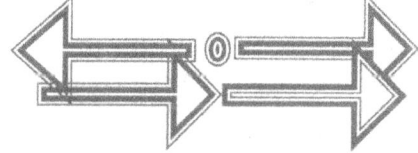

Einstein proved that the normal is not the case with light. Light travelling in opposing directions for one year will be one year from the source it came and the light will be one year apart. But Einstein's claim that this comes about because the light is equal to time did not make sense to me. If the light was time, then time was no factor to light. In that case light will travel through space in   a ratio of one meaning the moment it releases from the source it is on the other side of   space notwithstanding the distance the space has. That is not the case because   light is just a simple speed ratio like any cart or aircraft or spaceship. Light was distance during time duration and that comes down to being pretty fast, but it still remains speed and speed has any other relation to   time   than placing time as one to light.

After travelling for one year   light had a distance of C multiplied   by the seconds in one year to each side of the source but   the points of light travelling independent as light had that same distance being apart of the two points the light occupies at that moment. The two markers are just as far apart from each other as they are apart from the original starting point. With this in mind the use of C^2 by science might prove convenient, but it also proves with this mockery how big farce such innovative calculation is. There is no chance of anything going at C^2 because   there is no exceeding of C by light as such. If that were the case light would not be present in the explosion or antigravity. The speed of light is not a force it is a speed. It is a ratio putting space in relation to the time density the speed will establish. This whole argument pointed to Kepler holding 6tthe straight line in relevance to space and time.

From that point I concluded that the link must be the value of a straight line sharing a dimensional value with a half circle and the triangle. If we look at the line supposedly travelling straight we find that the straight flowing is equal to the square relating the triangle. This is completely Kepler indicating gravity being $a^3 = T^2 k$ Look at the dimension and not the number implicated. It is $^{1+2+3}$ and transfers that to the line being the ^{1} and the ^{2} being the square being equal to the triangle as ^{3}. But it diverts very much from normal mathematics and that is precisely what Kepler's formula also does. With Kepler $a^3 = T^2 k$ and with mathematics the volumetric size of space must either be according to the measure of normal mathematics if it is a cube then $a^3 = L \times B \times H$ and in the case of a sphere the measure will be $a^3\ 4/3\Pi r^3$.

This was a triangle in relation to the square we find in the half circle standing again related to the half circle. It is not standard mathematics and anyone drawing links between mathematics and the speed of light has no idea about what is involves. With that I have again antagonised millions of the most important people with which I have to share a view. I do share their view on the Cosmos but not with their view holding mathematics as a standard fit all and apply anywhere in the cosmos. It is about lines carrying dimensional properties and with that we have to consider the line once again.

Let us find the smallest possible line first. Reducing the line will eventually leave all sides on the same spot. Such a spot must be round in form. The line being the smallest line will start off as a dot. A line so small it has reached a point not dividable any more will have all sides literally on the precise same spot, and I have located singularity in just such a spot.

I came to the conclusion that the spot I found had to be singularity purely on the grounds that that spot holds only one side to serve as a start to the starting point of all directions possible. There in that side is only one spot is only one side applicable and one dimension present. With all the factors given one can only come to one conclusion and that is that there can be only singularity. In such a case more dividing by two will land further positions on the other side of the divide. That point serving as a position for all point and cannot allow further dividing is the smallest line or spot there may ever be. This spot is the result of a most basic process of reduction as the Hubble constant is a most basic process of doubling up during a matter of time. By reducing the line constantly the only value that will eventually remain without dispute from any party arguing about the facts is Π. By only having Π and a radius as one square (the radius effectively becomes one holding any and all sides on one point) of any significant measure as the radius it will be an evenly spaced dot. From the smallest ever possible dot will grow a line in every imaginable direction relating to a prospect of Π not favouring one direction that puts all directions at equilibrium meaning that any form of what ever might develop from such a spot will have the end and the start being in the same position, which will also have to be a sphere as the flow outward will be equal in all directions. Please think clearly, is that not precisely the commitment we find in gravity, where gravity is flowing from singularity outwards but never favouring any side?

This reasoning prompted me to look for singularity in such a spot because if the prime spot from which all came was a spot holding all, then the spot must hold the shortest line but more prominent it will hold the smallest form including the smallest circle or for that matter the smallest sphere. With gravity always being in the centre of a sphere where the space is least available in the entire structure (there is not even space left to fill) one finds a flow of gravity from that centre spot outwards in all possible direction even-handedly. The fact that the original gravity will begin as a circle or will be a circle is the direction it will take when being the first spot created. All progress will be evenly in all direction because no direction will stand out or be in favour above any other direction at first.

The spot forms a full circle, but the line running through the circle is forever present because that is the future radius of the circle that will one day develop the circle, which is equal to the present diameter. The fact of the presence of such a possible line in such a possible circle dividing the possible circle into two parts makes the centre line equal to the half circle. The line forms the half circle but not only that the line presents the half circle as much as the line is the half circle. The line then is 180^0 and the half circle is 180^0 because in singularity the two factors are the same. The same value is of course $\Pi^0 = 1$.

In this half circle of the future, which is no half circle as yet because of a lack of space there are three future points indicating the space less ness that will go on to become space filled with something. On top of such a circle to form must be a marker indicating an awaiting boundary or future border and at the bottom of the future circle there also must be a similar marker that is no marker as yet. Between the two possible points that are not there yet is a future line running that is not there yet. Then indicating the possibility of a position to come that will bring about the half circle being a future distance apart from the future line indicating a diameter that will one day be there a third such a marker must be established for the future. That forms a triangle with two more sides being connected by either a line being one or half pi being one. From singularity comes about that the line is the same as the half circle is the same as the triangle and all has one value being 180^0. From this come the most basic principles in as much as forming the ground rules of the law of Pythagoras.

●●●●●●●●●●●●●●●●●●●●●●●●● When drawing a line such a line then starts of with a dot serving the spot that holds all sides equal. That means the line serving as the future radius will be equal to the half circle which is then Π. The only aspect of the point that stands in for the end of the single line forming the radius of the circle is that we then mathematically reach the single dimension. We decreased the line to where a circle being Π formed on the single dimension. This dimension also hold the circle dividing line because from there the radius must once again generate a value and by such a gesture that the extending would form the circle that forms the sphere that eventually lead to the formation of particles. This leaves a problem to investigate.

With no line possible there had to be another dot that formed since the Universe has many dots that formed lines. But let us not to get confused and lost in the range of possible diversions but let us stick to two dots. One dot was next to the dot next to the dot, but as I said we stick to one dot

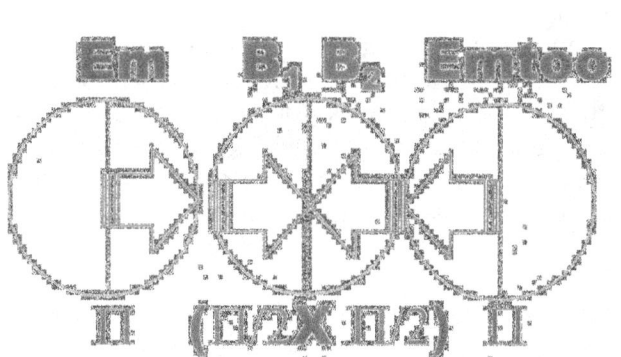

next too the second dot. $F = \dfrac{r^2}{M_1 M_2}$ is the first step gravity begun with. That leaves us with a huge problem in as much as when r = 0 then r^0 = 0 and 0 dividing any value will leave 0 as the answer. If the particles were inseparable at the start it must bring about that gravity would not be forming since the distance will not permit any dividing. By allowing the distance separating the particles to be zero, the particles melt into a unit. Again this is Mathematics and not my incoherency as some Academics dismissed my work.

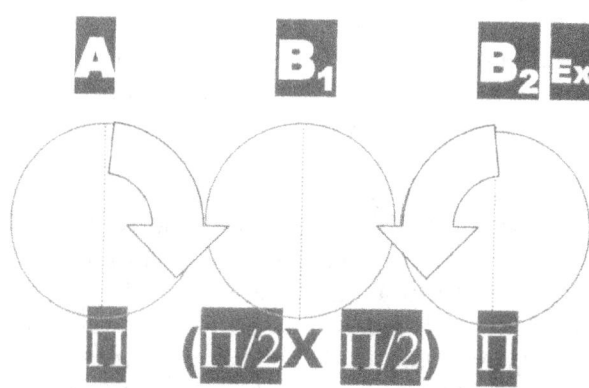

Extending into the distance

We return to the fact we established before that em and emtoo was divided by r and then r had to be one since r could not be zero. Such a centre would then carry the same value as em and emtoo. That means whatever value em and emtoo receives has to go in equal measure to r with em sharing half of the divide and emtoo sharing the other half of the divide.

Let me run through the argument one more time because I have been insulted by Academics in the past telling me I am bending mathematic rules with my applying double values to try and produce some argument. The two particles formed by an inseparable unit separated by a sharing of a spot. We know that at least two spots formed because there are many more than just two that remained to become part of the visual Universe. Let us name the spots because that is what humans do best if they do not know what to do with what they have to do. Let us call the on em and the other one spot next to em we then call emtoo. Between em and emtoo there was nothing because em and emtoo were inseparable. By they're being inseparable we would naturally be inclined to think that the separation value should be nothing or at least zero. But putting zero in that place is a mathematical excluding procedure leaving future mathematics excluded. With m multiplying m_2 and then dividing $\div$ r with zero (r=0) such a procedure will leave the lot at zero and with that nothing is going nowhere. That means although we think the space between the two parts are nothing the non-existing space has to be at least one to be a future factor.

Every part of the argument is sound but was never yet used. I repeat once more if my argument reflects on inconsistencies those inconsistencies are not about my work. In order to disprove my argument replace Mass one and Mass two with any number possible, then divide such a number with to the square being zero. If there was no space then the value of the particles had to be one. If there was no space between the particles the particles then had to form a unit. But if there is a mathematical possibility of reducing a line to the single dimension then there had to be a factor

representing r as a factor of one. Take $F = \dfrac{r^2}{M_1 M_2}$ and substitute any of the factors with zero

and the result coming about has to be zero. The factors in the equation have to have any and all the elements at a value of at least one. Only if r was a factor of one can gravity bring about any mathematical equation developing from this argument. That means the mass on both sides must have a factor of one being a limit, which does not allow such further reduction of r and any further reducing of r beyond the limit will not be tolerated. Only if r = 1 then r^2 can be 1 and mass can be apart. Like it or not but believing in the Big Bang must also bring about the accepting that the cosmos moved apart somewhat. The fact that r brought increase in the space separating the mass produces a problem that was solved already. About a century and a half ago Roche found just such a limit. Once again I was confronted by zero becoming growth. There is a huge hole that needs filling when bringing into a relation any forming of an alliance between a cosmos coming from nothing and filling with nothing and a cosmos growing spontaneously through balance shifting prominence. Mathematically the fact of applying nothing as a vale applying in the cosmos is not a strong and convincing argument. The minute one brings in zero as a multiplying factor forming a definite value working into the calculations of the cosmos, growth disappear. If growth was not a factor, the zero factors could be involved with some form of maintaining stability and where then further growth will accept the responsibility of zero.

Newton's claims about the principles he declared that is responsible for guiding physics carry no

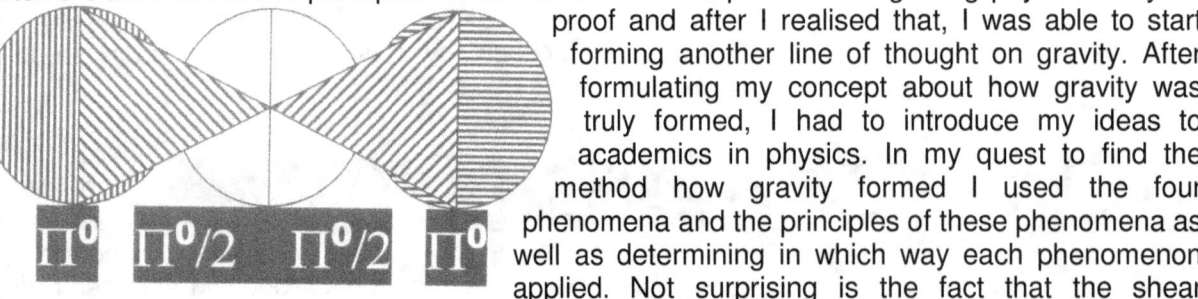

proof and after I realised that, I was able to start forming another line of thought on gravity. After formulating my concept about how gravity was truly formed, I had to introduce my ideas to academics in physics. In my quest to find the method how gravity formed I used the four phenomena and the principles of these phenomena as well as determining in which way each phenomenon applied. Not surprising is the fact that the shear existence and acceptance of the four cosmic phenomena is very securely hidden from normal investigation. Since Newtonians can't use Newton to explain the **Titius Bode law**, the **Roche limit**, the **Lagrangian positions** and the **Coanda Effect**, these phenomena are hidden so deep that it is very likely that you as the reader might never even have heard of these phenomena. It is because these phenomena portray Newton in a dim light and that no Newtonian would allow to happen! I use these phenomena to mathematically prove what gravity is, but the condition I apply is that gravity is not even close to what Newton suggested forms gravity. By my ability to prove gravity I had to place each one of the four in the way forming one part of what gravity is. Before I could manage that, I had to find a way as to explain what the working relation of the four cosmic regulating phenomena were and I had to figure out what was the known contributing facts to the cosmos that each was responsible in contributing to gravity in my effort in determining how they work and then implicated that specific formula's function mathematically in forming gravity in the cosmos. This was no easy task but I did it and by formula shows that my argument is logic and the mathematics prove that it works well. **I now wish to take you on the path I followed when I discovered thee real principles guiding gravity.**

The closest encounter worth noting we ever had with this law in the modern age of news and Television was the Shoemaker-Levy 9 incident during the previous decade. At the time and even in the present no one drew any similarities but after completing this book the reader should find why I could draw such similarities, which there is between this incident and the Roche limit. Even the phenomenon called the Sound Barrier became clear when applying the Roche factor with the laws governing the influence of singularity.

At the very first sign of any of the sides departing from the centre shared by all, all other points must also show signs of a willingness to depart. There will be one point where r still is one coming in as a factor but pi moves out from only being a factor of $\Pi^0 = 1$ and at that point pi will become a full factor of Π.

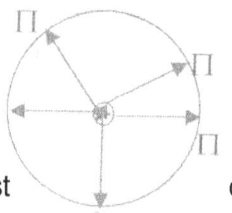

This point, which I now am appreciated value while the factor of one. His is the dawn of there but space was sparsely shifted to become Π for the very fist keep the Universe in the first referring to, is the point where Π a fully diameter D still remains a dimensional the second dimension where space was shared in some cases. It is when Π^0 time. But keeping Π as one ($\Pi^0 = 1$) we dimension.

The point without movement, the point holding singularity must have a value of Π being the eternal dot but since the dot has no dimension in having form the Π that indicates the dot must be Π^0. From such a point there has to be to the side of the centre point be a point where space do start. That point will then receive a diameter but that point will have form only in being a circle. In that point there is a shift from in relevance from Π to the centre Π^0 and for the first time it brought about two separate values for Π.

We have established the fact that em and emtoo was divided by r and then r had to be one since r could not be zero. Such a centre would then carry the same value as em and emtoo. That means whatever value em and emtoo receives has to go in equal measure to r with em sharing half of the divide and emtoo sharing the other half of the divide.

Because the three points existed on equal terms in singularity sharing a same spot the coming out of singularity will enforce that equal value comes to all. That means the circle gets to become Π, the diameter becomes Π and the distance setting the structures apart will also become Π. This is what the coming from one point brings along. Only when being part of the second dimension can there start being separate values.

While the form was still being in the single dimension from the one side of the form the dots had to establish identities apart but not separated yet. The one circle had a factor of $\Pi^0 = 1$ and the centre had to have a value of ($\Pi^0 / 2$) extending past the very next object but also cutting such an object into a square double half value that was going to come about as soon as the other dimensions came into form. In the relation at present em is extending toward emtoo by means of establishing a valid r and emtoo is establishing a valid extension to em be using r and this leads to two valid values for r being ((7+7) /10) and (10/7). The values I give here I shall explain later.

The only definite place one will locate zero is in between the starting point of the lines going in opposing direction in the position the lines hold before there was the least of directions applied, but that is only because there is no such a position, not because any line is coming from there. As I have indicated and positioned em and emtoo the two points may share a position but separation is forever a possibility and for that reason if there is no other reason we then have to put a dividing possibility at a value of one ($r^0 = 1$). The two lines are still one holding the opportunity of parting as an option but have not yet parted and therefore are on the very precise same spot. Being on the same spot does not mean being inseparable or being the same. It only means sharing a spot. The line coming from there is already there because it already has the choice of going in any and all opposing directions and when it starts running it will place filled space in that location not yet present but also holding a factor of one since it will become filled in the future. This is because the space at present is filled with a line. Where this space is now already filled with a line. The line had to have a start. The starting became the line running and by running the line is filling space. That means with the line there it filled the possibility that a line could form and not with a line not being in place at all. It is again taking the r separating em and emtoo on its factor value of one and not our human visual accepting value of zero. One may not discard any future possibilities of growth by giving those possibilities a value of zero. A line might form or space may form where the line later may form. We humans tend to dish out a value of zero where ever we do not visually are able to find a value at that precise second and in doing that we also place such a value as a running obstacle into the future. But our habit of doing that is proving to be a human shortfall because with our shortsightedness we think of the here and the now in excluding possibilities while we should think of the future by including possibilities. By disregarding a positional value as zero we exclude such a position from ever being possible. By giving it a factor of one we include such a position as a future possibility. When reversing a line we might find a better idea of what is in place and where it is in place. Gravity is officially a force without limits going past and through borders and has an unlimited reach. It seems to remain even and this is conflicting with the flow of perceptions about mathematics. In as much as showing that r serving in a factor value, as one has to form a limit where em and emtoo than that of r and aligning the discovered singularity produces the Roche limit as such a dominant factor in the cosmos. With my retracing a simple line helped me to find an explanation about the Roche limit, a feat not yet done in science. The Roche factor is next to singularity the second most basic foundation in cosmology and is the starting point where singularity spawned into dimensions. As it is fundamental in all cosmic development and with that it denounces

the gravity principle introduced by Newton as $F = G\dfrac{M_1M_2}{r^2}$. The formula

$F = \dfrac{r^2}{M_1M_2}$ is unable to explain the principle discovered by Titius and later by Bode and in

contrary to all statements to that effect made by Accepted Science policy makers the Titius Bode principle is not coincidental. In fact it is one of the four most adhered and important cosmic pillars holding the cosmos structural in place. From the two examples mentioned above comes gravity. In past few pages I proved how one could arrive at the facts that prove how the Titius Bode Principle leads us in the direction the origins of the solar system. But before we can accept the influence of the Titius Bode Principle we have to deal with "Nothing" and as such dismiss nothing from science. "Nothing" in the Universe is coincidental; "nothing" in the Universe does not apply. Where mathematics meat lines nothing disappears. Should any principle not match an accepted theory or change the accepted theory, the theory does not apply.

The content of my work holds a new view about Cosmology, which I have been working on for the past twenty-seven years and exclusively for the past six years. I always had a problem with the idea that space constituted of nothing, while I came to realise that lines mathematically couldn't start at zero because there is no evidence of zero as a factor in mathematics. Should you disagree with my statement the question in need of answering is this: What will the length of the shortest hypothetical line imaginable be and moreover, what would the total overall length be in that case? The shortest possible line (hypothetically) must be so short it must have an initial and ultimate point sharing the same spot. The two points must be one and only then can further reducing of any line not occur. If it used zero as a start, the zero part would not count, because the line will only start at a point past zero where the line then will start forming an infinitely small dot. I press this point in urging the understanding because there is such a point, but in my attempt to underline the fact, I have to per sway the reader to abolish four or five thousand years of accepted and practised mathematical culture and that is no easy feat. In applying the most basic method of taking the line back as far as possible brings a dot because of the equilibrium that will stem from such a position. The dot is in infinity, however small, it is not zero. Zero ultimately means not existing and then that point, as a start does not exist. The smallest line has a beginning and an end at the very same spot located in infinity, and infinity may be beyond human scope, though infinity is still not zero. Infinity may constitute of something we do not yet understand, but we may not define our human misunderstanding because infinity is not present in our minds and therefore by not sensing a value we disregard such a value as nothing whereas it is visibility nothing but in being potentially there it is one. It is the same as a person hearing a dog bark and investigate. When not sensing what the dog was barking at, the person turns around and disregards the barking as about nothing. The dog's reaction was not the indicator of the nothing the dog sensed something. The man's wits let him down and his wits produce the nothing. The dog will not bark about nothing because then the dog will not bark at all. The fact that the dog barked produces a possibility of something out there, which the dogs is getting annoyed about. The man's inability to detect what it is that the dog is sensing becomes the nothing but not the possibility of something worthwhile to investigate. We use nothing to avoid and not to valuate. In this aspect lies the difference there is between arithmetic and mathematical science where arithmetic can have position such as zero since arithmetic excludes the cosmos calculating numbers only. The nothing we see we made the nothing we find but the fact that distance is there, it is separating structures and by that is bringing in the factor of one which we cannot see. It is the way we try to disguise our inability to detect which produce the nothing we then use as a value, but still we substitute the nothing in applying arithmetic with the name as nothing and then the names used becomes the factor of one. Cosmology is not about numbers because no one can calculate the number of stars. Cosmology is all about lines and angles positioning objects and in those there features no zero. No line can be zero long and forming a position of zero degrees in relation to another object.

A man may have that many oxen or so many sheep and even this amount of wives, (in Africa) or not have any therefore having then a total of nothing, but there cannot be nothing between the Sun and its orbiting structures. The having and have-nots are part of arithmetic. Light will indicate a line flowing between the Sun and whatever planet, following dot after dot thereby proving the existing of the possibility of something going about by a straight line, and any straight line in relation to other straight lincs will be under the law of Pythagoras in as much as obeying the rules of trigonometry. There is no possibility of a straight line not forming in space. If there is space, there can be a straight line. The mere fact of two spots having different positions in space gives the two dots different values. If the line has the length of zero it is not present. If the triangle has one angle of zero it is no triangle because all other angles dismiss at the same time. Mathematics converts the values of integrating lines according to Pythagoras and arithmetic is about numbers to be added or subtracted.

By mathematically excluding zero from cosmology a new Universe opens to the human mind. With the distance between the Sun and Pluto being roughly one hundred times more than the distance between Mercury and the Sun, the distance must hold something more than pure vacuum filled with nothing except one atom hear and there occupying the vacuum between them and the Sun. If space supposedly comprises of nothing how can nothing then become plural forming more or be multiplied by a number as to indicate a growth in something not even existing. As the one becomes one hundred the one cannot substitute a value of nothing but then must be part of something. If the one substituted the nothing, all laws of mathematics will go in disarray because when one multiply any number by zero it becomes zero placing both planets in the Sun. If Pluto was one hundred times closer than it is at present was it one hundred times nothing closer? $100 \times 0 + 0 = 0$. That is mathematics!

By allowing the three hundred a value the nothing must form one making that which is between Pluto and the Sun not to be nothing but there has to be something. This argument follows mathematics to the letter and in precise detail. With Pluto and the Sun being apart that being apart has to have one of something in place forming the being apart from each other's cosmic position one time multiplied by the many ones we find in that space standing relative to other space regarding whatever the space becomes what is between the Sun and Pluto. That factor cannot stand in for not one, which is the same as nothing as that is because one cannot take the place in the position that zero secures. By excluding nothing from the equation space becomes something bringing in a value lying inside the realms of the infinite that must form singularity. As the zero becomes a dot, something else becomes clear about the dot. Looking at the night sky we find darkness overwhelming the space in relation to the stars bringing across light.

The Universe is about lines allowing light to flow from one point to another point and in following that line it has to continue in the line as the line has to represent something. The Universe is all in relation about lines indicating distances between cosmic structures. The cosmos is in short about lines connecting points in space being apart. It is about a line starting and continuing from such a start. But science advocates their opinion that such a start of a line flowing between any and all objects can hold zero because according to them the Universe are full of nothing. If the Universe in as much as outer space is a container filled with nothing at the present moment, and there is no place anything that was part of outer space previously could release to and there was no emptying of what ever filled it before, then it could not get rid of what was in the outer space when it started with what it started off with. We must then accept from what is not in the Universe was not in the Universe at the time during the start that at is present at present according to science because it then still must contain the same nothing and must have that same filling from the start to the present. If it was nothing it still must be nothing and that same substance being nothing is what it also used to grow using it as it grew because it filled outer space with nothing growing from and growing to nothing. Is that true? The filling of the Universe could not go anywhere so one has to presume it started off from nothing and from there it kept filling with nothing since what ever was in the Universe at the start had no place to escape to or no place through which to escape. That is only applying if it is nothing filling the Universe at large. Can nothing grow as much as a line is growing from a start of nothing?

The answer is that such lines not only indicate a distance but since the Universe came from such a small space as science propagate with the theory of the Big Bang then all particles in the Big Bang Universe were rather cramped for space when the Universe started from that small line between particles and is now the same line but is now so big. In the past it seemed being so small and showing the space between particles to be awfully short at one time. It was short but how short was it? Did it start off as nothing? Is the line starting at nothing as science wishes us to believe? If it does then all lines must start from nothing so we better investigate this trend with the start of a line. In this following I show my argument with which I hope to prove the counter part of what science believes. I am about to prove that which science sees as nothing in space and in material is the very location of singularity.

Lines mathematically cannot start at zero because there is no evidence of zero as a factor in mathematics. Should you disagree with my statement the question in need of answering is this: **What will the length of the shortest hypothetical line imaginable be and moreover, what would the total overall length be in that case?**

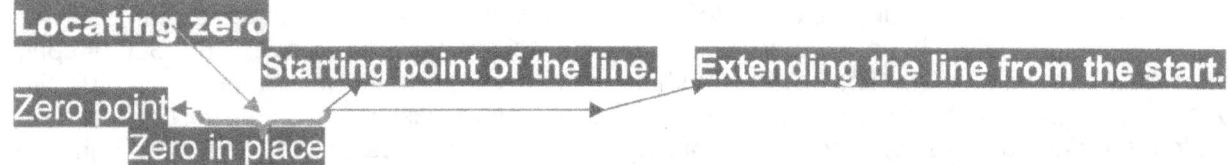

Let us duly test my statement by taking the line back as far as possible. The shortest possible line (hypothetically) must be so short it must have **an initial and ultimate point sharing the same spot.** The line that **cannot reduce** any **further** must be **so short** that **directions flowing away** from each other **are located** in the **same position.** Lines are responsible for everything connecting everything to anything. **Everything in the entire Universe holds relevancy to everything else in the entire Universe...the big question is ....how is that achieved. It is the same as this book being formed by the line up of words forming content through forming a concept. Every word in this book holds relevancy to every other word in this book...and that one can only see when one word is in contexts with all the other words forming the book. The Universe holds relevancy in the same manner as the graph...and therefore the graph can't use a line having a value of zero because the value of zero would remove the graph.**

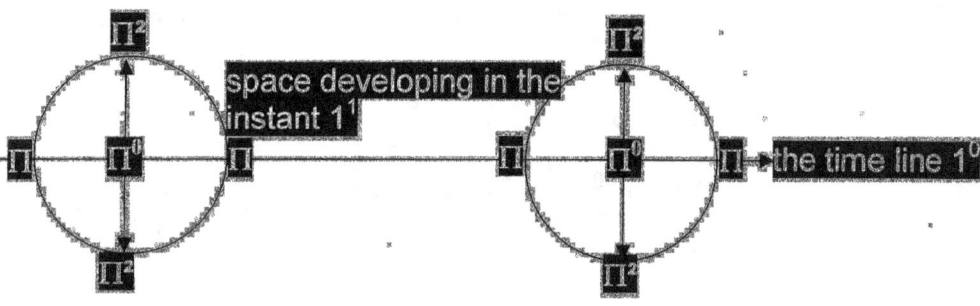

As one can read from information taken from the graph it is clear that the graph mimics the flow of the top taking only one sector in appreciation of movement. In fact one cans ee the graph portrays the top moving in a circle. Therefore what applies to space-time is depicted as the graph. The graph is the representation of time flowing by establishing space.

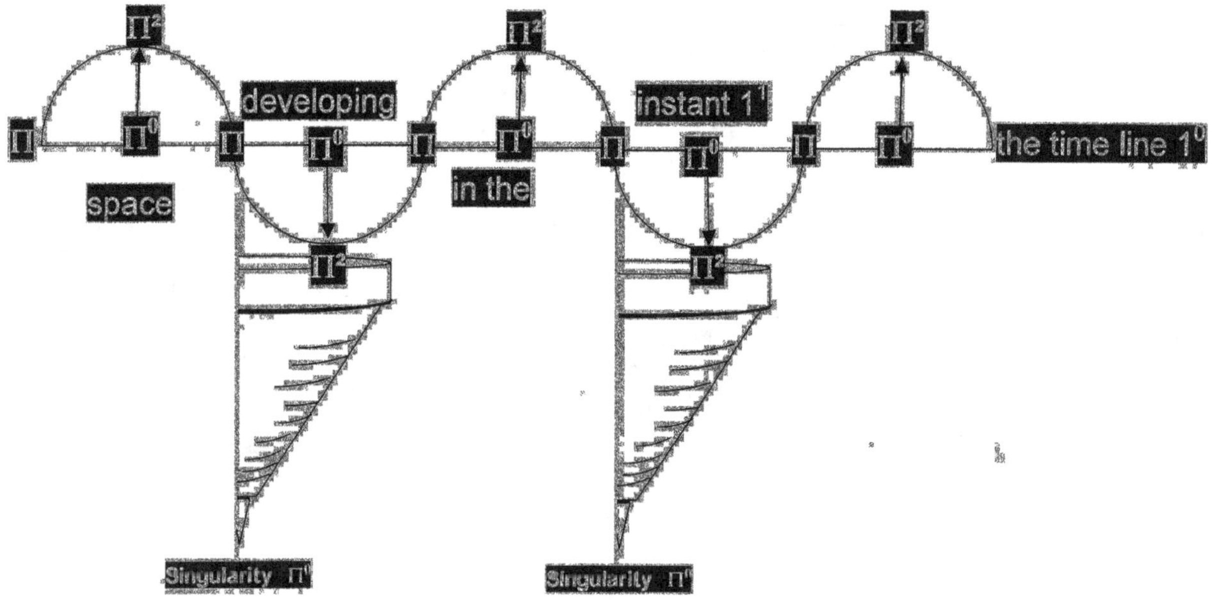

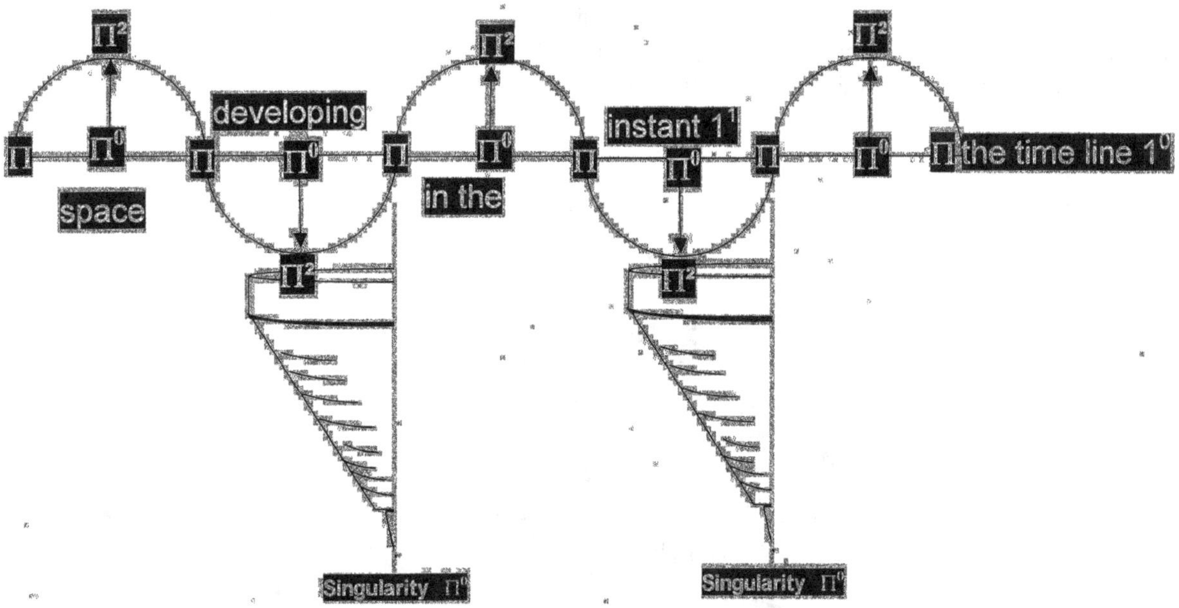

In the graph the line that science associates with zero is the time line 1^0 and crossing this line is space developing in the instant 1^1. The space developing is in relation to the movement of the object holding the time and the space relevant within the graph. The graph is representing anything that spins. This incorrect presumption dates back as far as mathematics go although I have this feeling the Egyptians and the Persians as well as the Hittites was aware f singularity and the role it had in mathematics.

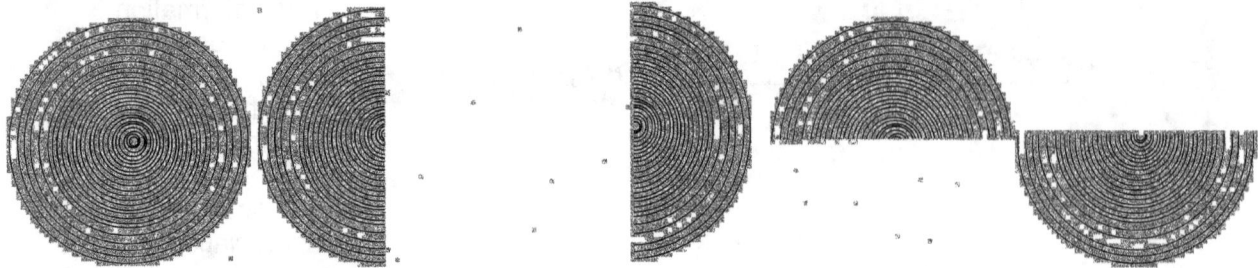

When viewing the top spinning from a human portray it seems to be a circle going nowhere as Newton saw it in his statement that the spin action nullifies the radius firmed in $\dfrac{dJ}{dt} = 0$. But when closer examined it seems the top divides into sectors halving the top from bottom to top as well as from left to right in any manner one wish to divide this.

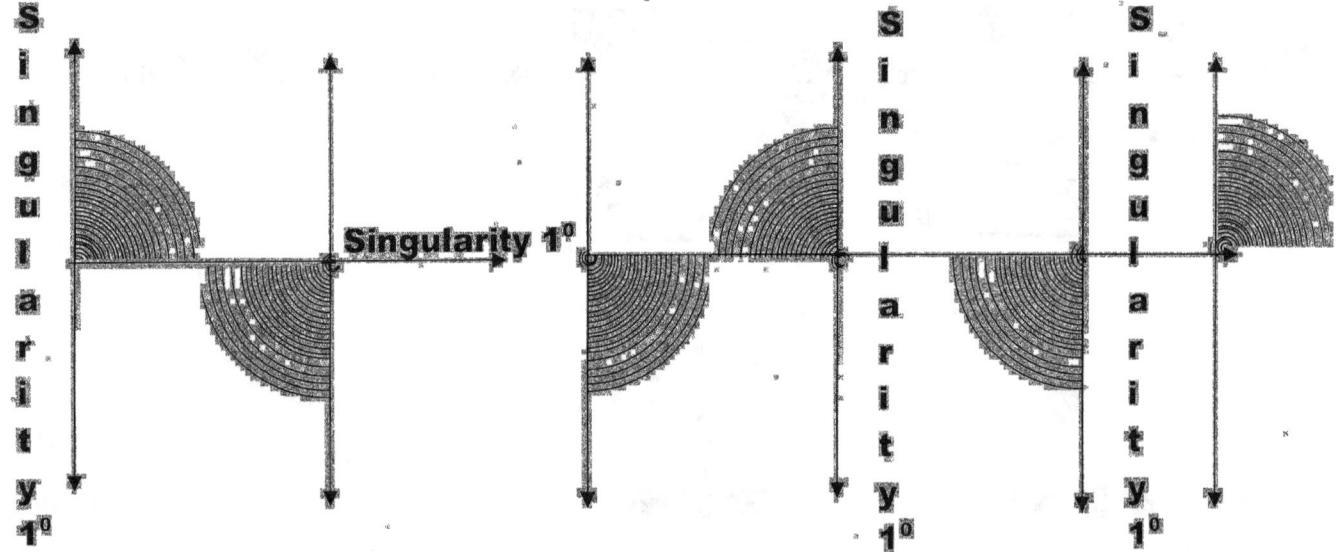

In closer study one can clearly see that the top spinning divides into four sectors, which supports, compliments as well as contradict and opposes each other. When gauging the situation in this manner we then get closer to assessing gravity more correctly.

$$T^2 = \frac{a^3}{k}$$

$$F \neq G \frac{M_1 M_2}{r^2} \frac{dJ}{dt} \neq 0$$

$$k = \frac{a^3}{T^2}$$

$$k^0 = \frac{a^3}{kT^2}$$

$$\frac{dJ}{dt} = 1^0$$

$$\frac{dJ}{dt} = 1^0$$

Earth surface and direction of rotation

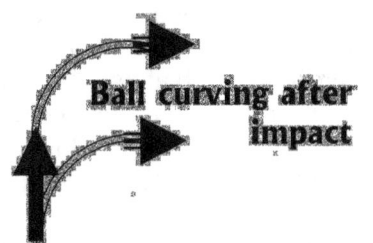

Any theoretical line being the shortest possible line cannot have the line holding the initial starting point at point zero and advance from there. Mathematics simply will not allow it. If the point had zero as all it had to offer, such a point is not present. Mathematics is not about numbers. Mathematics is about sizes, positions, and dimensions. Expressing three dimensions will be x^{1+1+1}. We tend to look at the dimension and not the number. We ignore the indicator as being the natural and only take the dimension the indicator points to in account.

We look at 1^0 and find the zero in space. In space there is space. In space there is 1^0 X 149 X 10^6 km of space between the Earth and the Sun but there cannot be 149 X 10^6 X 0 km between the Earth and the Sun. The zero means there is no such a position. If it used zero as a start, the zero part would not count, because the line will only start at a point past zero where the line then will start.

Zero ultimately means not existing and then that point, as a start does not exist and where the line then stars is a point in existing. When the line **has a beginning and an end at the very same spot** and it wishes to extend the position as to further the possibility it has, which direction should it favour. Extending the line in any one direction will favour one direction without any clear reason not extending in other directions. The fact of direction being present only proves and is proved by another point established, which is placed in relevance to such a second pointing a position already established by the relevancy of two point located in a direction to one another. But if one point starts one line there is no favour of direction since there is no established direction yet. The only mathematically sensible option about extending any line starting at a pre-designated point without any other point to establish a pre determined direction will be non-bias progress in all directions equally in order to give a meaningful flow of mathematical equilibrium. Not one direction stands superior to other directions and all directions are equal with no bias anywhere. Of this statement the Pythagoras mathematical principle is proof of and that I explain later.

Let us dissect nothing, as we find nothing in the presence of the cosmos. The distance between the Sun and Pluto is roughly one hundred times more and if the distance between Mercury and the Sun, but both has nothing between them and the Sun. The space filling the distance from the Sun to Mercury has nothing more than the space between Pluto and the Sun. That means the distance between the Sun and Pluto is as equal in relevancy as the distance from the Sun and Pluto since both is the measure of nothing. If the one substituted the nothing, all laws of mathematics will go in disarray because when one multiply any number by zero it becomes zero placing both planets in the Sun. The distance between the Sun and Pluto **is Pluto is 5900 X 10^6** kilometres of space, but in that statement we take it that the one of a kilometre is present in such a multiplication. The one constitutes the presence of fact being a statement of a value. By saying the distance constitutes of nothing we have to substitute the one factor with a factor of zero. Then the calculation must read **Pluto is 5900 X 10^6 X 0 = 0.** Including nothing as to state the presence of that part contained by the calculation delivers the total of zero. By excluding nothing from the equation space becomes something bringing in a value lying inside the realms of the infinite that must form singularity. Applying this logic to the Lagrangian system and interpreting that information to the law of Pythagoras a clear pattern come about.

The reaction responding from my argument is that it is silly, but should that be your personal opinion too then test where the silly part applies. Bring the zero into the calculation, the zero that science so eagerly place in outer space and see the mathematical result. By applying the distance one accepts automatically that the figure become calculated with one as it represent one in being a calculating part of the cosmos. The calculation as all calculations normally are is in order to calculate something and the something will at least stand in as one in relation to the rest being part of the calculation. But saying that the factor of one in fact represents nothing since nothing is so much the part in the calculation being calculated then the zero has to replace the one as the fact of being calculated.

The claim becomes obvious when observing the connection between the half circle, the straight line and the triangle, which could also promote all the qualities lurking behind the pyramid. Consider the connection between 180^0 sharing three different forms all part of mathematics where each is different in form, but equal in value and then one may realise in considering the very basic in mathematics being the Law of Pythagoras on which all mathematics are focused. The triangle stands in for one factor represented by one at a value of 180^0. So does the straight line become a factor of one and the half circle also becomes one where the factor of one equals all 180^0. All three are most seriously part of shapes in the cosmos. Revalue any one form to zero and the rest too must follow and share the same value. The Law of Pythagoras is about angles in relation to lines and not one angle can represent zero because that will reduce all the lines also to zero. The measure of angles between stars at a distance uses parsec as the indicator, but the parsec between the stars indicating an angle has to represent an angle whereby one may measure distance and such a distance cannot be zero because then the parsec will be equal to zero. Again it is multiplying the factor with the measure but if the measure is about a factor of zero, then the factor too becomes zero. That is as basic mathematics as I can present.

If the argument seems ridiculous it is not my mentioning such a fact that is ridiculous but the mere fact of the reasoning also becoming a recognising of an argument accepted by science making it as such ridiculous. If space is nothing then it has a number to use indicating just that value being zero or the capitol O indicating zero. Try and indicate what is measured and calculated in space, but not by simply "not thinking about the fact" and therefore simply ignoring that what is measured forming the sole value of space, but put the value of nothing as part of the distance in calculation because that is what is measured. When stating the distance between the Earth and the Sun place on paper what will allow the kilometres measured to represent the factor that is being measured. If represented by one being the total of one by hundred and forty nine million kilometres of nothing put that language in the International language of mathematics that spans all dialects spoken on Earth. Put it in mathematical terminology by saying there are 149 000 000 X 1 (multiplied by the kilometres) multiplied by what it is being measured which is 0 and what will the total come too… a full zero.

149 000 000 X 1 (km) x 0 (indicating what the km are made of) = 0

Mathematics says it. If there is something to be measured then the least value the measurement can have in relation to what is used in the measuring has to be one. It cannot be zero and be measured…and we do measure outer space! It sounds as if something here is at fault. It is not with my mentioning the inconsistency one should find fault but the fault is with the fact that it is there and no one noticed! I am not to blame just because I am mentioning it, but the blame must go where it belongs.

I think it is by now little understood although I imagine not nearly accepted that by adding a million of nothing to one nothing there will remain one nothing and that is still nothing. Nothing cannot accumulate therefore I cannot accept anything holding the vastness of space being able to constitute nothing as the major component.

Mercury has 58×10^6 km and Pluto is 5900×10^6 km space between the Sun and the planet. That indicates a distance and a distance comprises of something, for if was nothing then both would have equal nothing and be next to the Sun. I repeat, the distance indicates something because nothing would place them both in the Sun and moreover in the centre of the Sun. Having "nothing" between Mercury and the Sun and between Pluto and the Sun that then places Mercury and Pluto at a same position within the Sun. By saying Pluto has one hundred times more zero in between the Sun and Pluto makes such a statement laughable.

Being laughable it might be if I say it but if a learned Professor conducting a class does it. If I would say Mercury holds one hundred times less nothing, such a statement will make me an idiot, but used as science makes such a statement plausible. That means the more zero or the more nothing one find between cosmic structures puts such structures further apart. There can only be distance with something concrete applying the distance between it. The problem is identifying something from nothing that defines the difference there is in science. I cannot see how nothing can become plural or more sometimes. Realising this I went in search of that which nothing is substituting. The issue I went in search of is what to substitute the nothing with and fill the nothing with that something. Let us go on the interesting search of finding what prevents the Universe from tumbling in on itself. If the Universe was truly nothing, the nothing would not support the structure and the structure would disappear into the nothing not supporting it. The Universe is about lines forming angles and holding distance that much we established so far.

When reducing the circle in size one have to reduce the radius or the diameter because the pi is the indicator of the form as being a circle. Divide the r until there can be no dividing any further and that cannot in the end indicate zero because no matter how small, in that will forever be a value in place.

• r / 2 • r / 2 • r / 2 dividing r reduces r to infinity but not Π as Π remains stable, protected by the rotation of matter forming a circle around singularity

r or Π

Takin g that into account it is important to recognise that notwithstanding the size of a line, there is another line (or dot) eternally bigger as well as eternally smaller than the line in question. We can never grasp the size of a line that forms the utmost or the least of possibilities and therefore size belongs to the human mind forming conceptions of big and small, but it has no place in the cosmos at large. This

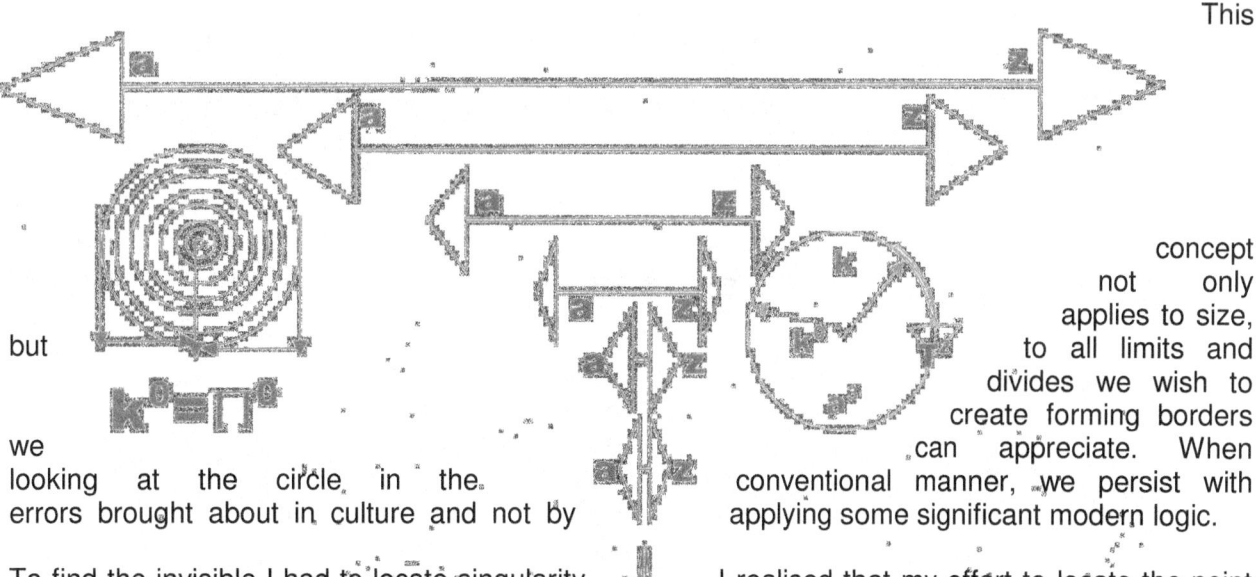

concept not only applies to size, to all limits and divides we wish to create forming borders can appreciate. When conventional manner, we persist with applying some significant modern logic.

but

we looking at the circle in the errors brought about in culture and not by

To find the invisible I had to locate singularity. holding singularity enabled me to backtrack the

I realised that my effort to locate the point exploding Universe to its origins.

By applying some basic effort I have located the the direction it took moving forward in time

position from where all movement came and

By reducing r or any line indefinitely to the tune of half each time, r would become infinitely small, beyond human calculating means, however as mentioned in the case of the smallest dot holding one spot, r would become insignificant beyond human comprehension even, but never reaching zero and still Π would remain intact and dictating form. I believe one can begin too see where my suspicions are heading because the flaw comes about in the manner mathematics are practised for thousands of years. But before coming to the mathematics I would first like to bring your attention to the practical side. I am promoting a theory in which I am able to prove there is as much contraction going on in the cosmic Universe as there is expansion and the contraction is as much part of the expansion. The Universe rides on a balance and we have to locate such a balance. To prove my theory I firstly had to locate the centre of the Universe. Even admitting to such a notion sounds like madness, but please allow me a chance to explain what I wish to explain in more detail for this is the crux of physics. If I wish to achieve success that would depend on my ability to convince all that outer space comprises of material and as such we can locate such material even if we are unable to see such material.

In the very centre there is a spot that changes everything radically without having a presence in the Universe. He point where the spot is reached, is also the spot where the spot is crossed. The instant of entering this spot is also the point of exiting this spot again. The spot has no point in 3D or claim any space of any sorts. The spot can't be halved or be reduced or go smaller by any margin and yet this spot holds the dynamics to alter everything circling about the spot. The spot commands everything attached to the spot and the spot controls gravity. Because the spot is never there, the spot can never disappear. Movement drawing a line through the spot can exit the spot and this gives the spot visibility. This spot is infinity. This spot is the smallest point thee could ever be in the Universe as the point holding infinity or singularity controls everything in relation to the spot. This is the spot in reference used by the clock pendulum Galileo devised.

The reversing of the circle radius is not alien to nature at all. An observation coming instinctively to mind one may recognise is that the form reminds rather explicitly of natural phenomenon as hurricanes, water whirls and even the shape most commonly favoured to express the cosmic object referred too as a Black Hole. The similarity may be more than coincidental. Let us consider the statement in the reverse. In our calculating of a circle we apply two formula methods. The one use an r to indicate the radius and the other use a D to indicate the diameter, which is double the radius and therefore needs to be divided by a four to eliminate the Newtonian inverse square law amounting to the difference there will be between the two. The one using the radius is Πr^2 and the other formula is using the diameter is $\Pi D^2 / 4$.

In any circle or sphere the size only depend on the fluctuation of r in the square as a component to the circle or sphere but that does not affect the form by indication of Π in any way there may be. The conclusion from this is that no line can start at zero because that will be a mathematical impossibility. This statement by itself excludes zero and with zero excluded one then begin to appreciate all the rest of the concepts governing corrected cosmology. A line or spot starting at zero would therefore be shorter than the shortest line possible. For obvious reasons can no line, or any line grow or extend from zero because such a line must then quit zero and become something, thus abandon its original value. That would mean the start of the line has a different value to the end and a line holds conformity through out. When any line is starting from point zero it can never leave zero because of the influence of being zero disqualifies any possibility of growth. If the line then had to grow in all directions at the same pace the line must therefore be a circle or being three-dimensional, a sphere.

Flowing from this fact is that in the Universe there can be no zero point or unfilled space. In the case of the growing sphere the value of the circle is Π, and that is where creation started. That gave me the clue where to start looking for singularity. One would find singularity in the value Π and the value Π will be in all things rotating in a circle. You might wonder how does that apply to the cosmos and moreover to gravity?

By accepting that there is some conductor (not the ether of old) between two cosmic structures can one accept that there are certain invisible undetectable influences on the edges outside the surrounding of material. One then can see how space conforms as it converts to liquid heat. In the same effort one can see how material confirms heat from space to material. By reducing and confining the heat drawn to the centre the space becomes more concentrated as the heat levels begins to rise.

 Take any bicycle pump and compress the plunger and the result will be that the heat created by such action will burn the finger you use to cover the valve hole. The heat comes about from concentrating the space, which holds the air. But the air does  not concentrate because one does not bring in more air than there was before. The relevancy changes as the space reduce to change the space back to heat. This action is the very opposite of an explosion. But reducing pace the action brings about that the space turns to heat. This is most crucial in accepting because this is the precondition about the understanding as much as accepting a new concept, which I try to introduce, and at the same time I try to produce a concerned effort in dismissing myths from cosmology.

The Spot

becoming

the **Dot**

Before the top comes in motion, motion is not yet defined b the line forming. When the top comes into motion, the first motion activates a line coming from a spot that was not even a dot. Space forms by parting from the spot that forms the dot by the motion of space. In that eternity parts from infinity. Motion parts from stagnation. Infinity breaks the laboriousness of eternity for the duration of infinity. The spot becomes dot and grows into the line that represents time.

Gravity is about turning space into hotter denser space as it reduces space and that is the reason why there is no visible or measurable stronger gravity in the centre of objects. The centre does not indicate more gravity because the gravity that is the measure of the accumulation of heat that the gravity produce in that space $a^3 = kT^2$. The heat increase as the gravity becomes more intense because the more intense heat is the gravity increase. By the reducing of the space coming down towards the smaller area such coming down leads to space reducing, which brings about heat increases. The stronger gravity personifies in the denser heat produced by the reduced space. As space increases heat dissipates and that we find is what happens in an explosion. The bigger the heat release the more space becomes available as winds (shock waves to use the name hiding the truth) blowing across fields. The winds come about as space multiplies through the release of heat creating new space that was not there before the time. In the explosion will the heat decrease be the decrease of gravity that was before the explosion bounding the heat into condensed space and the explosion is the release or antigravity reducing the heat as it increases the space. In my search I stumbled on two accepted but not intergraded laws and when I found and located singularity the two laws became very much plausible and factual.

Take a circle and reduce such a circle constantly to where it no longer can reduce. Reduce it to a point where only form remains part of the circle because the radius has gone beyond human measure and becomes so small it is not noticeable with what ever tools man may use, then what remains is pi since pi does not indicate size but indicate form, and form is all that then will remain.

I believe one can begin too see where my suspicions are heading because the flaw comes about in the manner mathematics are practised for thousands of years. Space is nothing because that means space is a standard fit all issued out before the time. Space cannot increase and winds are ghost blowing their breath. Winds are as much antigravity returning reduced space back to increased space. Before coming to the mathematics I would first like to bring your attention to the practical side. I am promoting a theory in which I am able to prove there is as much contraction going on in the cosmic Universe as there is expansion and the contraction is as much part of the expansion. The Universe rides on a balance and we have to locate such a balance. To prove my theory I firstly had to locate the centre of the Universe. Even admitting to such a notion sounds like madness, but please give me a chance to explain in more detail. I realised that my effort to locate the point holding singularity enabled me to backtrack the exploding Universe to its origins. By applying some basic effort I have located the position from where all movement came and the direction it took moving forward in time…and yes, even time as such.

Anything occupying space in the cube will apply r and by r I mean just a distance not using Π because Π serves as a form indication while the collective product of r will determine form as well as accumulative dimension total. Notwithstanding the name used confirming the shape or r named as length width or height, it is all just a straight line bringing about the cube with all its other names that may find attachment to specific form but nevertheless still remains only a six-sided cube with connecting lines applying different angles changing in some cases.

The normal perception is that any circle growing spontaneous would grow by the radius, which is r. In mathematics that may be true but it is not true in nature. In nature that cannot be the case because r is an indication of a straight line. By growing with the aid of a straight line from the centre to circle the influence that that would have on the circle would result in many circles following one another and not a continuous growth.

Gravity is the dimensional changing of space holding r as reference in the cube as to the sphere holding Π as the reference. In order to generate spin producing time in matter occupying space, therefore creating dimensional change, Π has to be a factor indicating the possibility of spin because implementing Π the circle sides will follow one another without establishing separation. The answer must be in finding Π, and thereby locating singularity. If singularity is in affect the original point of the cosmos birth, the reducing path we should follow will indicate the whereabouts such a point must be

In the normal applied mathematics there are two standard formulas used to calculate a circle. The one use an r to indicate the radius and the other use a D to indicate the diameter, which is double the radius and therefore needs to be divided by a four to eliminate the Newtonian inverse square law amounting to the difference there will be between the two. The one using the radius is Πr^2 and the other formula using the diameter is $\Pi D^2 / 4$. However one looks at the mathematical expressions and Kepler's formulating of space-time there is an exceptional difference between the two scientific uses. When investigating Kepler's formula one do find it appreciably differs from the normal Mathematical equation like $a^2 = r^2\Pi$ and $a^3 = 4/3\ \Pi r^3$. In the normally used mathematical expressions such equations tend to concentrate on the volumetric aspect. In the case pf Kepler's expression it is something else that wants to surface. It is another idea that is coming to mind. In Kepler's formula a^3 stands to symbolise the third dimension and such a third dimension becomes equal to two other dimensions grouping and sharing value to equal a^3 efforts. It is not the circle of the rotation because with such a normal circle the radius is in the square and Π evaluates form. Here there is no mention must of a factor Π, which one would suspect to be somewhere applying since the circle is Π and Π is the circle and the two are inseparable.

But not in Kepler's a^3, where there is no mention of Π at all. The fact that there is a radius of some sorts used to indicate a position cannot hold the square as it normally does in the case of the normal equations. In the mathematical equation the factor indicating the position of the circle edge has the square value being called the radius or in some cases the radius doubles and which then is the diameter, and the circle indicator is Π. But in this event the formula value will bring about a square value to the answer one receives. It will bring a value to the surface of the circle. In Kepler's formula it specifically does not. I am not the one that brought Newton into disrepute. Before me the cosmos did. The comets with they're not colliding did, and so did Roche and Lagrangian principles. Hubble was another one and it becomes apparent that every one that made a study about matters in the cosmos was in some disagreement about Newton. By Newton's effort to improvise on behalf of Kepler Newton made a statement that Kepler never made. In all honesty nature reacted strongly against the claims Newton made on behalf of Kepler and not about Kepler's work but about Newton's modifying of Kepler's work.

In short: how can a comet sail past the Sun time after time without colliding and still apply a contraction in the manner which Newton suggested by the one claiming a freezing grip on the other? This strongly contradicts $F = G\dfrac{M_1 M_2}{r^2}$. How can five structures as the LAGRANGIAN POINT form around a centre structure while the centre structure keeps the five in position at equilibrium? By rejecting Newton's improvising this strongly contradicts

$$F = G\frac{M_1 M_2}{r^2}$$

How can the Roche limit send structures spinning around an axis they establish at the time and either the lesser candidate destruct at a distance when there are in comparisons about size and gravity or they push each other into development beyond their individual means.

By rejecting Newton's improvising this strongly contradicts $F = G\dfrac{M_1 M_2}{r^2}$ How can Hubble's visual Universe expand while Newton's Universe must contract? I am not the one that started the dispute with Newton but it is the World of Physics that will not admit to such a dispute!

By rejecting Newton's improvising this strongly contradicts $F = G\dfrac{M_1 M_2}{r^2}$.

If the formula of $F = G\dfrac{M_1 M_2}{r^2}$ did apply, its validity would only apply in a very specific range, because the rest of the Cosmos sows no sign of adhering to $F = G\dfrac{M_1 M_2}{r^2}$ and at a very determinable point the formula does not affect objects in the air. After such a point one will find satellites able to orbit, be it art a definite pace that matches the rotation of the earth. Still...below

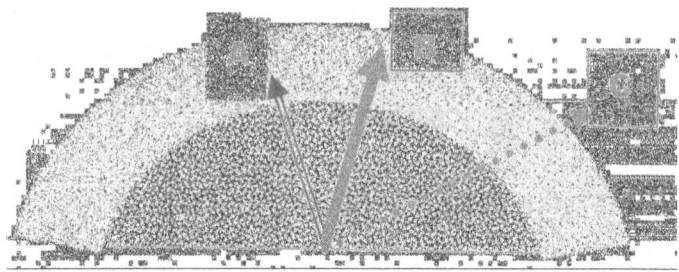

such a point (B) orbiting objects will come crushing down to the Earth.

From point (B) to the Earth Newton's formula apply and from point (B) upward Kepler's formula apply, but my pointing this out brings about all sorts of annoyance concerning academics. It must be clear to all persons that there are a big difference

between the applying of Newton's $F = G\dfrac{M_1 M_2}{r^2}$ and Kepler's $a^3 = T^2 k$. When the objects reach some point they will drop to the Earth and when that happens, mass do not play a part in the speeds they come to reach.

When examining the case where two balls drop vertically, gravity, as a force does not apply and therefore gravity does not come into effect because there is no difference in speed or duration.

With out any apparent reason the formula is substituted with the following formula:

g = G(M . m) /r^2 where:

G = the gravitational constant,

M = the mass of the body,

M = the mass of the lesser body

r^2 = the radius between the two bodies.

…And four hundreds of years this rubbish has never been questioned by the greatest minds that this far walked the planet or by those whom has the mental power to crack a safe…and they are unable to think the most simple thoughts such as finding singularity where Kepler placed singularity in **k^0 = a^3 ÷(T^{2}k).**

Let us take this formula back to the accepting of the Big Bang and find sensibility amongst a lot of confusion that I can see.

There was a beginning that saw a radius between objects so small the size will never again repeat. The diameter of the particles were also next to nothing but that should not be a contributing factor surely…the main focus point is that particles were as cramped as it shall never again be repeated.

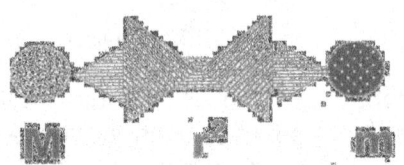
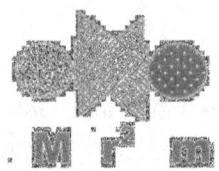

With the radius in the square dividing the shared and combined mass of the particles the relevant mass of the particles rises by the square as the radius reduces. If the radius becomes infinite, the relevant mass that the particles will produce goes up eternal. No force in the world would keep particles apart drawing on each other with an applying force but such a force is divided by an infinitely small separating radius. This is a recipe for joining and not dividing. Still according to the Universe I am able to witness the dividing became enormous and the joining practically irrelevant. The gravity was more than words can describe, the heat was able to melt it all in one structure, but that did not happen. It split into billions of individual atoms.

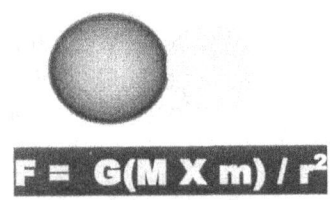

The two objects should have their own value of gravity and gravitons and in comparison with the gravitons of the earth; their value is insignificant. However, these two balls are in their own individual deuce to see who reaches the Earth first, and the iron ball's gravitons should give it a superior advantage. This comes about because the two objects are in a position where they compare in relation to one another and share a common second factor, which is the earth. In relation to the earth, the gravitons of the two balls do not come into consideration, but this do not play a part since the Earth is a common factor.

The balls, however, is put in a situation where they stand in relation to each other. When compared to one another, the gravitons should give the heavier ball a sizable advantage. The sensible example one can show to prove that where some matching structures in size come into conflict about occupied space sharing one of the structures are turned to heat in space by the other and larger structure. If the structure proves to large the superior structure turns the lesser compatriot into heat. Then being heat it will apply gravity and admit such heat into the ranks of its atmosphere, but not before it turned it into fragments good enough to be heat.

The Roche limit is the following principle, which also holds the name of the "curve ball" and the sound barrier. Since I am not going to explain the sound barrier in this letter I will again refer to the curve ball as an explanation. Considering Official Science policy the collision must be devastating and total destructive to one or both. Where r that is the radius between the two colliding structures disappear from the equation since the collision is already in progress the structure would be unable to maintain any viable distance

When an object has a condition of mass the object retains a position in relation to the centre of the Earth that holds a relation to the centre of the Sun, which holds a relation to the centre of the Milky Way, which holds a relation to …God only knows where this would ever end. The object is in a state of cosmic freeze and is where Newton in his stupidity tried to put any object with mass while spinning with the

Top **having mass** **Earth**

Earth in a state $\dfrac{dJ}{dt} = 1^0$. What this refers to is singularity which in terms of Kepler's

formula is $k^0 = a^3 \div (T^2k)$ or in Newton's corrupted version is $\dfrac{dJ}{dt} = 0$

When Newtonians are faced with explaining something of importance which they should understand but clearly does not understand and such explaining requires more than awarding mass or finding pressure to calculate and it can't be rewarded of having a status of nothing, Newtonians are then faced with a serious dilemma. To hide their insecurity they either award the most mind-blowing and devastating names by which they do not need to explain anything because the name alone makes any other member of the human race feel completely stupid, or they get so technical they cross over the line of sanity and get totally bizarre.

Ball curving after impact

Those Newtonians must have a lot of stupidity they feel they should hide behind a lot of complicated technicality. Let's examine the manner in senseless which our Super Educated Brainy Bunch Masters of Physics explain the "curve ball" phenomenon that is part of he Coanda effect. When a ball is hit or kicked or passed or whatever, the ball sometimes begins to rotate and when rotating the ball does not follow a line as it should go straight ahead but curves off centre in a direction leaning either left or right. I have been through this before in this book but now I wish to be more explicit and use this part to show how space-time by the measure of Π is involved in this phenomenon

Worldline of stationary ball

Time

World line of a ball travelling to the right.

space

The further the throw is

the more the curve the ball has

The point of impact

because this is not part of gravity…this is gravity as in being the Earth In

Ball curving after impact or purposeful kick

same as the path the orbits around the Sun.

their explaining there is a world line and a time line and in some or other peculiar circumstances the world splits from time because then the world gets a world line while the time lime is soldering on undeterred by the fact that time is leaving the world behind and while leaving the world high and dry and without time it is taking the ball on a journey to travel in curves. I am not

Where the should end

Where the ends after hard blow

Where the ends after soft blow

The position of the kicker / hitter of the ball

trying to belittle others but others has made a hell of a good job belittle me while those others that tell me my educating is so inferior they see no reason to meet me or that my schooling is so inferior in relation to their superior requirements that it has no purpose on their part to make an effort to read my work but meanwhile back at the ranch they are the ones underwriting this

to

shit. To them I say this, one can educate as long and as enduring as one wishes but stupidity can never be wiped away while the measure to think can be taught even by the effort of doing it on a self-help basis without proper official education. The answer to the curve ball is in realising Newton's mistake in assessing that $\dfrac{dJ}{dt} = 0$, which in mathematical principle could be nothing else than

$$\frac{dJ}{dt} = 1^0$$ and for that there is not much schooling required I would say!

The ball travels a path diverting from the true direction after being kicked. The diverting of direction does not depend on the force by which the ball is kicked because it could curve a lot while travelling a short distance or it could curve more gently while travelling much further while curving.

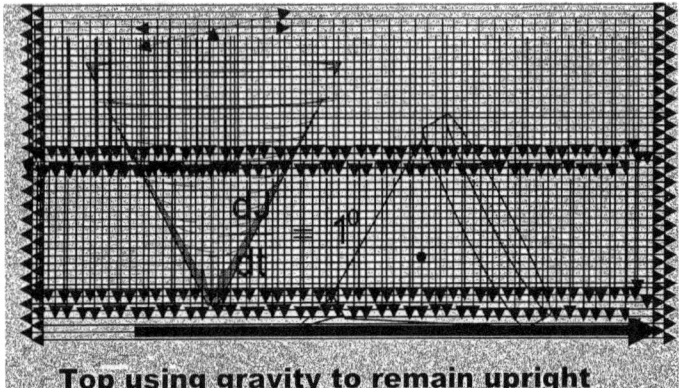

Top using gravity to remain upright

The value Newton awarded the top $\dfrac{dJ}{dt} = 0$ to remain upright is the same

value $\dfrac{dJ}{dt} = 1^0$ that the ball is awarded

to curve. It uses singularity, where as invisible, unseen and unnoticed lines are they are forming the grid that forms the entire Universe and every one is running all over and the lot is crossing one another while joining at ever cross because these lines are equal in measured value as $1^0 = 1^1$.

As indicated before the smallest any line could be is a dot and a dot is a circle where a circle is $r^2\Pi$. When we put the radius in relation to singularity we find space-time forming at Π because $\Pi r^{(0)2}$ is forming a space border at Π. That makes Π flat and the Universe forming space-time from singularity does not form a dimensional unit as Π because this shows Π is still very flat. Where Π^0 becomes Π the Universe becomes round but it still remains entirely flat. That means where time starts to produce space at a value of Π^0 going Π there is still a flat dimension forming. I wish to remind the reader that these positions still fall way outside the Universe we hold because at a point where Π^0 meets Π, only there does anything start that we can recognise but not yet measure. We know that time in progress is a constant renewal and repositioning of Π and by having Π flow we introduce space to the Universe as time delayed. Therefore time renews space by introducing Π and the significance of this is in the value Π has which so strongly resembles the Titius Bode value of seven in relation to ten.

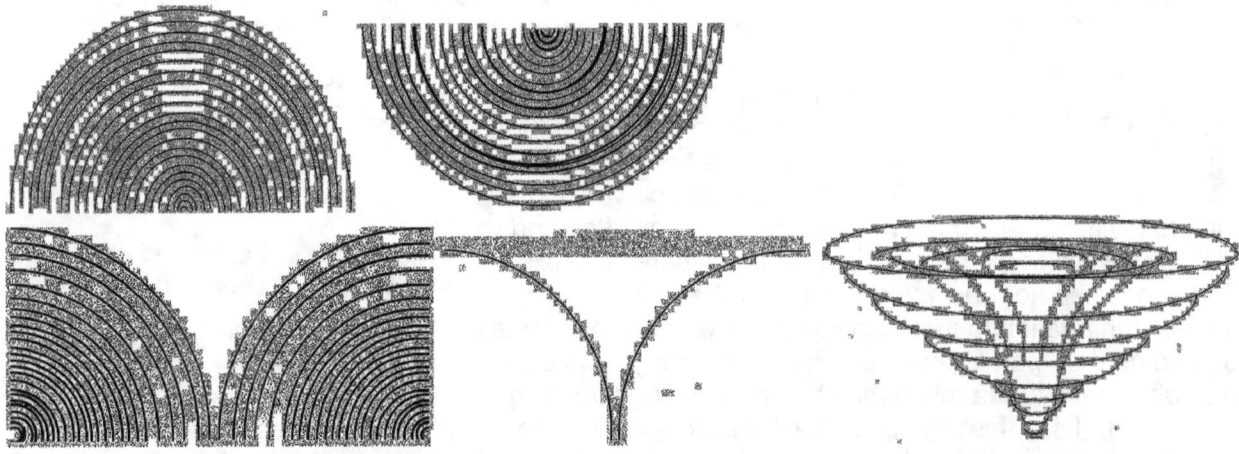

Looking at the interpretation of Π flowing and rough interpretations of what a Black hole represents there is some unmistakable alikeness in the resemblance of Π in comparison with the Black Hole. However, Π holds a double value or a two directional value of the Titius Bode law measurement. Where Π is (10 + 0.991 + 1 + 10) ÷ 7 we have the Titius bode measurement being only half the value or then the directional flow of time progressing into space. But I explain more about that a little later on in this book…

Studying the behaviour of the forming of Π teaches us more about the behaviour of time forming space in the cosmos.

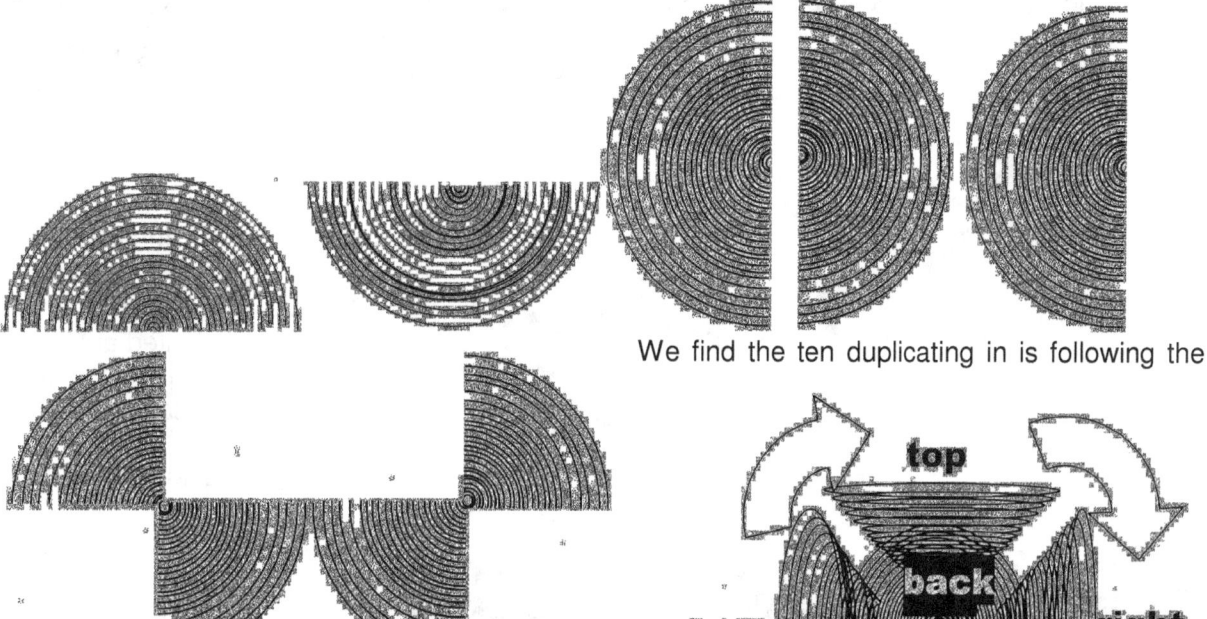

We find the ten duplicating in is following the left right intervals and the top bottom intervals and that makes the intervals not straightforward as it seems at first. The intervals show the directional change of four coming about and every time we find the cube in the directional four quadrants release one side to Π holding superiority. This means that time forming space as Π and then going onto more space ahs a change of status from being single dimensional or flat going onto being multi dimensional as one would expect being part of our Universe.

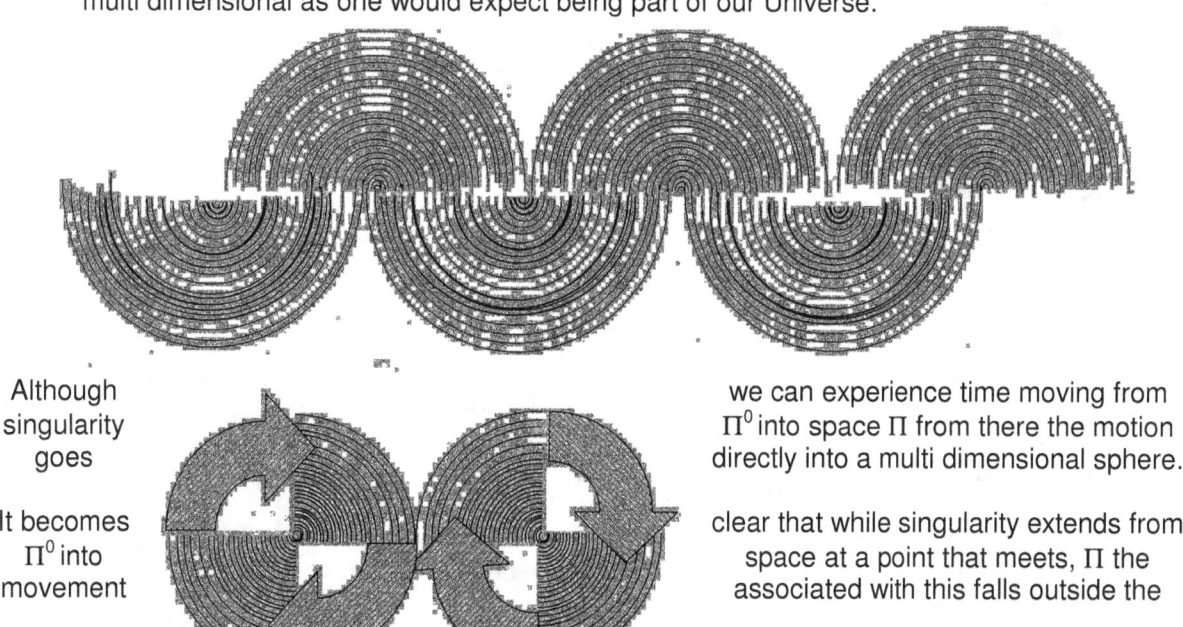

Although singularity goes

It becomes Π^0 into movement

we can experience time moving from Π^0 into space Π from there the motion directly into a multi dimensional sphere.

clear that while singularity extends from space at a point that meets, Π the associated with this falls outside the dimensional influence and layout of our Universe. Such extending of Π going to form $Π^2$ is a result of heat interacting where gravity or movement of time in space then is the consequence of the negative $(k^{-1} = T^2 / a^3)$ or the positive directional expanding $(k = a^3 / T^2)$ value of heat forming space.

It is when gravity becomes $Π^2$ that we have dimensional interaction in space. Before the gravity comes an issue whereby the gravity movement captures all movement we have the singularity expanding where this motion falling outside the realm of cosmic space we have $Π^0$ going onto touch space at Π and this action is still on the rim of the Universe on the inside where infinity stretches into space. Only when the space starts turning and in circling singularity does a dimensional Universe come about.

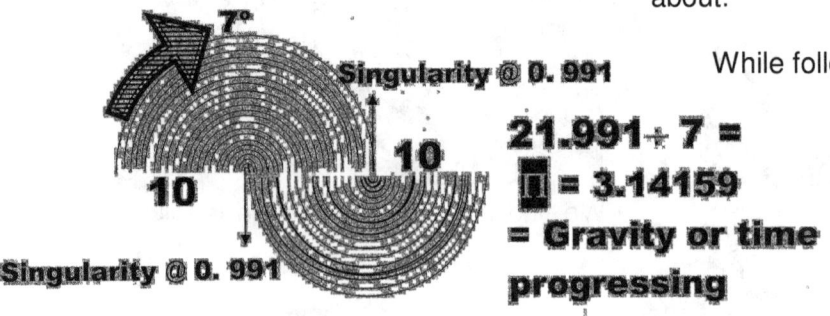

Singularity @ 0. 991

Singularity @ 0. 991

$21.991 \div 7 =$
$Π = 3.14159$
= Gravity or time progressing

While following the lead the Titius Bode law receives from extending Π from singularity $Π^0$ it is by measure of the Titius Bode law that gravity becomes valuated at $Π^2$. I am returning to prove this statement in a short while.

It is by introducing the five points the Lagrangian system holds in singularity forming the cube to cover the four quarters that Π provide, that Π then could receive a value of 20 + 1 + 0.991 in relation to the curving of 7 which then totals to Π. While the fives represent a different "box" each time as seven crosses into the next "box" of five holding sides to space, the seven remains the same but for alternating as it goes through singularity.

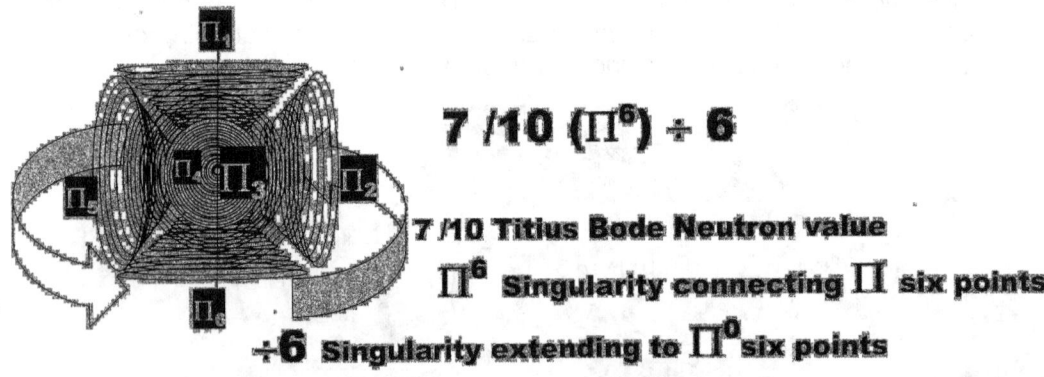

7 relating to 5+5 = 10

On both sides of the Universe

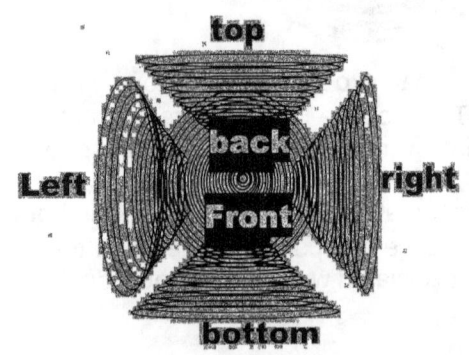

$7 / 10 (Π^6) \div 6$

7 /10 Titius Bode Neutron value
$Π^6$ **Singularity connecting Π six points**
$\div 6$ **Singularity extending to $Π^0$ six points**

top
back
Left   right
Front
bottom

By Implicating the Titius Bode law while still adopting Π does the value of continue to bring about spacetime. How, why and in what way I will prove further on in the book because at this point I wish to kill the ludicrous flat Universe madness that Newtonians

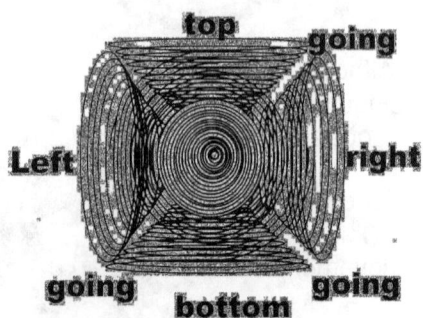

top   going
Left   right
going   going
bottom

underwrite. The value of Π goes six sided and three dimensional at $((7 \div 10)(\Pi^6) \div 6)$. We must read into this value more than what is represented on face value. In normal association would form space-time at $\Pi^3 = \Pi^2\Pi$ but where space meets time the dimensional limit stretches much further, in fact twice as far as the dynamics the third dimension brings along. Where space starts to spin we have the neutron value of space formed by the Titius Bode measure of half the value of Π being 7 /10 multiplied by the point where timeΠ^0 forms space at $\Pi^6 \div 6$ and at this point the relevancy holding the epitome of material forming a three dimensional concept that still involves time would be $3(\Pi^2 + \Pi^2)$ because just after that Π again draws time in space flat at $\Pi(\Pi^2 + \Pi^2)$. Having a six dimensional connotation to space-time this point$\Pi^6 \div 6$ incorporates both material holding space in control by movement $\Pi(\Pi^2 + \Pi^2)$ as well as space unoccupied $2(\Pi^3)$ we find a six dimensional value in relation with six sides which is the three dimensional aspect we are accustomed to.

It is from this value that space forming by dimensions in seven can be displaced to move to singularity once more and form a depleting of space-time. We know the movement as gravity where heat is rapidly condensed to a liquid in order to drive solids and to cool singularity by maintaining singularity..

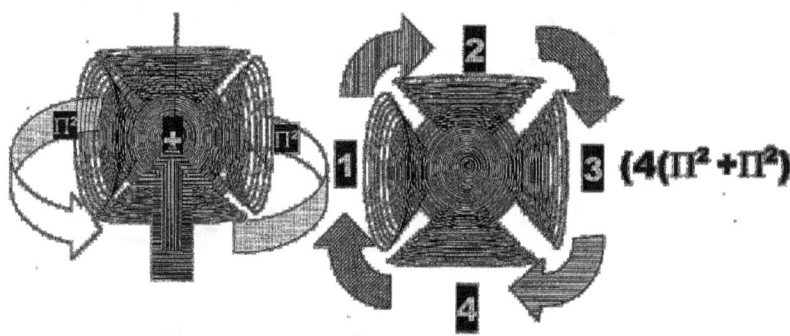

$\{4(\Pi^2 + \Pi^2)$

In a more controlled form we manage this movement while calling it electricity but it remains the same.

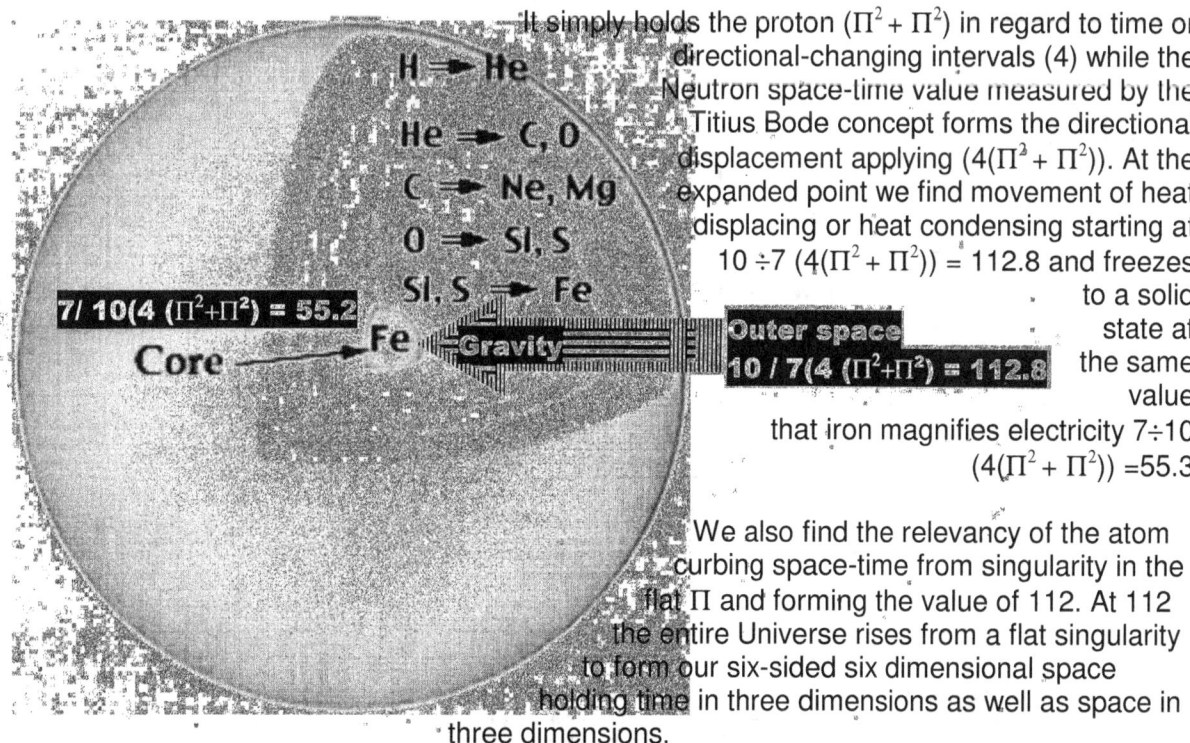

It simply holds the proton $(\Pi^2 + \Pi^2)$ in regard to time or directional-changing intervals (4) while the Neutron space-time value measured by the Titius Bode concept forms the directional displacement applying $(4(\Pi^2 + \Pi^2))$. At the expanded point we find movement of heat displacing or heat condensing starting at $10 \div 7 (4(\Pi^2 + \Pi^2)) = 112.8$ and freezes to a solid state at the same value that iron magnifies electricity $7 \div 10 (4(\Pi^2 + \Pi^2)) = 55.3$

We also find the relevancy of the atom curbing space-time from singularity in the flat Π and forming the value of 112. At 112 the entire Universe rises from a flat singularity to form our six-sided six dimensional space holding time in three dimensions as well as space in three dimensions.

The proton takes on a relative spin value of forming $(\Pi^2 + \Pi^2)$ while the neutron holds a value of $(\Pi^2\Pi)$ and the electron forms time (3) and the displacement of the atom is $(\Pi^2 + \Pi^2)(\Pi^2\Pi)(3) = 1836$

If I were allowed to make a guess about the Newtonian finding matter going anti matter then I would guess that the neutron became the anti-matter of the proton being the matter. By losing one dimension to going nuclear heat, it is conservable that the neutron became the liquid in the Coanda marriage where at that time the proton then became the solid. In that way one can explain to some certainty that the neutron lost one part to space allowing movement to take place and then because of the loss of that factor it became a form of liquid. This loss of one dimension to heat by the neutron allowed the four phenomena to consolidate their combined effort and form the marriage there are between the solid proton and the liquid neutron. The double movement value of the proton being $(\Pi^2+\Pi^2)$ changed to $(\Pi^2\Pi)$ where space entered the movement.

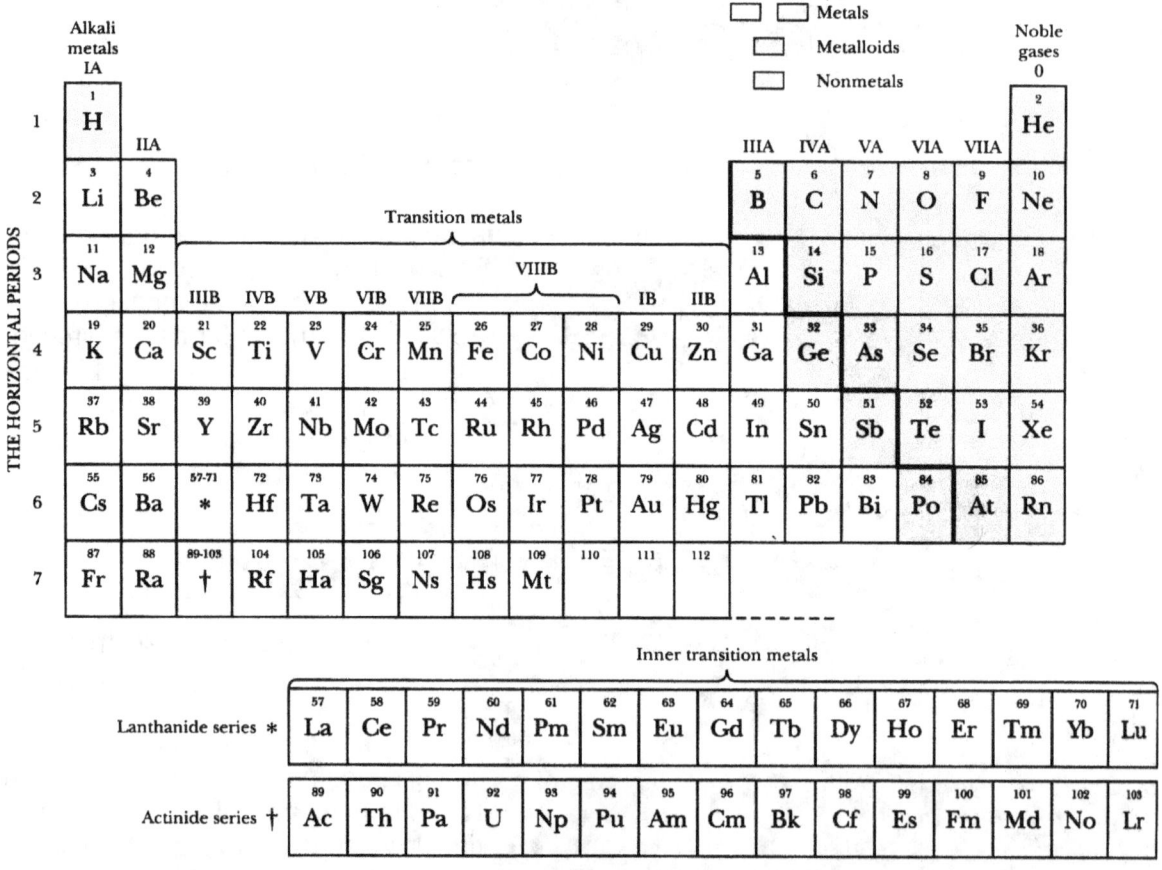

At the very top forming the true boundary of eternity as far as our Universe goes, we have the atom b ringing together the point where infinity meets eternity at a displacement value of 112.

When the Universe came into form which the Bible refers to as mighty winds blowing on the water the development of singularity in displacement was dividing in relation to the Titius Bode law of material moving in space $10/7\pi^2/2(\pi^2+\pi^2)=139$, material forming in spheres $7(\pi^2+\pi^2)=138$ and material spinning in space by gravity $7/10 \ \pi^2(\pi^2+\pi^2)=136$. I explain the layout or the meaning of these formulas in other more technically oriented books. In brief however I will mention the following.

At $7(\Pi^2+\Pi^2)=138.17$ the neutron developed with mighty winds running over the water as one factor (exploded by nuclear winds) turned the otherwise static and solid Universe to a liquid moving factor of the Universe. It did not run over the material because singularity can't move. It ran over the water or the fluid, that which became the then formless part $7/10(\Pi^2/2)(\Pi^2+\Pi^2)=138$ in relation to the form that came about (mentioned as the abyss, the waters and the wind) then solid part formed as a result of singularity extending $\Pi^0 \Rightarrow \Pi$ and form results from this by $7(\Pi^2+\Pi^2) = 138$ which then puts space-time in relevant movement of material $7/10(\Pi^2)$ which indicates the forming of gravity for the very first time and the movement of material in space $10/7(\Pi^2/2)$ formed as the solid component sporting gravity as material still does. This one can read from the factors in the formula $7/10(\Pi^2)(\Pi^2+\Pi^2) = 136$. This is what the Bible says happened. Then came the Big Bang where light became a factor and the Universe became three- dimensional at $7/10\Pi^6 / 6 =112$ and today we line in an atomic world inside the world of light with atoms nicely grouped.

We have a physics scenario whereby the faster anything moves, (say the top spins) the more the effect of gravity will be on the compacting of space and the increasing of heat. In order to compact 243 protons into one uranium atom we must spin it at the speed of light. This Newtonians call particle acceleration and in this manner Newtonians produce nuclear fuel. Yet this never dawned on Newtonians that this may come as a result of gravity increasing by reducing the space the proton takes up. When the uranium's atom once again slows down, this lot apparently is quite unstable and the volatility is associated with the release of heat. Again no Newtonian ever brings this association to mind. The reason why the Neutron has no mass is because the neutron forms a liquid and the proton forms a solid and when Uranium was nature's choice in stars, things back then spun much faster than everything spins today. The Neutron will never have mass because the neutron forms the water in relation to singularity being able to provide the solid. The Bible provides the adequate explaining Newtonian atheists completely miss about science! It is a pity we are stuck with such simple minded beings such as atheists are and then to have this lot live amongst us increases our miserable enduring we suffer with them being around. Everything mentioned in relation to gravity has the Roche limit as a measurement forming a strong relation as all can see with the Titius Bode value of either 7/10 or 10/ 7.

The Roche limit is:

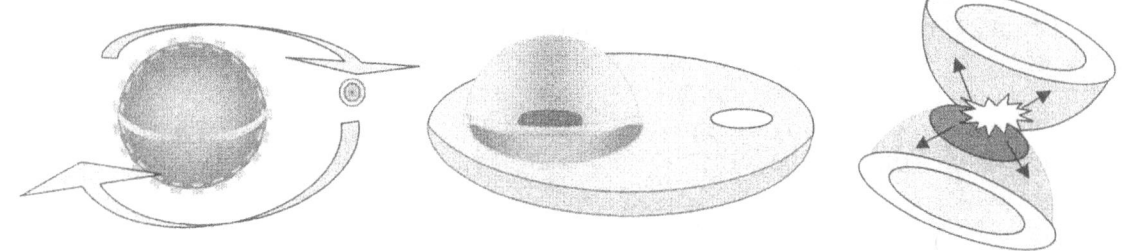

The region surrounding each star in a binary system, within which any material is gravitationally bound to that particular star. The boundary of the Roche lobes is an equipotential surface, and the lobes touch at the inner Lagrangian point, L_1, through which mass transfer may occur if one of the components expands to fill its lobe. It names after the French mathematician Edouard Albert Roche (1820-83).

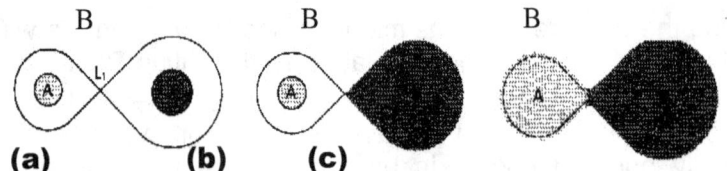

THE ROCHE LOBE: In a binary system, the Roche lobes of components A and B meet at the L_1 Lagrangian point. (a) In a detached system, neither star fills its Roche lobe. (b) In a semidetached system, one massive component, B, fills its Roche lobe. (c) In a contact binary, both components overfill their Roche lobes and share a common envelope. As with the graph I can see the two sides forming a connection therefore relevancy has to apply, all contradicting Newtonian claims of no connection but through mass attractions. The mass does not attract but one interferes with the other total influencing the space surroundings.

Even more astonishing is facts about the Binary star system that is seldom to never mentioned.

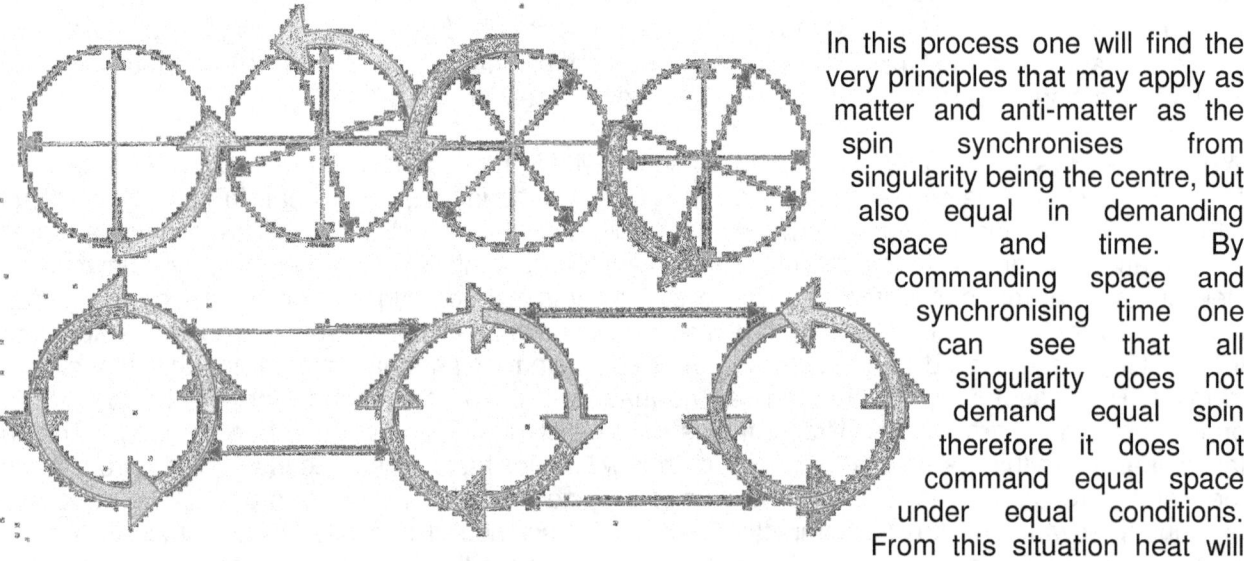

In this process one will find the very principles that may apply as matter and anti-matter as the spin synchronises from singularity being the centre, but also equal in demanding space and time. By commanding space and synchronising time one can see that all singularity does not demand equal spin therefore it does not command equal space under equal conditions. From this situation heat will arrive, as friction will devastate all material.

One should see why heat in our age comes about when produced. Heat is mainly a result of friction and exposing of the lack of space and billions are spent in industry to combat friction heat. That was precisely the conditions that were about in the pre or post Big bang second. Heat will always flow from the highest value to the lesser value. This is the concept I use for the basis of my entire theory. I base my theory on gravity producing cooling and contraction while heating produces motion by expanding and creating more space, which on the other hand also lead to cooling. By being without motion cosmic objects retain heat and expand. When overheating no amount of force can retain the container from becoming too little and with the heat coming about forming the expanding the space produced with this action will destroy the space any container may have not withstanding material used or the container size or even the force the container use to contain..

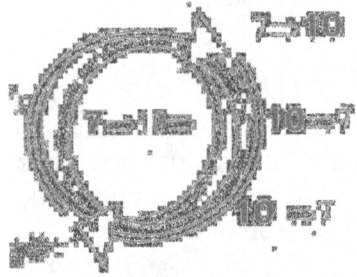

In this writing I am through out the time showing how the Titius Bode principles in conjunction with the Roche limit form gravity and the evidence of that we find as the jig formed in space between the different planets evolving. Gravity left its mark and first Titius and later Bode found the marks that gravity left for us to find. But since not even Newton could claim to know what gravity is and therefore nobody was able to read the markings of gravity this tendency was dismissed as coincidence.

During Creation before the Big Bang the compactness of the particles produced motion discrepancies between different particles bringing about friction where some particles overheated and formed heat. Heat then formed space in a process that became the Big Bang where heat produced space and formed the motion we see today in the Hubble constant. In space there is different dimensions that we call sides of which there are six opposing one another. Those sides that form from the heat that became space in motion but the sides play a much stronger role that is anticipated presently. The sides are dimensions and the dimensions come about since motion created the discrepancies. Let us think what the verdict at present is using current Mainstream Physics on the question why a water drop would float in a space capsule in space, in the presence of micro gravity. Why would a sphere always form when left on its own will to capture free form?

We all accept that the true cosmic form in Universe as a whole would be and most probably is the sphere…but why would the sphere form as the original form. If left to a natural outcome material will take on the form a sphere has. What is that special in the sphere as a cosmic choice? By merely blaming gravity pulling from the centre is rather avoiding the question with simplicity because the question arising from this answer is where is the centre of the Universe? The more growth or mass increase there comes about the more the motion applies by the centre will secure heat within the centre. The more heat there is secured in the centre the stronger the domination of such a heated centre will be on the surrounding space-time. By the securing of heat in the centre the reducing of k will produce less space occupied a^3 but much denser space occupied bringing about a much stronger T^2.

The gravity collects heat in the centre of material forming a sphere. It does so by concentrating space and as such produce heat. By producing more concentrated heat it secures independence and the independence will produce motion to move away from the singularity in control. Gravity has two factors influencing space, which are a straight-line k and a circle going around the centre T^2. It is a balance $a^3 = kT^2$ that forms $k^0 = a^3/k\, T^2$ Gravity T^2 increase as space a^3 reduces by the reducing of k. Much more to the point that implicates gravity within the star would be the heat the star concentrate within the centre of the star. Such heat shows much more relevancy than the mass it holds in the sphere of the star. The more space there is between the particles forming the mass within the sphere of the star the less the gravity output is because the less motion such a star can achieve and the less space the star will dismiss.

The piston in the exhaust/ intake position

Compression / ignition stroke

One must see the piston engine in the same manner as stars in operation or the Big Bang. By reducing the stroke k we find the space a^3 decreases as the temperature rises in ratio. Take this scenario and compare that to conditions applying during the period we think of as the Big Bang.

Compare the cylinder conditions to the Earth or any star. As k reduces from

the outside also reduces in about more heat must also reduce increase. That is sphere and with reversed. This is smaller it will increase T^2 as much as it decreases a^3

towards the centre the space represented by the declining k ratio and by equal measure. But a smaller distance k brings concentrated in the reduced space. If a^3 reduces then k but then T^2 holding the gravity or heat concentration must just what is happening inside the engine and inside the the Big Bang where with the Big Bang the situation is what Kepler said with his formula $T^2 = a^3 / k$. If k gets

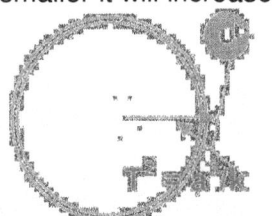

A smaller factor k will result in much more compact area. The smaller a^3 becomes the larger T^2. This will be compactness is the result of the distribution increasing and by increasing the mass or occupied space becomes denser and more intense. With that it is quite silly to measure the diameter of any star and from that try to determine the because the diameter could have reduced a billion achieved.

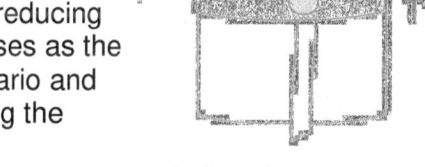

from the reducing of k. The density of the atom such

"mass"

times with the motion the star

The closer space gets to the centre of any star the hotter the star is. It is the same with the engine. The internal combustion engine teaches us the implication of gravity. By reducing the **k** factor the $\mathbf{a}^3$ factor also reduce but the heat in $\mathbf{T}^2$ surges.

The space reduced within the star centre is existing space that was the result of the left over of material, which was unable to establish gravity and reduce therefore established heat becoming space. The space that the star now moves into is that space which the star now contracts. It is existing space made redundant by motion applying as gravity. When Mainstream science investigate stars and find the true giants being very small as well as very massive every one acts surprised.

I wonder why…?

Has it got anything to do with the ignoring of Kepler as the cosmologist by favouring Newton as the mathematician? Are stars really about mass or are stars about reducing space to fit into space more particles and the more mass is the result of much denser particles holding more mass because of higher motion achieved? Is the increase of the density of the atoms reducing the space but at the same time increasing the mass claimed in space occupied by material? Can it be the result coming from stars with much higher motion potential producing much higher gravity? We know that the Black hole is spinning faster than the speed of light… and the gravity is contracting faster than the speed of light…

As gravity intensifies in true large stars the space vanishes from the equation where the star becomes forever smaller. That means gravity destroys space and does not pull matter. We all are very aware which star is the mighty gravity producer. So then has mass the least say when gravity is generated? It seems most likely to be true. Later findings proved Kepler as being astonishingly correct. The diminishing of **k** will reduce space $\mathbf{a}^3$ but it also will advance gravity $\mathbf{T}^{2.}$ It says so when reading mathematics correctly. $\mathbf{T^2 = a^3/\ k}$ shows that gravity grows where space demise. Every time the diameter diminishes the gravity produced by the star becomes immensely more. The space claimed decreases as we increase the motion with which the star holds relevance in the space it occupies and which it claims.

If mass was an issue the mass will play a major part in the time it take a body to release from the motion of the star. The difference in velocity that the star produce and the matching velocity outer space must produce to match the impact of the star motion on the outer space factor creates a velocity differentiation beyond human understanding. Again I must press the issue that since Galileo proved otherwise the concept of mass being the producing factor in gravity comes across as rather less thought through and more than a bit silly.

In Newtonian science terms where mass is recognised as that which is responsible for forming gravity it is thought that the bigger has to produce the most gravity. However, according to the latest information, the smaller stars are the true giants of gravity.

Having extensive gravity is not about being bulky and huge and big end enormous but spinning fast and being dense and condensed into a small space with a highly concentrated (cooled) environment.

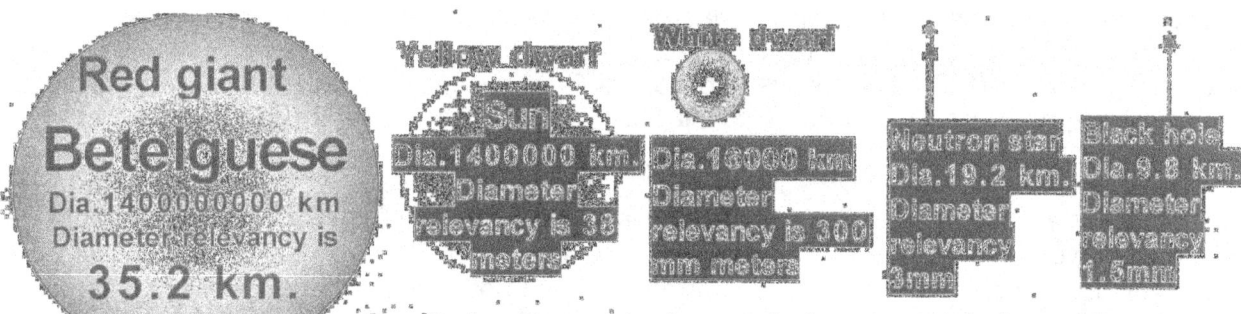

Gravity is strongest when and where space is least. Where space is least motion is producing most relevancy and change. Since motion produces gravity the Black hole contributes to producing the strongest gravity since the Black hole places all motion in space and no motion in the star. That forms the ultimate relevancy available and motion is all about relevancy. In the Black hole singularity controls matter and space apply all motion that is in fact the time factor to space occupied where the motion aspect is more commonly known as gravity. That is the location of strongest gravity. It is in the place that the heat is the most, which is in that centre area of any sphere. If any one does not believe me then test nature. Then it means that mass has the least say when gravity is generated? According to Kepler mass within space and gravity is the same thing. $a^3 = T^2 k$ Looking at evidence we find in the cosmos it seems most likely to be true but the radius k distinguishes the space a^3 required and the gravity T^2 produced. $k = a^3/T^2$. Later findings proved Kepler as being astonishingly correct. The diminishing of k will reduce space a^3 but it will also advance gravity T^2. $T^2 = a^3/k$ shows that gravity grows as space diminishes. Every time the diameter k diminishes and the space a^3 acquired by the star reduces the gravity T^2 produced by the star becomes immensely more. In Kepler's formula $k = a^3/T^2$ the smaller k becomes the smaller a^3 becomes and the bigger T^2 then gets. T^2 represents the gravity that positions the space a^3 at distance k from the centre capturing the structure through gravity applying T^2. We all are very aware which star is the mighty gravity producer. Has mass then the least say when gravity is generated? It seems most likely to be true. The idea is that gravity pulls material closer and more so in bigger stars. The gravity reducing space can only come from the accumulative effort of every individual atom according to proton mass (number) as a unifying effort of all the atoms in the star in accordance with mass applied. The idea is that mass is the same everywhere and is never changing. Why would there be such huge mass increases in the bigger or should I say smaller stars. What would entice the material inside such stars to grow more massive if mass comes about from the pulling of one particle closer to the next particle. If it was about pulling on each other the mass of the particles could not increase through that. Even by combining the mass of two individual atoms the increase is already in the equation.

By reducing space to the ultimate in the centre of the star where gravity peaks the heat sour to the ultimate and as heat rises and space reduces whereby more material fit into smaller space. This establishes conditions such as the conditions, which was present during the Big Bang and in the centre of the massive star the Big Bang is once more repeated. Within the space within the centre of the star being as little as possible the heat is as dense as possible but in that scenario more protons pack a smaller space and more protons in a smaller space accumulate more motion per space unit. This brings about more space diminished as more space transforms to more heat in less space. While the star is reducing space the Universe is expanding space to comply with cosmic equilibrium.

If we wish to find the future we should locate the past. If the cosmos is contracting, where to is it contracting? The direction of contracting must be in the opposing direction the direction of expanding. If we wish to locate the past from where the cosmos came and through that in what direction the cosmos came, it must take an effort to backtrack the direction it came. Should the argument come about that all came from nothing, then everything either still has to be at nothing, or our understanding of nothing leaves much to desire. Nothing means not existing, not being, never found and unable to produce any multiplication of any growth.

The above questions, but mostly the fact of what is more nothing and what is less nothing draw me to the realisation there can be no such a quantity in space as nothing because even space has to be something. Clearly as it is for any one to see one create space by nuclear explosions. The wind is shock waves, but what is the shock wave other than new space coming into prominence. In that way it is clear that releasing heat brings about the expanding of r as part of the sphere forming space. Hubble proved the Universe is expanding. Then by backtracking we have to set about reducing the sphere constituting the expanding Universe. If r in the circle is growing we have to reduce r to backtrack.

When the circle reduces, the value located to r will become implicated because r determines specific size. Not so in the case of Π, because Π in the true sense only indicate that the circle is a square without corners and therefore Π dictates form and not size. By reducing size only r comes into contest and will point to such reduction. By reducing the circle radius r by half continuously will lead to an infinite small circle but Π will remain because the circle as a form remains even being infinitely small.

In the past, and even in some quarters today, science is on the search for the 100 % efficiency machine. That theory runs on the surmising that a machine can drive as an output delivery without receiving input of energy. A few hundred years ago many Kings were fooled by such notion and some scientists truly spent a life in honest search of just such a device. Mostly the accomplishment came from cheats that very well new their machines were not up to the task, but in fooling a rich investor, brought about wealth to the inventor. As science progressed the no input giving all output machine became less and lesser a feature of the honest inventor. But the idea does not exclusively come from crooks finding a way to cheat the world. The practise of receiving without giving comes from science in the form of physics. It is physics taking the world on a wild goose chase in the way physics present the cosmic motion.

Physics propagates that the cosmos is all about running without input driving energy. The cosmos is all about wasting matter to a supply of motion. This idea prevails even after the world of science saw clearly in the past that there could be no such machine anywhere. Even the cosmos must be a machine driven by an input and an output. It is the input / output driving energy that must be located and the driving ability we have to locate. Science hold the mass drawing power to prominence, but what if it is not the drawing power of mass that holds prominence, but it is the reducing or contracting of space that is the driving motor behind the cosmos. All energy we humans at present use to accomplish matter motion, holds some form of heat redistribution. Even electricity is a form of pure heat. I say that in mind of what apply when the energy of electricity becomes over abundant and the machine overheats. By overheating it means that the motion the machine creates comes about from heat control and precisely planned heat distributing.

When I realised that it is not me that is drawn towards the earth, it is the space in which I find myself that reduces, and that produces the effort bringing me closer to the earth. The formula

$$F = G \frac{M_1 M_2}{r^2}$$ suggests driving, moving in a direction and contracting. It suggests the

reducing of space and not merely drawing or moving closer. When looking at any machine in practice, the machine draws power from space reducing whereby heat increases. Not releasing the heat to form space will lead to the destruction of the composition forming the machine. There is no form of matter, or element strong enough to resist matter deformation brought about by overheating. Having this in mind that matter does not resist heat, it is of importance to recognise that it is heat that is allowing space to give matter form. Looking at the manner in which energy is utilised it is space and heat forming matter allowing motion that allows work to achieve value.

At this moment science is all about a body falling where the two bodies are producing a force whereby the bodies draw one another closer. The bigger the mass, the bigger the drawing that comes about from the force unleashed by the mass of matter. The idea about this practise was phenomenal in 1705, it was impressive in 1805, but it is really ridiculous in 2005. Why would Boron form a solid having 5 protons weighing 10.811 g / mol and Argon a gas having 18 protons weighing 39.9 g / mol. But the "heavy" element with the biggest drawing power is a gas and the lightest element is a solid. That denounces the contracting force theory. The way we compile and use energy must be in a similar manner to the way the cosmos uses energy distribution. **We humans can create nothing, but nothing is all that we humans can create**. The rest of our achieving is by duplicating whatever nature provides.

To establish what drives the Universe except for blaming some medieval magical force coming from nowhere and is going nowhere we have to find what drives us. The energy we use in all forms is producing heat in space by either converting space to heat or heat to space. Explosions are about converting heat to space. Compressing is about reducing space to heat. That is all energy composing work and is the only method of producing energy notwithstanding the immeasurable many names we use to express the same function in different forms.

Arriving at the question about locating the space and time forming the centre the centre of the Universe one has to realise the centre of the Universe are in every singularity forming matter weather it is big or small, size carries no significance. It is the impartiality of singularity that is claiming the value and not the differentiation of matter. One must realise there are no big / small or hot /cold or near / far. It is all relevancies between matter claiming space and space is heat in a turnabout manner. Every aspect in the cosmos are locked-in Universes, sealed off from other Universes and inclusive or exclusive depending on singularity holding relevancies relating to one another. The relevancies rely on inter dependence and inter linking, but there are no differences according to human sizes or standards. Accepting that principle unlocks the "so called mysteries" of the Universe and brings about clear understanding. It is all about accepting, acknowledging and interpreting the role singularity maintains on matter.

One should not try to focus on an image of such a spot or dot because there is no image. The line dividing the cosmos and that run through every particle, no matter how large or small is beyond our vision. Such a small line, so small it is not even noticeable is large enough to part the cosmos into sectors. It splits the biggest there is into particles and we are not even able to notice the precise location of such a split. In truth there is no top or bottom that we living in 3D can see. We shall have to use a general conception brought about by intelligence. Your intellect tells you about such a spot, but that is all because that spot is on the other side of the Universe (quite latterly). From the centre of the dot there is a top and a bottom spot. From those points there is connection with four quarters. That produces six connecting points that are all aligning to the centre. Because it serves big and small, hot and cold equal and alike, and it is the smallest cutting the biggest into equality, size is of no issue. Size is what man makes of it. In the Universe there is no size in hot and cold, large and small. For the smallest there is, it is serving the largest there is equally.

Our instincts, our logic and our calculating process all indicate that the sphere holds a centre point from where six evenly positioned point's position matter to be. Using The formula $F = G \ (M_1.m_2)/r^2$ it indicates to a force pulling objects closer, where each force is coming from each centre point the body in question has. The contraction must commit the two bodies towards a point in each case being spot on in the middle, not withstanding what direction the force is applying, the body will draw to the centre.

If the Universe spins around a centre point holding singularity, and singularity confirms the centre of the Universe, then every particle holds the centre of the Universe making the number of universal centres immeasurable many, and every atom and sub atom particle presented outside the atom in smaller bits, are all not pieces of the Universe but they are a Universe surrounded by many Universes. If every atomic particle no matter how small is holding the centre of the Universe, then the gravity is coming about from that point because that is where the gravity applying in the Universe are applying contraction.

Then there is the other duplicating gravity principle where motion that brings about space is also putting distance between objects and this we humans understand to be growth of space or by a term much more commonly used as the Hubble constant but only in a balance where the space displaced and the space created match in time taken relating to space. This formula will apply as the second gravity $k/k^0 = a^3 / T^2 = k/= a^3 / T^2$ because the motion advances the distance between objects in relation but will only dominate when the first option broke the Roche law of $(\Pi \div 2)^2$. Then the relation will change to where the **k** factor holds a negative and total control. This means the space duplicated are not adequate to the space dismissed and a new space to time balance must be established. $k^0/k=T^2/a^3 = k^{-1} =T^2/a^3$ That is what Newton saw and that is what science recognises but that is not the primary gravity found in outer space at large. In outer space the Newtonian gravity only dominates where the bridging of singularity forcefully brought domination and control of a major singularity in capturing a minor singularity. Where that situation complies with cosmic rules the president set will be that the major contractor reduced the minor expander to heat and then capture the heat. Only heat applying motion brings particle separation as we may witness with the spinning top. Motion brings about space duplication, which revitalise space dismissing.

The Earth apply all motion Therefore $a^3 = T^2$ and to the top $k = k^0 = 1$

From the gravity and the opposing anti gravity applying came about two interacting translations of material where one formed gravity in securing a position and another by permitting expansion, releasing matter to establish space in relevancy bringing about antimatter in the form of matter performing antigravity. The one was filling space by giving away density and the other was applying density by giving up space. If it was true that matter was drawing closer matter, the securing of space would not stop the contacting and the regrouping of matter would start in the Universe even before expanding could start. The balance is in the minor particle finding commotion with enough heat that is allowing the particle to seek independence and a major particle removing space by seeking control of the heat that is in rebelling mode. When the minor particle gathers sufficient heat it will start an attempt to gain independence by placing as much space in relation to motion between the particles whereas the controller will demand the space in which the minor particle is moving. This relation allows stars to be freed from the galactica cocoons and move away from the galactica centre. It is all about capturing or releasing by establishing independence or surrendering independence. Since all the objects in motion in the Universe have not surrendered their independence they're in gravity that forms with **k** allowing the maintaining of independence and k^{-1} in a lesser but pivotal role securing stability. Only in the dynamics of the Earth that captured all

that is within the atmosphere of the Earth will k^{-1} be dominant as all within the Earth becomes part of the Earths motion of **k.** The control of space is about space duplicating and space dismissing. With motion comes space duplicating but the more the object grows the more will the object dismiss space and the less will the object duplicate space. By not duplicating space the object will increasingly destroy space because the value of space is fifty percent in the duplication of the space. By relenting the duplication of the space the motion of the star will tarnish and that will allow the star to dismiss much more space outside the star than merely the

$$T^2 > a^3 /k \; (k=k^0)$$

space, in which it is. By that the star gets an increasing ability to dismiss space that is not in the stars control but by putting such space to heat within the stars control. The star finally will achieve an ability to liquefy space into heat. The Sun has reached such a position. The essence of gravity is motion duplicating space and the dismissing of space. The Coanda effect is the best example and the spinning top illustrates the Coanda effect very well. The final step is to freeze space into material and become a dark star.

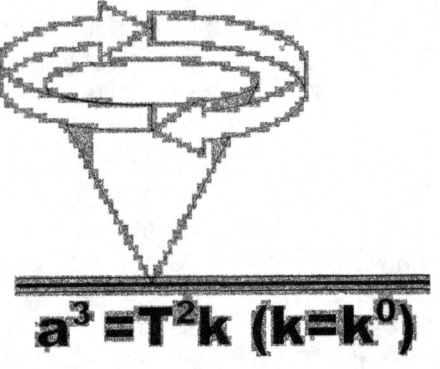

$$a^3 =T^2 k \; (k=k^0)$$

Before life intervenes and charge the top to bring about "cosmic Life" (a term I dare to use for the lack of a better term), the top is experiencing gravity just as we do by having gravity press the structure onto a surface where the top makes as much contact as possible with the surface that either is the Earth or stands in on behalf of the Earth. The top is resting without motion on the Earth in orbit of the Sun.

Then by artificially applying motion through the intervention of life the top acts in a manner that will only come about if the singularity carrying centre has generated massive heat centralised by gravity applying to dismiss space to the reducing of massive space with the aid of a massive number of protons or else by life generating some other artificial gravity we gave the name of electromagnetic induction. It is still applying the principle of heating singularity to produce centre heat charging motion. Although in this case the top is most likely spinning because of the action of human muscle creating motion. The important issue is that motion creates the spinning, which results in the top sustaining its own space by effectively enacting the Earths motion

When the top spins at a pace exceeding apparent barriers the top will begin to turn about the axis as if it is trying to extend the axis, which it is spinning about. The top jerks to the one side and jerks to the other side in an attempt to secure more space created by excess spin to use for the creating of additional space. Well that is precisely what it is doing in the attempt to turn around by extending the spinning axis which individual singularity provides. The space it created through spinning has

reached a point Earth granted the will find more better space-time mass does not the top is trying to

where that space is exceeding the space the object and by the object acquiring more space it space-time by which it then tries to secure a position for the singularity it is maintaining. The apply or draw the spinning top down and in fact undo the mass the Earth provides.

When the Earth reintroduces the effect of mass the top still try to create through motion sufficient space to use as a deterrent to the mass the Earth wishes to inflict on the top. The fight in the top is real and the determination of the top to prevent the declining of the space the top establishes almost seem as if it becomes a fight to the death where just the stronger will survive by killing off the weaker. Most times the fight lasts to the point where the top starts to spin vertically by running on the surface it used for spin. Only after the top showed the last act of defiance will the top again submit to the mass, which comes about from the tops inadequate spin tempo that will not any longer produce sustainable space to create he needed space and use the motion as antigravity

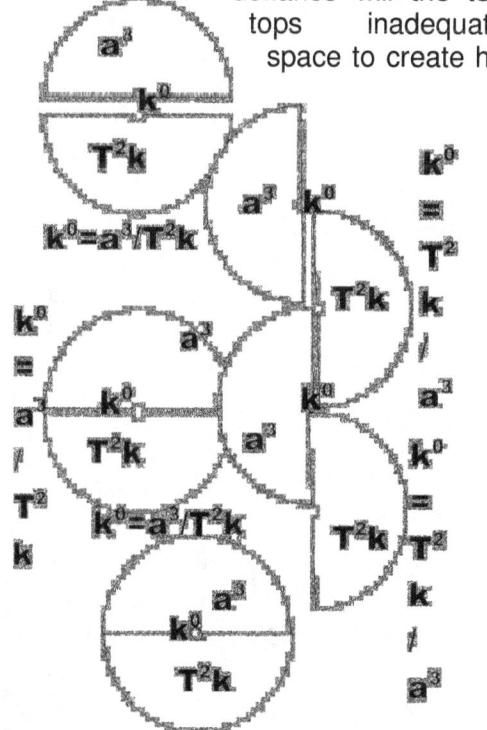

This must bring the evidence that mass does not bring about gravity but gravity creates mass when there is a lack of space brought about by motion duplicating space. In cases where motorised machines bring about motion the artificial energy is strong enough to apply motion that counteracts the mass factor completely

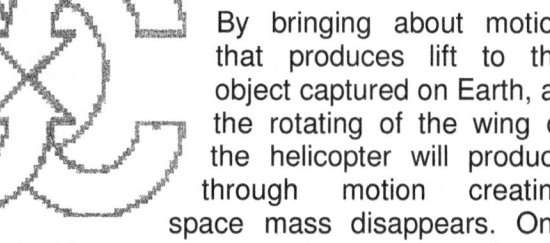

By bringing about motion that produces lift to the object captured on Earth, as the rotating of the wing of the helicopter will produce through motion creating space mass disappears. One must either decide to remain as stubborn as a mule and insist on it having mass just because of some ridiculous argument or admit that the action relieves the object of the mass the Earth produce.

The object will still sustain an effect from space duplication discrepancy but such discrepancy is then so much reduced the object is as light as air. In all of this I cannot find one shred of evidence where mass brought about any thing except being a result of the motion not producing independence to the top and therefore allowing the top to create an escape pass. Mass is the absence of adequate motion.

Gravity is the product of motion and with motion gravity comes naturedly in the process of motion. Due to the way the sphere is built it will always hold singularity in the centre of the structure. Singularity is a point within the centre of all spheres where no motion can be possible because the rotation pivots at that point and the pivoting changes direction precisely at that point without sides. The pivoting comes about where the line that forms the circle ends in the first dimension That we read into Kepler's formula about space-time originating from singularity $k^0 = a^3 / T^2 k$. At that point all space-time finds the relevance to return to the form of formlessness. The point that holds the value of Π^0 is located by dimension within any and all spheres but to top this Kepler showed that motion produce space and space is time through motion $a^3 = T^2 k$. Time forming the second basis for the entire Universe is the spin of heat in space. Without motion space-time collapse into singularity and within all spheres there is this point that cannot provide motion. Therefore through the form the sphere holds the sphere will diminish and destroy space in the centre by not providing motion within the very centre in the round structure.

All material has gravity built into the form it holds. Due to singularity forming Π^0 space collapse as a result of motion in the centre of all spinning structures. The faster the structure spin the more intense that space will be that collapse and therefore the more heat there will be to accumulate within that centre. The space ending there forces the motion to stop and the lack of motion produces space dissolving. But with space disappearing a flow of space forms that becomes necessary to replenish the space that went lost. This centre is also is within every atom where the proton forms such a centre and the atoms produce a unified effort of diminishing space where at the space less centre of the star the centre diminish space and the refilling of the space effect the centre where the space vanish as it accumulate in the centre of the sphere where no space ever can be. Towards that end all space will flow but will never be able to replenish the lost space because motion stropped at that point and the motion that brings about the space destroys the space just because the motion cannot be produced at such a point.

The fortunate part to us is the planning because the space that is dissolved is the space of the less successful spots that did not mange contracting that well but kept expanding with the friction to combat the heat coming about.

Therefore the spin involves space already established and the space contracting is space of other parts that overheated. Therefore the expanding becomes part of the growth as it becomes the contracted space. It is this effect that we gave the name the Coanda effect and is as much part of gravity as gravity is part of gravity. The spinning top again is as much proof that the Coanda effect is a product of gravity than flying is proof of the Coanda effect where it establishes antigravity through motion. As the top spins a centre is established with the motion of spin activating the centre not spinning. The centre comes about, as the centre remains motionless while the rest is spinning and the centre becomes an additional part being part of the motion of the object but within this a centre forms that is having no space and no motion. The motion travelling towards the centre must duplicate the following as well as the previous space crossing the centre to commit space-time to the   dimensional   form it holds.

Π^3

Singularity

$\overline{\Pi^2 + \Pi^2}$

Π^2

$\Pi^2\Pi$

Singularity extending the influence on flowing water

When there is a smaller object within such a space duplicating and that smaller space also serves as singularity and as an independent unit from the major singularity the space the smaller object holds is not duplicating to the same trend as the larger space. With the space being part of the larger space but also still being independent it will lack the capacity to match the larger space that is found more towards the outer rim of the sphere. Due to size restriction it will not match the required space duplication in using the matching motion.

As soon as the top starts spinning the top duplicate more space within that space of the containing object and by applying credible motion it suddenly find the required ability allowing the spinning top then to match the space reproducing. When spinning excessively it can overshadow the space duplication and in that it can seek the ability to find more space that is normally more to the outer edges of the sphere and therefore try to match the space reduplicating with space towards the outer edge. Gravity will always comply with two relevancies one is creating space flowing away and the other is space contracting.

Not only does this explain the energising of the top when spinning with excessive motion applied but also it proves that the gravity educed by contraction is part of the accepted form of space-time in the round securing of Π. The motion creates space duplicating within the parameters of the space containing the minor space but with the intensity of spinning the motion establish a concentration of heat at the centre that forms the motive for the independent spinning. The contraction is a result of singularity diminishing space by not providing motion at one specific point and therefore always creating space to flow by motion back to singularity to replenish the space that disappeared. When it comes down to the pulling of gravity that there is no pulling. The creating of motion establishes the duplicating of space but in the same instance the motion provide a point where motion does not apply and therefore it will destroy the space. The one accomplish the other and a relevancy is brought about because the motion creates space that creates a point of no motion destroying the space created. The stronger the motion is the more space will naturally displace and the more heat will then concentrate in the centre of the object applying motion. The minor object is part of the space and within the space of the major objects flow of space towards the major centre.

The Coanda effect is creating gravity. It is not replacing gravity it is not recreating gravity it is not enacting gravity it is forming gravity.

The rotating of the spinning top is as much part of the Coanda effect as flight is part of the Coanda effect and it all becomes the result of gravity provided by singularity coming as a result of motion creating space by providing a centre that will bring about contraction as space is rendered motionless within that singularity centre and therefore the centre is space less as much as it is motionless.

The contracting of space diminishing as a result of singularity established. The motion provides new space and the new space provides motion and in the midst of all of this a point forms through the motion where space will flow towards and disappear within that point of space less ness because of motion less ness.

The establishing of motion is creating space and providing a centre point of singularity. This proves that the atom links with singularity and that the value of singularity extending forms a relevance with the space-time it influences up to a point that it controls the space as much as it controls the motion of the space. That proves singularity extend way beyond the space of the material it holds and establish duplicating points of singularity within singularity by merely applying motion to space within space.

The Coanda effect is also the perfect example of the curvature of space-time brought about by the extending of singularity influencing due to the shape that imitates or duplicates the value of singularity and again conform Π. By establishing a new value of singularity as Π, singularity can once again take control and establish a new Π^2 as gravity in the new Π^3 forming space

By reducing the propeller in size and boxing in the airflow it is directing the flow to a centre where the spin will intensify as it accelerates. Gravity is created in such a way. The protons perform the spin creating the gravity and the proton number does not

$a^a = T^2 k \Rightarrow$ k^a $\mapsto a^a = T^2 k \Rightarrow$ k^a $\mapsto a^a = T^2 k$

k^a $\mapsto T^2 k = a^a \Rightarrow$ k^a $\mapsto T^2 k = a^a \Rightarrow$ k^a

Space duplicating is creating motion and from the motion space is created.

necessarily prove the strongest gravity.

The turbine engine is a star in the little. Massive space reducing brings on heat increase and with the accelerator of fuel added the heat increase generate a singularity enhancing equal to a star. It shows gravity come about by altering the space a^3, applying with heat added a new T^2 where that then produce the thrust to use the **k** coming about to elevate new movement, which is gravity

By reducing the propeller in size and boxing in the flow it is directing the flow the spin where it will intensify as it accelerates. Gravity is created in such a way. The protons perform the spin creating the gravity and the number does not necessarily prove the strongest gravity. The protons accelerate the moving of space whereby that spin will reduce the space volume. By accelerating the flow of space the volume per time unit decreases in relevancy.

In the reducing of the intake the gravity becomes artificially constructed because gravity is the reducing of space. By injecting a fuel the situation intensifies as the possibility of raising the heat levels become much more prudent.

The reduction of space will bring about heat. Injecting fuel in a place where such reducing of space already increases the heat the fuel will "spontaneously" ignite. The igniting created a heat level one can only find in the stars. Fuel will establish conditions that according to laws of cosmology only apply to stars where such heat levels will indicate the enormity of the gravity present that is generating the massive gravity accumulated in the concentration of heat in utmost reduced space a star is born in the gravity on Earth. In this example **k** increases as pushing space a^3, which creates to a new T^2 Bringing about a new will establish motion in absence of space. Gravity is the the thus

thrust the **k**

$$a^3 = T^2 k$$

$$a^3 = T^2 k$$

centre with more heat in the specific point relation to the heat intensifying at such a spot and the motion produces an energy that elevates such a spot to form a new Universe. It is the manner how stars start to be stars and apply gravity.

By reducing the space a^3 with the shortening of **k** a new standard comes about establishing a new T^2. Then moreover a new T^2 created by injecting fuel into a newly established **k** sets the groundwork for an a^3 coming about that can challenge the best star centres there is. With the ability of life to manipulate space-time man confuses nature in accepting there is a Tiger loose and this young star can challenge the space-time set by the gravity of the Earth. The Earth will allow such rebellion just to a point and a fight will ensure that will either release the spacecraft or down the aeroplane. But all this comes about by the artificial creation of a singularity miming singularity by presenting something that can respond and produce gravity. It concentrates space as gravity does, it concentrate space in huge quantity as gravity does and therefore nature takes the action as coming from well established singularity. But never forget that life established the manipulation of the cosmic phenomenon. The action itself is not cosmic. The action is artificially established by life's ability to manipulate the cosmos. The action becomes as result of cosmic life increasing but it is not normal and it is not cosmically natural. That stands in direct contrast to the very same application of the very same conditions but where the turbine is as artificial as money, the Roche factor is as natural as water.

Gravity is about a relation established when time begun between particles we know as material and particles we know as free or unoccupied space. Gravity reduces space to apply to fit the form of the sphere and later accept the form of the sphere. The fact of gravity is the producing of space by duplicating space just as Henri Coanda showed. Gravity produces space by mimicking space and producing motion that destroys space.

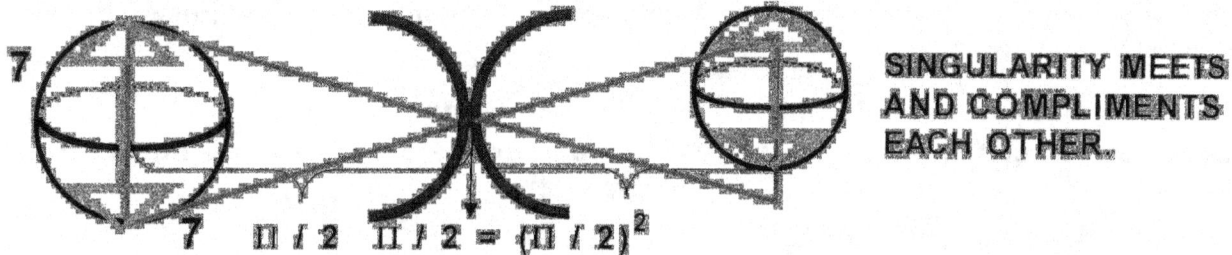

At this point the equality of the straight-line dimension to the triangle and the half circle holds prominence as a straight line, a half circle and a triangle is dimensionally equal. The common denominator will bolster all factors to an equivalent ratio.

The diameter of the cosmic structure holds the value of r and singularity holds the dimensional value of Π meaning that the radius or diameter (r) extends to become the diameter multiplying the value of singularity. But since r already consists of the square of space holding a definite positional relation with the value of singularity being Π the diameter comes into effect. Π extends each to an individual value to a point where the singularity on each side meets, bringing about a mutual Π^2 to the value dominance of the larger singularity control.

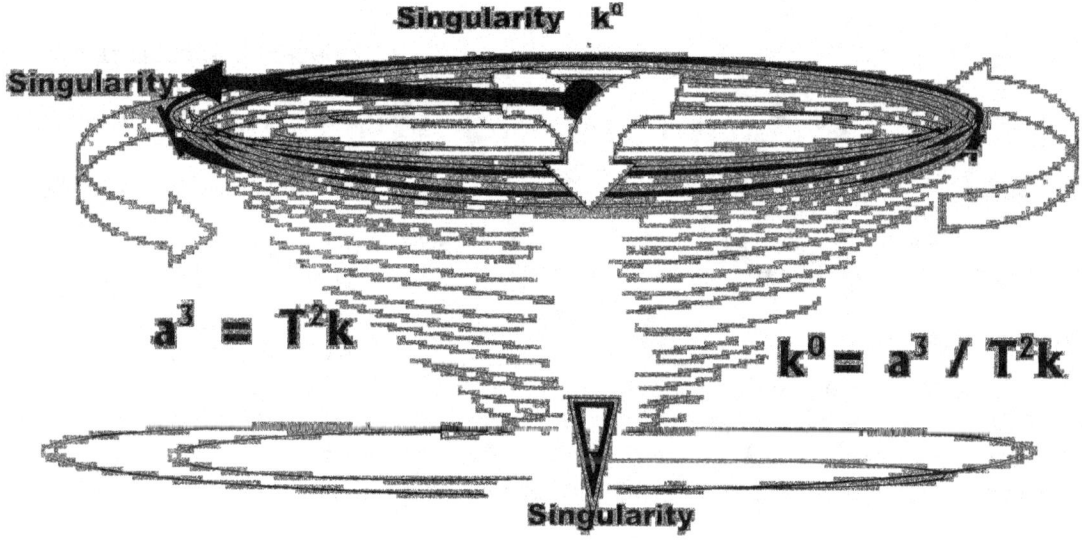

One must draw this statement of motion back to the point where singularity is getting sides. When there is singularity there can be no sides. It is 1 (one) from all angles there can be. That one fills a space. The space it fills does not really exist in the manner we humans see space to exist. It is a spot that is there without being there. It does not visually exist because it is not filling any substance and it cannot be recognised.

Once one accepts the fact of singularity that accepting of singularity then is contradicting all the things we know by not being any of the things we can recognise. There is no space. There can be no motion because there can be no space to have the motion within. It is a line that is so small it is not there and the only reason why we know it is there is because of the results it left as an imprint of its not being there. Not its absence because it is never absent. .It cannot be absent. It cannot go absent but it can never be there where it should be if I wish to locate it. If it was absent then it was zero or nothing but since it is there it is not there and that makes it present. The centre spot we cannot see and that we cannot detect has no sides to any side and has no place it fills because it fills all the places we cannot detect. The only way such a spot can fill space is by doubling the space it fills to become more than one place to fill. But the very instant that happens it halves the space it fill because it now cuts the space it has into two parts. That brings about that the point of not being is doubling the not being and by doubling the not being into being it also cut the not being that became present into half. We have to find this spot as we find religion. It is something that we can only know is there because we cannot disprove it is there but we can never prove it to be there. It is something seen through intellect and not through the eyes or light transfer. It is a point far beyond light. It is in our being and not in our vision. **Most important about this is to confirm what liquids are and what solids are in space.**

Flames are a **liquid heat**. **Smoke** is a **liquid heat**. **Vapour** forming compressed space is a **liquid heat**. **Rocket thrust** is **liquid heat**. **Air and atmospheric space** is **liquid heat**

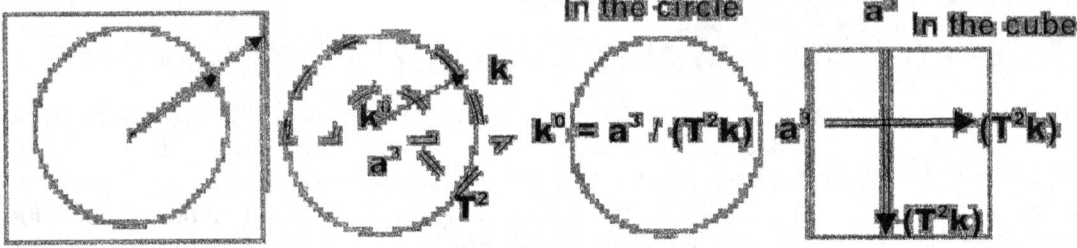

The big proof coming from the Coanda effect as explained above is that space flows through time as a product of the creation by motion op space-time. The opposing of a^3 by (T^2k) on the "other side" produces the **six sides** we now came to use. But that means gravity is where space disappears within spherical or circular structures and not in outer space.

$a^3 = T^2k$ therefore $1 = a^3 \backslash T^2k$ and is the same as $k^0 = a^3 / T^2k$ which means that
$k^0 = a^3 / a^3$ or $k^0 = T^2k / T^2k$

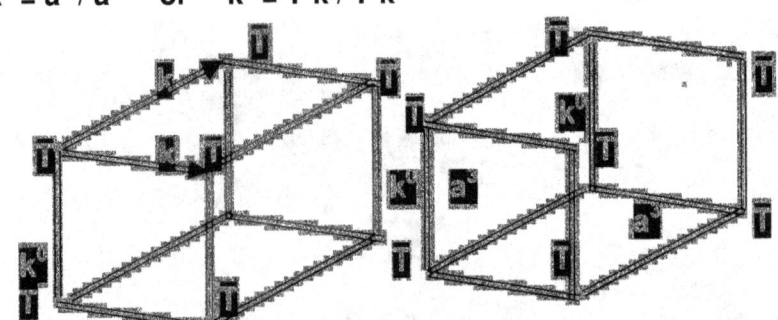

Space in the cube a^3 is the space in the square T^2 by motion through a straight-line **k**. That is mathematically what Kepler says. Space a^3 is formed by the square T^2 going double in creating a straight-line **k**. That means space a^3 is the doubling through motion T^2 by a straight line **k** and time T^2 is creating space a^3 by

implementing motion T^2 using straight- line **k**. By the motion T^2 of singularity k^0 space a^3 comes about in a straight-line **k**. Applying motion T^2 is the forming of gravity or anti gravity.

Gravity (cooling by contracting) and anti gravity (heat bringing expansion) is space created by the motion of the space in a line running between two points.

From this point that is there without being there that point now has to create a point that is there where we can see such a point start. The only way to do is to double on by adding and not by multiplying. It has to be in two places at one time but the time is filling the one place while the space is filling the next space. Therefore it cannot ever afterwards be in two places or on both sides of the Universe at one time because it is already filling both sides of the Universe and as such cannot again repeat such action. It is the time that fills the one side leaving the other side that is empty of space to form space that can move to the other side or it is the other side being in opposition without filling space that is forming space whereby moving from the space into the next space it is forming space. It is holding space that is not there and only by applying time or motion it can duplicate such a space and make that space valid. The one side is space not filled and on the other side it is space only filled by the space not filling one point duplicating such a point and then only filling the point through the motion that excludes the point from being on either the on or the other side of the Universe. Without the one side that is going to fill the other side and without the other side that is going to halve the one side not filled into two sectors of half the standing of one not one side is possible.

The only fact that the line is there is because the line is not there. By duplicating the line not being there the line establish the fact that it is there but by duplicating the line the line at the same instant becomes half of what it is at the time it is duplicating what it was not. This is coming straight from the horse's mouth. The Cosmos told Kepler that $a^3 = (T^2k)$. That is it. That is gravity. That is how the cosmos unveiled the cosmos. The cosmos said that from singularity k^0 came space a^3 with the motion establishing time T^2k. if it was this simple why make it so terribly complicated just to please the playing of the games by the mathematicians. The moment space realised **a³** such realising presented itself as motion. Translating Kepler's statement to verbally spoken English **a³ = (T²k)** can only be translated as "space moves" and "space is motion" and "space becomes motion". It is space will fill motion or motion will complete space. There is no other translation one can draw from this except by altering the concept through incorrect adding of facts the Universe never entered or stated in the formula the Universe unveiled to Kepler.

The line can only be if the line ⩔ fills the space it is not filling when the line is duplicating the filling of the space the line does not hold by applying motion to the filling of the space of the line. Then there will be either or but never both as we now wish to see the Universe. By the duplication it therefore insists on relevancy because without relevancy there can be no motion and no motion means no space. The strongest proof there is about this is the manner in which the Coanda principle applies the reproducing of space taking shape from a round object and involving motion to produce such duplication.

Conditions that prescribes the enactment of the Coanda effect is that the one surface has to duplicate singularity by establishing Π as a form. The round surface Π will bring about the shape of singularity Π that becomes enticed by the action of

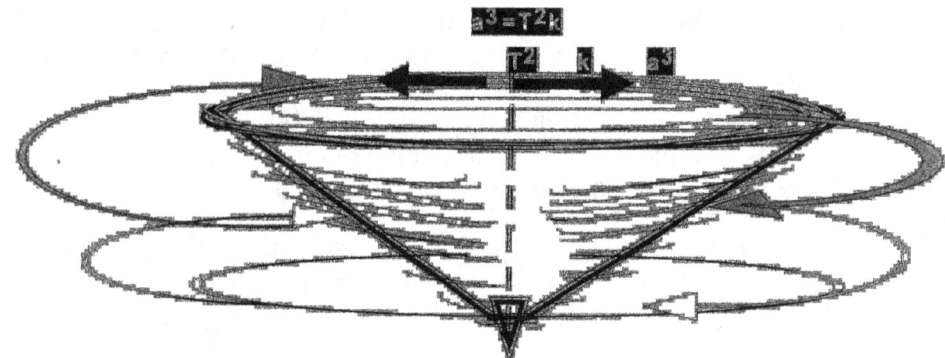

the motion of the liquid or of the solid or the motion of both around Π, which then establish and confirm singularity by form. The next factor is the presence of liquid. Air or atmosphere is liquid and water is liquid. Heat is liquid. The third factor being just as important is the motion establishing Π^2 by duplicating singularity as singularity becomes relevant through the applied motion that produces gravity from the singularity spot that provides the form.

7

Singularity establishing

Π

Π^2

The relation forming the duplication of singularity is a duplication but applies as a dimensional forming of Π and placing 7 in relation to 10 forming Π^2

The liquid applying motion forms the 10 disciplines. No motion leaves no Coanda as well as no gravity because gravity is motion that duplicate singularity

The Coanda principle which in fact should be seen as a law because it is that strong is the principle of gravity duplicating with the motion that provides such duplication a relation of the particles having the seven factor and such a factor of seven produces through motion another three dimension. This total that material fill while in motion is ten and when ten crosses the line of singularity too duplicate the seven in the other side of the Universe the crossing cuts singularity in two as much as it puts singularity in the square. But it involves the motion of concentrating space to be or hold fluids around solids that may or may not move. In this must be a solid, a round basis Π, fluids concentrated in space and motion applying to one or all of the factors.

$\Pi \Rightarrow \Pi^2$
$\Pi \Rightarrow \Pi^2$
$\Pi \Rightarrow \Pi^2$
$\Pi \Rightarrow \Pi^2$
$\Pi \Rightarrow \Pi^2$
$\Pi \Rightarrow \Pi^2$

Π Π Π Π
⇓ ⇓ ⇓ ⇓
$\Pi^2 \Pi^2 \Pi^2 \Pi^2$

That too forms the answer about the question concerning the Titius Bode gravity implicating of cosmology. The seven sides are linked by rotation nothing changes because there is a steady linking to the inside centre of the sphere. But it is to the outside that this rotation brings about dimensional complications. There are five T_1 points moving to five T_2 making contact with five moving points. The moving non fixed points is the point before reducing by five to the point after reducing by five that bring along the ten points in stead of the five to one point as it is the case with the Lagrangian system. Two points relating to seven points coming from by continuing in the same direction it is going to remaining seven points as going. That means in matter there are five times two points relating to seven in a moving constant and seven fixed rotating points

If it was not possible for space to use time in providing space for space to move too in duplicating the Coanda effect was not possible. But the Coanda effect is only applying gravity in a way we are not use to because we see our motion as being in a straight line following the curvature of the earth. In our way of thinking about gravity is that we go down wards by not going downwards. This then according to our misconception comes about a means of pulling is very incorrect because mass is the result of the lack of space we must duplicate to still remain in the cosmos.

By cosmos law we must move to fill the space we are moving towards and by our not applying we create the mass we are in. As we are part of the Earth space we are duplicating in time with the Earth because we have to comply too cosmic law. Instead we double the space by standing still of space as motion insist of replacing the space we have. By not being able to duplicate the space as we move onto the next space our motion creates we establish mass as a means to cheat gravity out of the space we should be duplicating. Mass is the effort we have to bring about since we cannot duplicate our space with motion. The duplication we are using to double our space stems from the protons up to the stars where singularity is providing such duplication.

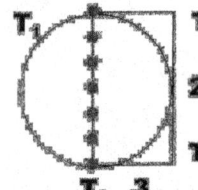

Time or spin or rotation, call it what you like but it is the moving from T_1 **to** T_2 using time that provides the dimension of depth to the dimension of distance between dimensions.

In the motionless Universe there will be on point in time and that point will represent k more than anything else. Every point being T_1 will only show the extending of **k** from singularity to that specific point. The fact that T_1 indicates no motion brings the Universe to a stand still and to a flat Universe.

It is that which give **k** the coming from the first dimension and by only extending from singularity it the forms two more positions becoming a^3. $k = a^3 / T^2$ but remove T_1 to T_2 from he equation and only **k** remains $a^3 / T^2 = k$. By the effort of spin or motion the Universe becomes the three-dimensional object it all seems to be.

Motion is what **gives space stability** and **security** in **six sides** where **three sides** are **opposing three sides**. This Kepler shows so very clearly in his statement $a^3 = T^2 k$. If there is anyone out there that cannot see this and miss the interpretation such a person cannot read mathematics

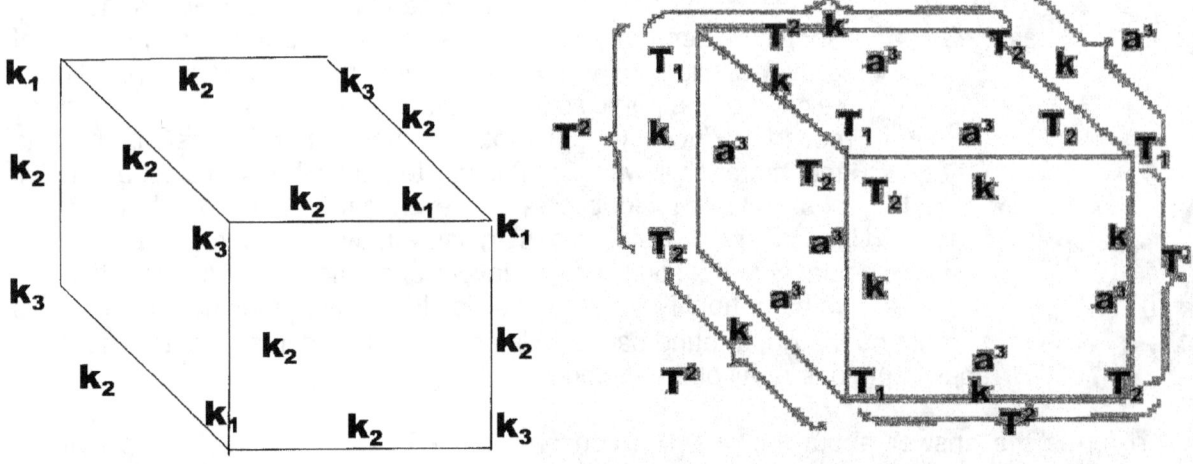

Einstein said gravity is the strongest where space disappears and even he, the Master misinterpreted his mathematics. The space mathematics show is not the Universe we see but space located at a point where space departs from singularity applying the value of k^0. It is a point within all atoms secured in a dimension smaller than the spin of the proton. It is where $k^0 = a^3 / k\, T^2$. It is at that very point that the one side is where space disappears and on the very other side motion gives space dimension stability holding size sides in form. It is beyond the micro cosmos at a point not mentioned yet or named yet.

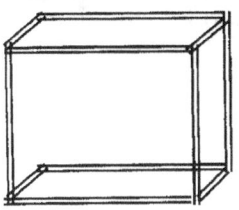

Our six-sided Universe is about space-time. It is space in motion and the motion is placing relevancies that are applying the time. Time even to human standards at present is the positioning of object after some other arrangement of positioning took place. One part of space is time and one part of time is space. If there were no space the next time cannot what was with what is going to be slightly differently arranged.

$$a^3 = a^{2+1}$$

$a^3 = k.\ T^2$ As I indicated previously it is all a dimensional differentiation.

$$T^2 = a^3/k^1$$

$$a^2 = a^{3-1}$$

$$T^2 = k^{3-1}$$

$$k^2 = k^{3-1}$$

The Universe we know shows space having six sides and even that Kepler proved as he proved space in spin $T^2 = T^{3-1}$ It is the same therefore it is dimensions repeating to form $a^6 = (a^3 \times a^{2+1})$

Space/ time in motion and movement are all the same things only separated by dimensions and dimensions are formed space where the dimensions become space being in motion and the space

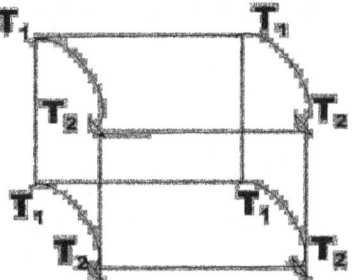

is motion by contraction or by expansion but because time is almost eternal at k^0 our perception of the Universe we are in is a stable and steady eternal structure. Gravity is motion and motion creates space to the third by the third in the third that interacts with one but establishes ten.

In this very manner is gravity the very same as speed where gravity is the moving

The reason why man can never fully create the complete 3D is because mans' inability to recreate motion that we find in time. The duration it takes any one point T_1 to move to any other point T_1 will have at the time of the arriving of T_2 produces the 3D Universe we are in. By such means does the Titius Bode changing five relating to seven to ten relating to seven bring about gravity or time or moving singularity. Use the name you like but it is all the same being Π^2

Gravity is the very same but it is the recalling of the space by creating motion in the space. Space is created from one position to another position and the duration it takes to complete the distance is time.

By recalling the space it is also reducing the space because it is counter acting the time expansion provides. That then is clarifying the reason why gravity will always on the limit be stronger than light. At a point it slows the time component down to such extend the space reduces faster in that time than what light can produce motion. But gravity and antigravity must be seen in the speed relations that come about from the process. Gravity is speed or velocity applying. It is space in volume in relation to time in motion. It is $k = a^3 / T^2$

When the object is stationary on the Earth a certain value to applies **k** in relation between the object and the Earth to the object. The space a^3 surrounding the craft influences the craft to maintain the time T^2 that the Earth applies. By departing to outer space the **k** influencing the space a^3 in the time T^2 changes as the time T^2. This is the result coming from the changing of **k** changes and since **k** changes all relevant factors change.

However the ratio of $k = a^3 / T^2$ does not end there. By shifting position **k** also produces an extension of increase on the formula.

The **k** that science sees that positions an object in relation to the centre of the Sun is not the only **k** applying. The line is **k** and there are many lines forming. There are the lines forming **k** by motion.

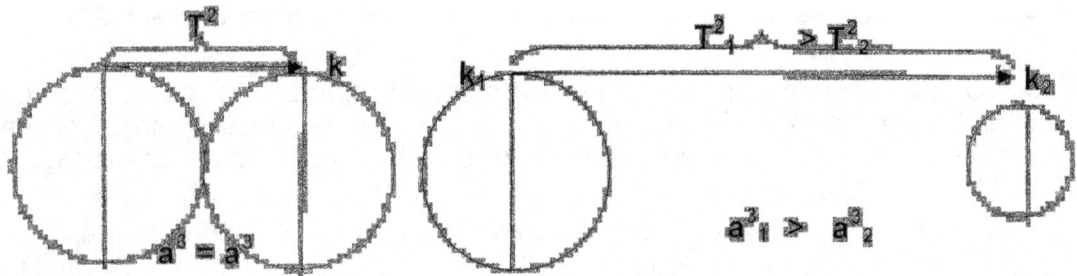

The only way **k** can extend is by revaluing T^2 to produce an individual gravity T^2 in the space a^3 that the occupier occupies. By increasing the heat that supplies the cosmic structure individuality, the structure then receives a new gravity T^2 because $k_1 < k_2$ in producing $k_1 = a^3 / T^2$ compared to $k_2 = a^3 / T^2$. It is not surprising that T^2 takes on such a different value because the surrounding space of a^3 that influences the ratio does change considerably. By changing the value of **k** brings a new cosmos about with a new gravity T^2 and a new value to space a^3.

By motion increasing or decreasing gravity changes values. This is very evident in the Coanda effect. It is also very apparent in the sound barrier as well as the re-entry of objects entering the atmosphere. An object entering brings about that the **k** of outer space changes to the **k** of the Earth. The Earth **k** secures a new time component T^2 but such a time component also bring about a revaluing of the space a^3 occupied by the structure. If the structure does not adhere to specific rules the Earth will reduce the material to heat as we can see with all objects coming through the atmospheric barrier. It is an interlinking formula $k = a^3 / T^2$ where any changes to one aspect brings about changes to all aspects.

This also come about and influence travelling in space. There is no unlimited speed that can be achieved once an object is in space. By increasing the time T^2 or decreasing the movement **k** such changes will bring changes on all factors and will demand revaluing of all aspects. This we find by using Kepler's formula $k = a^3 / T^2$. This is what happens to objects entering the atmosphere. There is no friction and such thoughts are senseless. There is not enough material in that area of space to bring about the friction required to unleash the heat we see develop from such an entry, The flames surrounding the structure is the area a^3 that reduces because of the changes brought on by the new time component and the new **k** factor changing the relevancies between the factors. Space travel is limited and the distance to time endured is gravity. For us to be able to leave the confinement of the solar system will require anti gravity equal to the gravity in the centre of the. It is the same requirement there is when we wish to leave the Earth centre.

By looking carefully at Kepler one can see that space comes about from matter moving by which space is created through the motion causing time to become an integral part of space. There is no chance of wiping out time or turning time in reverse, because by fiddling with time motion moves space back to space within singularity. Changing motion is taking gravity back to singularity and that reducing of k will lead eventually to conditions that prevailed before the Big Bang even.

Gravity starts where gravity started in the very first instant. The very place where **k** left singularity and stepped into the 3D **k** extended ever so slightly but never more influential. The point is where **k** formed a line outside singularity and went from k^0 to form **k.**

The length of the extending is so small it is beyond any manner of human conception or mathematical comprehension but so vital the Universe was the result from that action. With that action of extending k by the very utmost, utmost slightest of margins the Universe we know came into being what it is. The factor **k** went from **k^0** to **k.** By the extending of **k** space **a^3** came into place. But that space only came about by **T^2** producing motion. By **k** extending space came into place, but only through the motion of gravity and antigravity. The motion **T^2** produced brought **k** the independence which **a^3** secured. The factor **k = a^3 T^2**. Seen from the point singularity holds every aspect there is in the cosmos including other singularity is space-time. From where **k** starts expanding **k** produces space **a^3** through time or motion **T^2**. It is **k = a^3 T^2**. To one point holding singularity that point of singularity holds everything there is and there is only the space-time extending by **k** producing **k = a^3 T^2**. There cannot be space without motion. There cannot be motion without gravity and there cannot be space without gravity. Gravity is motion producing time forming space capturing gravity. It is a confined unit belonging to every singularity extending.

Accepting gravity to be motion is half the story and that is very much literally half the story.

The heat brings about expanding singularity from a one sided affair to filling a volumetric Universe. But all of it is a relevancy where ten positions will sacrifice individuality and compromise singularity in order to secure two positions in singularity.

By using the Titius Bode principle of seven on the one hand relating to ten and all ten coming about to sacrifice their position in order to save one vantage point material as well as space has to cross the border of singularity and fall in the other side of the Universe. The centre point holding singularity still forms the divide but all other points have a task to perform. Securing the centre singularity and maintaining the centre singularity brings about securing all singularity and maintaining all singularity. It is six relating to six where three is on the one side and three is on the other side. There is space formed and three is motion in space formed.

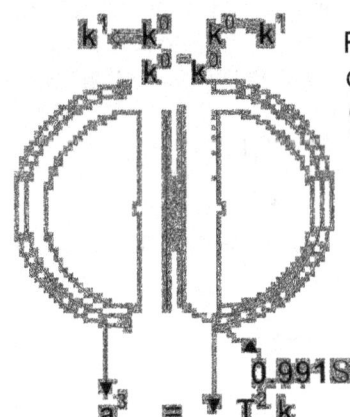

From the relevancy of the overheating which is bringing about space by creating motion there are positions taken in space occupied and controlled by singularity as well as positions influenced by singularity. There are those on the one side forming ten and then the eternal divide and there are ten on the other side also involving the one and the infinite.

On the one side there are **k^0 = a^3 / T^{2}k** and when that adds it is seven points. Then there is another three either being space or being motion creating space **a^3** or **T^2 k.**

On the one side of the Universe ten positions form a web with one and a range of dots smaller than one (0.9991) to form the protecting of singularity. On the other side there is the same number and that number form a unit 10 + 1 + .0991 + 10 = 21 that forms the space effected by the seven forming material and forms the motion of the Titius Bode principal. The factor of space surrounding material (always in the sphere to protect singularity by applying gravity) brings singularity back to the original value of Π by maintaining Π^2 in the motion and space required to maintain Π. This produces the dome covering the sphere forming the circle mimicking singularity. the atmosphere is as much part of the sphere as the sphere is part of gravity forming an extension of singularity being the eternal Π.

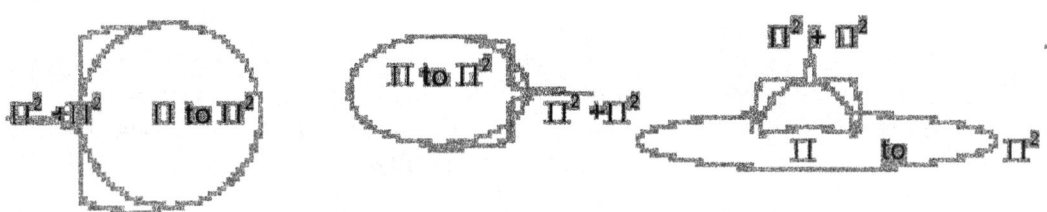

The atom once was as round as any sphere but with the enormous development of space-time and the massive favouring of space in relation to material the dome form grew flatter. This does not change the singularity forming Π **to** Π^2 in any way. We can see this very same tendency in almost all galactica of substance. The centre still holds a dome or a sphere while the edges grow flat with gravity growth reconstructing the original form. It is important to realise that because motion establish space as well as control space the ten factors representing the square of space will always grow much more than the seven factors does in the relevancy.

We can even follow how the proton relevancy reapplies in the shape that the galactica form or as the galactica takes on a "normal" neutron form. The motion producing time is essentially locked in singularity where time starts the motion with establishing space. But from our position outside singularity the motion creating space is the slowest time can be when not being eternal because it is periodic and not eternal which singularity is. Singularity cannot shift because singularity is the first dimension and as such immovable. Therefore anything not eternal is artificially created and beyond being permanent although we see the Universe as structurally solid. That only comes about as our perception. In stead singularity create space and time and by allowing space to reduce from singularity the reducing is placing the motion establishing space to reinforce as well as secure the immovability of singularity. It is providing singularity with permanency while singularity is removing the permanency from the creation sector. As the one proton moves into singularity and allow space to disappear the emotion brings about the other proton to move into the place of the proton that disappeared. At that point time is next too eternal because the very next spot time becomes eternal as it enters the immovable singularity, which is eternal. This is like watching your nails grow. After time one can see the nails showed growth but the growth is so time consuming we find the growth of the nails to be next to eternal. By the motion of the proton, the proton is doubling the next proton that disappeared as it connected to singularity. This motion at that point is 1836 times more time consuming than is the speed of light and at the point where the proton becomes singularity the motion is so much time consuming the time created to indicate the motion presents us with a permanently structured. It seems as if it might be 1836^3 times the speed of light but explaining this is far too time consuming. This period of space relative to motion being 1836^3 is the motion securing space in the Universe and is from our position eternally fixed. We see the proton as vibrating because the proton is securing the permanency of singularity. Any light shining permanent is actually flickering and the flickering that fragments when permanent and permanent. is so fast gives us the impression that it is Anything less permanent will seem broken into holding comparison to what we find to be at the same time anything flickering 1836^3 faster than light becomes solid.

$$T^2 = a^3 / k$$

$$2\,T^2 = 2a^3 / k$$

be on space as well as time because In the Sun on the very inside the displacing is three hundred thousand years to displace one kilometre of space. The space creates time and if the space is dense the time that creates the space must be slow moving. Einstein declared that time moves slower in larger gravity.

At the present moment science is looking at the speed of light as being three hundred thousand kilometres of space being displaced every second of time on Earth. This again is a speed. It places distance in relation to time but according to science they place the emphasis on space where it should space is duplicated by time and is the same thing..

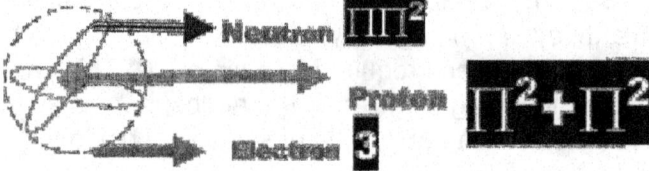

In this it takes light an effort of about ten million yeas to escape from the inner core of the Sun to the outer ridge. When space reduces time expands and when space expands time reduces. $k^0 = a^3 / T^2k$ therefore the bigger **k** will bring about a bigger T^2 since it reduces a^3 in the process. On the other hand will a bigger a^3 bring about a longer **k** and thereby shortening T^2.

From what we read into Kepler it says that time contacts space at a specific value and that value remains a unit that brings about a rule. The factor **k** moves one specific distance creating a very specific space $\mathbf{a}^3$ that takes a specific time $\mathbf{T}^2$ to duplicate. This ratio applies through out the Universe but the relevancies of the factors as the factors stands is in sequence with singularity extending. That is why the atomic conditions applying inside a Black hole is almost unrelated to atomic structures in the Sun or on Earth. The applying conditions demand different requirements on space-time brought about as the space-time is influenced by singularity at that point. The atom changes space-time to comply to standards the singularity allows prevailing within the borders singularity establishes. It is because of this reason exclusively that stars vary so much in the way they present and react to space-time.

It then is the atom in the most centre part where space and time meets singularity, that Einstein found a Universe collapsing to a single dimension, and every atom at a point post of the proton where gravity initiates in according with the proton dimensional colas of $(\Pi^2 + \Pi^2)(\Pi^2 \text{ X } \Pi \text{ X } 3) = $ **1836**

The Titius Bode law is an extending dynamic deriving from the law of the gravity dimensional factor where the space factor in a square of ten relates to a matter factor in the square by half (half since nothing can be in two places in the Universe simultaneously) of the matter factor of $\Pi^2 + \Pi^2$ or the square of space (10) relate to the matter factor of 7. From such a point every other point will be opposing any other point not pointing in the direction to which the first point is pointing, whereby it extends the direction it holds. No matter what the point is or where the point leads, such a point holding a specific direction will be unique in the direction it is rotating because at that or any other specific point wherever, it will be directing not in the direction it spins but in the direction flowing from the centre point outwards.

Gravity starts where gravity started in the very first instant. The very place where **k** left singularity and stepped into the 3D **k** extended ever so slightly but never more influential. The point is where k formed a line outside singularity and went from Π^0 to form Π.

The length of the extending is so small it is beyond any manner of human conception or mathematical comprehension but so vital the Universe was the result from that action. With that action of extending k by the very utmost slightest of margins the Universe we know came into being what it is. The factor **k** went from Π^0 to Π. By the extending of **k** space $\mathbf{a}^3$ came into place. But that space only came about by $\mathbf{T}^2$ producing motion. By **k** extending space came into place, but only through the motion of gravity and antigravity. The motion $\mathbf{T}^2$ produced brought **k** the independence which $\mathbf{a}^3$ secured. The factor **k** $= \mathbf{a}^3 \mathbf{T}^2$. Seen from the point singularity holds every aspect there is in the cosmos including other singularity is space-time. From where **k** starts expanding **k** produces space $\mathbf{a}^3$ through time or motion $\mathbf{T}^2$. It is **k** $= \mathbf{a}^3 \mathbf{T}^2$. To one point holding singularity that point of singularity holds everything there is and there is only the space-time extending by **k** producing **k** $= \mathbf{a}^3 \mathbf{T}^2$. There cannot be space without motion. There cannot be motion without gravity and there cannot be space without gravity. Gravity is motion producing time forming space capturing gravity. It is a confined unit belonging to every singularity extending.

That is the relation there is At the point of cosmic birth the cosmos is Π^3 as Π extends from Π^0 becoming 3D through spin. Where spin ends space ends and gravity ends. It is a unit undividable. Linking Π^3 to Π^0 is the motion Π^2 brought about by spin or rotary motion.

There are four time sectors in the Universe coming about from singularity and singularity fills every one. In the one sector there is a proton Π^2 connecting to singularity Π^3, which is connecting singularity in the form of Π to singularity in the form of Π^3. In that there is the Kepler formula Π^3 (**a**^{3}) = $\Pi^2\Pi^0$ (**k** **T**^{2}). Then there is the other proton filling the second opposing quarter of time implementing the same procedure with the same result coming about. In the third quarter is the first neutron connection following an identical path but linking to a proton forming the space-time. In the forth quarter the motion divided even further by splitting the motion as it conforms space-time from **3 to Π,** which in the end is the motion of Π^2. In all instances space **a**^{3} is confirming singularity **k**^{0} in time **T**^{2} except in the one that produced the Big Bang. The **k**^{0} confirms **T**$_1$ to **T**$_{2.}$ Singularity is and remains Π therefore Kepler takes on Π but the dimensional impact remains the same. This connecting happens both sides on the Universe Time has four parts to fill and singularity criss-cross fill it be connecting directly through the proton to proton ($\Pi^2 + \Pi^2$) or the proton to neutron (Π^2) or the link of the neutron to electron link ($\Pi3$) Adding this total confirms the mass difference between the proton and the electron at 1836 times. The proton has a mass of $1{,}673 \times 10^{-27}$ which is 1836,12 times greater than the electron's mass of $9{,}109 \times 10^{-31}$. ($\Pi^2 + \Pi^2$) X (Π^2) X ($\Pi3$) = **1836**.

It is the mass that space generates where the space has to reduce size by becoming more intense and concentrates 1836 time more when entering the point of singularity.

In single dimension seen from one aspect, with single dimension contacting the edges forming the sphere it will still keep the seven positions because the sphere remains a unified structure though apart because of singularity. In the core of the sphere connects the proton alliances $\Pi^2 + \Pi^2$ with the solidity of the neutron holding Π^2 as a second forming value. From the centre to the outside is a connecting of Π, in relation to the Π^2 that brings about the liquid or neutron form. In the centre of all the atoms space is relinquishing a position through the dissolving of heat by means of maintaining singularity. But through it all, another singularity forms in the very centre of the structure, claiming the position where space is the least available that bind the singularity of all the atoms sharing space as a unit. With that evidence I realised there are a connecting of singularity and that connection is electricity. In the cosmos all objects form a sphere. Some solids do not seem to be a sphere and space is no sphere, but the truth is hiding in the way of connecting. At the centre connects Π^2 forming the base of the solid. At any one specific given point forming the surface of the sphere is another marker holding the connecting relevancy of Π. When there is no sufficient heat to form space that will part Π from the other holding of Π, the two will combine in a solid joining connecting as 2Π that translates to Π^2.

In the way space and the sphere connects the sphere will have 7Π points holding a relation to 3Π points not within the sphere forming the 10Π that creation started with. This will mean there is a division forever, and such a division may run smaller everlasting. With fluids connecting it is simple to recognise the sphere as Π for the form will indicate Π as the form of the sphere. By gas forming the connection there are the three points of space being apart and not forming Π, but still holds a relevancy to Π^2 through the value of Π. The gravity applies as much to material as it applies to form installing form. In this way stars are spheres are just more cosmic atom and all rules apply as much to stars being just cosmic atoms as the rules apply to atoms being just individual stars. The Universe ends with singularity and starts with singularity where from this size is just space-time. The rules in the cosmos are the same applying to all in the same manner

To go back into our past we must take a circle that is representing the cosmos and reduce the circle to see what comes from there. Of all the models prepared by all the wise during the past this model was never made. That is what I did by following the suggestions arising from Kepler's formula and in that I uncovered a hornets nest. The hornet sting depends on the individual persons view of what a sting is. The sphere is any circle and all circles have two parts. One part is a line that indicates the distance between the two opposing sides being 180^0 apart. The second factor is the pi indicating the result of the square value of the other factor being the straight line which by the square produces the measure of the shape it indicates. But where as in the normal square the line matches angle touching lines that connects and in the square one will find the surface in the lines. With the circle the square falls inside the square but the lines cross at an angle and the edges hold the form value of the circle being **a**^{3} coming about as pi cuts corners literally. The circle holds a cross inside as the diameter doubles by the square and a square never crosses on the outside as in the case of the square where the lines touch at angles. With the square we refer to the lines indicating the borders of the square forming the sides that is the borders and indicate the final measure. The sides we face as in example length and breadth that becomes multiplied and form the square. In the case of the circle we use the lines crossing to the inside on the length they represent being a full line or half a line. In the case of the full line the name used for such a line is a diameter and in the case where only half the inside line of the circle apply we gave it the name of a radius. Being fully aware of the various names I prefer to use r to indicate any and all lines that forms a combination to indicate either surface or volume for either the square or the circle. If I refer to r I refer to any straight line.

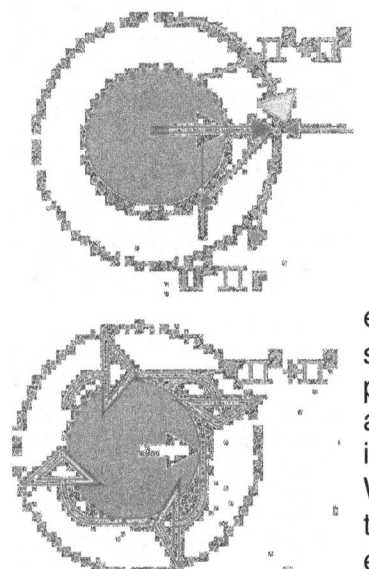

In the circle $\Pi^2\Pi$ which consists of the atmosphere the space surrounding the rotating object will also extend by Π as the concentration of the spinning motion draw or drag on past Π^2 extending the influence of Π^2 by the value of Π. Very clear evidence about this one can see in the Coanda effect. This extending of Π^2 to accommodate Π we refer to as the atmosphere, but physics apply to this extending in the normal fashion. The soil of the structure represents the solid proton being $\Pi^2+\Pi^2$. From the spinning motion Π does not stop at the end of the solid structure but the influence of Π extends and this then becomes the atmosphere. The influence of Π^2 stops at the end of the solid structure but the influence of Π extending plays a most dominant role in the cosmos, although not yet recognised and that factor is most crucial to a better understanding of the implications of laws governing the cosmos.

With the circle being $\Pi^2\Pi$ the Π^2 will reflect the circle in the square with Π forming the extending of Π^2. This is an extending of the six Π forming in alliance with the

centre Π. This produces that any extension of 6 forming material one further extending goes into space and relates to a seventh dimension. The extending of Π will not end immediately but will carry to the surrounding space the circle influence through rotation. The influence immediately above the circle will have the biggest influence and reduce gradually as the value of Π reduces in the leverage that the space has on Π and a gradual but definite change from Π to r will affect the extending of Π progressively more. The decline of Π will follow the same contour of the circle at 7^0. Every one of the dimensions indicates an individual significance as I shall show later and the increase into space runs by 7^0.

The man that to my humble opinion) took cosmology into a new dimension was HUBBLE, EDWIN POWELL (1899-1953), the American astronomer. He first studied nebulae, concluding in 1917 that the spiral-shaped ones (which we know as galaxies) were different in nature from diffuse nebulae, which he found to be gas clouds illuminated by stars. From 1923, using the 100-inch (2.5-m) telescope at Mount Wilson Observatory, he resolved the outer regions of the spiral nebulae M31 and M33 into star, identifying over 30 Cepheid variables in them. This proved that such 'nebulae' were truly independent star systems like our own – other galaxies. In 1925, he devised the so-called tuning-fork diagram of galaxies, dividing them into ellipticals, spirals, and barred spirals, which he believed to indicate an evolutionary sequence. By 1929, Hubble had good distance measurements for over twenty galaxies, including members of the Virgo Cluster.

By comparing distances with their velocities, as revealed by the redshifts in their spectra, he concluded that galaxies were receding with speeds that increased with their distance, a relationship known as the Hubble law. This was powerful evidence that the Universe is expanding. The dynamics of his work was so far reaching everybody (including Einstein had to revise their theories to accommodate his findings. His findings are the most disputed, undisputed observations in all of history.

The HUBBLE CLASSIFICATION is a widely used system for classifying galaxies according to their visual appearance, illustrated on the tuning-fork diagram. The sequence is based on three criteria: the relative sizes of the central bulge of stars and the flattened disk; the existence and character of spiral arms; and the resolution of the spiral arms and / or disk into stars and H II regions. The system was originated by E.P. Hubble. Is that which E.P. Hubble discovered the Universe? Is that which we see the Universe, because if we have our definitions are screwed up wrong, it is little wonder that "the Universe" doesn't make sense in the manner that we look at what we think of as the Universe.

Taking the outlook from the point the sphere is holding from that centre out into space there are ten points connecting to the centre. In that are the dimensions of singularity connecting to space where five connects to space in the second dimension of singularity, and five connect in the third dimension of singularity.

In the circle using $r^2\Pi$ the r has to have distinctive qualities placing it as a factor apart from Π. Where the growth shows no separate distinction but a continuous flow from the precise centre to the precise edge the flow would become in relation with Π depicting the circle and Π replacing r as reference to any point on the circle.

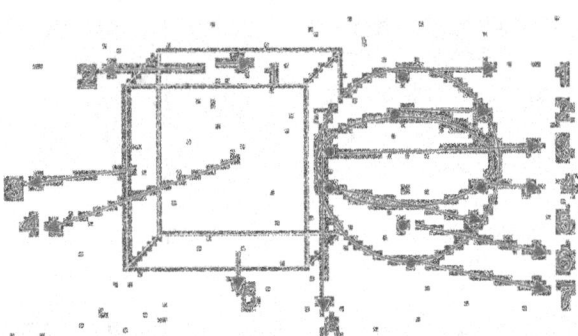

By using r as a distinction in the circle division is possible but by using Π there is no distinction possible making it a solid flow. Any object being in outer space floats and such floating is seemingly random with no specific detectable interfering favouring a movement in a particular direction. Such a devise is depending on influences not in our scope of detection. But then the object comes closer to the Earth and reaches on specific point where the six dimensions that influences the object in the cube suddenly changes. At one point one of the six dimensions fall away as it disappear and the object quite literally falls to the Earth. it changes a stance from floating to falling.

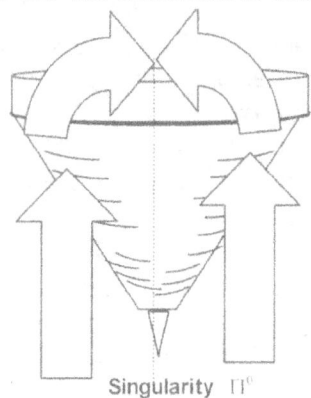

Singularity Π°

The support of one side disappeared and the centre point of the sphere took over the control. At that point the object is under the influence of one centre point in the sphere where the sphere in this case is the Earth and we also know that in such a centre point one will always find the strongest or the controlling gravity.

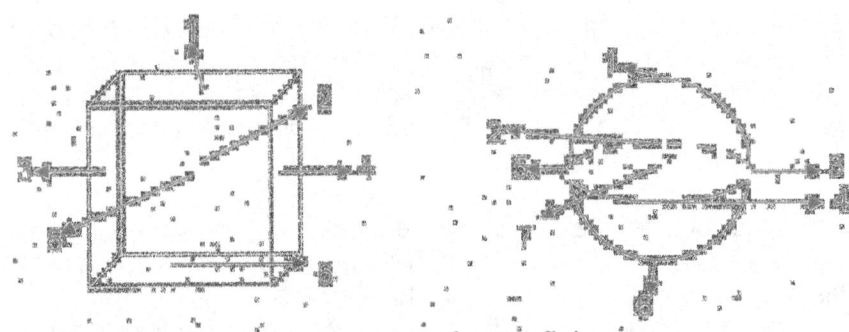

By rotational motion, the top creates a line confirming singularity running down the line and by generating the line the line charges gravity. The gravity is what drives the top as the top and as long as the top spins. There is an influence generated by the spin of the top that keeps the top upright while the top is spinning.

The line is generated but the line is far from magic. The line is where the centre of the Universe is which the Universe is then that what the top filled by particles from the line to the edge of the sphere. The particles in motion generate motion by electing a centre from the centre of every particle in the spinning top. Such an elected centre becomes the centre of the Universe as far as the top relates to a Universe because all the atoms in motion elect the centre of the Universe.

On the following explanation hinges all understanding of the entirety that we describe as physics. There are two valid forms we find in the Universe. The one supports a centre seventh point that is surrounded by six points and then there is the other one holding six points where not one of the six supports a conjoining seventh point.

Singularity split the Universe into two parts that under no circumstances can ever meet. The one side of the Universe perform a balancing act to the other side of the Universe that duplicates but never doubles. At the start, the dot began overheating while the dot remained cool by activating gravity. That is time and that is how time applies every instant time forms a Universe.

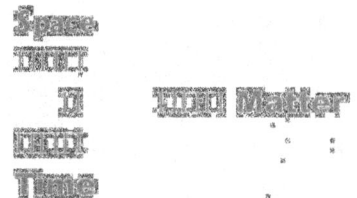

Where then the first dimension comes into place, space can't be but can form matter because matter holds space. From that point in singularity, time forms a position in which space is, but space also comes about when time places space in relation to more space and from time giving pace a relation, time comes into space by the fact that time splits the Universe in segments of matter relating to space filled with matter and time s the influencing that occurs when the spinning matter re-aligns. It is very important to realise that the Universe constitutes of one substance and that is singularity. Singularity is not part of the Universe but the relevance it holds to every other point holding singularity, which also is not part of the Universe, gives the lot relevance and that in relevance is what we call the Universe. However, there is no Universe as such because every point is a spot without space forming the tiniest sectors, which we call singularity, and that position holding singularity has no space and with that having no space, in itself that point therefore can't be part of the Universe. The Universe is space that is holding space in position as a result of time moving space.

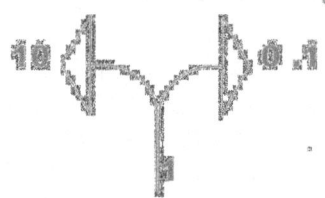

The Universe form mathematically and that gives mathematics a prominence. To understand the Universe, we have to reflect on the way numbers came about. It is most important to realise that it is the Universe that forms numbers and not numbers forming the Universe because the Universe placed numbers in the Universe and it is not numbers that is responsible for the Universe forming. By taking the queue from the numbers, lines that run in opposing directions puts in place Π^0 by having singularity increasing and with singularity adding time forming space is going larger as well as smaller. But the centre takes a value of one. Using 1^0 is a mathematical choice because it puts the correct value to singularity at $k^0=1$ or $\Pi^0 = 1$, but that also is responsible for that factor which splits the Universe into two parts, being smaller and being larger. That split is time forming space to the past and space to the future. Time is the adding of what is not part of space but in reflection the next time will be the history of time or therefore called space. From where there is no space, these points refers to other points also holding no space when seen as individual items, but it is in the referring of such points that space in deed comes into prominence and this concept puts in place an entire Universe. By this time comes about, resulting that space becomes the history of time.

There is no official Universe. The Universe does not exist. It comprises of innumerable spots that has no space and therefore is not there. The fact that there is a Universe is because the one spot that is not holds relevance to the next spot that is not. This is a mathematical reality. The number 1^0 has no meaning because 1^0 can have no value. However, the value of 2 can only be in place where 1^0 are holding a relation to another point holding singularity or 1^0. That is why mathematics is place.

It is a fact that I am where I am and where I am has no legality to exist for it can't form a part of the Universe in which I am since it holds no space, therefore I can't be where I think I am. That is the reason why the body life forms when such a body holding life is "being alive" disintegrates after life abandons the body, is that the body life holds is not a reality except for life putting it there while life has a reality to form the body that is in reality not there. The only reason why the spot I hold is there, is that it can only be attributed to the thought that it is not in place except for the reason that there is other points that also doesn't exist, except that it gives me a value, which I can have, only because of that I have a position to hold since there is another reality that I refer to which puts me in reality in another place that is confirmed by this point that is not part of the Universe.

That is physics and that is gravity and gravity is time putting space in as the history of time. The only reason I am in the Universe while in reality the Universe can't be a reality, is if I am in another reality that is not part of this Universe that has no reality. That is a mathematical fact and that is physics. There has to be a place that has no space where I am to give me legality to be where I think I now am because where I now am has no legality except that there are other points that also can't be, but for the space it forms in relation to me. This isn't religion, it is mathematical physics and that puts the atheist in the realms of stupidity and idiocy because they can only find foolishness in what in reality can't be because that which they put prominence too, can't exist in reality. I can only be in a place that at this point, holds no reality to me, but I still am in reference of that point holding reality just because of the fact that where I now think I am, can't hold reality. In other words, there has to be an "afterlife" in which I am, from where I never left to begin with and therefore to where I have to return since I am still in reference to that reality and that says I never left reality in the first place because there is no place forming reality where I now think of as reality which is where that life I call me then can be.

Only by being in reality somewhere else while filling a vacant spot that is not even present in the reality I think is reality which is in the here and now, can I establish a position where there is no reality to be placed and which is where life now thinks life is. It is time that the Newtonian atheist become wise and not to stay mentally underdeveloped. It is not the human body that keeps life as a guest because as soon as life departs from the body, the body fragments never to reform as it was before. It is a life that keeps the body as a host because life forms and develops the body to the satisfaction the requirements life imposes on the body. As soon as life has no need for the body, it

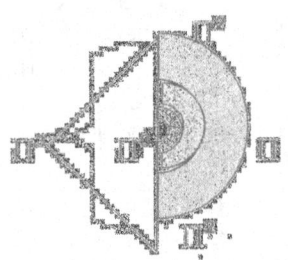

discards the body after which it is the body that degenerates in the absence of life that no longer maintain the structure the body has. The body is ridiculed and degenerated to form only cosmic atoms. But the simplicity of the Newtonian atheists understanding and underdevelopment of reason would not allow the Newtonian comprehension about things only the human mind may contemplate.

It is apparent that one cannot substitute the correct formula used to measure the area of a circle by using $a^3 = \Pi\ r^2$ because if k is the diameter then the formula must be $k^2 \Pi$. But k cannot be Π because in Kepler's formula k takes the value of the radius. In that case what will the value be of T^2? That places the formula outside the normal use of mathematics practised in the normal sense of $a^3 = \Pi\ r^2$.

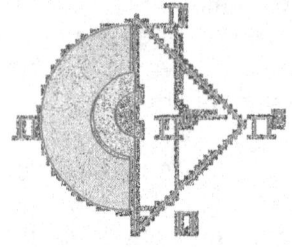

By using the Kepler Formula $a^3 = T^2 k$ it is good to change the values to Π and see what pans out. If $k = 1$, k at the same time would be k^0. By replacing $a^3 = \Pi^3$ then on the other side of the Universe $k = \Pi$ and $T^2 = \Pi^2$. But to secure this k in the centre must be 1 leaving $a^3 = 1(1\text{X}\ 1\text{X}1)=1$ and $T^2 = 1(1\text{X}1=1)$. That complies with Einstein's definition of space-time being: Space-time is a four dimensional position of the Universe where the position of an object is specified by three coordinates in space and one position in time.

If k is the middle being $k^0 =1$ then $a^3 = k^0 =1= T^2$ When time is in a shift freezing allocated positions forming space, then $a^3 / T^2 = k^0 =1$. In order not to overstep my limits by changing valid formulas I changed Kepler's formula to

On the one side of the Universe in relevance to all the dots that came before, three dots landed forming one side while three dots formed the second side and three dots formed the third side, all relating to a centre dot which in turn related to the original centre dot from which all the dots came and developed.

The Universe comes into position by deploying dots supporting other dots and some dots remains dots while other dots goes on to become dots of hybrids as it was supporting dots through claiming dots of lesser density and pass that on to dots with larger density.

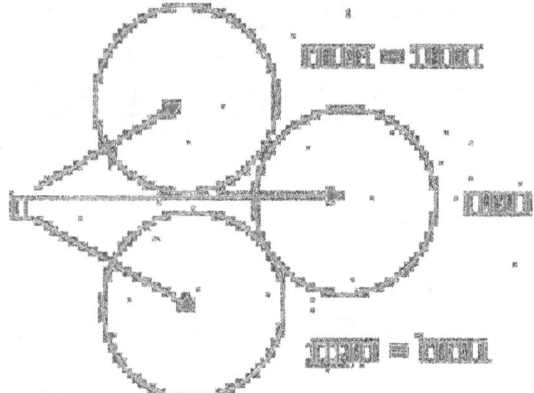

Matter form where matter has a position to time coming from Π^0 and forms space as
$$\Pi\Pi\Pi = \Pi\Pi\Pi$$

Space comes in place to form the history of time as space occupies a position since it has to be in some space $\Pi\Pi\Pi = \Pi\Pi\Pi$

...therefore $\Pi\Pi\Pi$ meets with $\Pi\Pi\Pi$ to form the atom holding the proton in $\Pi^2 + \Pi^2$ as well as the Neutron as $\Pi^2\Pi$ because the matter is within the space it holds and another Π^2 employs Π as a representative of singularity. This then placed the seven positions of singularity as the ending of matter and the three squares ($\Pi^2 +\Pi^2$ and Π^2) of singularity as the limit of material. The last $\Pi\Pi\Pi$ became $\Pi^0 \Pi^0 \Pi^0$ and that became the space producing heat without occupying matter in order to allow heat to be restrained inside the dome singularity provide.

When I refer to an atom it includes all unified cosmic structures holding an excluding formation such as an atom or a star or a galactica does. This is where **k** defines many but as a whole also one **a³ / T²** determining space within time holding space within time to the value of unifying the lot in one cosmos container. There is little difference in the cosmos to say a lead atom or a giant star or a large galactica It is all space confined to time extending singularity because in the cosmos there is no big or small.

From that the effect of gravity as a restraining on the exploding of space came into effect.

It is all about relevancies applying the relations gained

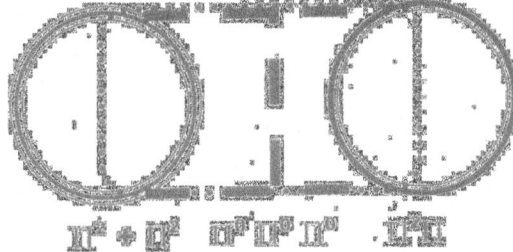

and lost through relations. If one place $\Pi^2 + \Pi^2$ on one side then $\Pi^2\Pi$ is related form where $\Pi^2 + \Pi^2$ is in the other side of the Universe being on the other side of the

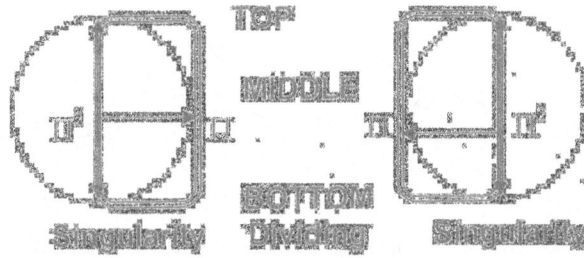

relevancy. Then $\Pi^0 \Pi^0 \Pi^0$ will again relate to the other two factors forming the "outside" of the other two being the "inside".

The Universe divides into two separate issues because of singularity. Nothing can be in two places at the same time where as all the rest in the Universe has to confine to the law applied by singularity. Objects can only be in one side of the Universe holding three parts or in the other side of the Universe holding three parts. From the totality three will be a double with six sides too shows, but that forms 3D. From singularity it is flat with three sides forming on either side of singularity.

But when the Universe was in the single dimension, all values were Π, therefore every value related to $\Pi\Pi\Pi$ forming three of the same that was very different because it was where Universes met and formed relations. Every dot formed an individual Universe and every dot

The names I use in TOP, MIDDLE and BOTTOM must not be viewed as sides but merely as terminology using names to implicate divisions. Direction depends on positions and positions form a value only when the observer forms part of the cosmos and not part of the observing.

At first when material presented one side of the Universe matter had three sides to show. Matter had to have space to keep matter somewhere in some part of some Universe and that made up three positions. Between the two Universes **k** and **T²** placed a value but since only singularity applied any values the value therefore was $\Pi^2\Pi$ where $\mathbf{T}^2 = \Pi^2$ indicated time coming from 7/10 in relation to 10/7 and $\Pi^2/2$ (proof of that is somewhere in the book) and $\mathbf{k} = \Pi$ valued by singularity. When space-time developed 3D the dimensions falling outside the sphere becoming space-heat formed as $\Pi^0 = 1$. The electron holds a relevancy of 3 relating to the Neutron being $\Pi^2\Pi$ and the three keeps the electrons in different Universes relating to separate or individual singularity.

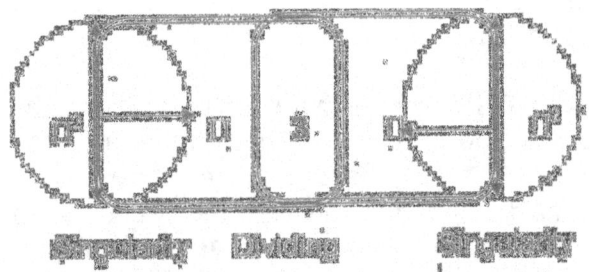

The relevancy between the two particles secures individual positioning between the opposing particles, which positions the material that sufficient space secures cooling and preventing overheating.

As the relevancy between the particles promote overheating or applying antigravity (overheating) to the responding cooling or applying of gravity, the one repels material into space-time while the other is collecting material into space-time. The one loses material and ensures a model of preventing overheating while the other gains material and sustain a model of overheating is prevented The one principle we named the Hubble constant where overheating produces space and the other one we called gravity where gravity is demolishing space, but both phenomenon is at present dominating the flow of time in the Universe and will do so until equilibrium again comes about.

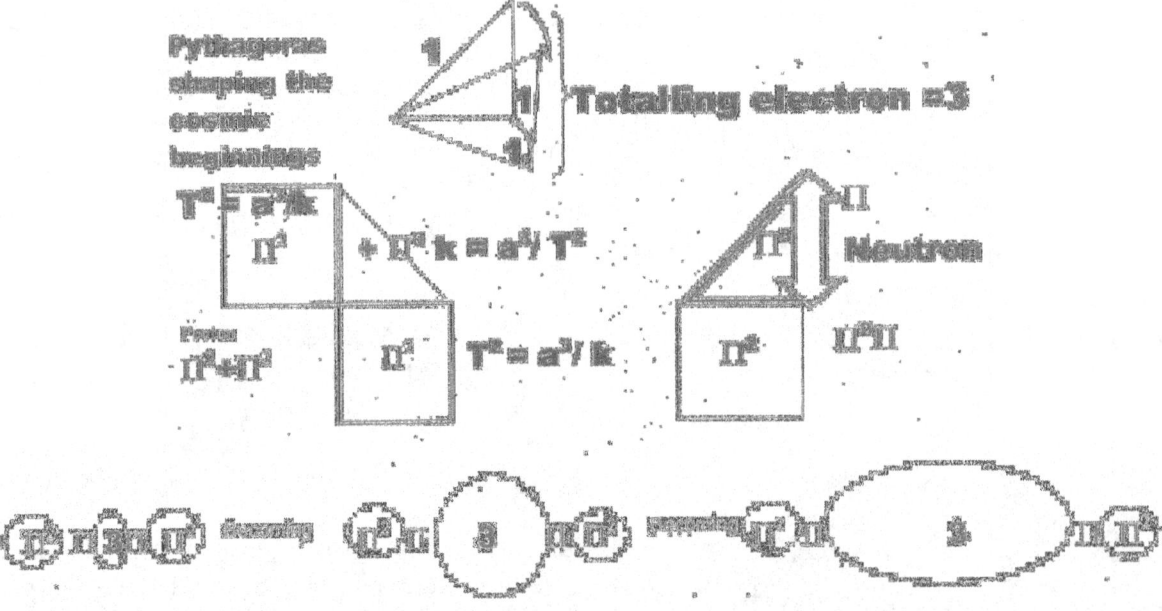

During the Big Bang two things happened. Particles all overheated. By overheating material enable the securing or claiming of more space.

Only by overheating or increasing heat can material claim more space. In order to supply **k** with any reason to grow into where **k** then become the fibre of material there had to be a way fitting natural processes to do such extending. The precondition of material to grow and claim space is to accumulate heat, and that must have happened because we can trace the excessive heat there was even today. If k grew, the temperature had to rise. But all temperature was the same in singularity. Singularity is homogeny in all areas including heat. Singularity can generate heat on one condition and that is that is by producing more spin. The Coanda effect is vivid proof of my statement. But to heat it must spin at a higher rate and by spinning singularity then produce more heat. That is in motion and all proof still exist that singularity had and has no movement. Spinning the top will bring the top to become more existed and then attempt to elevate the motion of the top to separate from the Earth gravity. In singularity there was no motion yet and there was no heat yet. Still to get k to extend the temperature had to rise. Let us have a good look at gravity.

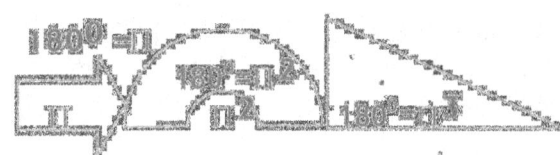
The triangle, the half circle and the straight –line has two things in common, they share 180^0 as a mutual value and they are part of singularity.

Using the concept that gravity applies Π as the circle factor Π as well as Π^2 replacing r^2 the replacing by Π brings two values as Π and Π^2. That I found is the case with gravity and will be apparent when explaining the sound barrier as well as the Four Cosmic Pillars. In order to create a distinction I

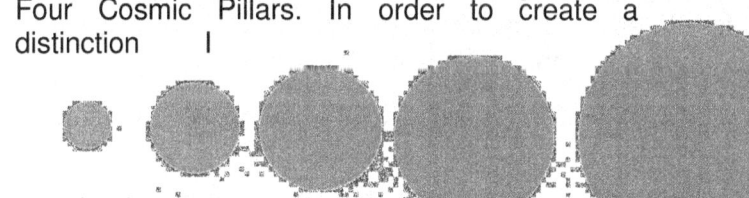

remained using r as the indicator of the cube or non-circle that has vacant space and by vacant space I refer to non-solid structures. In the solid structure I use Π as a value for reasons that will become apparent in due time.

Gravity does not apply mathematical equations to the letter as we would like, but rather use Kepler's thinking by enlisting an average gravity applying through out because it never favours and is equal every where. In gravity one finds the extending of Π implementing Π^2 on average as a unit and not the radius r as a specific.

Looking at the affect of gravity it shows the precise quality of no distinctive point, as gravity never seems to end at a point but flows all over affecting all that holds a position in its sphere of influence. The gravity coming from China meets the gravity coming from America at no particular spot but intermingles without distinction. This takes mathematics back to another fact beyond normal explaining. But the Lagrangian system proves much more than dimensional interlinking, it proves Pythagoras in principle.

No object can be in two spherical quarters in the same time, but has to alternate in aliens to the space in accordance to time rotation.

To alternate in aliens to the space the relation of time in space has to alternate relevancy to the cosmos.

Singularity holds five dimensions inside and five outside singularity as matter and space forming space-time. The ten dimensions I named the atomic relevancy is also showing the double value of singularity as singularity extends into as well as beyond space. The atomic relevancy is $(\Pi^2+\Pi^2)(\Pi^2 \text{ X } \Pi \text{ X } 3) = 1836$ that is the mass relation between the electron (3) and the proton. Proton = $(\Pi^2+\Pi^2)$ Neutron = $\Pi^2\,\Pi$. The atomic relevancy holds the dynamics of singularity control.

The network of individual singularity not only provide spinning through governing singularity in the sphere but also provide spinning in the geodesic through out the cosmos linking all matter to matter in a network no one will ever come to understand in full. In the sphere the four squares forming the triangles linking the lines to the half circles holds space in time maintaining singularity of different assortments. In view of the matter-to-matter Roche factor where the factor consists forming relation between particles occupying densified space-time of where $(\Pi / 2 \text{ X } \Pi / 2)$ relating to the foursquare triangle the value of gravity Π^2 comes in position as $\Pi^2 / 4 \text{ X } 4 = \Pi^2$.

By implication the Lagrangian five-point system is time formed with securing space displacement. Every four points form a securing proton relation of holding position to secure $(\Pi^2 + \Pi^2)$ in relation with centre singularity Π and the fifth is space filled with gravity Π^2. This is a relevancy applying all around and implicates any position to all other positions. This is gravity converting space to motion. It is space relating to time. It is the five of Pythagoras where the three of the triangle relate to the square of the line. It is $\Pi^{3+2=5}$ in relation to k or Π^0 being one as it refers to singularity. This is directly forming the basis of Mathematics or the basis of the Creation is the principle of space in motion. Our interpretation leads to out mathematical straying from facts.

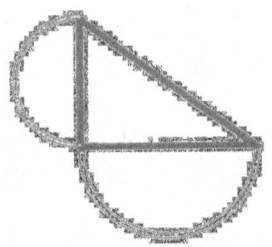 **The value of singularity stems directly from the law of Pythagoras** or **Pythagoras** is the result of **the average of singularity. With the shortest line being a dot, all lines must start from a position implicating** Π. A circle is a square without corners implementing Π and a half circle is therefore a triangle without corners. The corners are the factor that confused every one in the past. When replacing the value we normally attach to circle being r with Π, the law of Pythagoras becomes quite meaningful and mathematical.

By placing a connecting circle on the sides of the triangle half a circle forms. By implicating Π as a relevancy and not the straight-line r, two values of Π applies to each circle, and the straight line is no longer r, but is Π^2. This will bring about that each circle holds half the square value implicated to the allocated conditions applying to Π in that specific instance. By adding the two half squares forming the two half circles and then calculating the square root of the total that then forms the average diameter, an average of Π in the connecting line will come about. As both lines are the straight line forming singularity coming from one line being Π, the connecting line then must be the average of the two lines as Π^2.

A straight line a half circle and a triangle always have equality in dimensional capacity with of all sharing a calculated use 180^0. The reasons why this is the case is not apparent from the 3 dimensional positions we find ourselves using. Taking it back to single dimensional concepts it is very sensible indeed. Place the concept in line with Kepler within the boundaries of singularity where a line and a point and a position is inseparable but at the same time being the major divide. Leave out the symbol used and the dimensions are 1+2=3. The relevancy in Pythagoras is that the square plus the value of one line will produce the space within a square. It is two squares indicating one line but also a squire relating to a line giving an including space.

That is what **the law of Pythagoras says.**

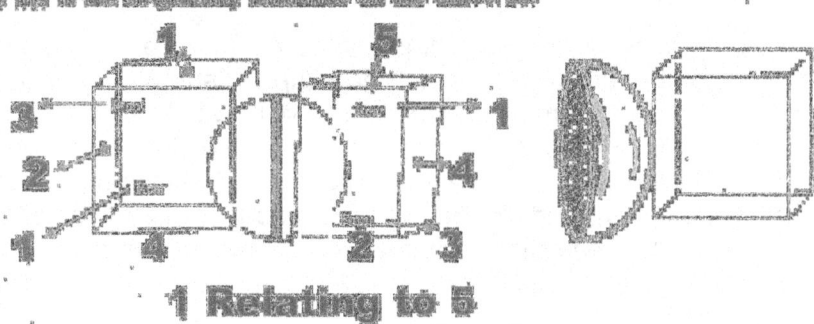

Once again that points back to Pythagoras and Kepler where $a^3 = k\,T^2$. The line forming the square proves the space included. $a^3 = k\,T^2$ presents $\Pi^3 = \Pi^2 (T^2)\,\Pi(k)$

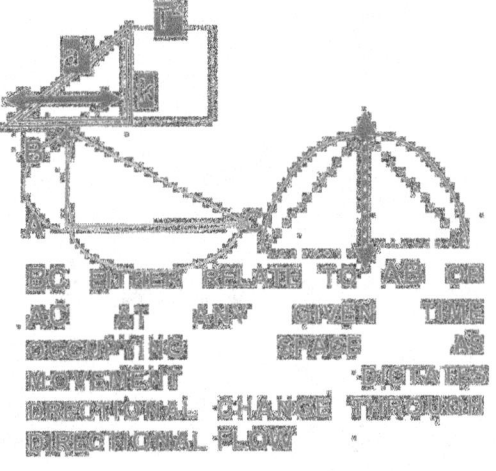

By having **k** in relation to any other side in the square T^2 will produce that area a^3 in three lines meeting committing three lines to space included. Mathematicians only see the two squares adding and when rooted it forms a line again. Yes that is the case but that is the proton part Pythagoras gives to the cosmos. There are so much more than that.

Placing singularity in the position of filling a centre it is two positions added to three positions will conclude the five points coming about. With the normal extending of singularity it will always form the triangle in a half circle whereby Π relates to the cube by 5 points to either side of the line singularity forms.

$(\Pi_{a2} \times \Pi_{a1}) + (\Pi_{b1} \times \Pi_{b2}) = (\Pi^2_a + \Pi^2_b)\,/\,2 = \Pi^2$ = gravity and that is proven by Pythagoras. Gravity is the average movement of matter through space in time determent from the position where matter in the sphere meets space in the cube from a point Π^2 in this 2(5) = 10 stands related to singularity as of Π to a point of the figures of (space) 7 from (matter)

From the star holding a dominant point or most valued point in singularity it affirm all three other structures, each holding singularity individually and in a compliment of 5.

Thus there are 10 standing related to seven and visa versa. By calculating the 4 squares in the circle with the dimensional changing of space (5) becomes the twenty. It is a triangle forming an area through two squares matching and it is a triangle receiving a three dimensional position in the flat Universe by connecting three adjoining lines, It is space coming from time coming from the centre of singularity. The normal flow will allow singularity extending to 10Π but when singularity blocks another sphere in singularity the two will form a joint value and by this joining the larger will dominate the space as well as the time of the lesser taking control of the surface and the atmosphere. Through this the Roche lobe comes about with all its other dynamics I describe farther on in the theses. The principle is the same, which we know as the conducting of lightning and Jupiter uses it extensively to implement this action.

In the sphere there are never only one direction implicated in movement. Movement are always in relation to the centre position because as a line goes up it also goes in or out. When a line goes north or south, it also comes towards the centre or going away from the centre. There is always relevancy present in movement. As this moving indicates direction it also apply Π^2 for indicating value forming the time factor. Because every moving line represents one quarter of the sphere in relation to the rest of the sphere and the line also indicate the relevant position between the point indicated and the point in the centre it is a relevancy of singularity in progress. By connecting the line, as Pythagoras will suggest the singularity within the sphere become a specific value indicated representing one half circle.

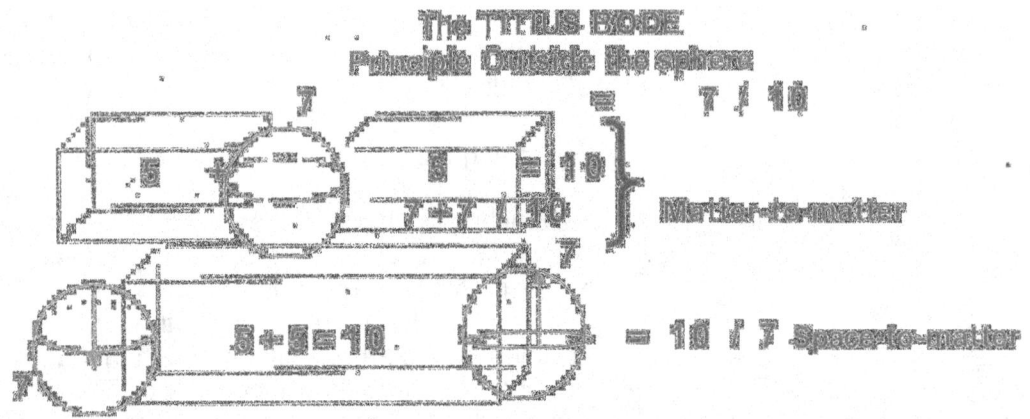

From the dimensional implication comes about, not only the Doppler's effect, but many more of phenomenon not yet understood. The dimensional relevancies formed between matter as six, matters end at seven and space at ten, comes the value of Π.

The process is all intermingled and stands in relevancy to one another. The relevancy compliment holds such attachment that none of the factors can even stand-alone. It is the way that science places every aspect in the cosmos as individual and not related to each other that launches the problems of miss understanding. The Value of singularity appreciates or demises by ten fold. For instance, the value of Π will increase by ten every time singularity applies another layer.

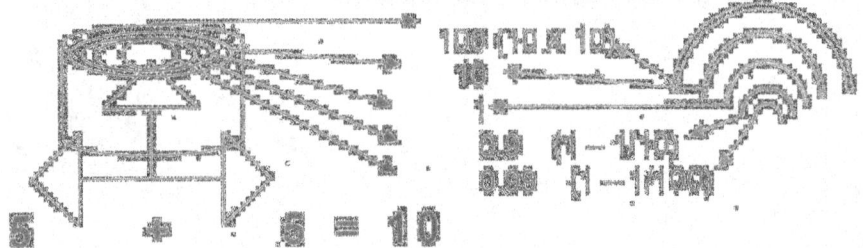

The normal flow will allow singularity extending to 10Π but when singularity blocks another sphere in singularity the two will form a joint value and by this joining the larger will dominate the space as well as the time of the lesser taking control of the surface and the atmosphere. Through this the Roche lobe comes about with all its other dynamics I describe farther on in the theses. The principle is the same, which we know as the conducting of lightning and Jupiter uses it extensively to implement this action. In the Roche limit the straight line forms part (1) and the half circle is part (2) and the triangle forms part (3) to singularity (4) Holding 5 points outside singularity. Every aspect connecting to the Universe changes everything it holds totally and becomes the anti-matter to which it was matter 180° previously.

Gravity is a relation not bound by borders and the influence stretches seemingly indefinite. As expansion comes about the expansion will have to follow what gravity produced initially.

In this, it is clear why the Titius Bode ([10 + 10 + 1 + .991] / 7) and the Lagrangian 5 \\ 1 systems part their ways when applying the different processes they hold. With all the differentiating, the observer must also consider the dual message that light uses in travelling through the vastness of universal space. The thought of nothing is just what it is, a thought of nothing and although it is in the human mind common nature to present nothing as a value in the recalling of something, nothing is a presentation of the figment in the human mind. There can be no number such as nothing and that was (possibly) Newton's biggest error. Nothing represents non-existing and that is just what nothing is, it is non-existing. However, with this statement the question arises as to how and why the Titius Bode law is implicated as one of the four pillars responsible for gravity?

To go back into the past one have to reduce the radius because the radius of the circle indicate the size while pi indicate form. The Big Bang is all about the cosmic radius expanding bringing about the filling of more space. By reducing the circle through the radius it becomes another matter of eliminating form because in such reducing of the radius it then becomes a matter of reducing a straight line. When reducing the circle in size one have to reduce the radius or the diameter because the pi factor is the indicator of the form as being a circle. It then becomes a process where it is just dividing the radius defined by the symbol r by halving the answer every time until there can be no dividing any further and such a reducing cannot end by becoming zero because in this process of reducing nothing disappeared. The lines just went smaller and smaller but if it was part of the Universe as it is still part of the Universe as it has nowhere to go but remain in and part of the Universe. One may divide by two, halving the result every time, which is a normal mathematical expression and by reducing by half can never reach zero. Allowing zero to be accomplished through any legal mathematical equation is not a mathematical fact and I challenge any person to prove such a feat mathematically. The numerical procedure may become tiresome or to small for any human to make sense of the outcome but never can such a dividing bring about zero in the ultimate answer. Zero cannot divide nor can zero multiply should I wish to retrace my steps to where I started. In using the method of the dividing by reducing the answer to half the size it will forever allow such a process to continue without ending in zero because no matter how small, the next value will be dividable by two in that will forever be a value in place. This stands as a mathematical fact and I do not have to prove my statement but those persons in academic positions, which portrait me as a person trying to per sway facts to fit my convenience, must explain how one can multiply distances filled with zero as a concept and get a valid mathematical answer other than zero. This alone is the biggest obstacle why there is so little exploring going about the space expansion present overall. In my dealings with Mainstream science in the past Mainstream science become truly all out aggressive with my trying to indicate that I am criticizing Newton and one such a fact is his way of producing zero as a result of a circle forming an end and at the same time starting with another beginning. I do not try to criticize Newton on any facts except on what he saw in Kepler's work which he tried to retranslate and the need he found to change Kepler's work. Do you as the reader realise that Johannes Kepler's introduced an astonishing presentation about space-time, singularity, gravity and the Big Bang in the years of his life between (1571-1630). By applying his formula we find some answers about questions yet unanswered. I raise the questions and with much study received new answers through dissecting what Kepler introduced to science. It was only possible when taken from the mathematics, which Kepler introduced from his cosmic findings. Kepler's answers were lost so many centuries ago when some Englishman saw him wise enough change Kepler's work and bring in the alterations as he saw fit. The changes were unnecessary because the changes were unasked for. Such forced changes came without the modern demand on proof because such proof was lacking in considering to what the accepted norm is today.

I have grown accustomed to outright rejection of members in mainstream physics because of my criticizing of science. Sure they can dismiss me in the light of my criticizing them on science but that they do without explaining why Mainstream Science makes such a performance about Einstein and his views about relativity, when Kepler said it far better and with more explaining and much less performing about four hundred years ago. By accepting Kepler one has to accept space-time, Kepler said $a^3 = T^2 k$. Kepler said there can be no space if there is no time $a^3 = T^2 k$. Kepler said space is the equivalent of time $T^2 = a^3 / k$. Kepler then also said when space came about so did time $k^0 = a^3 / T^2 k$. Kepler said space is time and time is space and the one cannot be without the other being there. Kepler said a^3 could only be present when it is equal (the same as) $T^2 k$ performing in the time aspect. Kepler said what he was told by the cosmos. Newton had no need in changing anything because by changing he brought about much misunderstanding as well. Kepler showed what time is but all Newtonians admit that not one Newtonian knows what time is. Time is the motion of space and space is motion filled with heat. Time is the spin or motion of heat in space. That means there is no greater all preserving Universe out there. Every point k^0 will establish space a^3 by applying motion $T^2 k$. There is space filled with heat and the heat applying motion at the point where space is. There is the space $a^3 = T^2 k$ in motion and the motion is time. There is only space-time. There is no liberated space without the restriction of time. There is no space restricting a Universe without the liberating of time through motion. That is what Edwin Hubble saw through his lens. But where there is motion there too is gravity or antigravity whatever relevancy one wish to apply. Motion is gravity, which is motion that is committing space to form in the presence of singularity and that too is antigravity.

Einstein received enormous admiration and / or huge criticism because Einstein declared that the Universe draws flat and then reforms. This got some that couldn't think very exited and got the rest that couldn't think very annoyed and then there was a third mindless mob that got much religiously consciously encumbered about the statement. How they get religion part of this goes above my thinking ability and sometimes when listening to some of these religious opinions and beliefs, even atheism makes more sense than these fools. It all comes down the way one thinks. It is a question of asking oneself to think how things move because time is what moves space. Time is the movement of everything in relation to one specific point. If an object moves from A to B that would mean everything that the object is made of, must displace. Everything that is the object that is shifting must break down and rebuild in a new location. To think the object comes from here and moves to there is so Newtonian as thinking of gravity by mass pulling…it is the Neanderthal's thoughts. To move one part has to be at the old location, and one part must be in the new location and there has to be a part that is nowhere because that part is in transit. Everything is made up of tiny parts and the tiny parts form big complex parts.

We are in time. We view the history of time. What I see, is the history of what was in relation to where I am when what I see occurred during the time when time placed the Universe in the order that I see it is n terms of me. The reason why things seem solid is because of time delay. I can see and I can feel because the replacing of what time places as space or as the history of time happens so often and in such a quick repeat that I observe and feel space that time replaced many times faster that what I am capable of perceiving. My powers of observation in terms of time displacing space is so slow that light might seem solid as it is in the case of a nuclear explosion. Time breaks down space and rebuilds space as Einstein said, but it happens in the Universe and not in the history of the Universe that is in repeat and is what our abilities allow us to observe.

Time is at a point where space breaks down just as Einstein said it is, but Einstein was confused about what forms the Universe. Time is where the triangle is equal to the straight line, which is equal

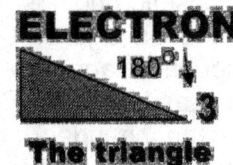

to the half circle and by duplicating the Universe (having a side on both sides of the Universe) form starts with a cube and a sphere. However, where we find form is where we find the delay or the history of time or there where it is what we call space. Space is the history of what was (the past) that time carries through (singularity) onto forming space (in the future) and this is an ongoing never-to-be interrupted process.

The vision about time that I try to portray can be matched by looking at a very fast spinning fan, but this fan is so fast spinning as nothing made of solid can ever spin. When the rotation exceeds a certain point, we can see rite through it as if the fan is not present although we can freeze the actions of the fan with a high-speed photograph to a point where every blade still stands still. In doing that we didn't freeze time but put light on paper to form a picture only intelligence can observe. We didn't freeze time because time can't freeze. The fan stands still because the photograph confirms this action. The fan moves so fast that the light moves through the fan undisturbed. However, the fan only forms a time base a little closer to time but doesn't even begin to be where time presents space. If I try to put my arm in the rotation area, I will lose my arm so fast that I will only observe the loss after the loss occurred. Yet, from the picture I can see that the fan blade stands still then moves to the next point where it stands still and this continues unabated which is the process we refer to as motion. Every molecule in the fan stood still. The photograph confirms that. At the instant where the picture is taken and put on paper, every atom is positioned in one place and therefore the atoms must move much quicker that the photo. However, the picture confirms what Einstein said when Einstein observed a Universe breaking down and rebuilding again. It is a matter of defining that Einstein what the Universe what time is referred to is and Space is only the and what space is. of time forming a repeating confirmation which light can delayed history from focus and I can observe the light to view what was at a specific point in relation to my point I have in time.

Time is when this lot is not and space is when this is disappearing into from which it is coming. This process is space-time. Space is the history of time, which is what came before time and which came after time and that makes space the history and the future history of time. That places time in space that can't be because that space is in singularity and singularity is where space can never be. Critical and all important to remember is that only motion can produce time by producing space coming from the past and going to the future and this process we think of as movement can put time in the present where singularity is and in which time is able to deform space and to reform space in another position according to the single point in singularity holding time. That is gravity and that has nothing to do with mass or attracting magical powers.

The sphere was the first to come about after space broke down the sphere into singularity and only after the sphere could not produce the gravity required to suppress the overheating forming the antigravity did the Universe try other options. Then the atom came in to form…did the atom clusters form…did the material clusters form…did the antigravity apply enough to allow overheating bring about expanding where the relevancy will produce one softer and one firmer structure…from which the antigravity expanded into space. When the heat turned to apply antigravity that eventually produced space, as we now know space to be the Big bang was happening. But before the Big Bang there were some mighty jerks, cracks and jerks announcing the Big Bang to come. It was the proton $\Pi^2 + \Pi^2$, which afterwards connected to other protons which became independent Π^2 connecting yet again to other Π^2 particle that with he help of further antigravity overheated and expanded to $\Pi3$. One can see there were many stages to produce what we now have. However most important is the fact that the laws that once implemented the stages of developing those laws still apply and still present our Universe to what it is.

How many dots was there is a question no person can answer because everything was un-dividable solid and yet it did group together to form every atom located in the 3D.

Individual singularity and governing singularity and group singularity is enhancing the gravity every time singularity finds an accumulation. We look at space and we wish to see a Universe. To the human mind the cosmos is the following.

With you're accepting of modern cosmology you're deciding of that then also will bring along a new perception about cosmology. Such a new perception must include a new perception about Kepler. In you're reading of this letter you will investigate Kepler and such investigations will introduce facts that no one before came to realise. It becomes clear when we go about analysing Kepler's formula somewhat differently that the Big Bang theory might be exceptionally correct in the way it is presented. But at the same time it is incredibly flawed when an overall view establishes a universal picture. For instance Science accepts that space expands but science does not permit any view about material expanding. The two are linked. Understanding the cosmos helps one to realise another part of Kepler for the first time. The realising involves issues that seem beyond answering given present facts because at present science is without such analysing of Kepler. By scrutinizing Kepler again we find that when re-aligning such investigating of Kepler and what Kepler really said, that which now is the unknown part of science becomes a new vision, such unknown becomes surprisingly clear. But it insists on a divorce separating Kepler's ideas from Newton's ideas about Kepler's ideas. It is necessary to give Kepler the recognition as a mathematician without Newton belittling Kepler's skills as a cosmologist and a mathematician. The difference is about finding what Kepler really brought to science in relation to what Newton saw what Kepler brought to science. It also puts a new appreciation on what Kepler said and diminishes Newton's effort to change what Kepler introduced.

The expanding of space **k** is the establishing of space **k** by the applying of T^2. The motion Hubble saw was space being established by time. It was space-time because without space there is no time as much as there is no time releasing space from singularity by means of forming space-time $a^3 = T^2 k$. Science was all impressed and some are still sceptical of Einstein committing space to time. Their scepticism runs so wide that they at this time in the liberated ages of lamination through wisdom reject Einstein's space-time to some situation that was in the beginning or that can be in a Black hole. Space-time is something we must rather use as some negotiating method to a situation that either was or will become part of the cosmos.

I have heard cosmologists declare that we must "somehow" not think of space but rather think of space-time, but while saying that they mildly refer to something to say to please Einstein and rather nothing else. This they say while the man that started cosmology some four hundred years ago introduced modern cosmology as space coming about through time. He declared space could only be if space was liberated and restricted at the same time by motion and the motion establishes time as a factor being as much part of space as space is space. But in the very same event the motion is the restricting of space by singularity forming T^2. The return of space to singularity and the returning becomes a rotation forming the second part of the time factor. While space is the liberation of singularity through motion **k** that is part of time, the restricting of the liberation of space from singularity is time T^2 in which singularity achieve the return of space to singularity. In this comes about four cosmic laws, which I named the four cosmic pillars

 Space is the forming of motion **k** because space is the liberating of f time from singularity where the hottest part of space will find a way to move away from the rest of the cosmos and that motion forms the time component **k**.

When accepting Kepler's work as the very basis of cosmology one has to accept that space is time and time is motion and motion is heat and heat is space. The one cannot be without the other because time and space is the very same thing, it is space-time and it is $a^3 = T^2 k$. Thinking of the cosmos must exclusively be in the form of space-time and in accepting there can be no space if there is no motion causing time. The proof that was presented then is accepted as unquestionable proven fact today given to scholars as accepted evidence. But such evidence was never produced in the past. If you do not agree with my statement…well test the following: The Sun at present is 1392 530 km in diameter. Let us only concentrate on the Earth in relation to the Sun, as that is what concerns us humans intimately. The Earth is roughly 150 million km from the Sun.

The Big Bang theory wants us to believe that everything once was so small all fitted on top of a needlepoint. The Sun is at present in diameter 1392 530. There then is a distance of 1.39 million km separating the one lot of atoms within the Sun from the other atoms on the other end of the Sun. Just as there is at the present 150 million kilometres between the particles that form the Earth and those forming the Sun. If we are to believe the Big Bang where the lot was less than one millimetre apart at one stage, we can appreciate that the distance there was between all the atoms that now forms the Sun and all the atoms that now forms the Earth was one meter apart. It seems rather trivial except that it also means the Sun in as much as being the space occupied by all the matter that now forms the Sun was some time back reduced to a speck of what now is present. The Earth too has to be a lot bigger at present than what it was some time back. It is rather unrealistic to believe that the structures remained the same size when the space placing them apart grew from millimetres to what is now applying. Kepler brought us the insight that the distance of **k** is at present 150×10^6 but believing the Big Bang then proves that **k** once was one meter and much less than one meter. With **k** at one meter it forces the size of the Sun and the size of the Earth to respond to that and shrink in sympathy. This also brings a realising that if they grow, then why is it that they grow because surely the idea manifesting as nothing can't grow. We have to bring all space being **k** in $\mathbf{a^3 = T^2\,k}$ in realising there is growth of particles as much as particles moving apart. The growth is everywhere and not selected to the nothing part. If the Sun and the Earth were that much closer back then, the Sun and the Earth was that much smaller. If a Sun with a diameter of 1392 530 km in diameter and an Earth with a diameter of 12756 km was one meter apart, then the distance separating the two was enough to join the two. As we can see they did not join. That brings across that by accepting the growth of space it then brings along the growth of all space. It includes the space holding material as much as the space not holding material. It includes all space. That leaves us with one fact being the principle issue of this very book: If space is nothing and space can grow, then how can nothing grow.

If space we think of as something because it holds material can grow as much and in relation to space we think of as nothing then the nothing has a lot of the something within the nothing because both is growing. It is an argument of complexity and it can only become apparent when Kepler becomes apparent and Kepler separates from what Newton saw Kepler to say. Kepler must come into his own as a mathematician, a cosmologist and a scientist because of his tremendous achievement. If we accept that Kepler's formula is mathematically $\mathbf{a^3 = T^2\,k}$, then it must also be $\mathbf{T^2 = a^3/\,k}$, $\mathbf{k = a^3/\,T^2}$ and $\mathbf{k^0 = a^3/\,(T^2 k)}$. What this then states is that Creation according to Kepler started at a point coming about from singularity $\mathbf{k^0}$ and grew out of singularity into space standing directly and undividable related to time where the time aspect is the motion created by and creating space.

Reading this letter will introduce facts that no one before came to realise during the past almost four centuries. This becomes clear when we go about analysing Kepler's formula somewhat differently. In a way for the first time it helps one come to realise another part of Kepler. But I must press the fact once more that this realising demands a divorce separating Kepler's ideas from Newton's ideas about Kepler's ideas. We must recognise the Master Kepler as an equal to Newton and not just someone with vague ideas and the ideas he had, later needed revising by a better Master. The difference is about finding what Kepler really brought to science in relation to what Newton saw what Kepler brought to science. It also puts a new appreciation on what Kepler said and diminishes Newton's effort to change what Kepler introduced.

Do you as the reader of this letter realise that Johannes Kepler's introduced an astonishing presentation about space-time, singularity and the Big Bang in the years of his life between (1571-1630). By my reinvestigating Kepler In the state Kepler's work was seen in, isolated from Newton I received new answers taken from the mathematics by which Kepler introduced his cosmic findings. Kepler's answers were lost so many centuries ago when some Englishman saw him wise enough change Kepler's work and bring in the alterations as he (the Englishman) saw fit. The changes were unnecessary because the changes were unasked for. It was the cosmos that directly that spoke to Kepler through mathematics and it is the what the cosmos that said things about itself to Kepler that Newton did not understand and then changed to what he (Newton) then understood. But now some three hundred and fifty years later in comes to our attention that Newton completely misunderstood what he thought he understood about Kepler.

The cosmos spoke to Kepler about space-time coming from singularity. Kepler gave us his findings. He translated what he cosmos told him (Kepler) as $a^3 = T^2k$. Translating Kepler's mathematical expression $a^3 = T^2k$ correctly to the verbal statement in English Kepler said that there is a space a^3 which is equal $=$ to the motion in the time duration T^2 thereof between two specific points which holds a relation to a centre where from there forms a straight line **k.** What is there mathematically not correct in Kepler's expression and why is any changing thereof necessary in any way? Test the following symbolic values in the mathematical expression and test the principal behind the expression in which Kepler stated them. Convince yourself about the evidence that Newton saw what Kepler saw where the translation thereof is mathematically incorrectly translated by Kepler's interpretation from mathematics to English:

a^3 The fact that any symbol uses a value to the third power indicates space or a volumetric established and separate unit. It is space because it is volume using the third dimension.

T^2 Is an indication of motion, moving from one point to another point or following a flat distance between two points. It is motion that is taking time in the second dimension.

k^1 Is the symbol used to indicate a straight line between two points with a definite beginning and a specific end position. It is Pythagoras by the triangle, half the square and the straight line sharing value in the 180^0 they represent. Kepler introduced this absolute basic mathematical principle.

What formed the grounds for any need by Newton to change Kepler's translations from the cosmic given to mathematics and then from mathematics to English? The space-time that the cosmos introduced was so brilliant it took the likes of a genius such as Einstein to realise the presence thereof many hundreds of years later.

What did I not translate correctly from the mathematical expressed to English? When I used Kepler's mathematics by my translating Kepler's work correctly I came upon answers not yet uncovered by Mainstream Science. Kepler gave the World mathematic translated cosmic answers that Kepler uncovered long before Newton, Einstein and others got wise about cosmology... Such is the advantage of recollecting Kepler facts that it does answer many questions, which went unnoticed and therefore not spoken about up to now and some were previously never even thought about. Mainstream Science never previously thought that through any examination of Kepler's work the scrutinizing would uncover these facts that I present. Subsequently Mainstream Science elected not to ask the correct questions and in the process Mainstream Science never found the correct answers. By not asking the questions Mainstream Science could not decipher any of the decoded mathematical messages, which Kepler received directly as a mathematical message spoken by the Universe and coming from the cosmos. We all know and appreciate that mathematics is just another language and the professional mathematicians have is responsibly translating mathematics to a verbally competent language.

They did not read nor recognise Kepler's mathematical translation and thereby was unable to translate Kepler's mathematics to the other communication forms being all verbally spoken dialects. In other cases human natural study methods brought along a cultural of Academics forcing students to comply whereby the students will accept the knowledge through our inherited past which when tested by modern standards is not that highly proven. We accept the answers as questions already fully answered because culture demands the accepting thereof.

Newton is institutionalised academic culture force fed by the generations of academics passing from the presence to the past whereby the academics in generation after generation introductive forces the accepting of Newton as correct above all else onto the following generation of future academics. The next generation of academics is conditioned mentally to accept Newton or die an academic death. The academic death comes by terminating all further study possibilities if Newton is incorrectly accounted in a students performing of an examination questionnaire. If and when the student will not accept such culture without reservations that his peers teach him that student will fail all further acceptance disallowing him the right to attend further classes in any of the classes given by the institution. I personally have experienced such bias from Academics in charge of institutions. By reading what Kepler said correctly so many centuries ago the effort brings all the answers to the questions academics at the present are incapable of answering. But it does not involve looking at what Newton said about what Kepler said… it is all about looking at what Kepler said. To understand Kepler one has to include the opinion of Kepler and remove the opinion of Newton about Kepler's findings. Why would a water droplet form a sphere when floating in a space capsule in outer space?

Kepler gave us the answer about the water drop forming a sphere even before Newton thought about the question. He gave us the answer three centuries before Einstein got the question wrong about his Universe going flat. Why would gravity always result in forming a sphere when gravity is left free and unhindered to capture form? Let us recollect Kepler's statements. He said that $a^3 = k/T^2$ and from that the mathematical relevancy guide one to the answer that $a^3/k\,T^2 = 1$ and $k^0 = 1$ bringing about that singularity is $k^0 = a^3/(T^2 k)$ which is the smallest space being in singularity produces gravity in forming space a^3 relating to time in motion $T^2\,k$. Indirectly Kepler said that mathematics prove that $k^0 = a^3/(T^2 k)$. $a^3 = k/T^2$: That is what Kepler brought into civilization for all time

a^3 symbolises in a mathematical interpretation the three-dimensional space.

T^2 is representing the period or time that Kepler suggested we should use to calculate time that holds the orbiting planet in direct contact with the space in relation to a very specific centre

k is the centre from which the planets must have grown if one accepts the Big Bang growth of particles and the affect of the Hubble constant on all cosmos matter. The specific value about the centre is most important because from the specific centre gravity always apply the strongest influence. From this I found that gravity is the strongest where space k^0 is the least. Gravity is about dismissing space to the advance of heat increasing. According to Kepler that is what he found to be true. Space a^3 will always be circling space around as T^2 in any position from the centre **k that** indicate the other space holding relevance. That is what Kepler said when he said $a^3 = T^2\,k$.

What in this is there to dispute...yet when I say these facts Academics find grounds to dispute my saying this about Kepler's saying that! Kepler gave us the answer but no one ever took notice! Kepler was the one that discovered **space / time** as $k = a^3/T^2$ Kepler was the one that discovered singularity as $k^0 = a^3/T^2 k$ Kepler was the one that discovered gravity holding space-time relative $k = a^3/T^2$ Newton claimed gravity as a force, but that is as vague as saying humans are life. If gravity is a force, then what is a force? If gravity is the product of the elusive graviton what is the graviton and where is the graviton. By using mathematic rules and laws correctly and investigating the formula that Kepler introduced intensely but without Newton interfering and telling Kepler what he (Kepler) should have found and whereas instead Newton should have been looking at what he (Kepler) found then he Newton would have seen what gravity is.

He (Kepler) said that the cosmos said that space is time being space-time. $a^3 = T^2\,k$. The space is held in check by motion from a centre and that is gravity. It becomes more than clear that space a^3 is time by dimension T^2 and time is space a^3 without dimension **k** Gravity is a^3/k but **k** is an addition of motion T^2. Motion T^2 of space a^3 being apart thereby forms k^1, which produces gravity. It is gravity that keeps the Sun and the planets at a specific distance and apart while the planet remains in motion around the Sun. It is $a^3 = T^2\,k$ that keeps the space in motion and at a distance whereby the space of the Sun is parted from the space of the Planet. That is gravity…what else can it be? After all it is space-time keeping the structures apart and space-time is a result of gravity. Once

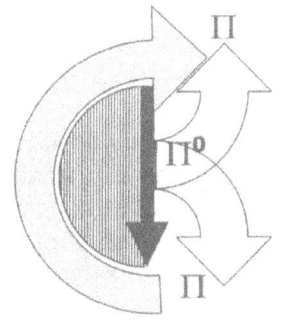

singularity was found the rest was simple but was finding singularity really that difficult. Not if one was guided by Kepler's formula. It is merely retracing **k** until **k** becomes k^0.

LAGRANGIAN POINT is another result flowing from the singularity position and most directly coming from Pythagoras principal and connecting that most basic mathematical in conformation of Kepler's formula.

Kepler brought in a formula that came through two lifetime studies and the formula read that space is equal to time in motion. Mathematically it reads that $a^3 = T^2 k$. This formula brought Newton's claims into dispute because what this formula said was that geodesic space is not confined space and while laws apply in the confined space of Earth it does not apply in geodesic space. From Kepler on can see the precise moment the Cosmos started. It is Mainstream science that is hindering the accepting of the formula that is dampening the human understanding of the Cosmos.

When did the cosmos start is the question every one is in search of since Biblical days. I would say it had to have started with space but science has the cosmic time linked with time. There could not have been space without time and there was no time without space.

The first moment came when **k** moved a way from singularity to establish space. It was when **k** introduced individual entities apart. It was when confinement was broken and particles appeared for the first time. Understanding this comes hand in hand with accepting that there is two forms of structures other that elements confined in an atom.
Through such motion gravity comes about because the motion is gravity and that produces the time aspect T^2 thus thereby changing the direction by rotary motion. The moment there is an area there is a measurable rotating brought about and no longer a non-interfering divide. Such a line holds space in a position that runs far beyond the boundaries and limits of the three-dimensional. Another factor of such a line would be that the radius (let us substitute the radius r with the using of Kepler's **k**), **k** would be immeasurably small. The factor **k** cannot be zero because infinitely close to that first **k** is the start of the third dimension where time plays the part as the fourth quarter. The presence of **k** is undeniable and recognisable yet it is not visible.

The fact that **k** is there albeit stripped of any influence disqualifies it from being zero and therefore not being there. With **k** already beyond any measurable space, leaving a^3 as a factor of one and not being able to pin any volume measure to that one **k** will have to be to the power of 0 being k^0. In Kepler's formula $a^3 = T^2 k$ the area a^3 would be one because of the dimensional non-existing of measured sides in any direction. If $k^0 = 1$ and $a^3 = 1$ the only alternative T^2 could possibly have is also one. The factor of T^2 identifies the time in the formula and when the formula indicates time as one, the time component must therefore be eternal. Only time in eternity does not change

The more commonly used mathematical formula applying when the calculation of the sphere comes about is $a^3 = 4/3 \, \Pi r^3$ where it places one third dimensional but lesser factor in direct relation to another third dimensional relation and all that in relation to the form that is applying. By using $a^3 = 4/3 \, \Pi r^3$ it is definitely not what Kepler said when he produced his formula. The square he allocated to time as well as the cube he allocated to space does not find representation in the mathematical expression of $a^3 = 4/3 \, \Pi r^3$. In mathematics we find a deliberate lack of the time factor when using the equation $a^3 = 4/3 \, \Pi r^3$. This must prove the differences in what Kepler found to express and what Newton thought Kepler tried to express. However using Kepler there is no criss-cross matching of dimensional accumulating. If I am reading the situation correctly Newton saw Kepler's mathematical skill somewhat below the dimension of Newton's genius and that spurred Newton on to bring changes to Kepler's formula but in doing so many things went missing through incorrect translations and miss interpretations of genius.

Kepler was not referring to a mathematical space on a flat sheet of paper, which was what Newton saw. Kepler produces a value linking space to gravity being time in as much as calculating space in the geodesic and measuring time in the geodesic and bringing about factors in formulas through figures that geodesic space informed Kepler about.
If Kepler was mistaken then the Cosmos was wrong about the cosmos and if Newton was incorrect it was about mathematics that Newton then was wrong about. Kepler places time in the square directly in relation to space in the cube in association where time shows two distinct qualities. The one factor is time in the circle rotating while the other is in the linear or the straight line implicating the position that the other would have. This proves that time or gravity (with time and gravity being the same thing)

In all instances of measuring the distance the orbit travels around the Sun as the space displaces or space covered by travelling in the time it is covered and dividing such a ratio one find the distance of the orbiting object from the Sun the in relation to the other factors form one or very close to one.

Planet	Period T years	T^2	Distance k	Space a^3	Ratio
Mercury	0.241	0.058	0.39	0.059	0.983
Venus	0.615	0.378	0.728	0.381	0.992
Earth	1.000	1.000	1.000	1.000	1.000
Mars	1.881	3.54	1.524	3.54	1.000
Jupiter	11.86	140.66	5.20	140.6	1.000
Saturn	29.46	867.9	9.54	868.25	0.999
Uranus	84.008	7069	19.19	7067	1.000
Neptune	164.8	27159	30.07	27189	0.999
Pluto	248.4	61708	39.46	61443	1.004

between the Sun and specific but different planets in the solar system.

At the first glance Kepler's formula seems to be numbers and positions applying

In that it is just about numbers but about that it is much, much more than just numbers. The numbers paint a picture and tell a story. Kepler produced a formula from the numbers and not numbers from a formula. The numbers brought about answers but we fail to ask the correct questions. By seeing a mathematical circle one miss the total picture and the story the numbers tell. The figures explain dimensions working in conjunction and together they combine dimension where the picture behind the story becomes a colour spectacle in comparison the mathematical grey that a mathematical circle produces. It is relevancies carried from the Sun and the Sun is the governing singularity representative for the entire solar system. This is about relevancies applying throughout the Universe. This balance is much, much more than what the figures say. It underlines and it explains gravity as a life form in the cosmos other than what we consider our life to be.

In their eagerness to calculate they calculated a formula to measure the circumference a^2 of a circle being Πr^2. I have seen an Astrophysics examination question paper where they use $4\Pi r^3 / 3$ as the formula to calculate the Sun and other stars volumetric space! They formulated the measuring procedure of the circle being in the third dimension that will show how big the volumetric space is of a sphere at a^3 being measured with the procedure being $4\Pi r^3 / 3$. This too was a fanciful devise allowing mathematicians to be much superior to the rest of the commoners and to dictate to the lowlife how and what they should think when they think and if they indeed can think of anything to think of. Then some Mathematician and an Englishman of Substance came onto the idea of gravity. Being a mathematician the Englishman placed the Universe at the feet of mathematicians. He saw circles where Kepler saw three dimensions. He saw three dimensions where Kepler saw nothing. He knew time had to be somewhere as something and then covered it by denouncing the circle cycle as nothing.

What then is it that Kepler saw as he formulated $a^3 = T^2 k$. At the normal flow of time it takes the electron a certain time to spin around the atom. The atom uses space a^3 and the atom is a certain length k that forces the distance the electron has to travel in one cycle period T^2. The atom a^3 connects the electrons travel k to gravity T^2. The relevance k produce to support a^3 is to point T^2 to two positions the electron will be in the duration of one specific time. The electron travel will be cyclic and periodic in relation to the space the atom holds. The space stands related to the gravity with which the Earth reduces space and with the space and speed with which the atom travels through space.

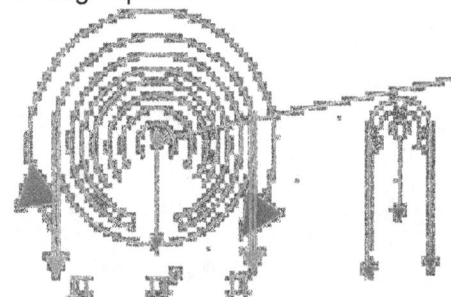

Should a person take time to responsibly study Kepler's formula with an assertiveness to draw a realistic conclusion about what Kepler's formula truly represent. Kepler showed that $a^3 = T^2 k$ when going back to Kepler's statements while ignoring Newton's devising of the formula. If we conclude that $a^3 = T^2 k$, then mathematically we can deduct that $k^0 =$

$a^3 \div (T^2 k)$ and putting the cosmos this way puts all there can be into perspective where the perspective is clearly that what ever can be can only be in regard to what circles around that centre holding singularity at $k^0 = a^3 \div (T^2 k)$

The perception is that light travels as fast as time flows but I disagree with that idea. Nevertheless the perception is there that the speed of light is as fast as travelling of any sorts can reach. We accept the electron's travelling speed imitates the speed of light as much as it is permitted by gravity to do so. By this imitation the electron come as close to time as it can ever come. The electron rotates around the atom nucleus indicating an atomic border of some sorts. From the centre one can draw a line pin pointing the position of the electron during the duration of a time period. In this time will indicate movement of the electron through space. The time indicated must be T^2 should space be in the third dimension a^3 and singularity will connect through k. By linking space a^3 to singularity, which produces time T^2 it will have to indicate the influence of singularity through the single dimension connecting of k. In the relation only k will be representing the single dimension factor since that places the Universe in space and time. Should one place the time factor but presenting time as t in a single dimension role the two dimensional time have to disappear with the three dimensional space that also will disappear. In that space and time must disappear. We all can accomplish that task by taking a photograph and print the image on paper and call it still photography. Then the paper will hold time in the square while the paper is in the cube all indicating individual and complimentary k pointing to individual and complimentary singularity producing space-time.

The one position in space will place the electron in time where the electron then will be below…behind and above the atom sub particles with which the electron shares frequency. The position k indicates implicate the electron T^2 in the environment a^3 establishes.

By back to where the reduce further $k^0 = 1$ value where k then single dimension. But equal to one and so is produce a line and we find that k represents a^3 to the full as well as T^2. The point where k forms the most slightly distance the area a^3 establishes a value outside the single dimension because T^2 adds a value. The fact that T^2 comes in as a factor in the presence of the first sign of k appearing indicates the start of motion taking a^3 from one location to another specific location. It indicates the travel of the planet during a month or a day or an hour. It does not indicate a circle except at the end when completing one cycle. T^2 is the distance in time a^3 will take k from indicating one point to indicating another point.

taking the line to k line or k cannot establishes such a finds a position in the in that case a^3 also is T^2. In fact k still has to

The formula points to a referring of the very time space was indicated by position location and time. The astonishing part is not as much the way Kepler formulated his formula to cover the movement and the position of the electron in relation with the rest of the atom, but the brilliant way the mathematicians neglected to see the fact. Kepler saw a three dimensional a^3 something in a specific position in time T^2 relating to a specific density k of the atom. With space in a cube as it cannot ever be otherwise the time too has to be in a square because placing time in the single dimension of t the time then becomes part of a single dimension such as one may find in a photograph picture. One can justly use the same formula to implicate the electron taking time to complete the distance between two points indicating the area from the centre of the atom.

The pulling away of the smaller space.

The double counter-acting returns.

The pulling towards within the larger space

I have explained infinity as well as where to locate the spot holding infinity. Then there is the space holding eternity and I would like to show why space-travel to the galactica and beyond is a farce. It is because of eternity. Kepler gave the answers, but then Newton decided to ignore those very answers on cosmic reality in favour of his personal perception about the prominence of a factor he devised called mass.

No line in the

Universe can ever move straight. That is the principle of gravity. Everything that moves straight is by the same movement diverted by 7 degrees to change direction and turn back towards where the singularity in control is located. The fact that the line will always diver by seven points to one, makes every straight line a circle in forming. That is why the comet doesn't collide with the Sun and planets orbit the Sun and Galactica forms circles. What ever is going away is also going to come back. There can never be an end to the Universe because going straight becomes a circle and a circle is the repeat of the same where only time allows the new perspective in relation to what was.

In this we find time. Time has to be infinity being in relation to eternity or that which can reduce no more standing relevant to that which will always expand because that which always expands is eternally progressing to grow bigger or become more. This state of affairs Mainstream Science called the Hubble constant. However, time is everything or eternity moving around infinity or everything moving around a spot holding something as 1^0. It is the movement of the top that produces a value of three to singularity 1^0 and that gives the movement of time a value of 3. It is time coming from the past $(+1^0)$ going to the present $(+1^0)$ and onto the future $(+1^0)$. The line that forms in relation to movement has three spots and in that we can detect the fact that time makes space three-dimensional. By going through the four quarters, the circle this produces brings about that movement to the four quarters relevancy of three and this produces seven spots where spots form a cube. The Universe is a cube, but not on the because the Universe has no outside. The Universe is a the inside where the Universe starts and this cube grows as the Universe expands into forming space.

action relates a seven outside cube on outwards In every linking r

sector the directional flow will provide a distinct meeting of Π to Π^2 and this allows the time component to be in the rotation.

As the meeting of r points to a very distinct different r in direction every instant of movement, such a point of meeting opposes the other points in meeting and will lead to destruction of the form Π in any the event of any value changes by Π changing Π^2 and r.

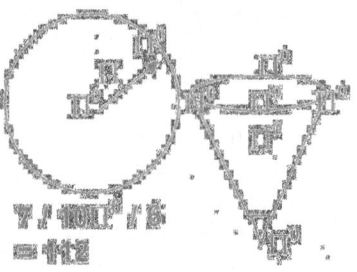

Keeping these factors in mind it is clear that Π^2 are the choice of gravity and not r^2.

Every quarter provides a distinct value that indicates the progress of the flow of time from the one point Π to the next point Π. Any changers occurring in Π will lead to a an unequal triangle providing two different values to r and will alternate the link between r and Π^2 bringing about different form (Π) and time (Π^2). When singularity (1^0) forming the lines of the triangle is not in equilibrium the triangle will destroy the matching of half circle. It is motion that brings about 3D as much as it is motion bringing about gravity Π^2.

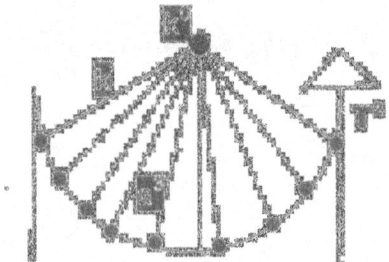

According to Einstein, the speed of light is a constant throughout the Universe. The speed of light results from two factors, being distance (kilometres) and time (seconds). This speed is accepted at 3×10^6 kilometres per second. Scientists know that it takes Sun light 10^6 years to reach the surface of the Sun, and we know the Sun is not thousands of billions of kilometres in diameter. The "Xepted scientific Newton Mistaken" explanation about this fact is that the Sunlight "bounce against matter" and this retard the Sun light all that dramatically. When light hits matter, (except in the case of glass), it joins singularity immediately. Therefore, the Sun holding matter on the inside has to be all- glass, or the "Xepted scientific" explanation is not very scientific at all. It all comes down to the density of matter in space valuing the time in that space away from the point maintaining singularity. Why can nobody but me see that?

The pendulum arm covers a specific distance per time unit, every instant it swings. This is because of Singularity in position a^3 during time T^2 in instant **k**

The space a^3 holds precise accordance to the time T^2 that it takes minus the compromise singularity claims from **k** by reducing space to the increase in heat.

The pendulum not only stops at a precise point that the Earth holds as singularity presenting the singularity the Earth dictates at that given time. The relation there is in what Kepler discovered and that which Galileo discovered has gone by without many (to my knowledge) seeing such an extreme direct link. Kepler formulated space-time and Galileo implemented space-time. The space the pendulum swing through is representative of the space captured by the anchor singularity formulated by Kepler as a^3, the pendulum arm becomes the indicator **k**, and the swing distance of the arm becomes the time T^2. This is the recopy with which man kept time and it is used as he principle method of time indication ever since coming into the light from the dark ages. It is a half circle indicated by a straight line forming two sides where each side is holding a triangle in relevancy. This is in sharp contrast to Newtonian claims that the cyclic repeat the Earth has with the Sun and all the numeral Equalities derived from that by using Kepler's formula still after one year comes to nothing.

It is quite shocking to realise the Science never consulted the pendulum for advise in their search on what forms time as well as what is time. The pendulum is not only synonymous with time but represents time in the mind of every civilised human.

If Galileo's pendulum is a devise used to measure time, then surely it should be possible to see from Galileo's pendulum what time truly is. Galileo's pendulum should be applied for just that purpose and it is worth noticing that Galileo's pendulum uses gravity as an indicator of time. With that in mind, whatever Newton had in mind about gravity, the true test is the pendulum. The pendulum puts eternity in relation to one point in singularity or in relation to infinity and that is time. Moreover, gravity drives the pendulum and from that we can deduct that time and gravity is very closely connected. It is so closely connected that it is the same thing because one cannot have the one without the other. Unfortunately science has not yet given this much thought, or so it seems.

With singularity placed in infinity within the centre of every rotating object every atom and its relation to its surroundings including

Π Π^0 Π

other atoms form space-time diverting from the point holding singularity as far as rotation goes because every object holds three relative positions in as far as where it was, where it is and where it will be in relation to singularity providing time. I elaborate on this else where.

In the action of the inseparable drawing closer and moving closer gravity finds the dual value of linear and circular gravity. There is no separation of the two factors acting as one but both have different application and values in the unit. This is the result of singularity having three parts acting as one but giving three distinctions in application. Gravity is as much part of dismissing space as it is about making contact with space in time. Since the connection comes about as a circle, the connecting points will relate to Π as the value. Due to the spinning nature of such a point with all surrounding the point will be alternating direction favouring change every second and in that the value to such a point can only be Π because of its constant changing. Using r would specifically oppose another r from every angle because the use of r will bring about a static relation to the previous and following instant and therefore it will cancel the constant spin flow.

$\Pi = r$ in constant directional change as time flows through rotation pinpoint positioning of singularity Π^0 with Π positioning space to either side forming the border set by singularity

The time frozen on paper in a single t is effective in remembering the viewer of an event but that is not the event in the present any longer. That was how the event occurred during the time from where the camera shutter opened T_1 to where the camera shutter closed T_2 and the time frame T^2 was then during the open period of the camera shutter. But afterward it represented **t** when looking at the picture and the looking of the picture became an event during a specific T^2 that went from where one is taking the first look to where one is looking away from the paper carrying the first dimensional image of an event gone by and that is at that stage a representation of **t** in another milieu of $a^3 = T^2 k$. The **t** in the single is when mathematically presented as only **t** indicating a mathematical single flat dimensional view of time and is then correctly applied because it represents a reminder of a four dimensional event $a^3 = T^2 k$ that went single dimensional because the moment in the fourth dimension was then frozen in a single dimension on paper while the fourth dimension $a^3 = T^2 k$ soldiered on and time will always be representing T^2 as Kepler stated in the square allocated to space having a cube $a^3 = T^2 k$ at a time even before gravity got a name. But reducing the dimension of time to a single **t** one will find the ability to mathematically design the paper on which the photo Image wlll be printed In time T^2 using space a^3 in the third dimension to apply the ink in the third dimension. Printed on the paper is an image that is not part of space-time while the ink used is space-time and the paper is space-time. The ingredient all hold different and **k** indicating different singularity connecting forming individual as well as group space-time.

In $a^3 = T^2 k$ each factor has an identifiable purpose and role it plays in fore filling a measurable format. One can't go and remove one factor as one pleases. Then the entire Universe becomes nullified, just as Newton foresaw it would happen. The only thing is that Newton claimed it does happen. That is one of several mistakes Newton made!

In $4 \pi^2 a^3 / T^2 = G (m + m_p)$
$a^3 = T^2 k$
$a^3 / k = T^2$ but at the same margin is
$k / a^3 = 1 / T^2$
$k = a^3 / T^2$
$k^0 = a^3 / T^2 / k = $ singularity
$a^3 / T^2 = G (m + m_p) / 4 \pi^2$
and $a^3 / T^2 = k$
then $k = G (m + m_p) / 4 \pi^2$
But I showed that $k = a^3 / T^2$ and Newton's claim is that $a^3 / T^2 = G (m + m_p) / 4 \pi^2$

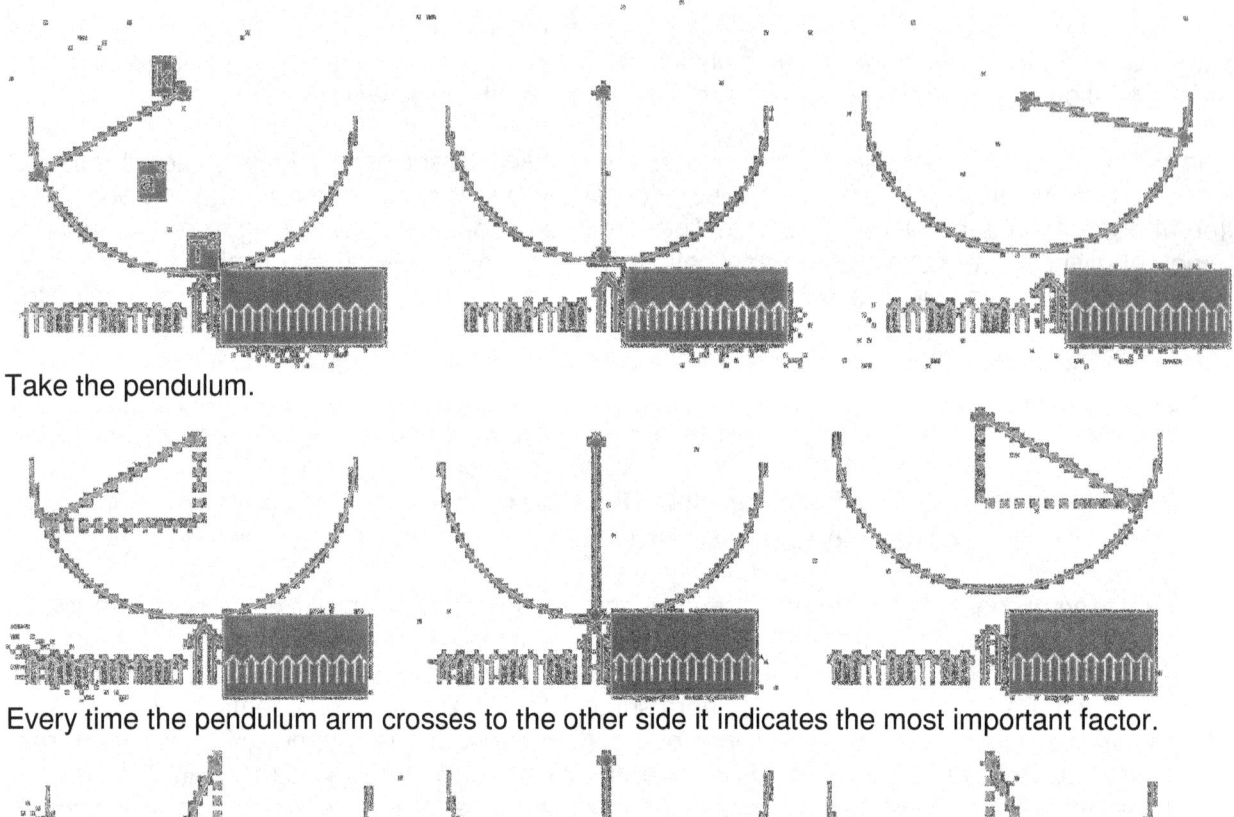

Take the pendulum.

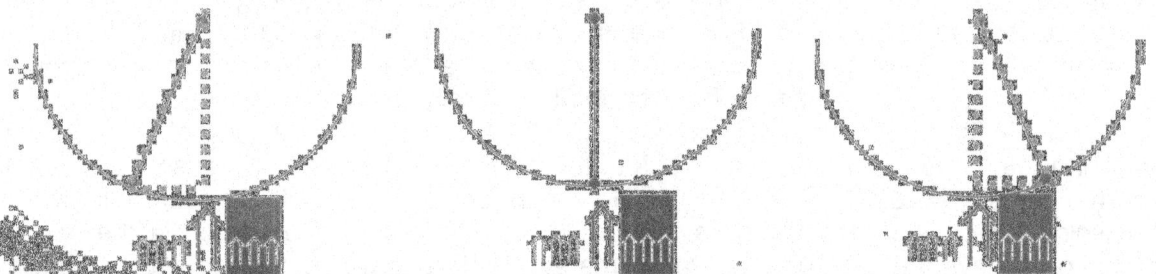

Every time the pendulum arm crosses to the other side it indicates the most important factor.

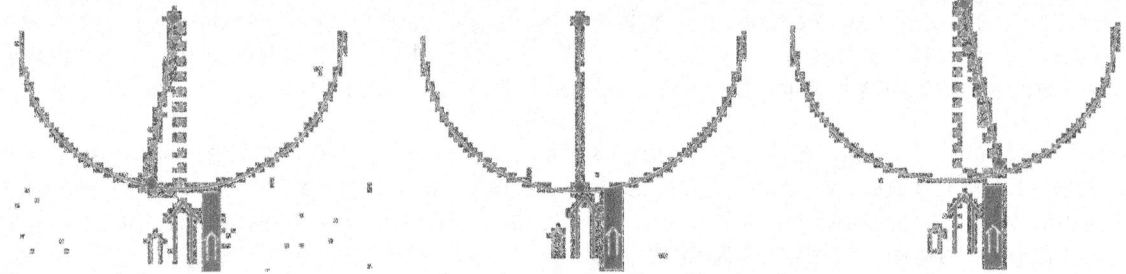

Every swing the pendulum arm does it brakes through the factor holding the Universe in place.

The pendulum not only crosses the singularity the Earth dictates at that given time and the pendulum not only points at the factor maintaining space-time on earth.
The only definite place one will locate zero is in between the starting point of the lines going in opposing direction in the position the lines hold before there was the least of directions applied, but that is only because there is no such a position, not because any line is coming from there. The two lines are still one holding the opportunity of parting as an option but have not yet parted and therefore are on the very precise same spot.

The line coming from there is already there because it already has the choice of going in any and all opposing directions and when it starts running it will place filled space in that location because the space was already filled with a line starting and not with a line not there at all. When reversing a line we might find a better idea of what is in place and where it is in place. Space-time depends on the relevancy of matter occupying space change position in accordance to all other matter relating or relevant or even only influenced by the space a^3 in the k duration of the time the matter changes position T^2 in the instant of changing. **Space-time is everything including singularity diverting from singularity** and that is what Galileo recognised without realising in his observation of the pendulum. Where Π^0 is singularity and Π^1 is the diversion from singularity forming $\Pi \times \Pi = \Pi^2$ being gravity or time.

The single dimension is a dimension covering everything into the dynamic of one. This brings about that **k = 1 a = 1** and **T = 1**. That is the first dimension and the first dimension is a dynamic of one being the result of a dimensional 0. $k^0 = 1$ $a^0 = 1$ and $T^0 = 1$. The factor k was at no stage zero. Only the dimensional factor is 0. The extending that k was capable of was zero. The factor k was never zero. The factor **k** can never indicate zero point from zero or point to zero because just one zero will dump the entire Universe into zero.

Then a moment arrived where **k** developed from k^0 to **k**. This brought about a revolution of cosmic proportions. Matter divided from singularity as matter claimed space. The growth of **k** had to produce a^3 which is an interpretation of the space **k** will bring in place. The tiniest and slightest of growth established **k** coming from k^0 to **k**. By establishing **k** that very second a^3 also came about. But through Kepler we can see how singularity achieved space to come about. It came through spin. Kepler said that $a^3 = T^2 k$. Space broke away from singularity by applying spin. Gravity is spin. When **k** extended it secured a^3 being space but through the spin of T^2 the space separated between particular singularity, to individual singularity. It secured individual space from space by rotating space. Through the rotation came boundaries, which we now consider to be particles in time. It is the time (from T_1 to T_2) that it takes **k** to swing into a different relevance to a^3 space. It established movement in the area holding the least space.

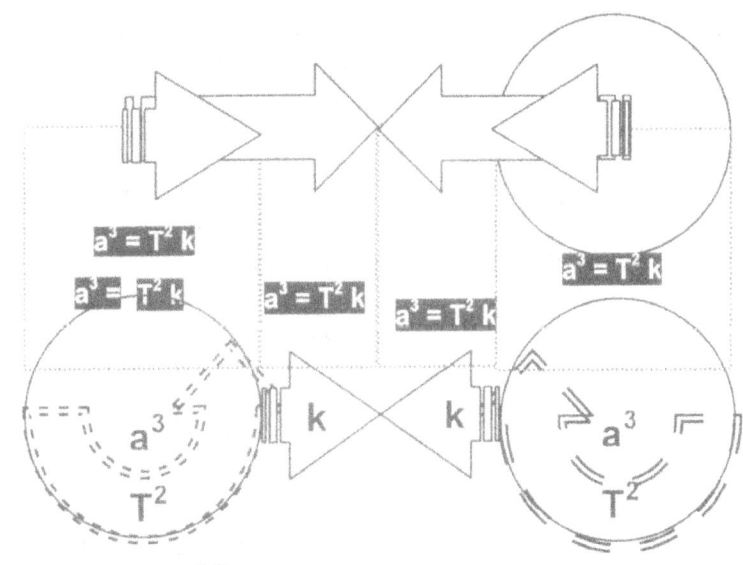

The relevancy started when k claimed space by motion from singularity. That is what Kepler claims. It does say how Kepler incidentally forgot to improvise for the claim of a circle but accidentally forgot about including $4\Pi^2$ on the one side and G (m + mp) on the other side. Kepler placed the growth **k** directly relating to the area a^3 separating as the spin T^2 provides the space. Kepler showed with his formula what gravity is. Gravity is the least space (singularity) **k** claiming space a^3 through spin T^2. Kepler announced space-time, the Hubble constant, the Big Bang theory and all other later cosmic developments.

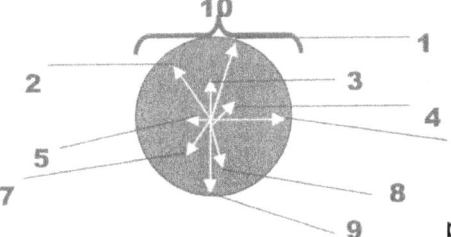

Space developed sectors through time applying differences as singularity changes the Universe from T_1 to T_2 through a^3 and T^2. The factor **k** positions the centre of the Universe and then sets rules applying in that Universe as far as setting space in time by applying space-time.

Space = a^3 and time T^2 coming about from singularity pointed by **k. Space-time $a^3 = T^2 k$**

Space developed sectors through time applying differences as singularity changes the Universe from **k** to **k** through a^3 and T^2. The factor **k** positions the centre of the Universe and then sets rules applying in that Universe as far as setting space in time by applying space-time.

Space = a^3 and time T^2 coming about from singularity pointed by **k. Space-time $a^3 = T^2 k$**

If the Sun held a relevancy of 10 relating to seven with one or two or three of the planets…well yes that might be coincidental, but when it shows such a relation with all of them where all of them includes planetary fragments being between the Earth and Mars making such a coincidental claim on ten structures perfectly distributed is more than ducking the truth. To honestly be honest about the finding of scientific truth and discarding such evidence as coincidental is being unfaithful to you.

While considering yourself as all scientists that should be to be level minded and not thinking you to be a participant of bias but in the same breath blow away such clear evidence should be honestly considered as gross clear dishonesty.

To go and dismiss this certainty as coincidental because it does not fit into a Newtonian Universe is stretching the truth to beyond the accepting norm. One should not try to focus on an image of such a spot or dot because there is no image. The line dividing the cosmos and that run through every particle, no matter how large or small is beyond our vision. Such a small line, so small it is not even noticeable to the cosmos in the 3D is large enough to part the cosmos into sectors. It splits the biggest there is into particles and we are not even able to notice the precise location of such a split. In truth there is no top or bottom that we living in 3D can see. We shall have to use a general conception brought about by intelligence. Your intellect tells you about such a spot, but that is all because that spot is on the other side of the Universe (quite literally). From the centre of the dot there is a top and a bottom spot. From those points there is connection with four quarters. That produces six connecting points that are all aligning to the centre.

In this singularity there can be no sides and without sides there can be no drawing showing the explanation by means of illustrations. Where I do just that, I ask for your forgiveness because being human means I have no capable means of performing an explanation. Yet I am forced to do just that and I have to allow the means of sketches, well knowing the implication that such act is not allowed.

There was the dot. The dot had no borders therefore there was no separation and still we know there were more than one in a group of one. The evidence of this is very present in the cosmos at present and one can find such evidence all around us.

With the cosmos still in a single dimension there were no limits as we know limits to form in the Universe we use and no borders indicating limits because after all it is the single dimension where there is only one dimension holding so much diversity. The borders were part of development because we can witness the legacy of such borders in the present day holding the 3D in place.

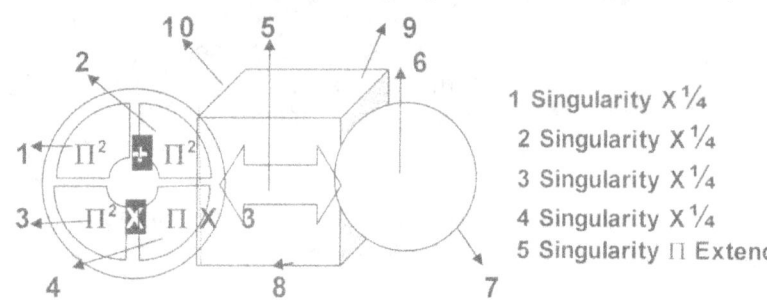

1 Singularity X ¼
2 Singularity X ¼
3 Singularity X ¼
4 Singularity X ¼
5 Singularity ∏ Extend

The overall picture resulted in a ring and all rings hold ∏ to secure the form. The only form that existed then was ∏ and therefore even today the borders use ∏ to indicate positions. But in the single dimension such definitions were far from clear and the only distinctions came from securing singularity in preserving the position of singularity to apply gravity and thereby absorb all anti-gravity. But anti gravity could not control expansion by counter acting contraction through gravity so the overheating continued forming non-existing borders in some thing infinitely solid just as Einstein predicted because this took place before light came about and therefore before the speed of light became part of the cosmos. The cosmos formed a partnership with one side overheating forming antigravity by expanding into space through the applying of the overheating and the other side formed gravity or contracting of space.

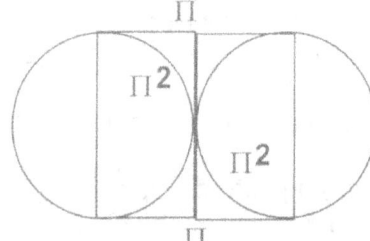

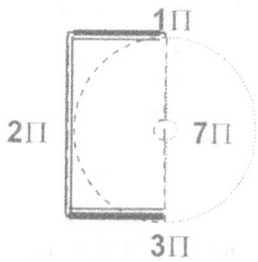

There are 6 points carrying ∏² and one ∏ forming 7 points locating 7∏.
On top of the 7 positions of ∏ included in the sphere there are 3 more ∏ outside the sphere. That makes a total of 10 ∏ or 10 dimensions.

A solid joining by double Π forming as Π²
Taking the sphere as a unit with 7 positions and

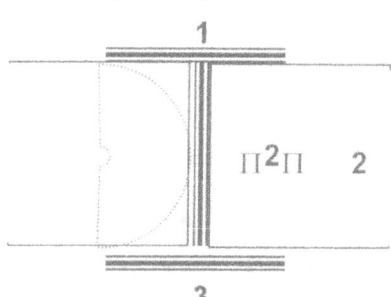

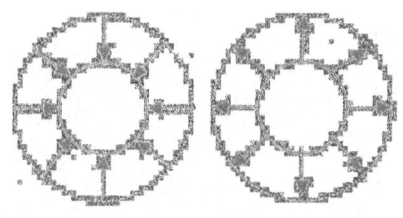

outside the 7 flanks 3
sides in the second
dimension = 10Π

Π²Π   2

Total connecting relevancy of the sphere forming matter
connecting to space = Π²Π 3

There are four locations indicating values inclusive of the

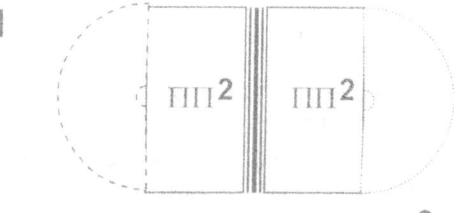

space of the atom. That is the base of gravity in the
atom and in the atom is the base of all gravity.

Gravity is about reducing space, which is all
about cooling heat through motion of occupied
space

A liquid connecting throughΠ²Π

Gravity is also about expanding, which is all about heating. Heating takes up more because of overheating unoccupied space and gravity or movement reduces space.

This says it all and yet every person with a position of influence in science is missing all there is to see in Cosmology! Greatness in Cosmological terms is not in size, but the measure goes by intensity of density and lack of space. A smaller (a^3) result in a larger T^2 where (a^3) is the space the object holds and T^2 is the sizable gravity the cosmic object has. It confirms Kepler and disagrees with Newton. According to Newton Betelguese should be formidable when applying gravity but the Black hole is the true undisputed giant! By taking the diameter, as the measure is clearly no solution in a method to calculate the gravity because it solve not one thing. The circumvent this failing the Academics changed the approached by applying the usual r measuring gravity but instead of having the radius holding the square they brought in the speed of light and further disrupted the truth by applying the square that should fall on the r in the normal calculations to the C that is supposedly there to bolster the gravity.

The speed of light is the worst or best form of antigravity depending on which way one look at matters. Light is the strongest antigravity there can be and to throw that into the Black hole by the square to hide the insufficiency of the methods applied to calculate gravity is once again another cover up to hide Newtonian not functioning in cosmology. By producing C^2 in an attempt to bolster the gravity figure they supposedly are able to calculate in the gravity of a Black hole and placing C^2 in conjunction with R symbolising the radius is just the way not to improve the incorrectness which their theory quit deliberately bring about as a measure to determine gravity.

Then what about the measuring of the gravity in the Neutron star and how will they explain the Neutron star because the Neutron star will either be stronger in gravity than the Black hole which is clearly not the case or it will be pathetically weak. If the Black hole has a diameter of 10 kilometres then the Neutron star has a diameter of twelve kilometres. By using C^2 in the Neutron star the Neutron star suddenly have a larger gravity than the Black hole has. It is either that or the neutron star is so weak it has less gravity than a comet. This ridiculous scenario developing just shows how little mathematicians have any grasp about cosmology. They should keep to building dams or skyscrapers where the mathematics of them is useful and is appreciated and stay out of cosmology. For this remark they will get back at me as they usually do.

Mass is the result of gravity and gravity is not the result of mass. Gravity brings about mass but mass does not produce gravity as Science wish to advocate. Gravity creates mass but mass does not establish gravity.

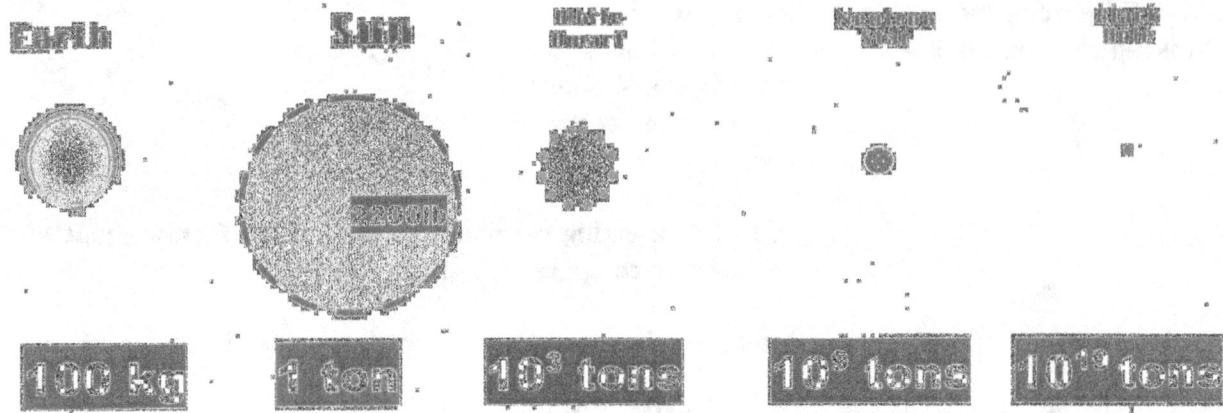

The idea is that gravity pulls material closer and more so in bigger stars. But that gravity pulling can only come from the accumulative effort of every individual atom according to proton mass (number) as a unifying effort of all the atoms in the star in accordance with mass applied. The idea is that mass is the same everywhere and is never changing. Why would there be such huge mass increases in the bigger or should I say smaller stars. What would entice the material inside such stars to grow more massive if mass comes about from the pulling of one particle closer to the next particle. If it was about pulling on each other the mass of the particles could not increase through that. Even by combining the mass of two individual atoms the increase is already in the equation.

By locking the two into a unit should not change the mathematics because 1+1 = 2 whether the two share one unit or two units cannot be any mass increase in a star because all material is within the unit. By fusing the star cannot become more massive using that specific method since it gains no further mass and two hydrogen atoms plus one oxygen atom can at best be equal to if not less massive than one neon atom. The mass cannot grow because the star does not produce mass, if we stick to the accepted views. By fusion protons will only join without further rising the mass they have apart or combined. If the particle has a mass as two units, and the units join in volumetric occupation, they still have the same mass. Some facts about science are too astonishing to be real!

The above facts are part of Accepted Science and accepted facts, but my theory about gravity dismissing space to the advancement of compacting matter is not accepted through all my trying to introduce my ideas to Accepted Science. One hundred pounds in mass will be equal to a mass of one ton in the Sun. One cubic meter being one ton on Earth will hold ten thousand tons of material in a star one class more developed than that what the Sun is. In more developed stars the figures rise above Human comprehension. But it is so clear that the space diminish as the mass becomes denser. By compacting matter the space reduces in the same process. I have been trying for years to get any professor to admit to that and the rest of my theory. Understanding star development must lead to understanding the Big Bang.

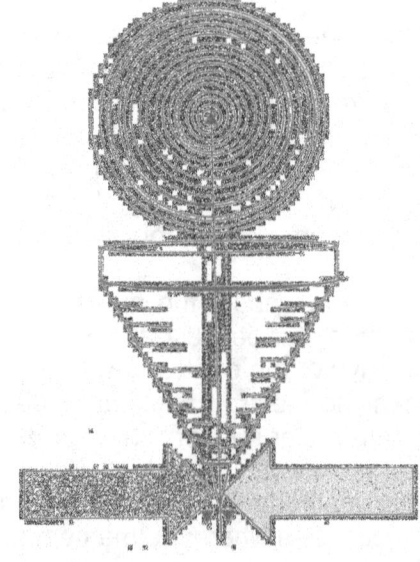

The heat available during the Big Bang explains the lack of space at the time. Space is heat expanded and heat is space contracted. The one is the reverse of the other. To expand is to bring about excess heat and to contract is to produce gravity by eliminating space. The Coanda effect backs my argument. Where gravity is applying the strongest heat is the most and space is the least.

Looking at the affect of gravity it shows the precise quality of no distinctive point, as gravity never seems to end at a point but flows all over affecting all that holds a position in its sphere of influence. The gravity coming from China meets the gravity coming from America at no particular spot but intermingles without distinction.

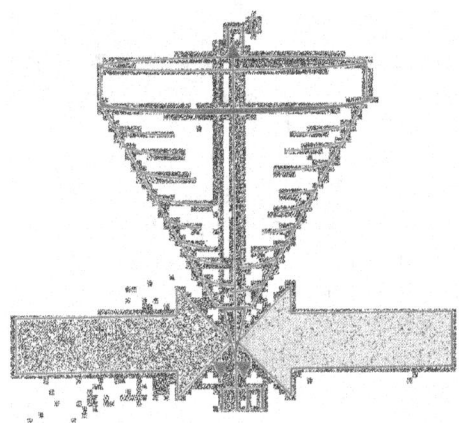

In the sketch, the circle to the right would come about from a straight line r growing influencing the appreciation of Π, but to influence Π would lead to a breakdown in r as Π and r are different entities. The circles to the left shows a continuous growth by extending Π every time and since Π is the same part as the previous Π, only extending that billionth of a millimetre or many times smaller each time, the circle will be truly continuous without any signs of a break.

In the context of dimensions one find coming from the centre Π^0 an established eternal flanking of Π to six positions since Π^0 forms the centre to the six sides and all six sides not having a diameter yet must apply Π to indicate specific value.

What I try to say in this elaborated effort is that where **k** for there very first time extended from Π^0 to the edge of 3D where Π begins it had a certain distance. We humans are incapable of ever measuring that extending growth but be as it may it is there. Such expanding is beyond is beyond human thought but it still had enough value to raise a Universe. By allowing this expanding of **k** personified as Π to continue uninterrupted as one flow of a continuing growth of Π a line of ever so small but still productive will follow one line upon another line producing a cover of the full area of $\mathbf{a}^3$ the time factor $\mathbf{T}^2$ or for that matter gravity Π^2 will be flowing constantly through out $\mathbf{a}^3$. The factor $\mathbf{a}^3$ will be improvised by the singularity measure of Π^3 and the time factors $\mathbf{T}^2$ and **k** will then be in singularity terms $\Pi^2\Pi$.

In the very centre, which I am referring too, rotation must end or start depending from what vantage point the relevance is placed.

The very centre form an eternal divide that will not allow what is on the one side to present an influence on the other side. It divides spin. It divides direction of spin. It divides all rotation from the outside that one may detect and such divide is there because at one point spin will run to the left coming from the right and just immediately next to that point must run a direction from left to right.

It cuts without contributing or participating in movement. It divides without any favour.

That is singularity not having a dimension of space and not having a dimension of time, or a radius connecting the rotating distance to Π. It is the point where $\mathbf{T}^2$ breaks down and $\mathbf{a}^3$ stops to exist. It is a point in all-rotating objects. Every rotating object holds a centre from where the rest of the rotating direction will differ at any and all given points. Not one point is exactly the same, but in the very middle, the centre no one can draw, measure or see is a point not in motion.

In the centre runs an axis line that forms the division of rotation. No one human will ever be able to indicate the precise line, but such a line must exist because of our logic telling us about such a line. In the centre one will always find one more line smaller than the outside but forever also always bigger as it is towards the inside.

This is what Kepler referred too when he revealed to the world what the cosmos revealed to him. This shows singularity being the only value that singularity could ever have…and that is 1. Having 1^0 is singularity and whether singularity is expressed as k^0 or is expressed as Π^0 or as 1^0, it will always indicate singularity is a measured value. Einstein can have as many bright theories about singularity, but the only value that can express singularity is having an exponential indicator of zero. It is not zero as Newton had it and it is nothing as marvellous as Einstein had it but it does form the basis of everything valid.

Why would a water drop floating in a space capsule in space, in micro gravity always form a sphere when left capture free form? We all accept that the true cosmic form would be and most probably will be the sphere…but why would the sphere form as the original form when matter is not pre-cast to have any specific form and therefore take on by cosmic pre-cast the sphere as form? We know gravity is there, but qualifying gravity as a force lets the process of investigating science a bit off the hook. Thing become rather simplistic in the modern age when all else is so highly investigated but gravity is merely defined as a force influencing matter. Why would we find in space, where there is supposedly nothing, something we named micro gravity and would bring about micro gravity when gravity is not present? What would cause gravity up to one point and from such a point in the area there supposedly is nothing there is micro gravity. By extending k froze to set the standard for heat. By freezing k came about and set the Universe into a concept other that one unified lump of not being anything of sorts. The freezing action brought on **k** and **k** brought on space through spin. By the extending off **k** did the Universe obtain a^3 / T^2. But it had to produce **k** by freezing **k** into existing from where k produced a^3 / T^2. This we have to understand about fusion. Fusion is about creating a freeze in the deepest of heat and where all is engulfing in heat surrounding all, from that freezing must come about to apply fusing. It is this first action ever that has to repeat to establish fusion. It is in gravity attempting to secure the most heat under the prevailing conditions where gravity eliminates the most space to establish a freezing centre in creating fusion. But that does not solve the indicating action. How did the Universe liberate material and heat/ space from singularity because with singularity comes eternal non-changing-everlasting in conditions remaining in absolute equilibrium. This equilibrium maintains because all development extends form equal equilibrium through out. What evoked change and that is the question the Atheist will never answer but that too is the most basic and ever-lasting fundamentals of the Universe. This split second start before the start, but there are other books I delve into this matter. From the deep freeze came the Hot Big Bang bringing about to Universal displacement relatives being $10 \div 7(4(\pi^2 + \pi^2)) = 112.795$ ands then a second one established 3D by introducing the six to seven sides Universe at a density point of measuring $7/10\pi^6/6 = 112.162$.

There is of course a lot more about this than what I mention at this point. But what made the Universe freeze to form the Universe in space and through time. It had to start with a specific reason applying. Once the process started there was no stop to it, but there is no chance that the initiation of the start was spontaneous by nature. With $k^0 = a^0/ T^0$ all stood still in singularity and that factor is still with us controlling and generating the cosmos. It is there for all not to see and for every one to establish. The effect coming from $k^0 = a^0/ T^0$ and about $k^0 = a^0/ T^0$ is beyond denial. That centre spot in the rotating of objects is up to this point of cosmic development as incapable of having space-time and can only secure space-time as it was capable of at the start and as it will be capable of   up to the very end. In only $k^0 = a^0/ T^0$ being present it was as stable as it still proves its stability. There were something external from the Universe that was outside controlling the $k^0 = a^0/ T^0$ Universe that set the lot into space being motion. That is the one aspect natural physics will never answer. What forced the cosmos out of the stability of singularity as $k^0 = a^0/T^0$. There are two controlling cosmic principles it control of all. The one is gravity and the other is the light performing as the example of the epitome of antigravity.

There has never been any explanation offered about gravity being able to bend light except mathematical calculations producing the factor prevailing and the principle produced. Einstein declared that large objects producing massive gravity could bend light. Human abilities are meaningless in the totality of the cosmos therefore what we perceive as massive and what we perceive as nothing is in cosmic standard amounting to about the same. It is our norm we create bringing on our true Human incompetence in realising limits prevailing in cosmic standards. If large gravity can bend light all gravity can bend light be affecting the flow of light. Gravity does not bend light because light is not a solid that can bend. By dismissing the space through which light travels such dismissing must lead to rerouting or redirecting the path light will displace space-time. The closer the light travel or pass the centre core of any object the closer it comes to the main gravity within the structure and that alone will bring on a greater effect the more the space conforms to density. Gravity is bringing space in relation with time through applied motion. It is about space becoming reduced or redundant because of motion either by space moving or by matter moving. The duration of the time in relation to the space affected bring about the gravity applied. Gravity is space in relation to the centre while at the same time it is the centre coming in new relevancies to space surrounding the material structure. There are always two components to gravity in space being $a^3 = k\,T^2$ It is k in the way the centre stands in the space changing in relation to a centre a^3 but at the same time it is a centre T^2 changing relation of that centre to another centre k applying control. It is a constant re-matching prevailing centre influences on space occupied by material and space not occupied directly by material. It is the time the space takes to bring about new positions in space occupied by motion and through motion that takes a certain duration time to move from point to point. Gravity is motion of space towards a centre and the time such motion takes while that centre is attempting motion away from a controlling centre and the time it takes to complete such attempt. Gravity is speed and speed in space in motion through time duration. Light is the attempt to establish motion not controlled by any centre and the time it takes to establish space between such a centre of space control and the light finding an ability to dispense the space by reactive motion.
Gravity produced space by allowing as much as producing overheating. But what is heat if there is no cold to set the standard for heat become the other end. How did singularity have parted principle of unifying? Singularity froze in applying gravity bringing about particle separation within singularity.
In the investigation of light and gravity and objects and gravity, the mathematical rule of the invert square law must apply without question. But according to the observation of Roche that is not the case. From what one gather through the Roche limit implicating two orbiting structure the opposite is applying. One must accept that although k proves as an indicator it is also much more when complying the thin influences brought about by singularity in the values carried on by singularity.

As a^3 increases, so does T^2 as well as k increase and with that the influence of gravity per space unit increases with the concentration demise of a^3. But why would that be and what are we missing? Light shows there is an influence out there in outer space, that redirects light's route through space when passing large gravity fields. It is about the relevancy of k influencing the a^3 to allow the T^2 of light to divert in route because of influences established by k on a^3 and slowing down or increasing the line diverting. In this measure one may also find the Roche limit applying, but to truly understand how the Roche limit comes in place and how the Roche limit work one have to replace Kepler's factors with singularity and singularity extending being Π^3 $\Pi^2\Pi$ and **3**.

There cannot be light without gravity and there cannot be gravity without light but that is on condition that the light we perceive to be light is only the symptom of what the Universe use as light confirmed. Light is not what we think light is but that explaining will too extended to explain in this part. Light is heat concentrated to a form almost material and space is heat dismissed. Therefore light, heat and space is the very same thing and all form the extending of antigravity applying. But gravity is more the redeeming and the re cooping of light where light then is antigravity. Light is space concentrated in motion at speed and gravity is reforming space by motion forming speed. Gravity is speed and light is space at speed. In drawing a most basic picture of light passing the gravity lines extending from any structure, I felt it was most insightful that the brains in cosmology was not able to see why light does *not bend* in the presence of increasing gravity. More surprising was that I found the mathematicians had to call on Einstein for advise on a most ordinary problem. Light does *not bend* when passing large objects. It is Kepler's formula applying, and the evidence is clearly in front of the searcher for truth. But one has to go back to Kepler to re-apply what Kepler formulised and change the significant from Newton's significance.

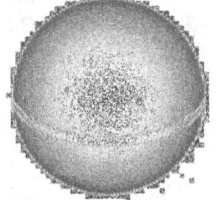

The sphere holds six sides in relation to form as unit and as does all other shapes and forms. All forms have to have at least six sides indicating different exposures to the Universe. But with gravity having a free choice, gravity always chooses the sphere. As I shall prove later on gravity is the strongest where the form produces the least evenly distributed space. The first condition for gravity is even-handedness through out the sphere holding the applying gravity and the second is to have most or the strongest gravity located where the space is least. That gravity then

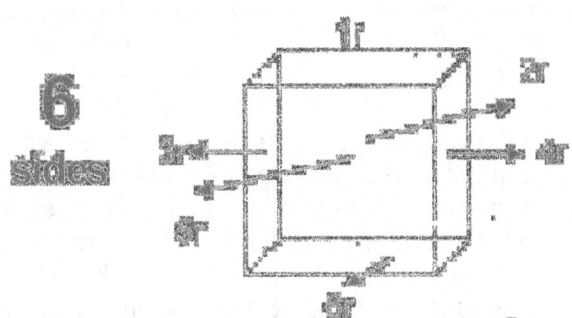

has a position in the very centre of the sphere and from that centre the gravity produces all the edges or borders that the sphere consist of. In the case of the sphere this factor makes the sphere much more dominating than any other form does. From the centre point controlling all sides is gravity and with gravity applying control the sphere has seven sides to the square in any other possible form having at least six sides.

The cube can come in whatever form there may be but the sphere adheres to precise measure and behind this principle is all that forms the Universe. The cube has six sides connected loosely and can change form just by changing the relevancy between one side (or more) in relation to the distance brought about by the other sides. The sphere being a complex circle stands related where the sides has to apply precise measure in equality. This becomes a law because in the precise middle one will find the strongest gravity as that gravity holds the object in form and true to form. If there is even gravity spread in all directions the form must be a sphere and the sphere insist on seven points relating to sides or borders.

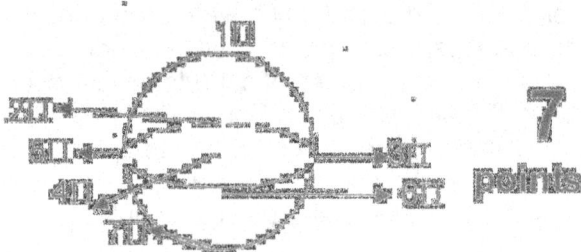

The understanding of this involves every aspect relating to the cosmos. This is the principle unlocking the four cosmic principles that are responsible for gravity and gravity is time.

In the sphere, which I am referring to there can be no radius but only the extending of Π from the centre Π in six opposing directions relating to one another by the square but remaining Π because of the unity the matter holds in relating to space. In the event of using r the individuality of r as a concept will bring about a specific ring at any specific point where every such a ring will define every time various on going circles. One cannot use such a definitive line because such a line will have to cut through atoms at some points. That is not the effect one get from gravity. Gravity includes and joins all aspects within the field but such definitive line will have to exclude some and include some because of the definite points and rings developing. The gravity influence we are in search of is something like a woven cloth covering the area and covering all in the area. It is like a silk blanket covering all the aspects.

That means the value Π is running into and past the entire surface as if the surface is one consistency without different particles. That is what motivated me to look for gravity as being heat because everything consists of heat and if gravity is condensing heat gravity then will be all including. It is not possible to draw a precise line that would form a precise ring and not cut some atoms in parts. Because where r is used there will always be an atom disallowing the precise positioning of the circle the circle continues on a solid basis holding Π as a positional reference and not r. In every sphere there then are the seven Π relating in precise dimensional and positional equality forming equilibrium to the centre Π as well as to one another by 90^0 and 180^0 implicating the dimensional positioning. Therefore the sphere holds $7_{by}{}^{\Pi}$ and the cube holds $6 \times r^2$. I use the symbol r to define the idea of a line weather the line might be used as a radius or to be indicating length breadth or width I do that to avoid the pitfall of using names to hide truths.

Where space comes into contact with the sphere the cube loses one of the six dimensions it has to the more dominating seven dimension of the sphere whereby the seven dimension in equilibrium will dominate the six dimension loosely connected by r bringing about that the cube then has 5 sides to the seven of the cube. Because the space surrounding the sphere takes on the shape of the sphere and not the other way round where the sphere resolves in accepting the form of the cube, one may presume the form of the sphere is the most dominant of the two choices.

In the centre of a sphere there is a definitive position where the strongest influence of gravity is located. It is the centre of the sphere where the space is the least. From that centre point gravity extends in keeping the edges of the sphere perfectly true to the form singularity has being Π in every aspect. But also such extending continues beyond the specific edges of the sphere as it influences the space surrounding the sphere. We know this by another name given to divert every one from realising no one has a clue why that phenomenon is applying. We call it the Coanda effect where the process shows that even liquids submit to change of form.

The condition for such a submission is that singularity carrying Π is in place and committing the influenced space. Only by applying singularity through using Π does the Coanda effect apply. With the sphere being defined by singularity in the centre as to confirm the point holding singularity with the use of Π as form, the sphere is influential enough to remove one side of the square cube, which is, loosely connecting sides to form the cube. The cube is very loosely formed and the sphere is very defiantly controlled by singularity therefore the sphere is able to dominate the space in the cube by removing the nearest side. With the removing of the side the cube in form loose one supporting dimension and therefore will not be able to secure what is in the sphere to the form of the sphere.

That is gravity.
It is reforming space to the requirements of singularity.
Gravity is The Roche limit,
   Gravity is The Lagrangian system
      Gravity is The Titius Bode law
         Gravity is The Coanda affect
Now we have located the allocated point from where gravity is controls the cosmos I now will take you on the earlier promised tour where this tour will lead you to gravity and the four phenomena behind gravity. It is about numbers...it is about form. Where we are heading space is so small that one dot represents an entire University coming into space. Please do not think of the cosmos as you would see the cosmos from your normal vantage point. Where we are heading is a place without space becoming space and it is allocated where you can only enter by intellect because only the mind's eye can go that small. It is where space is built block by block and that space is responsible for the expansion we witness as the Hubble expansion.

The location where we find the points I am referring to is so small that at that point the straight line, the half circle and the triangle is the precise same thing while holding the same value. Only direction is possible since space is still forming by movement. It is where singularity Π^0 meets space Π for the very first time and there can still be no radius r because forming a line requires space and space is not yet available.

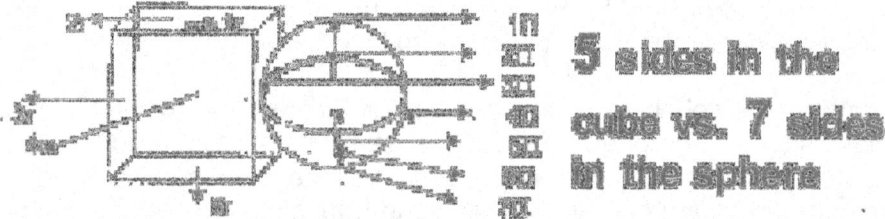

This means that in the cube at the point of contact between the cube and the sphere the cube experience such a contact point as if the "bottom falls out" of the cube and without a "bottom" to support objects they fall to the sphere as objects does fall to the earth. Remember that a body "floats" in space, but at one specific point it starts to "fall" to the earth. That is gravity and it is a dimension change much more than any force.

I shall explain this last remark later on.

That too is the Lagrangian system with five cosmic structures holding relevancy to the centre structure where the centre structure stands in for seven positions diverting from singularity and the orbiting structures standing in for five positions in space.

Gravity has its organ in movement brought about by for differences moving in relation to each other. Gravity is all to do with dimensional changing and reforming of forms to re-affirm alliances supporting singularity. It is the reforming of space converting space to more concentrated heat.

The Universe is in the three dimensions using twelve dimensions that is visible to us and indefinite number of stages in size differences ranging from the immeasurable small to the immeasurable large where mathematics become a short fall to the next and the previous dimension.

There are always 10 in positions running smaller and running larger. Even to us thinking we are at the edge, because we use light as an information source will find that the Universe are infinitely bigger than what we see and infinitely smaller than what we see.

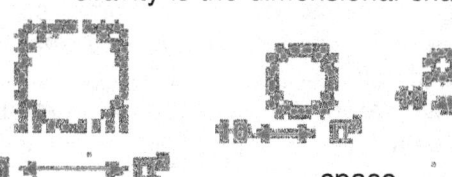

Gravity is the dimensional change of space taking space from 10 to Π^2. This happens by means of applying the Titius Bode configuration of space adapting form through the seven dimensions interlinking ten dimensions to reform the concentration of the space to heat.

The result of the five to one relation comes partly with the Titius Bode principle.

From this line of reasoning I dismissed the theory of the presence of a force being gravity but rather consider it as a dimensional changing contributed by the spin of the Earth and the spin comes from singularity located in the centre of the earth.

It is all about dimensional changing that influences space as a factor of ten to reduce to Π^2 on a continual basis from point forming new dimensions through billions of such points.

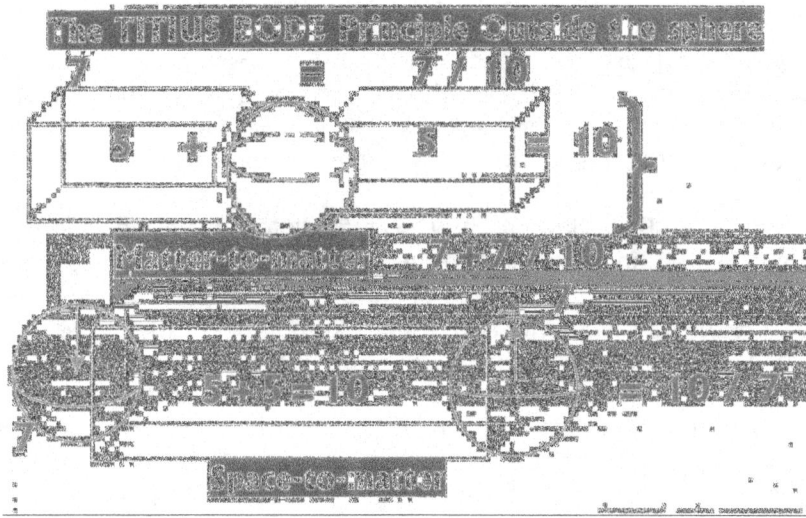

Space-time is a four dimensional position of the Universe where the position of an object is specified by three coordinates in space and one position in time. This evidence we find as matter grew into the dimension we now share with billions of stars in the cosmos.

With the dimensional change from space in the cube to space in the sphere a relation of 5 to 7 comes about depicting gravity on one side of the divided Universe. The principle of 5 sides in space relating to 7 in the sphere holding matter forms the basis of the Titius Bode and the Lagrangian principles.

The German mathematician and astronomer BODE, JOHANN ELERT (1747 – 1826), published a formula in 1772, now known as Bode's law, which yielded the approximate distances of the six known planets, from which he predicted the existence between Mars and Jupiter of an undiscovered planet. His major publication was Uranographia (1801), a comprehensive atlas of the entire sky showing over 17 000 stars and nebulae. For fifty years, he oversaw the publication of astronomical data in the Berlin Academy's yearbook.

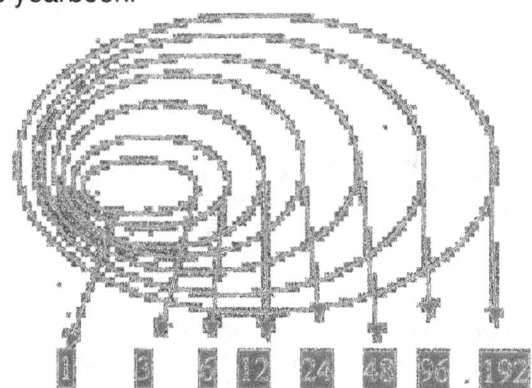

Planet	Mercury	Venus	Earth	Mars	Ceres	Jupiter	Saturn	Uranus
Bode's Law dist.	4	7	10	16	28	52	100	196
Actual dist.	3.9	7.2	10	15.2	28	52	95	192

BODE'S LAW

A numerical sequence announced by J.E. Bode in 1772, which matches the distances from the Sun of the six planets then known. It is also known as the Titus-Bode law, as it was first pointed out by the German mathematician Johann Daniel Titus (1729-96) in 1766. It is formed from the sequence

0, 3, 6, 12, 24, 48, 96, and 192 by adding 4 to each number. The planets were seen to fit this sequence quite well – as did Uranus, discovered in 1781. However, Neptune and Pluto do not conform to the 'law'. Bode's law stimulated the search for a planet orbiting between Mars and Jupiter that led to the discovery of the first asteroids. It is often said that the law has no theoretical basis, but it does show how orbital resonance can lead to commensurability.

It is also known as the Titus-Bode law, as it was first pointed out by the German mathematician Johann Daniel Titius (1729-96) in 1766. It is formed from the sequence 0,3,6,12,24,48,96, and 192 by adding 4 to each number. The planets were seen to fit this sequence quite well – as did Uranus, discovered in 1781. However, Neptune and Pluto do not conform to the 'law'. Bode's Law stimulated the search for a planet orbiting between Mars and Jupiter that led to the discovery of the first asteroids. It is often said that the law has no theoretical basis, but it does show how orbital resonance can lead to commensurability. The importance that becomes known is the sequence the Titius Bode law saw in the number arrangement of 3; 6; 12; 24; 48; 96 etc. The incorrect application of the Titus Bode law lies in subtracting the figure of 3 from 10 leaving 7. The other way of reasoning is to add four each time to the firs value of three starting with 3 and so on. The true significance of the Titus-Bode law is that it points directly to a circular growth of 7 stages. The 7 relating to 10 is a precise derogative of the Roche limit or the Roche limit is a precise derogative of the Titius Bode principle because the two systems interlink.

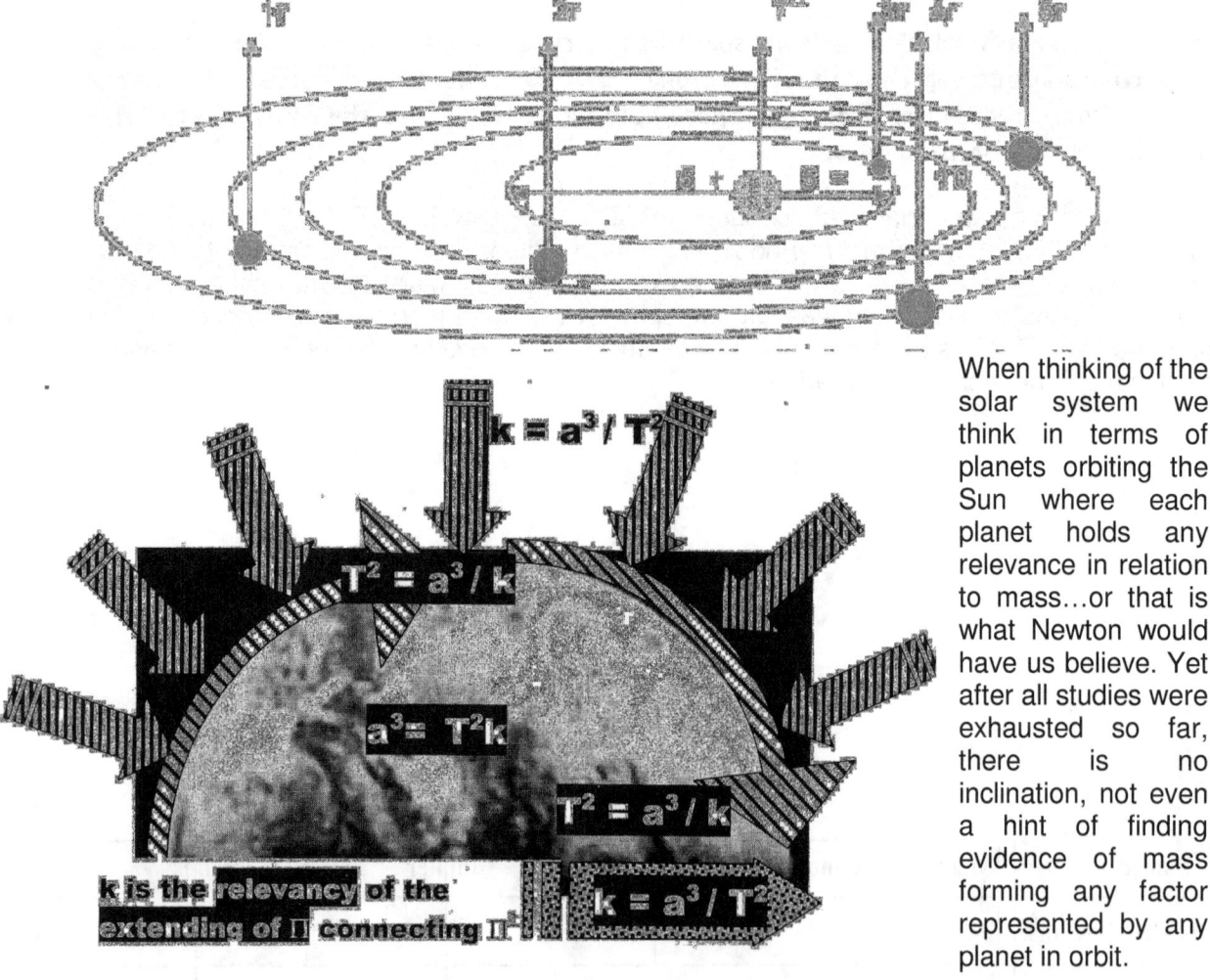

$$k = a^3 / T^2$$

$$T^2 = a^3 / k$$

$$a^3 = T^2 k$$

$$T^2 = a^3 / k$$

k is the relevancy of the extending of Π connecting Π

$$k = a^3 / T^2$$

When thinking of the solar system we think in terms of planets orbiting the Sun where each planet holds any relevance in relation to mass...or that is what Newton would have us believe. Yet after all studies were exhausted so far, there is no inclination, not even a hint of finding evidence of mass forming any factor represented by any planet in orbit.

The smallest two planets are both on the outside and on the inside. The densest planets is those which we would normally consider as having far more gravity if gravity was a pulling force, or how else could gravity compress the planets so solid? But with them in that respect being dense is meaningless because the densest of the lot are also the smallest of the lot. Then if one takes size as a measuring tool we find they are completely scattered in random. Not one would adhere to mass placing them in any specific order in relation to mass presenting prominence. Mass has no influence in any way at all in relation to the solar system evidence and it is most obvious that with Newton using mass as an indicator of value, then using mass in this manner made mass wishful thinking. No matter from which angle one approaches the views Newton takes on gravity, there is just no evidence in support of Newton's claims. However, there are cosmic phenomena that show a strong indication that there are influences which are very prominent.

…But what has the Titius Bode law got to do with the price of onions you might ask…well it is the way the space builds in relation to time. The relevancy of **k** is extending T^2 or then gravity as Π^2 and that is precisely the role that the Titius Bode law plays. The question we should ask is not what mass any object has, but why would the object find footing on the ground when movement would suggest all objects continue to move in a straight line. Newton said a body remains in a uniform state of rest or motion until a force is acted on it. That is Newton's first law and that is hogwash. Nothing in the Universe can ever stand still and everything has to constantly move. Every movement is following a straight line as much as it is going in a circle formation $a^3 = T^2k$ and that is the true law of Kepler. Everything is moving constantly and no force is ever acting in on movement by accelerating or decelerating except when life as a forceful influence may interfere with some aspects on Earth or its immediate surrounding in the most incredible minute way. Life is the only force and the Earth is the only place this force are located, notwithstanding the atheistic corruption of the truth and the mental instability of Newtonians creating make belief fairy tale science. Then further more, everything moving is moving by implementing the Titius Bode law of establishing space-time to the value of time as Π.

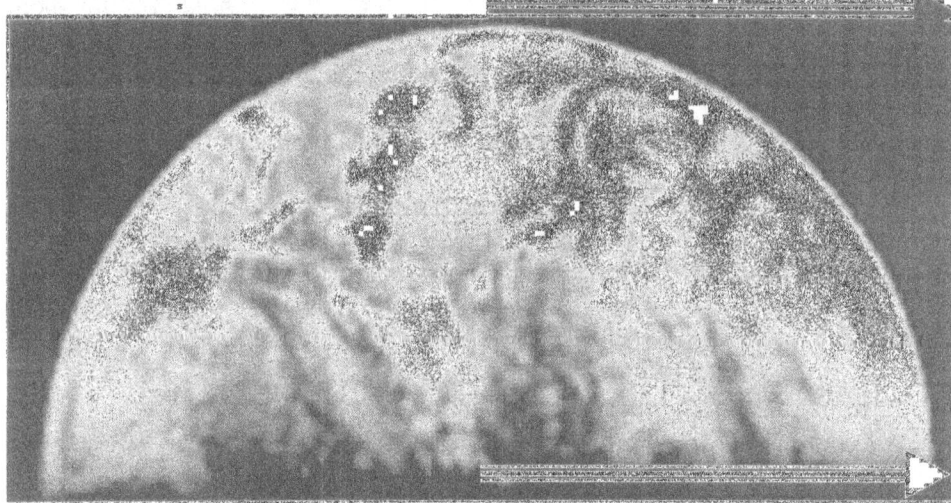

When an object with "mass" is moving it is standing still in relation to everything else on Earth. However, it is moving with the Earth. What does this statement entail? If it moved, it moved in a straight line. The object moved from one point to another point. Take my personal view I have in relation to my movement as a

typical example and I would have the idea that while I am standing still being motionless I am

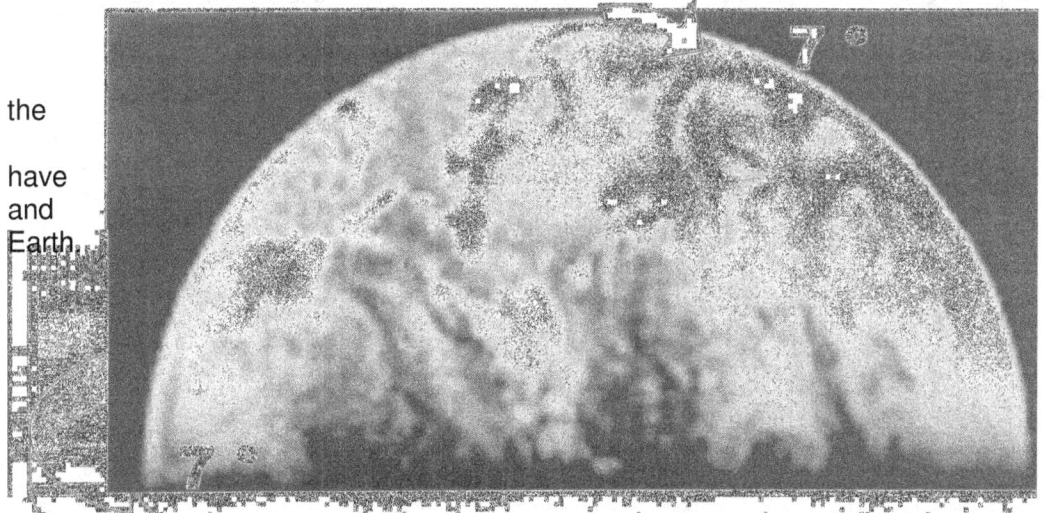

moving n a straight line because of perspective direction I about myself about the

the

have and Earth

Looking straight ahead towards the direction in which I am travelling I can see the sky coming

and then I follow the straight line that the Earth is heading while moving with the Earth. From my point of view it is clear that I am going in a direction straight ahead.

From my point of view the Earth is going straight and the Earth is taking me along as I am going straight also. This might be my human perception but my perception as a human totally fails me in a cosmic sense.

I might think I am going straight but in fact I am following the curvature of the Earth and in doing that I am redirecting my direction that I am going by seven degrees every time I move. I might think that I am going straight, but my travelling redirects seven degrees And Not only that, but I redirect another seven degrees at the same time. The Earth spins around its axis and the directional change coming from this spinning brings about seven degrees alteration to the direction that I seem to travel. However in the same event I am following the Earth spinning around the axis of the Sun and this brings about another seven degrees of change in direction.

If any object moved forward in relation to the Earth spinning around its axis or for that matter around the axis of the Sun, it would follow a straight line and "fall of the Earth". That it does not do so how does it remain attached to the Earth while the Earth is redirecting the object's travel and in that secures the object to the Earth.

One thing anyone can be sure of is that the magic of mystical powers of attraction that mass supposedly should display is as much fiction that sprang from Newton's mind as what the Emperor's magical clothes was that only the wise could see. There is no mass having unprovoked fits of attacks by compulsive attracting disorder and only the wanted-to-be-wise and other Newtonians were able to see the Emperor's clothes (or was it the magic of mass?)

There is something securing the object that is standing unattached or loose from the Earth's crust. Something is securing the object having mass to the Earth and we are taxed with the task to find out why this happens.

Travelling normally should lead to any loose object falling off the Earth because the Earth detours from its original direction and there is no reason why any object standing on Earth should follow this detour. Yet the truth is that the object changes direction at an angle of seven degrees just as the spinning Earth does. It means that going straight is diverting the direction of the course by seven degrees. There is never going straight just as there never can be standing still or never moving. This I have to acclimate: going ahead is changing direction by seven degrees. This is what gravity is …it is moving forward by the margin of changing direction to the value of seven. One may test the accuracy to find support in the results. The object being connected to the Earth has a direction change of 7 while the object holds a relative position in relation to singularity at a value of Π and all this is in ratio with gravity being Π^2.

This means that a body on Earth travels at $7\Pi\Pi^2 = 217$ km / h. If the object moves beyond $7\Pi\Pi^2 = 217$ km / h, at such a speed the object will begin to "fall off the Earth" or using other words the object will get "airborne" and start to "fly" That state of affairs has nothing to do with "air that is flowing underneath the car" and Lofting the car off the ground. The lifting at a speed of $7\Pi\Pi^2 = 217$ km / h is all about the limits of gravity applying to movement. To maintain movement in relation to the curving of the Earth the object has to maintain a speed below and not exceed $7\Pi\Pi^2 = 217$ km / h or the vehicle will lift from the Earth. This has nothing to do with air flowing under the car but is in line with the effect t gravity leaves.

This is about any object moving faster than $7\Pi\Pi^2 = 217$ km / h, which at that point will start to straiten the Earth's curvature and go straight as it no longer deems to have a poison in accordance with the time that the Earth dictates. By going for instance $7\Pi\Pi^2 X 2\Pi^0 = 434$ the object / vehicle / aircraft would be travelling double the speed gravity requires and the vehicle then would be an aircraft in the air.

By moving from 1^0 to 1^1 and from $1^0\Pi^0$ to $1^1\Pi$ requires space. Yet, when form came into form such moving did not leave the realm or the domain of singularity. That is still with us since the principle has nowhere to go but to remain in the Universe. The motion brought about Π as the motion brought about Π^2 using the same motion. It is the motion that moved 1^0 from 1^1 or $1^0\Pi^0$ to $1^1\Pi$ that became time in the square and the motion including time became space $1^0\Pi^1\Pi^2 = \Pi^3$

All this is happening while the crossing is all concerning singularity moving from one sector of singularity to the other sector of singularity which is $(\Pi/2) X (\Pi/2)$.

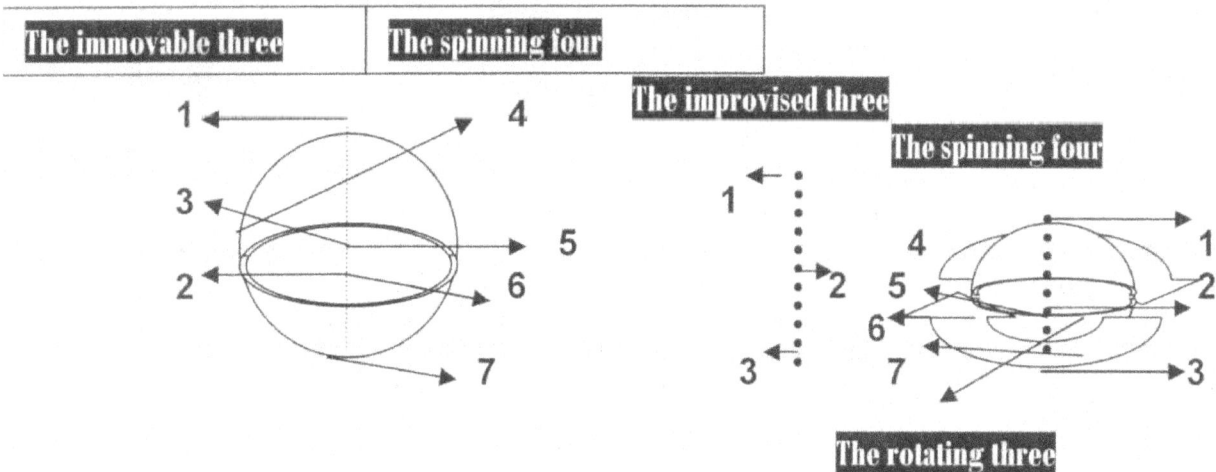

The relevancy forms part of the duplicating and dismissing displacement of space-time we call gravity. In that, we are looking at relevancies and no precise specifics. However, the Universe was built block by block in this manner. As it  was but is no longer only form that applies in the Universe but concrete measurements also come into play therefore even the relevancies may apply in different relations as they switch over to compensate for other factors alternating as they are coming into prominence. The lesser developing sphere orbits the 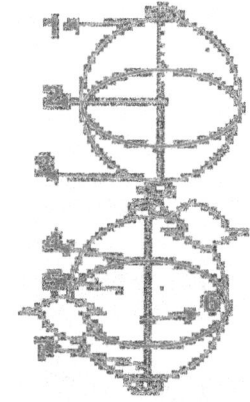 dominating sphere and between them, there are definitive relevancies. The centre circle singularity line of three is unaffected by spin which I shall call the immovable three. However, the immovable three holds such a stout position as far as the centre sphere is concerned. In relation to the orbiting circle the centre line is part of the building and destructing process that manifests as duplication as the centre singularity maintain domination and control over the orbiting structure. In addition, it has a major part in the motion building of the sphere in orbit and moreover building by generating the singularity line that generates the lesser and the orbiting sphere. In that relation the centre sphere reflects the centre line to serve the orbiting sphere by supplying the reference needed to establish motion in the orbiting sphere as one Unit. In that there is an undisputable reference of seven orbiting the centre and the centre providing three as a reflection of the seven which in all accounts for ten relating to the four which also is spinning as time and in total forms the seven taken in relation from the orbiting ten.

By not having motion the lines also have no space as the space extends to form space forms space and the line includes serving the three points to the outside. Where there is no motion, there is no space and where there is little motion, there is little space. The only space the line may relate to can be a point that is on the border of the sphere that is crossing singularity and connecting the two edges on either side of the sphere that is forming the sphere. That means the line from one point holding singularity to another point holding singularity that line will cross the centre line which gives the line in singularity valid space-time to control. Singularity does not have the ability of motion therefore singularity does not hold space. Singularity is also eternally indifferent to motion and motion can excite singularity but singularity cannot be shifted by motion. Three points form the line.

It is these three points coming from singularity as an extension to singularity that forms the line of time. Again I wish to remind that time is gravity is movement and while the top is motionless on the ground, the top adheres to the time restricted by the Earth. Once the top starts to spin, the activating of this centre line forming singularity gives the top a time line of self worth.

In everything that spins there has to be a line holding a centre that protects a point holding singularity. That is one point. Then there has to be a point at the top of this centre line and a point at the bottom of this centre line that connects to the centre. This form point two and point three. The line that forms can never move because that line forms with infinity that associates with time forming. This line holding three points connects by rotation to four points that has to spin. The four points also associate with time to the value of $4(\Pi^0)\Pi^2$. That holds time that connects the four points $4(\Pi^0)$ to rotation. This forms the movement aspect of time.

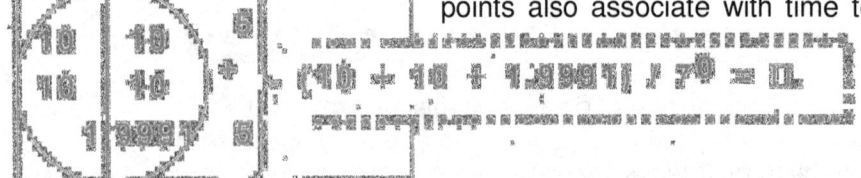

This value of four plus three points totalling seven is the seven that form the one part of the value of Π where Π is 21.9991 / 7. The value of Π is 21.9991 / 7 and there is a specific well defined reason why this value is specifically that.

The seven is the value of the rotation redirection that the straight line follows when it diverts the straight line into the circle by the measure of seven degrees. But that is one part of the value of Π which is the value of the curvature of space – time. That is the flow value but there is another part of Π that also forms the second part of gravity. This holds the value of 21.9991.

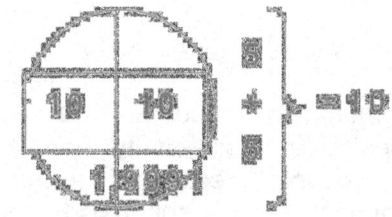

In order to solve this issue I had to locate and define gravity and believe me doing that is much more complicated than dismissing gravity to an inexplicable magical force compelled by a non-existing connection to mass. I had to find gravity and as Newton said, gravity comes from a centre point in any sphere. I had to find the centre of Π because there I could find the centre from which everything rotating is connected by gravity.

The Universe limits run from the Earth centre equal in all directions since the Earth is connected to singularity by gravity and when drawing the map that is in progress about the cosmos the allocated centre must be where the Earth now is.

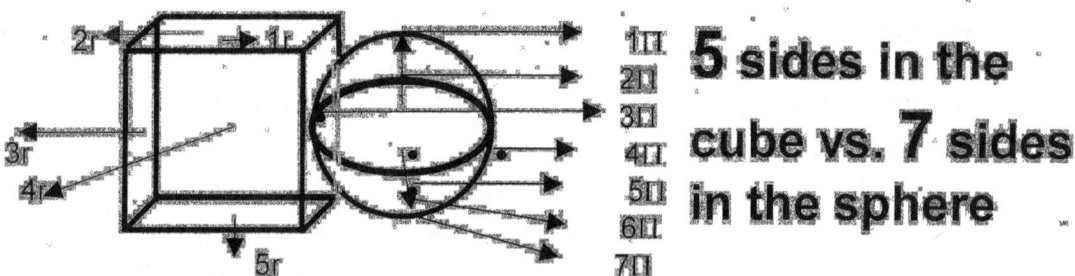

This places the value between the proton and time (possibly the Graviton?)

From Kepler's formula $a^3 = kT^2$

$a^3 = 7$ The single of matter relating to 10 (5+5=10) the square of space

$a^3 = \pi\pi^2\, 7/10$

Gravity dominating space

(10 / 7) \(7/ 10) = 2.04   which is equal to 1.4285   /   0.7 = 2.04

SPACE DIVIDED INTO TIME

(7/10) / (10/7) = 0.49   which is equal to .7   /   1.4285 = 0.49

SPACE MULTIPLIED WITH TIME

 7/10   **10 / 7 X 7/10 =2.04**

THE PROCESS PARTED USING THE ROCHE PRINCIPLE

A is   10 / 7 $(\Pi/2)^2$ 7/10   2.04 x ($\Pi/2$)^2 = 5.033

B is $(\Pi/2)^2$ X 2.04 =   5.033

A+ B is   10 / 7   10.066

SPACE DIVIDE INTO TIME

                    7/10/ 7/ 10 = 1 NO INFLUENCE

                    7/10 / 7/10 = 0.49

                    10 / 7   0.49

            10 / 7 /\\ 10 / 7   7/10=.49   7/10= .49

                .49 +.49 = .98

                .98 X 10.066 = 9.86468 = Π^2

TIME SPACE = Π^2 = 9.8696 = MATTER.

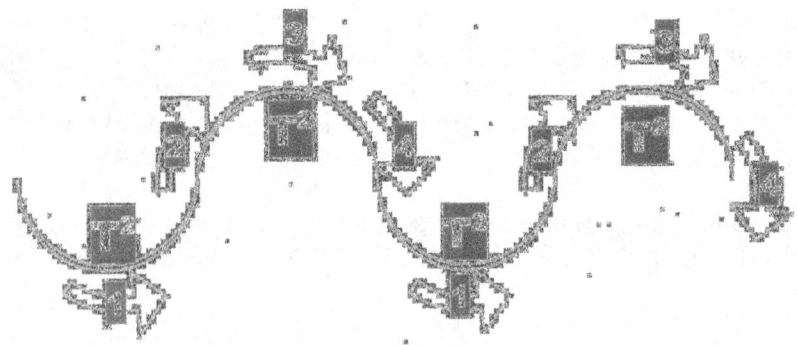

From this one can see that it proves that the relation of 7 and ten brings about a space decline from ten to Π^2 and that space decline becomes the value we think of as gravity. It comes about as the space decline in form to match the proton dismissing of space to heat and eventually to return the space to singularity.

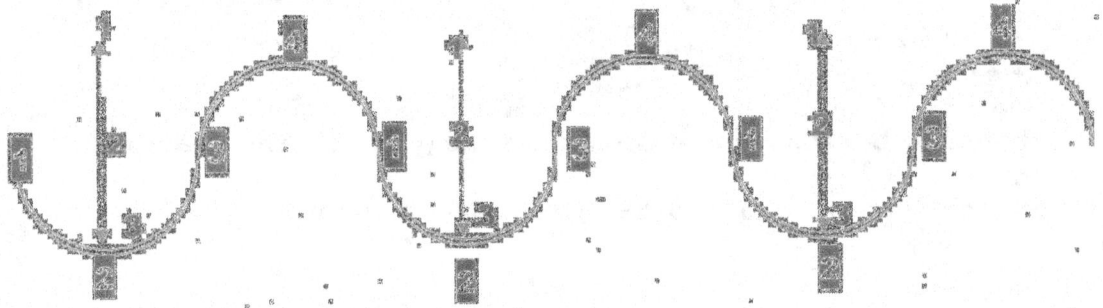

Taking the relevancy of $\Pi^2\Pi$ to the relation of material in forming **k** or $\Pi = 7$ and T^2 as $\Pi^2 = 10$,

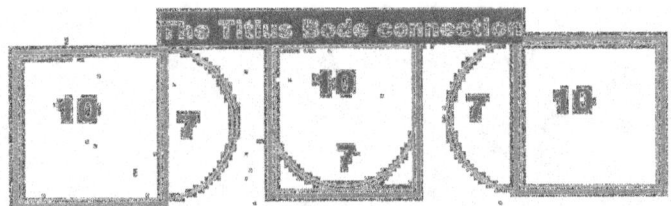

If Π is 7 in a single dimension then 10 is T^2 diverting from the single dimension a square in a natural state.

$7^2 / 10^2 = .49$ because 7 are matter square and 10^2 is the square of the

square of space. The square comes from the split by singularity of the Universe splitting the Universe in two compatible but never inclusive sectors and space having two squares because space is as it is, is already in a square. That is the result from Pythagoras placing space in a square.

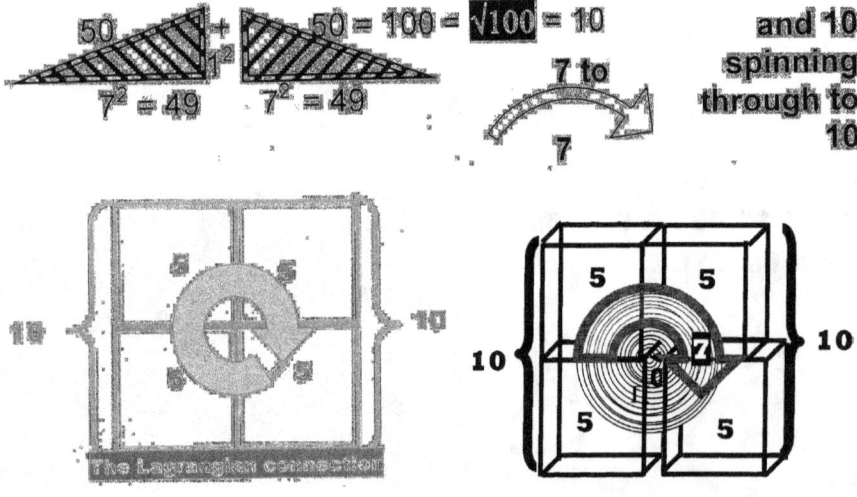

Space is already in the square but through time motion space goes from one point to another point in the square of motion.

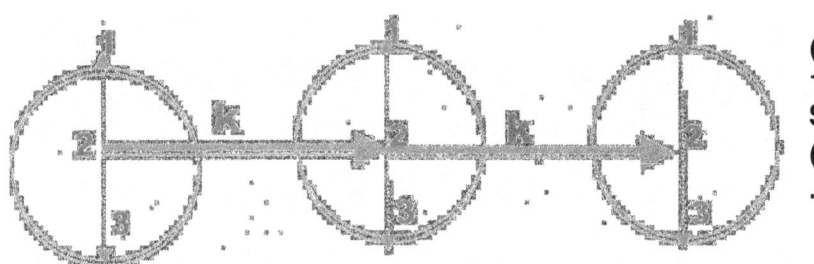

$(10 / 7) \setminus (7/ 10) = 2.04$
$1.4285 \quad / \quad 0.7 \quad = 2.04$
SPACE DIVIDED INTO TIME
$(7/10) / (10/7) \quad = 0.49$
$.7 \quad / \quad 1.4285 = 0.49$

In the Universe and even as far as hurricane and tornados singularity parts the circle around singularity into sectors of four as far as time goes.

Position 1 **(10 / 7)** divided by position 2 **(7/ 10)** placing two points of the four quarters of time. Position 3 is **(7/10)** and is divided by **(10/7)** positioning the other two positions of time in relation to positions occupied. This confirms the position material holds as a relevant.

Placing space in a position to time **10 / 7** X **7/10** Brings about the precise opposite of the first sector but not the opposite result because in this space dominates.

Since the square of space is directly involved with the dimensional depleting of space from ten to Π^2 and singularity divide the affect of gravity being Π^2 into two sectors forming one double of the four quadrants of time, the affect of Π^2 may not be ignored since that totally dominated gravity from even before the establishing of gravity in the form of the Roche limit.

$$A1 + A2 = .49 + .49 = .98$$

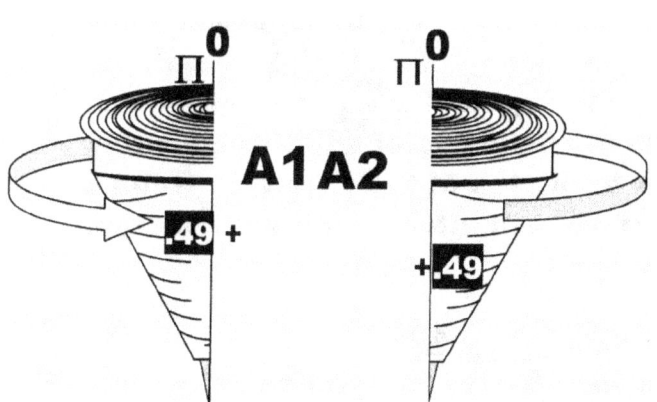

$$A1 + A2 = 5.033 + 5.033 = 10.066$$

The result is the dimensional dismissing of space from a relative 10 to Π^2.

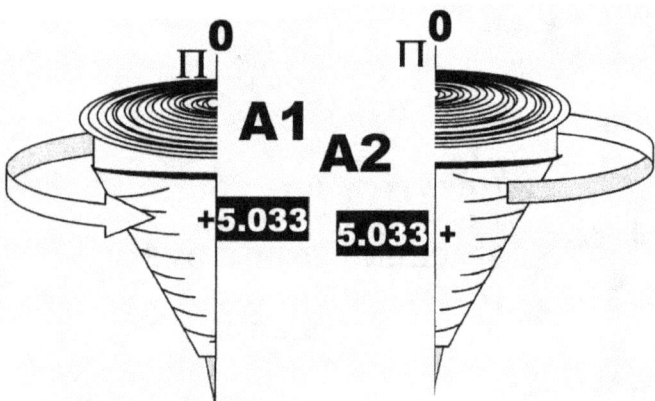

Matter in relation (part of) to the total dimension of space.
(10 / 7) \ (7/ 10) = 2.04
1.4285 / 0.7 = 2.04

Taking from both orbiting influences

SPACE DIVIDED INTO TIME
(7/10) / (10/7) = 0.49
.7 / 1.4285 = 0.49 **Taking from both orbiting influences**

SPACE MULTIPLIED WITH TIME
7/10 / 7/10 = 1 and   10 / 7 X 7/10 =1 **Therefore not influencing change**

THE PROCESS PARTED USING THE ROCHE PRINCIPLE

10 / 7	$(\Pi/2)^2$ **The Roche influence on Titius Bode**	
7/10	$2.04 \times (\Pi/2)^2 =$	5.033
$(\Pi/2)^2$	$2.04 \times (\Pi/2)^2 =$	5.033
10 / 7	5.033 +5.033 =	10.066 **from both objects**

In the case of the solid we have Π^0 by seven and in that we find that Π^0 is substituted by the number of $7\Pi^0$. When we find 7 stands related in terms of 7/10, which then holds the value of 7/10 (space-time on one side of the divide) the value of space-time concerning the solid part of Creation is then 7/10. That means on the solid side we find that Π is replaced by the Titius Bode value applying as 7/10 and in the square it is the square of material turning inside liquid which is 7/10 $= (0.7)^2 = 0.49$ on both sides of the divide $(7/10)^2 + (7/10)^2$ as material spins within the liquid we then have the solid moving from (7/10) to a new poison of (7/10) making (7/10) the value of Π and the gravity value of (7/10) becomes the square of (7/10), which is (0.49). This happens to the top of the sphere as well as the bottom of the sphere and combining to movement altogether gives a motion value of 0.49 X 2 =.98.

10 / 7= 0.49 + 0.49 on both sides of the divide

10 / 7= 0.49 on both sides of the of the rotation   0.49 X 2 =.98

10/7 :   10 / 7 = 1 also on both sides of the of the rotation

7/10 =.49   7/10= .49

.49   +   .49   = .98

.98 X 10.066 = 9.8 =Π^2 **TIME SPACE** = Π^2= 9.869 **TIME**

To prove my theory I firstly had to locate the centre of the Universe. The failure of Newtonian science to locate the centre of the Universe is an obstacle they never noticed. It should be a fact they have to predominately first establish where to locate the centre of the Universe.

With gravity pulling everything in that general direction and finding such a point should show where the lot is heading to where all contraction will eventually lead. Identifying that precise location is a far greater problem to investigate than is the critical mass density factors a devastating problem. This inconsistency to point where the contracting should be heading proves to be the Waterloo of science because science has no idea where to position such a centre. If we backtrack instead of fast track the contraction of the Universe we should be able to find the point of the beginning of everything. Its because of where science position the end of the Universe some thirty odd billion light years from where we now are that I concluded the centre of contraction must be allocated. Closer to home we must search for the point of gravity where the gravity is the strongest as it must be in the centre of the Earth.

The Universe limits run from the Earth centre equal in all directions since the Earth is connected to singularity by gravity and when drawing this map that is in progress about the cosmos the allocated centre must be where the Earth now is.

That was what inspired me to locate my centre of my Universe. Even admitting to such a notion sounds like madness, but please allow me a chance to explain in more detail. I realised that my effort to locate the point holding singularity enabled me to backtrack the exploding Universe to its origins. By applying some basic effort I have located the position from where all movement came and the direction it took moving forward in time...and yes, even time as such. Gravity is the dimensional changing of space holding r as reference in the cube as to the sphere holding Π as the reference. In order to generate spin that is producing time in matter occupying space, therefore creating dimensional change, Π has to be a factor indicating the possibility of spin because by implementing Π the circle sides will follow one another without establishing separation.

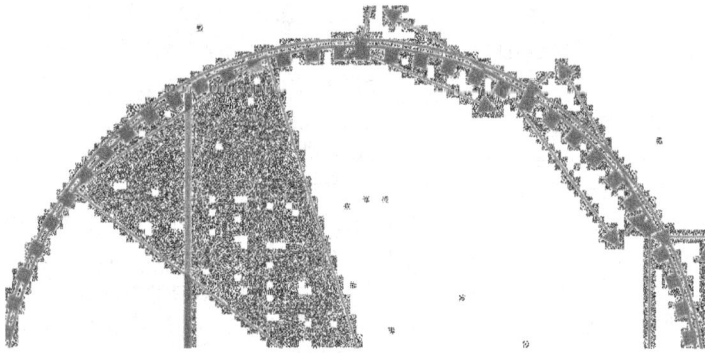

As soon as motion takes gravity straight, singularity will reposition the direction changing the direction of motion by 7^0. It is this turning of motion by redirecting the continuing of motion that sets the critical time within the proton connecting to singularity. Instead of r being specifically a straight line that gravity will inevitably be, Π which the form value of singularity is.

That is this 7^0 redirecting in the square of space of space, which is ten on both sides of singularity and time is that what we find to be the Titius Bode law of 7 / 10 and 10 / 7 in relation to the Roche limit of $\Pi^2/4$ which is producing the gravity of Π^2.

However the reducing in it is going from ten that is on one side and is crossing over the figure of 1.9991, (which is singularity on both sides of the Universe) and coming into contact with another 10 while turning 7^0 that we find to form Π. In all being the total forming on both sides of the Universe it is $(10 + 10 + 1.9991) / 7^0 = \Pi$.

The answer must be in finding Π, and thereby locating singularity. If singularity is in affect the original point of the cosmos birth, the reducing path we should follow will indicate the whereabouts such a point must be. That is where cosmology diverts from mathematics.

Again have to remind the reader that we discussing events happening where singularity meets space. It is way before space –time becomes three dimensional at $7/10(\Pi^0)^6/6 = 112$. At $7/10(\Pi^0)^6/6$ the element table ends (112) because the three-dimensional value holding six dots $(\Pi^0)^6$ in relation to 6 sides (/6) in movement (7/10) that then forms the Universe we use to live in would take singularity (Π^0) no further.

We are where the Roche factor is the singularity influenced half of the Lagrangian system, the Titius Bode is the dimensional duplication of the doubling of the Lagrangian system in space occupied and space not occupies by material.

Space-time is a **four dimensional position of the Universe where the position of an object is specified by three coordinates in space and one position in time**

By rotating around a centre that is standing still such a centre forms a divide that separate the unified unit. **Any point will be opposing itself** within the **rotating of 180°** where it **then change every aspect** of its **previous flowing** characteristics it had or **will once again have** in **360°** from there. While in rotation from the view point of a bystander it all may seem static and never changing

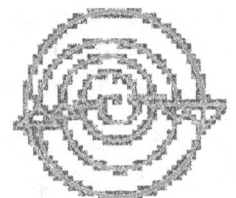

but to the object in spin every next instant in time will be diverting from every aspect it had every second passing, and the direction it held in relation to the direction it held the previous mille, mille second as it will totally be incompatible with the direction it holds the very next mille, mille second of rotation. This is why we can use degrees measuring the circle by (6^2) (forming the square relating to matter through singularity) X 10 (square if space) = 360^0 however it is always in motion. That proves no point can be static or constant, though it may seem that way to outsiders. Although matter is matter, matter can also be anti-matter and moreover form its own anti-matter at the same time. This degeneration of structure is very likely to occur with overheating. Revaluing Π to Π^2 will bring about a new contact point where Π meets **r** forming another relation in Π^2. Every time material swap sides it also qualifies as anti matter to matter because if it goes out of orbiting rotation frequency, it has the ability to collide with the same matter it forms union with but is located on the other part of the spin. It then becomes in a situation where Π **revalue to r. Time is** the **changes in relation** where Π **contacts a different r** not withstanding the many r points there may form because **every r constitutes a different value** to the Universe through other ratios and relevancies brought about **by heat and light. Time is the duration it takes Π to rotate between any two given points of r** and therefore must always amount to **a square (T^2)** moving from point to point through the **cube of space (a^3)** in that **duration of time (k)**. With that it proves **Kepler's a^3 (space) $=T^2$ k (time in the instant of motion)** but motion must continue through a specific value in space where the space-time is maintaining relevant equilibriums throughout singularity connecting**.**

When concerning a sphere and we establish the Laws of Pythagoras to provide a centre principle all lines running through the centre will be effectively related by groups forming 180^0 and 90^0.

We take a line running between two points as being 180^0 and the rest of the explaining is saved in the accepting part of mathematics. Any one of the two points the line start or ends at are a point in infinity. The start and the end depend on the viewer putting the relevance to favour the side of choice. That puts the point of end or beginning in the spectrum of choice and not fact. Any direction is as equal as all other directions.

A straight line, triangle and half a circle will always have equality in dimensional capacity providing equilibrium being 180^0 because each one shares a common denominator in singularity to the value of Π. As the straight line averts a zero it holds another straight line in place to set about such an averting where the two lines will always carry a relevancy in elation to progress (the triangle) and a common denominator in the start from singularity. This concept we apply as the graph or the vector. By going back to a line, any lines and all lines, the line is a connection of dots in infinity, running from one specific to another specific and avoiding zero or dots. At every point in infinity it dips into infinity coming out on the other side by choice of direction and the direction is unforced and change presents any angle including the straight line, which incidentally is just another angle.

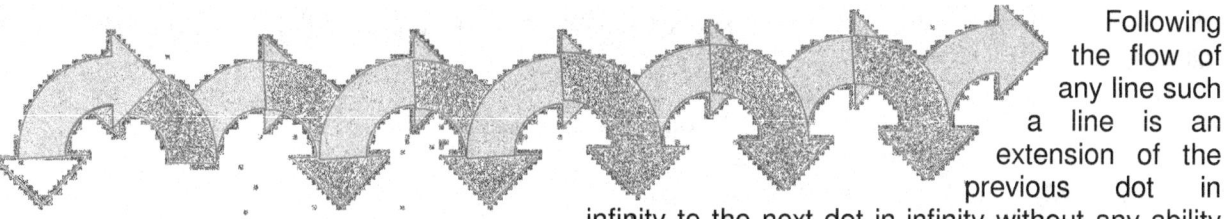

Following the flow of any line such a line is an extension of the previous dot in infinity to the next dot in infinity without any ability to skip or bypass any of the other dots in the connecting line. Any direction change including the remaining of travelling in the same direction is in relation to a line travelling all being the very same. Change does not affect the line.

When connecting to the dot representing infinity the flow can be in any and all possible directions, including in the same direction. We all live in a graph as the Universe with all in it is nothing less that a three-dimensional graph flowing according to time. That means in the case of Pythagoras the mere fact that the line shows changes in direction does not implicate or affect the line as a tool of mathematics. Whether the line changes into a half circle meeting at the other end again or meeting in a triangle in forming a half square by joining the point where it began, the result still indicate a line flowing between points.

going into progress linking from

The line dips into infinity every time it passes infinity when it cuts through Infinity. The line infinity comes natural as the line because all lines are infinite dots one point to another point. That brings about that coming from infinity might change in angle bit that directs the route and not the form. The form is all the same

In that way a circle is a straight line following a loop as it comes out of singularity at a different angle and a triangle is a straight line that dipped into singularity but at three stages changed the angle with which the line then left to follow different directions at specific points. From the point singularity observes it still remain a straight line because there is no direction alternation in the first dimension and in that dimension it still remains a straight line in which we on the outside may experience as three forms but are in fact one single line. Only when the direction changes completely in reverse, does the line become double in value but comes from multiplication for instance 2Π become Π^2.

Not long after the law of Pythagoras was understood where Pythagoras introduced mathematics Eratosthenes of Syene made as big a discovery as Pythagoras did. But in the one instance the world took notice because the world could see and understand and the other instance the world disregarded the findings because the world did not see what the implications was. The same apply to aircraft flying and when the aircraft wishes to escape the earth's singularity hold it has to comply with the laws laid down by the earth. The seven becomes as big a part of the concept as does Π as it all interacts.

So far I have shown that gravity is all to do with dimensional changing and reforming of forms to re-affirm alliances supporting the centre of the object asserting gravity. It is the reforming of space by converting from an un-attached cube in space to a confirming form such as the sphere bringing about more concentrated heat surrounding the object spinning. The Universe is in the three dimensions using twelve dimensions (including the two dimensions involving singularity) that is visible to us and indefinite number of stages in size differences ranging from the immeasurable small to the immeasurable large where mathematics become a short fall to the next and the previous dimension. I include the following presentation to give an idea of how I came to realise the integration of the four cosmic pillars in being responsible for forming gravity and how the blanket forming gravity is knitted. I am not going to elaborate using massively complicated detail as to how the lot fits together and what assembles where but when investigating the picture should give the reader a common idea of how the lot fits. I include the presentation to prove that I did use the four phenomena that Newtonian mass could never explain and therefore Newtonian science denied existing at all. Because the phenomena did not corroborate Newton's fraud Newtonians degenerated the phenomena to a level where it is being classified as coincidental where it occurred because the backwardness of Newtonian misconception could not figure out how the cosmos takes time to build space. The Newtonian reference to this method of Universe construction they call the Hubble constant, but in truth it is the way that time forms space by applying the four cosmic pillars.

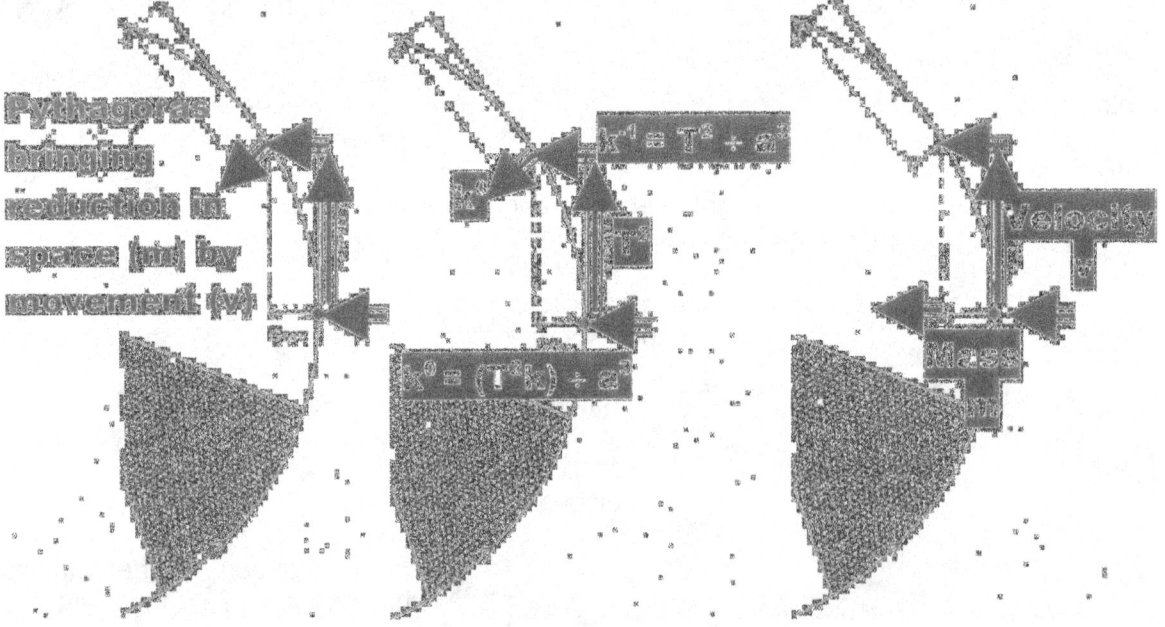

By moving in rotation such action places the reference of movement applying in relation to a centre holding singularity that produces gravity (and only the centre of a circle in rotation produces gravity where it is done by the committing of movement in relation to singularity) are we able to find the basic design of gravity. The movement should follow the lead of **k** but instead follow the spin of T^2. As the spin turns the direction of travel, such a change in direction reaffirms singularity in affirm the control of singularity by confirming the location of singularity. In order to assert time by employing singularity as Π there has to be a relation with ten coming from the past going through 1.991 which is singularity on both sides of the Universe and the entering the future by the measure of 10. This has all to do with the changing of direction by 7 and that puts the curvature of space-time to the value of Π. This is what Kepler's formula implies when it places $k^0 = a^3 / (T^2k)$ or as it should be when correcting one of Newton's many mistakes from $\dfrac{dJ}{dt} = 0$ to $\dfrac{dJ}{dt} = 1^0$. As the circle rotates the circle diverts from the straight line by 7^0. As the object should move in a straight line it would have to overcome the 7^0 diversions the circle produces. Time has a factor is 3. The reasons why this is true is far too involved to go into that at this point but there are other books in which I explain that

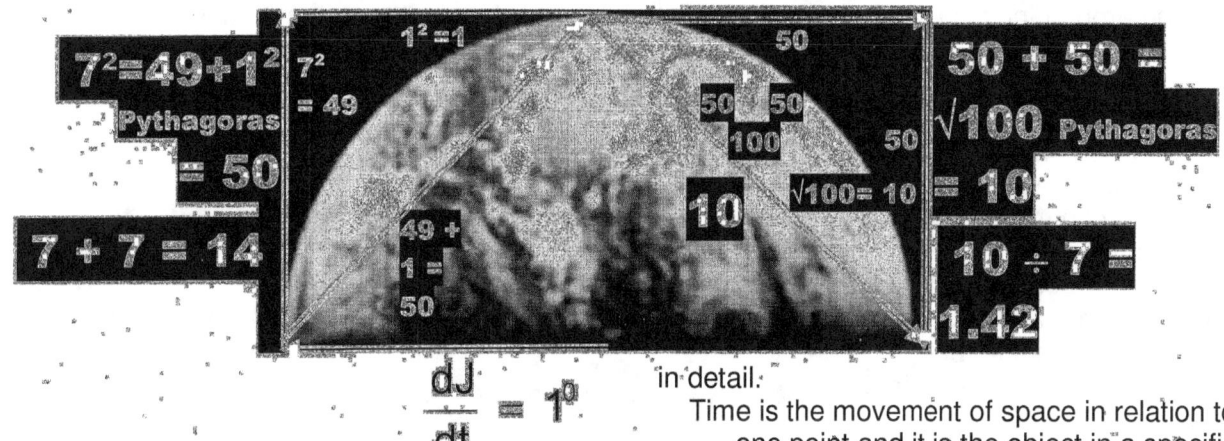

$$\frac{dJ}{dt} = 1^0$$

in detail.

Time is the movement of space in relation to one point and it is the object in a specific position coming from the past into the present and onto the future and this way we have movement of space from the past (**T**) through the present (**k**) onto the future (**T**). Therefore the object has space-time as $a^3 = T^2k$. By having $a^3 = T^2k$ we also have k^0 located at a point holding singularity $a^3 = T^2k$ which mathematically places $k^0 = a^3 \div (T^2k)$ and that indicates that the moving (T^2k) of space a^3 points to confirm the location of singularity k^0. The formula Kepler introduced shows clearly that the Universe is a sphere by the measure of k^0 (1) $= a^3 (3) \div (T^2k)(3) = (7)$.

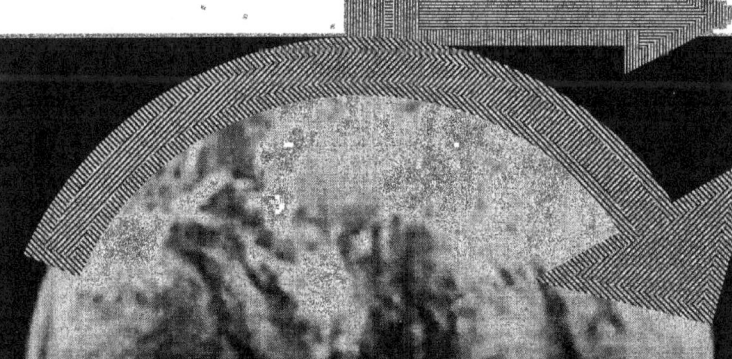

The compressing of the space surrounding object holding gravity is part of the gravity but view of

movement where 10 as a factor declines by seven as the law of Pythagoras prevails from was in times where space reached. The condensing of the surrounding the object forms an indicator of the development

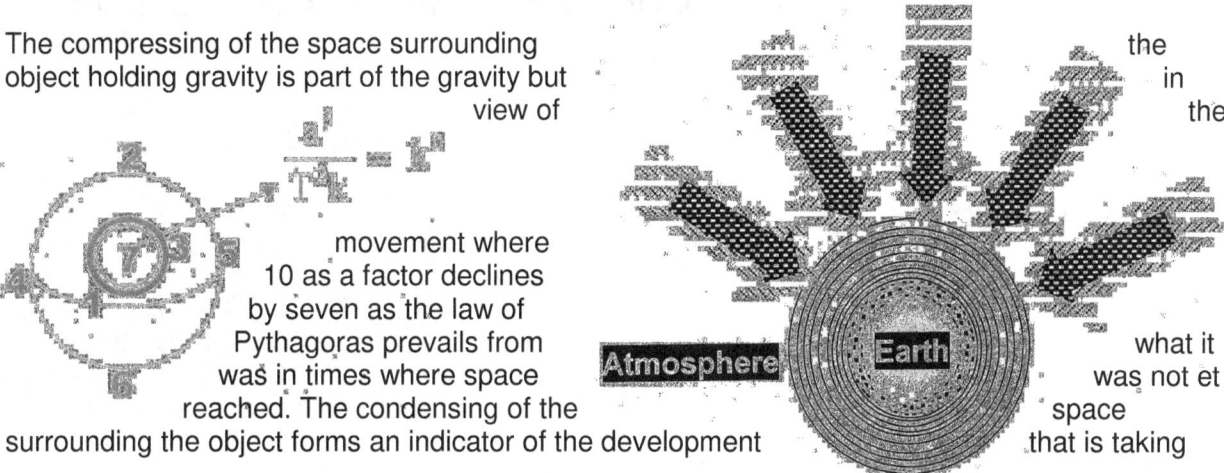

the
in
the

what it
was not et
space
that is taking

place within the star. In that light the movement will correspond in bringing about a colour that light forms as the movement condenses to a liquid around the star. The forming of liquid reducing the density of the space from a gas to a liquid and in much bigger starts to become the same solid the star is, is so much part of the process of explaining gravity that it is gravity. The movement of the star pumps liquid into the star in order to lubricate atoms and part atoms spinning. This movement condenses the space and it is the condensing or the cooling of the space that forms gravity. It is the fact that the Sun freezes its atmosphere to a cool 6500oC that forms the gravity the Sun has. All the space moving to the centre that Kepler found to be formulated as $k^0 = a^3 \div (T^2 k)$ is that liquid sloshing around the Sun. That yellow fluid is the evidence of the solar system forming an attachment $k^0 = a^3 \div (T^2 k)$.

However the value of k^0 is 1^0 and since $k^0 = 1^0$ anywhere in the entire Universe we have singularity 1^0 not only controlling the entirety but also we have singularity 1^0 confirming an entire Universe and that becomes the binding factor joining the entire Universe to the measure of 1^0. That places all gravity in relation to one point being equal everywhere, which means gravity must be unequal everywhere. In the Universe we find time moving by repositioning space as a unified item in relation to any single point while one point being any point remains unchanged but it is much more complex than one could gather from this remark. Only nothing filling the minds of Newtonians is as simple as mass pulling by forming gravity. The rest out there forming cosmic physics are much more meaningful than such simplicity.

A sphere can only be a sphere if the sphere spins and the spin activates one point in the centre being in a relation to six points affirming the boundaries of the sphere. The spin confirms the space holding the relevance and in that all atoms affirm the inside of all stars.

The Super Nova that has gravity "that has gone mad" (as Newtonians so scientifically explain the process of the misfortune gravity brought to the mega star in performing the super nova phenomenon and I still have to learn how gravity can go mad because where does gravity find intelligence to lose) has in fact expanded by increasing the relevancy **k** produces and this acquires an increase in occupied space a^3 as the spin T^2 tarnished in movement allowing a surge in heat levels inside the star that then in turn insist on larger space to be occupied by the star and this is all the result of reducing its much more rapid spin to become less moving heat spread over a larger area due to the less spin applying then as a result of the movement of the star (T^2) that has become too slow to confirm **k**. Gravity depends on the spin of a sphere. The spin of the sphere positions as

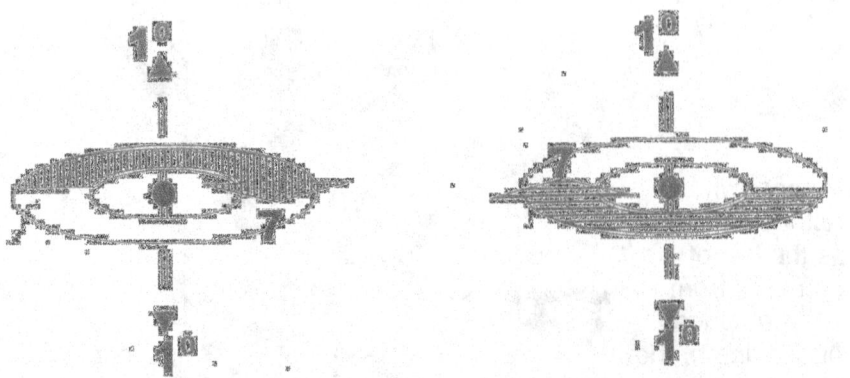

ell as allocates singularity. That is much more a law than it is a rule and this statement rules gravity by law. The deliberate rotation of five points across four sectors establishes a line with three points that is not a line at all. Although the line commits four sectors of total opposition, the four sectors define five points connected by a line holding singularity (1^0) as a beacon. This has the value of time where that which can never stop moving is parted from that which can never move, and that holds space occupied by material in place.

Quoted directly from the Oxford dictionary of Astronomy the following:
The definition of singularity is as follows:

Singularity: a mathematical point at which certain physical quantities reach infinite values for example, according to the general relativity the curvature of space-time becomes infinite in a black hole. In the Big Bang theory the Universe was born from singularity in which the density and temperature of matter were infinite. Let's hunt singularity down! Singularity can have but one mathematical value and that is 1. Let's find 1^0 and find the centre of the Universe, which is $\Pi^0 = 1^0$

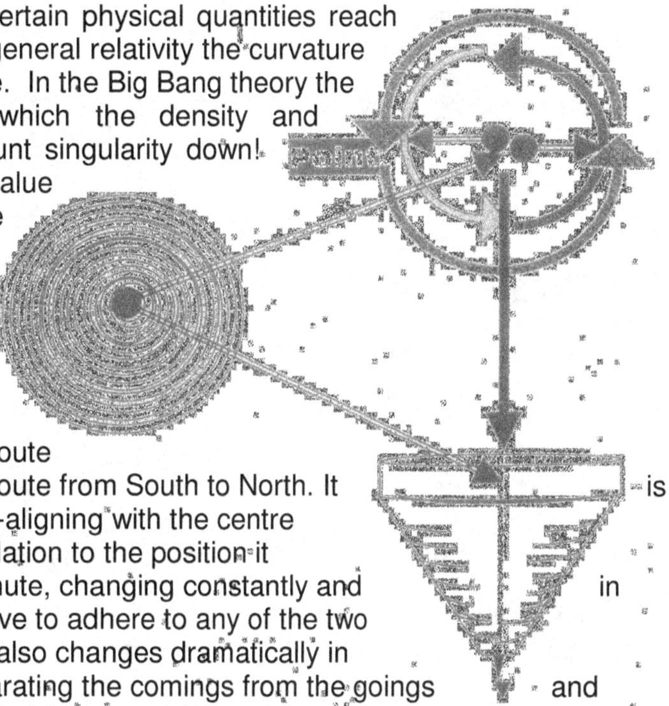

All atomic particles would have too referred to coming or going as much as circling around. This comes about as no atom can be ignoring any of the three positions it is aligning with in direction. Some would be on route from North to South and others would be on route from South to North. It is a response to an ever changing directional re-aligning with the centre holding the centre too a specific location in relation to the position it previously owned and would own the next minute, changing constantly and in according with time location. All matter will have to adhere to any of the two directions, which in fact is actually four, but it also changes dramatically in moment. In the centre a line MUST form separating the comings from the goings and again the goings from the comings, where no matter can be located. That line is too small too hold any atom, sub- atom particle or matter of any kind. All matter is either on the one side, or on the other side, but never can be neutral. It even gets more complicated because another line forms locating matter in groups. The one group is relating a poison from the "centre point" holding "back" and "front" running through "the centre" where the other line is relating from "side" to "side" running through the "centre point". The fact of the lines is that "they are there", but we cannot see them. Try as you may, no one will be able to calculate the very position that forms the lines, but as they change all particle characteristics, the lines are a reality as the spin of the matter is real. Being too small to hold atoms, they then therefore must become part of singularity, where singularity is a spot in the centre with two lines crossing the spot at an angle of 90^0. That is the basis of singularity, and since all the positions still relate too a centre of a circle, forming a part of a spinning circle, Π must form the basic value.

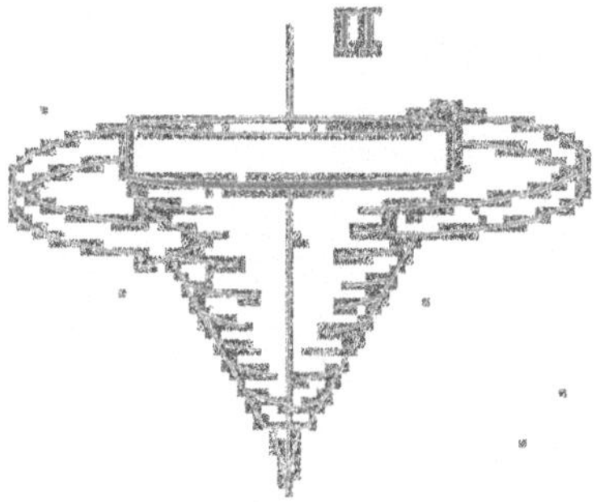

In that there are one specific group in relation to coming towards the centre while coming towards the front.

The following group is rushing away from the centre while coming to the front.

The third is heading for the back while heading for the centre and…

The fourth is rushing away from the centre while heading towards the rear. This part I just explained is one of the most crucial aspects of physics since this is gravity, as I am about to explain a bit further down the book.

To confirm singularity the spin of a line diverting by seven points from a point confirming singularity to a point confirming singularity changes nothing since singularity is equal. It is the movement by seven that distinguishes the gravity changing positions.

Time or spin or rotation, call it what you like, but it is the moving from T_1 to T_2 using time that provides the dimension of depth to the dimension of distance between dimensions. In the

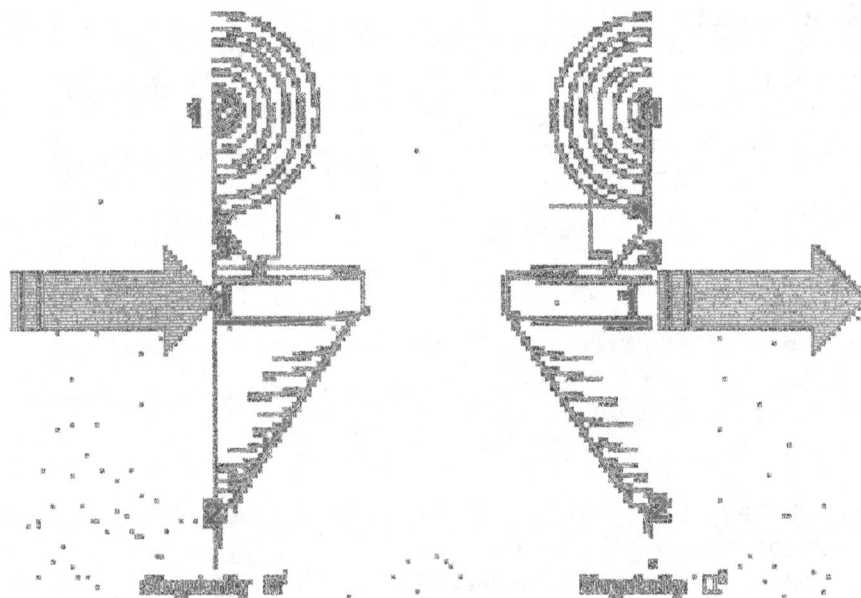

motionless Universe there will be on point in time and that point will represent k more than anything else. Every point being T_1 will only show the extending of **k** from singularity to that specific point. The fact that T_1 indicates no motion brings the Universe to a stand still and to a flat Universe.

It is that which give **k** the coming from the first dimension and by only extending from singularity it the forms two more positions becoming a^3. k =

a^3 / T^2 but remove T_1 to T_2 from he equation and only **k** remains $a^3 / T^2 = k$. By the effort of spin or motion the Universe becomes the three-dimensional object it all seems to be.

That too forms the answer about the gravity implicating of cosmology. linked

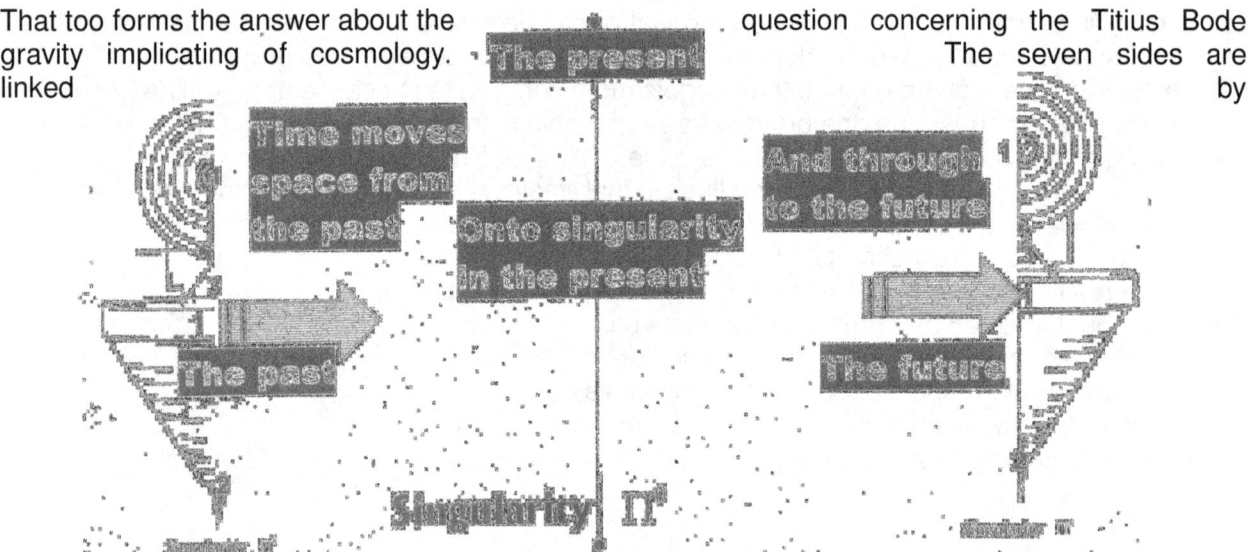

question concerning the Titius Bode The seven sides are by

rotation nothing changes because there is a steady linking to the inside centre of the sphere. But it is to the outside that dimensional five T_1 points moving to with five moving points. points is the point the point after reducing the ten points in stead it is the case with the points relating to seven continuing in the same

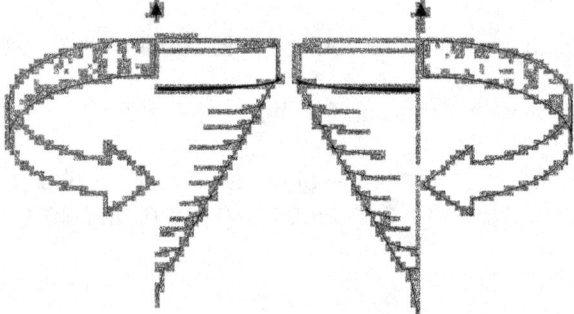

this rotation brings about complications. There are five T_2 making contact The moving non fixed before reducing by five to by five that bring along of the five to one point as Lagrangian system. Two points coming from by direction it is going to remaining seven points as going. That means in matter there are five times two points relating to seven in a moving constant and seven fixed rotating points

The line that forms in relation to singularity shifts every object in a direction while also repositioning every object that time shifts back and forward in a circular movement. The line moves the circle (T^2) by repositioning **k** and it repositions **k** by relocating (turning) T^2 and from this action we find space forming as (a^3) We have to realise that where space comes away from singularity there is a situation where singularity Π^0 extends and that extending becomes a value of Π. The fact that it is Π alone proves the Titius Bode law applying. When the value of Π generates the next 7 points the value coming about from this relevancy is 21.991. That I have explained and the reasons why that are that case, I also have explained. The value of Π is time coming from singularity.

Time moves on by the value of 3 and in the way of

proving this we find proof in the fact that a line is forming by 3, a top, a centre and a bottom, but holds no inside space. The lens can't be part of the Universe because it splits the Universe into seven clear sectors while the line by itself holds no space. On the one side of this line forming singularity by time we have the Universe opposing every virtue in relation to what applies just on the other side of this line of divide that holds no space. The line is the most crucial aspect of the Universe because it ties what is opposing in the Universe together without confusing what forms the Universe. Yet the line forms without showing any presence at all and the line can only be detected through intellect understanding the fact that the lines places six sides in three dimensions in relation to the line being present.

By the way the reason why the Universe will come to some end is because Π holds the value of 3.14159 moving through space is 3

Any spin of a round object positions seven points changing the straight line. To find gravity we have to find the value of the straight line because as the straight line defers, the direction of movement will change and this will reduce the movement in distance in relation to the law of Pythagoras.

By the rotating motion the circle redefines its movement direction in an angle that alters by seven degrees. It is aiming to go straight but then it has to turn by seven degrees because that is the redirection of rotation that any circle has.

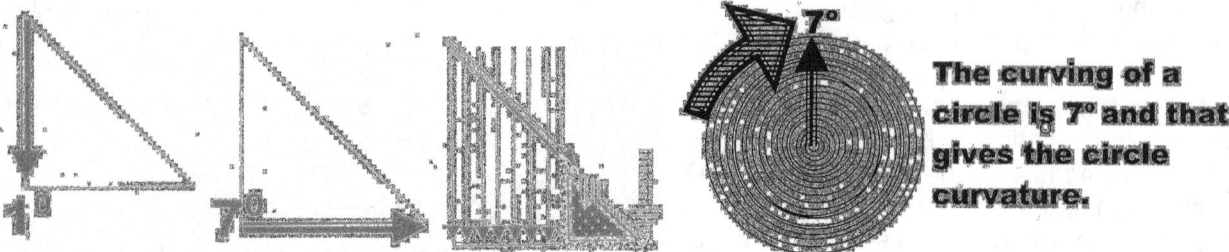

The curving of a circle is 7° and that gives the circle curvature.

When an object is descending down to Earth, the object is falling in relation to a continuous line of repeating triangles continuously diverting from the straight line of descending in relation to the Earth turning. The Earth gives a new point of reference and this references shifts by 7 degrees due to the circle alignment. In this the relation of the liquid become somewhat shorter due to the law of Pythagoras applying to the hypotenuse becoming shorter as it redirects the original direction and this compresses the space we call the atmosphere. Proof of this statement is the various layers all having different names but is still space compressing ever denser as it comes closer to the Earth. This is gravity. It is the liquid racing towards a spinning Earth because of a spinning Earth.

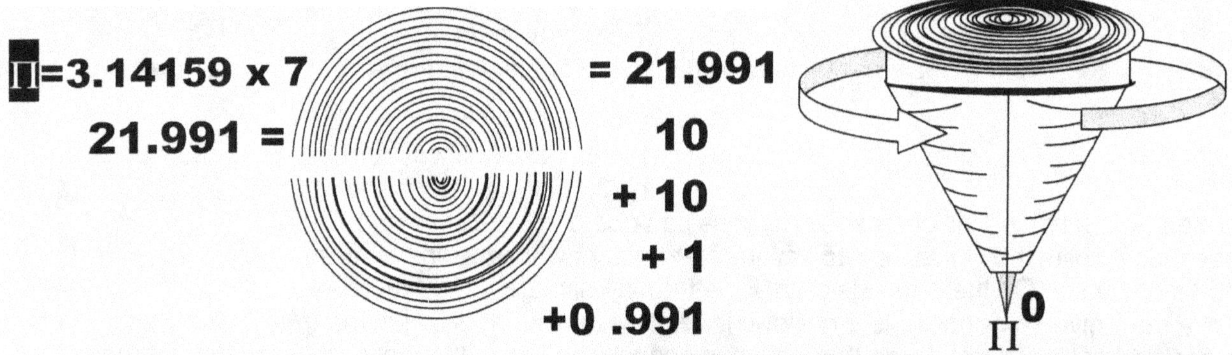

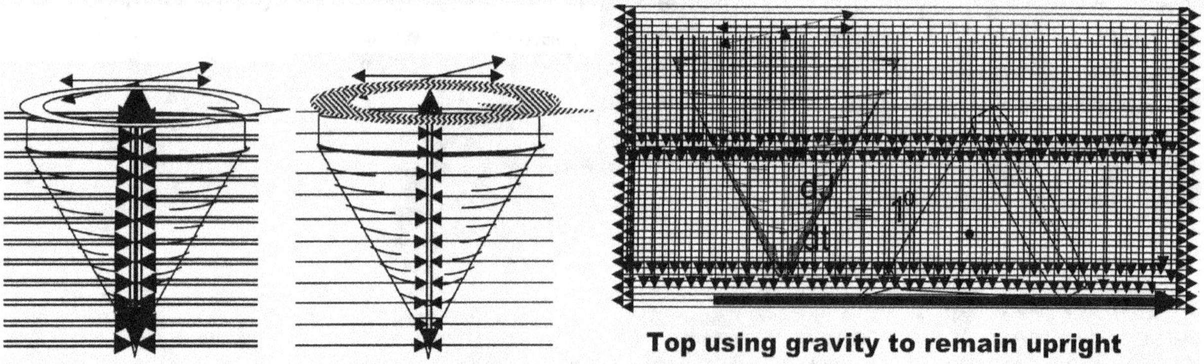

$$\Pi = 3.14159 \times 7 \qquad = 21.991$$

$$21.991 = \begin{array}{r} 10 \\ + 10 \\ + 1 \\ + 0.991 \end{array}$$

$$\Pi^0$$

Top using gravity to remain upright

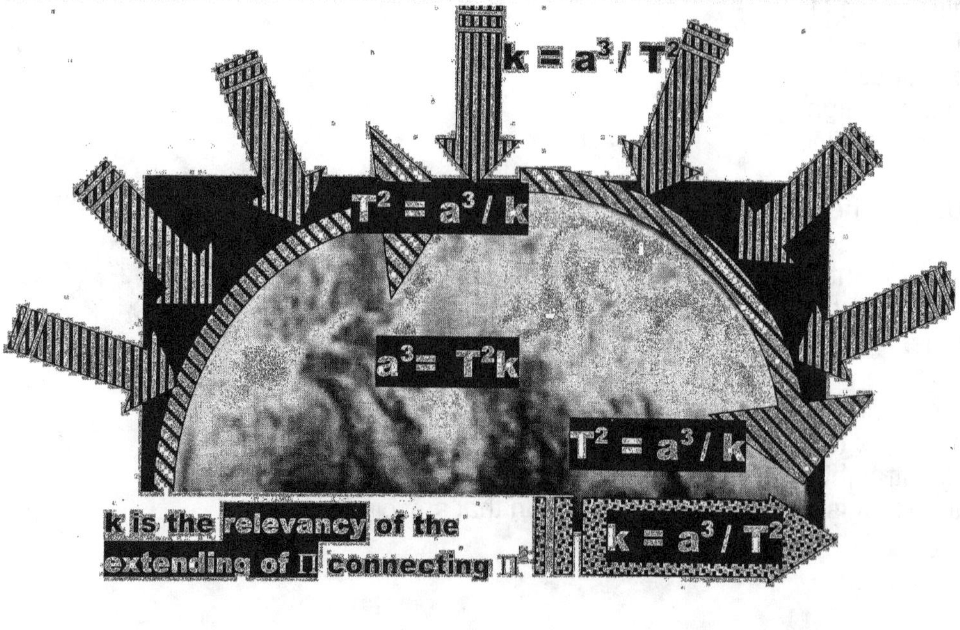

$$k = a^3 / T^2$$

$$T^2 = a^3 / k$$

$$a^3 = T^2 k$$

$$T^2 = a^3 / k$$

$$k = a^3 / T^2$$

k is the relevancy of the extending of Π connecting Π.

The relevancy of **k** interchanging remains applying in the gravity growth we find where space-time grows in relation to the Titius Bode law applying. The contraction we find indicated, as $k^{-1} = T^2 \div a^3$ becomes the same growth we find in **k** developing material using **k** as the growth factor $k = a^3 \div T^2$ but this growth extends outwards to influence time in

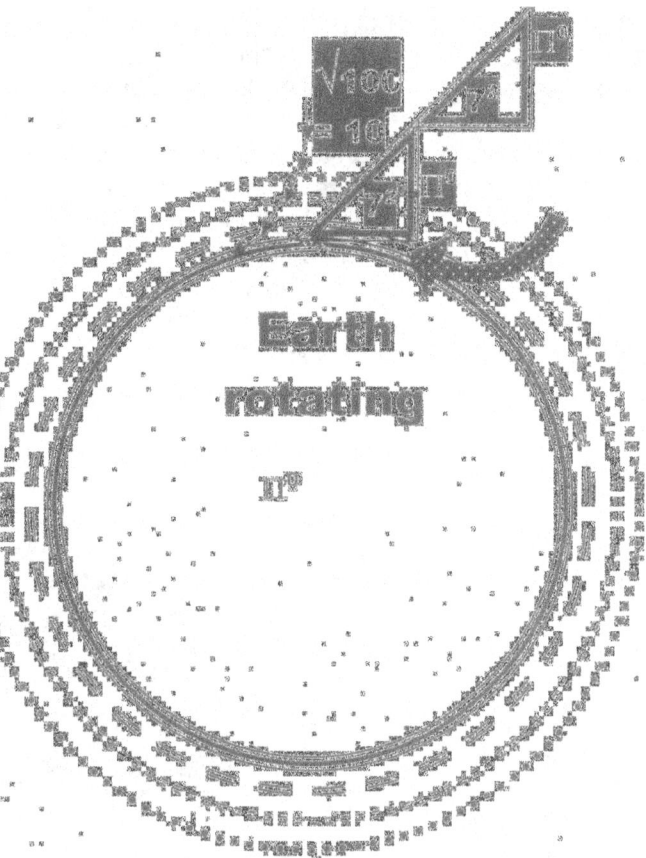

Atmospheric layers growing ever denser as space bends with the Earth's movement that diverts the directional flow and thus bending the straight line that space holds in relation to singularity

relation to the Titius Bode planetary allocation. The relevance of Π connects to 7 where space then becomes $5+5 = 10$. Where the planets are positioned we find the growth forming as $7+7$. The falling object continuously connects to Π by the measure of the Titius Bode taking time into forming space. There are the innumerable lines forming

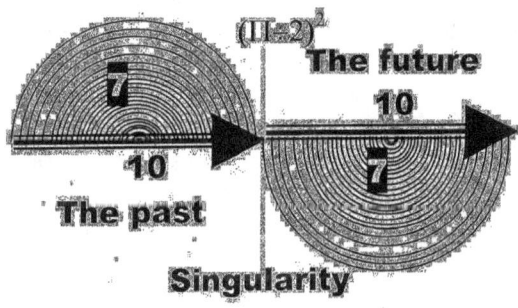

singularity by connecting to the curvature of the Earth and the lines are declining as the lines connect by compacting space available for occupation. This is where the relevance of **k** as a factor forms its important use. The factor forms Π in connecting to Π^2 as the space declines in measure but also forms Π as the Earth soldiers on orbiting the Sun. All movement is about forming the three- dimensional space status as singularity is converted from Π to $\Pi\Pi^2$ that then become Π^3.

The Titius Bode law represents the future half of Π and is the tool that takes Π from singularity onto (Π^2) through the Roche division of $(\Pi\div2)^2$ and taking it $(\Pi^2\Pi)$ into space –time (Π^3).

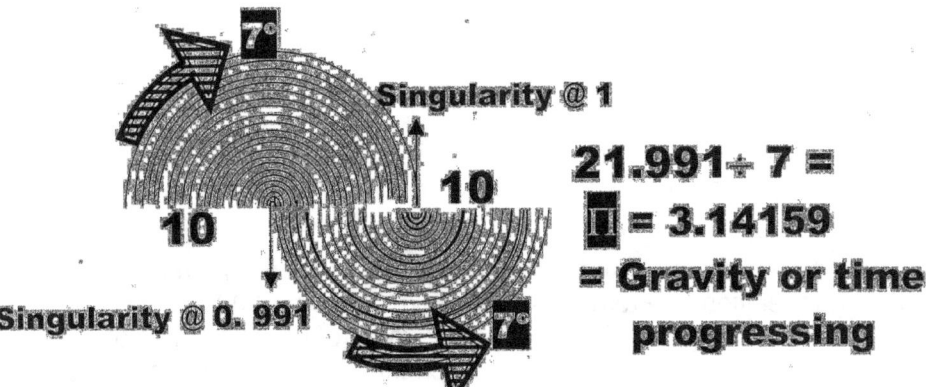

$$21.991 \div 7 =$$
$$\Pi = 3.14159$$
$$= \text{Gravity or time progressing}$$

One must understand time and time is a very complicated concept that I just have no room in this book to explain. I do try and explain time as a concept in **_an Open letter on Gravity_** but in that book there I do not explain gravity in this rather simplistic fashion. The problem with the work I present is that there are so many levels of understanding that I have to cater for that choosing the level of explaining in detail becomes problematic.

The movement of time is in relation to seven going from singularity to singularity but since singularity is equal the changing only holds relevance to the seven and the one forming singularity have only a mathematical function in the triangle of Pythagoras.

When considering the law of Pythagoras we find the square of the shift is added to the square of singularity and that brings the hypotenuse to a measured value of 50. However as indicated previously, the movement can never stand still because time is the movement of space in relation to one point and therefore we have the square in repeat.

The Titius bode is the progress (half of) Π as time moves into space and into space-time by the value of Π^2. In that there is 10 being related to 7.

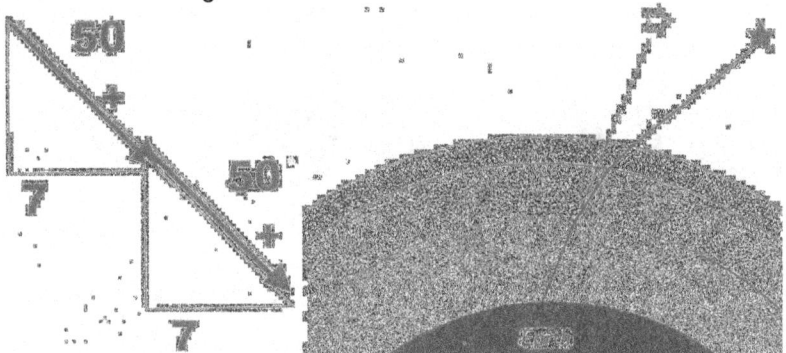

Also it is true that the seven points of the sphere reduces the effective sides of the cube from six points to having five. Since the seven points of the sphere rotates around 1 as 1 we have five points relocating at four positions. This constitutes to twenty positions in all. This again reaffirms the Pythagoras statement plying a part in forming gravity. By bending incoming space the vertical line is pushed by a vale of seven degree in a horizontal direction. Doing this calls for the triangular law of Pythagoras to come into effect. Instead of moving down directly as Newtonians surmise, the route down diverts by seven degrees to the one point it moves vertical. This is even evident in the way we observe light. Our light in the sky is blue not because it scatters because such a thought is so typical of Newtonian stupidity. Our sky is blue because the bending light is concentrated as it is bended into a denser atmosphere and this gives the blueness of the sky. This blue sky will be a factor in all planet-like structures. From realising this we can start to calculate gravity applying.

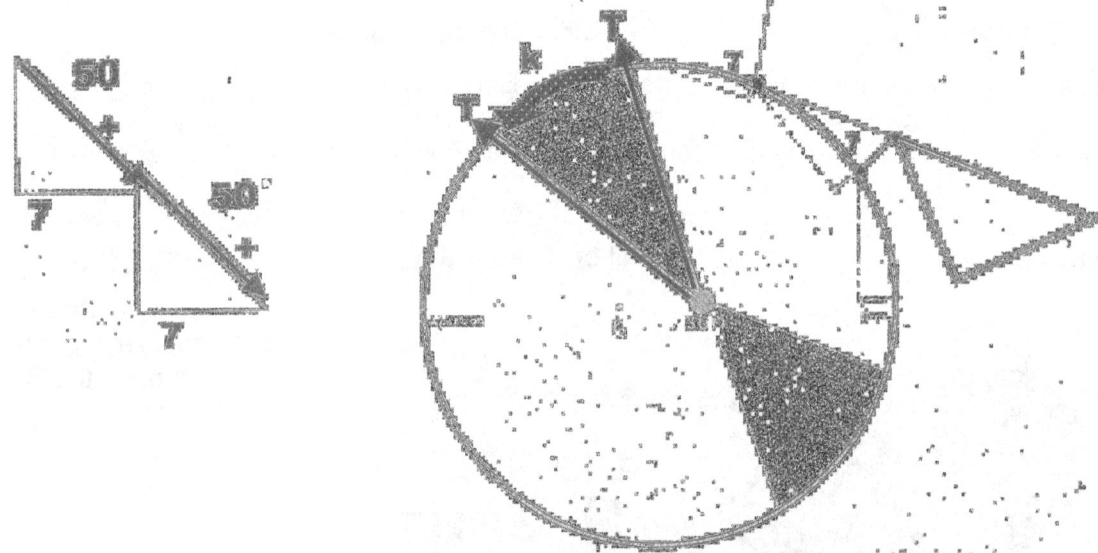

Gravity is the movement of space in relation to material and therefore the material moving constitutes 7 positions changing in relation to a total of space or liquid having 100 squared (since it is the hypotenuse of the law of Pythagoras. That puts a diversion of seven points constituting material that resembles singularity in relation to a total of ten positions that changes the space or the liquid factor.

In as brief as one can put it then the summarising will be as follows where gravity is having a cube (10) with a spinning sphere (7) in it and the mathematical solution to this calculations brings the

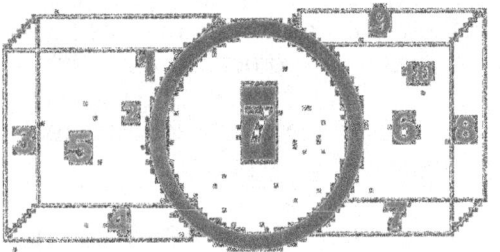
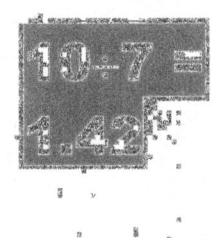

measured value of gravity formed as singularity (Π) in motion (Π^2). By the way and it ahs nothing to do with gravity, but the proof that everything will collapse in material where material represents infinity is the fact that infinity progresses in $\Pi = 3.1416$ while time in eternity moves by the measure of 3 and eventually material will accumulate space entirely by the measure of 0.1416, but this is just a thought worth mentioning.

However we can clearly see from this that the Titius Bode law is responsible for forming gravity and space is the remains of time forming layers by which a history called the Universe forms. We have 7 in relation to 10 as much as we have four in relation to seven where time is the component forming 3.

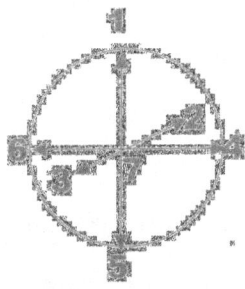

The movement of material forms a double in motion whereas this double motion relates to space by putting the diversion of the circle in relation to the space factor outer space provides. The second facto coming into use is putting seven in relation to ten.

In the fashion that we experience gravity as we form a part of the Earth and as we are taxed with mass, gravity to us depends on the location we are allocated at in relation to the line singularity forms.

When taking the movement of seven diverting the ten positions we have a change of 1.42. The movement went trough seven plus seven (7+7=14) which puts this movement of 14 in relation to 1.42 and that represents 9.86 or Π^2. Depending on where gravity is taken in relation to the axis of the Earth, this figure would change from 9.86 or Π^2 to $14 \div 1.42857 = 9.8$. Knowing the Newtonian mind the Brits and later the Yanks would believe that the Anglo American constitutes the centre of the Universe and they would therefore measure gravity along the lines of London and New York and find gravity to form a "constant" at 9.81. It helps to understand the simple ness of the Newtonian mind because them the "force" they measured is 9.81 Nm / s^2.

When we dissect the sphere to the bone of singularity we find Pythagoras at that bone. Because the sphere centre is the infinite point such a point will cover the infinite spot of the outside of such a spot where singularity produce matter in the centre extending. From the family of lines there will be three lines being close relatives crossing three lines at the centre and pointing to six edges at the birth point of material. That is then six points forming a^3. The factor of **k** is extending results in 3D lines with six points being material. From the centre of the sphere six but that six is standing in relation to the next six point which are they in a different location in time. From Kepler we know that space in 3D or a^3 is produced by time or motion in T^2. **k** =a^3 / T^2. Matter then holds six singularity positions where motion puts the six 6 effectively in 3D by the square of motion T^2. **a^3 = 6 and T^2 = 6^2**

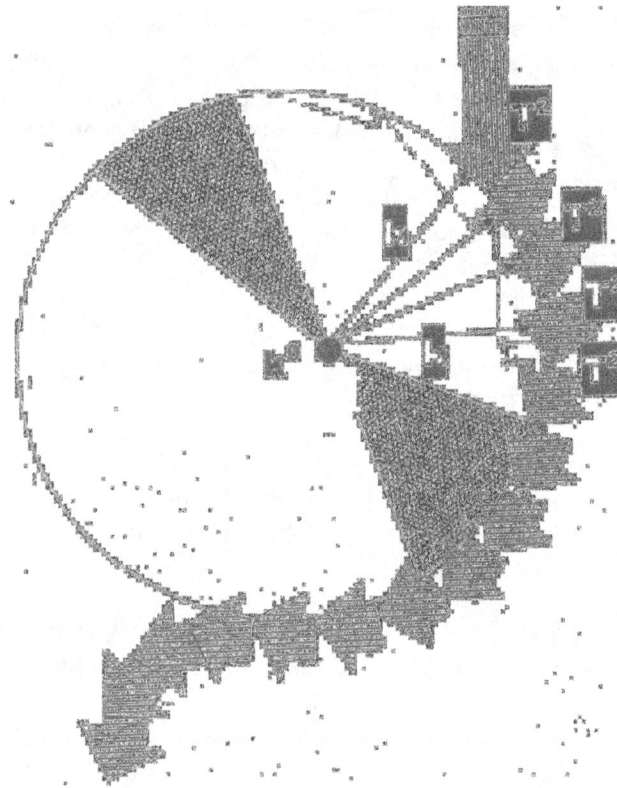

However, please allow me to warn any reader that the exercise do not remain as simple as this example may prove. From this point on the arguments about the way that time forms space (Newtonians call this process the Hubble constant due to their simplicity they show in understanding cosmology) and it really becomes complex when I prove by using the four pillars (the Titius Bode law, The Roche Limit, the Lagrangian points and gravity as the Coanda effect how science matches the explanation of the cosmic birth as the Bible tells is it took place. If one leaves out one word as it is used in the Bible, the entire exercise does not fit science or the Bible any more. However, explaining that makes cosmology really complicated! When looking at a sphere rotating one would look at sides constantly and eternally opposing the opposition as well as the direction the spin had just before.

The Coanda effect is singularity matching the Earth singularity and then through the motion between the liquid and the material gravity

forms by using the four cosmic pillars.

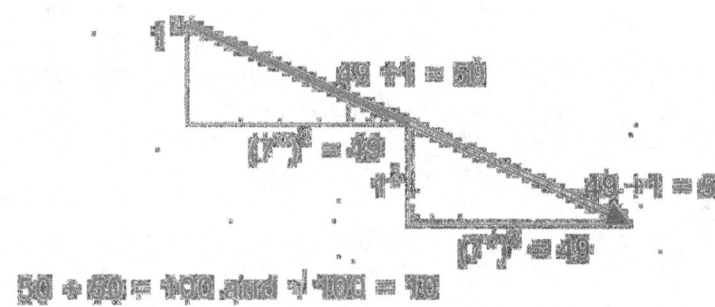

being also 7 + 3 = 10

6^2 **= 36.** Then space comes into play. Where six holds the material position the very next position is where space start and that then must be seven

50 + 50 = 100 but that is in both sides of the Universe.

$100^{1/2}$ **= 10. That places space unoccupied in the factor of 10**

That will bring about that a circle is an infinite number of six points in the square and in the square of double space. From that very simple mathematics come the proof of the Lagrangian system, the Roche principle and the Titius Bode principle.

In the Roche singularity apply all three components

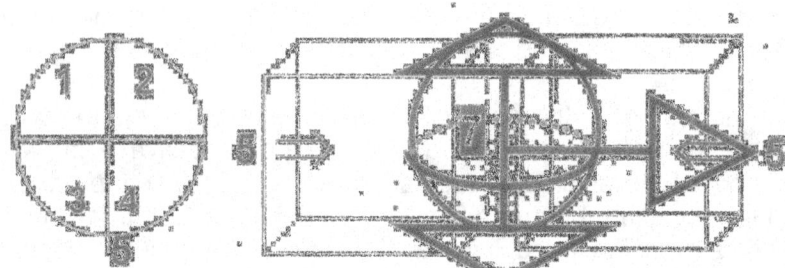

In the Roche limit the **straight line** forms part (1) and the **half circle** is part (2) and the **triangle** forms part (3) to singularity (4) Holding 5 points outside singularity

The influence of singularity as the extending of Π into space links Π^2 to r and forms 2(5)+2(5) =10+10=20
From the position of singularity there are different values in Π where each indicates a position. The value it represents being $\Pi\Pi\Pi$, Π^3, Π^2, Π and Π^0

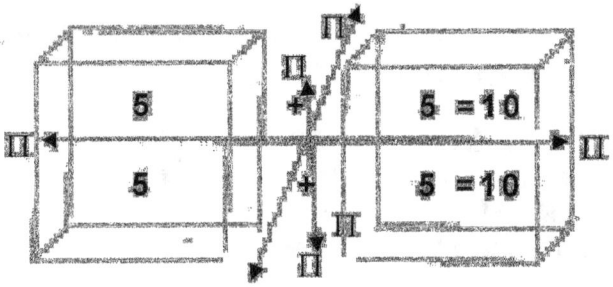

From there it influences singularity in the triangle flowing through to the half circle. It is an interaction between circular and linear motion as the value of Π continuous past Π^2 (at the end of the solid) and every cosmic structure holds an individual and specific singularity. The field where Π extends we call the atmosphere having a value of 21.991 / 7, which is Π.

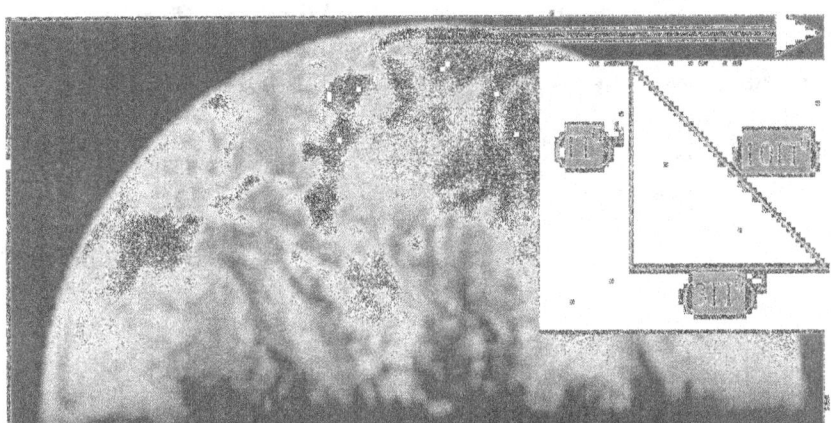
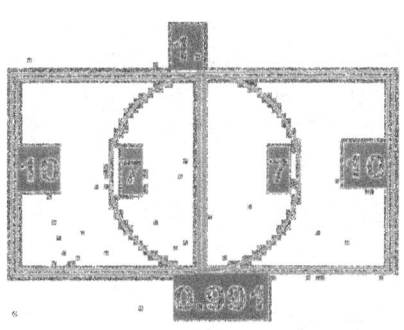

From singularity presenting the value of Π and where Π forms the relative point to gravity, we find the principle behind the Lagrangian Point System. Again we have to go to where singularity forms as a straight line that is equal to a triangle and that is equal to half circle because the lot is equal to 180°.

LAGRANGE (-TOURNIER), JOSEPH LOUIS DE (1736-1813)

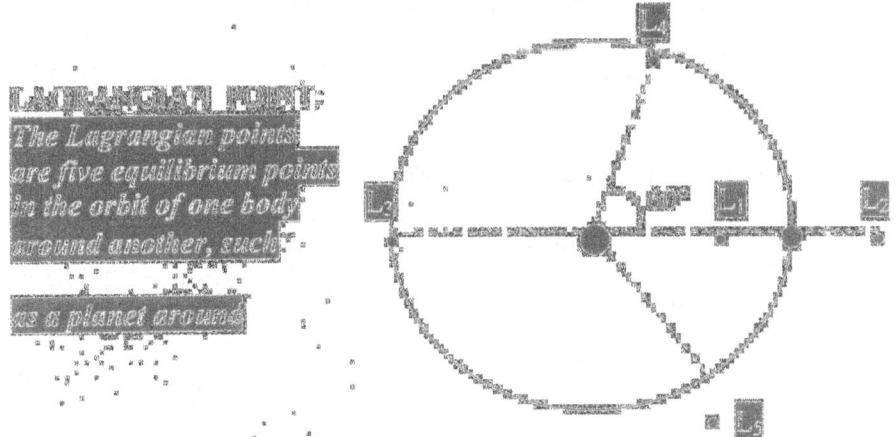

French mathematician, born in Italy. In celestial mechanics, he studied perturbations and stability in the Solar System. He examined the three-body problem for the Earth, Moon and Sun (1764) and the motion of Jupiter's satellites (1766). In 1772, he found the particular solutions to the problem that give rise to the equilibrium positions called Lagrangian points. Lagrange also studied the Moon's liberation. LAGRANGIAN POINT One of five points at which small bodies can remain the orbital plane of two massive bodies; also known as liberation points. Three of the points lie on the line joining the two massive bodies: L_1 lies between them, while L_2 and L_3 have the two bodies between them. These three points are unstable, slight displacements of a body from then resulting in its rapid departure. the fourth and fifth points (L_4 and L_5) each form an equilateral triangle with the two massive bodies, 60° ahead of and behind the smaller body in its orbit around the larger one. A well-known example of bodies flying at the L_4 and L_5 Lagrangian points are the Trojan asteroids in Jupiter's orbit. Among Saturn's satellites, Telesto and Calypso lie at the L_4 and L_5 Lagrangian points in the orbit of the much larger Tethys. In similar fashion, tiny Helene precedes Saturn's satellite Dione, keeping 60° ahead of Dione. The Lagrangian points are named after the French mathematician J.L. de Lagrange, who first calculated their existence.

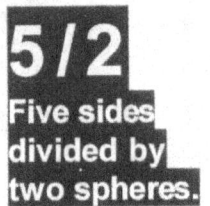

The Lagrangian System implicating the five positions extending from singularity

1 Half circle	$= 180^0$	L3 L4 L5	
2 Triangle 1	$= 180^0$	L3 L4 L5	
3 Triangle 2	$= 180^0$	L3 L4 L5	
4 Straight Line	$= 180^0$		

The half Circle $= 180^0$ combining as a sphere when comprising Singularity in the matching of the value of the straight line forming the half circle and combining as the triangle and all are equal 180^0

The Roche limit 5/2 becoming $= (\Pi/2 \text{X} \Pi/2) = 2.4674$ in relation to singularity as movement or gravity interferes with the forming of space.

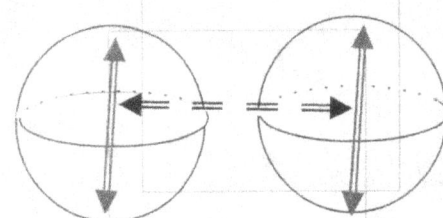

In the Roche limit the space factor provides space to a solid structure and therefore the value of r is replaced by the value of Π bringing about a square in half of Π. The cube holding 5 to either side removes allowing the extending of Π to indicate position to space. Where Π extends to lock onto the next sphere's extending indicator, Π has to connect to Π forming the square of space and translating that to the half of Π being $(\Pi/2)^2$.

The space between the spheres divide in half, but because of the extending of Π and not applying r as ordinary mathematics will suggest where Π replaces r the singularity extending from Π^0 will be half of Π in the square of $\Pi = (\Pi/2)^2 = $ **2.4674.** In this lies the dynamics why planets have a positional (be it rather a dimensional) relation of 7/10. Half of the five of the Lagrangian points is the Roche in conjunction with singularity. With singularity coming involved singularity will enforce the value change to fit Π.

The Lagrangian ratio is normal dimensional particle layout brought with development through time as a result of the Roche principle. One can see the Lagrangian system by looking at the elements developing clusters of five forming such characteristics of similarity in groups. Looking at how singularity applied connection in 3D there will always be five relating to a centre where that centre carries half of the value. The will be singularity taking a centre with one point to the top of the centre and one point to the side of the centre. The centre then becomes the one side in corresponding to the side the centre takes. From the centre another connecting will be to the bottom the front and the back. Every point will form a position where it will support the centre in displacing space by providing heat in an attempt to secure the prevention of overheating. The function of the liking of five to a centre on as a group effort is with the group work individually to secure the survival of the individuals forming the group and being the group but as a group as well as the securing the group as a unit by a mutual concentrating of space. By halving this effort in the Roche proves the motive behind the forming of the five in the Lagrangian points because of the destruction or development that space less ness then in Roche bring about. The five to one forming the centre of the governing gravity by establishing a principle singularity such a centre improves the gravity effort by all taking part to dismiss heat and create a concentrating heat flow to the centre whereby the group as individuals and as a structure will survive through mutual gravity. This is the effort of the whole cluster in underwriting the centre spot control. It is plugging five spots same as electrons do and in that accelerate the space flow by reducing the space flow as a motorcar carburettor does. In producing mutual support it underlines the individual support required by all participants to prevent future overheating or antigravity. The extending of space collecting and the extending of space reducing prevents overheating by establishing matter in six by connecting matter in six to a centre one. That centre one has the factor of less than one forming the Alfa singularity to the cluster forming the gravity.

As I indicated previously there are many stages where singularity connects either through space-time implicated or more directly. Π ConnectsΠ whereby Π will join > by 1 /10 and 10 /7 standing related by the linking of matter and space and space and time with singularity applying motion to space creating gravity in the process. The fact that $(\Pi^2 + \Pi^2)$ being the proton forming time is because the gravity created is the proton in motion. It must be clearly understood that only by the motion of the proton in the square is 3D created and not by space. The contact that the first motion makes with the rest of the Universe forms time in the splitting to fast to repeat. By the proton applying time that action of spin creates the duration of time being so close to eternal that all other time following will be much faster but seem much slower. By motion duplicating space can 3D establish space and by connecting to space through time can 3D in slow motion become reality. The motion of heat in space is a result of the motion of heat bringing space. But since that is so close to singularity that is hardly a human concern. By alternating alliances with singularity and on the other side of the Universe by aligning with space created can space establish the time component within the realms of time-eternal. One should remember that such time in motion at that point where the proton is meeting singularity such time is 1836 times longer in duration than the speed of light represented by the speed of the proton. The mass is a result of gravity and not the cause of gravity. The gravity is a time discrepancy coming about from space depleted in size. I essence it is a time zone that much slower. The proton is filling the four quarters of time and the gravity the proton produces fill the space next to time or if you wish the gravity filling the space. If we go back to the Roche with the second singularity being to clues we find gravity Π^2 being halved in thee square of the half by space filled to close to singularity being confirmed.. That means the gravity being Π^2 is filling the space with gravity. That filling by gravity will extend to a marker pointing singularity as the extension thereof which, then will be Π. The gravity committed in the space immediately following time $(\Pi^2 + \Pi^2)$ the proton commit once again to other time or gravity factor applying to other space-time. That means the space of material in the atom or in the star which is just another cosmic atom holds the seven positions of which six is in material $(\Pi^2 + \Pi^2)$ and (Π^2) and one indicating space within the atom being the seventh marker as (Π). This fills the atom that was about before the Big Bang. Then comes the part that concludes what came about during the Big Bang. Three dimensions being part of the atom falls outside the atom but is part of the atom as it produces a relevancy attached to the atom forming a value of space in between particles. That concludes the atom as $(\Pi^2 + \Pi^2)(\Pi^2)(\Pi 3)$ = 1836 time difference. Let's start with time in progress to explain the atom because the atom is the Universe and I am going to prove that in due time.

The Roche limit is:

The region surrounding each star in a binary system, within which any material is gravitationally bound to that particular star. The boundary of the Roche lobes is an equipotential surface, and the lobes touch at the inner Lagrangian point, L_1, through which mass transfer may occur if one of the components expands to fill its lobe. It names after the French mathematician Edouard Albert Roche (1820-83).

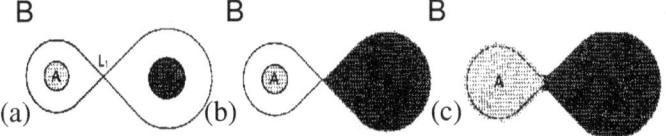

THE ROCHE LOBE: **In a binary system, the Roche lobes of components A and B meet at the L_1 Lagrangian point. (a) In a detached system, neither star fills its Roche lobe. (b) In a semidetached system, one massive component, B, fills its Roche lobe. (c) In a contact binary, both components overfill their Roche lobes and share a common envelope.**

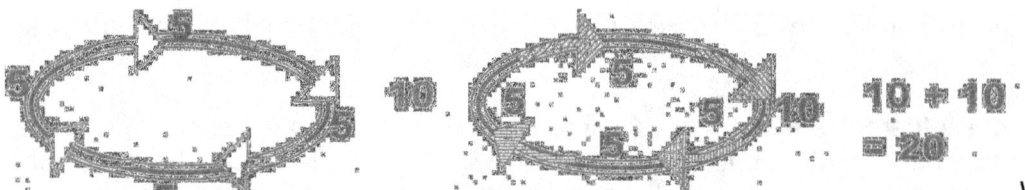

We must consider this in terms of building the Universe because from sand crystals a skyscraper rises. When looking at the sky scraper we do not see individual sand grains but eventually it is the multiplication of the combination when all the and mortise adds to form a unit that the constructed sky scraper comes about. The sand accumulates although it is very small in relation to the building.

The same way the Universe build by implementing new relevancies and from the new relevancies form the components that is one Universe. In implementing the forms available the Big Bang builds a Universe not far off in a distant part but in our very presence by using what is available to build with. Newtonians look at the city and try and make conclusions from that. It is the same as when Newtonians are looking at a city where they do not see streets parting buildings and buildings standing as units that are made up immeasurable many things much smaller.

They wish to give that which they see a measurable mass and a constant so that future playing with mathematics is simpler to achieve. The cosmos comprises of the most micro and tiniest particles that assemble in fragments forming clusters, that group as atoms spinning as stars that combine as galactica.

The Lagrangian point system is part of what represents one side of the form the cosmos uses while the Titius Bode law combines the Lagrangian points by a double and associates that total with the seven points that the sphere holds. That forms the building blocks of the Universe in relation to the Roche limit.

The Lagrangian points system doesn't only resemble the Coanda gravity but it is the Coanda gravity. The Lagrangian points resemble every aspect that singularity forms and the Lagrangian points system implements gravity in such a close duplication thereof that one can hardly miss the formula.

Again it implements the Coanda effect in order to keep the spinning satellites in allocated positions spinning with the liquid while singularity through the Roche limit positions the satellites spinning around the centre.

Only when the Roche limit is intergrading in the way it applies in relation to the formula Kepler introduced as space-time serving gravity does the Coanda effect make any sense. But in that event cosmology becomes something that makes sense. In the manner that Newtonians now present cosmology it is the most ludicrous subject God invented and nothing is realistic. I mean the concept that nothing can actually be is realistic.

Gravity is the mixing of two cosmic phenomena using the third to manifest as the fourth phenomenon. On the one side we find seven and this is interacting with ten in relation to the Roche lobe, which is $\Pi^2 \div 4$ resulting in gravity forming.

Whenever any circle is spinning opposing sides develop and even if it is the same body circling what could have been the one characteristic applying will be opposing the previous set of rules by embodying the very opposite set of rules since the direction of spin changes the entire Universe that the spinning body is facing. Everything is forever changing in the face of singularity at the centre of the Universe that is never changing. Remember the centre of the Universe is a centre formed in the centre wherever anything is spinning around such a centre.

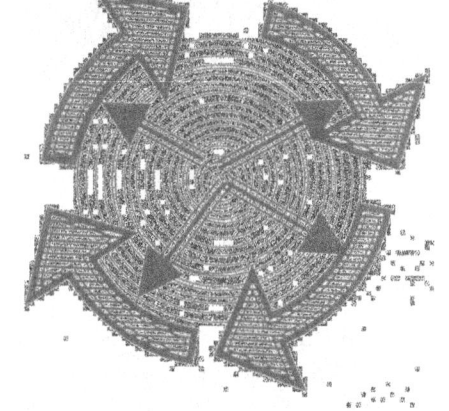

In one rotation we find the very same object in self-confrontation at four points each serving a total different reality to what previously applied. If we have the northern hemisphere favouring $\frac{10}{7}$ then on the other side the value of favour would change to $\frac{7}{10}$ and the mixing of this will put spinning in cross reference to each other.

Time is the building block of space and space is the history of time in development. Those that was looking for a white hole from where material was suppose to flow should have seen time forming a hole from where space progresses in forming. The way it forms is by initiating the Titius Bode principle and that is why the entire solar system and in fact the entire Universe is using the Titius Bode sequence to develop space.

Southern Hemisphere   Northern Hemisphere

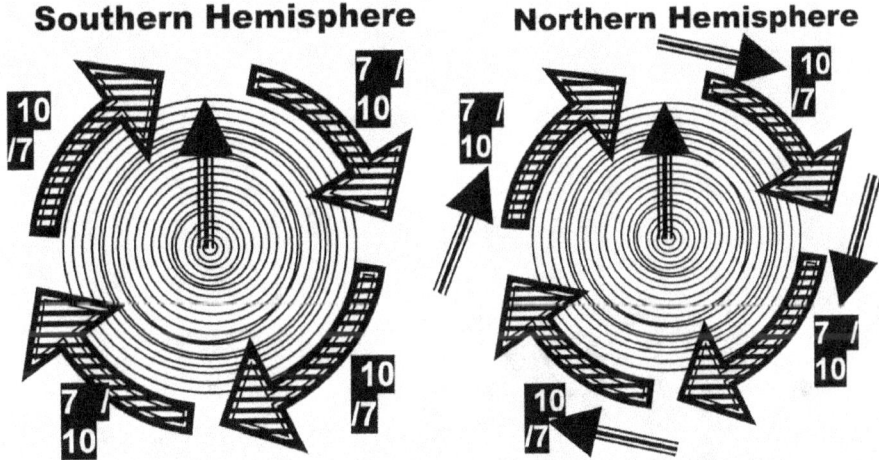

There are borders and regions of limits where this interaction of space becoming time takes place. Such a limit is the Roche limit and this circle limit is mark by singularity crossing influence borders.

The Roche limit is putting a divide to points in singularity by the implementing of $\left(\frac{\Pi}{2}\right)^2$ that is mixing the process claiming space.

The manner in which the Roche lobe manifests its presence is to reflect on space. When one star holds singularity closer than the distance of 2.4674 times the diameter of the star, the domineering singularity will liquefy the subordinate singularity. This manifestation exemplifies and it proves the Coanda effect establishing gravity by a solid forming a liquid in relation to singularity. In the Universe we have space became the crucial aspect and therefore we have **k** as a connecting between T^2, which is the movement and k^0, which is singularity. As k^0 relating to **k** represents Π^0 that is singularity extending to space, which then becomes Π we have Π in motion forming Π^2.

But in order to extend singularity in ratio to the 3D aspect we have Π which is the diameter of the star extending by movement singularity in motion as gravity $\Pi^2 \div 4$. The closest moving of another point holding singularity can one be at $\Pi^2 \div 4$ and anything closer is doomed by becoming liquefied as it then is overheated and heated to as state of turning liquid. In other books I show that gravity is the cooling of material and by spinning the star maintain temperature control. Gravity is the maintaining of the control of heat within the star. This however, is only worth mentioning as that is far to complex t become an issue in a simple book such as that. This reflects onto gravity as follows

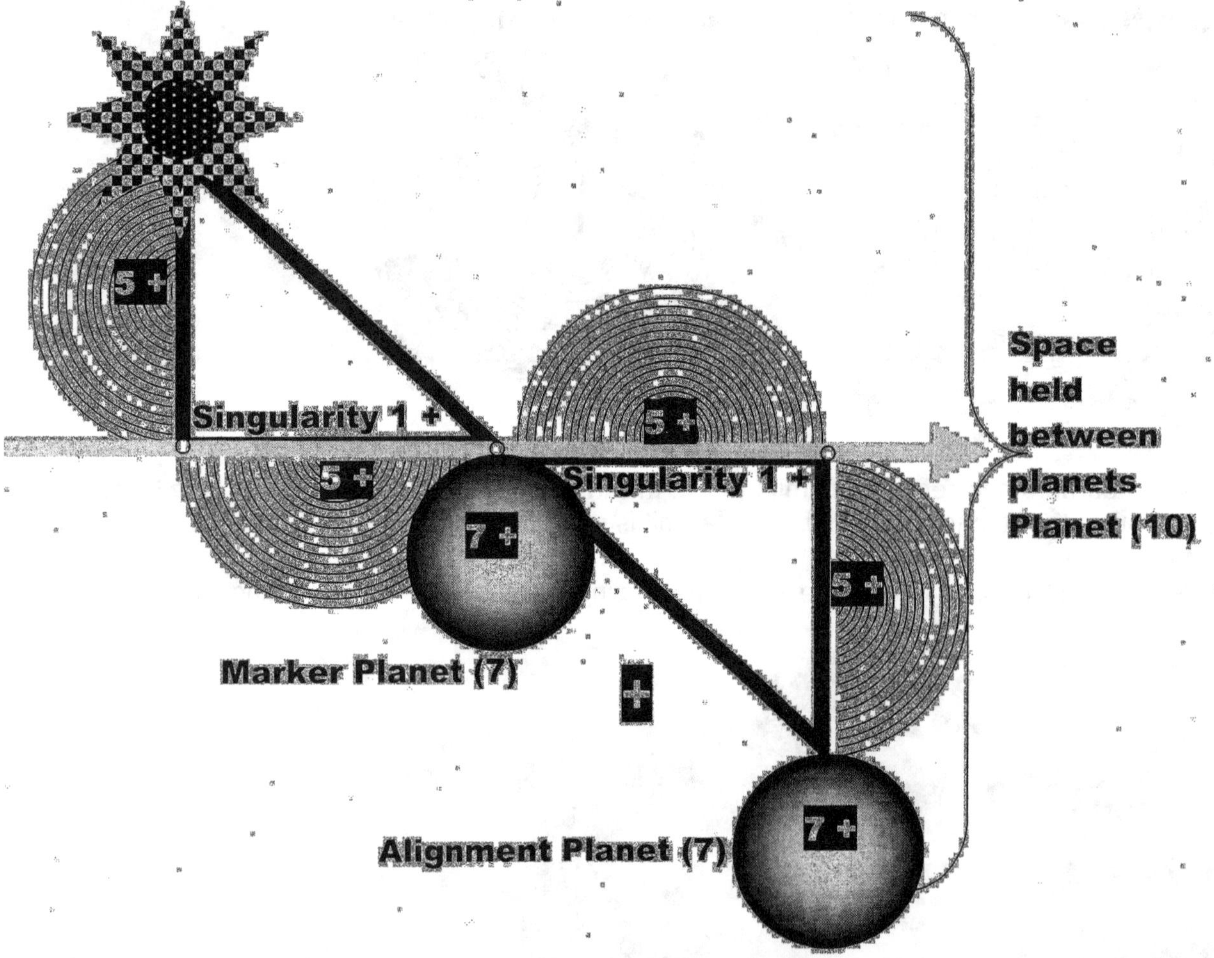

Gravity is motion in space and motion through space. Gravity is a relevancy between space travelling and the time it takes the space to travel. Gravity produces or reduces space during a certain period of synchronized spin of material in motion. Gravity is $a^3 = k\,T^2$ where it then becomes $k = a^3 / T^2$ In the light of this all other explaining fails the test of accuracy. It is no force because mass depend on gravity and gravity does not depend on mass.

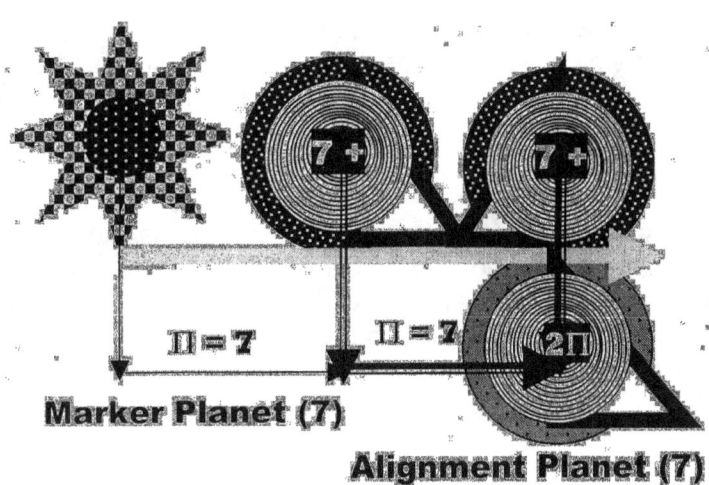

Marker Planet (7)

Alignment Planet (7)

The aligning of the planets goes according to what the Roche limit indicates. The planet uses the first point serving singularity as a marker to indicate the orbit space-time of the immediate planet next to the marker and being just on the outside of the marker. It is for that reason that Mercury has such a peculiar orbit schedule that doesn't make much sense from the perspective we have on Earth. The planet on the outside confirms the orbit space-time applying to it in accordance to the first planet orbiting just to its inside. The two planets align according to the Titius Bode space-time value of 7 + 7 relating to 10 and in both cases it is in accordance to **k** that provides the relevancy forming space-time applying to that specific planet $k = a^3 / T^2$.

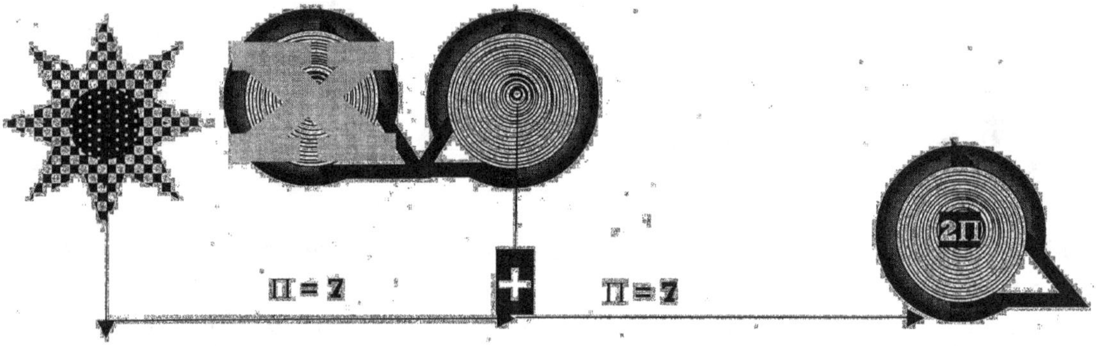

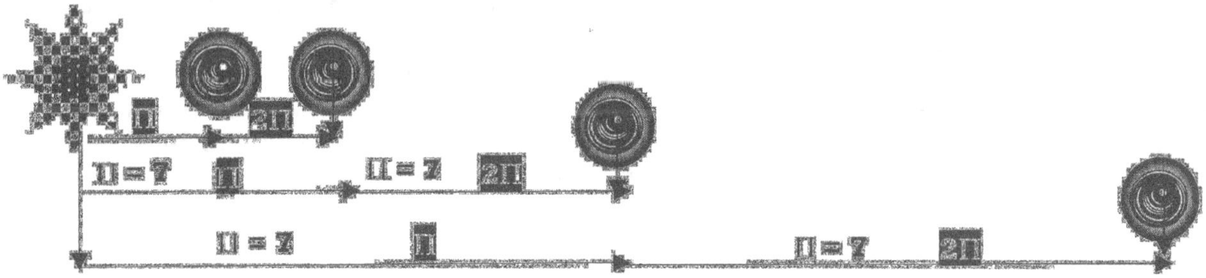

	1	3	6	12	24	48	96	192
Planet	Mercury	Venus	Earth	Mars	Ceres	Jupiter	Saturn	Uranus
Bode's Law distance	4	7	10	16	28	52	100	196
Actual distance	3.9	7.2	10	15.2	28	52	95	192

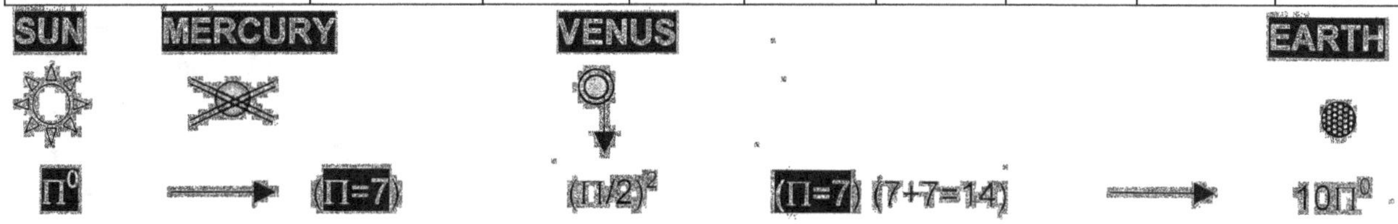

SUN   **MERCURY**   **VENUS**   **EARTH**

Π^0 ⟶ $(\Pi=7)$ $(\Pi/2)^2$ $(\Pi=7)$ $(7+7=14)$ ⟶ $10\Pi^0$

(This is what I try to explain)
From our alignment with singularity the values confirming the Titius Bode law of 7 and 10 will only be confirmed as a true value because we are the only ones according to our position that confirms singularity as being in the centre of the Universe. At every point a planet serves singularity, a new point lining up with and confirming of the forming of the centre of the Universe is confirmed. The point serving Venus or Mars would never line up correctly in our view but the point still serves as 7 + 7 in relation to 10 from that point serving singularity as forming the centre of the Universe. The confirming of the point holding 7 takes the immediate first planet as the guide to the distance the

first seven would be and from there it uses another seven which then is the precise same length as the first seven. All the inside planets rotating as well as all the outsides planets rotating holds no influence on any particular planet orbiting the Sun except the most immediate inside planet that confirms the first distance relating to 7 or to Π. It is only the Sun and the first inside planet that has any relevance to singularity $(\Pi \div 2)^2$.

The Titius Bode principle is a relation where space is the ten factors and material is the seven factors. By space being diminished by material one relation comes about and where material dismisses space another relation of seven to ten comes about. The Titius Bode Principle is equal to gravity @ $= \Pi^2 = 9.8696$ Proving that the Titius Bode Principle is a product flowing Directly from the growth of singularity forming space-time The Titius Bode principle directly valuating TIME to SPACE $= \Pi^2 = 9.8696 =$ MATTER HOLDING THE SECOND PROTON COUPLING

The spherical positioning layout forming the Titius Bode Principle

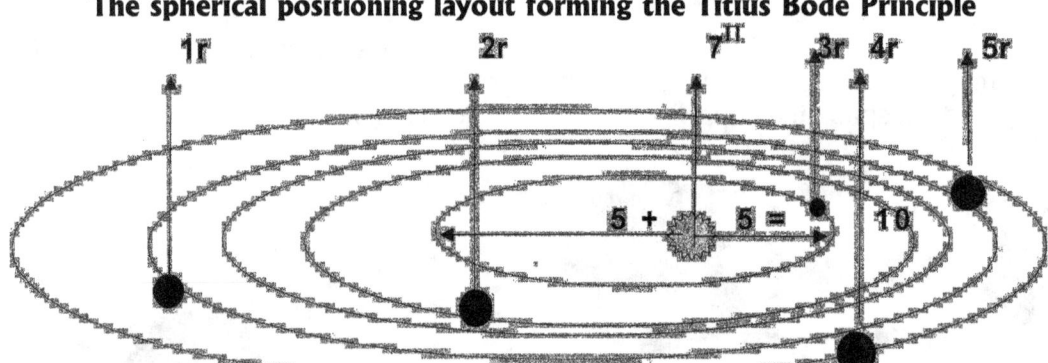

The $5 + 5 = 10$ is a position of dimensions as space loses value to singularity. The 7 that matter diverts in points from singularity may seem, as coincidental but is valid. Still in accordance to our perception valuing the number in degrees, it seems coincidental but if it is coincidental, it is nevertheless a figure of diverting proven as accountable in all other calculations and plays a most dynamic role.

The Lagrangian 5 point system results as much from the Curvature of space-time as does the form the Black Hole holds. The Galactica is the opposing equivalent of the Black Hole and has identical but opposing similarities being the five points positioned to singularity. The galactica is generating space and the Black hole is degenerating space.

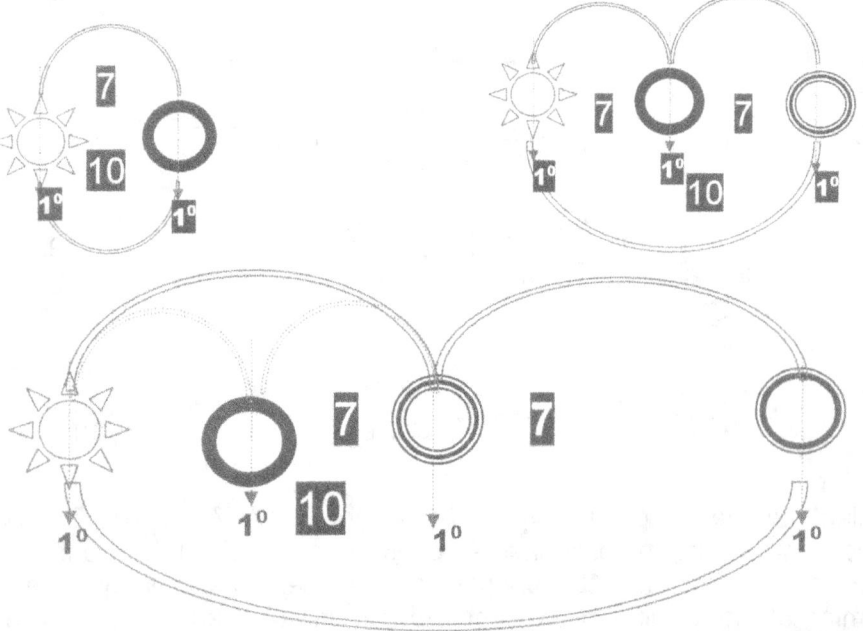

From the matter-to-matter relation in the Titius Bode configuration there are 7 / 10 + 7 / 10 = .7 + .7 = 1.4
From the space-to-matter relation in the Titius Bode configuration there is 10 / 7 = 1.42

In this maintaining of cross referencing of singularity located in individual atoms providing spin to the governing singularity that maintain structural form in solids, many factors of singularity all form a close knit network and being inseparable as one unit, by the same margin it also is strictly individual to a point of destructing. From the inner or governing singularity outward all is concerned as spade-heat. From the dividing singularity only one reference holds a matter value forming the position next to the governing singularity and therefore 7+7 becomes a factor and not all the dividing singularity between the point of reference and the governing singularity. That way the star to the outside takes a position doubling the distance every time. In balance everything in space to the outside of the governing singularity is space be it space or matter that makes no difference therefore that is 10.

The extension of Π is well received as a dimensional implication to matter holding seven positions from singularity and space having four quarters through out the rotation of singularity forming the centre to the five dimensions (one side lost to the cube's six sides connecting to the five remaining sides) making the total sides facing space from the point holding singularity at any given instant at a value of twenty (4 X 5 = 20). Then adding the singularity cross of Π being (1+0.991) = 1.991 the relation never becomes 22/7 as s often is used. This is crude because in more precise calculations it becomes .91 + 1 = 21.91/7 = Π and in that way one can define the precise measure of Π.

The sectors provide individual singularity as a means in sustaining governing singularity by which provision comes through maintaining governing singularity the required spin in maintaining cooling. If this process did not apply, there would be no connecting individual singularity to major singularity. The sectors provide individual singularity a means in sustaining governing singularity by which provision comes through maintaining governing singularity the required spin in maintaining cooling. If this process did not apply, there would be no connecting individual singularity to major singularity.

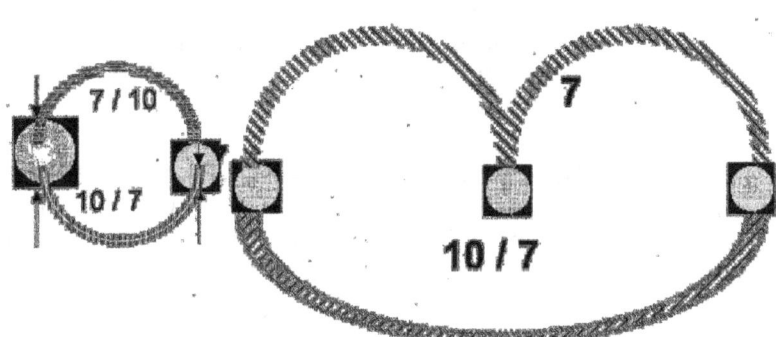

SINGULARITY BY DIVIDING SPACE INTO MATTER AND MATTER INTO SPACE, ANG ALL OF THIS ACCORDING TO THE TITIUS BODE LAW OF 10 / 7 AND 7 / 10 IN CONJUNCTION WITH THE ROCHE PRINCIPLE OF $(\Pi/2)^2$
Time started at a point where infinity parted from eternity, a point not having a name so you can cal that point whatever you wish to say, as long as you say time did not move at all. Then the command came and time overheated for the first Π^2 in time. That brought space into play.

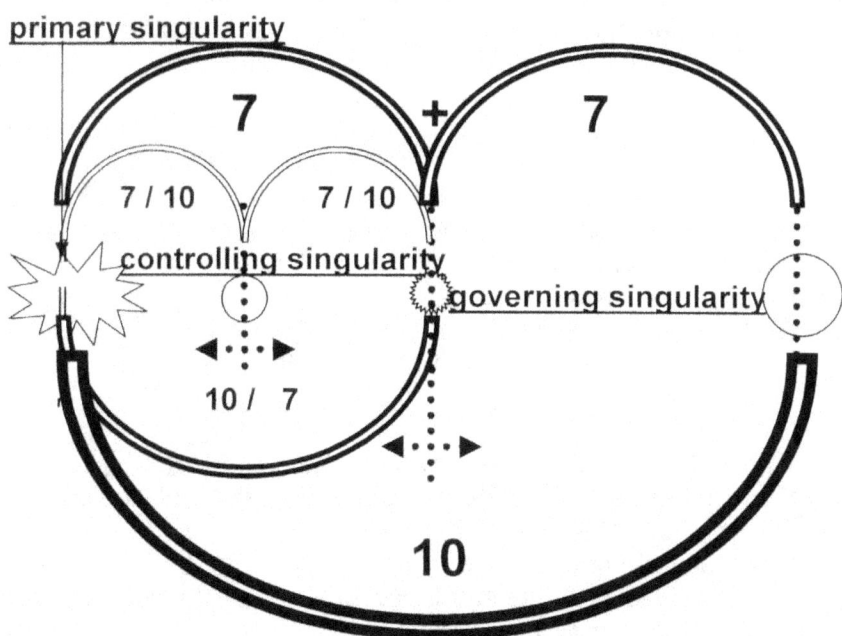

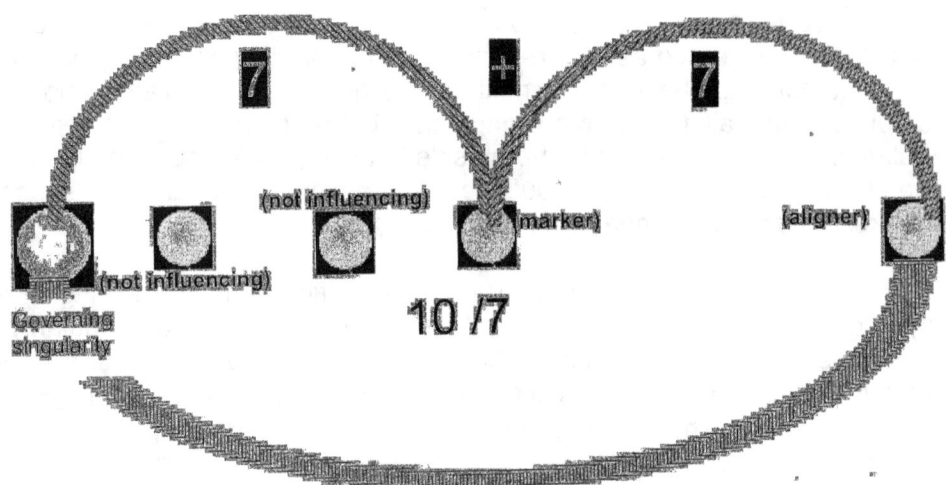

From the orbiting structure (planet) aligning singularity only one structure the very inside singularity applies as a position of reference and that is reference to the distance applied between the governing singularity. From the Sun (governing singularity) the matter marker is 7/10 = 0.7 with the only one other forming a marker Π= 21.991/7 = and from that there is the seven forming the main component of the roundness of Π. Adding the double 7 implicating time allows the two components of 7 to form 14. From the Sun (governing singularity) the outer planet forming the marker in search of position holds space in the square 10 / 7 = 1.42 in aligning with the 7 forming material of the Sun. Therefore there are two sevens relating to ten forming the material positioning of the structure in orbit and from the governing singularity all outside the Sun is the square of space (ten) aligning with one particle (seven) and not one of the other structure to the inside or the outside holds any value. Because 7 + 7 = 14 and 10 / 7 = 1.42 the distance doubles every time there is an aligning of three orbiting planets.

In this there is definite proof of influences coming about between particles sharing gravity. But then again the entire Universe shares gravity and as such then all will influence everything. It is the position singularity holds in relation to the Universe and the Milky Way forming currents and seasons moreover than the Sun shining brighter or not. The Sun in size over dominating the earths in comparison disqualifies any positional influence that can alter the earths heat standings. Through shear size the Sun can shine at the top and the bottom of the Earth simultaneously without effort from all normal possible angles. I show a relation between singularity in different positions maintaining seasons and north/south polarity, not only as far as concerning the Earth but also outside influencing polarization. This has to do with the second position singularity holds in accordance to matter and space and is an "*electromagnetic*" (used for the lack of a better word) sustained positional opposing derived precisely from the graph in the manner when calculating electricity.

In this it is clear why the Titius Bode ([10 + 10 + 1 + .991] / 7) and the Lagrangian 5 \\ 7 systems part their ways when applying the different processes they hold. With all the differentiating, the observer must also consider the dual massage that light uses in travelling through the vastness of universal space. The thought of nothing is just what it is, a thought of nothing and although it is in the human mind common nature to present nothing as a value in the recalling of something, nothing is a presentation of the figment in the human mind. There can be no number such as nothing and that was (possibly) Newton's biggest error. Nothing represent non-existing and that is just what nothing is, it is non-existing.

The Titius Bode influence in a manner that on the one side holds the matter-to-matter relation of 7+7/10 whilst on the other side during the same time holds the space-to-matter relation of 10/7 forming equal and opposing values. From this the orbits of cosmic structures are always oval favouring the singularity dynamics of the one structure at one point and switching the favouring to the other structure on the opposing side. Because the structures can never be equal in size (singularity will not permit that where the Roche principle will intervene) the shape is always "off centre" as well. This influences coming about as the Titius Bode principal manifest in other ways proving Kepler's time relation with space through distance from singularity controlling the factors.

Once again the following proves that mass is a result of gravity and gravity does not come about through mass, because by using a new a³ it can establish a new k, which will convert that gravity T² to apply to the new a³. However it fluids, which smoke and dense heat also are.

On that and other grounds I maintain that the Sun on the inside is liquid. Gravity and the establishing thereof is not a God given write of birth bestowed on all the heaviest to create.

The gravity it develops is a "*cosmic life*" not to be confused with carbon life we find and have and are on Earth, but that which makes the Universe alive. That establishing or creating or exciting of singularity by applying new heat in space by separating distance by spin is a new cosmos entity standing apart from the rest of the Cosmos. Every singularity is a Universe that can apply new values and rules by changing any of the Kepler Factors.

By altering any of the three Kepler factors space-time can establish a new significant gravity in the midst of gravity applying. By creating a new spin in the presence of the Earth gravity, the spin creates a gravity that will encourage in example the spinning top to try (it can never happen but it is trying all the same) through a newly charged singularity to develop a gravity that will produce such vigorous movement T^2 that will take the top to a position apart from the rotation it normally has with the Earth. When the speed of the rotation exceeds the limitation Earth the Earth allow the spinning top will start wobbling from side to side indicating a maximum effort to create lift and go in a separate spec at a separate distance from the Earth.

When the top slows down the wobble will become present again, as the gravity established through the spin will fight to stay alive and apart from that of the Earth. But it shows that in Kepler's formula new space comes about from establishing a new T^2, which the spinning then forms in the alliance of space created through the manifestation of a new $a^3 = k\,T^2$ in the boundaries of the Earth. Make no mistake about the fact that the spin is new gravity that comes about in the area the top occupies and the **k** is now the rotation coming about from the centre of the new spin.

The wobbling at the bottom and at the peak of the spin effort of the rotation is a gravity struggling for independence either to maintain independence or at the top to establish ultimate independence.

The ten dimensions I named the atomic relevancy is also showing the double value of singularity as singularity extends into as well as beyond space. The atomic relevancy is $(\Pi^2+\Pi^2)(\Pi^2 \times \Pi \times 3) =$ **1836** that is the mass relation between the electron (3) and the proton. Proton = $(\Pi^2+\Pi^2)$ Neutron $=\Pi^2\,\Pi$. The atomic relevancy holds the dynamics of singularity control. In the ratio and dimensions we find in the atom, all space-time derives from the atom, whatever the atom is.
Matter in relation (part of) with the total dimension of space.

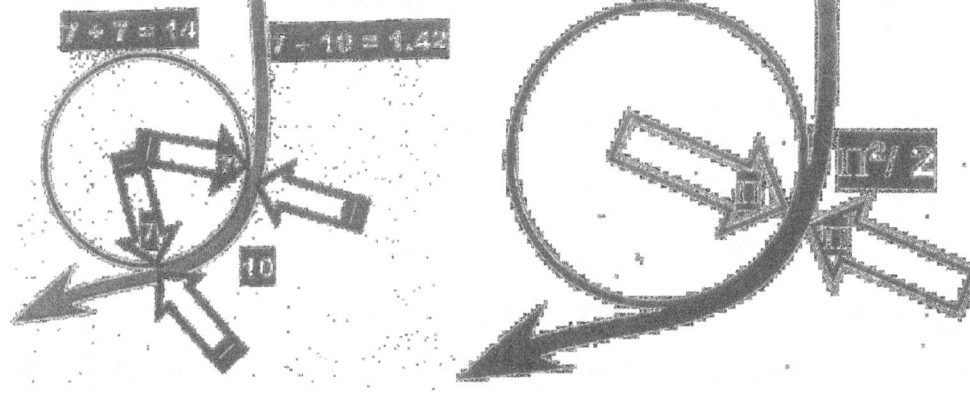

The Titius Bode implemented $\left(\dfrac{10}{7} \div \dfrac{7}{10}\right) = 2.04$

$\dfrac{1.4285}{0.7} = 2.04$ Taking from both orbiting influences

$2.04 \times \left(\dfrac{\Pi}{2}\right)^2 = 5.033$

$2.04 \times \left(\dfrac{\Pi}{2}\right)^2 = 5.033$

$5.033 + 5.033 = 10.066$ from both objects

SPACE DIVIDED INTO TIME

$\left(\dfrac{7}{10}\right) \div \left(\dfrac{10}{7}\right) = 0.49$

$\dfrac{0.7}{1.4285} = 0.49$ Taking from both orbiting influences

SPACE MULTIPLIED WITH TIME

$\dfrac{7}{10} \div \dfrac{7}{10} = 1$ and $\dfrac{10}{7} \times \dfrac{7}{10} = 1$ Therefore not influencing change

THE PROCESS PARTED USING THE ROCHE PRINCIPLE

$\dfrac{10}{7}$ The Titius Bode $\left(\dfrac{\Pi}{2}\right)^2$ The Roche influence on

$\dfrac{7}{10}$ The Titius Bode

$\left(\dfrac{\Pi}{2}\right)^2$ $\dfrac{10}{7}$

SPACE DIVIDED INTO TIME

$\dfrac{7}{10}$ $\left(\dfrac{7}{10}\right) \div \left(\dfrac{10}{7}\right) = 0.49$

$\dfrac{10}{7}$

$\left(\dfrac{10}{7} \div \dfrac{7}{10}\right) = .49$ $\left(\dfrac{10}{7} \div \dfrac{7}{10}\right) = .49$

.49   +   .49 = .98

$.98 \times 10.066 = 9.86468 = \Pi^2$

TIME SPACE $= \Pi^2 = 9.8696$

TIME SPACE $= \Pi^2 = 9.8696$ = Space and time in a dimensional implication

From the manner that gravity forms where the movement is centred on singularity by applying the Roche limit in conjunction with the Coanda effect we have gravity. Take the circle away and replace that with the earth. Take the fluid away and replace that with the atmosphere, which is, a liquid and one find the gravity of the Earth in place and the Coanda effect permits. In gravity there is no mass as a factor but mass results from gravity as a secondary factor.

The Roche limit forms an intergrading part that connects the Titius Bode law standing in relation to time with the Titius Bode law standing in relation to material. In **An Open Letter on Gravity Part** 1 Volume 1 + 2 and **An Open Letter on Gravity Part** 2

I explain these aspects of gravity extensively.

I also prove the formula there is between the Titius Bode law in relation to the Roche limit as well as the Lagrangian points mathematically as thee three combine in the unit called the Coanda effect.

From using the formula I prove the value of gravity as it is used as g = 9.81 Nm/s^2 and with that evidence there can be no doubt as to what gravity is and why gravity is what it is.

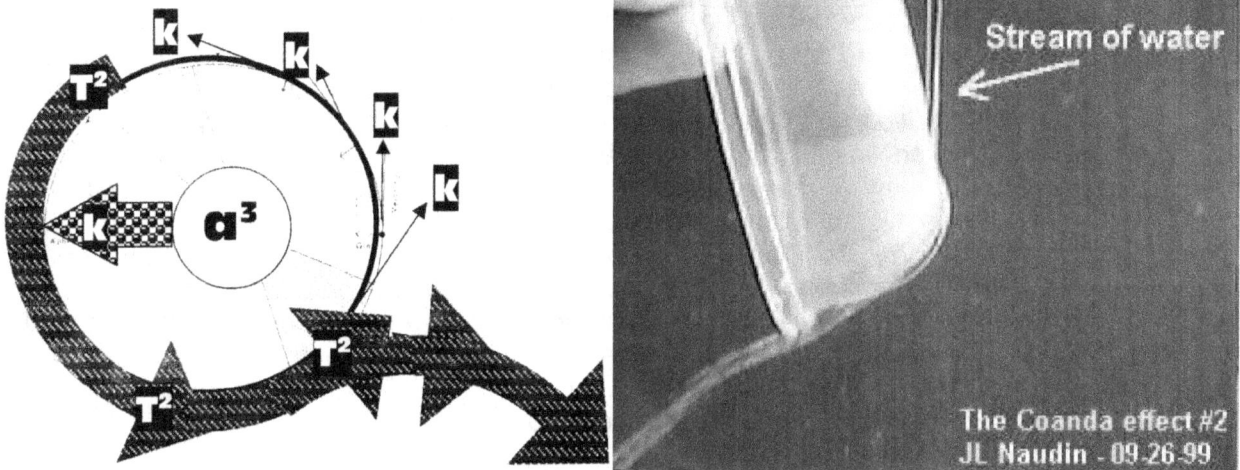

The Coanda effect #2
JL Naudin - 09.26.99

It is for that reason that size is not what influences mass or that gravity is exponentially more in correspondence with mass increasing However this fact does not relate to size increases but on the contrary directly links to size reductions.

The smaller the structure maintains a much higher velocity and a much higher velocity will as such reduce the size the object maintains. With a much higher velocity it will compress more space into any confined unit by which it will increase the density as it decreases the structural size the object holds. By diversely increasing the density it indicates that the space is collapsing much more due to the higher velocity coming from the increased spin and this spin holds a far greater value of movement. The smaller the structure is, the faster it spins, the more it crams material into a reduced pace and the more density will bring about more mass.

The fact that smaller cosmic objects generate more "mass" in smaller space proves that the Coanda effect is gravity. The faster the spin is, the higher the space reducing is and the more "mass" the higher the rotating movement will generate.

It must be very clearly understood that although singularity links all there is by equal value ($\Pi^0 = 1$), it is in the relation to space –time $k^0 = a^3 / T^2k$ or $\Pi^0 = \Pi^3 / \Pi^2\Pi$ that an entire Universe is different in all aspects thereof. Every star has Π^2 as gravity, but depending on $\Pi^0 = \Pi^3 / \Pi^2\Pi$ the value of $\Pi^2 = \Pi^3 /\Pi$ is completely different, even on the same structure. In all of this the founding value of the Universe is the relevancy applying to the atom in as much as being $(\Pi^2 + \Pi^2)(\Pi^2\Pi)(3) = 1836$ as the displacement reducing value of time forming space.

Professor Strauss, the first book I wrote that was still written in Afrikaans, I brought to you and you asked me whom do I intend would read it, and I said everyone because I am just anyone. Then I walked a long route while enticing all groups to read my writings as to trying to get my principle across by establishing my ideas and get my concepts introduced and this was the path I followed since our then meeting we had when this venture started. That was seven years ago and since then I followed your advice to contact academics all over and try to find one academic seeing my point of view where then I would be able to find a channel through him by which I could introduce my ideas to the academic world at large. Further more I tried and failed to get numerous publishers interested but with no academic officially backing my view that road was even more costly and proved even more unsuccessful. While I was going about I went on to write many other books with which I was trying to approach science from various angles Sung a more explicit fashion every time and even with later more progressive books confirming my concepts more as to underwrite what I initially believe to be true, it did me no good in the end. My concepts remained what it was at the start and every book only confirmed what I believed in as being in support of what my view initially was. Details I presented might seem more elaborate now, nevertheless it is still the very same arguments that I initially had and I have not changed one sentence expressing one idea since that first book, although I have added around (I guess from the top of my head without checking) about twenty four books to the list, in which I have not retracted even the least statement or the readdressed the longest argument that I then made in my first book that was in Afrikaans.

During the seven years of incredible unsuccessful drought finding no interested party since then when we met with my Afrikaans book, I have tried to interest countless academics in reading my work, but no sooner did they reach page three where I say I don't agree with Newton than they reject my work on the basis that I reject Newton and with no further investigation on their part in reading my work, they showed no further concern in finding out what I have to say on this matter. My disagreeing with Newton was so disgusting to all the astute members of the physics world that they treated my books as if my views was scientific porn would and because it filled them with such detest they had no incentive to read anything more, because any one questioning Newton's concepts was then indulging in science porn, or so it seems to me is what they think is the case. I have been rejected by academics on countless occasions because I reject Newton. The rejection was never because my work was at fault and they were able to show where I stray from valid arguing, or that any person would find an argument (even one argument) on which to show my arguments were not sane and therefore incorrect. All this is now history because as much as it concerned me a while ago, it now at this stage couldn't bother me less.

Then in the meanwhile I tried to get some reviews from more astute members of society than the group where I belong with my tradesman-mentality I tried to establish some opinion about my work from sectors I thought could help me find some form of censure or review. I handed the books to medical doctors, professional engineers, one accountant, several lawyers, one dentist and some teachers, which among them was Head Masters. Also there were other tradesmen with my intellectual mentality that asked me to have a copy as to satisfy their curiosity, just to check on me about my ability to write about such matters as I claimed I wrote about. There was very little response other than being surprised about gravity attracting as a force and yet no comets collide with the Sun but from their ranks I think their interest came because they suspected my abilities in writing in the first place just because I am a motor mechanic, and to their thinking no motor mechanic can fathom concepts Einstein introduced. The so called professionals had one response which was actually very insulting when I ponder on the response in hindsight: everyone of the so-called professional educated person's were annoyed with me because I informed them about things (the four cosmic principles no less) they never heard of before and between the lines they all told me that if those were as important as I valued those to be, everyone would have heard about them and while no one has heard of the Titius Bode law, The Roche limit, The Lagrangian points and the Coanda effect, therefore my inability to use facts of a more common knowledge nature is proving to be a handicap of my ability and making them understand with the writing on my part. There were those that said in so many words that I ought to use information that they should know about and their lack of understanding was just because of my lack of studies in being University educated and in that were my clear inability to write about things they have to be able to understand and enjoy. They accused me of not having an ability to inform them to a more satisfactory level with the work I introduce as a new principle by which I underwrite gravity. They were not informed about issues I put under scrutiny and therefore they couldn't follow my line of arguing because I was using principles and terminology they, with their superior status and obvious better intellect resulting from their far more advanced education, had never heard of and with them never encountering the facts I use that then proved to them I had no ability to inform. Their inability to understand was my fault because I could have used more commonly known and more widely accepted information that everyone knows about instead of proving my arguments using facts they have never heard of before. To their most informed opinion they were unable to understand what I was saying because I was illiterate and mentioned information that they never heard of and although this was solely contributed by their lack of intensive reading widely (the Afrikaans word I am looking for is "onbelesenheid" or also meaning not a well read background or not having a broad and widely field of reading but unfortunately I was unable to find the equivalent specific word in English), the audacity that a mere motor mechanic with no University education whatsoever used arguments and terminology unknown to them and spoke of concepts they could not follow, notwithstanding their superior intellectual status and advanced schooling, this shortfall could only be as a result coming from what my inabilities produced and as I am less educated, therefore such an enormous shortfall in their ability to become informed as to what I was saying was totally my fault. This made me realise whatever I do, the fact that no one ever heard of my arguments and the fact that only I can't see how anything cannot have mass or pressure in the cosmos or could not be filled with nothing, their failure to understand I alone am to blame for that.

With the Afrikaans book back then, you asked me whom do I intend would form the audience that would read my book and I replied every person that can read and had an interest in reading what I had to say, which was as broad an audience as I could muster. Should you now ask me whom I have in mind that should read this book then this time I will tell you I wrote it with only you in mind and no other person. I have run out of potential readers. At the same time it should also be said that I couldn't care the worth of one small piece of shit if no one ever opens the books I wrote because I know I am correct and I know the other lot on Earth are all incorrect and as mad as that may sound when saying everyone is wrong and I am the only one that is correct, it is not half as mad as the concepts put forward that the Universe works on the principles of mass and pressure and where mass and pressure is absent, there one would find nothing filling everything, while ignoring true cosmic principles evident in the Universe on the grounds that they are coincidental because they don't prove to use mass. All those that were not interested in reading my work because of my inability to perform to their liking now can fly off to hell for all I care. I just couldn't care less about the fact that no one this far would look past my status or not being able to look past the given status of Newton and which Newton does not deserve and look at the correctness of my concepts compared to the ridiculous mythology backing the background that Newton holds.

The book I wish to send is a book I named ***An Open Letter on the Veracity of Gravity***. By reading the book to its final pages any well-schooled academic such as you yourself are, should be able to gauge (in a much brooder manner) the outline of the entirety of work because this touches on aspects that I explain deeper in other books. It should take a highly qualified academic as you are, to form a brooder view of the concept I put forward. I hope that my personal drawings, which are mostly a form of embroidering my concepts, would make the reading easier and thereby should allow you to form a better opinion about my concepts. As you may imagine, I regard the work as rather sensitive, although I do not mind other senior academics reading my work, but still I do not wish to have the information (all new) strewn on campus by every novice. I sincerely hope you may find the time to read the book, (which consist mostly drawings and sketches in any case).

I would have preferred to make an appointment and meet you in person, but due to an open-heart operation I recently had, this will be the only means of contact I am allowed to have since I may not travel for the following foreseeable time and when saying that I can hardly walk to answer a telephone. Should you be kind enough to agree to my request, I shall mail a copy of the book. You can either contact me personally by telephone or through a friend using the e-mail address as given, which is as reliable a way of contacting me.

Thank you very much for the time and effort put into ad hearing to my request.

Peet Schutte.